Cryptosporidium: parasite and disease

Simone M. Cacciò • Giovanni Widmer
Editors

Cryptosporidium: parasite and disease

Springer

Editors
Simone M. Cacciò
Department of Infectious, Parasitic and Immunomediated Diseases
Istituto Superiore di Sanità
Rome, Italy

Giovanni Widmer
Tufts Cummings School of Veterinary Medicine
Division of Infectious Diseases
North Grafton, Massachusetts
USA

ISBN 978-3-7091-1561-9 ISBN 978-3-7091-1562-6 (eBook)
DOI 10.1007/978-3-7091-1562-6
Springer Wien Heidelberg New York Dordrecht London

Library of Congress Control Number: 2013954572

Printed on acid-free paper

Springer is part of Springer Science+Business Media (www.springer.com)

Preface

It is more than a 100 years since *Cryptosporidium* parasites were first described by Edward Tyzzer.[1] In addition to the initial discovery, two events stand out for having significantly impacted our awareness of these parasites and the diseases they cause. The HIV epidemic and the emergence of cryptosporidiosis as a potentially severe opportunistic infection in people living with AIDS was significant, not only because of the clinical implications but because it motivated a substantial research effort and a desire to better understand these parasites. A second notable event was the waterborne cryptosporidiosis outbreak which occurred in Milwaukee in 1993. The importance of this outbreak is illustrated by the fact that the article[2] reporting on the epidemiological investigation of the outbreak was cited over 600 times, more than any other paper with the term *Cryptosporidium* in the title listed in PubMed. Subsequently, the application of genotyping methods based on the polymerase chain reaction revealed the common occurrence of cryptosporidiosis in animals, particularly in young livestock, and the importance of zoonotic transmission of this parasite. Whereas waterborne outbreaks have provided incentives for improving drinking water quality, progress in treatment and prevention of cryptosporidiosis has been disappointing. As described in this book, basic research has generated a wealth of information on many aspects of *Cryptosporidium* biology, but this knowledge has not had a decisive impact on the progress towards the production of effective treatments or vaccines. Cryptosporidiosis thus remains a serious infection, not only for immunocompromised individuals but also for children living in underdeveloped countries. In recognition of the disease burden, and to underline its link with poverty, the World Health Organization (WHO) has included cryptosporidiosis in the Neglected Diseases Initiative since 2004.

---

[1] Tyzzer, E.E. (1907) A sporozoon found in the peptic glands of the common mouse. Proc. Soc. Exp. Biol. Med. 5,12–13.

[2] Mac Kenzie WR, Hoxie NJ, Proctor ME, Gradus MS, Blair KA, Peterson DE,Kazmierczak JJ, Addiss DG, Fox KR, Rose JB, et al. (1994) A massive outbreak in Milwaukee of *Cryptosporidium* infection transmitted through the public water supply. N Engl J Med. 331, 161–167.

The content of this book reflects the extent to which our knowledge of *Cryptosporidium* parasites has expanded in recent years. The 13 chapters are written by scientists, clinicians and veterinarians having many years of experience with these parasites and who together have published hundreds of research papers. This practical experience and scholarly activity is reflected in the quality of the contributions. The book covers a wide range of subjects, ranging from clinical cryptosporidiosis to the epidemiology, taxonomy, host-parasite interaction and molecular biology. Recent progress in the field of *Cryptosporidium* "omics" is also covered.

We have grouped the chapters in four parts. Part I covers taxonomy and epidemiology. It includes four chapters on the molecular taxonomy, epidemiology, evolution and ecology of *Cryptosporidium* species infecting humans, livestock and other vertebrates. Part II covers "omics"-related subjects (genomics, transcriptomics and proteomics), as well as an overview of *Cryptosporidium* metabolism and its many unique features. Host-parasite interaction is the focus of Part III. Clinical cryptosporidiosis, immunology and the current state of drug development are covered. The last part is devoted to waterborne cryptosporidiosis. Two chapters review the implications of *Cryptosporidium* oocysts in drinking and recreational water and give an overview of treatments to remove and inactivate oocysts in drinking water.

This book is evidence of the devotion of 34 authors residing in seven countries who together have invested a significant effort in reviewing and interpreting recent progress in their respective fields. The editors would like to express their gratitude to the authors for taking on this task.

Contents

Part I Taxonomy and Epidemiology of Cryptosporidium

1 **Taxonomy and Molecular Taxonomy** 3
Una Ryan and Lihua Xiao

2 **Epidemiology of Human Cryptosporidiosis** 43
Simone M. Cacciò and Lorenza Putignani

3 **Molecular Epidemiology of Human Cryptosporidiosis** 81
Gordon L. Nichols, Rachel M. Chalmers, and Stephen J. Hadfield

4 **Cryptosporidiosis in Farmed Animals** 149
Lucy J. Robertson, Camilla Björkman, Charlotte Axén, and Ronald Fayer

5 **Cryptosporidiosis in Other Vertebrates** 237
Martin Kváč, John McEvoy, Brianna Stenger, and Mark Clark

Part II Molecular Biology

6 ***Cryptosporidium*: Current State of Genomics and Systems Biological Research** 327
Aaron R. Jex and Robin B. Gasser

7 **From Genome to Proteome: Transcriptional and Proteomic Analysis of *Cryptosporidium* Parasites** 345
Jonathan M. Wastling and Nadine P. Randle

8 ***Cryptosporidium* Metabolism** 361
Guan Zhu and Fengguang Guo

Part III Host-parasite Interaction

9 **Human Cryptosporidiosis: A Clinical Perspective** 383
Henry Shikani and Louis M. Weiss

10 **Immunology of Cryptosporidiosis** 423
Guoku Hu, Yaoyu Feng, Steven P. O'Hara, and Xian-Ming Chen

11 **Treatment of Cryptosporidiosis** 455
Jan R. Mead and Michael J. Arrowood

Part IV *Cryptosporidium* and Water

12 ***Cryptosporidium* Oocysts in Drinking Water and Recreational Water** 489
Paul A. Rochelle and George D. Di Giovanni

13 **Removal and Inactivation of *Cryptosporidium* from Water** 515
Paul Monis, Brendon King, and Alexandra Keegan

Index 553

Contributors

Michael J. Arrowood Centers for Disease Control and Prevention, Division of Foodborne, National Center for Emerging and Zoonotic Infectious Diseases, Waterborne and Environmental Diseases, Atlanta, USA

Charlotte Axén Department of Animal Health and Antimicrobial Strategies, National Veterinary Institute, Uppsala, Sweden

Camilla Björkman Division of Ruminant Medicine and Veterinary Epidemiology, Department of Clinical Sciences, Swedish University of Agricultural Sciences, Uppsala, Sweden

Simone M. Cacciò Department of Infectious, Parasitic and Immunomediated Diseases, Istituto Superiore di Sanità, Rome, Italy

Rachel M. Chalmers Cryptosporidium Reference Unit, Public Health Wales Microbiology, Singleton Hospital, Swansea, United Kingdom

Xian-Ming Chen Department of Medical Microbiology and Immunology, Creighton University School of Medicine, Omaha, USA

Mark Clark Department of Biological Sciences, North Dakota State University Department 2715, Fargo, USA

Ronald Fayer Agricultural Research Service, United States Department of Agriculture, Environmental Microbial and Food Safety Laboratory, Beltsville, USA

Yaoyu Feng School of Resources and Environmental Engineering, East China University of Science and Technology, Shanghai, China

Robin B. Gasser The University of Melbourne, Parkville, Australia

George D. Di Giovanni University of Texas School of Public Health, El Paso, USA

Fengguang Guo Department of Veterinary Pathobiology, College of Veterinary Medicine & Biomedical Sciences, Texas A&M University, College Station, USA

Stephen J Hadfield Cryptosporidium Reference Unit, Public Health Wales Microbiology, Singleton Hospital, Swansea, United Kingdom

Guoku Hu Department of Medical Microbiology and Immunology, Creighton University School of Medicine, Omaha, USA

Aaron R. Jex The University of Melbourne, Werribee, Australia

Alexandra Keegan Australian Water Quality Centre, South Australian Water Corporation, Adelaide, Australia

Brendon King Australian Water Quality Centre, South Australian Water Corporation, Adelaide, Australia

Martin Kváč Institute of Parasitology, Biology Centre of the Academy of Sciences of the Czech Republic, České Budějovice, Czech Republic

University of South Bohemia in České Budějovice, České Budějovice, Czech Republic

John McEvoy Department of Veterinary and Microbiological Sciences, North Dakota State University Department 7690, Fargo, USA

Jan R. Mead Atlanta Veterans Medical Center and Department of Pediatrics, Emory University, Atlanta, Decatur, USA

Paul Monis Australian Water Quality Centre, South Australian Water Corporation, Adelaide, Australia

Gordon Nichols European Centre for Disease Prevention and Control, Solna, Stockholm, Sweden

Steven P. O'Hara Division of Gastroenterology and Hepatology, Department of Internal Medicine, Mayo Clinic College of Medicine, Rochester, USA

Lorenza Putignani Unit of Parasitology and Unit of Microbiology, Bambino Gesù Children's Hospital, IRCCS, Rome, Italy

Nadine P. Randle Department of Infection Biology, Institute of Infection and Global Health, University of Liverpool, Liverpool, UK

Lucy J. Robertson Parasitology Laboratory, Section for Microbiology, Immunology and Parasitology, Department of Food Safety and Infection Biology, Norwegian School of Veterinary Science, Oslo, Norway

Paul A. Rochelle Metropolitan Water District of Southern California, Water Quality Laboratory, La Verne, USA

Una Ryan School of Veterinary and Life Sciences, Murdoch University, Murdoch, Australia

Henry Shikani Department of Pathology, Albert Einstein College of Medicine, Bronx, USA

Brianna Stenger Department of Veterinary and Microbiological Sciences, North Dakota State University Department 7690, Fargo, USA

Department of Biological Sciences, North Dakota State University Department 2715, Fargo, USA

Jonathan M. Wastling Department of Infection Biology, Institute of Infection and Global Health, University of Liverpool, Liverpool, United Kingdom

Louis M. Weiss Department of Pathology, Division of Parasitology, Department of Medicine, Division of Infectious Diseases, Albert Einstein College of Medicine, Bronx, USA

Lihua Xiao Division of Foodborne, Centers for Disease Control and Prevention, National Center for Emerging and Zoonotic Infectious Diseases, Waterborne and Environmental Diseases, Atlanta, USA

Guan Zhu Department of Veterinary Pathobiology, College of Veterinary Medicine & Biomedical Sciences, Texas A&M University, College Station, USA

Part I
Taxonomy and Epidemiology of Cryptosporidium

Chapter 1
Taxonomy and Molecular Taxonomy

Una Ryan and Lihua Xiao

Abstract *Cryptosporidium* parasites belong to the phylum Apicomplexa and possess features of both the coccidia and gregarines. Currently, 25 species of *Cryptosporidium* are recognized in fish, amphibians, reptiles, birds and mammals. All 25 species have been confirmed by morphological, biological, and molecular data. *Cryptosporidium duismarci* and *C. scophthalmi* lack sufficient biological and/or molecular data to be considered valid species. In addition to the named species, more than 40 genotypes from various vertebrate hosts have been described. For these genotypes to receive taxonomic status, sufficient morphological, biological, and molecular data are required and names must comply with the rules of the International Code for Zoological Nomenclature (ICZN). A different interpretation of the ICZN led to the proposal that *Cryptosporidium parvum* be renamed as *Cryptosporidium pestis* and that *C. parvum* be retained for *C. tyzzeri*. However, this proposal violates the guiding ICZN principle of maintaining taxonomic stability and avoiding confusion. In addition, *C. pestis* lacks a full taxonomic description and therefore is not a valid species. The taxonomic status of *Cryptosporidium* spp. is rapidly evolving and many genotypes are likely to be formally described as species in the future.

U. Ryan (✉)
School of Veterinary and Life Sciences, Murdoch University, Murdoch, WA 6150, Australia
e-mail: una.ryan@murdoch.edu.au

L. Xiao
Division of Foodborne, Waterborne and Environmental Diseases, Centers for Disease Control and Prevention, National Center for Emerging and Zoonotic Infectious Diseases, 1600 Clifton Road, Mailstop D66, Atlanta, GA 30329, USA
e-mail: lax0@cdc.gov

S.M. Cacciò and G. Widmer (eds.), *Cryptosporidium: parasite and disease*,
DOI 10.1007/978-3-7091-1562-6_1, © Springer-Verlag Wien 2014

1.1 Introduction

The genus *Cryptosporidium* is composed of protozoan parasites that infect epithelial cells in the microvillus border of the gastrointestinal tract of all classes of vertebrates. The parasite causes self-limiting diarrhoea in immunocompetent individuals but the infection may be chronic and life-threatening to those that are immunocompromised (Hunter et al. 2007). The incubation period for illness symptoms is approximately 7 days (range 1–14 days) and illness is usually self-limiting, with a mean duration of 6–9 days (Hunter et al. 2007), although longer times (mean duration 19–22 days, maximum up to 100–120 days) were reported in Australian cases (Robertson et al. 2002). Relapses are common; reports indicate 1–5 additional episodes in 40–70 % of patients (Hunter et al. 2004). The predominant symptom is watery diarrhoea, sometimes profuse (1–2 L/day in a small minority of cases, usually very young or old), sometimes mucous but rarely bloody.

The oocyst, the environmental stage of *Cryptosporidium*, is ubiquitous and therefore cryptosporidiosis can be acquired through a variety of transmission routes including direct contact with infected persons (person-to-person transmission), contact with animals (zoonotic transmission) and ingestion of contaminated food (foodborne transmission) or water (waterborne transmission) (Xiao 2010) (Chap. 2).

The advent of molecular tools for characterization and phylogenetic analysis of *Cryptosporidium* has revolutionized our understanding of *Cryptosporidium* taxonomy, which in turn underlies our ability to understand the biology, epidemiology and health related importance of various *Cryptosporidium* species (Xiao et al. 1999a, 2004a; Fayer 2010).

1.2 Cryptosporidium's Taxonomic Relationship to Gregarine Parasites

The genus *Cryptosporidium* belongs to the phylum Apicomplexa, which comprises parasitic eukaryotes possessing an apical complex at some stage in their life cycle. Within the Apicomplexa, *Cryptosporidium* species have been traditionally considered to be intestinal coccidian parasites, based on the possession of life cycle features and morphological characteristics similar to other enteric coccidian parasites (Levine 1988). Despite similarities, *Cryptosporidium* demonstrates several peculiarities that separate it from any other coccidian. These include (1) the location of *Cryptosporidium* within the host cell, where the endogenous developmental stages are confined to the apical surfaces of the host cell (intracellular, but extracytoplasmic); (2) the attachment of the parasite to the host cell, where a multi-membranous attachment or feeder organelle is formed at the base of the parasitiphorous vacuole to facilitate the uptake of nutrients from the host cell; (3) the presence of two morpho-functional types of oocysts, thick-walled and thin-walled, with the latter responsible for the initiation of the auto-infective cycle in the infected host; (4) the small size of the oocyst (4.9 × 4.4 μm for *C. parvum*) which lacks morphological structures such as sporocyst, micropyle and polar granules (Tzipori and Widmer 2000; Petry 2004); (5) the insensitivity to all anti-coccidial agents tested so far (Blagburn and Soave 1997; Cabada and White 2010); and (6) the presence of a novel gamont-like extracellular stage similar to those found in gregarine life cycles (Hijjawi et al. 2002; Rosales et al. 2005).

Gregarines are a diverse group of apicomplexan parasites that consist of large single-celled parasites that inhabit the intestines and other extracellular spaces of invertebrates and lower vertebrates, which are abundant in natural water sources (Barta and Thompson 2006; Levine 1988). Molecular phylogenies suggest that *Cryptosporidium* is evolutionarily divergent from other coccidia but related to gregarines and is an early branch at the base of the apicomplexan phylum (Zhu et al. 2000; Barta and Thompson 2006). Genomic and biochemical data also indicates that *Cryptosporidium* differs from other apicomplexans in that it has lost the apicoplast organelle (Zhu et al. 2000; Abrahamsen et al. 2004). In addition, *Cryptosporidium* appears to have lost many *de novo* biosynthetic pathways, such as the capacity to synthesize amino acids and nucleotides. A recent whole-genome-sequence survey for the gregarine, *Ascogregarina taiwanensis* supports the phylogenetic affinity of *Ascogregarina* with *Cryptosporidium* at the base of the apicomplexan clade (Templeton et al. 2010). More recently, genome sequencing of *C. muris* has identified the presence of mitochondrial structure and proteins that are absent from *C. parvum* and *C. hominis* but present in gregarines (Widmer and Sullivan 2012). *Ascogregarina* and *Cryptosporidium* however also possess features that unite them with the Coccidia, including an environmental oocyst stage, metabolic pathways such as the Type I fatty acid and polyketide synthetic enzymes, and a number of conserved extracellular-protein-domain architectures (Templeton et al. 2010). Future genome studies of other gregarine parasites will hopefully provide a clearer understanding of the correct taxonomic placement of the genus *Cryptosporidium*.

1.3 Standards for Taxonomic Status in the Genus *Cryptosporidium*

Traditionally, species delimitation within the Apicomplexa has been based on combinations of morphological features detected by light microscopy or electron microscopy (EM), unique life cycles and host specificity. Because oocysts of *Cryptosporidium* are among the smallest exogenous stages of any apicomplexan and lack distinctive morphological features to clearly differentiate *Cryptosporidium* species, morphology alone cannot be used to differentiate *Cryptosporidium* species (Fall et al. 2003).

To provide clarity, the following minimum requirements have been proposed for the naming of new species of *Cryptosporidium*: (i) morphometric studies of oocysts; (ii) genetic characterization with sequence information deposited in GenBank; (iii) demonstration of natural and, whenever feasible, at least some experimental host specificity; and (iv) compliance with International Code of Zoological Nomenclature (ICZN) (Egyed et al. 2003; Xiao et al. 2004a; Jirků et al. 2008).

1.4 Valid Species Within the Genus *Cryptosporidium*

To date a total of 25 species of *Cryptosporidium* have been formally described and are considered valid, including the recently described *C. viatorum* in humans and *C. scrofarum* (previously pig genotype II) in pigs (Tables 1.1, 1.2, 1.3, and 1.4 and Fig. 1.1). There are also over 40 genotypes, with a high probability that many of these will eventually be given species status with increased biological and molecular characterisation. Understanding the transmission dynamics of *Cryptosporidium* has traditionally been difficult because most species of *Cryptosporidium* are morphologically identical. Therefore, alternative molecular characterization tools such as PCR and DNA sequence analysis have been required to reliably differentiate/identify species and genotypes of *Cryptosporidium*. The 18S ribosomal RNA (rRNA) gene and the hypervariable 60-kDa glycoprotein (gp60) gene have been widely used as targets to identify species and track transmission (Xiao 2010; Plutzer and Karanis 2009; Ng et al. 2011) (Chap. 3).

1.5 Species in Marine Mammals and Fish

Very little is known about the prevalence and genetic diversity of species of *Cryptosporidium* in marine environments and the role that marine animals play in transmission of these parasites to humans. Molecular research on marine mammals has identified *C. muris* and seal genotypes 1 and 2, in ringed seals in Canada (Santín et al. 2005). Seal genotype 3 was identified in a harp seal (*Pagophilus groenlandicus*) from the Gulf of Maine (Bass et al. 2012) and a genotype similar to the *Cryptosporidium* sp. skunk genotype was identified in an Antarctic Southern elephant seal (*Mirounga leonina*) (Rengifo-Herrera et al. 2011). More recently, seal genotype 4 was identified in Weddell seals (*Leptonychotes weddellii*) (Rengifo-Herrera et al. 2013) (Table 1.1).

Cryptosporidium has been described in both fresh and marine water piscine species with parasitic stages located either on the stomach or intestinal surface, or at both sites (Table 1.1). The first account of *Cryptosporidium* in a piscine host was *Cryptosporidium nasorum*, identified in a Naso tang, a tropical fish species (Hoover et al. 1981). Hoover and colleagues also noted a similar infection in an unnamed

Table 1.1 *Cryptosporidium* sp. reported in marine mammals and fish

Species	Host	Site of Infection	Size (μm) L × W	Reference	GenBank accession number (18SrRNA)
Cryptosporidium sp.seal genotype 1 and 2, *C. muris*	Ringedseals (*Phoca hispida*), harbor seals (*Phoca vituline*), Hooded seal (*Cystophora cristata*)			Santín et al. 2005; Bass et al. 2012	AY731234 (AY731236-hsp70), AY731235 (AY731238-*hsp*70),
Cryptosporidium sp.seal genotype 3	Harp seal (*Pagophilus groenlandicus*)	–	–	Bass et al. 2012	JN858909 (JN860884-*hsp*70)
Cryptosporidium sp. (similar to skunk genotype)	Southern elephant seal (*Mirounga leonina*)	–	–	Rengifo-Herrera et al. 2011, 2013	GQ421425 (GQ421426-*hsp*70)
Cryptosporidium sp. seal genotype 4	Weddell seals (*Leptonychotes weddellii*)			Rengifo-Herrera et al. 2013	JQ740103 (JQ740105-*hsp*70)
C. nasorum	Naso Tang(*Naso literatus*)	Intestine	3.6 × 3.6	Hoover et al. 1981	–
Cryptosporidium sp.	Carp (*Cyprinus carpio*)	Intestine	–	Pavlásek 1983	–
Cryptosporidium sp.	Cichlid (*Oreochromis* sp.)	Stomach	–	Landsberg and Paperna 1986	–
Cryptosporidium sp.	Brown trout (*Salmo trutta*)	Intestine	–	Rush et al. 1987	–
Cryptosporidium sp.	Barramundi (*Lates calcarifer*)	Intestine	–	Glazebrook and Campbell 1987	–

(continued)

Table 1.1 (continued)

Species	Host	Site of Infection	Size (μm) L × W	Reference	GenBank accession number (18SrRNA)
Cryptosporidium sp.	Rainbow trout (*Oncorhynchus mykiss*)	Stomach	5–7	Freire-Santos et al. 1998	–
Cryptosporidium sp.	Red drum(*Sciaenops ocellatus*)	Stomach	7 × 4	Camus and Lopez 1996	–
Cryptosporidium sp.	Pleco (*Plecostomus* sp.)	Intestine and Stomach	–	Muench and White 1997	–
Piscicryptosporidium reinchenbachklinkei	Gourami (*Trichogaster leeri*)	Stomach	2.4–3.18 × 2.4–3.0	Paperna and Vilenkin 1996	–
Piscicryptosporidium-cichlidis (previously *Cryptocystidium villithecum*)	Cichlid (*Oreochromis* sp.)	Stomach	4.0–4.7 × 2.5–3.5	Paperna and Vilenkin 1996	–
Piscicryptosporidium sp. (previously *Chloromyxum-like*)	Gilthead sea bream (*Sparus aurata*)	Stomach	–	Paperna and Vilenkin 1996	–
C. molnari	Gilthead sea bream (*Sparus aurata*), European sea bass (*Dicentrarchus labrax*), Murray cod (*Maccullochella peelii peelii*)	Stomach (and intestine)	4.72 (3.23–5.45) × 4.47 (3.02–5.04)	Alvarez-Pellitero and Sitjà-Bobadilla 2002; Palenzuela et al. 2010; Barugahare et al. 2011	HM243548, HM243550, HQ585890
Cryptosporidium sp. (*C. scophthalmi*)	Turbot (*Scophthalmus maximus*)	Intestine	4.44 (3.7–5.03) × 3.91 (3.03–4.69)	Alvarez-Pellitero et al. 2004	–

Piscine genotype 1	Guppy (*Poecilia reticulata*)	Stomach	4.6 × 4.4	Ryan et al. 2004a	AY524773
Cryptosporidium sp.	Alewife (*Alosa pseudoharengus*)	–	–	Ziegler et al. 2007	–
Piscine genotype 2	Angelfish (*Pterophyllum scalare*)	Stomach	3.4 × 4.1	Murphy et al. 2009	
Piscine genotype 3	Mullet (*Mugil cephalus*)	Intestine	–	Reid et al. 2010	FJ769050
Piscine genotype 4	Golden algae eater (*Crossocheilus aymonieri*), Kupang damsel (*Chrysiptera hemicyanes*) Oscar fish (*Astronatus ocellatis*), Neon tetra (*Paracheirodon innesi*)	Intestine	–	Reid et al. 2010; Morine et al. 2012	HM989833
Piscine genotype 5	*Angelfish* (*Pterophyllum scalare*), Butter bream (*Monodactylidae*), Golden algae eater (*Crossocheilus aymonieri*)	–	–	Zanguee et al. 2010	HM989834
Piscine genotype 6, piscine genotype 6-like	Guppy (*Poecilia reticulata*), Gourami (*Trichogaster trichopterus*)	–	–	Zanguee et al. 2010; Morine et al. 2012	HM991857, JQ995776

(continued)

Table 1.1 (continued)

Species	Host	Site of Infection	Size (μm) L × W	Reference	GenBank accession number (18SrRNA)
C. parvum, *C. parvum-like* *C. xiaoi* and *C. scrofarum*	Whiting (*Sillago vittata*), Barramundi (*Lates calcarifer*)			Reid et al. 2010; Gibson-Kueh et al. 2011	–
Cryptosporidium sp.	Barramundi (*Late scalcarifer*)	Distal stomach and proximal small intestine		Gabor et al. 2011	–
Piscine genotype 7	Red eye tetra (*Moenkhausia sanctaefilomenae*)	–	–	Morine et al. 2012	JQ995773
Piscine genotype 8	Oblong silver biddy (*Gerres oblongus*)	–	–	Koinari et al. 2013	KC807985

Table 1.2 Avian species and genotypes of *Cryptosporidium* and hosts confirmed using molecular analysis

Species/ genotype	Avian host species	Site of infection	Size (μm) L × W	References	GenBank accession no. 18S
C. meleagridis	Turkey (*Meleagris gallopavo*) (type host), Indian ring-necked parrot (*Psittacula krameri*), Red-legged partridge (*Alectoris rufa*), Cockatiels (*Nymphicus hollandicus*); Bohemian waxwing (*Bombycilla garrulus*), Rufousturle dove (*Streptopelia orientalis*), and fan-tailed pigeon (*Columba livia*), Chicken (*Gallus gallus*); Quails (Coturnixcoturnix japonica), Pekin ducks (*Anas platyrhynchos*)	Intestine	4.5–6.0 × 4.2–5.3	Slavin 1955; Lindsay et al. 1989; Morgan et al. 2000a, 2001; Glaberman et al. 2001; Abe and Iseki 2004; Pagès-Manté et al. 2007; Abe and Makino 2010; Wang et al. 2010; Qi et al. 2011; Berrilli et al. 2012; Wang et al. 2012	AF112574
C. baileyi	Chicken (*Gallus gallus*)(type host); Brown quail (*Synoicus australis*) Cockatiels (*Nymphicus hollandicus*), Whooping crane (*Grus vipio*), Gray-bellied bulbul (*Pycnonotus* spp.), Black vulture (*Coragyps atratus*), Saffron finch (*Sicalis flaveola*), mixed-bred falcons (*Falco rusticolus* x *Falco cherrug*) Ruddy Shelduck (*Tadornaferruginea*), Red-billed leiothrixes (*Leiothrix lutea*), Pekin ducks	Cloaca, bursa, trachea	4.8 (4.8–5.7) × 6.4 (5.6–7.5)	Current et al. 1986; Morgan et al. 2001; Abe and Iseki 2004; Ng et al. 2006; Huber et al. 2007; Kimura et al. 2004; van Zeeland et al. 2008; Nakamura et al. 2009; Abe and Makino 2010; Amer et al. 2010; Wang et al. 2010; Qi et al. 2011; Sevá et al. 2011a; Wang et al. 2011; Schulze et al. 2012; Wang et al. 2012	L19068

(continued)

Table 1.2 (continued)

Species/ genotype	Avian host species	Site of infection	Size (μm) L × W	References	GenBank accession no. 18S
	(*Anas platyrhynchos*), Buffy-fronted seedeater (*Sporophila frontalis*), Java sparrows (*Padda oryzivora*), Mynas (*Acridotheres tristis*), Zebra finches (*Taeniopygia guttata*), Crested Lark (*Galerida cristata*), Gouldian finch (*Chloebia gouldiae*), Black-billed magpie (*Pica pica*), Ostriches (*Struthio camelus*), Red-breasted Merganser (*Mergus serrator*), Quails (*Coturnix coturnix japonica*)				
C. galli	Chicken (*Gallus gallus*)(type host), Finches (*Spermestidae* and *Fringillidae*), Capercaille (*Tetrao urogallus*), Pine grosbeak (*Pinicola enucleator*), Turqoise parrots (*Neophema pulchella*), Cuban flamingo (*Phoenicopterus ruber ruber*), Rhinocerous hornbill (*Buceros rhinoceros*), Red-cowled cardinal (*Paroaria dominicana*), Zebra finches (*Taeniopygia guttata*), Chocolate parson finches (*Peophila cincta*), Chestnut finches (*Lonchura castaneothorax*), Painted	Preventriculus	8.25 (8.0–8.5) × 6.3 (6.2–6.4)	Pavlásek 1999, 2001; Ryan et al. 2003a; Ng et al. 2006; Antunes et al. 2008; Nakamura et al. 2009; da Silva et al. 2010; Qi et al. 2011; Sevá et al. 2011a	AF316624, AY168847

	firetail finches (*Emblema picta*), Canaries (*Serinus* sp.), Glosters (*Serinus canaria*), Hazel hen (*Tetrastes bonasia rupestris*), Lesser seed-finches (*Oryzoborus angolensis*), Rufousbellied thrush (*Turdus rufiventris*), Green-winged saltators (*Saltator similis*), Slate-coloured seedeater (*Sporophila schistacea*), Gold-finch (*Carduelis carduelis*), Great-billed seed-finch (*Oryzoborus maximiliani*), Ultramarine grosbeak (*Cyanocompsa brissonii*), Rusty-collared seedeater (*Sporophila collaris*), Bohe-mian waxwings (*Bombycilla garrulus*), silver-eared Mesia (*Leiothrix argentauris*)				
Avian Genotype I	Red Factor Canary (*Serinus canaria*), Canary (*S. canaria*). Indian peafowl (*Pavo cristatus*)	–		Ng et al. 2006; Nakamura et al. 2009	DQ650339
Avian Genotype II	Eclectus (*Eclectus roratus*), Galah (*Eolophus roseicapilla*), Cock-atiel (*Nymphicus hollandicus*), Major Mitchell Cockatoo (*Cacatua leadbeateri*, *Ostriches* (*Struthio camelus*), White-eyed parakeet (*Aratinga leucophthalma*)		6.0–6.5 × 4.8–6.6	Meireles et al. 2006; Ng et al. 2006; Nakamura et al. 2009; Sevá et al. 2011a; Nguyen et al. 2013	DQ650340

(continued)

Table 1.2 (continued)

Species/ genotype	Avian host species	Site of infection	Size (μm) L × W	References	GenBank accession no. 18S
Avian Genotype III	Galah (*Eolophus roseicapilla*), Cockatiel (*Nymphicus hollandicus*), Sun Conure (*Aratinga solstitialis*), Peach-Faced Lovebirds (*Agapornis roseicollis*), Java sparrow (*Padda oryzivora*), cockatiel (*Nymphicus hollandicus*)	Proventriculus	7.0 × 6.0	Ng et al. 2006; Nakamura et al. 2009; Makino et al. 2010; Gomes et al. 2012	DQ650342
			7.4 × 5.8	Ng et al. 2006	DQ650344
			8.25 × 6.3		
Avian Genotype IV	Japaneese whiteeye (*Zosterops japonica*)				
Avian Genotype V	Cockatiels (*Nymphicus hollandicus*)	Intestine	5.8 × 4.5	Abe and Makino 2010; Qi et al. 2011	HM116381
Eurasian wood-cock Genotype	Eurasian woodcock (*Scolopax rusticola*)		8.5 × 6.4	Ryan et al. 2003b; Ng et al. 2006	AY273769
				Morgan et al. 2001, Zhou et al. 2004; Jellison et al. 2004	AF316630
Duck Genotype	Black duck (*Anas rubripes*), Canada Geese (*Branta canadensis*)				
Goose Genotype I	Canada Geese (*Branta canadensis*)			Xiao et al. 2002a; Zhou et al. 2004; Jellison et al. 2004	AY120912
Goose Genotype II	Canada Geese (*Branta canadensis*)			Zhou et al. 2004, Jellison et al. 2004	AY504512
Goose Genotype III	Canada Geese (*Branta canadensis*)			Jellison et al. 2004	AY324638
Goose Genotype IV	Canada Geese (*Branta canadensis*) - isolate KLJ-3b				
Goose Genotype V	Canada Geese (*Branta canadensis*)- isolate KLJ-7			Jellison et al. 2004	AY324641

Table 1.3 Amphibian and reptile *Cryptosporidium* species and genotypes and their hosts confirmed using molecular analyses

Species/ genotype	Amphibian/reptile host species	Site of infection	Size (μm) L × W	Reference	GenBank accession number (18SrRNA)
C. fragile	Black-spined toads (*Duttaphrynus melanostictus*)	Stomach	6.2 (5.5–7.0) × 5.5 (5.0–6.5)	Jirků et al. 2008	EU162751–EU162754
C. serpentis	Amazon tree boa (*Corallus hortulanus*), Black rat snake (*Elaphe obsoleta obsolete*), Bornmueller's viper (*Vipera bornmuelleri*), Bull snake (*Pituophis melanoleucus melanoleucus*), California kingsnake (*Lampropeltis getulus californiae*), Cornsnake (*Elaphe guttata guttata*), Common death adder (*Acanthophis antarticus*), Desert monitor (*Varanus griseus*), Eastern/Mainland Tiger snake (*Notechis scutatus*), Frilled lizard (*Chlamydosaurus kingui*), Giant madagascar or Oustalet's chameleon	Stomach	6.2 (5.6–6.6) × 5.3 (4.8–5.6)	Levine 1980; Kimbell et al. 1999; Morgan et al. 1999a; Xiao et al. 2004b; Hajdusek et al. 2004; Pedraza-Díaz et al. 2009; Richter et al. 2011; Sevá et al. 2011b; Rinaldi et al. 2012	AF151376

(continued)

Table 1.3 (continued)

Species/ genotype	Amphibian/reptile host species	Site of infection	Size (μm) L × W	Reference	GenBank accession number (18SrRNA)
	(*Chamaeleo oustaleti*), Leopard gecko (*Eublepharis macularius*), Mexican black kingsnake (*Lampropeltis getulus nigritus*), Milk snake (*Lampropeltis triangulum*), *Mountain viper* (*Vipera wagneri*), *Python* (*Python molurus*), Savannah monitor (*Varanus exanthematicus*), Skink (*Mabuya perrotetii*), Taipan (*Oxyuranus scutellatus*)				
C. varanii	African fat-tailed gecko (*Hemitheconyx caudicinctus*) *Leopard gecko* (*Eublepharis macularius*), Boa constrictor, Cornsnake (*Elaphe guttata guttata*), Leopard gecko (*Eublepharis macularius*), Desert monitor (*Varanus griseus*), Gecko	Intestine and Cloaca	4.7 (4.2–5.2) × 5 (4.4–5.6)	Pavlásek et al. 1995; Koudela and Modry 1998; Morgan et al. 1999a; Xiao et al. 2004b; Hajdusek et al. 2004; Plutzer and Karanis 2007; Pedraza-Díaz et al. 2009; Richter et al. 2011	AF112573

	(*Gekkoninae* sp.), Green iguana (*Iguana iguana*), *Lampropeltis* sp.; Louisiana pine snake (*Pituophis ruthveni*), Plated lizard (*Gerrhosaurus* sp.), Schneider's Skink (*Eumeces schneideri*), *Taipan* (*Oxyuranus scutellatus*)				
Lizard genotype/ C. *serpentis*-like	Leopard gecko (*Eublepharis macularius*), Corn snake (*Pantherophis guttatus*)			Xiao et al. 2004b; Richter et al. 2011	AY120915
Tortoise genotype I	Indian star tortoises (*Geochelone elegans*), Herman's tortoise (*Testudo hermanii*), Ball python (*Python regius*), Russian tortoise (*Agrionemys* [*Testudo*] *horsfieldii*)	Stomach		Xiao et al. 2002a, b; Alves et al. 2005; Pedraza-Díaz et al. 2009; Griffin et al. 2010; Richter et al. 2012	AY120914
Tortoise genotype II (*C. duismarci*)	Marginated tortoise (*Testudo marginata*), Ball python (*Python regius*) Veiled chameleon (*Chamaeleo calyptratus*), Pancake tortoise (*Malacochersus*	Intestine		Traversa et al. 2008; Pedraza-Díaz et al. 2009; Griffin et al. 2010; Traversa 2010; Richter et al. 2012	GQ504270

(continued)

Table 1.3 (continued)

Species/genotype	Amphibian/reptile host species	Site of infection	Size (μm) L × W	Reference	GenBank accession number (18SrRNA)
	tornieri), Russian tortoise (*Agrionemys* [*Testudo*] *horsfieldii*)				
Snake genotype I	New Guinea Viper boa (*Candoia asper*), Japanese grass snakes (*Rhabdophis tigris*)			Xiao et al. 2002a; Kuroki et al. 2008	AB222185
Snake genotype II	Boa constrictor (*Boa constrictor ortoni*)			Xiao et al. 2004b	AY268584

Table 1.4 Mammalian *Cryptosporidium* species

Species	Major host(s)	Size of oocysts (in mm)	Site of infection	References	GenBank accession number (18SrRNA)
C. muris	Rodents	6.1 (5.6–6.4) × 8.4 (8.0–9.0), 8.1 (7.0–9.0) × 5.9 (5.0–6.5)	Stomach	Tyzzer 1907	AB089284
C. parvum	Cattle/sheep/ humans	4.9 (4.5–5.4) × 4.4 (4.2–5.2)	Small intestine	Tyzzer 1912	AF308600
C. wrairi	Guinea pigs	4.6 (4.0–5.0) × 5.4 (4.8–5.6)	Small intestine	Vetterling et al. 1971	AF115378
C. felis	Cats	4.5 (5.0–4.5) × 5.0 (6.0–5.0)	Small intestine	Iseki 1979	AF108862
C. andersoni	Cattle	7.4 (6.0–8.1) × 5.5 (5.0–6.5)	Abomasum	Lindsay et al. 2000	AF093496
C. canis	Dogs	4.95 × 4.75	Small intestine	Fayer et al. 2001	AF112576
C. hominis	Humans	4.8 (64.4–5.4) × 5.2 (4.4–5.9)	Small intestine	Morgan-Ryan et al. 2002	AF108865
C. suis	Pigs	4.6 (4.9–4.4) × 4.2 (4.0–4.3)	Small and large intestine	Ryan et al. 2004b	AF115377
C. bovis	Cattle	4.89 (4.76–5.35) × 4.63 (4.17–4.76)	Small intestine	Fayer et al. 2005	AY741305
C. fayeri	Marsupials	4.9 (4.5–5.1) × 4.3 (3.8–5.0)	Small intestine	Ryan et al. 2008	AF159112
C. macropodum	Marsupials	4.9 (4.5–6.0) × 5.4 (5.0–6.0)	Small intestine	Power and Ryan 2008	AF513227
C. ryanae	Cattle	3.16 (2.94–4.41) × 3.73 (2.94–3.68)	Small Intestine	Fayer et al. 2008	AY587166
C. xiaoi	Sheep	3.94 (2.94–4.41) × 3.44 (2.94–4.41) μm	Small intestine	Fayer and Santín 2009	EU408314
C. ubiquitum	Sheep/wildlife	5.19 (4.92–5.61) × 4.87 (4.45–5.44)	Small intestine	Fayer et al. 2010	AF262328
C. cuniculus	Rabbits	5.98 (5.55–6.40) × 5.38 (5.02–5.92)	Small intestine	Robinson et al. 2010; (Inman and Takeuchi 1979)	FJ262725
C. tyzzeri	Mice	4.64±0.05 μm × 4.19±0.06	Small intestine	Ren et al. 2012	AF112571
C. viatorum	Humans	5.35 (4.87–5.87) × 4.72 (4.15–5.20)	Small intestine	Elwin et al. 2012b	HM485434
C. scrofarum	Pigs	5.16 (4.81–5.96) × 4.83 (4.23–5.29)	Small Intestine	Kváč et al. 2013	EU331243

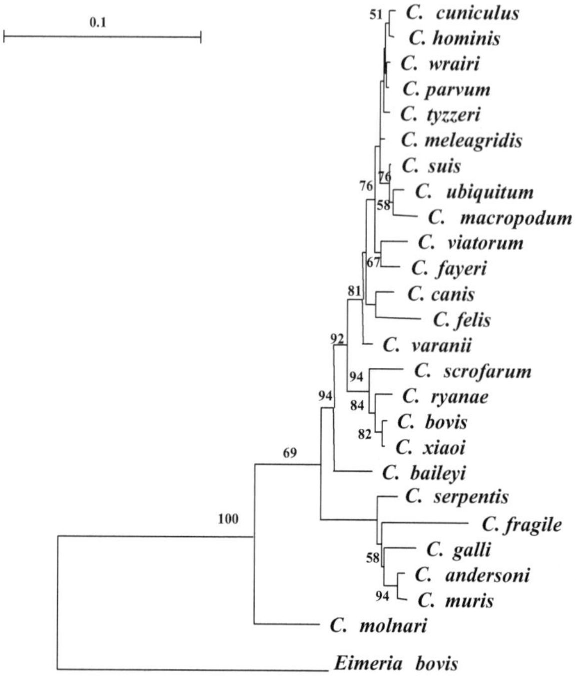

Fig. 1.1 Phylogenetic relationships of currently accepted *Cryptosporidium* species at the 18S ribosomal RNA locus inferred by neighbour-joining analysis based on genetic distances calculated using Kimura-2 parameters. Bootstrap values (>60 %) from 1,000 pseudoreplicates are shown

species of marine fish (Hoover et al. 1981). Levine (1984) suggested that *C. nasorum* should be given the status of species. However, lack of viable oocyst measurements and taxonomically valid diagnostic features and the fact that only developmental stages on the intestinal microvillous surface were described, has resulted in *C. nasorum* being considered a nomen nudem (i.e., a name that is invalid because an insufficient description was published) (Ryan et al. 2004a; Xiao et al. 2004a).

Cryptosporidium molnari was described in gilthead sea bream (*Sparus aurata*) and European sea bass (*Dicentrarchus labrax*) in 2002 (Alvarez-Pellitero and Sitjà-Bobadilla 2002) (Table 1.1). *Cryptosporidium molnari* primarily infects the epithelium of the stomach and seldom the intestine (Alvarez-Pellitero and Sitjà-Bobadilla 2002). In 2004 *Cryptosporidium scophthalmi* was described in turbot (*Psetta maxima, syn. Scophthalmus maximus*) and mainly infects the epithelium of the intestine and very seldom the stomach (Alvarez-Pellitero et al. 2004). An unusual feature of both parasites is that while merogonial and gamogonial stages appeared in the typical extracytoplasmic position, oogonial and sporogonial stages were located deeply within the epithelium (Alvarez-Pellitero and Sitjà-Bobadilla 2002; Alvarez-Pellitero et al. 2004). *Cryptosporidium molnari* has recently been characterised genetically (Palenzuela et al. 2010), but no DNA sequences are available for *C. scophthalmi*. Until genetic sequences are provided for *C. scophthalmi*, it cannot be considered a valid species because of the high genetic heterogeneity and morphological similarity among *Cryptosporidium* species and genotypes in fish.

A total of 13 additional species/genotypes have been identified in fish using molecular tools; piscine genotype 1 from a guppy (*Poecilia reticulate*) (Ryan et al. 2004a); piscine genotype 2 from a freshwater angelfish (*Pterophyllum scalare*) (Murphy et al. 2009), piscine genotype 3 from a sea mullet (*Mugil cephalus*) (Reid et al. 2010); piscine genotype 4 from a golden algae eater (*Crossocheilus aymonieri*), a kupang damsel (*Chrysiptera hemicyanes*) and an oscar fish (*Astronatus ocellatis*); piscine genotype 5 from an angelfish (*Pterophyllum scalare*), a butter bream (*Monodactylidae*) and a golden algae eater (*Crossocheilus aymonieri*); piscine genotype 6 from a guppy (*Poecilia reticulate*) (Zanguee et al. 2010); piscine genotype 7 from red eye tetra (*Moenkhausia sanctaefilomenae*) (Morine et al. 2012), piscine genotype 8 from Oblong silver biddy (*Gerres oblongus*) (Koinari et al. 2013); *C. parvum*, *C. xiaoi* and *C. scrofarum* in whiting (*Sillago vittata*) (Reid et al. 2010); *C. parvum* in Nile tilapias (*Oreochromis niloticus*) and a Silver barb (*Puntius gonionotus*) and *C. parvum* and *C. hominis* in mackerel scad (*Decapterus macarellus*) from Papua New Guinea (Koinari et al. 2013) and rat genotype III in a goldfish (*Carassius auratus*) (Morine et al. 2012).

The prevalence of *Cryptosporidium* in fish ranges from 0.8 % to up to 100 % mostly among juvenile fish (Landsberg and Paperna 1986; Sitjà-Bobadilla et al. 2005; Alvarez-Pellitero et al. 2004, 2009; Murphy et al. 2009; Reid et al. 2010; Zanguee et al. 2010; Barugahare et al. 2011). The pathogenesis of *Cryptosporidium* in fish has not been extensively studied; however *C. molnari* has been associated with clinical signs in gilthead sea bream, consisting of whitish feces, abdominal swelling and ascites which resulted in mortalities in fingerlings (Alvarez-Pellitero and Sitjà-Bobadilla 2002). Another study, which examined interactions between bacteria and *C. molnari* in gilthead sea bream under farm and laboratory conditions reported that the ability of gilthead sea bream to survive an additional bacterial infection may be compromised where a previous mixed infection of bacteria and *C. molnari* existed (Sitjà-Bobadilla et al. 2006). Piscine genotype 1 was associated with high mortalities amongst guppies and was detected in the stomach, with oogonial and sporogonial stages observed deep within the epithelium, similar to *C. molnari* (Ryan et al. 2004a). Previous studies on piscine genotype 2, which was identified in a hatchery, revealed that infected fish exhibited variable levels of emaciation, poor growth rates, swollen coelomic cavities, anorexia, listlessness and increased mortality (Murphy et al. 2009). In affected fish, large numbers of protozoa were identified both histologically and ultrastructurally associated with the gastric mucosa. A high prevalence of *Cryptosporidium* was associated with high mortalities in immature hatchery barramundi (*Lates calcarifer*) in Australia (Gabor et al. 2011).

In 1996, Paperna and Vilenkin proposed a new genus, designated *Piscicryptosporidium*, for *Cryptosporidium*-like species infecting a number of piscine hosts. The genus included two species, *P. reichenbachklinkei* and *P. cichlidaris* previously described as *Cryptosporidium* sp. from cichlid fishes of the genus

Oreochromis (Paperna and Vilenkin 1996). Several unique features were cited to support the genus including the covering of the surface of the parasitophorous sac by rudimentary microvilli and the localisation of the oocysts deep within the gastric mucosa (Paperna and Vilenkin 1996). However, these apparently differential features have also been described in some mammalian *Cryptosporidium* spp. For example, *C. parvum* has been occasionally found within some cells (Beyer et al. 2000; Marcial and Madara 1986) and microvilli are usually retained in different mammalian species (Alvarez-Pellitero and Sitjà-Bobadilla 2002). Recently, *P. reichenbachklinkei* and *P. cichlidaris* were published as *Cryptosporidium reichenbachklinkei* and *C. cichlidaris* (Jirků et al. 2008), with the comment that the genus was considered a synonym of *Cryptosporidium*.

No molecular data was provided by Paperna and Vilenkin, (1996) to support the genus or species, however more recent characterization of *C. molnari* and piscine genotypes 1–7 indicate that piscine-derived species and genotypes of *Cryptosporidium* are genetically very distinct and primitive to all other species (Ryan et al. 2004a; Palenzuela et al. 2010; Reid et al. 2010; Zanguee et al. 2010; Barugahare et al. 2011; Morine et al. 2012). However, further studies at additional loci are required to validate whether or not *Piscicryptosporidium* is a valid genus for piscine species.

1.6 Species in Birds

Currently only three avian *Cryptosporidium* spp. are recognised; *Cryptosporidium meleagridis*, *Cryptosporidium baileyi* and *Cryptosporidium galli*. Two other species of *Cryptosporidium* have been named from birds: *Cryptosporidium anserinum* from a domestic goose (*Anser domesticus*) (Proctor and Kemp 1974) and *Cryptosporidium tyzzeri* from chickens (*Gallus gallus domesticus*) (Levine 1961). Neither of these reports gave adequate description of oocysts or provided other useful information and are therefore not considered valid species (Lindsay and Blagburn 1990). Naturally occurring cryptosporidiosis in birds manifests itself in three clinical forms; respiratory disease, enteritis and renal disease. Usually only one form of the disease is present in an outbreak (Lindsay and Blagburn 1990).

Tyzzer first described avian cryptosporidiosis in 1929 (Tyzzer 1929) but it was not until 1955 that Slavin found a structurally similar parasite in the ileum of turkey poults and named the parasite *C. meleagridis* (Slavin 1955) (Table 1.2). Since then *C. meleagridis* has been characterized genetically and has been detected in many avian hosts (Table 1.2). It is the third most prevalent species infecting humans (Leoni et al. 2006; Elwin et al. 2012a). In some studies, *C. meleagridis* prevalence is similar to that of *C. parvum* (Gatei et al. 2002a; Cama et al. 2007). The ability of *C. meleagridis* to infect humans and other mammals, and its close relationship to *C. parvum* and *C. hominis* at the 18S rRNA, actin and 70 kDa heat shock protein (HSP70) loci, has led to the suggestion that mammals actually were the original hosts, and that the species has later adapted to birds (Xiao et al. 2002a, 2004a).

Recently, sequence analysis of the 18S gene and HSP70 loci, has been used to provide evidence of zoonotic transmission of *C. meleagridis* from chickens to a human on a Swedish farm (Silverlås et al. 2012).

A second species of avian *Cryptosporidium*, originally isolated from commercial broiler chickens, was named *C. baileyi* based on its life cycle and morphologic features (Current et al. 1986). Natural *C. baileyi* infections have been reported in many anatomic sites in avian hosts including the conjunctiva, nasopharynx, trachea, bronchi, air sacs, small intestine, large intestine, ceca, cloaca, bursa of fabricius, kidneys and urinary tract, but is more frequently associated with respiratory cryptosporidiosis (Lindsay and Blagburn 1990). *Cryptosporidium baileyi* is probably the most common avian *Cryptosporidium* species and has been reported in a wide range of avian hosts (Table 1.2). The relationship between *C. baileyi* and the rest of the intestinal and gastric parasites is unclear as phylogenetic analysis at the SSU rRNA, actin and COWP loci group *C. baileyi* with the intestinal parasites but analysis at the HSP70 locus placed *C. baileyi* in a cluster that contained all gastric *Cryptosporidium* instead of the intestinal parasite cluster (Xiao et al. 2002a).

A third species of avian *Cryptosporidium* was first described by Pavlásek (Pavlásek 1999, 2001) in hens on the basis of biological differences. More recently, the parasite was re-described on the basis of both molecular and biological differences (Ryan et al. 2003a). Unlike other avian species, life cycle stages of *C. galli* developed in epithelial cells of the proventriculus and not the respiratory tract or small and large intestines (Pavlásek 1999, 2001). *Cryptosporidium galli* is the most prevalent *Cryptosporidium* species in Passeriformes where it frequently causes chronic infection (da Silva et al. 2010). Natural *C. galli* infections have been reported in a wide range of species (Table 1.2).

Blagburn et al. (1990) may also have detected *C. galli* in birds when they used light and electron microscopy to characterize *Cryptosporidium* sp. in the proventriculus of an Australian diamond firetail finch that died of acute diarrhea. A subsequent publication also identified a species of *Cryptosporidium* infecting the proventriculus in finches and inadvertently proposed the name *Cryptosporidium blagburni* in Table 1.1 of the paper (Morgan et al. 2001). However, Pavlásek (Pavlásek 1999, 2001) had provided a detailed description of what appeared to be the same parasite and named it *C. galli* and molecular analyses have revealed *C. galli* and *C. blagburni* to be the same species (Ryan et al. 2003a). Distance, parsimony and maximum likelihood analysis of 3 loci (18S rRNA, HSP70 and actin) identified *C. galli* as a distinct species (Ryan et al. 2003a). The gastric location of *C. galli* in the host and its large size suggest that it is most closely related to the other gastric *Cryptosporidium* species and this has been supported by molecular analysis.

In addition to the three recognized species of *Cryptosporidium*, 11 genotypes: the avian genotypes I–V, the black duck genotype, the Eurasian woodcock genotype and goose genotypes I–IV have been reported (Table 1.2). These genotypes may be renamed as distinct species in the future once more biological and molecular data become available. In addition, *C. hominis*, *C. parvum*, *C. serpentis*, *C. muris*, *C. andersoni* and muskrat genotype I have also been identified in a small number of birds, most of which were probably the results of accidental ingestion of oocysts of these organisms (Ryan et al. 2003a; Zhou et al. 2004; Ng et al. 2006; Chatterjee 2007; Zylan et al. 2008; Jellison et al. 2007, 2009; McEvoy and Giddings 2009; Quah et al. 2011; Gomes et al. 2012).

1.7 Species in Amphibians

The first report of *Cryptosporidium* in amphibians was the description of oocysts in the feces of a captive mice-fed ornate horned frog, *Ceratophrys ornata* (Crawshaw and Mehren 1987). The identity of the species is unknown and, as mice can be infected with several species and genotypes of *Cryptosporidium*, it remains unknown if the frog was naturally infected or passing oocysts from infected mice. Currently the only accepted species is *Cryptosporidium fragile*, which was described from the stomach of naturally infected black-spined toads (*Duttaphrynus melanostictus*) from the Malay peninsula (Jirků et al. 2008). Developmental stages were confined to the surface of gastric epithelial cells. In transmission experiments, *C. fragile* was not infective for one fish species (*Poecilia reticulate*), four amphibian species (*Bufo bufo*, *Rana temporaria*, *Litoria caerulea*, *Xenopus laevis*), one species of reptile (*Pantherophis guttatus*) and SCID mice. Phylogenetic analysis of the full length 18S rRNA revealed *C. fragile* to be genetically distinct and grouped with the gastric species.

1.8 Species in Reptiles

Cryptosporidium infections are common in reptiles and have been reported in more than 57 reptilian species (O'Donoghue 1995; Xiao et al. 2004b). Unlike in other animals in which *Cryptosporidium* infection is usually self-limiting in immunocompetent individuals, cryptosporidiosis in reptiles is frequently chronic and sometimes lethal in snakes. *Cryptosporidium serpentis* was named by Levine (1980) based on a report of hypertrophic gastritis in four species of snakes by Brownstein et al. (1977), but was a nomen nudum until Tilley et al. (1990) provided supporting morphologic and biological data. Since then it has been reported in lizards and other reptiles (Table 1.3). *Cryptosporidium serpentis* has also been identified in cattle (Azami et al. 2007; Chen and Qiu 2012).

Cryptosporidium varanii was described by Pavlásek et al. (1995) from an Emerald Monitor (*Varanus prasinus*) in the Prague Zoo, based on oocyst morphology, histology of endogenous stages in the intestine, and failure of oocysts to infect mice. It was subsequently genetically characterized and identified in other lizards and in snakes (Pavlásek and Ryan 2008) (Table 1.3). *Cryptosporidium saurophilum* was named by Koudela and Modry (1998) from a Schneider's skink (*Eumeces schneideri*). Cross-transmission experiments showed that the species was transmissible to lizards but not to snakes and mice. Molecular characterization revealed *C. saurophilum* and *C. varanii* were the same species (Xiao et al. 1999b, 2004b; Pavlásek and Ryan 2008) and therefore *C. varanii* took precedence over *C. saurophilum* based on its earlier publication date (Pavlásek and Ryan 2008).

Tortoise genotype I has been identified in the feces of Indian star tortoises (*Geochelone elegans*), Hermann's tortoises (*Testudo hermanii*) and a ball python (*Python regius*) (Xiao et al. 2004b; Alves et al. 2005; Pedraza-Díaz et al. 2009; Richter et al. 2012). In 2008, Traversa et al. characterized a second tortoise genotype from the feces of a marginated tortoise (*Testudo marginata*) (*Cryptosporidium* sp. ex *Testudo marginata* CrIT-20). The same genotype was subsequently identified in a chameleon (*Chamaeleo calyptratus*), a python (*Python regius*), a Pancake tortoise (*Malacochersus tornieri*), a Russian tortoise (*Agrionemys* [*Testudo*] *horsfieldii*) and Hermann's tortoises (Pedraza-Díaz et al. 2009; Griffin et al. 2010; Richter et al. 2012).

In 2010, Traversa proposed the name *Cryptosporidium ducismarci* for this second tortoise genotype (Traversa 2010), however the manuscript lacks the traditional formal descriptions associated with naming a new parasite species including a description of the oocysts. While genetic evidence suggests that is likely to be a separate species, *C. ducismarci* cannot be considered a valid species until it is formally described and should therefore be referred to as tortoise genotype II. Tortoise genotype I has been identified in gastric mucosa in a Russian tortoise, while tortoise genotype II has been identified in intestinal lesions in a Russian tortoise and a Pancake tortoise (Griffin et al. 2010). Phylogenetically, tortoise genotype II is genetically closer to *C. varanii* whereas the gastric tortoise genotype I is closer to *C. serpentis* (Griffin et al. 2010).

Other reptile-associated genotypes include snake genotype 1 which was identified in a New Guinea Viper boa (*Candoia asper*) and Japanese grass snakes (*Rhabdophis tigris*) (Xiao et al. 2002b; Kuroki et al. 2008). Snake genotype II has been identified in a Boa constrictor (*Boa constrictor ortoni*) (Xiao et al. 2004b). *Cryptosporidium parvum*, *C. tyzzeri* (mouse genotype I) and *C. muris* have also been identified in snake and lizard feces. It is unclear whether these species are infecting the animals or passing through from ingested prey (Upton et al. 1989; Xiao et al. 2004b; Morgan et al. 1998, 1999a and Traversa et al. 2008). Avian genotype V has also been detected in two green iguanas (*Iguana iguana*) (Kik et al. 2011). In that study, *Cryptosporidium* developmental stages were identified on the intestinal epithelium and the infection was associated with cloacal prolapse and cystitis (Kik et al. 2011).

1.9 Species in Mammals

Over 150 species belonging to 12 mammalian Orders have been reported as hosts for *Cryptosporidium* species (Fayer 2008, 2010). However, the majority of these reports were based on microscopic observations of oocysts and lack data necessary for species designation. Currently, 18 species of *Cryptosporidium* are recognized in mammals, *C. muris*, *C. parvum*, *C. wrairi*, *C. felis*, *C. andersoni*, *C. canis*, *C. hominis*, *C. suis*, *C. bovis*, *C. fayeri*, *C. macropodum*, *C. ryanae*, *C. xiaoi*, *C. ubiquitum*, *C. cuniculus*, *C. tyzzeri*, *C. viatorum* and *C. scrofarum*. All of these species have been identified using combinations of morphologic, biological, and molecular data. Many of these species will be discussed in greater depth in subsequent chapters and therefore they are briefly discussed here in chronological order of their dates of publication (Table 1.4). In addition to named species, more than 28 *Cryptosporidium* genotypes have been described in mammals including horse, hamster, ferret, skunk, squirrel, bear, deer, fox, mongoose, hedgehog, wildebeest, muskrat I and II, opossum I and II, chipmunk I to III, rat I to V and deer mouse I to IV (Fayer 2010; Xiao et al. 2004a). For these genotypes to receive taxonomic status, sufficient morphological, biological, and molecular data are required and names must comply with the rules of the International Code for Zoological Nomenclature (ICZN).

Cryptosporidium muris, the type species of *Cryptosporidium*, was described as a protozoan parasite in the gastric glands of the stomach of the common mouse (Tyzzer 1907, 1910). Transmission was successful to other mice but not to a rat. Nearly 80 years later (Iseki et al. 1989), reported that oocysts from rats (*C. muris* strain RN66) were infectious to SPF laboratory rats, mice, guinea pigs, rabbits, dogs, and cats. Stages developed in the stomach and all hosts excreted oocysts (Iseki et al. 1989). Molecular data at multiple loci have confirmed the genetic distinctness of *C. muris*, which groups phylogenetically with the other gastric parasites (*C. andersoni*, *C. galli*, *C. serpentis*) (Xiao et al. 1999a; Morgan et al. 2000b). *Cryptosporidium muris* is well known to have broad host specificity. In addition to various rodent species, natural *C. muris* infections have been documented for pigs, Bactrian camels, giraffes, dogs, cats, nonhuman primates, seals, bilbies, and tawny frogmouth (Fayer 2010; Feng et al. 2011). It has also been reported in humans (Katsumata et al. 2000; Guyot et al. 2001; Gatei et al. 2002b, 2003, 2006; Tiangtip and Jongwutiwes 2002; Palmer et al. 2003; Muthusamy et al. 2006; Al-Brikan et al. 2008).

Cryptosporidium parvum was described as a coccidium in the small intestine of the common mouse (Tyzzer 1912). Subsequently, over 150 mammals were reported hosts of *Cryptosporidium* based on finding oocysts resembling *C. parvum* in their feces (Fayer 2010). The advent of molecular typing, however, has revealed that many *Cryptosporidium* species and genotypes have oocysts that are indistinguishable from *C. parvum* and therefore molecular tools are necessary to confirm the identity of a species. The terms *C. parvum* bovine genotype, genotype II, and genotype B have been used in the past to distinguish this species from *C. hominis*. The *C. parvum* genome project (Abrahamsen et al. 2004) and numerous phylogenetic analyses at >15 loci have confirmed the genetic uniqueness of *C. parvum* (Fayer 2010). *Cryptosporidium parvum* primarily infects cattle, sheep and humans but has been reported in a wide range of hosts (Xiao and Ryan 2008; Fayer 2010).

Cryptosporidium wrairi, first described as a coccidian in guinea pigs (*Cavia porcellus*) (Jervis et al. 1966), was named by Vetterling et al. (1971) as an acronym for the Walter Reed Army Institute of Research. Endogenous stages, but not oocysts, were described from guinea pigs with enteritis. Molecular studies at numerous loci have confirmed that it is genetically distinct (Xiao et al. 2004b). Infections in nature, found only in guinea pigs, strongly suggest that *C. wrairi* is host specific.

Cryptosporidium felis oocysts were first described from the feces of cats (Iseki 1979). Oocysts failed to infect mice, rats, guinea pigs and dogs (Asahi et al. 1991). *Cryptosporidium felis* has also been detected in cattle and in humans (Bornay-Llinares et al. 1999; Raccurt 2007; Lucio-Forster et al. 2010; Cieloszyk et al. 2012; Elwin et al. 2012a; Gherasim et al. 2012; Insulander et al. 2012). Molecular analysis confirmed the genetic distinctness *C. felis* (Xiao et al. 2004b).

Cryptosporidium andersoni infects the abomasum of cattle (*Bos taurus*) and produces oocysts that are morphologically similar to, but slightly smaller than those of *C. muris* and was originally mistakenly identified in cattle as *C. muris* based on its oocyst size. In 2000, it was described as a new species based on location of endogenous stages in the abomasum, its host range, and genetic distinctness at multiple loci (Lindsay et al. 2000). Oocysts were not infectious to outbred, inbred immunocompetent, or immunodeficient mice, nor to chickens or goats (Lindsay et al. 2000). More recently, a novel *C. andersoni* genotype from Japan has been reported to infect SCID mice (Satoh et al. 2003; Koyama et al. 2005; Matsubayashi et al. 2005). In contrast to *C. muris*, *C. andersoni* is mostly a parasite of cattle, having been found only occasionally in other animals such as Bactrian camels, sheep, and goats (Fayer 2010; Feng et al. 2011). It is occasionally detected in humans (Leoni et al. 2006; Morse et al. 2007; Waldron et al. 2011).

Cryptosporidium canis (previously dog genotype 1) was first identified as the dog genotype by Xiao et al. (1999b) and described as a species in 2001 (Fayer et al. 2001) on the basis that *C. canis* oocysts were infectious for calves but not mice and were genetically distinct from all other species. *Cryptosporidium canis* and its sub-genotypes (*C. canis* fox genotype and *C. canis* coyote genotype) have been reported in dogs, foxes and coyotes (Fayer 2010). *Cryptosporidium canis* has also been reported worldwide in humans (Lucio-Forster et al. 2010; Fayer 2010; Elwin et al. 2012a).

Cryptosporidium hominis, a pathogenic species found primarily in humans and originally referred to as the *C. parvum* human genotype, genotype 1, and genotype H was named a new species based on molecular (16 genetic loci) and biological differences between isolates from human and bovine sources (Morgan-Ryan et al. 2002). Under experimental conditions *C. hominis* was non-infective for mice, rats, cats, dogs, and cattle but infective for calves, lambs, and piglets (Morgan-Ryan et al. 2002; Xiao et al. 2002b). Natural infections have been reported in a dugong, cattle, goats and marsupials (Morgan et al. 2000c; Park et al. 2006; Fayer 2010; Abeywardena et al. 2012; Ryan and Power 2012).

Cryptosporidium suis, previously recognized as pig genotype I (Morgan et al. 1998), was named a new species based on analysis of the 18S rDNA, HSP70, and actin loci (Ryan et al. 2004b). It is not infectious for mice (Morgan et al. 1999b) and poorly infectious for cattle (Enemark et al. 2003). It has been reported in naturally infected pigs worldwide, but causes only mild or no clinical signs in pigs (Enemark et al. 2003; Zintl et al. 2007). It has been reported in four humans (Xiao et al. 2002a; Leoni et al. 2006; Cama et al. 2007; Wang et al. 2013) and a *C. suis*-like genotype has been reported in cattle (Geurden et al. 2006; Fayer et al. 2006; Khan et al. 2010; Ng et al. 2012) and rodents (Ng-Hublin et al. 2013).

Cryptosporidium bovis from calves, identified as *Cryptosporidium* genotype bovine B (Xiao et al. 2002b) was named a new species based on the fact that it was not transmissible to neonatal BALB/c mice or lambs in transmission experiments and based on its genetic distinctness at the 18S rRNA, HSP70, and actin loci (Fayer et al. 2005). This species primarily infects young cattle and has a wide geographic distribution (Feng et al. 2007). It is closely related to *C. ryanae* (Fayer et al. 2005).

Cryptosporidium fayeri (previously marsupial genotype I) was first described in a red kangaroo (*Macropus rufus*) in 1988 (Morgan et al. 1998) and was formally described as a species in 2008 (Ryan et al. 2008). *Cryptosporidium fayeri* oocysts are indistinguishable from *C. parvum* but were found to be infectious for neonatal ARC Swiss mice (Ryan et al. 2008). Its host range is largely confined to marsupials with the exception of one report in sheep (Ryan et al. 2005) and a recent report in a 29-year-old immunocompetent woman who suffered prolonged gastrointestinal illness in Sydney (Waldron et al. 2010). Identical subtypes were found in marsupials in the area (Waldron et al. 2010). *Cryptosporidium fayeri* is genetically very closely related to the opossum genotype I identified in North American opossums (*Didelphis virginiana*), which is considered a sub-genotype of *C. fayeri* (Ryan et al. 2008).

Cryptosporidium macropodum (previously marsupial genotype II or EGK3) was first identified in eastern grey kangaroos (*Macropus giganteus*) in 2003 (Power et al. 2004) and was named a new species in 2008 (Power and Ryan 2008). It has only been reported in marsupial hosts. It is genetically distinct and not closely related to *C. fayeri* but rather more closely related to *C. suis* (Power and Ryan 2008). Both *C. fayeri* and *C. macropodum* appear to be host adapted to marsupial species as demonstrated by absence of clinical signs in marsupial hosts despite excretion of high oocyst numbers (Power and Ryan 2008).

Cryptosporidium ryanae (previously deer-like genotype) was named a new species based on transmission and genetic data in 2008 (Fayer et al. 2008). The deer-like genotype has never been found in deer but was called "deer-like" because its 18S rRNA sequence was very similar to the deer genotype (Xiao et al. 2002a). Oocysts of *C. ryanae* are not infectious for BALB/c mice or lambs (Fayer et al. 2008). Together with *C. bovis*, *C. ryanae* is responsible for the majority of *Cryptosporidium* infections in post-weaned calves. It was the only species identified in zebu cattle and water buffaloes in a recent study in Nepal (Feng et al. 2012).

Cryptosporidium xiaoi (previously *C. bovis*-like or *C. bovis*) from sheep was initially identified by Chalmers et al. in 2002 and was formally described as a species in 2009 (Fayer and Santín 2009). *Cryptosporidium xiaoi* mainly infects sheep and appears to be asymptomatic but has been reported in yaks, goats, fish and kangaroos (Fayer and Santín 2009; Reid et al. 2010; Yang et al. 2011; Rieux et al. 2013). *Cryptosporidium xiaoi* is genetically distinct but very closely related to *C. bovis* (Fayer and Santín 2009).

Cryptosporidium ubiquitum (previously cervine genotype, cervid, W4 or genotype 3) was first identified by Xiao et al. (2000) in storm water samples in lower New York State (storm water isolate W4, GenBank accession no. AF262328). Subsequently, Perez and Le Blancq (2001) identified this genotype in white tailed deer-derived isolates from lower New York State and referred to it as genotype 3. Since then it has been described in a wide variety of hosts worldwide including humans and was formally described as a species in 2010 (Fayer et al. 2010). *Cryptosporidium ubiquitum* infects the greatest number of host species of any species of *Cryptosporidium* that has been substantiated by molecular testing and is one of the main *Cryptosporidium* species infecting sheep (along with *C. xiaoi*). It is also commonly found in humans (Fayer et al. 2010; Cieloszyk et al. 2012; Elwin et al. 2012a). It is genetically very distinct and is most closely related to *C. suis* (Fayer et al. 2010).

Cryptosporidium cuniculus (previously rabbit genotype) was first described in rabbits by Inman and Takeuchi (1979), who described the microscopic detection and ultra-structure of endogenous *Cryptosporidium* parasites in the ileum of an asymptomatic female rabbit. The rabbit genotype was first identified in rabbits from the Czech Republic (Ryan et al. 2003a) and *C. cuniculus* was formally re-described as a species in 2010 (Robinson et al. 2010). *Cryptosporidium cuniculus* oocysts were infectious for weanling rabbits, immunosuppressed Mongolian gerbils and immunosuppressed Porton mice but not neonatal mice (Robinson et al. 2010). *Cryptosporidium cuniculus* has a close genetic relationship with *C. hominis* with limited differences at the 18S rRNA, HSP70 and actin genes (0.51 %, 0.25 % and 0.12 %, difference, respectively) and is known to infect humans. In 2008, it was responsible for a human cryptosporidiosis outbreak in the UK (Chalmers et al. 2009), which has raised considerable awareness about the importance of investigating rabbits in drinking water catchments as a source of *Cryptosporidium* transmissible to humans. A recent study in the UK reported that *C. cuniculus* was the third most commonly identified *Cryptosporidium* species in patients with diarrhea (Chalmers et al. 2011). *Cryptosporidium cuniculus* has also been identified in a human patient in France and in children in Nigeria (Anon 2010; Molloy et al. 2010).

Cryptosporidium tyzzeri (previously mouse genotype I) was first reported in mice in Australia (Morgan et al. 1998) and was named a new species in 2012 (Ren et al. 2012). It has a wide geographic distribution and has been found in various countries, including the United States, Australia, the United Kingdom, Spain, Portugal, Poland, China, and Japan (cf. Morgan et al. 1999b; Ren et al. 2012). *Cryptosporidium tyzzeri* has smaller oocysts than *Cryptosporidium parvum* (see Table 1.4) and is not infectious for calves and lambs (Ren et al. 2012). It is also not infectious for pigs (Kváč et al. 2012). Unlike *C. parvum* in mice, *C. tyzzeri* does not appear to infect the colon (Ren et al. 2012). Although phylogenetically related to *C. parvum* (which infects mainly ruminants and humans), *C. tyzzeri* is genetically distinct from *C. parvum* and mostly infects domestic mice and small rodents but has been recently been reported in humans (Rasková et al. 2013).

Cryptosporidium viatorum was recently described from human travelers returning to Great Britain from India, Nepal, Bangladesh and Pakistan (Elwin et al. 2012b). Natural infection results in diarrhoea, abdominal pain, fever and occasionally nausea and/or vomiting. As with most intestinal *Cryptosporidium* species, *C. viatorum* oocysts are indistinguishable from *C. parvum* and measure 5.35 × 4.72 μm with a length to width ratio of 1.14 (Elwin et al. 2012b). *Cryptosporidium viatorum* is phylogenetically most similar to *Cryptosporidium fayeri* (<98 % similarity) at the 18S rRNA, HSP70 and actin genes (Elwin et al. 2012b). It has been also identified in two human patients in Sweden (Insulander et al. 2012). No other natural hosts of *C. viatorum* have been reported and potential reservoirs remain unknown.

Cryptosporidium scrofarum (previously pig genotype II) was first reported in pigs in Australia (Ryan et al. 2003c) and was formally described as a species in pigs in 2013 (Kváč et al. 2013). *Cryptosporidium scrofarum* oocysts were not infectious for adult SCID mice, adult BALB/c mice, Mongolian gerbils (*Meriones unguiculatus*), Southern multimammate mice (*Mastomys coucha*), yellow-necked mice (*Apodemus flavicollis*), or guinea pigs (*Cavia porcellus*). Phylogenetic analysis indicates that it is most closely related to *C. ryanae*, *C. xiaoi*, and *C. bovis* (Kváč et al. 2013). It has been reported in domestic pigs in eight countries and is prevalent in adult pigs, generally asymptomatic and with a generally low infection intensity (Kváč et al. 2013). In addition, wild boars, calves, marine fish, rodents and humans have been reported as natural hosts of this species (Kváč et al. 2009a, b, c; Němejc et al. 2012; Ng et al. 2011; Reid et al. 2010; Ng-Hublin et al. 2013). Evidence suggests that *C. suis* infects pigs of all ages, whereas *C. scrofarum* only infects pigs >6 week of age (Jeníková et al. 2011).

1.9.1 The C. tyzzeri *Versus* C. pestis *Debate*

In 2006, Slapeta proposed in an opinion article to rename *C. parvum* as *Cryptosporidium pestis* and retain *C. parvum* for mouse genotype I (now *C. tyzzeri*) (Slapeta 2006). This was largely based on a rigid interpretation of the Principle of Priority of the ICZN as Slapeta maintained that when Tyzzer described *C. parvum* in 1912 in mice (Tyzzer 1912), it is probable that he was in fact describing *C. tyzzeri*. This led to a debate in the literature as to the validity of *C. pestis* (Xiao et al. 2007). The debate was re-ignited in 2012, when Ren et al. re-named mouse genotype I as *C. tyzzeri*, the validity of which was disputed by Slapeta (Ren et al. 2012; Slapeta 2011; Xiao et al. 2012).

The dispute on the validity of *C. pestis* and *C. tyzzeri* originates from the uncertainty of the identity of *C. parvum* described by Tyzzer in 1912 and on the interpretation of the Principle of Priority of the ICZN. Because most animals are infected with multiple species and genotypes of *Cryptosporidium* spp., we will never know what Tyzzer originally described. It may have been *C. tyzzeri*, but may also have been any of the other *Cryptosporidium* species and genotypes that have been found in naturally infected domestic mice (Ren et al. 2012), including mouse genotype II and *C. parvum*. In addition, the ICZN clearly states that the Principle of Priority is to be used to promote stability and is not intended to upset a long-accepted name (Ride et al. 1999). Importantly, in 1985, Upton et al. re-described *C. parvum* specifically for bovine and human isolates (Upton and Current 1985). Because mice are known to be naturally infected with *C. parvum* and have been used widely as a laboratory model for *C. parvum*, the 1985 taxonomic re-description of *C. parvum* was done following ICZN rules. Prior to this re-description, the name was seldom used but since then the name *C. parvum* has been widely accepted by almost all researchers in the community (a recent PubMed search of *C. parvum* found no publications before 1985 and over 3,240 publications since 1985).

The designation of *C. tyzzeri* for the mouse genotype I, brings further clarity to the taxonomy of *Cryptosporidium* spp. in humans, cattle, and domestic mice (Ren et al. 2012). Slapeta objected to *C. tyzzeri* (Slapeta 2011), because the name was used by Levine (1961) for a *Cryptosporidium* sp. in chickens originally described by Tyzzer (1929), while the *C. tyzzeri* described by Levine (1961) is considered a synonym of *C. meleagridis* Slavin 1955. On the basis of this, Slapeta suggested that the name *C. tyzzeri* by Ren et al. (2012), was both a primary homonym and a junior synonym. However, homonymy and synonymy refer to the application of the same name to different taxa and different names to the same taxon, respectively. Because the taxon *C. tyzzeri* Levine 1961 is not an established one, the homonymy and synonymy suggested by Slapeta do not exist. This is because the nomen nudum nature of *C. tyzzeri* Levine 1961 was previously pointed out in several reviews of *Cryptosporidium* taxonomy (Upton 2003; Fayer 2008), and was stated clearly in the Etymology section of *C. tyzzeri* (Ren et al. 2012).

Cryptosporidium pestis has never been formally described and therefore is not a valid species as it lacks a full taxonomic description, and the name has only been used in the literature in four publications since 2006 (three of which were by Slapeta). To re-name *C. parvum* as *C. pestis* would be confusing to not only *Cryptosporidium* researchers but also the wider veterinary and medical community and water industry, who struggle to keep up with the taxonomy as it is. It would also be in violation of the underlying principles of the ICZN. Therefore, the prevailing name *C. parvum* for the species infective to calves and humans must be retained to avoid confusion.

1.10 Conclusions

To date 25 *Cryptosporidium* species are recognized as valid. Undoubtedly many *Cryptosporidium* genotypes will be formally described as species in the future. As new species of *Cryptosporidium* are named, recommendations from the "Code" in achieving stability and ensuring the uniqueness of each species should be borne in mind.

References

Abe N, Iseki M (2004) Identification of *Cryptosporidium* isolates from cockatiels by direct sequencing of the PCR-amplified small subunit ribosomal RNA gene. Parasitol Res 92:523–526

Abe N, Makino I (2010) Multilocus genotypic analysis of *Cryptosporidium* isolates from cockatiels, Japan. Parasitol Res 106:1491–1497

Abeywardena H, Jex AR et al (2012) Genetic characterisation of *Cryptosporidium* and *Giardia* from dairy calves: discovery of species/genotypes consistent with those found in humans. Infect Genet Evol 12:1984–1993

Abrahamsen MS, Templeton TJ et al (2004) Complete genome sequence of the apicomplexan, *Cryptosporidium parvum*. Science 304:441–445

Al-Brikan FA, Salem HS et al (2008) Multilocus genetic analysis of *Cryptosporidium* isolates from Saudi Arabia. J Egypt Soc Parasitol 38:645–658

Alvarez-Pellitero P, Sitjà-Bobadilla A (2002) *Cryptosporidium molnari n. sp.* (*Apicomplexa: Cryptosporidiidae*) infecting two marine fish species, *Sparus aurata L.* and *Dicentrarchus labrax L.* Int J Parasitol 32:1007–1021

Alvarez-Pellitero P, Quiroga MI et al (2004) *Cryptosporidium scophthalmi n. sp. (Apicomplexa: Cryptosporidiidae*) from cultured turbot *Scophthalmus maximus*. Light and electron microscope description and histopathological study. Dis Aquat Organ 62:33–45

Alvarez-Pellitero P, Perez A et al (2009) Host and environmental risk factors associated with Cryptosporidium scophthalmi (Apicomplexa) infection in cultured turbot, *Psetta maxima* (L.) (Pisces, Teleostei). Vet Parasitol 165:207–215

Alves M, Xiao L et al (2005) Occurrence and molecular characterization of *Cryptosporidium* spp. in mammals and reptiles at the Lisbon Zoo. Parasitol Res 97:108–112

Amer S, Wang C et al (2010) First detection of *Cryptosporidium baileyi* in Ruddy Shelduck (*Tadorna ferruginea*) in China. J Vet Med Sci 72:935–938

Anon (2010) ANOFEL Cryptosporidium National Network. Laboratory-based surveillance for *Cryptosporidium* in France, 2006–2009. Euro Surveill 15(33):19642

Antunes RG, Simões DC et al (2008) Natural infection with *Cryptosporidium galli* in canaries (*Serinus canaria*), in a cockatiel (*Nymphicus hollandicus*), and in lesser seed-finches (*Oryzoborus angolensis*) from Brazil. Avian Dis 52:702–705

Asahi H, Koyama T et al (1991) Biological nature of *Cryptosporidium* sp. isolated from a cat. Parasitol Res 77:237–240

Azami M, Moghaddam DD et al (2007) The identification of *Cryptosporidium* species (protozoa) in Ifsahan, Iran by PCR-RFLP analysis of the 18S rRNA gene. Mol Biol 41:934–939

Barta JR, Thompson RC (2006) What is *Cryptosporidium*? Reappraising its biology and phylogenetic affinities. Trends Parasitol 22:463–468

Barugahare R, Dennis MM et al (2011) Detection of *Cryptosporidium molnari* oocysts from fish by fluorescent-antibody staining assays for *Cryptosporidium* spp. affecting humans. Appl Environ Microbiol 77:1878–1880

Bass AL, Wallace CC et al (2012) Detection of *Cryptosporidium* sp. in two new seal species, *Phoca vitulina* and *Cystophora cristata*, and a novel *Cryptosporidium* genotype in a third seal species, *Pagophilus groenlandicus*, from the Gulf of Maine. J Parasitol 98:316–322

Berrilli F, D'Alfonso R et al (2012) Giardia duodenalis genotypes and *Cryptosporidium* species in humans and domestic animals in Côte d'Ivoire: occurrence and evidence for environmental contamination. Trans R Soc Trop Med Hyg 106:191–195

Beyer TV, Svezhova NV et al (2000) *Cryptosporidium parvum* (Coccidia, Apicomplexa): some new ultrastructural observations on its endogenous development. Eur J Protistol 36:151–159

Blagburn BL, Soave R (1997) Prophylaxis and chemotherapy: human and animal. In: Fayer R (ed) Cryptosporidium and cryptosporidiosis. CRC Press, Florida, pp 111–128

Blagburn BL, Lindsay DS et al (1990) *Cryptosporidium* sp infection in the proventriculus of an Australian diamond firetail finch (*Staganoplura bella* Passeriformes, Estrildidae). Avian Dis 34:1027–1030

Bornay-Llinares FJ, da Silva AJ et al (1999) Identification of *Cryptosporidium felis* in a cow by morphologic and molecular methods. Appl Environ Microbiol 65:1455–1458

Brownstein DG, Strandberg JD et al (1977) *Cryptosporidium* in snakes with hypertrophic gastritis. Vet Pathol 14:606–617

Cabada MM, White AC (2010) Treatment of cryptosporidiosis: do we know what we think we know? Curr Opin Infect Dis 23:494–499

Cama VA, Ross JM et al (2007) Differences in clinical manifestations among *Cryptosporidium* species and subtypes in HIV-infected persons. J Infect Dis 196:684–691

Camus AC, Lopez MK (1996) Gastric cryptosporidiosis in juvenile red drum. J Aquat Anim Health 8:167–172

Chalmers RM, Elwin K et al (2002) *Cryptosporidium* in farmed animals: the detection of a novel isolate in sheep. Int J Parasitol 32:21–26

Chalmers RM, Robinson G et al (2009) *Cryptosporidium* sp. rabbit genotype, a newly identified human pathogen. Emerg Infect Dis 15:829–830

Chalmers RM, Elwin K et al (2011) Sporadic human cryptosporidiosis caused by *Cryptosporidium cuniculus*, United Kingdom, 2007–2008. Emerg Infect Dis 3:536–538

Chatterjee R (2007) A potential new crypto source. Environ Sci Technol 41:3399–3400

Chen F, Qiu H (2012) Identification and characterization of a Chinese isolate of *Cryptosporidium serpentis* from dairy cattle. Parasitol Res 111:1785–1791

Cieloszyk J, Goñi P et al (2012) Two cases of zoonotic cryptosporidiosis in Spain by the unusual species *Cryptosporidium ubiquitum* and *Cryptosporidium felis*. Enferm Infecc Microbiol Clin 30:549–551

Crawshaw GJ, Mehren KG (1987) Cryptosporidiosis in zoo and wild animals. In: Erkrankungen der Zootiere. Verhandlungsbericht des 29. Internationalen symposiums Über die Erkrankungen der Zootiere, Cardiff. Akademie-Verlag, Berlin, pp 353–362

Current WL, Upton SJ et al (1986) The life cycle of *Cryptosporidium baileyi* n. sp. (Apicomplexa, Cryptosporidiidae) infecting chickens. J Protozool 33:289–296

da Silva DC, Homem CG et al (2010) Physical, epidemiological, and molecular evaluation of infection by *Cryptosporidium galli* in Passeriformes. Parasitol Res 107:271–277

Egyed Z, Sreter T et al (2003) Characterization of *Cryptosporidium* spp.- recent developments and future needs. Vet Parasitol 111:103–114

Elwin K, Hadfield SJ et al (2012a) The epidemiology of sporadic human infections with unusual cryptosporidia detected during routine typing in England and Wales, 2000–2008. Epidemiol Infect 140:673–683

Elwin K, Hadfield SJ et al (2012b) *Cryptosporidium viatorum* n. sp. (Apicomplexa: Cryptosporidiidae) among travellers returning to Great Britain from the Indian subcontinent, 2007–2011. Int J Parasitol 42:675–682

Enemark HL, Ahrens P et al (2003) *Cryptosporidium parvum*: infectivity and pathogenicity of the 'porcine' genotype. Parasitology 126:407–416

Fall A, Thompson RC et al (2003) Morphology is not a reliable tool for delineating species within *Cryptosporidium*. J Parasitol 89:399–402

Fayer R (2008) General biology. In: Fayer R, Xiao L (eds) Cryptosporidium and cryptosporidiosis, 2nd edn. CRC Press, Boca Raton, pp 1–42

Fayer R (2010) Taxonomy and species delimitation in *Cryptosporidium*. Exp Parasitol 124:90–97

Fayer R, Santín M (2009) *Cryptosporidium xiaoi* n. sp. (Apicomplexa: Cryptosporidiidae) in sheep (*Ovis aries*). Vet Parasitol 164:192–200

Fayer R, Trout JM et al (2001) *Cryptosporidium canis* n. sp. from domestic dogs. J Parasitol 87:1415–1422

Fayer R, Santín M et al (2005) *Cryptosporidium bovis* n. sp. (Apicomplexa: Cryptosporidiidae) in cattle (*Bos taurus*). J Parasitol 91:624–629

Fayer R, Santín M et al (2006) Prevalence of species and genotypes of *Cryptosporidium* found in 1–2 year-old dairy cattle in the eastern United States. Vet Parasitol 135:105–112

Fayer R, Santín M et al (2008) *Cryptosporidium ryanae* n. sp. (Apicomplexa: Cryptosporidiidae) in cattle (*Bos taurus*). Vet Parasitol 156:191–198
Fayer R, Santín M et al (2010) *Cryptosporidium ubiquitum* n. sp. in animals and humans. Vet Parasitol 172:23–32
Feng Y, Ortega Y et al (2007) Wide geographic distribution of *Cryptosporidium bovis* and the deer-like genotype in bovines. Vet Parasitol 144:1–9
Feng Y, Yang W et al (2011) Development of a multilocus sequence tool for typing *Cryptosporidium muris* and *Cryptosporidium andersoni*. J Clin Microbiol 49:34–41
Feng Y, Karna SR et al (2012) Common occurrence of a unique *Cryptosporidium ryanae* variant in zebu cattle and water buffaloes in the buffer zone of the Chitwan National Park, Nepal. Vet Parasitol 185:309–314
Freire-Santos F, Vergara-Castiblanco CA et al (1998) *Cryptosporidium parvum*: an attempt at experimental infection in rainbow trout *Oncorhynchus mykiss*. J Parasitol 84:935–938
Gabor LJ, Srivastava M et al (2011) Cryptosporidiosis in intensively reared Barramundi (Lates calcarifer). J Vet Diagn Invest 23:383–386
Gatei W, Suputtamongkol et al (2002a) Zoonotic species of *Cryptosporidium* are as prevalent as the anthroponotic in HIV-infected patients in Thailand. Ann Trop Med Parasitol 96:797–802
Gatei W, Ashford RW et al (2002b) *Cryptosporidium muris* infection in an HIV-infected adult, Kenya. Emerg Infect Dis 8:204–206
Gatei W, Greensill J et al (2003) Molecular analysis of the 18S rRNA gene of *Cryptosporidium* parasites from patients with or without human immunodeficiency virus infections living in Kenya, Malawi, Brazil, the United Kingdom, and Vietnam. J Clin Microbiol 41:1458–1462
Gatei W, Wamae CN et al (2006) Cryptosporidiosis: prevalence, genotype analysis, and symptoms associated with infections in children in Kenya. Am J Trop Med Hyg 75:78–82
Geurden T, Goma FY et al (2006) Prevalence and genotyping of *Cryptosporidium* in three cattle husbandry systems in Zambia. Vet Parasitol 138:217–222
Gherasim A, Lebbad M et al (2012) Two geographically separated food-borne outbreaks in Sweden linked by an unusual *Cryptosporidium parvum* subtype, October 2010. Euro Surveill 17(46):20318
Gibson-Kueh S, Yang R et al (2011) The molecular characterization of an Eimeria and *Cryptosporidium* detected in Asian seabass (Lates calcarifer) cultured in Vietnam. Vet Parasitol 181:91–96
Glaberman S, Sulaiman IM et al (2001) A multilocus genotypic analysis of *Cryptosporidium meleagridis*. J Eukaryot Microbiol 48:19S–22S
Glazebrook JS, Campbell SR (1987) Diseases of Barramundi (Lates calcarifer) in Australia: a review. In: Copland JW, Grey DL (eds) Management of wild and cultured sea bass/barramundi (Lates calcarifer): proceedings of an international workshop, Darwin, N.T. Australia, 24–30 September 1986. Australian Centre for International Agricultural Research (ACIAR) Proceedings No. 20. ACIAR, Canberra, pp 204–206
Gomes RS, Huber F et al (2012) Cryptosporidium spp. parasitize exotic birds that are commercialized in markets, commercial aviaries, and pet shops. Parasitol Res 110:1363–1370
Griffin C, Reavill DR et al (2010) Cryptosporidiosis caused by two distinct species in Russian tortoises and a pancake tortoise. Vet Parasitol 170:14–19
Guyot K, Follet-Dumoulin A et al (2001) Molecular characterization of *Cryptosporidium* isolates obtained from humans in France. J Clin Microbiol 39:3472–3480
Hajdusek O, Ditrich O et al (2004) Molecular identification of *Cryptosporidium* spp. in animal and human hosts from the Czech Republic. Vet Parasitol 122:183–192
Hijjawi NS, Meloni BP et al (2002) Successful in vitro cultivation of *Cryptosporidium andersoni*: evidence for the existence of novel extracellular stages in the life cycle and implications for the classification of *Cryptosporidium*. Int J Parasitol 32:1719–1726
Hoover DM, Hoerr FJ et al (1981) Enteric cryptosporidiosis in a naso tang, *Naso lituratus* Block and Schneider. J Fish Dis 4:425–428

Huber F, da Silva S et al (2007) Genotypic characterization and phylogenetic analysis of *Cryptosporidium* sp. from domestic animals in Brazil. Vet Parasitol 150:65–74

Hunter PR, Hughes S et al (2004) Sporadic cryptosporidiosis case-control study with genotyping. Emerg Infect Dis 10:1241–1249

Hunter PR, Hadfield SJ et al (2007) Subtypes of *Cryptosporidium parvum* in humans and disease risk. Emerg Infect Dis 13:82–88

Inman LR, Takeuchi A (1979) Spontaneous cryptosporidiosis in an adult female rabbit. Vet Pathol 16:89–95

Insulander M, Silverlås C et al (2012) Molecular epidemiology and clinical manifestations of human cryptosporidiosis in Sweden. Epidemiol Infect 141:1009–1020

Iseki M (1979) *Cryptosporidium felis* sp. from the domestic cat. Jpn J Parasitol 28:13–35

Iseki M, Maekawa T et al (1989) Infectivity of *Cryptosporidium muris* (strain RN 66) in various laboratory animals. Parasitol Res 75:218–222

Jellison KL, Distel DL et al (2004) Phylogenetic analysis of the hypervariable region of the 18S rRNA gene of *Cryptosporidium* oocysts in feces of Canada geese (*Branta canadensis*): evidence for five novel genotypes. Appl Environ Microbiol 70:452–458

Jellison KL, Distel DL et al (2007) Phylogenetic analysis implicates birds as a source of *Cryptosporidium* spp. oocysts in agricultural watersheds. Environ Sci Technol 41:3620–3625

Jellison KL, Lynch AE et al (2009) Source tracking identifies deer and geese as vectors of human-infectious *Cryptosporidium* genotypes in an urban/suburban watershed. Environ Sci Technol 43:4267–4272

Jeníková M, Němejc K et al (2011) New view on the age-specificity of pig *Cryptosporidium* by species-specific primers for distinguishing *Cryptosporidium suis* and *cryptosporidium* pig genotype II. Vet Parasitol 176:120–125

Jervis HR, Merrill TG et al (1966) Coccidiosis in the guinea pig small intestine due to a *Cryptosporidium*. Am J Vet Res 27:408–414

Jirků M, Valigurová A et al (2008) New species of *Cryptosporidium* Tyzzer, 1907 (Apicomplexa) from amphibian host: morphology, biology and phylogeny. Folia Parasitol (Praha) 55:81–94

Katsumata TD, Hosea IG et al (2000) Possible *Cryptosporidium muris* infection in humans. Am J Trop Med Hyg 62:70–72

Khan SM, Debnath C et al (2010) Molecular characterization and assessment of zoonotic transmission of *Cryptosporidium* from dairy cattle in West Bengal, India. Vet Parasitol 171:41–47

Kik MJ, van Asten AJ et al (2011) Cloaca prolapse and cystitis in green iguana (*Iguana iguana*) caused by a novel *Cryptosporidium* species. Vet Parasitol 175:165–167

Kimbell LM, Miller DL et al (1999) Molecular analysis of the 18S rRNA gene of *Cryptosporidium serpentis* in a wild-caught corn snake (*Elaphe guttata guttata*) and a five-species restriction fragment length polymorphism- based assay that can additionally discern *C. parvum* from *C. wrairi*. Appl Environ Microbiol 65:5345–5349

Kimura A, Suzuki Y et al (2004) Identification of the *Cryptosporidium* isolate from chickens in Japan by sequence analyses. J Vet Med Sci 66:879–881

Koinari et al (2013) Still in press. doi: 10.1016/j.vetpar.2013.08.031. http://authors.elsevier.com/sd/article/S0304401713004974

Koudela B, Modry D (1998) New species of *Cryptosporidium* (Apicomplexa, Cryptosporidiidae) from lizards. Folia Parasitol 45:93–100

Koyama Y, Satoh M et al (2005) Isolation of *Cryptosporidium andersoni* Kawatabi type in a slaughterhouse in the northern island of Japan. Vet Parasitol 130:323–326

Kuroki T, Izumiyama S et al (2008) Occurrence of *Cryptosporidium* sp. in snakes in Japan. Parasitol Res 103:801–805

Kvác M, Hanzlíková D et al (2009a) Prevalence and age-related infection of *Cryptosporidium suis*, *C. muris* and *Cryptosporidium* pig genotype II in pigs on a farm complex in the Czech Republic. Vet Parasitol 160:319–322

Kvác M, Kvetonová D et al (2009b) *Cryptosporidium* pig genotype II in immunocompetent man. Emerg Infect Dis 15:982–983

Kvác M, Sak B et al (2009c) Molecular characterization of *Cryptosporidium* isolates from pigs at slaughterhouses in South Bohemia, Czech Republic. Parasitol Res 104:425–428

Kváč M, Kestřánová M et al (2012) *Cryptosporidium tyzzeri* and *Cryptosporidium muris* originated from wild West-European house mice (*Mus musculus domesticus*) and East-European house mice (*Mus musculus musculus*) are non-infectious for pigs. Exp Parasitol 131:107–110

Kváč M, Kestřánová M et al (2013) *Cryptosporidium scrofarum* n. sp. (Apicomplexa: Cryptosporidiidae) in domestic pigs (*Sus scrofa*). Vet Parasitol 191:218–227

Landsberg JH, Paperna I (1986) Ultrastructural study of the coccidian *Cryptosporidium* sp. from stomachs of juvenile cichlid fish. Dis Aquat Organ 2:13–20

Leoni F, Amar C et al (2006) Genetic analysis of *Cryptosporidium* from 2414 humans with diarrhoea in England between 1985 and 2000. J Med Microbiol 55:703–707

Levine ND (1961) Protozoan parasites of domestic animals and of man. Burgess Publishing Company, Minneapolis

Levine ND (1980) Some corrections of coccidian (Apicomplexa: Protozoa) nomenclature. J Parasitol 66:830–834

Levine ND (1984) Taxonomy and review of the coccidian genus *Cryptosporidium* (Protozoa, Apicomplexa). J Protozool 31:94–98

Levine ND (1988) Progress in taxonomy of the Apicomplexan protozoa. J Protozool 35:518–520

Lindsay DS, Blagburn BL (1990) Cryptosporidiosis in birds. In: Dubey JP, Speer CA, Fayer R (eds) Cryptosporidiosis in man and animals. CRC Press, Boca Raton, pp 133–148

Lindsay DS, Blagburn BL et al (1989) Morphometric comparison of the oocysts of *Cryptosporidium meleagridis* and *Cryptosporidium baileyi* from birds. Proc Helminthol Soc Wash 56:91–92

Lindsay DS, Upton SJ et al (2000) *Cryptosporidium andersoni* n. sp. (Apicomplexa: Cryptosporiidae) from cattle, *Bos taurus*. J Eukaryot Microbiol 47:91–95

Lucio-Forster A, Griffiths JK et al (2010) Minimal zoonotic risk of cryptosporidiosis from pet dogs and cats. Trends Parasitol 26:174–179

Makino I, Abe N et al (2010) *Cryptosporidium* avian genotype III as a possible causative agent of chronic vomiting in peach-faced lovebirds (*Agapornis roseicollis*). Avian Dis 54:1102–1107

Marcial MA, Madara JL (1986) *Cryptosporidium*: cellular localization, structural analysis of absorptive cell parasite membrane-membrane interactions in guinea pigs, and suggestion of protozoan transport by M cells. Gastroenterology 90:583–594

Matsubayashi M, Kimata I et al (2005) Infectivity of a novel type of *Cryptosporidium andersoni* to laboratory mice. Vet Parasitol 129:165–168

McEvoy JM, Giddings CW (2009) *Cryptosporidium* in commercially produced turkeys on-farm and posts-laughter. Lett Appl Microbiol 48:302–306

Meireles MV, Soares RM et al (2006) Biological studies and molecular characterization of a *Cryptosporidium* isolate from ostriches (*Struthio camelus*). J Parasitol 92:623–626

Molloy SF, Smith HV et al (2010) Identification of a high diversity of *Cryptosporidium* species genotypes and subtypes in a pediatric population in Nigeria. Am J Trop Med Hyg 82:608–613

Morgan UM, Sargent KD et al (1998) Molecular characterization of *Cryptosporidium* from various hosts. Parasitology 117:31–37

Morgan UM, Xiao L et al (1999a) Phylogenetic analysis of *Cryptosporidium* isolates from captive reptiles using 18S rDNA sequence data and random amplified polymorphic DNA analysis. J Parasitol 85:525–530

Morgan UM, Sturdee AP et al (1999b) The *Cryptosporidium* "mouse" genotype is conserved across geographic areas. J Clin Microbiol 37:1302–1305

Morgan UM, Xiao L et al (2000a) *Cryptosporidium meleagridis* in an Indian ring-necked parrot (*Psittacula krameri*). Aust Vet J 78:182–183

Morgan UM, Xiao L et al (2000b) Molecular and phylogenetic analysis of *Cryptosporidium muris* from various hosts. Parasitology 120:457–464

Morgan UM, Xiao L et al (2000c) Detection of the *Cryptosporidium parvum* "human" genotype in a dugong (*Dugong dugon*). J Parasitol 86:1352–1354

Morgan UM, Monis PT et al (2001) Molecular and phylogenetic characterisation of *Cryptosporidium* from birds. Int J Parasitol 31:289–296
Morgan-Ryan UM, Fall A et al (2002) *Cryptosporidium hominis* n. sp. (Apicomplexa: Cryptosporidiidae) from *Homo sapiens*. J Eukaryot Microbiol 49:433–440
Morine M, Yang R et al (2012) Additional novel *Cryptosporidium* genotypes in ornamental fishes. Vet Parasitol 190:578–582
Morse TD, Nichols RA et al (2007) Incidence of cryptosporidiosis species in paediatric patients in Malawi. Epidemiol Infect 135:1307–1315
Muench TR, White MR (1997) Cryptosporidiosis in a tropical freshwater catfish (*Plecostomus* spp.). J Vet Diagn Invest 9:87–90
Murphy BG, Bradway D et al (2009) Gastric cryptosporidiosis in freshwater angelfish (Pterophyllum scalare). J Vet Diagn Invest 21:722–727
Muthusamy D, Rao SS et al (2006) Multilocus genotyping of *Cryptosporidium* sp. isolates from human immunodeficiency virus-infected individuals in South India. J Clin Microbiol 44:632–634
Nakamura AA, Simões DC et al (2009) Molecular characterization of *cryptosporidium* spp. from fecal samples of birds kept in captivity in Brazil. Vet Parasitol 166:47–51
Němejc K, Sak B et al (2012) The first report on *Cryptosporidium suis* and *Cryptosporidium* pig genotype II in Eurasian wild boars (*Sus scrofa*) (Czech Republic). Vet Parasitol 184:122–125
Ng J, Pavlásek I et al (2006) Identification of novel *Cryptosporidium* genotypes from avian hosts. Appl Environ Microbiol 72:7548–7553
Ng J, Yang R et al (2011) Molecular characterization of *Cryptosporidium* and *Giardia* in pre-weaned calves in Western Australia and New South Wales. Vet Parasitol 176:145–150
Ng J, Eastwood K et al (2012) Evidence of *Cryptosporidium* transmission between cattle and humans in northern New South Wales. Exp Parasitol 130:437–441
Ng-Hublin JS, Singleton GR et al (2013) Molecular characterization of *Cryptosporidium* spp. from wild rats and mice from rural communities in the Philippines. Infect Genet Evol 16:5–12
Nguyen ST, Fukuda Y et al (2013) Prevalence and molecular characterization of *Cryptosporidium* in ostriches (*Struthio camelus*) on a farm in central Vietnam. Exp Parasitol 133:8–11
O'Donoghue PJ (1995) *Cryptosporidium* and cryptosporidiosis in man and animals. Int J Parasitol 25:139–195
Pagès-Manté A, Pagès-Bosch M et al (2007) An outbreak of disease associated with cryptosporidia on a red-legged partridge (*Alectoris rufa*) game farm. Avian Pathol 36:275–278
Palenzuela O, Alvarez-Pellitero P et al (2010) Molecular characterization of *Cryptosporidium molnari* reveals a distinct piscine clade. Appl Environ Microbiol 76:7646–7649
Palmer CJ, Xiao L et al (2003) *Cryptosporidium muris*, a rodent pathogen, recovered from a human in Peru. Emerg Infect Dis 9:1174–1176
Paperna I, Vilenkin M (1996) Cryptosporidiosis in the gourami *Trichogaster leeri*: description of a new species and a proposal for a new genus, *Piscicryptosporidium*, for species infecting fish. Dis Aquat Organ 27:95–101
Park JH, Guk SM et al (2006) Genotype analysis of *Cryptosporidium* spp. prevalent in a rural village in Hwasun-gun, Republic of Korea. Korean J Parasitol 44:27–33
Pavlásek I (1983) *Cryptosporidium* sp. in *Cyprinus carpio* Linne, 1758 in Czechoslovkia. Folia Parasitol 30:248
Pavlásek I (1999) *Cryptosporidia*: biology, diagnosis, host spectrum, specificity, and the environment. Rem Klin Mikrobiol 3:290–301
Pavlásek I (2001) Findings of *Cryptosporidia* in the stomach of chickens and of exotic and wild birds. Veterinarstvi 51:103–108
Pavlasek I, Ryan U (2008) *Cryptosporidium varanii* takes precedence over *C. saurophilum*. Exp Parasitol 118:434–437
Pavlásek I et al (1995) *Cryptosporidium varanii* n. sp. (Apicomplexa: Cryptosporidiidae) in Emerald monitor (*Varanus prasinus* Schlegal, 1893) in captivity in Prague Zoo. Gazella 22:99–108

Pedraza-Díaz S, Ortega-Mora LM et al (2009) Molecular characterisation of *Cryptosporidium* isolates from pet reptiles. Vet Parasitol 160:204–210

Perez JF, Le Blancq SM (2001) *Cryptosporidium parvum* infection involving novel genotypes in wildlife from lower New York State. Appl Environ Microbiol 67:1154–1162

Petry F (2004) Structural analysis of *Cryptosporidium parvum*. Microsc Microanal 10:586–601

Plutzer J, Karanis P (2007) Molecular identification of a *Cryptosporidium saurophilum* from corn snake (*Elaphe guttata guttata*). Parasitol Res 101:1141–1145

Plutzer J, Karanis P (2009) Genetic polymorphism in *Cryptosporidium* species: an update. Vet Parasitol 165:187–199

Power ML, Ryan UM (2008) A new species of *Cryptosporidium* (Apicomplexa: Cryptosporidiidae) from eastern grey kangaroos (*Macropus giganteus*). J Parasitol 94:1114–1117

Power ML, Slade MB et al (2004) Genetic characterisation of *Cryptosporidium* from a wild population of eastern grey kangaroos *Macropus giganteus* inhabiting a water catchment. Infect Genet Evol 4:59–67

Proctor SJ, Kemp RL (1974) *Cryptosporidium anserinum* sp. n. (Sporozoa) in a domestic goose *Anser anser* L., from Iowa. J Protozool 21:664–666

Qi M, Wang R et al (2011) *Cryptosporidium* spp. in pet birds: genetic diversity and potential public health significance. Exp Parasitol 128:336–340

Quah JX, Ambu S et al (2011) Molecular identification of *Cryptosporidium parvum* from avian hosts. Parasitology 138:573–577

Raccurt CP (2007) Worldwide human zoonotic cryptosporidiosis caused by *Cryptosporidium felis*. Parasite 14:15–20

Rasková V, Kvetonová D et al (2013) Human cryptosporidiosis caused by *Cryptosporidium tyzzeri* and *C. parvum* isolates presumably transmitted from wild mice. J Clin Microbiol 51:360–362

Reid A, Lymbery A et al (2010) Identification of novel and zoonotic *Cryptosporidium* species in marine fish. Vet Parasitol 168:190–195

Ren X, Zhao J et al (2012) *Cryptosporidium tyzzeri* n. sp. (Apicomplexa: Cryptosporidiidae) in domestic mice (*Mus musculus*). Exp Parasitol 130:274–281

Rengifo-Herrera C, Ortega-Mora LM et al (2011) Detection and characterization of a *Cryptosporidium* isolate from a southern elephant seal (*Mirounga leonina*) from the Antarctic peninsula. Appl Environ Microbiol 77:1524–1527

Rengifo-Herrera C, Ortega-Mora LM et al (2013) Detection of a novel genotype of *Cryptosporidium* in Antarctic pinnipeds. Vet Parasitol 191:112–118

Richter B, Nedorost N et al (2011) Detection of *Cryptosporidium* species in feces or gastric contents from snakes and lizards as determined by polymerase chain reaction analysis and partial sequencing of the 18S ribosomal RNA gene. J Vet Diagn Invest 23:430–435

Richter B, Rasim R et al (2012) Cryptosporidiosis outbreak in captive chelonians (*Testudo hermanni*) with identification of two *Cryptosporidium* genotypes. J Vet Diagn Invest 24:591–595

Ride WDL, Cogger HG et al (1999) International code of zoological nomenclature, 4th edn. International Trust for Zoological Nomenclature/The Natural History Museum, London

Rieux A, Paraud C et al (2013) Molecular characterization of *Cryptosporidium* spp. in pre-weaned kids in a dairy goat farm in western France. Vet Parasitol 192:268–272

Rinaldi L, Capasso M et al (2012) Prevalence and molecular identification of *Cryptosporidium* isolates from pet lizards and snakes in Italy. Parasite 19:437–440

Robertson B, Sinclair MI et al (2002) Case-control studies of sporadic cryptosporidiosis in Melbourne and Adelaide, Australia. Epidemiol Infect 128:419–431

Robinson G, Wright S et al (2010) Re-description of *Cryptosporidium cuniculus* Inman and Takeuchi, 1979 (Apicomplexa: Cryptosporidiidae): morphology, biology and phylogeny. Int J Parasitol 40:1539–1548

Rosales MJ, Cordón GP et al (2005) Extracellular like gregarine stages of *Cryptosporidium parvum*. Acta Trop 95:74–78

Rush BA, Chapman PA et al (1987) *Cryptosporidium* and drinking water. Lancet 2:632–633
Ryan U, Power M (2012) *Cryptosporidium* species in Australian wildlife and domestic animals. Parasitology 139:1673–1688
Ryan UM, Xiao L et al (2003a) A redescription of *Cryptosporidium galli* Pavlasek 1999, 2001 (*Apicomplexa*: *Cryptospodiidae*) from birds. J Parasitol 89:809–813
Ryan UM, Xiao L et al (2003b) Identification of novel *Cryptosporidium* genotypes from the Czech Republic. Appl Environ Microbiol 69:4302–4307
Ryan UM, Samarasinghe B et al (2003c) Identification of a novel *Cryptosporidium* genotype in pigs. Appl Environ Microbiol 69:3970–3974
Ryan UM, O'Hara A et al (2004a) Molecular and biological characterization of *Cryptosporidium molnari*-like isolate from a guppy (*Poecilia reticulata*). Appl Environ Microbiol 70:3761–3765
Ryan UM, Monis P et al (2004b) *Cryptosporidium suis* n. sp. (Apicomplexa: Cryptosporidiidae) in pigs (*Sus scrofa*). J Parasitol 90:769–773
Ryan UM, Bath C et al (2005) Sheep may not be an important zoonotic reservoir for *Cryptosporidium* and *Giardia* parasites. Appl Environ Microbiol 71:4992–4997
Ryan UM, Power M et al (2008) *Cryptosporidium fayeri* n. sp. (Apicomplexa: Cryptosporidiidae) from the Red Kangaroo (*Macropus rufus*). J Eukaryot Microbiol 55:22–26
Santín M, Dixon BR et al (2005) Genetic characterization of *Cryptosporidium* isolates from ringed seals (*Phoca hispida*) in Northern Quebec, Canada. J Parasitol 91:712–716
Satoh M, Hikosaka K et al (2003) Characteristics of a novel type of bovine *Cryptosporidium andersoni*. Appl Environ Microbiol 69:691–692
Schulze C, Kämmerling J et al (2012) *Cryptosporidium baileyi*–infection in Red-breasted Merganser (*Mergus serrator*) ducklings from a zoological garden. Berl Munch Tierarztl Wochenschr 125:428–431
Sevá AP, Funada MR et al (2011a) Genotyping of *Cryptosporidium* spp. from free-living wild birds from Brazil. Vet Parasitol 175:27–32
Sevá AP, Sercundes MK et al (2011b) Occurrence and molecular diagnosis of *Cryptosporidium serpentis* in captive snakes in São Paulo, Brazil. J Zoo Wildl Med 42:326–329
Silverlås C, Mattsson JG et al (2012) Zoonotic transmission of *Cryptosporidium meleagridis* on an organic Swedish farm. Int J Parasitol 42:963–967
Sitjà-Bobadilla A, Padrós F et al (2005) Epidemiology of *Cryptosporidium molnari* in Spanish gilthead sea bream (*Sparus aurata* L.) and European sea bass (*Dicentrarchus labrax* L.) cultures: from hatchery to market size. Appl Environ Microbiol 71:131–139
Sitjà-Bobadilla A, Pujalte MJ et al (2006) Interactions between bacteria and *Cryptosporidium molnari* in gilthead sea bream (*Sparus aurata*) under farm and laboratory conditions. Vet Parasitol 142:248–259
Slapeta J (2006) *Cryptosporidium* species found in cattle: a proposal for a new species. Trends Parasitol 22:469–474
Slapeta J (2011) Naming of *Cryptosporidium pestis* is in accordance with the ICZN Code and the name is available for this taxon previously recognized as *C. parvum* 'bovine genotype'. Vet Parasitol 177:1–5
Slavin D (1955) *Cryptosporidium meleagridis* (sp. nov.). J Comp Pathol 65:262–266
Templeton TJ, Enomoto S et al (2010) A genome-sequence survey for *Ascogregarina taiwanensis* supports evolutionary affiliation but metabolic diversity between a Gregarine and *Cryptosporidium*. Mol Biol Evol 27:235–248
Tiangtip R, Jongwutiwes S (2002) Molecular analysis of *Cryptosporidium* species isolated from HIV-infected patients in Thailand. Trop Med Int Health 7:357–364
Tilley M, Upton SJ et al (1990) A comparative study of the biology of *Cryptosporidium serpentis* and *Cryptosporidium parvum* (Apicomplexa:Cryptosporidiidae). J Zoo Wildlife Med 21:463–467
Traversa D (2010) Evidence for a new species of *Cryptosporidium* infecting tortoises: *Cryptosporidium ducismarci*. Parasit Vectors 3:21–23

Traversa D, Iorio R et al (2008) *Cryptosporidium* from tortoises: genetic characterisation, phylogeny and zoonotic implications. Mol Cell Probes 22:122–128

Tyzzer EE (1907) A sporozoan found in the peptic glands of the common mouse. Proc Soc Exp Biol Med 5:12–13

Tyzzer EE (1910) An extracellular coccidium, *Cryptosporidium muris* (gen. et sp. nov.), of the gastric glands of the common mouse. J Med Res 23:487–511

Tyzzer EE (1912) *Cryptosporidium parvum* (sp. nov.), a coccidium found in the small intestine of the common mouse. Arch Protisenk 26:394–412

Tyzzer EE (1929) Coccidiosis in gallinaceous birds. Am J Hyg 10:269–383

Tzipori S, Widmer G (2000) The biology of *Cryptosporidium*. Contrib Microbiol 6:1–32

Upton SJ (2003) *Cryptosporidium*: they probably taste like chicken. In: Thomson RCA, Armson A et al (eds) *Cryptosporidium*: from molecules to disease. Elsevier, Amsterdam, pp 3–10

Upton SJ, Current WL (1985) The species of *Cryptosporidium* (Apicomplexa: Cryptosporidiidae) infecting mammals. J Parasitol 71:625–629

Upton SJ, McAllister CT et al (1989) *Cryptosporidium* spp. in wild and captive reptiles. J Wildl Dis 25:20–30

van Zeeland YR, Schoemaker NJ et al (2008) Upper respiratory tract infection caused by *Cryptosporidium baileyi* in three mixed-bred falcons (*Falco rusticolus x Falco cherrug*). Avian Dis 52:357–363

Vetterling JM, Jervis HR et al (1971) *Cryptosporidium wrairi* sp. n. from the guinea pig *Cavia porcellus*, with an emendation of the genus. J Protozool 18:243–247

Waldron LS, Cheung-Kwok-Sang C et al (2010) Wildlife-associated *Cryptosporidium fayeri* in human, Australia. Emerg Infect Dis 16:2006–2007

Waldron LS, Dimeski B et al (2011) Molecular epidemiology, spatiotemporal analysis, and ecology of sporadic human cryptosporidiosis in Australia. Appl Environ Microbiol 77:7757–7765

Wang R, Jian F et al (2010) Large-scale survey of *Cryptosporidium* spp. in chickens and Pekin ducks (*Anas platyrhynchos*) in Henan, China: prevalence and molecular characterization. Avian Pathol 39:447–451

Wang R, Qi M et al (2011) Prevalence of *Cryptosporidium baileyi* in ostriches (*Struthio camelus*) in Zhengzhou, China. Vet Parasitol 175:151–154

Wang R, Wang F et al (2012) *Cryptosporidium* spp. in quails (*Coturnix coturnix japonica*) in Henan, China: molecular characterization and public health significance. Vet Parasitol 187:534–537

Wang L, Zhang H et al (2013) Zoonotic *Cryptosporidium* species and *Enterocytozoon bieneusi* genotypes in HIV-positive patients on antiretroviral therapy. J Clin Microbiol 51:557–563

Widmer G, Sullivan S (2012) Genomics and population biology of *Cryptosporidium* species. Parasite Immunol 34:61–71

Xiao L (2010) Molecular epidemiology of cryptosporidiosis: an update. Exp Parasitol 124:80–89

Xiao L, Ryan UM (2008) Molecular epidemiology. In: Fayer R, Xiao L (eds) *Cryptosporidium* and cryptosporidiosis, 2nd edn. CRC Press, Boca Raton, pp 119–172

Xiao L, Morgan UM et al (1999a) Genetic diversity within *Cryptosporidium parvum* and related *Cryptosporidium* species. Appl Environ Microbiol 65:3386–3391

Xiao L, Escalante H et al (1999b) Phylogenetic analysis of *Cryptosporidium* parasites based on the small-subunit rRNA gene locus. Appl Environ Microbiol 65:1578–1583

Xiao L, Alderisio K et al (2000) Identification of species and sources of *Cryptosporidium* oocysts in storm waters with a small-subunit rRNA-based diagnostic and genotyping tool. Appl Environ Microbiol 66:5492–5498

Xiao L, Sulaiman IM et al (2002a) Host adaptation and host–parasite co-evolution in *Cryptosporidium*: implications for taxonomy and public health. Int J Parasitol 32:1773–1785

Xiao L, Bern et al (2002b) Identification of the *Cryptosporidium* pig genotype in a human patient. J Infect Dis 185:1846–1848

Xiao L, Fayer R et al (2004a) *Cryptosporidium* taxonomy: recent advances and implications for public health. Clin Microbiol Rev 17:72–97
Xiao L, Ryan UM et al (2004b) Genetic diversity of *Cryptosporidium* spp. in captive reptiles. Appl Environ Microbiol 70:891–899
Xiao L, Fayer R et al (2007) Response to the newly proposed species *Cryptosporidium pestis*. Trends Parasitol 23:41–42
Xiao L, Ryan UM et al (2012) *Cryptosporidium tyzzeri* and *Cryptosporidium pestis*: which name is valid? Exp Parasitol 130:308–309
Yang R, Fenwick S et al (2011) Identification of novel *Cryptosporidium* genotypes in kangaroos from Western Australia. Vet Parasitol 179:22–27
Zanguee N, Lymbery JA et al (2010) Identification of novel *Cryptosporidium* species in aquarium fish. Vet Parasitol 174:43–48
Zhou L, Kassa H et al (2004) Host-adapted *Cryptosporidium* spp. in Canada geese (*Branta canadensis*). Appl Environ Microbiol 70:4211–4215
Zhu G, Keithly JS et al (2000) What is the phylogenetic position of *Cryptosporidium*? Int J Syst Evol Microbiol 50(Pt 4):1673–1681
Ziegler PE, Wade SE et al (2007) Prevalence of *Cryptosporidium* species in wildlife populations within a watershed landscape in southeastern New York State. Vet Parasitol 147:176–184
Zintl A, Neville D et al (2007) Prevalence of *Cryptosporidium* species in intensively farmed pigs in Ireland. Parasitology 134:1575–1582
Zylan K, Bailey T et al (2008) An outbreak of cryptosporidiosis in a collection of stone curlews (*Burhinus oedicnemus*) in Dubai. Avian Pathol 37:521–526

Chapter 2
Epidemiology of Human Cryptosporidiosis

Simone M. Cacciò and Lorenza Putignani

Abstract *Cryptosporidium* species are protozoan parasites that infect the epithelial cells of the gastrointestinal tract of vertebrates. In humans, cryptosporidiosis is usually a self-limiting infection in immunocompetent individuals, but severe diarrhoea and dissemination to extra-intestinal sites can occur in high-risk individuals, such as the very young, the elderly and immunosuppressed individuals, particularly those with HIV infection. The oocyst, the infectious stage of *Cryptosporidium*, is immediately infectious upon excretion with the host faeces, which favours direct transmission. Oocysts have the capacity to persist in the environment and to withstand standard water treatment and some species of *Cryptosporidium*, particularly *C. parvum*, have a wide host range and can be transmitted to humans by direct contact with animals or through ingestion of water and food contaminated with oocysts. Due to the presence of multiple transmission routes, the epidemiology of cryptosporidiosis is complex. The investigation of sporadic cases and outbreaks of cryptosporidiosis has contributed to a better understanding of risk factors and infection sources. Genotyping techniques have enabled a better understanding of the epidemiology of cryptosporidiosis in different geographical, seasonal and socioeconomic context.

S.M. Cacciò (✉)
Department of Infectious, Parasitic and Immunomediated Diseases, Istituto Superiore di Sanità, Viale Regina Elena, 299, Rome 00161, Italy
e-mail: simone.caccio@iss.it

L. Putignani
Unit of Parasitology and Unit of Metagenomics, Bambino Gesù Children's Hospital, IRCCS, Rome, Italy

Piazza Sant'Onofrio, 4, Rome, 00165, Italy
e-mail: lorenza.putignani@opbg.net

S.M. Cacciò and G. Widmer (eds.), *Cryptosporidium: parasite and disease*,
DOI 10.1007/978-3-7091-1562-6_2, © Springer-Verlag Wien 2014

2.1 Introduction

Cryptosporidia are obligate intracellular parasites of many species from all vertebrate classes. First described in laboratory mice by Tyzzer in 1912, *Cryptosporidium* was recognized as a cause of diarrheal disease in animals and then in humans during the 1970s. In the following decade, cryptosporidiosis emerged worldwide as a common cause of severe or life-threatening infection in immunocompromised patients, especially those with AIDS, and of acute, self-limiting gastroenteritis in otherwise healthy subjects, especially children (see Chap. 9).

In humans, infection is caused mainly by the zoonotic species *Cryptosporidium parvum*, which is highly prevalent in young livestock, and *Cryptosporidium hominis*, which is essentially a human parasite. With the development of improved genotyping methods, an increasing number of species and genotypes have been recognized as human pathogens, albeit with low prevalence (Putignani and Menichella 2010). The epidemiology of the infection involves both direct transmission from animals to humans or from person to person, as well as indirect transmission through ingestion of water and food contaminated with infectious oocysts (Cacciò et al. 2005; Smith et al. 2006b).

This chapter focuses on the epidemiology of human infections. The molecular epidemiology of human cryptosporidiosis is presented in Chap. 3.

2.2 Life Cycle of *Cryptosporidium*

The *Cryptosporidium* life cycle is represented in Fig. 2.1, and the major phases are briefly described here. The life cycle of the parasite is completed within a single host (i.e., a monoxenous cycle), and involves both asexual and sexual replication. For a more detailed account, the readers are referred to previously published articles (Chen et al. 2002; Smith et al. 2005; Fayer 2008).

Excystation: after ingestion of oocysts by a susceptible host, the first step towards infection is excystation, the process by which the oocyst wall opens along a suture to allow the release of four infectious sporozoites. This process has been studied under in vitro conditions and several factors that mimic the transit through the acidic stomach to the alkaline small intestine have been shown to enhance excystation. In particular, temperature and pH appear to be the most important triggers. *Cryptosporidium* species that infect the stomach of the host (like *Cryptosporidium muris*) respond more rapidly to those triggers compared to species that infect the intestine of the host, indicating the need of gastric species to rapidly excyst and release the sporozoites upon ingestion (Widmer et al. 2007). Therefore, the role of host derived triggers during the excystation process varies for different *Cryptosporidium* species (Smith et al. 2005).

Attachment and invasion: like other apicomplexans, *Cryptosporidium* possesses an apical complex formed by the apical ring, the conoid and secretory

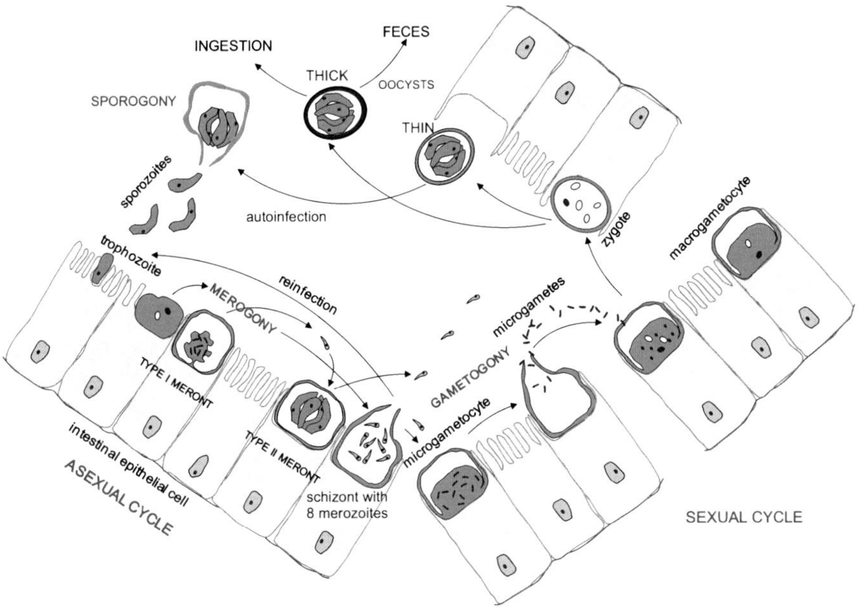

Fig. 2.1 Life cycle of parasite. The infection is acquired through the ingestion of sporulated oocysts. Motile sporozoites, from opened oocyst, attach to intestinal epithelial cells. The trophozoite undergoes an asexual replication (merogony), resulting in the production of eight merozoites (type I meronts). Merozoites, released into the intestinal lumen, infect new intestinal epithelial cells, and originate type II meronts, characterized by four merozoites. The merozoites can undergo a sexual cycle (gametogony) and develop into macrogametocytes. The microgametocyte produces numerous microgametes which are released into the intestinal lumen. A microgamete will fuse with a macrogamete and the resulting zygote undergoes sporogony. Fully sporulated thick and thin oocysts are shed into the intestinal lumen at the completion of sporogony. The infectious thick oocysts are excreted in the feces, thus completing the life cycle. An autoinfection in which excystation takes place within the same host may also be possible and is mediated by 'thin-walled' oocysts

organelles (a single rhoptry and micronemes). The apical complex is intimately involved in the process of attachment and invasion of host cells (Borowski et al. 2008). When the sporozoite contacts the host cell membrane, the rhoptry extends to the attachment site, while micronemes and dense granules move to the apical region. The content of secretory organelles is released, which triggers the process of recognition and attachment to the host cell, invasion, and formation of the parasitophorous vacuole in which the parasite replicates. The sporozoites within the parasitophorous vacuole are not directly in contact with the host cell, and occupy a unique intracellular but extracytoplasmic niche, typical of *Cryptosporidium* parasites (also named parasitophorous sac, Valigurová et al. 2008). A complex process at the host-parasite interface leads to the formation of the so-called feeder organelle, which is thought to be essential in salvaging nutrients and directly separates the parasite cytoplasm from the affected host cell (Umemiya et al. 2005). As the internalization process progresses, the sporozoite differentiates into a spherical trophozoite.

Asexual reproduction: merogony is characterized by the division of the trophozoite nucleus and cytoplasm to generate meronts. In *C. parvum*, there are two types of meronts, type I and type II. Type I meronts have six or eight nuclei. Each gives rise to a merozoite, which is structurally similar to the sporozoite. Mature merozoites leave the meront and invade other host cells, where they can develop into another Type I or into a Type II meront. Type II meronts have four nuclei and generate four merozoites.

Sexual reproduction: upon infection of new host cells, merozoites from Type II meronts can initiate the sexual phase by differentiating into either a microgamont or a macrogamont. Nuclear division in the microgamont leads to the production of numerous microgametes (equivalent to sperm cells) that are released from the parasitophorous vacuole. Each macrogamont (equivalent to an ovum) may be fertilized by a microgamete. The product of fertilization, the zygote, develops into an oocyst.

Sporogony: the zygote differentiates into four sporozoites (sporogony) within the oocyst. It is thought that fully sporulated oocysts are released into the lumen of the intestine and pass out of the body with the faeces, where they are immediately infectious for other susceptible hosts. Oocysts that sporulate in the respiratory tract are found in nasal secretions and sputum (Mor et al. 2010). *Cryptosporidium* oocysts are spherical, measuring only 3–6 μm in diameter. Each contains four haploid sporozoites and possesses a tick wall.

It is believed that some of the oocysts possess a thin wall and can cause autoinfection in the same parasitized host by liberating their sporozoites in the gut lumen. The process of autoinfection is believed to occur only in species of *Cryptosporidium* and *Caryospora*. Upon their release, sporozoites undergo the developmental processes of schizogony, gametogony and sporogony in enterocytes of the same infected host. Therefore, the life cycle of *Cryptosporidium* ensures the production of very large numbers of infective oocysts, due to the recycling of merozoites to produce further type I generations of schizonts, and the endogenous re-infection from thin-walled oocysts.

2.3 *Cryptosporidium* Species Infecting Humans

The first human cases of cryptosporidiosis were reported in 1976 (Nime et al. 1976; Meisel et al. 1976). The patients were a 3-year-old child and a 39-year old individual who had severe bullous pemphigoid (a skin disease characterized by blisters) and received treatment with cyclophosphamide and prednisolone. Both patients lived on a farm with cattle and had a dog. They presented with severe watery diarrhea. Diagnosis was based on microscopic examination (including electron microscopy) of rectum and jejunal biopsy specimens. At that time, identification at the level of species was not possible, and infections were attributed to

Cryptosporidium spp. In 1978, it was shown that oocysts are shed with the feces of infected hosts (Pohlenz et al. 1978) and since then, the diagnosis of cryptosporidiosis has been based on the demonstration of oocysts in feces.

A series of experiments performed using parasite isolates from different hosts (a calf, a lamb, a human and a deer) showed that *Cryptosporidium* could be transmitted to newborn animals (mice, rats, guinea pigs, piglets, calves and lambs), thus demonstrating a lack of host specificity and underlining a zoonotic potential (Tzipori et al. 1980). It was therefore argued that a single species, *C. parvum*, was the causative agent of cryptosporidiosis in mammals, including man.

The role of *Cryptosporidium* as a serious human pathogen was firmly demonstrated in the 1980s in individuals with HIV/AIDS, who experienced a persistent and life threatening infection, often involving parasite dissemination to the hepatobiliary and the respiratory tracts in addition to the entire gastrointestinal (GI) tract (Ma and Soave 1983; Navin and Juranek 1984). The next major event that attracted worldwide interest towards the parasite was the 1993 massive waterborne outbreak in Milwaukee, Wisconsin, that involved an estimated 403,000 person (MacKenzie et al. 1995). This event demonstrated the ability of *Cryptosporidium* to resist water treatment and be transmitted through drinking water. Following these major events, molecular methods for the detection and identification of *Cryptosporidium* species on different matrices were actively developed and their application has dramatically changed our understanding of the taxonomy and epidemiology of *Cryptosporidium*.

In the early 1990s, the application of Southern blotting (Ortega et al. 1991), Western blotting (Nina et al. 1992) and isoenzymes profiles (Ogunkolade et al. 1993) provided the first evidence of genetic heterogeneity among *C. parvum* isolated from humans and livestock. These studies demonstrated for the first time that humans were infected with two types of *Cryptosporidium* parasites, one being apparently the same as found in cattle and the other exclusively found in humans. However, due to the large amount of biological material needed, these techniques have not been used to characterize field isolates, and have been largely replaced by DNA amplification techniques (PCR). Different assays, including PCR followed by restriction length fragment polymorphism (PCR-RFLP) (Ortega et al. 1991), PCR followed by sequencing (Morgan et al. 1997), Random Amplification of Polymorphic DNA (RAPD) (Morgan et al. 1995), and length polymorphisms of simple DNA repeats (Cacciò et al. 2000), were developed to investigate genetic heterogeneity among isolates.

A number of studies, based on the analysis of single or multiple genetic markers (e.g., Peng et al. 1997; Spano et al. 1998), confirmed the presence of two genetically distinct subgroups within *C. parvum*, which were referred in the literature to as 'human' and 'cattle,' H and C, or Type 1 and Type 2. This result was subsequently confirmed in many laboratories and demonstrated the existence of two distinct transmission cycles of human cryptosporidiosis, one comprising ruminants and humans (potentially zoonotic cycle) and the other exclusively comprising humans (solely anthroponotic cycle).

In 2002, on the basis of accumulating observations of genotypic and biological differences, a new species, *C. hominis*, was finally proposed for *C. parvum* parasites exclusively infecting humans (Morgan-Ryan et al. 2002; see Chap. 1 for the taxonomy of *Cryptosporidium*).

The increasing application of genotyping techniques in epidemiological surveys has demonstrated that the large majority (>90 %) of human cases of cryptosporidiosis are due to *C. hominis* and C. *parvum* (Xiao 2010). However, it is now clear that other species have the potential to cause infection in humans, including *Cryptosporidium meleagridis* and, more occasionally, *C. canis*, *C. felis*, *C. ubiquitum* and *C. viatorum*, and other *Cryptosporidium* genotypes of unknown taxonomic status. The prevalence of these less common species and genotypes varied geographically: for example, *C. meleagridis* was found to be as prevalent as *C. parvum* in children from Peru and in Thailand (Cama et al. 2008; see Chap. 3).

2.4 Incubation Period

Based on evidence from experimental infections, it has been estimated that the incubation period is between 5 and 7 days. Data from experimental infection of healthy volunteers (Okhuysen et al. 1996; Chappel et al. 2006, 2011), and investigations of waterborne and foodborne outbreaks are the main source of information.

Experimental infections of healthy volunteers have been performed using the three main human pathogens, namely *C. parvum*, *C. hominis* and *C. meleagridis*. For *C. parvum*, volunteers were challenged with different inocula (10 to >10,000 oocysts) of three strains of animal origin (Iowa, TAMU and UCP, MD, Okhuysen et al. 1996). Results showed that 12 (86 %) of 14 volunteers who received the TAMU isolate developed diarrhoea, as did 15 (52 %) of 29 who received the Iowa isolate and ten (59 %) of 17 who received the UCP isolate (Okhuysen et al. 1996). The mean incubation period was 7.7 days (for IOWA), 7 days (for UCP) and 4 days (for TAMU).

In the study with *C. hominis*, 21 healthy adults were challenged with 10–500 oocysts of the TU502 (Xu et al. 2004) isolate, and 13 developed diarrhoea, while nine had oocysts detected in faecal samples (Chappel et al. 2006). The mean incubation period was 5.4 days (range, 2–10 days). Finally, in the case of *C. meleagridis*, the study involved five volunteers, who were challenged with 10^5 oocysts of the isolate TU1867 (Akiyoshi et al. 2003). The incubation period was of 5.3 days (range, 4–7 days) (Chappel et al. 2011).

During the large waterborne outbreak of cryptosporidiosis in Milwaukee, the mean incubation period was estimated to range from 3 to 7 days, but it appeared to be shorter in the elderly (5–6 days) compared to either children (7 days) or adults (8 days) (Naumova et al. 2003).

In 1996, another large waterborne outbreak occurred in Japan, and was caused by contamination of the town's potable water (Yamamoto et al. 2000). An estimated number of 9,140 individuals were affected. The median incubation period for the 14 persons for whom this calculation was possible was 6.4 days (range, 5–8 days).

In 2002, an outbreak of cryptosporidiosis occurred among visitors to a public swimming pool in Sweden and affected an estimated with about 800–1,000 individuals (Insulander et al. 2005). The median incubation period was estimated at 5 days (range, 2–13 days).

Three investigations of foodborne cryptosporidiosis have also provided data on incubation period. In 1993, an outbreak linked to consumption of contaminated apple cider occurred in central Maine (Millard et al. 1994). The median incubation period was 6 days (range, 10 h to 13 days).

In 1997, an outbreak of acute gastroenteritis occurred among members of a group attending a dinner banquet catered by a restaurant in Spokane (Anon 1998). Foodborne transmission was implicated through a contaminated ingredient in multiple menu items. The incubation period was estimated between 3 and 9 days. In 2006, acute gastroenteritis caused by *C. parvum* was reported by four members from the same company who had eaten a raw meat dish called "Yukke: Korean-style beef tartar" and raw liver at a rotisserie in Sakai City, Japan (Yoshida et al. 2007). Based on information from interviews, the median incubation period was 5.5 (range, 5–7 days).

Further data comes from other investigations. A study of an outbreak of GI illness among a class of 96 undergraduate veterinary students in New Zealand (Grinberg et al. 2011) indicated a median incubation period of 5 days (range, 0–11 days).

Variability in the estimated duration of the latent period can be attributed to inaccuracies in the estimation of the time from exposure to the onset of symptoms, or to difference in inoculum size (Chappel et al. 2006), but may also reflect alterations in the incubation period due to partial immunity derived from prior exposure to the parasite. Phenotypic differences among parasite isolates may also explain the observed range.

2.5 Asymptomatic Infection

Little information is available on asymptomatic carriage of *Cryptosporidium* in humans. In studies of experimental infection of healthy volunteers with *C. parvum*, it was noted that some individuals passed oocysts in their stools in the lack of overt symptoms (diarhoea), indicating that asymptomatic shedding of the parasite occurs (Okhuysen et al. 1999). In contrast, asymptomatic shedding was not seen in any of the volunteers experimentally challenged with *C. hominis* oocysts (Chappel et al. 2006).

A prospective study in the US found asymptomatic cryptosporidiosis in 12 of 78 (6.4 %) immunocompetent and 11 of 50 (22 %) immunodeficient children (Pettoello-Mantovani et al. 1995). By comparison, *Cryptosporidium* was found in

4.4 % of immunocompetent and 4.8 % of immunodeficient children of a control symptomatic population. A recent study in the UK screened 230 asymptomatic children in preschool day care centres, using a highly sensitive detection protocol based on immunomagnetic separation and DNA typing (Davies et al. 2009). The observed prevalence (1.3 %) was lower than that from the study of Pettoello-Mantovani et al. (1995), that was based on the less sensitive microscopic technique. However, the detection of *C. ubiquitum* and of the skunk genotype in those asymptomatic children raises the possibility that some species/genotypes may have low pathogenicity and be more common than previously thought.

In Scandinavian countries (Denmark, Sweden, Norway and Finland) a meta-analysis of asymptomatic and symptomatic cryptosporidiosis in adults showed a lower prevalence (0.99 %) in the former compared to the latter (2.99 %). In a community-based study in Melbourne, Australia, 1,091 faecal specimens from asymptomatic individuals were screened for the presence of bacteria, viruses and parasites (Hellard et al. 2000), and *Cryptosporidium* was found in four samples (0.4 %). The role of carriers in the transmission of *Cryptosporidium* was also suggested by a large-scale case–control study of sporadic cases in the United Kingdom (Hunter et al. 2004), that identified changing diapers as an independent risk factor for infection with *C. hominis*, even if the child did not have diarrhoea.

The prevalence of asymptomatic infection in other countries is only partially known. A study of 377 Aymara school students (5–19 years of age) from villages of the northern Bolivian Altiplano (Esteban et al. 1998) found a very high prevalence of *Cryptosporidium* (31.6 %) based on microscopy. The authors argued that the mild or asymptomatic infections observed in children were due to some level of immunity that develops after continuous exposure to the parasite.

In Jeddah, Saudi Arabia, a study of asymptomatic children from nurseries found that 9 of 190 (4.7 %) were positive for *Cryptosporidium*, as compared to 20 of 63 (32 %) children with diarrhoea from paediatric clinics of the same city (Al-Braiken et al. 2003). In Makkah, Saudi Arabia, *Cryptosporidium* was detected in 4 % of 589 stool samples from asymptomatic school children aged 7–12 years from 13 primary schools (Al-Harthi 2004). Finally, in a study of 276 asymptomatic aboriginal (Orang Asli) children (141 boys and 135 girls, aged 2–15 years) living in villages in the Malaysia (Al-Mekhlafi et al. 2011), *Cryptosporidium* was detected in 20 (7.2 %) children.

An interesting study aimed at determining the contribution of asymptomatic immigrants in the spreading of the disease in the Kashmir region (India), where cryptosporidiosis is not considered to be endemic. Analysis of stool samples from 45 non-diarrheic and nine diarrheic HIV-infected individuals revealed that all were carriers of *Cryptosporidium* spp. (Masarat et al. 2012). Remarkably, epidemiological traits revealed that the asymptomatic individuals were non-Kashmiri army personals and travellers (immigrants), while local emigrant merchants represented symptomatic cases. This suggests that the non-diarrheic HIV positive population may be a potential source of endemic spread of cryptosporidiosis, and that obligatory laboratory testing in HIV positive immigrant population, like merchants and travellers, regardless of symptoms, should be mandatory to understand patterns of transmission.

2.6 Symptomatic Infection

The symptomatic period of infection is characterized by diarrhoea, abdominal pain, nausea or vomiting, mild fever, anorexia, malaise, fatigue and weight loss (Fayer and Ungar 1986; Casemore 1990). Diarrhoea can be of sudden onset and is generally watery and voluminous; between three and six stools (but sometimes many more) may be passed each day, which are sometimes offensive and may contain mucus. Symptoms usually last up to 3 weeks, but some patients experience chronic diarrhoea of a month or longer. Oocysts may continue to be shed for a mean period of 7 days (range 1–15 days) after symptoms have ceased, although exceptionally for up to 2 months (Jokipii and Jokipii 1986) (see Chap. 9).

2.7 Risk Factors

Our knowledge of risk factors for acquiring cryptosporidiosis is mainly derived from investigations of outbreaks, albeit the majority of human infection is sporadic. Outbreak investigations have demonstrated the existence of multiple transmission routes, including contact with infected animals, person-to-person transmission in households and care settings, consumption of contaminated foods and drinks, consumption of water from private and public supplies, exposure to recreational water in swimming pools or water parks, and travel to endemic countries (Nichols et al. 2009) (Fig. 2.2). It is important to stress that risk factors are likely to differ for *C. parvum* and *C. hominis* and in different geographical settings, due to the difference in host range, transmission cycles and prevalence. This underscores the importance of identifying species or genotype in epidemiologic studies. However, molecular typing has been used only in a few case–control studies, both from developed (Hunter et al. 2004; Lake et al. 2007) and developing countries (Cama et al. 2008; Molloy et al. 2011).

In industrialised countries, case–control studies have been conducted in the US (Roy et al. 2004), the UK (Hunter et al. 2004; Lake et al. 2007), and Australia (Robertson et al. 2002). In all these studies, statistically significant risk factors, identified by multivariate analysis, included contact with persons with diarrhoea, particularly young children, and contact with cattle, especially calves. Further, travel abroad was significantly associated with an increased risk in the US and UK studies (Roy et al. 2004; Hunter et al. 2004). History of travel abroad was excluded in the Australian study to allow focus on endemic disease (Robertson et al. 2002). In addition, the US study identified swimming in fresh water as a risk factor and the Australian study found swimming in a chlorinated swimming pool as

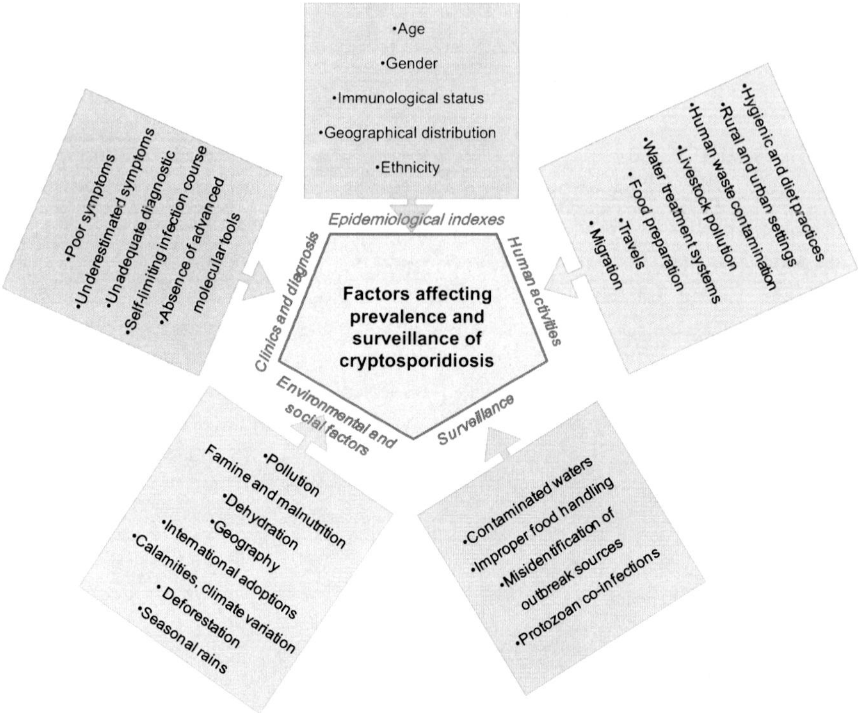

Fig. 2.2 Factors affecting prevalence and adequate surveillance of cryptosporidiosis

a risk factor. Both the UK and Australian studies identified a dose dependent risk associated with drinking unboiled water, which was also reported in a regional study in the UK (Goh et al. 2004). Negative associations with eating ice cream and raw vegetables was found in the UK and US studies.

A more recent study from the UK (Lake et al. 2007) investigated the role of wider environmental and socioeconomic factors (e.g. water supply, socioeconomic status, land use, livestock densities and healthcare accessibility) upon human cryptosporidiosis. By comparing 3,368 laboratory-confirmed cases to an equal number of controls, the authors concluded that risk factors for *C. hominis* and *C. parvum* must be considered separately. Indeed, for *C. hominis* cases the strongest risks factors were living in areas with many higher socioeconomic status individuals, living in areas with a high proportion of young children and living in urban areas. In contrast, agricultural land use surrounding the place of residence and the water supply were significant risk factors for *C. parvum* illness (Lake et al. 2007).

Despite the higher prevalence of cryptosporidiosis in developing and tropical countries, few studies have been conducted to identify risk factors in those areas. In Peru, a total of 156 cases of cryptosporidiosis were found in 109 of 553 children during a 4-year longitudinal birth cohort study (Cama et al. 2008). Investigation into risk factors did not identify statistically significant associations between *Cryptosporidium* spp. and any of the variables considered (basic aspects of sanitation and zoonotic, foodborne, and waterborne transmission), possibly because children are exposed to these parasites through different transmission routes, which makes single exposure variables difficult to demonstrate (Cama et al. 2008) (Fig. 2.2).

In Indonesia, a study of 917 patients with acute diarrhea (715 in-patients and 202 out-patients from the Hospital of University of Airlanaga in Surabaya) found *C. parvum* oocysts in 26 (2.8 %) of the patients and in 15 (1.4 %) of 1,043 control patients. Investigation of risk factors by multiple logistic regression indicated that contact with pets (cats), rainy season, occurrence of a flood, and crowded living conditions were significant risk factors for cryptosporidiosis (Katsumata et al. 1998).

A 2-year study in the Nile River Delta in Egypt examined risk factors for cryptosporidiosis in children with diarrhea (Abdel-Messih et al. 2005). A prevalence of 17 % (241 of 1,275) was observed, and clinical findings included vomiting, persistent diarrhoea and the need for hospitalization. Children <12 months of age were 2.4 times more likely to be infected with *Cryptosporidium* ($p<0.01$) and children 12–23 months were 1.9 ($p<0.05$) times more likely to be infected with the organism as compared to older children. Breastfeeding had a trend towards protection against *Cryptosporidium*-associated diarrhoea ($p = 0.07$).

Finally, a study in Nigeria found that 134 of 692 children (19.4 %) were infected with *Cryptosporidium* spp. The study also provided information on the species for 49 isolates (Molloy et al. 2011). Using generalized linear mixed-effects models, risk factors were identified for all *Cryptosporidium* infections, as well as for *C. hominis* and *C. parvum* both together and separately. Malaria and absence of *Ascaris* infection were risk factors for all *Cryptosporidium* infections, whereas stunting and younger age were highlighted as risk factors for *C. hominis* infections. Stunting and malaria were identified as risk factors for *C. parvum* infection (Molloy et al. 2011).

Together, studies in developing countries indentified several risk factors: age <2 years, absence of breastfeeding, contact with pets, living in overcrowded conditions, low birth weight, male gender, malnourishment and co-infections as significant risk factors for cryptosporidiosis (Putignani and Menichella 2010) (Fig. 2.2).

The role of host genetics in susceptibility to infection was studied in a cohort of 226 Bangladeshi children aged 2–5 years, who were prospectively followed for >3 years (Kirkpatrick et al. 2008). Ninety-six children (42.5 %) were diagnosed with *Cryptosporidium* infection. A total of 51 (22.6 %) had asymptomatic infection, whereas 58 (25.7 %) had symptomatic cryptosporidiosis, of whom 17 (29.3 %) had recurrent disease. Infected children, both asymptomatic and symptomatic, were more likely to carry the human leukocyte antigen (HLA) class II DQB1*0301 allele ($P = 0.009$), and a strong association was found between the DQB1*0301/DRB1*1101 haplotype and the development of both asymptomatic and symptomatic infection ($P = 0.009$). Infected children were also more likely to carry the

B*15 HLA class I allele. This was the first description of a genetic component of the immune response to *Cryptosporidium* infection, which included HLA class I and II alleles (Kirkpatrick et al. 2008).

2.8 Post-Infectious Sequelae

Little is known on the long-term consequences of *Cryptosporidium* infection in humans. The gut epithelium generally recovers after resolution of symptoms, but there are some indications that long-term sequelae may arise (Cacciò et al. 2009).

In immunocompetent individuals, the medium-term health effects of cryptosporidiosis is characterized by the recurrence of loss of appetite, vomiting, abdominal pain, and diarrhoea, independently if the patients have been infected by *C. parvum* or by *C. hominis* (see Chap. 9).

Interestingly, significant differences in the occurrence of extra-intestinal symptoms as sequelae of cryptosporidiosis following infections with *C. hominis* and *C. parvum* have been reported in immunocompetent persons (Hunter et al. 2004). Indeed, eye pain and recurrent headache are associated with *C. hominis* infection but not with *C. parvum* infection, whereas other symptoms, such as fatigue and joint pains, are present after infections with both species, but are significantly more common after *C. hominis* infection (Hunter et al. 2004). The relatively small number of case patients who reported joint pains (13 control subjects versus 36 case patients) means that the establishment of firm conclusions about the nature and distribution of joint symptoms will require further investigations (Hunter et al. 2004).

The potential association of *Cryptosporidium* with the inflammatory bowel syndrome (IBS), a common GI disorder characterized by abdominal pain and alterations in bowel habits, is still uncertain. Indeed, a study of intestinal mucosal biopsies and serology from patients with IBS did not support a major role for the parasite in its pathogenesis (Chen et al. 2001). However, experimental *C. parvum* infection in a rat model resulted in jejunal hypersensitivity to distension, which was also associated with activated mast cell accumulation at 50 days post-infection (Khaldi et al. 2009). These findings are consistent with the observations that IBS patients have a marked increase in mast cell numbers and higher tryptase concentrations in jejunal fluid. Thus, further studies are needed to understand the role that *Cryptosporidium* infection may have in the establishment of IBS.

A seronegative reactive arthritis secondary to cryptosporidiosis has been reported in adults (Hay et al. 1987; Ozgül et al. 1999; Collins and Highton 2004) and children (Shepherd et al. 1989; Cron and Sherry 1995), including one report of Reiter's syndrome (arthritis, conjunctivitis and urethritis) (Cron and Sherry 1995).

2.9 Burden of Disease

2.9.1 Overall Prevalence of Infection

In the developed countries, diarrhoea is the most common reason for missing work, while in the developing world, it is a leading cause of death. Internationally, the mortality rate is 5–10 million deaths each year (Nemes 2009). In this scenario, *Cryptosporidium* is a major cause of diarrheal disease, globally (Shirley et al. 2012). Unlike many common causes of infectious enteritis, control and treatment of this infection are still problematic. Indeed, control is focused mainly on prevention and no widely effective vaccine or drug-based intervention strategies are available (see Chap. 11). Furthermore, control strategies are particularly deficient for infections of severely immunocompromised individuals, the elderly, children or malnourished people, especially in developing countries. Cryptosporidiosis also presents a significant burden on immunocompetent individuals, and can have permanent effects on physical and mental development of children infected at an early age (Jex et al. 2011). Therefore, a tight monitoring of global infection is nowadays even more essential than in the past, because of the need to include control strategies for different categories of individuals, also in the absence of symptomatic evidence.

Moreover, *Cryptosporidium* is considered an emergent pathogen, often underappreciated for limitations or absence of appropriate diagnostics tools (e.g., microscopy often fails to detect low parasite load) and/or for only indirect association to severe medical growth faltering, malnutrition, and diarrheal mortality.

Recently, significant advances in molecular typing and subtyping analyses have yielded new insights into the epidemiology of cryptosporidiosis; however, in developing countries, point-of-care sites remain crucial for control of enteritis, especially in large areas such as the Indian subcontinent, where they are still lacking in many territories, while highly active antiretroviral therapy (HAART) is not equally distributed. In industrialized countries, outbreaks due to food-borne and water-borne transmission routes, growing transplantation surgery in routine medical practice and related long-term steroid treatments given in combination, are strongly contributing to the increase in disease incidence (Putignani and Menichella 2010; Shirley et al. 2012).

2.9.2 Seasonality

The incidence of cryptosporidiosis exhibits strong seasonality, with low endemic levels followed by pronounced seasonal outbursts (McLauchlin et al. 2000). In a recent meta-analysis on the seasonality of cryptosporidiosis, which was based on 61 published studies, increases in temperature and precipitation were associated with an increase in the incidence of cryptosporidiosis (Jagai et al. 2009).

Precipitation was found to be a strong seasonal driver for cryptosporidiosis in moist tropical climates. On the other hand, in temperate climates, the incidence of cryptosporidiosis peaked with an increase in temperature. The study further shows that while climatic conditions typically define a pathogen habitat area, meteorological factors affect timing and intensity of seasonal outbreaks (Jagai et al. 2009).

Clearly, the seasonal patterns tend to vary with location. For example, in India, the incidence of cryptosporidial diarrhea among children residing in the more temperate northern part of India correlated positively with temperature and negatively with humidity, but this correlation is not observed for children residing in the more tropical southern region. In another study from North-Eastern India, the highest prevalence of cryptosporidiosis was observed during the rainy months, and symptomatic as well as asymptomatic cryptosporidiosis in children was found to increase with increasing rainfall in Kolkata (Desai et al. 2012). In Kuwait, peak incidence occurred during the months of March and April, with no cases during the hottest months of July and August (Daoud et al. 1990).

Due to difference in transmission routes, it is likely that seasonal patterns may vary for different *Cryptosporidium* species. Indeed, genotyping studies conducted in the United Kingdom and New Zealand have found that human cases due to *C. parvum* peak in the late spring whereas those caused by *C. hominis* peak in the fall (McLauchlin et al. 2000; Learmonth et al. 2004). This was interpreted as increased exposure to animal oocysts following the calving and lambing season for *C. parvum*, and to increased travel, exposure to water and attendance to day care centres for *C. hominis*. The dramatic decline in human cases due to *C. parvum* observed in the United Kingdom after the large food-and-mouth outbreak of 2001, was initially explained by a reduced exposure of people to animals and, therefore, reduced zoonotic transmission (Smerdon et al. 2003). However, as the decline has continued, intervention measures, such as the introduction of water regulations and major structural changes in public water supply, are now believed to have played a major role (Sopwith et al. 2005).

2.9.3 Geographic Distribution

Cryptosporidiosis has a worldwide distribution, but the prevalence of infection is assumed to be higher in developing countries (Putignani and Menichella 2010). It should be noted, however, that in many developed and developing countries, surveillance systems for routine detection of cryptosporidiosis are not in place, and few studies have been conducted to estimate how prevalence can vary over time (Nichols 2008).

The distribution of the major *Cryptosporidium* species infecting humans varies geographically. Previous studies have shown that *C. parvum* and *C. hominis* are responsible for >90 % of human cases of cryptosporidiosis in most areas (Xiao and Ryan 2008). In the United Kingdom, in other European countries and in New

Zealand, *C. parvum* is responsible for slightly more infections than *C. hominis* (Xiao 2010). In the Middle East, *C. parvum* is the dominant species in humans. In contrast, *C. hominis* is responsible for more infections than *C. parvum* in the United States, Australia, China and Japan, as well as in most developing countries. Notably, the prevalence of *C. meleagridis* can be as high as that of *C. parvum* in certain areas of the world (Cama et al. 2008). Major differences in transmission routes may account for the observed differences in the distribution of *Cryptosporidium* species (Xiao 2010). The distribution of *C. parvum* and *C. hominis* can also vary within a single country; for example, *C. parvum* is more common than *C. hominis* in rural states in the United States, Ireland and New Zealand (Feltus et al. 2006; Zintl et al. 2009; Snel et al. 2009). Further geographic differences can be observed in the distribution of *C. parvum* and *C. hominis* subtypes (see Chap. 3).

2.9.4 Infection in Children

Cryptosporidiosis occurs more frequently in infants and children than in adults, both in developed and developing countries (Snelling et al. 2007; Xiao 2010; Putignani and Menichella 2010). This is likely to reflect both exposure and immunity. In the United States, endemic parasitic infections are more frequent than commonly perceived. Cryptosporidiosis occurs mainly in children aged 1–9 years, with the onset of infection peaking in the summer in association with communal swimming venues and recreational water use (Barry et al. 2013).

Needless to say, however, the impact of cryptosporidiosis is much higher in the poorest regions of the world. Globally, one in ten child deaths result from diarrhoeal disease during the first 5 years of life, and most occur in sub-Saharan Africa and south Asia (Liu et al. 2012). The role of various pathogens have been very recently studied in 9,439 children with moderate-to-severe diarrhoea and 13,129 control children without diarrhoea (Kotloff et al. 2013). Remarkably, *Cryptosporidium* was identified as a significant pathogen regardless of HIV prevalence and site of collection, and was the second most common pathogen in infants, and was associated with an increased risk of death in toddlers aged 12–23 months (Kotloff et al. 2013).

Children in developing countries are uniquely vulnerable to persistent infection because of the independent and synergistic effects of immune naiveté, malnutrition, and HIV infection (Mor and Tzipori 2008). In these areas, cryptosporidiosis is most prevalent during early childhood, with as many as 45 % of children experiencing the disease before the age of 2 years (Valentiner-Branth et al. 2003). *Cryptosporidium* also plays a causal role in childhood malnutrition and has been linked to impaired physical fitness in late childhood. Studies in Sub-Saharan Africa (reviewed by Mor and Tzipori 2008) have documented significantly higher cryptosporidiosis prevalence among malnourished children. It is difficult to ascertain the direction of this association, i.e., if malnutrition predisposes to infection or if infection actually impairs nutrient absorption and therefore causes weight loss and growth stunting. Similar findings have been reported from longitudinal studies in Peru (Checkley et al. 1998) and Brazil (Bushen et al. 2007).

The prevalence and predictors of *Cryptosporidium* infection, and its effect on nutritional status, have recently been explored among 276 children (aged 2–15 years) in aboriginal villages in the Malaysian state of Selangor (Al-Mekhlafi et al. 2011). Faecal smears were examined by microscopy while socio-economic data were collected using a standardized questionnaire. Nutritional status was assessed by anthropometric measurements. *Cryptosporidium* infection was detected in 7.2 % of the children, and was found to be significantly associated with low birth weight ($\leq$2.5 kg), being part of a large household, and prolonged breast feeding ($>$2 years).

The impact of *Cryptosporidium* on children has been demonstrated also in Arab countries, such as Egypt, Jordan, Kuwait, Libya, Palestine, Saudi Arabia and Tunisia. Prevalence rates of 1–43 % (mean 8.7 %) in diarrheic immunocompetent children and of 1–82 % (mean 41 %) in immunocompromised children and adults were reported (Ghenghesh et al. 2012). Higher infection rates were found in children living in rural and semi-urban areas than in those residing in urban areas. *Cryptosporidium*-associated diarrhoea occurred mainly in children aged 1 year or less and was inversely correlated with age (Ghenghesh et al. 2012).

Access to quality drinking water in poor regions of the world is thought to be important in limiting gastro-intestinal infections. A study was conducted to determine whether or not bottled drinking water, intended such as "protected" water supply, could prevent or delay cryptosporidiosis among children in an endemic semi-urban community in Southern India (Sarkar et al. 2013). A total of 176 children were enrolled and received either bottled ($n = 90$) or municipal ($n = 86$) drinking water. Weekly surveillance visits were conducted until children reached their second birthday. Stools were collected every month and during diarrheal episodes, and tested for the presence of *Cryptosporidium* spp. by PCR. Cryptosporidiosis, mostly in an asymptomatic form, was observed in 118 of 176 (67 %) children during the follow-up period at a rate of 0.59 episodes/child-year. Diarrhea associated with *Cryptosporidium* spp. tended to be longer in duration and more severe. Stunting at 6 months and a higher disease burden were associated with a higher risk of cryptosporidiosis, but interestingly drinking bottled water was not associated with a reduced risk of cryptosporidiosis. The lack of association between drinking bottled water and cryptosporidiosis suggests possible spread from asymptomatically infected individuals involving multiple transmission pathways.

2.9.5 *Infection in Immunocompromised Individuals*

Cryptosporidiosis is a leading cause of severe diarrhoea and extraintestinal infection in immunocompromised individuals (see Chap. 9). Besides HIV-infected patients, individuals at high risk include those with X-linked hyperimmunoglobulin

M syndrome (XHIM), CD40 ligand or gamma-interferon deficiency, children with leukemia, and organ transplant recipients (Cacciò et al. 2009).

The importance of cryptosporidiosis in HIV-infected people is well established and the subject has been extensively reviewed from the point of view of the epidemiology and clinical features (Hunter and Nichols 2002; Del Chierico et al. 2011), as well as from the perspective of treatment and control (Abubakar et al. 2007). In short, the risk of faecal carriage, the severity of illness and the development of unusual complications of cryptosporidiosis are related to the CD4 cell count (Pozio et al. 1997). Indeed, patients with CD4 counts of less than 50 are at greatest risk for both the severity of the disease and prolonged carriage. In severely immunocompromised persons, the parasite can also colonize extra-intestinal sites, particularly the gall bladder, biliary tract, pancreas, and lungs (Hunter and Nichols 2002).

The introduction of the highly active antiretroviral therapy (HAART) has had a remarkable impact on many opportunistic viral, bacterial and parasitic infections, resulting in a marked reduction in their occurrence and clinical course, at least in developed countries (Pozio and Gomez Morales 2005). The HAART therapy is based on a combination of nucleoside and non-nucleoside reverse transcriptase inhibitors and HIV protease inhibitors, and results in immune restoration, characterized by an increase in memory and naïve $CD4^{+}C$ T cells and the recovery of $CD4^{+}C$ lymphocyte reactivity against opportunistic pathogens. Cryptosporidiosis, however, remains a major problem for patients failing HAART, for most individuals living with AIDS in developing countries without access to HAART, and for severely malnourished children.

An extended study on intestinal parasitic infections, including cryptosporidiosis, was carried out in Congo by enrolling hospitalized AIDS patients (Wumba et al. 2010). Stool samples were collected from 175 patients older than 15 years. Parasites were detected by microscopy, immunofluorescence antibody tests and PCR (for diagnosis of microsporidia). At baseline, 19 patients (10.8 %) were under HAART and 156 (89.2 %) were eligible for it. Hospitalization was essentially due to intestinal infection associated with diarrhoea (49.7 % of the patients). A parasite was found in 47 of 175 (26.9 %) patients, and 27 out of 175 (15.4 %) were infected with at least one opportunistic parasite. The overall prevalence rate for *Cryptosporidium* sp. was 9.7 %, and increased to 12.6 % when only patients with diarrhoea were considered. A number of other protozoan and helminths were observed, but no significant relationship was established between any individual parasite and diarrhoea. These results underline the importance of opportunistic infections in symptomatic AIDS patients regardless of diarrhoea at the time of the hospitalisation, and showed that routine microscopic examination for *Cryptosporidium* spp. should be considered due to the absence of clinical markers.

In HIV patients, hepatic parenchymal and biliary tract diseases are common. A recent study focused on clinical aspects of AIDS-related cholangiopathy (De Angelis et al. 2009). Although the etiology is unclear, several opportunistic infections (including *Cryptosporidium* and cytomegalovirus) are suspected to cause it. The most common finding after endoscopic retrograde cholangiopancreatography is diffuse sclerosing cholangitis in combination with papillary stenosis. Clinically, the most common manifestations are right upper quadrant pain and fever accompanied by an elevated serum alkaline phosphatase level. In vitro experiments have shown that concurrent active HIV replication and *C. parvum* infection synergistically increase cholangiocyte apoptosis and thus jointly contribute to AIDS-related cholangiopathies (De Angelis et al. 2009).

Primary immunodeficiencies are rare inherited disorders of the innate, cellular, and/or humoral immune system. A particular susceptibility to infection with *Cryptosporidium* is observed in children with X-linked hyper-immunoglobulin M syndrome (XHIM), resulting from CD40 ligand deficiency (CD40L), hyper-IgM syndrome type 3 caused by CD40 deficiency, primary CD4 lymphopenia, severe combined immunodeficiency syndrome, and gamma interferon deficiency (Cacciò et al. 2009). These patients are unable to clear the infection and extraintestinal infection, particularly of the bile tract, may result in chronic liver inflammation or even lead to liver cirrhosis. Colonization of the biliary system may also predispose to the development of sclerosing cholangitis (SC) and cholangiocarcinoma.

In a study of the association between XHIM and tumors of the pancreas, liver, and biliary tree, 14 of 20 boys (70 %) were found to be infected with *Cryptosporidium* (Hayward et al. 1997). In these patients, cholangiopathy and/or cirrhosis preceded the development of the tumors, suggesting that infection or inflammation of bile ducts caused by *Cryptosporidium* may play an important role in the development of malignancy. Another study in Poland found chronic cryptosporidiosis in three out of five patients with XHIM and in a single patient with primary CD4 lymphopenia, and reported SC in these patients (Wolska-Kusnierz et al. 2007). Further support for the association between *Cryptosporidium* infection and SC was found in a study in the United Kingdom that enrolled 35 children with clinical evidence of liver disease and found 12 of 27 children (44 %) infected with the parasite, among whom nine had SC (Rodrigues et al. 2004).

In a recent analysis of 126 patients with XHIM syndrome reported to the European Society for Immunodeficiency registry, approximately one-sixth developed liver disease, and in more than 50 % of cases this was associated with *Cryptosporidium* species infection (Toniati et al. 2002). These figures may even be an underestimate, because more-sensitive molecular techniques reveal that a number of patients are colonized by *Cryptosporidium* without evidence of its presence on conventional microbiology screening (Rodrigues et al. 2004; Wolska-Kusnierz et al. 2007). Thus, unrecognized cases of cryptosporidiosis in children with primary immunodeficiencies may lead to serious consequences, with development of sclerosing cholangitis, liver cirrhosis, and cholangiocarcinoma.

Data on cryptosporidiosis in solid-organ transplant recipients are limited, but those available illustrate the need of a high index of suspicion in any transplant patient who presents with severe diarrhea (Cacciò et al. 2009; Krause et al. 2012). Intestinal infections with *Cryptosporidium* species have been reported in renal transplant patients. Some cases resulted in either mild disease or asymptomatic carriage, but severe cryptosporidiosis can occur in these patients, including biliary involvement treated with reduction of immunosuppression and a short course of antiparasitic agents (Abdo et al. 2003; Hong et al. 2007). In the recent study of Krause et al. (2012), *Cryptosporidium* was detected as the cause of gastroenteritis in six children (four kidney recipients, one liver and kidney recipient, and one heart transplant recipient). All patients were hospitalized due to prolonged diarrhoea, fever, abdominal pain and weight loss, and most presented deterioration of kidney functions and abnormal values of liver enzymes.

About ten cases of cryptosporidosis has also been reported in liver transplant recipients. In a study from Belgium on 461 children following liver transplantation, three (0.65 %) developed diffuse cholangitis associated with intestinal *Cryptosporidium* species carriage (Campos et al. 2000). All three recipients required reoperation on the bile duct anastomosis, but biliary cirrhosis developed in one patient, requiring retransplantation. In a retrospective study from Pittsburgh, four (0.34 %) pediatric cases of cryptosporidiosis were identified among 1,160 nonrenal, abdominal organ transplant recipients (Gerber et al. 2000). Three of these four cases occurred in patients receiving liver transplants, and one occurred following a small bowel transplantation. All four patients spontaneously resolved their infections. Manz and Steuerwald (2007) reported a case of cryptosporidiosis in an adult patient treated with interferon and ribavirin for recurrent hepatitis C after liver transplantation. The patient did not have HIV infection or immunoglobulin deficiency and recovered after the treatment was stopped and the dosage of immunosuppressant was lowered. A case of multiple infections with distinct *Cryptosporidium* species has been described for a transplanted ileum (Pozio et al. 2004). In Iran, among 44 liver transplant children in Shiraz Nemazee hospital, *C. parvum* and *C. meleagridis* were detected in 11.36 % of the children (Agholi et al. 2013).

The study by Sulżyc-Bielicka et al. (2012) evaluated the prevalence of *Cryptosporidium* spp. in 87 patients with diagnosed colorectal cancer, by using a ProSpecT *Cryptosporidium* Microplate Assay. *Cryptosporidium* sp. was found in 12.6 % of the patients, with a prevalence comparable to patients with immune deficiency; however, no specific correlation was found between *Cryptosporidium* spp. infection and gender, age, neoplasm differentiation grade, or neoplastic tumour localisation.

2.9.6 Infection in Travellers

In 2011, approximately 980 million people travelled internationally. More than 522 million people from developed countries travelled overseas; an estimated 50–100 million people travelled to developing countries (Ross et al. 2013). Approximately 8 % of travellers to the developing world require medical care during or after travel and more than a quarter of those who seek medical assistance present with GI symptoms. Traveller diarrhoea (TD) occurs in 20–60 % of European or North American travellers in inter-tropical areas (Cavallo and Garrabé 2007). Bacteria, viruses and parasites all may contribute to TD, but their relative importance is still uncertain; however, protozoan infections with *Giardia* and *Cryptosporidium* are frequently identified as the cause of GI complaints in returning travellers

(Freedman et al. 2006; Thielman and Guerrant 1998; Okhuysen 2001). Travelling represents an important risk factor for acquiring infection also with spore-forming protozoa such as *Cyclospora*, Microsporidia, and *Isospora* (Goodgame 2003). Several studies have shown that a large proportion of travellers and immigrants from tropical and subtropical countries are affected by GI disorders and harbour intestinal pathogens in the absence of evident GI problems (Saiman et al. 2001; Freedman et al. 2006; Ansart et al. 2005; Caruana et al. 2006; Whitty et al. 2000; Fotedar et al. 2007).

One of the first studies of travellers returning from developing countries was performed in Germany in 1997 (Jelinek et al. 1997). To estimate the prevalence of *C. parvum* and *Cyclospora cayetanensis*, 978 stool samples were taken from 795 patients (469 suffering from diarrhoea) returning from developing countries. Of the 795 patients, infection with *C. cayetanensis* was detected in five subjects (1.1 %), while 13 patients (2.8 %) were infected with *C. parvum*. All patients with either *C. parvum* or *C. cayetanensis* infection suffered from watery diarrhoea, suggesting that in cases of persistent watery diarrhoea these pathogens should be always considered in the differential diagnosis (Jelinek et al. 1997).

In the Netherlands, a new diagnostic strategy was recently implemented for the routine diagnosis of intestinal parasites in returning travellers and in immigrants (ten Hove et al. 2009). Over a period of 13 months, unpreserved stool samples, patient characteristics and clinical data were collected from those attending a travel clinic. Stool samples were analysed on a daily basis by microscopic examination and antigen detection, and compared with a weekly performed multiplex real-time PCR analysis for *Entamoeba histolytica*, *Giardia*, *Cryptosporidium* and *Strongyloides stercoralis*. Microscopy and antigen detection screening of 2,591 stool samples showed *E. histolytica*, *Giardia*, *Cryptosporidium* and *S. stercoralis* in 0.3 %, 4.7 %, 0.5 %, and 0.1 % of the cases. These detection rates were lower than those obtained with real-time PCR. PCR positivity was 0.5 %, 6.0 %, 1.3 %, and 0.8 % (ten Hove et al. 2009), showing that high-throughput molecular screening could provide more accurate estimates of the prevalence of these pathogens.

It has been hypothesized that travellers may be exposed to parasite species/genotypes that do not circulate in their home countries, and that symptomatic infection with these parasites occurs because of the lack of or insufficient cross-protection resulting from previous exposures. In agreement with this hypothesis, Tanriverdi et al. (2008) used microsatellite typing to show that the *C. hominis* isolates infecting patients that had travelled to the UK from Pakistan less than 2 weeks prior to the isolation of the parasites were significantly different from the large cluster of autochthonous cases from the UK. Furthermore, a novel *Cryptosporidium* species was recently identified among travellers with gastro-intestinal symptoms returning to the UK from the Indian subcontinent (Elwin et al. 2012). Based on morphological and molecular data, this parasite was designated a new species, *Cryptosporidium viatorum*. The name was chosen to underscore its link to foreign travel (Elwin et al. 2012). The epidemiology of *C. viatorum* cases was found to be different from that of individuals infected with *C. parvum* and *C. hominis*: most *C. viatorum* cases occurred in the first months of the year, vomiting was reported less often, but the duration of symptoms was longer. The cases of *C. viatorum* were all travellers to the Indian subcontinent, whereas cases of *C. hominis* and *C. parvum* were more likely associated to travel elsewhere (Elwin et al. 2012).

Another study on *Cryptosporidium* infections contracted whilst travelling abroad used a PCR-coupled single-strand conformation polymorphism analysis, followed by targeted sequencing of the gp60 gene (Jex and Gasser 2008). The study investigated *C. hominis* and *C. parvum* isolates ($n = 115$) from UK citizens inferred to have been infected while travelling abroad (to 25 countries) or in the UK. Isolates were classified to the genotype and sub-genotype levels, leading to the identification of five *C. hominis* and four *C. parvum* gp60 genotypes. A particular *C. hominis* subgenotype (IbA10G2R2) was found in the majority (71 %) of the isolates, collected from individuals who have travelled to 14 different countries, while other subgenotypes appeared to be quite rare (Jex and Gasser 2008).

2.10 Transmission Routes

There are multiple transmission routes through which humans can acquire *Cryptosporidium* infections. The main routes of transmissions are shared by the two major human pathogens, *C. hominis* (Fig. 2.3) and *C. parvum* (Fig. 2.4); however animal pollution and environmental contamination particularly affect zoonotic transmission of *C. parvum* (Fig. 2.4).

2.10.1 Person-to-Person Transmission

Cryptosporidium is easily transmitted among children and staff members in nurseries (Hannah and Riordan 1988), day care centres (Heijbel et al. 1987), and schools (Lee and Greig 2010).

Nosocomial infection is also well documented, and both direct (Baxby et al. 1983; Koch et al. 1985) and indirect (Martino et al. 1988; Navarrete et al. 1991) person-to-person transmission (via contaminated hands) has been incriminated as the likely route. Nosocomial infection can cause secondary cases among roommates (Bruce et al. 2000) and family members (Pandak et al. 2006), further supporting the highly infectious capacity of the parasite.

Direct transmission among HIV-positive men who have sex with men occurs more frequently than in HIV-positive drug users (Pedersen et al. 1996), and a case–control study in Australia (Hellard et al. 2003) has shown that men having more than one sexual partner are more likely to have *Cryptosporidium* diarrhoea, therefore indicating that sexual contacts represent a risk factor for faecal-oral parasite transmission.

In developing countries, the high prevalence of *C. hominis* and of anthroponotic subtypes of *C. parvum*, particularly in children, has also been

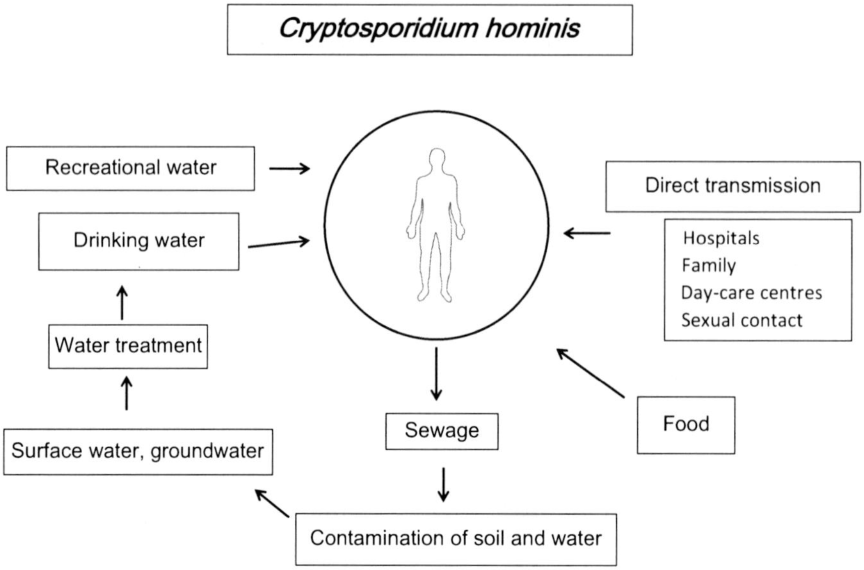

Fig. 2.3 Main transmission routes of *Cryptosporidium hominis*

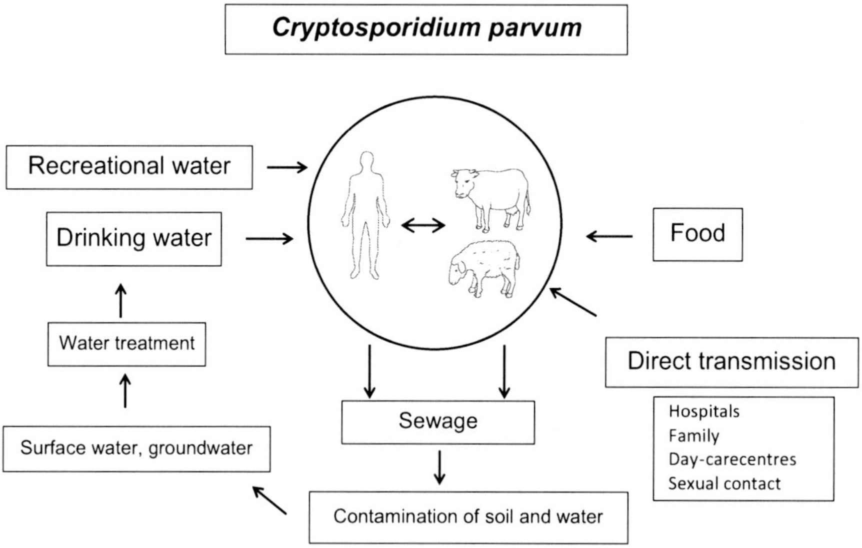

Fig. 2.4 Main transmission routes of *Cryptosporidium parvum*

taken as an indication of the importance of person-to-person transmission (Xiao 2009) (Figs. 2.3 and 2.4).

2.10.2 Zoonotic Transmission

Cryptosporidium parvum is the most important zoonotic agent of cryptosporidiosis, with a large range and abundance of animal reservoirs, mainly in young farmed animals (see Chap. 4). Therefore, individuals who come in contact with those animals, either for occupational or recreational reasons, may be at risk. Outbreaks of cryptosporidiosis have been reported among veterinarians and veterinary students (Pohjola et al. 1986; Preiser et al. 2003; Gait et al. 2008), other people exposed to agricultural animals (Stantic-Pavlinic et al. 2003) and children visiting farms (Shield et al. 1990; Stefanogiannis et al. 2001; Hoek et al. 2008). Contact with farmed animals was identified as a significant risk factor for sporadic cases of human cryptosporidiosis in the UK (Hunter et al. 2004; Goh et al. 2004).

Due to the high prevalence of *C. parvum* infection and the high numbers of oocysts shed in faeces (up to $>5 \times 10^6$ oocysts per gram), calves are considered to pose the most significant threat to environmental contamination and transmission to humans (Current et al. 1983). Initially, it was believed that human infection with *C. parvum* were all of zoonotic origin and calves have been implicated as the main source of infectious oocysts, but further studies based on highly polymorphic markers have shown that certain *C. parvum* subtypes are found in humans but not in animals, and are likely to be transmitted through an anthroponotic cycle (Mallon et al. 2003). Thus, a significant fraction of human *C. parvum* infections may not originate from livestock reservoirs (Grinberg et al. 2008). Nevertheless, calves are frequently infected with a *C. parvum* subtype that is commonly found in humans in the same geographic areas (Xiao 2009) and epidemiologic studies have supported the occurrence of zoonotic transmission (Hunter et al. 2007).

Epidemiologic investigations have demonstrated the role of sheep in human cryptosporidiosis more than 20 years ago (Casemore 1989), and this has been further supported by molecular studies, particularly in the UK, where five human outbreaks have been linked to contacts with lambs (Chalmers and Giles 2010). Recently, a case of zoonotic transmission of a rare *C. parvum* subtype from infected lambs to a children has been reported in Italy (Cacciò et al. 2013).

On the contrary, little is known about the role of goats in zoonotic cryptosporidiosis. Goats can be infected with *C. xiaoi*, which is a non-zoonotic species, but also with *C. parvum* (Rieux et al. 2013).

The zoonotic potential of canine and feline cryptosporidiosis has been a major concern to both veterinarians and physicians, but the actual risk of transmission from pets to human appears to be minimal, at least in develop countries. To date, there have been only 26 *C. canis* and 97 *C. felis* cases reported in people, and the majority was from immunocompromised individuals (Lucio-Forster et al. 2010). Epidemiological investigations in the UK and USA (Goh et al. 2004; Glaser et al. 1998) have failed to establish a significant relationship between owning a dog and infection with *Cryptosporidium*, whereas owning a dog or a cat was found to be a risk factor for cryptosporidiosis in Guinea Bissau and Indonesia (Miron et al. 1991; Katsumata et al. 1998). There is only one report of possible transmission of *C. canis* between a dog and two siblings living in a household in Peru (Xiao 2009). It is likely that only immunocompromised individuals are at risk of acquiring cryptosporidiosis from pets.

Rabbits are the natural host of *Cryptosporidium cuniculus* (Robinson et al. 2010), a species that is now known to infect humans (Chalmers et al. 2009). First identified as a human pathogen during a waterborne outbreak, *C. cuniculus* appears to be the third most commonly identified species in patients with diarrhea in the UK, after *C. parvum* and *C. hominis* (Chalmers et al. 2011). The existence of identical *C. cuniculus* subtypes in humans and pets or wild rabbits suggest zoonotic potential (Zhang et al. 2012).

Zoonotic transmission of *C. meleagridis*, an avian parasite that can infect humans, has been recently reported in Sweden (Silverlås et al. 2012). Interestingly, results of molecular characterization suggest laying hens or broiler chickens as the source of infectious oocysts.

Fish, reptiles and amphibians appear not to pose a risk for human cryptosporidiosis (see Chap. 5).

2.10.3 Waterborne Cryptosporidiosis

Waterborne infectious diseases are a globally emerging public health issue. Various community outbreaks due to contamination of water have highlighted the importance of intestinal protozoa in public health. Among these important pathogens are *Giardia duodenalis*, *E. histolytica*, *C. cayetanensis*, *Isospora belli*, Microsporidia and, of greater relevance, *Cryptosporidium* (Karanis et al. 2007).

The ubiquitous presence of *Cryptosporidium* spp. in the aquatic environment is explained by the large number of hosts, the extremely high number of oocysts shed by these hosts, and the remarkable stability of oocysts (Smith et al. 2006b). Thus, water represents a very important vehicle of infection for the population, and waterborne cryptosporidiosis is a serious public health concern, particularly for populations at risk of severe infection (pregnant women, children, HIV-positive and transplanted patients) (Chap. 12).

Indeed, out of the 71 *Cryptosporidium*-linked outbreaks described in the last decade, 40 (56.3 %) appear to be correlated to waterborne transmission. Geographically, the outbreaks seem to be concentrated in the USA, Canada, Australia and Europe, especially in the UK and Ireland, and affect both adults and children (Putignani and Menichella 2010; Chalmers 2012). Surveillance data has revealed the presence of *Cryptosporidium* spp. in the entire water treatment system (see Chap. 12), which represents an unacceptable health risk, particularly for at risk populations (pregnant women, children, HIV-positive and transplanted patients). Such evidence suggests that focus ought to be placed on prevention of human and animal waste contamination especially in authorized recreational waters. Remarkably, cryptosporidiosis is the most frequently reported gastrointestinal illness in outbreaks associated with treated (disinfected) recreational water venues in USA (Yoder and Beach 2007; Chap. 12).

Following several large outbreaks linked to drinking water in UK and USA, most notably the 1993 Milwaukee outbreak that involved an estimated 403,000 cases of cryptosporidiosis (MacKenzie et al. 1995), emphasis on monitoring and intervention of water supplies, and greater awareness and investigation of water as a transmission vehicle, has led to a decrease in the number of outbreaks due to drinking water (Chap. 12).

Remarkably, however, cryptosporidiosis remains the most frequently reported GI illness in outbreaks associated with treated (disinfected) recreational water venues in the UK and the USA (Hlavsa et al. 2011; Smith et al. 2006a; Yoder and Beach 2007; Yoder et al. 2010). Outbreaks have occurred in swimming pools, paddling or wading pools, water parks ad fountains, and were caused by *C. hominis* and *C. parvum* (see Chap. 12). During the summer of 2007, Utah experienced a statewide outbreak of GI illness caused by *Cryptosporidium* (CDC, MMWR Report 2012). Of 1,506 interviewed patients with laboratory-confirmed cryptosporidiosis, 1,209 (80 %) reported swimming in at least one of approximately 450 recreational water venues during their potential 14-day incubation period. Because swimmers were the primary source of *Cryptosporidium* contamination, healthy swimming campaigns are needed to increase awareness and practice of healthy swimming behaviours, especially not swimming while ill with diarrhoea. The healthy swimming campaign, as part of a multipronged prevention effort, might have helped prevent recreational water-associated outbreaks of cryptosporidiosis in Utah (CDC, MMWR Report 2012). Local and state health departments can use cryptosporidiosis surveillance data to better understand the epidemiologic characteristics and the disease burden of cryptosporidiosis in the USA, to design prevention strategies that reduce disease spread and to establish research priorities. The role of water in the transmission of *Cryptosporidium* in developing countries is less known. Clearly, the potential for transmission is enhanced by the absence of sanitary and parasitological drinking water monitoring, and the burden of the infection is surely underestimated due to the scarcity of appropriate surveillance programs and the relative inadequacy of laboratory diagnosis (Mak 2004). A Quantitative Microbiological Risk Assessment (QMRA) conducted in Africa (Hunter et al. 2009) demonstrated that interruptions in water supplies that forced people to revert to drinking raw water caused a greater risk of infection, particularly in young children. Thus, poor reliability of drinking water supplies has an impact on the achievement of health improvement targets (Hunter et al. 2009).

The importance of stability of the treatment process and the importance of watershed protection has been stressed by a comprehensive QMRA performed on a source water survey from 66 waterworks in 33 major cities across China (Xiao et al. 2012). The annual diarrhoea morbidity caused by *Cryptosporidium* in drinking water was estimated to be 2,701 cases per 10^5 immunocompromised persons and 148 cases per 10^5 immunocompetent persons, giving an overall rate of 149 cases per 100,000 population. The burden of cryptosporidiosis associated with drinking water treated with the conventional process was higher than the reference risk level suggested by the WHO, but lower than that suggested by the United States Environmental Protection Agency (Xiao et al. 2012).

A survey of *Giardia* and *Cryptosporidium* conducted in 206 samples of surface waters used as drinking water sources by public water systems in four densely urbanized regions of Sao Paulo State, Brazil and a QMRA, showed 102 samples positive for *Giardia* and 19 for *Cryptosporidium*, with maximum concentrations of 97.0 cysts/L and 6.0 oocysts/L, respectively. The probability of *Giardia* infection was close to the rates of acute diarrheic disease for adults (1–3 %) but lower for children (2–7 %). The daily consumption of drinking water was an important contributing factor for these differences (Sato et al. 2013).

2.10.4 Foodborne Cryptosporidiosis

Contamination of different types of food with *Cryptosporidium* oocysts has been demonstrated in studies from different regions of the world. Those studies have mainly focused on fruits and vegetables, because these foods are prone to contamination and are often consumed raw or after minimal thermal treatment, therefore increasing the possibility of transmission (Robertson and Chalmers 2013). Due to the highly variable, and usually low, rates of recovery of oocysts from food matrices, improved methods have been recently developed and validated in the UK (Cook et al. 2006).

Studies in Costa Rica and Peru (Monge and Arias 1996; Ortega et al. 1997) have shown contamination of numerous raw vegetables, including basil, cabbage, celery, cilantro, green onions, leeks, lettuce, parsley, and yerba buena. A more recent study in Costa Rica (Calvo et al. 2004) investigated the presence of *Cryptosporidium* spp., *Cyclospora* spp., and Microsporidia on lettuce, parsley, cilantro, strawberries and blackberries collected from five local markets. Fifty different samples of each product, 25 taken in the dry season and 25 in the rainy season, were evaluated. All products were found contaminated with *Cryptosporidium* spp., *Cyclospora* spp., and/or Microsporidia. *Cryptosporidium* was not detected in strawberries, microsporidia were absent on blackberries and *Cyclospora* was only isolated from lettuce during the dry season. These results show the importance of introducing good agricultural practices, especially due to the resistance of *Cryptosporidium* and *Cyclospora* to disinfecting agents (Putignani and Menichella 2010).

Food contamination with *Cryptosporidium* oocysts, however, is not limited to developing countries. In Norway, a search for parasites in fruits and vegetables was undertaken in the period from 1999 to 2001 (Robertson and Gjerde 2001). Of the 475 samples, 29 were found to be positive for *Cryptosporidium* oocysts and *Giardia* cysts, of which 19 only for *Cryptosporidium* (lettuce and mung bean sprouts). Mung bean sprouts were significantly more likely to be contaminated with *Cryptosporidium* oocysts or *Giardia* cysts than the other fruits and vegetables, even if concentrations were generally low (approximately 3 (oo)cysts per 100 g product). There was no association between imported produce and detection of parasites.

Cryptosporidium oocysts and *Giardia* cysts were also detected in water samples concerned with field irrigation and production of bean sprouts (Robertson and Gjerde 2001).

In Poland, 163 samples comprising Peking cabbages, leeks, white cabbages, red cabbages, lettuces, spring onions, celery, cauliflowers, broccoli, spinach, Brussels sprouts, raspberries, strawberries, all from local markets, were tested (Rzezutka et al. 2010). *Cryptosporidium* oocysts were detected in one leek sample, one celery sample, four cabbage samples, including all cabbage types tested. Of note, *Cryptosporidium*-positive samples came from districts with the highest number of cattle herds. In Spain, 19 fresh produce samples from local markets were tested, and oocysts were found in 33 % of Chinese cabbage, 75 % of Lollo rosso lettuce, and 78 % of Romaine lettuce (Amorós et al. 2010).

These studies demonstrated that contamination of fruits and vegetables occurs also in developed countries.

A few outbreaks have been linked to the consumption of contaminated vegetables. In 2008, a *C. parvum* outbreak in Sweden was linked to chanterelle sauce with fresh parsley added after the preparation of the sauce (Insulander et al. 2008), while in Finland a salad mixture was the suspected vehicle for a *C. parvum* outbreak (Ponka et al. 2009).

Contamination of dairy products and fruit juices has also been linked to outbreaks of cryptosporidiosis. The consumption of unpasteurized cow milk has been suggested as the cause of outbreaks in the UK and Australia (Gelletlie et al. 1997; Harper et al. 2002). Outbreaks associated with consumption of fruit juice is becoming an emergent public health problem since the early 1990s, when the first outbreak associated with apple cider was described (Millard et al. 1994). However, in the period from September to November 2003, 12 local residents in Northern Ohio were diagnosed with cryptosporidiosis for having drunk ozonated apple cider (Blackburn et al. 2006). In response to epidemiologic investigations of outbreaks in which juice is implicated, the USA Food and Drug Administration has implemented process control measures to regulate the production of fruit juice, according with the Hazard Analysis Critical Control Point (HACCP) plan. However juice operations that are exempt from processing requirements or do not comply with the regulation, continue to be implicated in outbreaks of illness. The CDC receives reports of food-associated outbreaks of illness (FoodNet, http://www.cdc.gov/FoodNet/) and its *Foodborne Outbreak Reporting System* has reviewed, from 1995 through 2005, ten implicating apple juice or cider, eight linked to orange juice, and three involving other types of fruit juice-associated outbreaks (Putignani and Menichella 2010). Among the 13 outbreaks of known aetiology, two were caused by *Cryptosporidium* and one by Shiga toxin-producing *E. coli* O111 and *Cryptosporidium* (Vojdani et al. 2008). The incidence of foodborne disease outbreaks caused by contaminated low pH fruit juices is increasing (Lynch et al. 2006). The association of *Cryptosporidium* with fruit juice is a raising safety concern in food industries. In 1998, CDC implemented enhanced surveillance for foodborne-diseases outbreaks by increasing communication with state, local, and territorial health departments and revising the outbreak report form. Since 2001, reports are submitted through a web internet application called electronic Foodborne Outbreak Reporting System (Putignani and Menichella 2010).

Outbreak investigations have also put into focus the role of food handlers as a source of food contamination and subsequent transmission of cryptosporidiosis (Robertson and Chalmers 2013). In 1998, a large outbreak of gastroenteritis occurred in >100 persons at a University campus in Washington DC. A case–control study of 88 case patients and 67 control subjects showed that eating in one of two cafeterias was associated with diarrheal illness. Further epidemiologic and molecular evidence indicate that an ill foodhandler was the likely outbreak source, and *C. hominis* as identified as the etiological cause of gastroenteritis (Quiroz et al. 2000).

In 2005 an outbreak of diarrhoea, affecting a group of 99 company employees, was described near Copenhagen (Ethelberg et al. 2009). All people were ill and 13 tested positive for *C. hominis*. Disease was associated with eating from the canteen salad bar on one, possibly two, specific weekdays. Three separate salad bar ingredients were found to be likely sources: peeled whole carrots served in a bowl of water, grated carrots, and red peppers. The likely source of infection was an infected food-handler, who may have contaminated food served at the buffet (Ethelberg et al. 2009).

Recently, the role of the food handlers has been investigated in Venezuela, where cryptosporidiosis is an important public health problem (Freites-Martinez et al. 2009). Despite a basic investigation approach, 14 out of 119 fecal samples from food workers were found positive for *Cryptosporidium* spp. in association with other protozoa, the most frequently detected being *Endolimax nana*, *Blastocystis hominis*, *Entamoeba coli*, *Giardia*, and *E. histolytica*/*Entamoeba dispar*.

Finally, because of their capacity to filter large volumes of water potentially contaminated with oocysts from human and animal faecal waste, and because they are usually consumed raw, molluscs have been postulated as a route of transmission (Robertson 2007). Indeed many species, including mussels, oysters and clams, have been found to harbour *Cryptosporidium* oocysts in their digestive tract. However, there are no evidence of infections or outbreaks linked to consumption of molluscs.

2.11 Conclusions

Cryptosporidium is an important cause of diarrhoea, worldwide. In developed countries, large waterborne outbreaks continue to occur, emphasizing the need for better regulation and for improvements of drinking water treatment processes. Also, the increasing number of outbreaks linked to the use of recreational water (swimming pools, water parks) indicate the need for better control measures and guidelines. Immunocompromised individuals are particularly susceptible to *Cryptosporidium*, and may develop severe infections and extra-intestinal dissemination, yet an effective therapy to eradicate the parasite is not available. In developing countries, the parasite is endemic and significantly associated with moderate-to-severe diarrhoea in infants, a finding that highlights the need to develop resources to diagnose, treat, and prevent cryptosporidiosis in resource-poor settings. Under this situation, routine diagnosis and effective reporting of *Cryptosporidium* to local and national surveillance organizations remain of key importance in understanding the epidemiology of this important, but often underestimated, pathogen.

Acknowledgments This work was supported by the Ministry of Health, Ricerca Corrente 2013, Bambino Gesù Children's Hospital, IRCCS, to LP. The authors thank Federica Del Chierico for help with the preparation of figures.

References

Abdel-Messih IA, Wierzba TF, Abu-Elyazeed R et al (2005) Diarrhea associated with *Cryptosporidium parvum* among young children of the Nile River Delta in Egypt. J Trop Pediatr 51:154–159

Abdo A, Klassen J, Urbanski S et al (2003) Reversible sclerosing cholangitis secondary to cryptosporidiosis in a renal transplant patient. J Hepatol 38:688–691

Abubakar I, Aliyu SH, Arumugam C et al (2007) Treatment of cryptosporidiosis in immunocompromised individuals: systematic review and meta-analysis. Br J Clin Pharmacol 63 (4):387–393

Agholi M, Hatam GR, Motazedian MH (2013) Microsporidia and coccidia as causes of persistence diarrhea among liver transplant children: incidence rate and species/genotypes. Pediatr Infect Dis J 32(2):185–187

Akiyoshi DE, Dilo J, Pearson C et al (2003) Characterization of *Cryptosporidium meleagridis* of human origin passaged through different host species. Infect Immun 71(4):1828–1832

Al-Braiken FA, Amin A, Beeching NJ et al (2003) Detection of *Cryptosporidium* amongst diarrhoeic and asymptomatic children in Jeddah, Saudi Arabia. Ann Trop Med Parasitol 97 (5):505–510

Al-Harthi SA (2004) Prevalence of intestinal parasites in schoolchildren in Makkah, Saudi Arabia. New Egypt J Med 31:37–43

Al-Mekhlafi HM, Mahdy MA, 'Azlin MY et al (2011) Childhood *Cryptosporidium* infection among aboriginal communities in Peninsular Malaysia. Ann Trop Med Parasitol 105 (2):135–143

Amorós I, Alonso JL, Cuesta G (2010) *Cryptosporidium* oocysts and *Giardia* cysts on salad products irrigated with contaminated water. J Food Prot 73(6):1138–1140

Anon (1998) Foodborne outbreak of cryptosporidiosis–Spokane, Washington, 1997. MWR Morb Mortal Wkly Rep 47(27):565–567

Ansart S, Perez L, Vergely O et al (2005) Illnesses in travelers returning from the tropics: a prospective study of 622 patients. J Travel Med 12(6):312–318

Barry MA, Weatherhead JE, Hotez PJ et al (2013) Childhood parasitic infections endemic to the United States. Pediatr Clin North Am 60:471–485

Baxby D, Hart CA, Taylor C (1983) Human cryptosporidiosis: a possible case of hospital cross infection. BMJ 287:1760–1761

Blackburn BG, Mazurek JM, Hlavsa M et al (2006) Cryptosporidiosis associated with ozonated apple cider. Emerg Infect Dis 12(4):684–686

Borowski H, Clode PL, Thompson RC (2008) Active invasion and/or encapsulation? A reappraisal of host-cell parasitism by *Cryptosporidium*. Trends Parasitol 24(11):509–516. doi:10.1016/j.pt.2008.08.002

Bruce BB, Blass MA, Blumberg HM et al (2000) Risk of *Cryptosporidium parvum* transmission between hospital roommates. Clin Infect Dis 31(4):947–950

Bushen OY, Kohli A, Pinkerton RC et al (2007) Heavy cryptosporidial infections in children in northeast Brazil: comparison of *Cryptosporidium hominis* and *Cryptosporidium parvum*. Trans R Soc Trop Med Hyg 101:378–384

Cacciò S, Homan W, Camilli R et al (2000) A microsatellite marker reveals population heterogeneity within human and animal genotypes of *Cryptosporidium parvum*. Parasitology 120:237–244

Cacciò SM, Thompson RC, McLauchlin J et al (2005) Unravelling *Cryptosporidium* and *Giardia* epidemiology. Trends Parasitol 21(9):430–437

Cacciò SM, Pozio E, Guarino A et al (2009) Long-term consequences of *Cryptosporidium* infections in immunocompetent and immunodeficient individuals. In: Fratamico PM, Smith JL, Brogden KA (eds) Sequelae and long-term consequences of infectious diseases. ASM Press, Washington, DC, pp 245–257

Cacciò SM, Sannella AR, Mariano V et al (2013) A rare *Cryptosporidium parvum* genotype associated with infection of lambs and zoonotic transmission in Italy. Vet Parasitol 191 (1–2):128–131. doi:10.1016/j.vetpar.2012.08.010

Calvo M, Carazo M, Arias ML et al (2004) Prevalence of *Cyclospora* sp., *Cryptosporidium* sp, microsporidia and fecal coliform determination in fresh fruit and vegetables consumed in Costa Rica. Arch Latinoam Nutr 54(4):428–432

Cama VA, Bern C, Roberts J et al (2008) *Cryptosporidium* species and subtypes and clinical manifestations in children, Peru. Emerg Infect Dis 14:1567–1574

Campos M, Jouzdani E, Sempoux C et al (2000) Sclerosing cholangitis associated to cryptosporidiosis in liver-transplanted children. Eur J Pediatr 159:113–115

Caruana SR, Kelly HA, Ngeow JYY et al (2006) Undiagnosed and potentially lethal parasite infections among immigrants and refugees in Australia. J Travel Med 13(4):233–239

Casemore DP (1989) Sheep as a source of human cryptosporidiosis. J Infect 19:101–104

Casemore DP (1990) Epidemiological aspects of human cryptosporidiosis. Epidemiol Infect 104:1–28

Cavallo JD, Garrabé E (2007) Infectious aetiologies of travelers' diarrhoea. Med Mal Infect 37 (11):722–727

Centers for Disease Control and Prevention (CDC) (2012) Promotion of healthy swimming after a statewide outbreak of cryptosporidiosis associated with recreational water venues–Utah, 2008–2009. MMWR 61(19):348–352

Chalmers RM (2012) Waterborne outbreaks of cryptosporidiosis. Ann Ist Super Sanita 48:429–446

Chalmers RM, Elwin K, Hadfield SJ et al (2011) Sporadic human cryptosporidiosis caused by *Cryptosporidium cuniculus*, United Kingdom, 2007–2008. Emerg Infect Dis 17(3):536–568

Chalmers RM, Giles M (2010) Zoonotic cryptosporidiosis in the UK – challenges for control. J Appl Microbiol 109:1487–1497

Chalmers RM, Elwin K, Thomas AL et al (2009) Long-term *Cryptosporidium* typing reveals the aetiology and species-specific epidemiology of human cryptosporidiosis in England and Wales, 2000 to 2003. Euro Surveill 14:19086

Chappel C, Okhuysen PC, Langer-Curry RC et al (2006) *Cryptosporidium hominis*: experimental challenge of healthy adults. Am J Trop Med Hyg 75(5):851–857

Chappel C, Okhuysen PC, Langer-Curry RC et al (2011) *Cryptosporidium meleagridis*: infectivity in healthy adult volunteers. Am J Trop Med Hyg 85(2):238–242. doi:10.4269/ajtmh.2011.10-0664

Checkley W, Epstein LD, Gilman RH et al (1998) Effects of *Cryptosporidium parvum* infection in Peruvian children: growth faltering and subsequent catch-up growth. Am J Epidemiol 148:497–506

Chen W, Chadwick V, Tie A et al (2001) *Cryptosporidium parvum* in intestinal mucosal biopsies from patients with inflammatory bowel disease. Am J Gastroenterol 96:3463–3464

Chen XM, Keithly JS, Paya CV et al (2002) Cryptosporidiosis. N Engl J Med 346:1723–1731
Collings S, Highton J (2004) *Cryptosporidium* reactive arthritis. N Z Med J 117(1200):1, following U1023
Cook N, Paton CA, Wilkinson N et al (2006) Towards standard methods for the detection of *Cryptosporidium parvum* on lettuce and raspberries. Part 2: validation. Int J Food Microbiol 109(3):222–228
Cron RQ, Sherry DD (1995) Reiter's syndrome associated with cryptosporidial gastroenteritis. J Rheumatol 22(10):1962–1963
Current WL, Reese NC, Ernst JV et al (1983) Human cryptosporidiosis in immunocompetent and immunodeficient persons. Studies of an outbreak and experimental transmission. N Engl J Med 308(21):1252–1257
Daoud AS, Zaki M, Pugh RN et al (1990) *Cryptosporidium* gastroenteritis in immunocompetent children from Kuwait. Trop Geogr Med 42:113–118
Davies AP, Campbell B, Evans MR et al (2009) Asymptomatic carriage of protozoan parasites in children in day care centers in the United Kingdom. Pediatr Infect Dis J 28(9):838–840
De Angelis C, Mangone M, Bianchi M et al (2009) An update on AIDS-related cholangiopathy. Minerva Gastroenterol Dietol 55(1):79–82
Del Chierico F, Onori M, Di Bella S et al (2011) Cases of cryptosporidiosis co-infections in AIDS patients: a correlation between clinical presentation and GP60 subgenotype lineages from aged formalin-fixed stool samples. Ann Trop Med Parasitol 105(5):339–349
Desai NT, Sarkar R, Kang G (2012) Cryptosporidiosis: an under-recognized public health problem. Trop Parasitol 2(2):91–98. doi:10.4103/2229-5070.105173
Elwin K, Hadfield SJ, Robinson G et al (2012) *Cryptosporidium viatorum* n. sp. (Apicomplexa: Cryptosporidiidae) among travellers returning to Great Britain from the Indian subcontinent, 2007–2011. Int J Parasitol 42(7):675–682
Esteban JG, Aguirre C, Flores A et al (1998) *Cryptosporidium* prevalences in healthy Aymara children from the northern Bolivian Altiplano. Am J Trop Med Hyg 58(1):50–55
Ethelberg S, Lisby M, Vestergaard LS et al (2009) A foodborne outbreak of *Cryptosporidium hominis* infection. Epidemiol Infect 137(3):348–356
Fayer R (2008) General biology. In: Fayer R, Xiao L (eds) *Cryptosporidium* and cryptosporidiosis. CRC Press, Boca Raton, pp 1–42
Fayer R, Ungar BL (1986) *Cryptosporidium* spp. and cryptosporidiosis. Microbiol Rev 50 (4):458–483
Feltus DC, Giddings CW, Schneck BL et al (2006) Evidence supporting zoonotic transmission of *Cryptosporidium* in Wisconsin. J Clin Microbiol 44:4303–4308
Fotedar R, Stark D, Beebe N et al (2007) Laboratory diagnostic techniques for *Entamoeba* species. Clin Microbiol Rev 20(3):511–532
Freedman DO, Weld LH, Kozarsky PE et al (2006) Spectrum of disease and relation to place of exposure among ill returned travellers. N Engl J Med 354(2):119–130
Freites-Martinez AD, Colmenares D, Perez M et al (2009) *Cryptosporidium* sp infections and other intestinal parasites in food handlers from Zulia state, Venezuela. Invest Clin 50(1):13–21
Gait R, Soutar RH, Hanson M et al (2008) Outbreak of cryptosporidiosis among veterinary students. Vet Rec 162:843–845
Gelletlie R, Stuart J, Soltanpoor N et al (1997) Cryptosporidiosis associated with school milk. Lancet 350(9083):1005–1006
Gerber DA, Green M, Jaffe R et al (2000) Cryptosporidial infections after solid organ transplantation in children. Pediatr Transplant 4:50–55
Ghenghesh KS, Ghanghish K, El-Mohammady H et al (2012) *Cryptosporidium* in countries of the Arab world: the past decade (2002–2011). Libyan J Med 7. doi: 10.3402/ljm.v7i0.19852. Epub 2012 Nov 27
Glaser CA, Safrin S, Reingold A et al (1998) Association between *Cryptosporidium* infection and animal exposure in HIV-infected individuals. J Acquir Immune Defic Syndr Hum Retrovirol 17:79–82

Goh S, Reacher M, Casemore DP et al (2004) Sporadic cryptosporidiosis, North Cumbria, England, 1996–2000. Emerg Infect Dis 10:1007–1015

Goodgame (2003) Emerging causes of traveler's diarrhea: *Cryptosporidium*, *Cyclospora*, *Isospora*, and Microsporidia. Curr Infect Dis Rep 5(1):66–73

Grinberg A, Lopez-Villalobos N, Pomroy H et al (2008) Host-shaped segregation of the *Cryptosporidium parvum* multilocus genotype repertoire. Epidemiol Infect 136:273–278

Grinberg A, Pomroy WE, Squires RA et al (2011) Retrospective cohort study of an outbreak of cryptosporidiosis caused by a rare *Cryptosporidium parvum* subgenotype. Epidemiol Infect 139(10):1542–1550

Hannah J, Riordan T (1988) Case to case spread of cryptosporidiosis; evidence from a day nursery outbreak. Public Health 102(6):539–544

Harper CM, Cowell NA, Adams BC et al (2002) Outbreak of *Cryptosporidium* linked to drinking unpasteurised milk. Commun Dis Intell Q Rep 26(3):449–450

Hay EM, Winfield J, McKendrick MW (1987) Reactive arthritis associated with *Cryptosporidium* enteritis. Br Med J 295(6592):248

Hayward AR, Levy J, Facchetti F et al (1997) Cholangiopathy and tumors of the pancreas, liver and biliary tree in boys with hyper IgM. J Immunol 158:977–983

Heijbel H, Slaine K, Seigel B et al (1987) Outbreak of diarrhea in a day care center with spread to household members: the role of *Cryptosporidium*. Pediatr Infect Dis J 6(6):532–535

Hellard M, Hocking J, Willis J et al (2003) Risk factors leading to *Cryptosporidium* infection in men who have sex with men. Sex Transm Infect 79(5):412–414

Hellard ME, Sinclair MI, Hogg GG et al (2000) Prevalence of enteric pathogens among community based asymptomatic individuals. J Gastroenterol Hepatol 15(3):290–293

Hlavsa MC, Roberts VA, Anderson AR, CDC et al (2011) Surveillance for waterborne disease outbreaks and other health events associated with recreational water-United States, 2007–2008. MMWR Surveill Summ 60(12):1–32

Hoek MR, Oliver I, Barlow M et al (2008) Outbreak of *Cryptosporidium parvum* among children after a school excursion to an adventure farm, south west England. J Water Health 6 (3):333–338

Hong DK, Wong CJ, Gutierrez K (2007) Severe cryptosporidiosis in a seven-year-old renal transplant recipient. Case report and review of the literature. Pediatr Transplant 11:94–100

Hunter PR, Nichols G (2002) Epidemiology and clinical features of *Cryptosporidium* infection in immunocompromised patients. Clin Microbiol Rev 15(1):145–154

Hunter PR, Hughes S, Woodhouse N et al (2004) Health sequelae of human cryptosporidiosis in immunocompetent patients. Clin Infect Dis 39:504–510

Hunter PR, Hadfield SJ, Wilkinson D et al (2007) Subtypes of *Cryptosporidium parvum* in humans and disease risk. Emerg Infect Dis 13:82–88

Hunter PR, Zmirou-Navier D, Hartemann P (2009) Estimating the impact on health of poor reliability of drinking water interventions in developing countries. Sci Total Environ 407 (8):2621–2624. doi:10.1016/j.scitotenv.2009.01.018

Insulander M, de Jong B, Svenungsson B (2008) A foodborne outbreak of cryptosporidiosis among guests and staff at a hotel restaurant in Stockholm County, Sweden. Euro Surveill 13(51), 19071

Insulander M, Lebbad M, Stenström TA et al (2005) An outbreak of cryptosporidiosis associated with exposure to swimming pool water. Scand J Infect Dis 37(5):354–360

Jagai JS, Castronovo DA, Monchak J et al (2009) Seasonality of cryptosporidiosis: a meta-analysis approach. Environ Res 109(4):465–478. doi:10.1016/j.envres.2009.02.008

Jelinek T, Lotze M, Eichenlaub S et al (1997) Prevalence of infection with *Cryptosporidium parvum* and *Cyclospora cayetanensis* among international travellers. Gut 41(6):801–804

Jex AR, Gasser RB (2008) Analysis of the genetic diversity within *Cryptosporidium hominis* and *Cryptosporidium parvum* from imported and autochtonous cases of human cryptosporidiosis by mutation scanning. Electrophoresis 29(20):4119–4129

Jex AR, Smith HV, Nolan MJ et al (2011) Cryptic parasite revealed improved prospects for treatment and control of human cryptosporidiosis through advanced technologies. Adv Parasitol 77:141–173

Jokipii L, Jokipii AM (1986) Timing of symptoms and oocyst excretion in human cryptosporidiosis. N Engl J Med 315(26):1643–1647

Karanis P, Kourenti C, Smith H (2007) Waterborne transmission of protozoan parasites: a worldwide review of outbreaks and lessons learnt. J Water Health 5(1):1–38

Katsumata T, Hosea D, Wasito EB et al (1998) Cryptosporidiosis in Indonesia: a hospital-based study and a community-based survey. Am J Trop Med Hyg 59:628–632

Khaldi S, Gargala G, Le Goff L et al (2009) *Cryptosporidium parvum* isolate-dependent post-infectious jejunal hypersensitivity and mast cell accumulation in an immunocompetent rat model. Infect Immun 77:5163–5169

Kirkpatrick BD, Haque R, Duggal P et al (2008) Association between *Cryptosporidium* infection and human leukocyte antigen class I and class II alleles. J Infect Dis 197(3):474–478

Koch KL, Phillips DJ, Aber RC et al (1985) Cryptosporidiosis in hospital personnel: evidence for person-to-person transmission. Ann Intern Med 102:593–59

Kotloff KL, Nataro JP, Blackwelder WC et al (2013) Burden and aetiology of diarrhoeal disease in infants and young children in developing countries (the Global Enteric Multicenter Study, GEMS): a prospective, case–control study. Lancet 382(9888):209–222. doi:10.1016/S0140-6736(13)60844-2

Krause I, Amir J, Cleper R et al (2012) Cryptosporidiosis in children following solid organ transplantation. Pediatr Infect Dis J 31(11):1135–1138

Lake IR, Harrison FCD, Chalmers RM et al (2007) Case–control study of environmental and social factors influencing cryptosporidiosis. Eur J Epidemiol 22:805–811

Learmonth JJ, Ionas G, Ebbett KA et al (2004) Genetic characterization and transmission cycles of *Cryptosporidium* species isolated from humans in New Zealand. Appl Environ Microbiol 70 (7):3973–3978

Lee MB, Greig JD (2010) A review of gastrointestinal outbreaks in schools: effective infection control interventions. J Sch Health 80(12):588–598. doi:10.1111/j.1746-1561.2010.00546.x

Liu L, Johnson HL, Cousens S et al (2012) Global, regional, and national causes of child mortality: an updated systematic analysis for 2010 with time trends since 2000. Lancet 379:2151–2161

Lucio-Forster A, Griffiths JK, Cama VA et al (2010) Minimal zoonotic risk of cryptosporidiosis from pet dogs and cats. Trends Parasitol 26(4):174–179. doi:10.1016/j.pt.2010.01.004

Lynch M, Woodruf R, Painter J et al (2006) Surveillance for foodborne-disease outbreaks-United States, 1998–2002. MMWR 55(10):1–42

Ma P, Soave R (1983) Three-step stool examination for cryptosporidiosis in 10 homosexual men with protracted watery diarrhea. J Infect Dis 147:824–828

MacKenzie WR, Schell WL, Blair KA et al (1995) Massive outbreak of waterborne *Cryptosporidium* infection in Milwaukee, Wisconsin: recurrence of illness and risk of secondary transmission. Clin Infect Dis 21(1):57–62

Mak JW (2004) Important zoonotic intestinal protozoan parasites in Asia. Trop Biomed 21 (2):39–50

Mallon ME, MacLeod A, Wastling JM et al (2003) Multilocus genotyping of *Cryptosporidium parvum* type 2: population genetics and sub-structuring. Infect Genet Evol 3(3):207–218

Manz M, Steuerwald M (2007) Cryptosporidiosis in a patient on PEG interferon and ribavirin for recurrent hepatitis C after living donor liver transplantation. Transpl Infect Dis 9:60–61

Martino P, Gentile G, Caprioli A et al (1988) Hospital-acquired cryptosporidiosis in a bone marrow transplantation unit. J Infect Dis 158:647–648

Masarat S, Ahmad F, Chisti M et al (2012) Prevalence of *Cryptosporidium* species among HIV positive asymptomatic and symptomatic immigrant population in Kashmir, India. Iran J Microbiol 4(1):35–39

McLauchlin J, Amar C, Pedraza-Díaz S et al (2000) Molecular epidemiological analysis of *Cryptosporidium* spp. in the United Kingdom: results of genotyping *Cryptosporidium* spp. in

1,705 fecal samples from humans and 105 fecal samples from livestock animals. J Clin Microbiol 38(11):3984–3990
Meisel JL, Perera DR, Meligro C et al (1976) Overwhelming watery diarrhea associated with a *Cryptosporidium* in an immunosuppressed patient. Gastroenterology 70(6):1156–1160
Millard PS, Gensheimer KF, Addiss DG et al (1994) An outbreak of cryptosporidiosis from fresh-pressed apple cider. JAMA 272(20):1592–1596
Miron D, Kenes J, Dagan R (1991) Calves as a source of an outbreak of cryptosporidiosis among young children in an agricultural closed community. Pediatr Infect Dis J 10:438–441
Molloy SF, Tanner CJ, Kirwan P et al (2011) Sporadic *Cryptosporidium* infection in Nigerian children: risk factors with species identification. Epidemiol Infect 139(6):946–954
Monge R, Arias ML (1996) Presence of various pathogenic microorganisms in fresh vegetable in Costa Rica. Arch Latinoam Nutr 46:192–294
Mor SM, Tzipori S (2008) Cryptosporidiosis in children in Sub-Saharan Africa: a lingering challenge. Clin Infect Dis 47(7):915–921. doi:10.1086/591539
Mor SM, Tumwine JK, Ndeezi G et al (2010) Respiratory cryptosporidiosis in HIV-seronegative children in Uganda: potential for respiratory transmission. Clin Infect Dis 50(10):1366–1372. doi:10.1086/652140
Morgan UM, Constantine CC, O'Donoghue et al (1995) Molecular characterization of *Cryptosporidium* isolates from humans and other animals using random amplified polymorphic DNA analysis. Am J Trop Med Hyg 52(6):559–564
Morgan UM, Constantine CC, Forbes DA et al (1997) Differentiation between human and animal isolates of *Cryptosporidium parvum* using rDNA sequencing and direct PCR analysis. J Parasitol 83(5):825–830
Morgan-Ryan UM, Fall A, Ward LA et al (2002) *Cryptosporidium hominis* n. sp. (Apicomplexa: Cryptosporidiidae) from *Homo sapiens*. J Eukaryot Microbiol 49:433–440
Naumova EN, Egorov AI, Morris RD et al (2003) The elderly and waterborne *Cryptosporidium* infection: gastroenteritis hospitalizations before and during the 1993 Milwaukee outbreak. Emerg Infect Dis 9(4):418–425
Navarrete S, Stetler HC, Avila C et al (1991) An outbreak of *Cryptosporidium* diarrhea in a pediatric hospital. Pediatr Infect Dis J 10:248–250
Navin TR, Juranek DD (1984) Cryptosporidiosis: clinical, epidemiologic, and parasitologic review. Rev Infect Dis 6(3):313–327
Nemes Z (2009) Diarrhea from the infectologist's point of view. Orv Hetil 150(8):353–361
Nichols G, Lane C, Asgari N et al (2009) Rainfall and outbreaks of drinking water related disease and in England and Wales. J Water Health 7(1):1–8
Nichols G (2008) Epidemiology. In: Fayer R, Xiao L (eds) *Cryptosporidium* and cryptosporidiosis. CRC Press, Boca Raton, pp 79–118
Nime FA, Burek JD, Page DL et al (1976) Acute enterocolitis in a human being infected with the protozoan *Cryptosporidium*. Gastroenterology 70(4):592–598
Nina JM, McDonald V, Deer RM et al (1992) Comparative study of the antigenic composition of oocyst isolates of *Cryptosporidium parvum* from different hosts. Parasite Immunol 14:227–232
Ogunkolade BW, Robinson HA, McDonald V et al (1993) Isoenzyme variation within the genus *Cryptosporidium*. Parasitol Res 79:385–388
Okhuysen PC (2001) Traveler's diarrhea due to intestinal protozoa. Clin Infect Dis 33(1):110–114
Okhuysen PC, Chappell CL, Crabb JH et al (1999) Virulence of three distinct *Cryptosporidium parvum* isolates for healthy adults. J Infect Dis 180(4):1275–1281
Okhuysen PC, Chappell CL, Kettner C et al (1996) *Cryptosporidium parvum* metalloaminopeptidase inhibitors prevent in vitro excystation. Antimicrob Agents Chemother 40 (12):2781–2784
Ortega YR, Roxas CR, Gilman RH et al (1997) Isolation of *Cryptosporidium parvum* and *Cyclospora cayetanensis* from vegetables collected in markets of an endemic region in Peru. Am J Trop Med Hyg 57:683–686

Ortega YR, Sheehy RR, Cama VA et al (1991) Restriction fragment length polymorphism analysis of *Cryptosporidium parvum* isolates of bovine and human origin. J Protozool 38:40S–41S

Ozgül A, Tanyüksel M, Yazicioglu K, Arpacioglu O (1999) Sacroiliitis associated with Cryptosporidium parvum in an HLA-B27-negative patient. Rheumatology (Oxford) 38(3):288–289

Pandak N, Zeljka K, Cvitkovic A (2006) A family outbreak of cryptosporidiosis: probable nosocomial infection and person-to-person transmission. Wien Klin Wochenschr 118 (15–16):485–487

Pedersen C, Danner S, Lazzarin A et al (1996) Epidemiology of cryptosporidiosis among European AIDS patients. Genitourin Med 72:128–131

Peng MM, Xiao L, Freeman AR et al (1997) Genetic polymorphism among *Cryptosporidium parvum* isolates: evidence of two distinct human transmission cycles. Emerg Infect Dis 3 (4):567–573

Pettoello-Mantovani M, Di Martino L, Dettori G et al (1995) Asymptomatic carriage of intestinal *Cryptosporidium* in immunocompetent and immunodeficient children: a prospective study. Pediatr Infect Dis J 14(12):1042–1047

Pohjola S, Oksanen H, Jokipii L et al (1986) Outbreak of cryptosporidiosis among veterinary students. Scand J Infect Dis 18(2):173–178

Pohlenz J, Bemrick WJ, Moon HW et al (1978) Bovine cryptosporidiosis: a transmission and scanning electron microscopic study of some stages in the life cycle and of the host-parasite relationship. Vet Pathol 15(3):417–427

Ponka A, Kotilainen H, Rimhanen-Finne R et al (2009) A foodborne outbreak due to *Cryptosporidium parvum* in Helsinki, November 2008. Euro Surveill 14(28):19269

Pozio E, Gomez Morales MA (2005) The impact of HIV-protease inhibitors on opportunistic parasites. Trends Parasitol 21(2):58–63

Pozio E, Rezza G, Boschini A et al (1997) Clinical cryptosporidiosis and human immunodeficiency virus (HIV)-induced immunosuppression: findings from a longitudinal study of HIV-positive and HIV-negative former injection drug users. J Infect Dis 176:969–975

Pozio E, Rivasi F, Cacciò SM (2004) Infection with *Cryptosporidium hominis* and reinfection with *Cryptosporidium parvum* in a transplanted ileum. APMIS 112:309–313

Preiser G, Preiser L, Madeo L (2003) An outbreak of cryptosporidiosis among veterinary science students who work with calves. J Am Coll Health 51(5):213–215

Putignani L, Menichella D (2010) Global distribution, public health and clinical impact of the protozoan pathogen *Cryptosporidium*. Interdiscip Perspect Infect Dis. doi:10.1155/2010/753512, pii: 753512

Quiroz ES, Bern C, MacArthur JR et al (2000) An outbreak of cryptosporidiosis linked to a foodhandler. J Infect Dis 181(2):695–700

Rieux A, Paraud C, Pors I et al (2013) Molecular characterization of *Cryptosporidium* spp. in pre-weaned kids in a dairy goat farm in western France. Vet Parasitol 192(1–3):268–272. doi:10.1016/j.vetpar.2012.11.008

Robertson B, Sinclair MI, Forbes AB et al (2002) Case–control studies of sporadic cryptosporidiosis in Melbourne and Adelaide, Australia. Epidemiol Infect 128(3):419–431

Robertson LJ (2007) The potential for marine bivalve shellfish to act as transmission vehicles for outbreaks of protozoan infections in humans: a review. Int J Food Microbiol 120(3):201–216

Robertson LJ, Chalmers RM (2013) Foodborne cryptosporidiosis: is there really more in Nordic countries? Trends Parasitol 29(1):3–9. doi:10.1016/j.pt.2012.10.003

Robertson LJ, Gjerde B (2001) Occurrence of parasites on fruits and vegetables in Norway. J Food Prot 64(11):1793–1798

Robinson G, Wright S, Elwin K et al (2010) Re-description of *Cryptosporidium cuniculus* Inman and Takeuchi, 1979 (Apicomplexa: Cryptosporidiidae); morphology, biology and phylogeny. Int J Parasitol 40:1539–1548

Rodrigues F, Davies EG, Harrison P et al (2004) Liver disease in children with primary immunodeficiencies. J Pediatr 145:333–339

Ross AG, Olds GR, Cripps AW et al (2013) Enteropathogens and chronic illness in returning travelers. N Engl J Med 368(19):1817–1825. doi:10.1056/NEJMra1207777

Roy SL, DeLong SM, Stenzel SA et al (2004) Risk factors for sporadic cryptosporidiosis among immunocompetent persons in the United States from 1999 to 2001. J Clin Microbiol 42 (7):2944–2951

Rzezutka A, Nichols RA, Connelly L et al (2010) *Cryptosporidium* oocysts on fresh produce from areas of high livestock production in Poland. Int J Food Microbiol 39(1–2):96–101. doi:10.1016/j.ijfoodmicro.2010.01.027

Saiman L, Aronson ZJ et al (2001) Prevalence of infectious diseases among internationally adopted children. Pediatrics 108(3):608–612

Sarkar R, Ajjampur SS, Prabakaran AD et al (2013) Cryptosporidiosis among children in an endemic semi-urban community in Southern India: does a protected drinking water source decrease infection? Clin Infect Dis 57:398–406

Sato MI, Galvani AT, Padula JA et al (2013) Assessing the infection risk of Giardia and Cryptosporidium in public drinking water delivered by surface water systems in Sao Paulo State, Brazil. Sci Total Environ 442:389–396. doi:10.1016/j.scitotenv.2012.09.077

Shepherd RC, Smail PJ, Sinha GP (1989) Reactive arthritis complicating cryptosporidial infection. Arch Dis Child 64(5):743–774

Shield J, Baumer JH, Dawson JA et al (1990) Cryptosporidiosis–an educational experience. J Infect 21(3):297–301

Shirley DA, Moonah SN, Kotloff KL (2012) Burden of disease from cryptosporidiosis. Curr Opin Infect Dis 25(5):555–563

Silverlås C, Mattsson JG, Insulander M et al (2012) Zoonotic transmission of *Cryptosporidium meleagridis* on an organic Swedish farm. Int J Parasitol 42(11):963–967. doi:10.1016/j.ijpara.2012.08.008

Smerdon WJ, Nichols T, Chalmers RM et al (2003) Foot and mouth disease in livestock and reduced cryptosporidiosis in humans, England and Wales. Emerg Infect Dis 9(1):22–28

Smith A, Reacher M, Smerdon W et al (2006a) Outbreaks of waterborne infectious intestinal disease in England and Wales, 1992–2003. Epidemiol Infect 134(6):1141–1149

Smith HV, Cacciò SM, Tait A et al (2006b) Tools for investigating the environmental transmission of *Cryptosporidium* and *Giardia* infections in humans. Trends Parasitol 22(4):160–167

Smith HV, Nichols RA, Grimason AM (2005) *Cryptosporidium* excystation and invasion: getting to the guts of the matter. Trends Parasitol 21(3):133–142

Snel SJ, Baker MG, Venugopal K (2009) The epidemiology of cryptosporidiosis in New Zealand, 1997–2006. N Z Med J 122(1290):47–61

Snelling WJ, Xiao L, Ortega-Pierres G et al (2007) Cryptosporidiosis in developing countries. J Infect Dev Ctries 1(3):242–256

Sopwith W, Osborn K, Chalmers R et al (2005) The changing epidemiology of cryptosporidiosis in North West England. Epidemiol Infect 133(5):785–793

Spano F, Putignani L, Crisanti A et al (1998) Multilocus genotypic analysis of *Cryptosporidium parvum* isolates from different hosts and geographical origins. J Clin Microbiol 36 (11):3255–3259

Stantic-Pavlinic M, Xiao L, Glaberman S et al (2003) Cryptosporidiosis associated with animal contacts. Wien Klin Wochenschr 115(3–4):125–127

Stefanogiannis N, McLean M, Van Mil H (2001) Outbreak of cryptosporidiosis linked with a farm event. N Z Med J 114(1144):519–521

Sulżyc-Bielicka V, Kołodziejczyk L, Jaczewska S et al (2012) Prevalence of *Cryptosporidium* sp. in patients with colorectal cancer. Pol Przegl Chir 84(7):348–351

Tanriverdi S, Grinberg A, Chalmers RM et al (2008) Inferences about the global population structures of *Cryptosporidium parvum* and *Cryptosporidium hominis*. Appl Environ Microbiol 74(23):7227–7234

ten Hove RJ, van Esbroeck M, Vervoort T et al (2009) Molecular diagnostics of intestinal parasites in returning travellers. Eur J Clin Microbiol Infect Dis 28(9):1045–1053

Thielman NM, Guerrant RL (1998) Persistent diarrhea in the returned traveller. Infect Dis Clin North Am 12(2):489–501

Toniati P, Giliani S, Jones A et al (2002) Report of the ESID collaborative study on clinical features and molecular analysis of X-linked hyper-IgM syndrome. Eur Soc Immunodeficiencies Newsl F9(Suppl):40

Tzipori S, Angus KW, Campbell I et al (1980) *Cryptosporidium*: evidence for a single-species genus. Infect Immun 30:884–886

Umemiya R, Fukuda M, Fujisaki K et al (2005) Electron microscopic observation of the invasion process of *Cryptosporidium parvum* in severe combined immunodeficiency mice. J Parasitol 91(5):1034–1039

Valentiner-Branth P, Steinsland H, Fischer TK et al (2003) Cohort study of Guinean children: incidence, pathogenicity, conferred protection, and attributable risk for enteropathogens during the first 2 years of life. J Clin Microbiol 41:4238–4245

Valigurová A, Jirků M, Koudela B et al (2008) Cryptosporidia: epicellular parasites embraced by the host cell membrane. Int J Parasitol 38(8–9):913–922

Vojdani JD, Beuchat LR, Tauxe RV (2008) Juice associated outbreaks of human illness in the United States, 1995 through 2005. J Food Prot 71(2):356–364

Whitty CJM, Carroll B, Armstrong M et al (2000) Utility of history, examination and laboratory tests in screening those returning to Europe from the tropics for parasitic infection. Trop Med Int Health 5(11):818–823

Widmer G, Klein P, Bonilla R (2007) Adaptation of *Cryptosporidium* oocysts to different excystation conditions. Parasitology 134:1583–1588

Wolska-Kusnierz B, Bajer A, Cacciò SM et al (2007) *Cryptosporidium* infection in patients with primary immunodeficiencies. J Pediatr Gastroenterol Nutr 45:458–464

Wumba R, Longo-Mbenza B, Mandina M et al (2010) Intestinal parasites infections in hospitalized AIDS patients in Kinshasa, Democratic Republic of Congo. Parasite 17(4):321–328

Xiao L (2010) Molecular epidemiology of cryptosporidiosis: an update. Exp Parasitol 124:80–89

Xiao L, Ryan UM (2008) Molecular epidemiology. In: Fayer R, Xiao L (eds) *Cryptosporidium* and cryptosporidiosis. CRC Press/IWA Publishing, Boca Raton, pp 119–171

Xiao L (2009) Overview of *Cryptosporidium* presentations at the 10th international workshops on opportunistic protists. Eukaryot Cell 8(4):429–436. doi:10.1128/EC.00295-08

Xiao S, An W, Chen Z et al (2012) The burden of drinking water-associated cryptosporidiosis in China: the large contribution of the immunodeficient population identified by quantitative microbial risk assessment. Water Res 46(13):4272–4280. doi:10.1016/j.watres.2012.05.012

Xu P, Widmer G, Wang Y et al (2004) The genome of *Cryptosporidium hominis*. Nature 432 (7015):1107–1112, Erratum in: Nature 432(7015):415

Yamamoto N, Urabe K, Takaoka M et al (2000) Outbreak of cryptosporidiosis after contamination of the public water supply in Saitama Prefecture, Japan, in 1996. Kansenshogaku Zasshi 74(6):518–526

Yoder JS, Beach MJ (2007) Cryptosporidiosis surveillance-United States, 2003–2005. MMWR 56(7):1–10

Yoder JS, Harral C, Beach MJ, Centers for Disease Control and Prevention (CDC) (2010) Cryptosporidiosis surveillance – United States, 2006–2008. MMWR Surveill Summ 59(6):1–14

Yoshida H, Matsuo M, Miyoshi T et al (2007) An outbreak of cryptosporidiosis suspected to be related to contaminated food, October 2006, Sakai City, Japan. Jpn J Infect Dis 60(6):405–407

Zhang W, Shen Y, Wang R et al (2012) *Cryptosporidium cuniculus* and *Giardia duodenalis* in rabbits: genetic diversity and possible zoonotic transmission. PLoS One 7(2):e31262

Zintl A, Proctor AF, Read C et al (2009) The prevalence of *Cryptosporidium* species and subtypes in human faecal samples in Ireland. Epidemiol Infect 137:270–277

Chapter 3
Molecular Epidemiology of Human Cryptosporidiosis

Gordon L. Nichols, Rachel M. Chalmers, and Stephen J. Hadfield

Abstract Most human cryptosporidiosis is caused by the species *Cryptosporidium hominis* and *C. parvum*, with less common species globally including *C. meleagridis*, *C. cuniculus*, *C. ubiquitum*, *C. viatorum*, *C. canis* and *C. felis*, although the distribution and prevalence may vary. Minor species that are only very rarely found in symptomatic people include *C. andersoni*, *C. suis*, *C. scrofarum*, *C. bovis*, and *C. muris*. The use of molecular methods to characterise isolates has contributed to our understanding of the overall epidemiology of cryptosporidiosis and its transmission, the biology and host range of each species and the evolution from a common ancestor. The development of more discriminatory typing methods has shown that outbreaks may be caused by subtypes identified at particular loci. The 60 kDa glycoprotein (gp60) gene has been the most widely used locus due to its high sequence variation. A gp60 naming system has been proposed and widely adopted which has greatly assisted with global epidemiological analysis. However, limitations of a single locus scheme have been identified including low gp60 subtype diversity in certain countries. A plethora of multilocus genotyping (MLG) methods have been developed to provide increased discrimination and are beginning to deepen our understanding of parasite population structures and transmission dynamics. However, there is a need for a standard MLG scheme that, when applied to routine surveillance datasets, will distinguish between outbreak and sporadic cases. This will allow investigation of geographically and temporally dispersed outbreaks which is not currently possible in areas of low gp60 sequence variation.

G.L. Nichols (✉)
European Centre for Disease Prevention and Control, Tomtebodavägen 11A, Stockholm, Solna 17183, Sweden

University of East Anglia, University of Thessaly, Public Health, England
e-mail: gordon.nichols@ecdc.europa.eu

R.M. Chalmers • S.J. Hadfield
Cryptosporidium Reference Unit, Public Health Wales Microbiology, Singleton Hospital, Swansea SA2 8QA, UK
e-mail: Rachel.Chalmers@wales.nhs.uk; Stephen.Hadfield@wales.nhs.uk

S.M. Cacciò and G. Widmer (eds.), *Cryptosporidium: parasite and disease*,
DOI 10.1007/978-3-7091-1562-6_3, © Springer-Verlag Wien 2014

3.1 Introduction to Molecular Epidemiology

Cryptosporidiosis is a disease that can be controlled by interventions; these are currently directed at limiting exposure to the parasite. Although vaccination has been attempted (Roche et al. 2013) there are no commercial vaccines and specific treatment options are limited (McDonald 2011). In order to intervene and prevent disease, epidemiological investigations should focus on identifying those at most risk, determining sources, transmission pathways, hosts and vectors and risk factors that are linked with infection. The evidence to date suggests clear human and animal sources and transmission routes that commonly involve water as a vehicle. However, there remain unanswered epidemiological questions and the role of molecular typing, both to the species level and further, in addressing these questions needs to be well documented. A review of the molecular epidemiology of *Cryptosporidium* infections undertaken in 2007 (Nichols 2008) informed us of a number of things. First, that not all cryptosporidiosis is caused by the same organism (Nichols et al. 1991), a concept proposed in 1980 (Tzipori et al. 1980). Second, it showed that the host range for *Cryptosporidium* species and genotypes varies (Xiao and Ryan 2004); third, that there is clear evidence of distinct human transmission routes and sources (Peng et al. 1997), with more than one species responsible for human disease; and fourth, that *C. parvum* and *C. hominis* are generally the most common species causing human cryptosporidiosis (McLauchlin et al. 2000; Morgan-Ryan et al. 2002; Peng et al. 1997). There was evidence that the distribution of these and other species differs geographically, with *C. meleagridis* being more common in some settings (for example in a shanty town in Peru) (Cama et al. 2003; Xiao et al. 2001b). It was shown that outbreaks can sometimes involve more than one *Cryptosporidium* species and individuals can be co-infected (McLauchlin et al. 2000; Peng et al. 1997) and that contaminated water can contain multiple species and genotypes (Xiao et al. 2006). Differences in the descriptive epidemiology for sporadic disease between *C. hominis* and *C. parvum* (Hunter et al. 2004; Nichols 2008) were demonstrated and though unrelated, the degree of genetic variation in different species (panmictic vs. clonal) can vary (Mallon et al. 2003b). The term genotype is used to refer to a taxonomic rank below the species level using a defined set of genetic markers, which for *Cryptosporidium* includes isolates with sequences sufficiently different at the SSU rRNA gene (similarity usually close to 99.5 %) but that don't have enough other data for species status. The term subtype is used when defined loci are investigated to detect variation within species. The term isolate is used for a specific population of organisms from a human or animal, and the term strain should be restricted to isolates that are maintained in culture, although these may not necessarily be genetically homogeneous. In one multi-locus study, some subtypes of *C. parvum* from human patients were found to be uncommon in agricultural animals (Mallon et al. 2003b), while natural infection with *C. hominis* in sheep and cattle can occur occasionally (Smith et al. 2005). The potential for significant disease reduction was demonstrated with interventions to reduce contamination of drinking water

(Sopwith et al. 2005). Molecular epidemiology has also demonstrated the potential for identifying new transmission routes (McLauchlin et al. 2000). These themes will be explored below.

3.1.1 Phenotyping

Early in the development of *Cryptosporidium* epidemiology, phenotypic methods were used to try to delineate species and provide investigation tools (McLauchlin et al. 1998; Nichols 1992; Nichols et al. 1991). Approaches used included monoclonal antibody and polyacrylamide gel electrophoresis or isoenzyme analysis to examine differences between *Cryptosporidium* species (Nina et al. 1992a, b; Ogunkolade et al. 1993). However, such approaches did not have the sensitivity of genetic approaches and have not proved to be very useful in the practice of molecular epidemiology, other than highlighting the presence of several *Cryptosporidium* species present in different hosts.

3.1.2 Genotyping to Species Level

Application of molecular methods for typing (genotyping) *Cryptosporidium* isolates has provided improved sensitivity and specificity over phenotyping and enabled more widespread investigation of variation in clinical and subclinical infections and in food, water and environmental samples. One of the important achievements of the application of molecular epidemiology has been the identification of infection sources (that is, the hosts providing a source of human or animal infection), by delineating the species of *Cryptosporidium*, their natural hosts and the evidence for infectivity and pathogenicity of these species for humans. The rationale for species designation was reviewed in 2010 by Fayer (2010) and is discussed, along with the current taxonomy of the genus, in Chap. 1. The most current nomenclature is used here, assuming in places that the prior name correctly represents the organism. *Cryptosporidium* species and their host distribution with some molecular epidemiological information are summarised in Table 3.1.

To date human cryptosporidiosis has been strongly linked to *C. hominis*, *C. parvum*, *C. meleagridis* and *C. cuniculus*. Experimental infections conducted in human volunteers and outbreak investigations has provided further evidence (Table 3.1). In some settings, *C. canis and C. felis* are also associated with human cryptosporidiosis, and *C. ubiquitum* and *C. viatorum* are recognised as emerging human pathogens. In the case of many of the minor species (those only rarely found in humans, see Table 3.1) there is merely the occasional finding of oocysts in faeces of symptomatic patients, including *C. andersoni*, *C. suis*, *C. scrofarum*, *C. bovis*, and *C. muris*. Some species have not been described in humans. This may reflect the greater adaptation of these species to their hosts, especially those found in fish, reptiles and amphibians, that is reflected in the taxonomy (Palenzuela et al. 2010).

Table 3.1 *Cryptosporidium* species and their involvement in human infections

Cryptosporidium species	Involvement in human cryptosporidiosis	Animal reservoirs of human infection	Molecular epidemiology
Species strongly linked to human infections			
C. hominis (Morgan-Ryan et al. 2002)	The species most commonly linked to human- derived infection, either directly person-to-person (especially in daycare centres, household contacts, toileting or nappy changing), or indirectly via contaminated drinking water, recreational water, food or fomites A common cause of sporadic cases and outbreaks of cryptosporidiosis Infectivity data from experimental infections in adults indicates that ingestion of a single oocyst carries a discrete probability of infection	There is no defined animal host for this species, although low density shedding of oocysts has been reported occasionally from cattle, sheep and goats	High parasite genetic heterogeneity has been reported in non-industrialised countries, compared with apparently less variation in some industrialised countries (Tanriverdi et al. 2008). A genome sequence for isolate TU502, has been published (Xu et al. 2004)
C. parvum (Xiao et al. 2012)	The species commonly transmitted from young ruminants to humans, and sometimes between humans, either directly through farm animal contact or person-to-person, or indirectly through contaminated drinking water, recreational water, environmental contact, food or fomites A common cause of sporadic cases and outbreaks of cryptosporidiosis Infectivity data from experimental infections in adults indicates variation between isolates, although virulence factors are as yet undefined. Ingestion of a single oocyst of some strains carries a discrete probability of infection	A common cause of diarrhoea in pre-weaned calves, lambs and goat kids, and also reported in foals, alpaca, llama (see Chaps. 4 and 5). Infection may be asymptomatic especially in older animals and sometimes in lambs. High oocyst output from pre-weaned animals increases risk of direct and indirect transmission to humans	High heterogeneity in isolates from humans has been reported from many countries; some subtypes (e.g. *gp60* IIc) appear host-adapted to humans with no defined animal host A full genome sequence for the zoonotic IOWA isolate (gp60 subtype IIaA15G2R1a) has been published (Abrahamsen et al. 2004). A three-way comparison with anthroponotic *C. parvum* TU114 (gp60 IIc) and *C. hominis* identified loci where the anthroponotic *C. parvum* sequence is more similar to *C. hominis* than to the zoonotic *C. parvum*, suggesting that proteins encoded by these genes may influence the host range (Widmer et al. 2012)

C. meleagridis (Slavin 1955)	One of the more common non-*C. parvum*, non-*C. hominis Cryptosporidium* species infecting humans Mainly sporadic cases, more frequent in some populations, for example in Bangkok, Thailand and Lima, Peru. Cases in UK have been linked to travel to endemic countries A possible farm-related outbreak has been reported in Sweden linked to poultry (Silverlas et al. 2012) Infectivity data from experimental infections in adults indicates mild illness	Associated with enteritis, diarrhoea and death in birds. Readily transmissible between birds and mammals	Subtyping *C. meleagridis* in Japan has identified several genotypes infecting birds (Abe 2010; Abe and Makino 2010)
C. cuniculus (Robinson et al. 2010)	Caused a drinking waterborne outbreak in the UK when a wild rabbit contaminated treated water (Chalmers et al. 2011a). Has also been demonstrated to occur in sporadic disease in the UK (Chalmers et al. 2011a; Elwin et al. 2012a). Individual reports from France (Network 2010) and children in Nigeria (Molloy et al. 2010)	Asymptomatic shedding of oocysts in natural and experimentally infected rabbits (Robinson and Chalmers 2010; Robinson et al. 2010)	May be overlooked by commonly used typing tools due to genetic similarity to *C. hominis* (Chalmers et al. 2011a). *C. cuniculus* gp60 subtype Va is found in humans and only occasionally in rabbits (including the rabbit that caused the outbreak in the UK) and Vb is found in both hosts (Zhang et al. 2012). Seasonal and sex distribution of *gp60* subtypes has been reported in humans in UK (Chalmers et al. 2011a)
C. canis (Fayer et al. 2001)	The species found in dogs that has been isolated from some infected humans (Elwin et al. 2012a), and epidemiologically linked to diarrhoea in children in a shanty town in Lima, Peru (Priest et al. 2006)	Diarrhoea usually seen in puppies and younger dogs; asymptomatic shedding of oocysts has been reported in older and immune- competent dogs	

(continued)

Table 3.1 (continued)

Cryptosporidium species	Involvement in human cryptosporidiosis	Animal reservoirs of human infection	Molecular epidemiology
C. felis (Iseki 1979)	The species found in cats and has been detected in infected humans in the UK (Chalmers et al. 2011b), France (Network 2010) and Spain (Cieloszyk et al. 2012). Having a compromised immune system and being in contact with a cat were risk factors for human infection (Elwin et al. 2012a). *C. felis* has been associated with diarrhoea in children in a shanty town in Lima, Peru (Priest et al. 2006)	More common in kittens and younger cats and has been associated with diarrhoea. *Cryptosporidium* was not detected in 57 healthy kittens in the UK (Gow et al. 2009) and uncommon (4 %) in 149 kittens in New York (Spain et al. 2001), although transmission from cat to human has been reported (Egger et al. 1990)	
Species occasionally found in human infections			
C. andersoni (Lindsay et al. 2000)	This species has been reported in a few human infections in the UK, Australia, and Malawi (Leoni et al. 2006; Morse et al. 2007; Waldron et al. 2011a), including in HIV positive people (Agholi et al. 2013)	Occurs in the stomach of cattle (Lindsay et al. 2000), and has been reported in one study in bird faeces (Ng et al. 2006). This species has also been detected in wastewaters (Ben Ayed et al. 2012; Liu et al. 2011; Nichols et al. 2006a; Ruecker et al. 2013; Xiao et al. 2001a), molluscs (Miller et al. 2005; Nichols et al. 2006a) and tapwater (Castro-Hermida et al. 2008; Feng et al. 2011c)	*C. andersoni* subtyping of cattle strains has identified at least five MLST types (Zhao et al. 2013)
C. baileyi (Current et al. 1986)	There is a single unsubstantiated report of human infection (Ditrich et al. 1991) with this bird species, made without molecular analysis	Many bird species can be affected, particularly chickens (Abbassi et al. 2000)	

C. bovis (Fayer et al. 2005)	This species is usually found in cattle (Fayer et al. 2005); small numbers of human infections have been reported of people stating cattle contact in Australia and India (Helmy et al. 2013; Ng et al. 2012)	Mainly shed by weaned calves with no or few clinical signs	
C. fayeri (Ryan et al. 2008)	This kangaroo species been detected in a individual infected person in Australia who had contact with marsuipials (Waldron et al. 2010)	Found in kangaroos but no clinical signs are described in animals (Ryan et al. 2008; Yang et al. 2011)	
C. muris (Tyzzer 1910)	The first *Cryptosporidium* species to be described (Tyzzer 1910). It occurs, as its name suggests, in mice. *C. muris* has larger oocysts than most species and is reported to infect humans only occasionally (Tosini et al. 2010)	Infection is of the stomach of mice; clinical signs are not usually reported; infection and oocyst shedding has been reported in other animal species	A genome sequence has been published (Widmer and Sullivan 2012)
C. scrofarum (Kvac et al. 2013)	This species is found in farmed pigs worldwide (Kvac et al. 2013), in wild pigs (Nemejc et al. 2012) and has also been identified in fish (Reid et al. 2010) and a calf (Ng et al. 2011). To date this species has been identified in a single human immunocompetent case in Czech Republic who had contact wild pigs (Kvac et al. 2009)	Infection is seen in pigs older than 6 weeks resulting in low levels of oocyst shedding and no association with diarrhoea	
C. suis (Ryan et al. 2004a)	This species is found in domestic pigs (Ryan et al. 2004a) and has been detected in a small number of immunocompetent and in HIV positive patients with diarrhoea (Cama et al. 2007; Leoni et al. 2006; Wang	Infection is usually seen in pre-weaned pigs but is not associated with diarrhoea	The frequent detection of this organism in drinking water in China without human illness being commonly identified (Feng et al. 2011c) suggests the organism is poorly infectious or lacks pathogenicity for most people

(continued)

Table 3.1 (continued)

Cryptosporidium species	Involvement in human cryptosporidiosis	Animal reservoirs of human infection	Molecular epidemiology
	et al. 2013; Xiao et al. 2002) in UK and Peru who had contact with pigs		
C. tyzzeri (Ren et al. 2012)	This species is considered to be the species first found in mice by Tyzzer in 1912 (Yang et al. 2011; Tyzzer 1910). There is one individual report from Czech Republic involving contact with wild mice (Raskova et al. 2013)	Infection is usually most intense in the mouse ileum but no clinical signs were recorded in experimental infections	
C. ubiquitum (Fayer et al. 2010)	Has been demonstrated in infected humans in the UK (Chalmers et al. 2011b; Elwin et al. 2012a), Canada (Wong and Ong 2006), Spain (Cieloszyk et al. 2012) and Nigeria (Molloy et al. 2010)	A common parasite of weaned lambs (Fiuza et al. 2011; Shen et al. 2011; Sweeny et al. 2011) and has been found in many other mammals indicating a broad host range encompassing ruminants, rodents and primates. Contamination of water sources, stormwaters and wastewaters in many geographical locations	Newly developed *gp60* primers have been used to compare human and animal isolates in different geographical locations revealing differences in transmission (Lihua paper submitted – check we can state this)
C. viatorum (Elwin et al. 2012c)	A species identified from infected people returning to the UK from the Indian subcontinent (Jellison et al. 2009), sporadic cases have also been reported in Sweden, following visits to South America and Kenya (Insulander et al. 2013)	It is not clear whether this species is found only in humans, as an animal host has not been identified yet	
Species not reported in humans			
C. fragile (Jirku et al. 2008)	This species has been identified in toads (Wang et al. 2013)	Infection seems host specific as infections of four amphibian and one fish species failed	

C. galli (Ryan et al. 2003)	*C. galli* has been isolated from wild and pet birds (Antunes et al. 2008; da Silva et al. 2010; Nakamura et al. 2009; Qi et al. 2011; Seva Ada et al. 2011)	Infection of the proventriculus can lead to diarrhoea and fatalities, shedding can be prolonged and is not necessarily associated with diarrhoea
C. macropodum (Power and Ryan 2008)	This species is found in marsupials including kangaroos (Yang et al. 2011)	Infection is limited to marsupials, without any clinical signs
C. molnari (Alvarez-Pellitero and Sitja-Bobadilla 2002)	A fish species (Alvarez-Pellitero and Sitja-Bobadilla 2002), first found in farmed bream and bass and a *C. molnari*-like species has been detected in guppies (Ryan et al. 2004b) (Ng et al. 2011)	Infection is seasonal, mostly in younger fish in the spring
C. ryanae (Fayer et al. 2008)	This species is found in buffalos (Amer et al. 2013; Feng et al. 2012) and cattle (Fayer et al. 2008; Helmy et al. 2013; Liu et al. 2009; Santin and Zarlenga 2009)	Infection is host specific and shedding is seen in weaned calves but is thought to be asymptomatic
C. scophthalmi (Alvarez-Pellitero et al. 2004)	*C. scophthalmi* (Alvarez-Pellitero et al. 2004) are fish species found in turbot	Infection occurs in the intestine and is seen seasonally in younger fish; has been linked to reduced growth rates
C. serpentis (Levine 1980)	This species is common in snakes and lizards (Pavlasek and Ryan 2008; Richter et al. 2011)	Infection is common. In snakes, infection manifests as anorexia, persistent post-prandial regurgitation, lethargy, mid-body swelling and chronic weight loss, which is usually protracted and almost always fatal. In lizards infection is usually asymptomatic
C. varanii (Pavlasek and Ryan 2008)	This species is found in lizards and snakes (Pavlasek and Ryan 2008)	Infection is seen in lizards and snakes. Clinical signs include anorexia, progressive weight loss, abdominal swelling and death, particularly in young animals

(continued)

Table 3.1 (continued)

Cryptosporidium species	Involvement in human cryptosporidiosis	Animal reservoirs of human infection	Molecular epidemiology
C. wrairi (Vetterling et al. 1971)	*C. wrairi* (Vetterling et al. 1971) was identified from guinea pigs	Disease not described	
C. xiaoi (Fayer and Santin 2009)	*This species*) is found in lambs and sheep (Fayer and Santin 2009) in many countries (Imre et al. 2013; Sweeny et al. 2011), in goats (Diaz et al. 2010), as well as in kangaroos (Yang et al. 2011) and fish (Reid et al. 2010)	In sheep, asymptomatic carriage; no association of oocyst shedding and clinical symptoms in experimentally infected lambs, while infection in goat kids can be associated with diarrhoea	Closely related to *C. bovis*

However, some zoonotic species have been detected, for example *C. parvum*, *C. scrofarum* and *C. xiaoi*, in fish (Reid et al. 2010).

The genotypes of *Cryptosporidium* that have been reported comprise those isolates that have yet to be given species status, those that are sub-species of one of the recognised species and those that have been shown to be different from recognised species but are not fully characterised. Most are named after the host in which they are first reported. It is likely, based on previous experience, that several of these will, in the fullness of time, be described as separate species. Some of these genotypes have also been found in humans, including the *C. hominis* monkey genotype (Xiao et al. 1998) detected in humans in UK and Malawi (Elwin et al. 2012a; Mallon et al. 2003b), a chipmunk genotype I identified in chipmunks (Lv et al. 2009), squirrels, geese (Jellison et al. 2009) and reported in humans in the USA, France and Sweden (Insulander et al. 2013; Network 2010), a skunk genotype found in skunks, racoons (Chavez et al. 2012), geese (Jellison et al. 2009), an elephant seal (Rengifo-Herrera et al. 2011) and reported in human infections in the UK (Chalmers et al. 2009; Davies et al. 2009; Elwin et al. 2012a) (interestingly there are no skunks outside of zoos in the UK) and a horse genotype identified in foals (Burton et al. 2010) and reported in humans in the UK and USA (Elwin et al. 2012a).

Many other genotypes (over 50) have not been detected in humans. For example, a seal genotype (Bass et al. 2012); two muskrat genotypes (Perz and Le Blancq 2001; Zhou et al. 2004) that have also been found in samples from geese (Jellison et al. 2009); a fox genotype (Zhou et al. 2004) and three different rat genotypes (Paparini et al. 2012). This list of genotypes not found in humans is not definitive, and does not mean these types cannot cause human infection. The species and genotype host range data suggest that over time there has been selection and speciation of *Cryptosporidium* lineages not only in individual hosts, but also in host phyla. However the host range of some species appears to be wider than for others. In addition it remains likely that some "hosts" are merely transport vectors of oocysts from their environment rather than being infected themselves (e.g. *C. parvum* found in molluscs, fish, or geese).

3.1.3 Subtyping

Efforts to investigate genetic diversity within and between *Cryptosporidium* species, aimed at identifying infection sources and investigating transmission dynamics, have mainly been focussed on *C. parvum* and *C. hominis* as the most important human pathogens. *C. cuniculus*, *C. meleagridis*, *C. andersoni* and *C. muris* have also been investigated (Feng et al. 2011a; Glaberman et al. 2001; Wang et al. 2012; Zhang et al. 2012). Caution must be expressed for "subtypes" based on sequence analysis of the ssu rRNA gene alone, arising from the polymorphic nature of different copies of the gene (Xiao et al. 1999). Because of a lack of identified genetic markers, other species have been less investigated. The 60 kDa glycoprotein (*gp60*, also called *gp40/15*) gene (Strong et al. 2000) is the most common marker that has been used, either alone or as part of a multi-locus panel. However, as a protozoan with a sexual cycle, *Cryptosporidium* species diversity is best investigated at multiple loci, although there is, as yet, no standardised multilocus subtyping scheme. There is experimental evidence that gp60 subtypes of *C. parvum* and *C. hominis* do not segregate with a multi-locus subtyping method (Widmer and Lee 2010), confirming the need for a standardised multi-locus approach to isolate identification.

3.2 Factors Influencing the Molecular Epidemiology of Human Infections

The same factors that impact the prevalence of *Cryptosporidium* in humans (Chap. 2) also influence the molecular epidemiology, i.e.: host-related factors of age, sex, immunological status, geographical location and ethnic group; exposure including hygiene, diet, water consumption, rural and urban settings, contact with human waste and livestock, food and water protection, travel and immigration; and environmental and socio-economic factors including animal pollution, malnutrition and dehydration, geography, calamities (e.g. floods), weather and climate (Nichols 2008; Nichols et al. 2006; Putignani et al. 2010). Exposure factors will also be influenced by the distribution of *Cryptosporidium* species in various hosts and in vehicles of transmission such as water (see Chaps. 4, 5, and 12). In addition, parasite-related infectivity and pathogenicity factors also influence the molecular epidemiology of human infections (Bouzid et al. 2013; Nichols 2008; Xiao 2008). Although 25 putative virulence factors have been proposed, none of them has yet been confirmed (Bouzid et al. 2013), yet various studies have shown differences in outcome of infection between *C. parvum* and *C. hominis*. For example, a study of sporadic cryptosporidiosis in the UK showed significantly longer duration of diarrhoea in *C. hominis* patients than in those with *C. parvum* infection (Hunter et al. 2004). Comparison of three *C. parvum* strains in adult volunteer studies found differences in ID50, attack rate and duration of diarrhoea (Okhuysen et al. 1999). The association between gp60 subtypes and outcome of infection is discussed later. When host and parasite factors are coupled, they help explain the molecular epidemiology and potential risks posed by different exposures and *Cryptosporidium* species to human health. For example, the microbiological evidence from the rare occurrence of *C. felis* and *C. canis* in humans in most settings is that dog and cat faeces are probably not a major source of human infection (Bowman and Lucio-Forster 2010). This is supported by many analytical epidemiological studies, none of which identify contact with companion animals as a risk factor for human cryptosporidiosis (Bowman and Lucio-Forster 2010). Nevertheless, some individuals are susceptible and it is sensible to advise hand washing after contact with any animal, although there does not seem to be any particular susceptibility to infection with these species in immunocompromised patients.

The common occurrence and low infectious dose of *C. parvum* and *C. hominis* in human infections (Chap. 2), and the rare occurrence of many of the other *Cryptosporidium* species, suggests that there is some species barrier that prevents many of these rarer species infecting people. Because *C. parvum* and *C. hominis* are common in the environment as a result of animal and human faecal contamination from animal waste and sewage inputs, there may also be a relationship between infection and exposure to oocysts that are common in the environment. It has been shown that in some countries a broader range of species can be detected in all patients (Morgan-Ryan et al. 2002). This contrasts to the view that some *Cryptosporidium* species are more commonly found in immunocompromised patients than in immunocompetent ones. However, with improved protocols for detecting different species it should be recognised that the ability to detect some species has improved. A recent example of this is *C. cuniculus* that was identified as *C. hominis* using routine PCR screening methods but has been detected as a small but significant component of the UK *Cryptosporidium* cases following more detailed typing.

3.3 Genotyping and Subtyping Methods

Molecular testing or screening stools for *Cryptosporidium* using PCR protocols are much more sensitive than microscopic methods (Hadfield et al. 2011), allowing larger numbers of cases of cryptosporidiosis to be diagnosed, and improving estimates of carriage. However, assays for the detection of presence/absence of the genus may not be designed to identify species, for which conventional (single round and nested) and real-time PCRs have been described. PCR primers and conditions need to be selected carefully to amplify all *Cryptosporidium* species of interest and to avoid non-specific amplification. Any PCR-based method will be dependent, among other things, on efficient DNA extraction. This is also discussed in this section.

3.3.1 DNA Extraction from Clinical Samples

There is no recommended method, but efficient DNA extraction underpins the application of molecular assays (Elwin et al. 2012b). The most commonly investigated human samples for epidemiological purposes are stools, and extraction can be undertaken readily either directly from fresh or frozen stools or from semi-purified oocyst suspensions or sediments, including immuno-magnetically separated oocysts, providing the oocysts are disrupted so that sporozoite DNA can be made available (Elwin et al. 2012b).

Options for oocyst disruption processes include boiling (Elwin et al. 2001), freeze–thaw cycles (Bialek et al. 2002; Nichols et al. 2006b) and chemical, enzymatic or mechanical treatments (Balatbat et al. 1996; Lindergard et al. 2003; McLauchlin et al. 1999; Pieniazek et al. 1999) or a combination such as bead-beating, followed by guanidine thiocyanate extraction and silica purification (Amar et al. 2007; Boom et al. 1990; McLauchlin et al. 1999; Patel et al. 1998; Zintl et al. 2009).

Options for DNA extraction include commercial spin columns, glassmilk, and chelex resin. One study reported that direct extraction by bead-beating, guanidine thiocyanate and silica purification provided slightly lower PCR positivity and higher threshold cycle (Ct) values in real-time PCR, indicating lower DNA yield, than spin-column extraction from boiled, semi-purified oocyst suspensions, although direct extraction is cheaper, and amenable to automation (Elwin et al. 2012b). *Cryptosporidium* DNA extraction has been demonstrated successfully in an automated, universal DNA extraction method for enteric bacteria, viruses and protozoan based on bead-beating (MagNa Lyser instrument, Roche Diagnostics Ltd) and lysis under denaturing conditions in the presence of proteinase-K followed by nucleic acid binding to the silica surface of magnetic particles (QIAsymphony® SP, Qiagen) (Halstead et al. 2013). Contaminants are removed by washing, and pure nucleic acid is eluted in either modified TE buffer or water.

If the original stool sample is not available, the stained microscopy slides left over from the original diagnosis can be used; the smeared material can be scraped off into lysis buffer and extracted by a suitable method (Amar et al. 2001).

Some preservatives and fixatives applied to stool samples, such as 10 % formalin, sodium acetate–acetic acid–formalin (SAF) and 2.5 % potassium dichromate, may inhibit downstream PCR. If they have not penetrated the oocysts, they can be removed by washing. In one study, *Cryptosporidium* DNA was successfully amplified from "aged" stool samples that had been collected and processed in 10 % formalin from HIV-AIDS patients by including washing steps in PBS before multiple freeze-thaw cycles and preceding stool extraction kit by 95 °C incubation in lysis buffer (QIAamp DNA stool mini kit; Qiagen (Del Chierico et al. 2011). The repeated PBS washings were deemed to remove excess formalin. Further, the selection of a relatively small size DNA target (358 bp of the gp60 gene) minimized the effect of formalin-induced DNA damage, including fragmentation and cross-linking.

Stools can contain many PCR inhibitors, including heme, bilirubin, bile salts, and complex polysaccharides. Boiling in 10 % polyvinylpolypyrrolidone before oocyst disruption and DNA extraction can reduce inhibition, alternatively spin columns appear to be effective. The use of reagents such as 400 ng/μl of non-acetylated bovine serum albumin in the PCR is recommended.

The choice of disruption and extraction processes will depend on the types and consistency of samples being tested, the normal procedures in the laboratory, the number of parasites present, whether other pathogens are also being sought, the laboratory resources (expertise, time, consumables and equipment available), and the requirement for any long-term archiving of material.

3.3.2 Species Identification by Conventional PCR

Conventional PCRs which either produce amplicon sizes characteristic of individual species or requiring downstream analysis usually by restriction enzyme digestion identifying restriction fragment length polymorphisms (RFLP), or by sequencing, have been described in detail previously (Chalmers et al. 2011c). There is no single recommended method, although sequencing provides the definitive identification, as some species/genotypes have RFLP patterns that are indistinguishable.

Conventional PCR assays have targeted various genes including small subunit (SSU) rRNA, *Cryptosporidium* oocyst wall protein (COWP), thrombospondin-related adhesive proteins, 70-kDa heat shock protein (HSP70), and actin genes (Chalmers et al. 2011c) and were described in detail (Xiao 2008). Sequencing the SSU rRNA gene is regarded as the benchmark and has been widely used in many studies, because its analysis offers improved sensitivity due to the multiple copies of the gene and because it has both conserved and polymorphic regions. Use of other gene targets may lead to under-estimation of the presence of some species, as the designed primers don't amplify DNA from all species. However, conventional PCR is time-consuming and resource-heavy.

An alternative method for species identification is capillary electrophoresis single strand conformation polymorphism (CE-SSCP). One study found that species could be readily differentiated based on the SSCP mobility of amplified ssu rRNA gene molecules, and clones that differed by single-nucleotide polymorphisms could be distinguished in this way (Power et al. 2011).

3.3.3 Species Identification by Real-Time PCR

Real-time PCR methods for *Cryptosporidium* were first developed over 12 years ago (Amar et al. 2003; Fontaine and Guillot 2002; Higgins et al. 2001; MacDonald et al. 2002) and can provide a more rapid result than conventional PCR. When coupled with modern filtration sampling methods and immunomagnetic separation, real-time PCR can also be used for examining *Cryptosporidium* isolates from water (Fontaine and Guillot 2003). Real-time PCR offers advantages of improved work flow and efficiency for large-scale molecular surveillance, as well as reduced handling avoiding cross contamination and carryover of PCR products, amplification performance monitoring, and lower consumables costs and staff time (Hadfield et al. 2011).

There is no recommended method and various real-time PCR assays have been described for the identification of *Cryptosporidium* species in human clinical samples, especially where many years of base-line work has established the usual prevalence of each species (Table 3.2). In the UK, where the vast majority of clinical cases are caused by *C. parvum* or *C. hominis* (Chalmers et al. 2009, 2011c), typing has evolved to streamline the process: Hadfield et al. (2011) initially developed and applied a TaqMan minor groove binding (MGB) probe-based assay that provides specific identification of *C. hominis* and *C. parvum* by targeting a locus of unknown function, LIB13 (Tanriverdi et al. 2003), while simultaneously detecting all known *Cryptosporidium* species and genotypes by amplification of a region of the SSU rRNA gene (Morgan et al. 1997) with identification by direct sequence analysis. Designed for investigation of human samples, semi-purification of oocysts prior to DNA extraction (Elwin et al. 2001) effectively screens out many potential cross-reacting genera which can be problematic when targeting this gene which is ubiquitous and conserved among eukaryotes. This assay improved typability of microscopy-positive stools to 99.5 % compared with 96.9 % by

Table 3.2 Some real-time PCR approaches used in investigation of human samples for *Cryptosporidium*

Cryptosporidium amplification target	Detection platform and chemistry	Purpose of application	Most suitable application	Reference
SSU rRNA gene	Light cycler (Roche) Hybridisation probes	Species identification	Differentiation of intestinal species	Limor et al. (2002)
β-tubulin gene	Light cycler (Roche) SYBR Green/Hybridisation probes	Species identification	Differentiation of *C. parvum* and *C. hominis*	Tanriverdi et al. (2002)
Lib13 locus	Light cycler (Roche) and SYBR Green	Species identification	Differentiation of *C. parvum* and *C. hominis*	Tanriverdi et al. (2002)
COWP gene	Light cycler (Roche) SYBR Green	Species identification	Real-time, then RFLP as melt curves couldn't differentiate	Amar et al. (2004)
DNA-J like protein gene	iCycler (BioRad) MGB TaqMan	Genus detection in diagnostic parasitology	Used in a multiplex assay for simultaneous detection of *Entamoeba histolytica*, *Giardia duodenalis*, and *C. parvum*	Verweij et al. (2004)
SSU rRNA gene	iCycler (BioRad) Scorpion probes	Species identification	Detection of *Cryptosporidium* spp. Specific detection of *C. parvum* and *C. meleagridis*	Stroup et al. (2006)
SSU rRNA gene	iCycler (BioRad) SybrGreen	Genus detection	Detection of *Cryptosporidium* spp. in cell culture and human faecal samples	Parr et al. (2007)
COWP gene	iCycler (BioRad) TaqMan	Genus detection in diagnostic parasitology	Used in a multiplex assay for simultaneous detection of *Cryptosporidium* spp. *E. histolytica* and *G. duodenalis*	Haque et al. (2007)
SSU rRNA gene and Lib13 locus	iCycler (BioRad) + Mx3000p (Stratagene)	Genus detection and species identification	Detection of *Cryptosporidium* spp. Differentiation of *C. parvum* and *C. hominis*	Jothikumar et al. (2008)
SSU rRNA gene	Rotor-Gene 6000 (Qiagen) Melt curve (dye not stated)	Genus detection	Multiplexed detection of *Cryptosporidium* spp., *Dientamoeba fragilis*, *Entamoeba histolytica* and *G. duodenalis*	Stark et al. (2011)

SSU rRNA gene, and Lib13 locus,	Rotor-Gene 6000 (Corbett Research) MGB TaqMan	Species identification	Differentiation of *C. parvum* and *C. hominis*. Detection and identification by sequencing of other *Cryptosporidium* spp.	Hadfield et al. (2011)
DNA-J like protein gene	ABI PRISM 7300 (Life Technologies) Dual-labelled probe	Genus detection	Detection of *Cryptosporidium* spp.	Calderaro et al. (2011)
gp60 Va and Vb alleles	Rotor-Gene 6000 (Corbett Research) Melt curve analysis (Sybr Green)	Species detection and subtype family identification	Detection of *C. cuniculus*. Identification of Va and Vb subtype families	Hadfield and Chalmers (2012)
SSU rRNA gene	ViiA7 (Life Technologies) TaqMan singleplex array card	Genus detection	Detection of *Cryptosporidium* spp., with parallel detection of enterotoxigenic *E. coli*, enteropathogenic *E. coli*, enteroaggregative *E. coli*, and Shiga-toxigenic *E. coli*, *Shigella*/enteroinvasive *E. coli*, *Salmonella*, *Campylobacter jejuni*/*Campylobacter coli*, *Vibrio cholerae*, and *Clostridium difficile*, *G. duodenalis*, *E. histolytica*, *Ascaris lumbricoides*, *Trichuris trichiura*, rotavirus, norovirus group 2, adenovirus, astrovirus, and sapovirus	Liu et al. (2013)
DNA-J like protein gene	ABI 7500 (Life Technologies) Probe not stated	Genus detection	Detection of *Cryptosporidium* spp. multiplexed *E. histolytica* and *G. duodenalis*	McAuliffe et al. 2013 J Infect 67:122–129
SSU rRNA gene	Platform not stated TaqMan and hybridisation probes	Genus detection and species identification	Single-tube detection of *Cryptosporidium* spp. and identification of *C. hominis*/*C. cuniculus* and *C. parvum*	Mary et al. (2013)
DNA J-like protein	ABI 7900HT Fast/ABI 7500 (Applied Biosystems) TaqMan MGB	Genus detection in diagnostic parasitology	Detection of *Cryptosporidium* spp., with parallel detection of *A lumbricoides*, *Necator americanus*, *Ancylostoma duodenale*, *G. duodenalis*, *E. histolytica*, *T. trichiura*, and *Strongyloides stercoralis*	Mejia et al. (2013)

PCR-RFLP COWP gene and increased the detection of unusual species (Chalmers et al. 2013). The assay originally incorporated an exogenous internal control (IC) for the identification of inhibited reactions. The IC was considered superfluous for reference genotyping of clinical samples as no inhibition was identified in analysis of over 2,000 samples, largely due to the semi-purification step in sample preparation (Elwin et al. 2012b), but should be considered when using PCR for primary testing. Because the vast majority of samples are *C. parvum* or *C. hominis*, and mixed infections were detected in $<1\ \%$ of samples, the assay has since been re-formatted (Hadfield et al., in preparation) as a single tube, multiplex PCR using re-designed *C. hominis* primers and probe based on the CpA135 gene (Tosini et al. 2010) that avoid amplification of *C. cuniculus* which is co-detected in most assays (Chalmers et al. 2009). The reference test algorithm based on this single multiplex real-time PCR includes re-testing the small number of samples that do not amplify in the *C. parvum*/*C. hominis* assay by amplification of a region of the SSU rRNA gene with identification by direct sequence analysis.

Alternative molecular algorithms need to be applied for primary testing incorporating genotyping, where it is not known whether *Cryptosporidium* are present or absent. For example, one paper describes a real-time PCR assay for detection and quantification of all *Cryptosporidium* species with a conserved sequence TaqMan MGB probe that is then followed by specific identification of *C. hominis* and *C. parvum* in positive samples using a duplex assay with specific TaqMan MGB probes, performed on the same amplicon (Mary et al. 2013). However, one limitation of the assay pointed out by the authors is that other *Cryptosporidium* species cannot be characterized. Other papers describe multiplex assays for primary testing of stools for a variety of parasites (Table 3.2).

A TaqMan array 384 well real time PCR technique has been used for screening for a range of 19 enteric pathogens including *Cryptosporidium* (Liu et al. 2013). Similar approaches have been used by others. If applied uniformly this sort of approach could give a much better uniformity of surveillance than current targeted approaches. The singleplex real-time PCR format has advantages over multiplex approaches (Haque et al. 2007; Jothikumar et al. 2009; Nazeer et al. 2013; Stark et al. 2011; Taniuchi et al. 2011), which have predominantly been focussed on parasites, in covering a broader spectrum of pathogens and possibly being amenable to use as a single screening method for all the common enteric pathogens. Re-formatting the assay targets would permit identification of *Cryptosporidium* species/genotypes.

Further developments in PCR-based detection and quantitation include digital PCR which has the potential to have utility in food and water microbiology, where more precise quantitation is needed (Morisset et al. 2013).

3.3.4 *Characterisation of* Cryptosporidium *in Stools by Alternative Methods*

An alternative to PCR-based methods is loop-mediated isothermal amplification (LAMP), based on autocycling strand-displacement DNA synthesis by *Bst* polymerase which may overcome the inhibitory potential of some samples; one study reported a high prevalence of *Cryptosporidium* in animal faeces that were negative by PCR (Bakheit et al. 2008), and another demonstrated improved detection over microscopy (Nago et al. 2010) although improved specificity is required for species characterisation and independent evaluation is required.

The presence of two extra-chromosomal linear virus-like double stranded RNA elements in some *Cryptosporidium* species has been utilised in a heteroduplex mobility assay (HMA) where heterogeneity between two segments of DNA, one being a reference sequence and the other a test sample, is identified; heating to denature the DNA into single strands and slow cooling encourages annealing of both the original homoduplex and the formation of a heteroduplex comprising the two different start strands. These can be differentiated by the slower PAGE migration by the heteroduplexes. One group in the UK developed a HMA for epidemiological investigations (Leoni et al. 2003). Although unusual strains identified by HMA were confirmed by sequencing, and epidemiological utility identified, this method has not been widely adopted.

3.3.5 *GP60 Subtyping Methods*

Sequencing the hypervariable gp60 gene has been used to further classify some *Cryptosporidium* species and genotypes, most commonly *C. parvum* and *C. hominis*, demonstrating extensive intra-species diversity. This analysis has been used to increase our understanding of the epidemiology of human cryptosporidiosis and to assist in outbreak investigations and source tracking of contamination of food and water.

The gp60 gene encodes two surface glycoproteins, gp15 and gp40, and contains hyper-variable regions and a microsatellite sequence of tandem tri-nucleotide repeats coding for the amino acid serine at the 5′ (gp40) end of the gene (Strong et al. 2000). GP60 subtype nomenclature is based on that proposed for *C. parvum* by Sulaiman et al. (2005) and *C. hominis* by Cama et al. (2007). A hyper-variable region towards the 3′ end of the gene is used to classify subtype families and provide the first part of the nomenclature: for *C. hominis* termed Ia, Ib, Id, etc. and for *C. parvum* IIa, IIb, IIc etc. Note there is now no Ic as the isolate to which it was ascribed was actually *C. parvum* and is re-classified as IIc; this family is regarded as human-restricted *C. parvum* as there is no defined animal host, although IIc has been reported in a hedgehog (Dyachenko et al. 2010), and a IIc isolate has been shown to be infectious to immunosuppressed mice (Tanriverdi et al. 2007). The second part of the nomenclature, used to designate the "subtype", is based on the number and combination of various serine repeats, whether TCA (represented by the letter A), TCG (represented by the letter G), or TCT (represented by the letter T). A third expression may be used for some isolates which have one or more characteristic sequences repeated immediately after the serine microsatellite region, and is represented by the letter R and a number. Finally, where isolates diverge at the 3′ end of the gene but are insufficiently different to form a new subtype family, lower case letters are used to indicate the difference. Readers are referred to standard nomenclature as shown in Table 3.3. Typical examples are *C. hominis* IbA10G2 (family Ib with ten TCA and two TCG serine repeats) and *C. parvum* IIaA15G2R1 (family IIa with 15 TCA and two TCG serine repeats and one characteristic repeat sequence).

The most commonly used gp60 assay is a nested PCR using external primers AL3531 and AL3535 which amplify ~850 to ~900 bp region of the gene and the internal primers AL3532 and AL3534 which amplify a ~810 to ~870 bp region (Alves et al. 2003). The sensitivity of the assay is not as good as that described for nested SSU rRNA (Chalmers et al. 2005), and an alternative, amplifying a shorter product of ~358 to ~400 bp, can be helpful (Sulaiman et al. 2005). Commonly used gp60 primers do not amplify species other than *C. parvum* and *C. hominis*, and different primers have been developed, for example for *C. meleagridis* (Glaberman et al. 2001) and *C. fayeri* (Power et al. 2009).

Sequencing the gp60 gene has identified a large number of distinct "subtypes" within *C. hominis* and *C. parvum* from humans, which suggested genetic 'richness' (i.e. high number of distinct types) but low diversity (i.e. the relationship between the number and relative prevalence of each those distinct types) (Jex and Gasser 2010). However, at a more local level, this is dependent on the population studied (see Sect. 3.4 and Table 3.4). As indicated earlier there is the need for a standardised multi-locus approach for subtyping *C. parvum* and *C. hominis.*

Alternatives to sequencing the *gp60* gene have been investigated for comparing isolates. PCR-based restriction endonuclease fingerprinting was used by one group to measure mutations in the gene, allowing single-nucleotide substitutions for amplicons of approximately 1 kb in size to be detected (Pangasa et al. 2010). Another group used a dual fluorescent terminal-restriction fragment length polymorphism (T-RFLP) analysis of the gp60 gene to investigate the genetic diversity of *Cryptosporidium* subtype populations in a single host infection. Variation within one human *Cryptosporidium* sample and mouse samples from seven consecutive passages with *C. parvum* was revealed, showing that differences in the ratio of subtype populations occur between infections (Waldron and Power 2011). Single strand conformational polymorphism analysis has also been used (Jex et al. 2007b; Wu et al. 2003) and can provide a rapid, cost-effective alternative for differentiation in population-based studies but sequencing would be required to confirm subtypes. Fragment sizing of gp60 amplicons has also been used in multilocus subtyping, which is discussed below.

As a subtyping tool, investigation of the gp60 gene offers both advantages and disadvantages. It is the most variable gene identified so far, and there is a large amount of sequence data available. The data are readily generated, although there is no single reference repository for comparison against reference subtypes and

Table 3.3 Major gp60 families of Cryptosporidium spp. and representative sequences (Updated from Feng et al. (2011b))

Species and family	Dominant trinucleotide repeat	Other repeat (R)	GenBank accession no.
C. hominis			
Ia	TCA	AA/GGACGGTGGTAAGG	AF164502 (IaA23R4)
Ib	TCA, TCG, TCT	–	AY262031 (IbA10G2), DQ665688 (IbA9G3)
Id	TCA, TCG	–	DQ665692 (IdA16)
Ie	TCA, TCG, TCT	–	AY738184 (IeA11G3T3)
If	TCA, TCG	–	AF440638 (IfA19G1)
Ig	TCA	–	EF208067 (IgA24)
Ih	TCA, TCG	–	FJ971716 (IhA14G1)
Ii (*C. hominis* monkey genotype)	TCA	–	HM234173 (IiA17)
Ij[a]	TCA	–	JF681174 (IjA14)
C. parvum			
IIa	TCA, TCG	ACATCA	AY262034 (IIaA15G2R1), DQ192501 (IIaA15G2R2)
IIb	TCA	–	AF402285 (IIbA14)
IIc	TCA, TCG	–	AF164491 (IIcA5G3a)
	TCA, TCG	–	AF164501 (IIcA5G3b)
	TCA, TCG	–	EU095267 (IIcA5G3c)
	TCA, TCG	–	AF440636 (IIcA5G3d)
	TCA, TCG	–	HM234172 (IIcA5G3e)
	TCA, TCG	–	HM234171 (IIcA5G3f)
	TCA, TCG	–	Accession no. not available; IIcA5G3g (REF)
	TCA, TCG	–	Accession no. not available; IIcA5G3h (REF)
	TCA, TCG	–	AM947935 (IIcA5G3i)
	TCA, TCG	–	HQ005749 (IIcA5G3j)
	TCA, TCG	–	JN867336 (IIcA5G3k)
IId	TCA, TCG	–	AY738194 (IIdA18G1)
IIe	TCA, TCG	–	AY382675 (IIeA12G1)
IIf	TCA	–	AY738188 (IIfA6)

(continued)

Table 3.3 (continued)

Species and family	Dominant trinucleotide repeat	Other repeat (R)	GenBank accession no.
IIg	TCA	–	AY873780 (IIgA9)
IIh	TCA, TCG	–	AY873781 (IIhA7G4)
IIi	TCA	–	AY873782 (IIiA10)
IIk	TCA	–	AB237137 (IIkA14)
IIl	TCA	–	AM937006 (IIlA18)
IIm	TCA, TCG	–	AY700401 (IImA7G1)
IIn	TCA	–	FJ897787 (IInA8)
IIo	TCA, TCG	–	JN867335 (IIoA16G1)
C. meleagridis			
IIIa	TCA, TCG	–	AF401499 (IIIaA24G3)
IIIb	TCA, TCG	–	AF401501 (IIIbA13G1), AB539720 (IIIbA20G1)
IIIc	TCA	–	AF401497 (IIIcA6)
IIId	TCA	–	DQ067570 (IIIdA6)
IIIe	TCA, TCG	–	AB539721 (IIIeA20G1)
C. fayeri			
IVa	TCA, TCG, TCT	–	FJ490060 (IVaA11G3T1)
IVb	TCA, TCG, TCT	–	FJ490087 (IVbA9G1T1)
IVc	TCA, TCG, TCT	–	FJ490069 (IVcA8G1T1)
IVd	TCA, TCG, TCT	–	FJ490058 (IVdA7G1T1)
IVe	TCA, TCG, TCT	–	FJ490071(IVeA7G1T1)
IVf	TCA, TCG, TCT	–	FJ490076 (IVfA12G1T1)
C. cuniculus			
Va	TCA	–	FJ262730 (VaA18)
Vb	TCA	–	FJ262734 (VbA29)
Horse genotype			
VIa[b]	TCA, TCG	–	FJ435960 (VIaA11G3), DQ648547 (IIjA15G4)
VIb	TCA	–	FJ435961 (VIbA13)

C. wrairi			
VIIa	TCA, TCT	–	GQ121020 (VIIaA17T1)
Ferret genotype			
VIIIa	TCA, TCG	–	GQ121029 (VIIIaA5G2)
C. tyzzeri			
IXa	TCA	A/GTTCTGGTACTGAAGATA	GQ121030 (IXaA6R3), AY378188 (IIfA6R2)
IXb	TCA	–	HM234177 (IXbA6R2)
Mink genotype			
Xa	TCA, TCG	–	HM234174 (XaA5G1)
Opossum genotype			
XIa	TCA, TCG, TCT	–	HM234181 (XIaA4G1T1)
C. ubiquitum			
XII[c]	–	–	Not available
Hedgehog genotype			
XIII	TCA	ACATCA	KC305643 (XIIIaA22R9)

[a]Baboon isolate, originally referred to as Ii (ref)
[b]Originally referred to as IIj
[c]Lihua Xiao, personal communication

Table 3.4 Distribution of *C. parvum* and *C. hominis* gp60 families in humans (Updated from Xiao (2010))

Geographical region (UN) and country	Patient group	*C. parvum* cases subtyped							*C. hominis* subtyped								
		N	IIa	IIb	IIc	IId	IIe	Other	N	Ia	Ib	Id	Ie	If	Ig	Ih	
Europe																	
Belgium	Sporadic cases	6	4	0	1	1	0	0	13	0	13	0	0	0	0	0	Geurden et al. (2009)
England, North	Sporadic cases	Not investigated							8	0	8	0	0	0	0		Glaberman et al., (2002)
France	Waterborne outbreak	1	1	0	0	0	0	0	22	0	21	1	0	0	0		Cohen et al., (2006)
Ireland	Sporadic cases	79	78	0	0	1	0	0	25	0	25	0	0	0	0	0	Zintl et al. (2009)
Ireland	Sporadic cases	249	170	0	0	0	0	0	Not investigated								Zintl et al. (2011)
Italy	Unspecified patients	None identified							5	0	5	0	0	0	0		Wu et al. (2003)
Italy	AIDS patients	8	4	0	4	0	0	0	1	1	0	0	0	0	0	0	Del Chierico et al. (2011)
Netherlands	Sporadic cases[a]	13	9	0	1	3	0	0	64	0	63	1	0	0	0	0	Wielinga et al. (2008)
Northern Ireland	Waterborne outbreak, sporadic cases	52	52	0	0	0	0	0	68	1	67	0	0	0	0		Glaberman et al. (2002)
Portugal	Unspecified	4	1	0	3	0	0	0	3	2	0	0	0	1	0		Peng et al. (2001)
Portugal	HIV + adults[a]	25	9	1	7	8	0	0	15	1	10	1	2	1	0	0	Alves et al. (2006)
Slovak Republic	Family members	Not investigated							1	0	1	0	0	0	0	0	Ondriska et al. (2013)
Slovenia	Children and adults	31	29	0	1	0	0	1	2	1	1	0	0	0	0	0	Soba and Logar (008)
Slovenia	Patients with gastro-intestinal disorders	21	21	0	0	0	0	0	71	0	53	12	0	1	5	0	Koehler et al. (2013)
Slovenia	Unspecified	3	3	0	0	0	0	0	None identified								Peng et al. (2001)
Sweden	2 Foodborne outbreaks	27	4	0	0	23	0	0	2	0	1	1	0	0	0	0	Gherasim et al. (2010)
Sweden	Sporadic cases	107	69	0	11	24	1	2xIIo	63	5	44	10	1	3	0	0	Insulander et al. (2013)
Switzerland	HIV+	2	1	0	0	1	0	0	3	0	2	1	0	0	0	0	O'Brien et al. (2008)
UK	Unspecified	None identified							3	0	3	0	0	0	0	0	O'Brien et al. (2008)
UK	Sporadic cases	69	56	0	1	9	0	3	Not investigated								Chalmers et al. (2011b)
UK	Patients from farms[a]	11	11	0	0	0	0	0	Not investigated								Smith et al. (2010b)
UK	Sporadic cases	Not investigated							101	5	93	0	0	1	2	0	Chalmers et al. (2008)
UK	Outbreak cases[a]	22	22	0	0	0	0	0	Not investigated								Chalmers et al. (2005)

Oceania																
Australia	Sporadic cases	6	6	0	0	0	0	0	12	0	10	0	2	0	0	0	Chalmers et al. (2005)
Australia	Unspecified	24	23	0	1	0	0	0	38	3	34	0	0	1	0	0	Jex et al. (2008)
Australia	Unspecified	23	18	0	5	0	0	0	74	0	73	1	0	0	0	0	Jex et al. (2007a)
Australia	Children[a]	9	8	0	1	0	0	0	41	1	24	13	0	3	0	0	O'Brien et al. (2008)
Australia	Unspecified	32	30	0	1	1	0	0	37	2	28	5	1	1	0	0	Waldron et al. (2009)
Australia	Farm dwellers[a]	7	5	0	0	1	0	0	None identified								Ng et al. (2012)
Australia	Sporadic cases[a]	80	79	0	0	1	0	0	164	4	156	0	0	4	0	0	Waldron et al. (2011a)
Australia	Outbreak	None identified							48	0	37	6	1	1	3	0	Ng et al. (2010b)
Australia	Outbreak	Not investigated							8	0	8	0	0	0	0	0	Mayne et al. (2011)
Australia	Sporadic cases	24	23	0	1	0	0	0	38	3	34	0	0	1	0		Jex et al. (2008)
Australia, Western	Sporadic cases	49	48	0	0	1	0	0	195	0	39	88	4	3	60		Ng et al. (2010b)
Australia, Western	Unspecified[a]	4	4	0	0	0	0	0	3	0	2	1	0	0	0	0	Ng et al. (2008)
New Zealand	Outbreak[a]	4	4	0	0	0	0	0	0								Grinberg et al. (2011)
New Zealand	Sporadic cases[a]	41	41	0	0	0	0	0	Not investigated								Grinberg et al. (2008a)
Australia	Outbreak	3	3	0	0	0	0	0	21	0	21	0	0	0	0	0	Ng et al. (2010b)
Northern America																	
Canada	Outbreak	None identified							4	0	0	4	0	0	0	0	Ong et al. (2008)
Canada	Sporadic cases[a]	5	5	0	0	0	0	0	Not investigated								Budu-Amoako et al. (2012)
Canada	Adults[a]	6	4	0	0	0	0	2	4	2	0	1	1	0	0	0	Trotz-Williams et al. (2006)
USA	Outbreak	7	7	0	0	0	0	0	0								
USA, 4 states	Cases	5	5	0	0	0	0	0	51	43	8	0	0	0	0	0	Xiao et al. (2009)
USA	Various	Not investigated							55	11	30	8	6	0	0		Sulaiman et al. (2001)
USA, New Orleans	HIV + adults	6	1	0	5	0	0	0									Xiao and Ryan (2004)
USA, Upper Midwest	Cases	166							Not investigated								Herges et al. (2012); Gatei et al. (2006)
USA, Wisconsin	Unspecified	30	30	0	0	0	0	0	1	0	1	0	0	0	0	0	Feltus et al. (2006)
Southern Asia																	
Bangladesh	Children	4	0	0	0	0	0	4xIIm	48	8	10	6	13	11	0	0	Hira et al. (2011)
India	Children	7	0	0	7	0	0	0	47	35	1	8	3	0	0	0	Ajjampur et al. (2007)

(continued)

Table 3.4 (continued)

Geographical region (UN) and country	Patient group	*C. parvum* cases subtyped							*C. hominis* subtyped								
		N	IIa	IIb	IIc	IId	IIe	Other	N	Ia	Ib	Id	Ie	If	Ig	Ih	
India	HIV + adults	9	0	5	3	0	0	1	28	6	8	5	0	8	0	0	Muthusamy et al. (2006)
India		14	0	0	6	5	3	0	39	12	1	8	13	5	0	0	
India	Children	6	0	0	1	1	0	2xIIm, 2xIIn	59	16	11	7	16	5	0	0	Ajjampur et al. (2010)
India	Unspecified	0	0	0	0	0	0	0	7	5	1	1	0	0	0		Peng et al. (2001)
India, Kolkata	Children with diarrhoea	Not investigated							43	7	15	16	5	0	0		Gatei et al. (2007)
Iran	Children[a]	22	7	0	0	15	0	0	3	0	0	1	0	2	0		Nazemalhosseini-Mojarad et al. (2011)
Nepal	Unspecified patients	None identified							6	6	0	0	0	0	0		Wu et al. (2003)
Tehran, Iran	Children	17	6	0	0	11	0	0	2	0	0	0	0	2	0		Taghipour et al. (2011)
Sub-Saharan Africa																	
Ethiopia	Patients with diarrhoea	13	13	0	0	0	0	0	1	0	1	0	0	0	0	0	Adamu et al. (2010)
Kenya	Paediatric population	23	2		17			2xIIi, 2xIIm	28	10	10	4	3	0	0	1	Molloy et al. (2010)
Kenya	Unspecified	4	0	4	0	0	0	0	2	2	0	0	0	0	0		Peng et al. (2001)
Madagascar	Children	1	0	0	1	0	0	0	10	3	0	4	1	0	0		Areeshi et al. (2008)
Malawi	Children	2	0	0	1	0	1	0	26	4	5	11	6	0	0	0	Peng et al. (2003)
Nigeria	Patients	1	0	0	0	0	1	0	2	2	0	0	0	0	0	0	Maikai et al. (2012)
Nigeria	Children	2	0	0	2	0	0	0	3	2	1	0	0	0	0	0	Ayinmode et al. (2012)
South Africa	HIV + children	5	0	0	5	0	0	0	15	1	4	5	5	0	0	0	Leav et al. (2002)
Uganda	Children	15	0	0	10	0	0	5	13	4	1	1	7	0	0	0	Akiyoshi et al. (2006)
Northern Africa																	
Egypt Ismailia province	Children[a]	14	7	0	0	7	0	0	Did not amplify								Helmy et al. (2013)
Western Asia																	
Jordan	Children	13	3	0	2	8	0	0	15	0	8	7	0	0	0		Hijjawi et al. (2010)
Kuwait	Children	59	28	0	2	29	0	1	4	0	2	1	1	0	0	0	Sulaiman et al. (2005)
Kuwait	Children	61	29	0	12	20	0	0	22	8	0	12	2	0	0	0	Iqbal et al. (2011)
Yemen	Patients attending clinic	32	7	0	0	0	0	0	1	0	0	0	1	0	0		Alyousefi et al. (2013)

Southeast Asia																	
Malaysia	HIV/AIDS patients	6	2	0	0	2	1	1	3	0	3	0	0	0	0		Lim et al. (2011)
Malaysia	HIV + patients	13	12	0	0	1	0	0	5	3	2	0	0	0	0		Iqbal et al. (2012)
Eastern Asia																	
China		None identified							9	1	6	2	0	0	0		Wang et al. (2011)
China	Unspecified	1	0	1	0	0	0	0	2	0	1	1	0	0	0		Peng et al. (2001)
Japan	Unspecified[a]	2	1	0	1	0	0	0	3	1	1	0	1	0	0	0	Abe et al. (2010)
Japan	Unspecified patients	1	1	0	0	0	0	0	2	1	0	0	1	0	0		Wu et al. (2003)
Latin America and Caribbean																	
Cuba	Children	None identified							6	1	3	2	0	0	0		Pelayo et al. (2008)
Guatemala	Unspecified	1	0	0	1	0	0	0	3	0	0	0	3	0	0		Peng et al. (2001)
Jamaica	HIV + adults	7	0	0	7	0	0	0	25	0	22	0	3	0	0		Gatei et al. (2008)
Peru	HIV + adults	22	0	0	22	0	0	0	141	35	39	40	13	0	0		Cama et al. (2007)
Peru	Children	15	0	0	15	0	0	0	56	15	16	7	15	3	Mixed		Cama et al. (2008)

[a]Studies included testing of animals

non-standardised nomenclature has been used by some researchers. Investigation of samples from animals and humans has provided information about the distribution of families and subtypes useful in supporting (or otherwise) zoonotic transmission (see below). However, the sequence similarity between isolates and the relationship to similarity at other loci is not extensively known. The proteins encoded by the gene have a functional role in invasion of host cells and initiation of disease. As such, gene products are exposed to the host immune response, and the gene is likely to be under considerable selective pressure, reflected by its diversity. Interestingly, when Widmer (2009) compared the distribution of serine repeat length in *C. parvum* isolates from humans with those from animals, the former showed a bi-modal distribution with more shorter repeats than occur in animals, indicative of host-driven selective pressure.

Recombinant progeny of *C. parvum* strains have been used to examine co-infections in immunosuppressed mice (Tanriverdi et al. 2007). The recombinants were obtained by crossing two genetically distinct isolates co-infected in immunosuppressed mice and then by targeted propagation in mice of the progeny oocysts from populations lacking one parental allele at one or more loci. Sixteen progeny clones were genotyped at 40 loci located on each of the eight chromosomes and a range for the meiotic crossover frequency was determined. Like other Apicomplexan parasites, *C. parvum* exhibited a high rate of recombination. Evolution is therefore likely to be rapid as recombination events between different strains can occur; new subtypes are constantly being placed on GenBank. More compatible with the biology of *Cryptosporidium*, and the presence of a sexual phase in the life cycle, is a multilocus subtyping scheme, for both epidemiological and population biology investigations.

3.3.6 Multilocus Typing Methods

Multilocus genotyping (MLG) methods have become more widely used to improve on the discrimination offered by gp60 single locus analysis. Most often this involves investigation of amplified PCR products from multiple loci to produce either a multilocus fragment type (MLFT) or multilocus sequence type (MLST). MLFT identifies variations in the length of microsatellite (repeating unit lengths up to 4 bp) or minisatellite (>4 bp) markers, driven by the number of tandem repeats and is measured cost-effectively by gel or capillary electrophoresis. MLST, as the name suggests, is based on the heterogeneity identified by DNA sequencing providing not only the number but also variation in the markers, which may also include those with only single nucleotide polymorphisms (SNPs), providing improved discrimination between *Cryptosporidium* isolates (Gatei et al. 2006).

One of the most striking features of multi-locus investigations of *C. parvum* and *C. hominis* is the diversity of loci that have been investigated. There is no standardised MLFT or MLST scheme, and a variety of loci have been used in global, regional and local parasite population studies and in epidemiological investigations (Table 3.5). Robinson and Chalmers (2012) attempted to use

Table 3.5 Informative multi-locus investigations of *Cryptosporidium parvum* and *Cryptosporidium hominis* involving human cases (Adapted and updated from Robinson and Chalmers (Strong et al. 2000))

Location	Samples subtyped	Loci investigated	Number of MLGs identified: *C. hominis*	Number of MLGs identified: *C. parvum*	Key findings	Reference
Global comparison studies						
Investigation of population structure in *C. hominis* from US, Kenya, Australia, Guatamala	62 *C. hominis* samples	gp60 hsp70 Poly-T SSR-1 SSU rDNA TRAP-C2	9	–	The same number of MLGs and gp60 subtypes were identified. Did not identify recombination, supporting clonal population but acknowledged the need for subsequent studies involving the use of more polymorphic genes	Sulaiman (2001)
Investigation of geographically distinct isolates from Kolkata, India, Nairobi, Kenya, Lima, Peru and New Orleans, USA	69 *C. hominis* sporadic human cases	cp47 cp56 dzhrgp gp60 hsp70 msc6-7 Mucin1 rpgr	34	–	Geographical clustering was observed when gp60 excluded	Gatei et al. (2006)
Investigation of genetic variability in isolates from various countries (USA, Scotland, Israel, Uganda, Australia)	Reference isolates of human and animal origin: 10 *C. hominis*, 16 *C. parvum*	1887 5B12 Cp492	10	16	Differential rate of repeat lengths inferred for *C. hominis* and *C. parvum*	Tanriverdi (2006)

(continued)

Table 3.5 (continued)

Location	Samples subtyped	Number of MLGs identified: Loci investigated	C. hominis	C. parvum	Key findings	Reference
		MS5—Mallon MS9—Mallon MSA MSB MSC MSE MSF MSG MSI MSK TP14				
Investigation of population structures in France and Port-au-Prince, Haiti	*C. hominis*: 51 human cases	ML1 ML2	8	17 in humans	Most *C. hominis* isolates (Tanriverdi et al. 2008) were of one MLG	Ngouanesavanh et al. (2006)
	C. parvum: 42 human cases and 14 animal samples (7 calf; 7 goat)	MS5—Mallon MS9—Mallon		10 in calves 5 in goats	From analysis of human isolates in France a basic clonal population structure was observed and in *C. hominis* basic or epidemic clonal population structure. In Haiti, the low degree of polymorphism suggested epidemic clonality	

Investigation of global partitioning in *C. parvum* and *C. hominis* populations and comparison of the population structures among isolates from Uganda, USA, England and Wales (including foreign travellers), Israel, New Zealand, Serbia and Turkey	*C. hominis*: 290 sporadic and outbreak human cases *C. parvum*: 227 human cases and animal isolates (calves, horse, goat)	1887 MS9—Mallon MSA MSB MSC MSE MSG MSK TP14	81	74 (not differentiated by host in the analysis)	For both parasite species, a quasi-complete phylogenetic segregation was observed among the countries. Statistical tests, linkage disequilibrium (LD), and eBURST analysis indicated both clonal and panmictic population structures. Gene flow was not sufficient to erase genetic divergence among geographically separated populations, which basically remained allopatric	Tanriverdi et al. (2008)
Area-specific studies						
Investigation of genetic variability in northern, central and southern Italy	59 *C. parvum* (9 HIV + ve humans, 29 calves, 11 kids, 10 lambs)	ML1 ML2	–	6	Some alleles were found in different hosts while some were host or geographically restricted	Caccio (2001)
Investigation of genetic variability in England	*C. hominis*: 104 human cases (28 sporadic, 60 linked to 3 waterborne outbreaks, 16 from 8 familial outbreaks) *C. parvum*: 126 human (73 sporadic, 40 linked to 3 waterborne outbreaks, 13 from 6 familial outbreaks), 17 sporadic livestock samples (lambs or calves)	gp60 ML1 MS5—Mallon	6	25 human 5 animal	Isolates of both *C. hominis* and *C. parvum* from separate waterborne outbreaks were genetically homogeneous, suggesting preferential or point source transmission	Leoni et al. (2007)

(continued)

Table 3.5 (continued)

Location	Samples subtyped	Loci investigated	Number of MLGs identified: *C. hominis*	Number of MLGs identified: *C. parvum*	Key findings	Reference
Investigation of genetic variability and risk factors in North West England and Wales	*C. hominis*: 118 sporadic human cases *C. parvum*: 63 sporadic human cases	gp60 ML1 ML2	9	31	Analysis with epidemiologic data found an association between reported contact with farm animals and individual *C. parvum* microsatellite alleles	Hunter et al. (2007)
Investigation of population structure in Aberdeenshire, Scotland	*C. hominis*: 71 sporadic human cases *C. parvum*: 64 sporadic human cases; 44 bovine cases	gp60 hsp70 ML1 MS5—Mallon MS9—Mallon MS12 TP14	7	26 human 15 bovine	Overall cluster analysis identified three *C. parvum* and one *C. hominis* cluster. *C. hominis* isolates were essentially clonal. Human *C. parvum* isolates were in LD suggesting clonality, whereas bovine isolates were not, suggesting panmixia	Mallon et al. (2003a)
Investigation of population structure in Orkney and Thurso, Dumfriesshire, Scotland	*C. parvum*: 40 human cases, 190 bovine cases, 11 ovine cases	gp60 ML1 hsp70 MS12 MS5—Mallon MS9—Mallon TP14	7	36 human 24 bovine 6 ovine	Found no evidence to support geographic or temporal sub-structuring of the *C. parvum* populations, but host-substructuring was identified by cluster analysis, with human *C. parvum* isolates falling into two human-only groups and a human/animal group	Mallon (2003b)

Re-analysis of *C. parvum* cases from Aberdeenshire, Orkney and Dumfriesshire, Scotland (Mallon et al. 2003a, b)	297 *C. parvum* samples (human, cattle, sheep)	gp60 hsp70 ML1 MM5 MM18 MM19 MS5—Mallon MS9—Mallon MS12 TP14	–	95 (not differentiated by host in the analysis)	Limited sub-structuring found within a small geographical area, but substantial sub-structuring over larger geographical distances	Morrison et al. (2008) C. *parvum* data from Mallon (2003b)
Investigation of population structure in Kolkata, India	*C. hominis*: 47 paediatric patients	Chrom3T CP47 CP56 GP60 HSP70 MSC6-7 Mucin1	–	25	Four major groups identified. Significant intra- and inter-genic LD observed with minimum recombination or expansion of limited subtypes, indicative of a mostly clonal population structure	Gatei (2007)
Study of genetic diversity in Kingston, Jamaica	HIV + ve patients: 25 *C. hominis* 7 *C. parvum*	CP47 DZ-HRGP HSP70 MSC6-7 RPGR	2	4	Presence of homogeneous *C. hominis* and *C. parvum* populations which were distinct from isolates from other countries	Gatei (2008)
Investigation of population structure in Malawi	41 HIV + ve and –ve *C. hominis* patients (MLGs obtained for 18 only)	GP60 HSP70	–	10	LD analysis showed possible recombination. Cryptosporidiosis in the study area was largely caused by anthroponotic transmission	Peng (2003)

(continued)

Table 3.5 (continued)

Location	Samples subtyped	Loci investigated	Number of MLGs identified: *C. hominis*	Number of MLGs identified: *C. parvum*	Key findings	Reference
Investigation of population structure in northern, central, and southern Italy	*C. parvum*: 9 human, 122 bovine, 21 ovine, 21 caprine *C. hominis*: 9 human	MS1 MS9 TP14 MM5 MM18 MM19 GP60	2	102 (9 human)	Predominantly clonal population structure, but also evidence that part of the diversity could be explained by genetic exchange	Drumo et al. (2012)
Investigation of population structure in Minnesota and Wisconsin, USA	212 *C. parvum*: 166 human, 46 cattle	TP14 CP47 MS5 MS9 MSC6-7 DZ-HRGP GRH gp60	–	94 (90 human, 13 bovine)	Among the 94 MLGs identified, 60 were represented by a single isolate. Approximately 20 % of isolates belonged to MLT 2, a group that included both human and cattle isolates. Population analyses revealed a predominantly panmictic population with no apparent geographic or host substructuring	Herges et al. (2012)
Outbreak investigations						
Various locations in Wales and England	*C. parvum*: 25 human cases 4 lambs 2 calves 1 effluent sample (also contained *C. ubiquitum*)	hsp70 ITS-2 (SSCP) gp60 ML1 ML2	–	4	Single MLGs identified in each outbreak. Outbreak cases and suspected sources were linked	Chalmers et al. (2005)

Stockholm, Sweden	Outbreak cases associated with a swimming pool	gp60 hsp70 TP14	–	2	MLGs were clustered according to pool use: one MLG was found among outdoor pool users and secondary cases, and another MLG from users of indoor pool	Mattsson (2008)
North West England	88 human samples from drinking waterborne outbreak: 10 *C. hominis* 78 *C. parvum*	gp60 ML1 ML2	1	17	Clustering of one *C. parvum* MLG was observed (statistically significant) in patients living in the area most dependent on one water supply	Hunter et al. (2008)

published data to identify the most informative markers as candidates for the development of a standardised multi-locus fragment size-based typing (MLFT) scheme for epidemiological purposes. They identified that in 31 MLFT studies, 55 markers had been used, of which 45 were applied to both *C. parvum* and *C. hominis*. Further analysis of 11 selected studies that provided sufficient raw data from three or more markers of >2 bp repeats, and a sampling frame containing at least 50 samples, revealed that some markers were present in all schemes at an average redundancy (i.e. markers which did not significantly enhance discrimination) of 40 % for *C. hominis* and 27 % for *C. parvum*. They ranked markers based on the most discriminatory combinations, which revealed two different sets of potentially most informative candidate markers, one for each species. It is likely, however, that wider application of whole genome sequencing will enable the identification of more informative markers.

Accurate analysis of microsatellite and minisatellite markers can be problematic due to the difficulties encountered by DNA polymerases during PCR amplification resulting in multiple fragment sizes evidenced by multiple "stutter" bands on electrophoresis gels or mixed sequence chromatograms (Diaz et al. 2012). Shorter microsatellites appear to be more problematic in this respect as well as being more technically challenging for size assignment by MLFT methods. Careful selection of marker and analysis method with validation with sequenced isolates is therefore essential. It should also be noted that sizes given by different MLFT methods are not necessarily equivalent.

Although sequence analysis is perhaps most desirable as it can potentially give the most accurate fragment size measurement and greater discrimination, if MLG is to be used during outbreak investigations fragment sizing provides more rapid assessment of relationship between isolates. Either way, there is a need for technical evaluation including typability (percentage of samples generating a complete multi-locus type) and discriminatory power by direct fragment size analysis and analysed for correlation with epidemiological data in suitable sampling frames.

An alternative sequence based approach to analysis of micro- and minisatellites and the gp60 gene was reported by Bouzid et al. (2010a) who reported SNPs in ten slower mutating loci identified in published genomes. As more whole genome data becomes available it is likely that more such SNPs will be discovered, allowing rapid and more extensive analysis using array-based techniques making genome-wide association studies possible for investigation of aspects of the parasite's biology e.g. virulence.

3.4 Strain Differences in Epidemiology and Outcome

More than 80 studies have been published investigating gp60 sequences in human-derived *C. parvum* and *C. hominis* isolates (Table 3.4). Broadly, these show some differences in distribution of gp60 families between studies, and it is clear that some sampled populations have greater diversity than others (Table 3.4). For example, *C. hominis* family Ib appears to predominate in most studies in Europe, North America and high income countries in Oceania. There are two common Ib subtypes: IbA10G2 which is especially prevalent in northern Europe, and IbA9G3 which is common in Australia, India and some countries in Africa, but more rare in northern Europe (Chalmers et al. 2008; Wielinga et al. 2008). However, one study in Western Australia found Id and Ig to be more common than Ib, especially in young children (Ng et al. 2010a). All *C. hominis* outbreaks in Europe, and nearly half of those in the USA have been caused by IbA10G2 (Xiao 2010). In Peru and China studies have revealed greater diversity within Ib than elsewhere (Cama et al. 2007; Feng et al. 2009; Peng et al. 2001).

In most high-income countries, *C. hominis* Id and Ia occur only occasionally and other gp60 families more rarely. There are some notable exceptions, in addition to the study in Western Australia mentioned above, another study in Australia reported C. *hominis* Id as the second most common after Ib (O'Brien et al. 2008). During the investigation of increased numbers of reported cases in 2007 in four states in the USA, Xiao et al. (2009) found that *C. hominis* Ia was most prevalent, and identified as rare subtype IaA28R4. This subtype had been found in cases linked to recreational water outbreaks in that year and in previous years, and the authors speculate that the cases may have been part of a larger, multistate outbreak for which typing was not widely undertaken (Xiao et al. 2009). In Sweden and Slovenia, Id is reported second to Ib, and is also common in reports from the Indian sub-continent, Peru, Kenya, Malawi and in one study in the US (Table 3.4). In a study of sporadic cases in the UK, having a non-IbA10G2 subtype was significantly linked to recent travel outside Europe (Chalmers et al. 2008).

There is greater diversity of *C. hominis* subtypes in less industrialised nations, and in studies in Peru these have found linkage between subtype and clinical outcomes that appear to vary with the particular host population under investigation, producing different spectra in children and HIV + adults. In HIV + patients in Lima, having Id was significantly associated with diarrhea in general and chronic diarrhea in particular, while infections with Ia and Ie were more likely asymptomatic (Cama et al. 2007). In a longitudinal cohort study of children in the same area of Lima, all *C. hominis* infections were associated with diarrhea, although Ib was significantly associated with diarrhea, nausea, vomiting and general malaise during both first and subsequent infections and thus appeared more virulent (Cama et al. 2008). In first episodes of infection, Ia was also very pathogenic being significantly associated with diarrhea, nausea and vomiting (Cama et al. 2008).

Among *C. parvum*, IIc is considered to be anthroponotic, having no defined animal host. It is widespread, especially where *C. hominis* is prevalent, like in Africa, Western Asia, Latin American and the Caribbean (Table 3.4), and the lack of a defined animal host for IIc supports its anthropomorphic transmission. Although *C. hominis* IbA10G2 has been found on rare occasions in animals, there is as yet no evidence for animals as a reservoir of human infection.

The other major gp60 families common in humans, IIa and IId, have been found in animals, especially ruminants, and this supports the zoonotic potential of *C. parvum* IIa and IId. One study in Spain found that family IIa is particularly associated with bovines, and IId with ovines and caprines (Quilez et al. 2008), although human outbreaks associated with sheep (linked to bottle feeding lambs) have occurred with IIa in the UK (Chalmers and Giles 2010). In most high income countries the anthroponotic *C. parvum* family IIc is generally rare, and *C. parvum* isolates are mainly IIa, although in Sweden IId is more common. In contrast, with a single exception, there have been few or no reports of IIa in sub-Saharan Africa, the Indian sub-continent, South and Central America and the Caribbean, where IIc is commonly found; coupled with the relatively high proportion of *C. hominis* cases in these countries, the importance of anthroponotic sources and transmission is highlighted here. The exception is one study in Ethiopia where IIa predominated and IIc was not found; zoonotic transmission here is supported by the predominance of *C. parvum* over *C. hominis* (Adamu et al. 2010), possibly reflecting the importance of cattle in that country. In North Africa, Western and South Eastern Asia, zoonotic transmission is also indicated by gp60 family distribution.

Any distribution analysis also needs to take into account the population investigated e.g. diarrhoea status, age, sex, HIV status, urban versus rural location, season, etc. Although inference about zoonotic potential can be made from prior knowledge of the host distribution of subtypes, further evidence for interpretation of the distribution of gp60 families and subtypes has been provided in some studies where humans were sampled alongside animals in the same area. For example, studies in Canada, Australia, the Netherlands and Egypt found a high proportion of human cases with the same gp60 subtypes as those found in livestock animals (Budu-Amoako et al. 2012; Helmy et al. 2013; Ng et al. 2012; Wielinga et al. 2008) indicating they may be a reservoir for infection. In one study in New Zealand, all foal, human, and bovine *C. parvum* isolates investigated had the gp60 IIaA18G3R1 subtype (Grinberg et al. 2008b). A high prevalence of *C. parvum* IIaA15G2R1 was found in both humans and cattle in the Netherlands (Wielinga et al. 2008), in 9/15 HIV + patients and all bovine, one of two sheep and all zoo ruminants tested in Portugal (Alves et al. 2006), but in only 2/9 Australian patients, 2/7 Australian cattle samples and all Canadian, European and US cattle sampled (O'Brien et al. 2008). Other Australian patients in that study had IIaA18G3R1, IIaA19G2R1, IIaA19G3R1, IIaA19G4R1, IIcA5G3a1. In fact, IIaA18G3R1 was found in farm dwellers and calves on the same farms in New South Wales (Ng et al. 2012). A greater diversity of subtypes was detected previously in human and cattle samples across the state (Ng et al. 2008). In Canada, a study on Prince Edward Island revealed that of three subtypes found in beef cattle,IIaA16G2RI, IIaA16G3RI and IIaA15G2RI, two, IIaA16G2RI and IIaA15G2RI, were also identified in the human isolates (Budu-Amoako et al. 2012). In Ontario, greater diversity was identified in both humans and dairy claves in the province of Ontario where three isolates from humans were the same subtypes as isolates from the calves (Trotz-Williams et al. 2006). In Iran, cattle were deemed to be a possible source of IIa subtypes (IIaA15G2R1, IIaA16G3R1), and one IId subtype (IId A15G1) in children, but there were more IId subtypes in children than cattle, the source of which was not known (Nazemalhosseini-Mojarad et al. 2011). Another study in Iran, also found that IId was common in children in Tehran, this time IIdA20G1a (Taghipour et al. 2011). In a study of bovines and humans in Egypt, both the bovine and the human infections were with IIaA15G1R1 and IIdA20G1, although the humans also had IIaA15G2R1 (Helmy et al. 2013). In England and Wales IIaA17G1R1 and IIaA15G2R1 predominate in cattle and lambs (Brook et al. 2009; Mueller-Doblies et al. 2008; Smith et al. 2010b) and in sporadic human *C. parvum* cases (Chalmers et al. 2011c). However, in Ireland, as in New South Wales Australia, IIaA18G3R1 predominates in both humans (Zintl et al. 2011) and cattle (Thompson et al. 2007), but is rare in England and Wales.

Data arising from an on-farm study in England and Wales, showing the proportion of human cases that had the same subtype as those found in suspected source animal contacts (Smith et al. 2010b), were used to estimate, from national surveillance data, the proportion of all laboratory confirmed cases that might have been acquired from direct contact with farm animals (Chalmers et al. 2011a). The proportion was 25 % for *C. parvum* cases, and 10 % of all reported *Cryptosporidium* cases.

Some studies have combined analytical epidemiology with analysis of gp60 families and subtypes. For example, in a study of *C. parvum* cases in the UK, cryptosporidiosis cases with IIa were more likely than IId to have visited a farm, or had contact with farm animals or with their faeces in the 2 weeks prior to illness (Chalmers et al. 2011c). Within gp60 family IIa, genotypes IIaA15G2R1 and IIaA17G1R1 predominated (22 cases each); nine other IIa genotypes accounted for 12 cases. The IId genotypes were mainly IIdA17G1R1 and IIdA18G1 (three each). Cases with IIaA17G1R1 were particularly linked to zoonotic exposures: visiting a farm or having farm animal contact in the 2 weeks prior to illness.

The analysis of a single locus is not sufficient to infer geographical diversity of subtypes. Global gp60 subtype distribution revealed limited microbiological evidence for host specificity (Jex and Gasser 2008). Although many gp60 subtypes have only been found in humans, many are rare findings and we do not know whether they have an animal host or not.

Some studies have used MLG to make inferences about the global population structures of *C. hominis* and *C. parvum* (Table 3.5). For example, Gatei et al. (2006) compared fragment sizing and sequence-based analysis of *C. hominis* subtypes at multiple loci, both including and excluding gp60. When gp60 was excluded from the analysis, all but three of 27 MLGs identified were found in samples from single countries. They concluded that geographical isolation has led to the emergence of distinct *C. hominis* subtypes. Targets other than gp60 were identified as suitable to discriminate geographically distinct subtypes, including the MSC6–7, DZHRGP, CP47 and HSP70 loci which showed genetic differentiation by region, and their inclusion in a multi-locus scheme was deemed to provide a superior subtyping tool for *C. hominis*.

A fragment sizing approach used by Tanriverdi et al. (2008), indicated that gene exchanges do not seem sufficient to erase genetic divergence among geographically separated populations, and that multilocus genotyping enables the tracking of *Cryptosporidium* isolates to their country of origin. They suggested that, rather than conforming to a strict paradigm of either clonal (deriving from a common ancestor without interbreeding with other types within the species) or panmictic (interbreeding freely within the species) species, their data are consistent with the co-occurrence of both modes of propagation, with the relative contribution of each mode appearing to vary according to the prevailing ecological determinants of transmission. This seems to be important as apparently conflicting conclusions have been drawn from studies in different countries and different regions (Table 3.5). For example, in another study in a smaller geographical area, Scotland, lack of partitioning may be explained by the high frequency of cattle movements, typical in the UK, preventing the establishment of settled and epidemiologically isolated herds (Morrison et al. 2008).

It should also be recognised that population studies based on clinical isolates tend to be biased towards analysis of the most virulent strains and may not be representative of overall population structure (Gupta and Maiden 2001). Another limitation of these studies is that the presence of mixed MLGs is often not accounted for in analyses and the contribution of mixed infections to population diversity is not well understood.

Interestingly, in the interim since Gatei's work in 2006, although DZHRGP and CP47 featured in the preferred marker list generated by the analysis of Robinson and Chalmers (2012), MSC6-7 and hsp70 did not appear in the top half of the most informative rankings, other markers being deemed more useful.

Only a small number of publications have reported on epidemiological investigations applying MLG to both sporadic and outbreak cases. In outbreaks, MLG has been demonstrated to be useful in identifying related cases. In the UK, Hunter et al. (2008) demonstrated that although 17 different MLGs were detected among cases, clustering of one MLG was statistically significant for patients resident in the area most dependent on one water supply. In a swimming pool-related outbreak in Sweden, MLGs were clustered according to pool use: one MLG was found among outdoor pool users and secondary cases, and another MLG from users of an indoor pool (Mattsson et al. 2008).

Overall there is a need for the application of a standard multi-locus subtyping scheme applied to more host species and most especially in non-industrialised countries where the burden of illness and long-term sequelae from *Cryptosporidium* infection are greatest.

3.5 Application of Molecular Methods in Sporadic and Outbreak Related Cryptosporidiosis

3.5.1 Outbreak Detection, Investigation, Management, Control and Surveillance

Molecular typing and subtyping have proved valuable in the investigation of outbreaks of cryptosporidiosis, with subtyping of *C. parvum* and *C. hominis* allowing a degree of confidence that isolates can be recognised as an "outbreak strain". Early investigations indicated that outbreaks could be caused by *C. hominis* or *C. parvum*, and that drinking water and recreational water outbreaks could be caused by either species (McLauchlin et al. 2000). However, the development of more specific and sensitive methods of species identification and subtyping has resulted in an ability to more sensitively examine outbreaks that are not from these sources.

3.5.2 Foodborne Outbreaks

Many outbreaks linked to food have been reported (Centers for Disease C, Prevention 1995, 1996, 1997; Djuretic et al. 1997; Millard et al. 1994; Robertson and Chalmers 2013), but the investigation of such sources has been limited because of the lack of systematic subtyping and reporting of strains to national surveillance (Chap. 2). Outbreaks in Sweden in Stockholm and Umea caused by *C. parvum* subtype IIdA24G1 and another one in Orebro by *C. parvum* IIdA20G1e illustrate the value in applying subtyping to the investigation of *Cryptosporidium* cases. *C. parvum* seems to predominate in foodborne outbreaks suggesting animal sources of contamination. However, some foodborne outbreaks have been linked to contamination by food handlers (Ethelberg et al. 2009; Harper et al. 2002; Lee and Greig 2010; Ponka et al. 2009; Quiroz et al. 2000), where the local nature of the outbreak makes subtyping less essential as the route of transmission is clear and interventions exist to prevent future outbreaks (improvement of hygiene). It remains possible that foodborne cryptosporidiosis is more common than we currently recognise, particularly through salad items and soft fruits as has been common for *Cyclospora* infections (Chalmers et al. 2000b). A large *Cryptosporidium* outbreak across England and Scotland in 2012 was attributed epidemiologically to a bagged salad product, partly as a result of the identification of an outbreak strain based on gp60 subtyping of *C. parvum* (unpublished data). However, there is a need for routine subtyping of all clinical *Cryptosporidium* isolates using even more specific methods, along with these results feeding into national surveillance, so that other clusters of cases can be detected and links to sources of contamination investigated. This was particularly the case in the late summer of 2012 when there was a simultaneous increase in *C. hominis* infections in three EU countries the cause of which is not known (Fournet et al. 2012).

One of the difficulties in investigating foodborne cryptosporidiosis is that the implicated foods are generally short-shelf life products which will have been discarded before the cases of illness have been identified, leaving no relevant food samples for testing. Further, issues of sampling from large batches of products are raised where the distribution of the contamination may be uneven and missed by sampling procedures. There are as yet no standard methods for the detection of *Cryptosporidium* in food, although an ISO method for soft fruit and leafy green vegetables is in preparation. This does not include molecular tests but they could be incorporated in the process in a similar way to that applied to water, by either testing part of the pellet by immunomagnetic separation (IMS) or by scraping material from *Cryptosporidium*-positive microscopy slides (Ruecker et al. 2013). Although this procedure has been informative in investigations (see below) and gp60 subtyping has been applied, multi-locus methods have not, and careful preparation will be needed to ensure that sufficient DNA is extracted to permit the sequencing of multiple markers.

3.5.3 Waterborne Outbreaks

Typing has been useful in the analysis and follow-up of waterborne outbreaks, some of which are mentioned above (see also Chap. 12). Microsatellite typing was used in the investigation of cases that may have been linked to a UK drinking water related outbreak and was useful in identifying cases most likely to have been involved (Hunter et al. 2008). A unique drinking water related outbreak of cryptosporidiosis in a city in England was linked to water contamination from a rabbit, the *Cryptosporidium* species, *C. cuniculus*, only identified by typing (Robinson and Chalmers 2010; Smith et al. 2010a) (Pulleston et al., submitted for publication). A state-wide outbreak of cryptosporidiosis throughout New South Wales, Australia used genetic characterization to identify *C. hominis* IbA10G2 subtype as the causative parasite acquired from swimming pools (Waldron et al. 2011b). A drinking water related outbreak in Wales was attributed to *C. hominis* gp60 subtype IbA10G2, which was also found in source waters serving the drinking water supply (Chalmers et al. 2010).

3.5.4 Source Tracking (Isolates from Water) and Source Attribution (Animal Sources of Human Infections)

Comparing isolates from water and host animals with those from humans can be used to track the levels of exposure to humans through source waters used for drinking water and to attribute human infection to particular source animals (Ruecker et al. 2012). For example, a rare gp60 subtype IIaA20G2R1 that caused an outbreak of cryptosporidiosis among lambs but not calves in a mixed sheep/cattle farm in central Italy, was also found in the faeces of an 18 month old boy who lived on the farm and had been hospitalised due to acute gastrointestinal symptoms demonstrating transmission from lambs to humans (Caccio et al. 2013). This route of transmission is a common cause of outbreaks in the spring in the UK (Gormley et al. 2011), and typing has been applied in the investigation of farm-related outbreaks (Chalmers and Giles 2010).

Cryptosporidium has been one of the most successfully studied organisms with respect to source attribution. The common occurrence of *C. parvum* in agricultural animals and the almost complete absence of *C. hominis* has been key in focussing the attention of water authorities on the differential contributions of human and animal sources of contamination of drinking water (Nichols 2008). The identification of a variety of *Cryptosporidium* species that have a narrow host specificity has also been useful in ruling out some potential transmission routes, for example dogs and cats, as being common (Bowman and Lucio-Forster 2010). Comparing isolates from water and host animals with those from humans can be used to track the levels of exposure to humans through source waters used for drinking water (Nichols et al. 2003, 2006a, 2010) and to attribute human infection to particular source animals (Ruecker et al. 2012). Some studies have identified most of the *Cryptosporidium* isolates to be not strains that commonly cause human disease (Yang et al. 2008).

The value of this work, in addition to identifying transmission pathways and sources, has been to extend our understanding of the host range of *Cryptosporidium* spp. present in the natural environment, and to reassure us that most of these isolates do not seem to be linked with human disease.

3.5.5 Outbreak Detection, Investigation, Management and Control

There have been numerous reviews and analyses of outbreaks of cryptosporidiosis, some general enteric ones and others exclusively due to *Cryptosporidium* (Baldursson and Karanis 2011; Barwick et al. 2000; Beaudeau et al. 2008; Chalmers 2012; Chalmers and Giles 2010; Craun et al. 2005; Curriero et al. 2001; Djuretic et al. 1997; Dziuban et al. 2006; Furtado et al. 1998; Glaberman et al. 2002; Hlavsa et al. 2011; Karanis et al. 2007; Kramer et al. 1996; Lee and Greig 2010; Lee et al. 2002; LeJeune and Davis 2004; Levine et al. 1990; Levy et al. 1998; McLauchlin et al. 2000; Nichols et al. 2006, 2009; Risebro et al. 2007; Rooney et al. 2004; Schuster et al. 2005; Semenza and Nichols 2007; Smith et al. 2006; Vojdani et al. 2008; Xiao 2008; Yoder et al. 2004, 2008). In 2008, Xiao and Morgan reviewed outbreaks where gp60 genotypes were identified (Xiao 2008). In reviewing the molecular epidemiology it is always useful to assess what using modern molecular methods has brought to the investigation and control of such outbreaks.

The approaches to outbreak investigation of protozoan infections have been reviewed recently (Pollock and Nichols 2012). The detection and investigation of outbreaks is crucially dependent on having good microbiological diagnostic regimens in place in primary diagnostic laboratories. In the absence of this only the largest of outbreaks will be detected. It is not generally possible for primary laboratories to type strains to species or sub-species level, and therefore having a reliable reference service, or a research centre of excellence is fundamental to providing the detail derived from subtyping. These reference services also interact with each other to provide a common nomenclature of species and types so that strain comparison can be understood internationally. Recent reviews have addressed taxonomy and nomenclature (Xiao et al. 2004). The typing is useful in real time to focus investigations on linked cases and eliminate background cases that are not part of the outbreak. There needs to be communication between public health teams and local and national governments, drinking water suppliers and regulators as well as swimming pool operators, in order that these groups understand the specific problems resulting from the chlorine resistance of *Cryptosporidium* oocysts, so that when an outbreak is detected there is some agreement about the science. In particular, what testing is required, what results from water testing mean, what levels of oocyst contamination are acceptable, if any, what are the best approaches to intervention to limit the outbreak, who is in charge of the investigation, what approach should be used for the investigation and also so that the people involved know each other and therefore can have some prior experience of trusting each other.

Detection of a cluster of cases locally should result in a problem assessment group being established, upgraded to an outbreak control team should the evidence warrant this. Because cryptosporidiosis is commonly waterborne, cases usually present as a defined local cluster. However, outbreaks linked to overseas travel or to widely distributed food items, mean that outbreaks can arise that are not local. It is here, particularly, where speciation and subtyping are important in identifying an outbreak strain. For outbreaks where faecally contaminated water is the likely source there is a greater chance of a mixed outbreak caused by more than one strain, because the contamination may have come from several infected people or animals, compared to an outbreak caused by a water contaminated from a single individual. In the case of swimming pools there is good evidence for mixed outbreaks caused by *C. parvum* and C. *hominis*. The same is true for drinking water outbreaks. Outbreaks of foodborne illness where the infection derives from a single human case are thought to be predominantly derived from a single outbreak strain. The consequences of these difficulties is that in the absence of typing and subtyping some outbreaks can be difficult to investigate. In many of these cases older outbreaks reported there was a delay between when cases occured and when the results were available.

The control and prevention of outbreaks of cryptosporidiosis relies on the use of outbreak data to identify the sorts of combinations of events that can contribute to outbreaks. Some of these were outlined in a previous review (Nichols et al. 2006). For drinking water there are specific issues related to backwash water re-cycling, catchment management, water safety plans, source and treated water monitoring, animal management around treatment facilities, and emergency planning (Chaps. 12 and 13). For swimming pools the management approaches are established, but while reliance on the optimal operation of coagulation and filtration is essential, there is an increasing recognition that secondary disinfection using UV is necessary to deal with viable oocysts that break through conventional filtration at times.

3.6 Cryptosporidiosis in Immunocompromised People

Examinations of *Cryptosporidium* infections in people with some specific defects in immunity (Hunter and Nichols 2002) has been greatly assisted by molecular epidemiology (Wang et al. 2013). The occurrence of different *Cryptosporidium* species in immunocompetent and immunocompromised people was reviewed by Xiao and Morgan (2008). Both groups had infections with a range of species, and there do not appear to be *Cryptosporidium* species that are infectious only for any of the immunocompromised groups, although infected patients with compromised immunity will excrete the parasite for extended periods.

3.7 Serological Markers

The detection of serum antibodies in response to *Cryptosporidium* infection has been useful in the examination of pathogenicity of individual *Cryptosporidium* species for humans (Chappell et al. 2006, 2011; DuPont et al. 1995; Moss et al. 1998). It has also been of use in comparing the incidence of cryptosporidiosis in different populations, particularly those served by different water supplies (Egorov et al. 2004; Frost et al. 2000a, b, 2001, 2003, 2005; Kozisek et al. 2008). For detailed molecular epidemiology however, the current tools for examining the response to infection will not clearly differentiate between different *Cryptosporidium* species, and are unlikely to be able in the future to differentiate between different genetic subtypes without specific cloned proteins screened in microarrays.

3.8 Conclusions and Future Developments

This chapter started by reviewing what past molecular epidemiology studies have shown us. The question that remains is what issues are still left to address. For us, one critical missing piece of the story at present is what outbreaks and sources of contamination and infection we are missing using current methods and surveillance algorithms. In particular, there seems to be the possibility of outbreaks occurring that are not being detected using current surveillance and/or typing and subtyping methods. When an outbreak strain is identified, it is commonly one that is also found in sporadic cases, and therefore being definitive about its role in the outbreak is difficult. This problem in common bacterial causes of diarrhoea like *Salmonella* is being addressed through whole genome sequencing of isolates to allow greater differentiation of strains (Singh et al. 2013; Zankari et al. 2013). For *Cryptosporidium* the problem is a bit harder because strains cannot readily be cultured to yield high numbers of parasite for genome sequencing. However, it is likely that an approach to whole genome sequencing larger numbers of strains using oocyst purification and whole genome amplification may yield approaches that are applicable to use in disease surveillance. The advantages of genome sequencing are that the dramatic and continuing reduction in the costs make this a feasible tool for use in molecular epidemiology and the finer differentiation of strain differences will lead to a much more proactive approach to the surveillance and control of cryptosporidiosis.

Using single sporozoite sequencing of the SSU rRNA gene from *C. andersoni* shows individual haploid sporozoites can contain two different copies of the gene (Ikarashi et al. 2013) (Le Blancq et al. 1997). More systematic whole genome sequencing will undoubtedly shed light on the recombination events that give rise to this and to distinctive features of *Cryptosporidium* genetics.

All diagnostic laboratories that test faecal samples for enteric and foodborne infections should be screening for *Cryptosporidium* using the recognised standard methods. While the infection is more common in young children, we also find that testing people of all ages is useful. Experience suggests that systematic subtyping of positive samples obtained through this approach can be enormously helpful in both understanding the causes of both cases and outbreaks, but also in response to outbreaks when they occur. Despite the widespread occurrence of large outbreaks across the World, it remains likely that there are more cases occurring in small outbreaks, most of which are not being detected. This is particularly the case for swimming pool related outbreaks where the exposed population may not be as large as occurs when a water catchment is contaminated. Food related outbreaks resulting from salad or fruit with a wide distribution will probably not be picked up by traditional epidemiological methods and we think that there is still a need for more discriminatory genetic methods to separate out strains associated with outbreaks from background sporadic cases.

The mapping (Piper et al. 1998) and sequencing of the complete genomes (Abrahamsen 2001; Abrahamsen et al. 2004; Liu et al. 1999; Widmer and Sullivan 2012; Xu et al. 2004) of three species of *Cryptosporidium* and an anthroponotic strain of *C. parvum* (Widmer et al. 2012) has lead to an increased understanding of the biochemistry and genetics of this genus (Chap. 6). Approaches to amplifying *Cryptosporidium* whole genome fragments have been examined and the commercial kits available have been compared (Bouzid et al. 2010b). *Cryptosporidium* genes can be examined through a Library of Apicomplexan Metabolic Pathways at http://www.llamp.net (Shanmugasundram et al. 2013), the Cryptosporidium Genomics Resource http://cryptodb.org/cryptodb/ (Heiges et al. 2006), the EPICDB relational database that organizes and displays experimental, high throughput proteomics data for *T. gondii* and *C. parvum* http://toro.aecom.yu.edu/cgi-bin/biodefense/main.cgi (Madrid-Aliste et al. 2009) and an extended database of protein domains for several eukaryotic pathogens http://www.atgc-montpellier.fr/EuPathDomains/ (Ghouila et al. 2011). However, there are still a considerable number of genes that are not annotated and assigned to a function, and there remains a strong need for additional work on individual proteins (Rider and Zhu 2010). Comparison of zoonotic and anthroponotic strains of *C. parvum* and *C. hominis* suggest these genes may influence the host range (Widmer et al. 2012). However, more gene sequences are needed for variants within these species and of other species, both human pathogenic and non pathogenic for comparison and further investigation.

Rather than an approach that passively collects *Cryptosporidium* surveillance data and then conducts questionnaires on cases that appear to be part of a cluster, the use of enhanced surveillance, where a questionnaire is administered to all patients, allows epidemiological, demographic and risk information to be collected systematically. Such an approach, conducted over an extended time period, can be useful in conducting ad-hoc case-case comparisons across years. Such an approach proved useful in the investigation of a multi-region outbreak in 2012 (unpublished) that early case-case examination suggested the outbreak might be linked to salad items from a national retailer and which was subsequently shown to be from this source by a national case–control study.

Other issues that remain to be definitively answered follow. Does the rare occurrence of some species in humans reflect greater susceptibility in some human phenotypes, some immunocompromised people or some age groups? Is the detection of oocysts in the faeces always correlated with human infection and disease, and can some species infect but not produce oocysts in humans? Are any oocyst detections in humans a result of the consumption of heavily contaminated food with species that are not infectious to humans?

References

Abbassi H, Wyers M, Cabaret J, Naciri M (2000) Rapid detection and quantification of *Cryptosporidium baileyi* oocysts in feces and organs of chickens using a microscopic slide flotation method. Parasitol Res 86(3):179–187, PubMed PMID: 10726987

Abe N (2010) Genotype analysis of *Cryptosporidium meleagridis* isolates from humans in Japan. Jpn J Infect Dis 63(3):214–215, PubMed PMID: 20495279

Abe N, Makino I (2010) Multilocus genotypic analysis of Cryptosporidium isolates from cockatiels. Jpn Parasitol Res 106(6):1491–1497, PubMed PMID: 20339870

Abrahamsen MS (2001) *Cryptosporidium parvum* genome project. Comp Funct Genom 2(1): 19–21, PubMed PMID: 18628893. Pubmed Central PMCID: 2447187

Abrahamsen MS, Templeton TJ, Enomoto S, Abrahante JE, Zhu G, Lancto CA et al (2004) Complete genome sequence of the apicomplexan, *Cryptosporidium parvum*. Science 304 (5669):441–445, PubMed PMID: 15044751

Adamu H, Petros B, Hailu A, Petry F (2010) Molecular characterization of Cryptosporidium isolates from humans in Ethiopia. Acta Trop 115(1–2):77–83, PubMed PMID: 20206592

Agholi M, Hatam GR, Motazedian MH (2013) HIV/AIDS-associated opportunistic protozoal diarrhea. AIDS Res Hum Retrovir 29(1):35–41, PubMed PMID: 22873400. Pubmed Central PMCID: 3537293

Ajjampur SS, Gladstone BP, Selvapandian D, Muliyil JP, Ward H, Kang G (2007) Molecular and spatial epidemiology of cryptosporidiosis in children in a semiurban community in South India. J Clin Microbiol 45(3):915–920, PubMed PMID: 17251402. Pubmed Central PMCID: 1829120

Ajjampur SS, Sarkar R, Sankaran P, Kannan A, Menon VK, Muliyil J et al (2010) Symptomatic and asymptomatic Cryptosporidium infections in children in a semi-urban slum community in southern India. Am J Trop Med Hyg 83(5):1110–1115, PubMed PMID: 21036847. Pubmed Central PMCID: 2963979

Akiyoshi DE, Tumwine JK, Bakeera-Kitaka S, Tzipori S (2006) Subtype analysis of Cryptosporidium isolates from children in Uganda. J Parasitol 92(5):1097–1100, PubMed PMID: 17152957

Alvarez-Pellitero P, Sitja-Bobadilla A (2002) *Cryptosporidium molnari* n. sp. (Apicomplexa: Cryptosporidiidae) infecting two marine fish species, *Sparus aurata* L. and *Dicentrarchus labrax* L. Int J Parasitol 32(8):1007–1021

Alvarez-Pellitero P, Quiroga MI, Sitja-Bobadilla A, Redondo MJ, Palenzuela O, Padros F et al (2004) *Cryptosporidium scophthalmi* n. sp. (Apicomplexa: Cryptosporidiidae) from cultured turbot Scophthalmus maximus. Light and electron microscope description and histopathological study. Dis Aquat Org 62(1–2):133–145

Alves M, Xiao L, Sulaiman I, Lal AA, Matos O, Antunes F (2003) Subgenotype analysis of Cryptosporidium isolates from humans, cattle, and zoo ruminants in Portugal. J Clin Microbiol 41(6):2744–2747, PubMed PMID: 12791920. Pubmed Central PMCID: 156540

Alves M, Xiao L, Antunes F, Matos O (2006) Distribution of Cryptosporidium subtypes in humans and domestic and wild ruminants in Portugal. Parasitol Res 99(3):287–292, PubMed PMID: 16552512

Alyousefi NA, Mahdy MA, Lim YA, Xiao L, Mahmud R (2013) First molecular characterization of Cryptosporidium in Yemen. Parasitology 140(6):729–734, PubMed PMID: 23369243

Amar C, Pedraza-Diaz S, McLauchlin J (2001) Extraction and genotyping of *Cryptosporidium parvum* DNA from fecal smears on glass slides stained conventionally for direct microscope examination. J Clin Microbiol 39(1):401–403, PubMed PMID: 11136813. Pubmed Central PMCID: 87744

Amar CF, Dear PH, McLauchlin J (2003) Detection and genotyping by real-time PCR/RFLP analyses of Giardia duodenalis from human faeces. J Med Microbiol 52(Pt 8):681–683, PubMed PMID: 12867562

Amar CF, Dear PH, McLauchlin J (2004) Detection and identification by real time PCR/RFLP analyses of *Cryptosporidium* species from human faeces. Lett Appl Microbiol 38(3):217–222, PubMed PMID: 14962043

Amar CF, East CL, Gray J, Iturriza-Gomara M, Maclure EA, McLauchlin J (2007) Detection by PCR of eight groups of enteric pathogens in 4,627 faecal samples: re-examination of the English case-control infectious intestinal disease study (1993–1996). Eur J Clin Microbiol Infect Dis 26(5):311–323, PubMed PMID: 17447091

Amer S, Zidan S, Feng Y, Adamu H, Li N, Xiao L (2013) Identity and public health potential of *Cryptosporidium* spp. in water buffalo calves in Egypt. Vet Parasitol 191(1–2):123–127, PubMed PMID: 22963712

Antunes RG, Simoes DC, Nakamura AA, Meireles MV (2008) Natural infection with *Cryptosporidium galli* in canaries (*Serinus canaria*), in a cockatiel (*Nymphicus hollandicus*), and in lesser seed-finches (*Oryzoborus angolensis*) from Brazil. Avian Dis 52(4):702–705, PubMed PMID: 19166068

Areeshi M, Dove W, Papaventsis D, Gatei W, Combe P, Grosjean P et al (2008) *Cryptosporidium* species causing acute diarrhoea in children in Antananarivo, Madagascar. Ann Trop Med Parasitol 102(4):309–315, PubMed PMID: 18510811

Ayinmode AB, Fagbemi BO, Xiao L (2012) Molecular characterization of Cryptosporidium in children in Oyo State, Nigeria: implications for infection sources. Parasitol Res 110(1): 479–481, PubMed PMID: 21744017

Bakheit MA, Torra D, Palomino LA, Thekisoe OM, Mbati PA, Ongerth J et al (2008) Sensitive and specific detection of *Cryptosporidium* species in PCR-negative samples by loop-mediated isothermal DNA amplification and confirmation of generated LAMP products by sequencing. Vet Parasitol 158(1–2):11–22, PubMed PMID: 18940521

Balatbat AB, Jordan GW, Tang YJ, Silva J Jr (1996) Detection of *Cryptosporidium parvum* DNA in human feces by nested PCR. J Clin Microbiol 34(7):1769–1772, PubMed PMID: 8784586. Pubmed Central PMCID: 229111

Baldursson S, Karanis P (2011) Waterborne transmission of protozoan parasites: review of worldwide outbreaks—an update 2004–2010. Water Res 45(20):6603–6614, PubMed PMID: 22048017

Barwick RS, Levy DA, Craun GF, Beach MJ, Calderon RL (2000) Surveillance for waterborne-disease outbreaks–United States, 1997–1998. MMWR CDC Surveill Summ 49(4):1–21, PubMed PMID: 10843502

Bass AL, Wallace CC, Yund PO, Ford TE (2012) Detection of *Cryptosporidium* sp. in two new seal species, Phoca vitulina and Cystophora cristata, and a novel Cryptosporidium genotype in a third seal species, Pagophilus groenlandicus, from the Gulf of Maine. J Parasitol 98(2): 316–322

Beaudeau P, de Valk H, Vaillant V, Mannschott C, Tillier C, Mouly D et al (2008) Lessons learned from ten investigations of waterborne gastroenteritis outbreaks, France, 1998–2006. J Water Health 6(4):491–503, PubMed PMID: 18401114

Ben Ayed L, Yang W, Widmer G, Cama V, Ortega Y, Xiao L (2012) Survey and genetic characterization of wastewater in Tunisia for *Cryptosporidium* spp., Giardia duodenalis, Enterocytozoon bieneusi, *Cyclospora* cayetanensis and Eimeria spp. J Water Health 10(3): 431–444, PubMed PMID: 22960487

Bialek R, Binder N, Dietz K, Joachim A, Knobloch J, Zelck UE (2002) Comparison of fluorescence, antigen and PCR assays to detect *Cryptosporidium parvum* in fecal specimens. Diagn Microbiol Infect Dis 43(4):283–288, PubMed PMID: 12151188

Boom R, Sol CJ, Salimans MM, Jansen CL, Wertheim-van Dillen PM, van der Noordaa J (1990) Rapid and simple method for purification of nucleic acids. J Clin Microbiol 28(3):495–503, PubMed PMID: 1691208. Pubmed Central PMCID: 269651

Bouzid M, Tyler KM, Christen R, Chalmers RM, Elwin K, Hunter PR (2010a) Multi-locus analysis of human infective *Cryptosporidium* species and subtypes using ten novel genetic loci. BMC Microbiol 10:213, PubMed PMID: 20696051. Pubmed Central PMCID: 2928199

Bouzid M, Heavens D, Elwin K, Chalmers RM, Hadfield SJ, Hunter PR et al (2010b) Whole genome amplification (WGA) for archiving and genotyping of clinical isolates of *Cryptosporidium* species. Parasitology 137(1):27–36, PubMed PMID: 19765343

Bouzid M, Hunter PR, Chalmers RM, Tyler KM (2013) Cryptosporidium pathogenicity and virulence. Clin Microbiol Rev 26(1):115–134, PubMed PMID: 23297262. Pubmed Central PMCID: 3553671

Bowman DD, Lucio-Forster A (2010) Cryptosporidiosis and giardiasis in dogs and cats: veterinary and public health importance. Exp Parasitol 124(1):121–127, PubMed PMID: 19545532

Brook EJ, Anthony Hart C, French NP, Christley RM (2009) Molecular epidemiology of Cryptosporidium subtypes in cattle in England. Vet J 179(3):378–382, PubMed PMID: 18083583

Budu-Amoako E, Greenwood SJ, Dixon BR, Sweet L, Ang L, Barkema HW et al (2012) Molecular epidemiology of cryptosporidium and giardia in humans on Prince Edward Island, Canada: evidence of zoonotic transmission from cattle. Zoonoses Pub health 59(6):424–433, PubMed PMID: 22390418

Burton AJ, Nydam DV, Dearen TK, Mitchell K, Bowman DD, Xiao L (2010) The prevalence of Cryptosporidium, and identification of the Cryptosporidium horse genotype in foals in New York State. Vet Parasitol 174(1–2):139–144, PubMed PMID: 20932647

Caccio S, Spano F, Pozio E (2001) Large sequence variation at two microsatellite loci among zoonotic (genotype C) isolates of *Cryptosporidium parvum*. Int J Parasitol 31(10):1082–1086, PubMed PMID: 11429171

Caccio SM, Sannella AR, Mariano V, Valentini S, Berti F, Tosini F et al (2013) A rare *Cryptosporidium parvum* genotype associated with infection of lambs and zoonotic transmission in Italy. Vet Parasitol 191(1–2):128–131, PubMed PMID: 22954678

Calderaro A, Montecchini S, Gorrini C, Dettori G, Chezzi C (2011) Similar diagnostic performances of antigen detection and nucleic acid detection of *Cryptosporidium* spp. in a low-prevalence setting. Diagn Microbiol Infect Dis 70(1):72–77, PubMed PMID: 21513845

Cama VA, Bern C, Sulaiman IM, Gilman RH, Ticona E, Vivar A et al (2003) *Cryptosporidium* species and genotypes in HIV-positive patients in Lima, Peru. J Eukaryot Microbiol 50(Suppl): 531–533, PubMed PMID: 14736153

Cama VA, Ross JM, Crawford S, Kawai V, Chavez-Valdez R, Vargas D et al (2007) Differences in clinical manifestations among *Cryptosporidium* species and subtypes in HIV-infected persons. J Infect Dis 196(5):684–691, PubMed PMID: 17674309

Cama VA, Bern C, Roberts J, Cabrera L, Sterling CR, Ortega Y et al (2008) *Cryptosporidium* species and subtypes and clinical manifestations in children, Peru. Emerg Infect Dis 14(10): 1567–1574, PubMed PMID: 18826821. Pubmed Central PMCID: 2609889

Castro-Hermida JA, Garcia-Presedo I, Almeida A, Gonzalez-Warleta M, Correia Da Costa JM, Mezo M (2008) Presence of *Cryptosporidium* spp. and Giardia duodenalis through drinking water. Sci Total Environ 405(1–3):45–53, PubMed PMID: 18684490

Centers for Disease C, Prevention (1998) Foodborne outbreak of cryptosporidiosis—Spokane, Washington, 1997. MMWR Morb Mortal Wkly Rep 47(27):565–567, PubMed PMID: 9694641

Centers for Disease C, Prevention (1996) Foodborne outbreak of diarrheal illness associated with *Cryptosporidium parvum*—Minnesota, 1995. MMWR Morb Mortal Wkly Rep 45(36): 783–784, PubMed PMID: 8801445

Centers for Disease C, Prevention (1997) Outbreaks of Escherichia coli O157:H7 infection and cryptosporidiosis associated with drinking unpasteurized apple cider—Connecticut and New York, October 1996. MMWR Morb Mortal Wkly Rep 46(1):4–8, PubMed PMID: 9011776

Chalmers RM (2012) Waterborne outbreaks of cryptosporidiosis. Ann Ist Super Sanita 48(4): 429–446, PubMed PMID: 23247139

Chalmers RMHS, Elwin K, Robinson G (2013) Improved performance of real-time PCR for typing Cryptosporidium for epidemiological purposes. IV International Cryptosporidium and Giardia conference. Welington, pp 122–129

Chalmers RM, Giles M (2010) Zoonotic cryptosporidiosis in the UK—challenges for control. J Appl Microbiol 109(5):1487–1497, PubMed PMID: 20497274

Chalmers RM, Nichols G, Rooney R (2000) Foodborne outbreaks of cyclosporiasis have arisen in North America. Is the United Kingdom at risk? Comm Dis Pub Health PHLS 3(1):50–55, PubMed PMID: 10743320

Chalmers RM, Ferguson C, Caccio S, Gasser RB, Abs ELOYG, Heijnen L et al (2005) Direct comparison of selected methods for genetic categorisation of *Cryptosporidium parvum* and *Cryptosporidium hominis* species. Int J Parasitol 35(4):397–410, PubMed PMID: 15777916

Chalmers RM, Hadfield SJ, Jackson CJ, Elwin K, Xiao L, Hunter P (2008) Geographic linkage and variation in *Cryptosporidium hominis*. Emerg Infect Dis 14(3):496–498, PubMed PMID: 18325272. Pubmed Central PMCID: 2570818

Chalmers RM, Elwin K, Thomas AL, Guy EC, Mason B (2009a) Long-term Cryptosporidium typing reveals the aetiology and species-specific epidemiology of human cryptosporidiosis in England and Wales, 2000 to 2003. Eur Surveill Bull Eur Mal Transm Eur Commun Dis Bull 14 (2):1–9. PubMed PMID: 19161717

Chalmers RM, Robinson G, Elwin K, Hadfield SJ, Thomas E, Watkins J et al (2010) Detection of *Cryptosporidium* species and sources of contamination with *Cryptosporidium hominis* during a waterborne outbreak in North West Wales. J Water Health 8(2):311–325, PubMed PMID: 20154394

Chalmers RM, Elwin K, Hadfield SJ, Robinson G (2011a) Sporadic human cryptosporidiosis caused by *Cryptosporidium cuniculus*, United Kingdom, 2007–2008. Emerg Infect Dis 17(3): 536–538, PubMed PMID: 21392453. Pubmed Central PMCID: 3165992

Chalmers RM, Smith R, Elwin K, Clifton-Hadley FA, Giles M (2011b) Epidemiology of anthroponotic and zoonotic human cryptosporidiosis in England and Wales, 2004–2006. Epidemiol Infect 139(5):700–712, PubMed PMID: 20619076

Chalmers RM, Smith RP, Hadfield SJ, Elwin K, Giles M (2011c) Zoonotic linkage and variation in *Cryptosporidium parvum* from patients in the United Kingdom. Parasitol Res 108(5): 1321–1325, PubMed PMID: 21193928

Chappell CL, Okhuysen PC, Langer-Curry R, Widmer G, Akiyoshi DE, Tanriverdi S et al (2006) *Cryptosporidium hominis*: experimental challenge of healthy adults. Am J Trop Med Hyg 75 (5):851–857, PubMed PMID: 17123976

Chappell CL, Okhuysen PC, Langer-Curry RC, Akiyoshi DE, Widmer G, Tzipori S (2011) Cryptosporidium meleagridis: infectivity in healthy adult volunteers. Am J Trop Med Hyg 85(2):238–242, PubMed PMID: 21813841. Pubmed Central PMCID: 3144819

Chavez DJ, LeVan IK, Miller MW, Ballweber LR (2012) Baylisascaris procyonis in raccoons (Procyon lotor) from eastern Colorado, an area of undefined prevalence. Vet Parasitol 185 (2–4):330–334, PubMed PMID: 22119387

Cieloszyk J, Goni P, Garcia A, Remacha MA, Sanchez E, Clavel A (2012) Two cases of zoonotic cryptosporidiosis in Spain by the unusual species Cryptosporidium ubiquitum and Cryptosporidium felis. Enferm Infecc Microbiol Clin 30(9):549–551, PubMed PMID: 22728073

Cohen S, Dalle F, Gallay A, Di Palma M, Bonnin A, Ward HD (2006) Identification of Cpgp40/15 Type Ib as the predominant allele in isolates of Cryptosporidium spp. from a waterborne outbreak of gastroenteritis in South Burgundy, France. J Clin Microbiol 44(2):589–591

Craun GF, Calderon RL, Craun MF (2005) Outbreaks associated with recreational water in the United States. Int J Environ Health Res 15(4):243–262, PubMed PMID: 16175741

Current WL, Upton SJ, Haynes TB (1986) The life cycle of Cryptosporidium baileyi n. sp. (Apicomplexa, Cryptosporidiidae) infecting chickens. J Protozool 33(2):289–296

Curriero FC, Patz JA, Rose JB, Lele S (2001) The association between extreme precipitation and waterborne disease outbreaks in the United States, 1948–1994. Am J Pub Health 91(8): 1194–1199, PubMed PMID: 11499103. Pubmed Central PMCID: 1446745

da Silva DC, Homem CG, Nakamura AA, Teixeira WF, Perri SH, Meireles MV (2010) Physical, epidemiological, and molecular evaluation of infection by Cryptosporidium galli in Passeriformes. Parasitol Res 107(2):271–277, PubMed PMID: 20407911

Davies AP, Campbell B, Evans MR, Bone A, Roche A, Chalmers RM (2009) Asymptomatic carriage of protozoan parasites in children in day care centers in the United Kingdom. Pediatr Infect Dis J 28(9):838–840, PubMed PMID: 19684527

Del Chierico F, Onori M, Di Bella S, Bordi E, Petrosillo N, Menichella D et al (2011) Cases of cryptosporidiosis co-infections in AIDS patients: a correlation between clinical presentation and GP60 subgenotype lineages from aged formalin-fixed stool samples. Ann Trop Med Parasitol 105(5):339–349, PubMed PMID: 21929875. Pubmed Central PMCID: 3176465

Diaz P, Quilez J, Robinson G, Chalmers RM, Diez-Banos P, Morrondo P (2010) Identification of Cryptosporidium xiaoi in diarrhoeic goat kids (Capra hircus) in Spain. Vet Parasitol 172(1–2): 132–134, PubMed PMID: 20537797

Diaz P, Hadfield SJ, Quilez J, Soilan M, Lopez C, Panadero R et al (2012) Assessment of three methods for multilocus fragment typing of Cryptosporidium parvum from domestic ruminants in North West Spain. Vet Parasitol 186(3–4):188–195, PubMed PMID: 22154970

Ditrich O, Palkovic L, Sterba J, Prokopic J, Loudova J, Giboda M (1991) The first finding of Cryptosporidium baileyi in man. Parasitol Res 77(1):44–47, PubMed PMID: 1825238

Djuretic T, Wall PG, Nichols G (1997) General outbreaks of infectious intestinal disease associated with milk and dairy products in England and Wales: 1992 to 1996. Comm Dis Rep CDR Rev 7(3):R41–R45, PubMed PMID: 9080728

Drumo R, Widmer G, Morrison LJ, Tait A, Grelloni V, D'Avino N et al (2012) Evidence of host-associated populations of Cryptosporidium parvum in Italy. Appl Environ Microbiol 78(10): 3523–3529, PubMed PMID: 22389374. Pubmed Central PMCID: 3346357

DuPont HL, Chappell CL, Sterling CR, Okhuysen PC, Rose JB, Jakubowski W (1995) The infectivity of Cryptosporidium parvum in healthy volunteers. N Engl J Med 332(13): 855–859, PubMed PMID: 7870140

Dyachenko V, Kuhnert Y, Schmaeschke R, Etzold M, Pantchev N, Daugschies A (2010) Occurrence and molecular characterization of Cryptosporidium spp. genotypes in European hedgehogs (Erinaceus europaeus L.) in Germany. Parasitology 137(2):205–216

Dziuban EJ, Liang JL, Craun GF, Hill V, Yu PA, Painter J et al (2006) Surveillance for waterborne disease and outbreaks associated with recreational water–United States, 2003–2004. Morb Mortal Wkly Rep Surveill Summ 55(12):1–30, PubMed PMID: 17183230

Egger M, Nguyen XM, Schaad UB, Krech T (1990) Intestinal cryptosporidiosis acquired from a cat. Infection 18(3):177–178, PubMed PMID: 2365471

Egorov A, Frost F, Muller T, Naumova E, Tereschenko A, Ford T (2004) Serological evidence of Cryptosporidium infections in a Russian city and evaluation of risk factors for infections. Ann Epidemiol 14(2):129–136, PubMed PMID: 15018886

Elwin K, Chalmers RM, Roberts R, Guy EC, Casemore DP (2001) Modification of a rapid method for the identification of gene-specific polymorphisms in Cryptosporidium parvum and its application to clinical and epidemiological investigations. Appl Environ Microbiol 67(12): 5581–5584, PubMed PMID: 11722909. Pubmed Central PMCID: 93346

Elwin K, Hadfield SJ, Robinson G, Chalmers RM (2012a) The epidemiology of sporadic human infections with unusual cryptosporidia detected during routine typing in England and Wales, 2000–2008. Epidemiol Infect 140(4):673–683, PubMed PMID: 21733255

Elwin K, Robinson G, Hadfield SJ, Fairclough HV, Iturriza-Gomara M, Chalmers RM (2012b) A comparison of two approaches to extracting Cryptosporidium DNA from human stools as measured by a real-time PCR assay. J Microbiol Methods 89(1):38–40, PubMed PMID: 22366300

Elwin K, Hadfield SJ, Robinson G, Crouch ND, Chalmers RM (2012c) Cryptosporidium viatorum n. sp. (Apicomplexa: Cryptosporidiidae) among travellers returning to Great Britain from the Indian subcontinent, 2007–2011. Int J Parasitol 42(7):675–682

Ethelberg S, Lisby M, Vestergaard LS, Enemark HL, Olsen KE, Stensvold CR et al (2009) A foodborne outbreak of *Cryptosporidium hominis* infection. Epidemiol Infect 137(3):348–356, PubMed PMID: 19134228

Fayer R (2010) Taxonomy and species delimitation in Cryptosporidium. Exp Parasitol 124(1): 90–97, PubMed PMID: 19303009

Fayer R, Santin M (2009) Cryptosporidium xiaoi n. sp. (Apicomplexa: Cryptosporidiidae) in sheep (Ovis aries). Vet Parasitol 164(2–4):192–200

Fayer R, Trout JM, Xiao L, Morgan UM, Lai AA, Dubey JP (2001) Cryptosporidium canis n. sp. from domestic dogs. J Parasitol 87(6):1415–1422, PubMed PMID: 11780831

Fayer R, Santin M, Xiao L (2005) Cryptosporidium bovis n. sp. (Apicomplexa: Cryptosporidiidae) in cattle (Bos taurus). J Parasitol 91(3):624–629

Fayer R, Santin M, Trout JM (2008) Cryptosporidium ryanae n. sp. (Apicomplexa: Cryptosporidiidae) in cattle (Bos taurus). Vet Parasitol 156(3–4):191–198

Fayer R, Santin M, Macarisin D (2010) Cryptosporidium ubiquitum n. sp. in animals and humans. Vet Parasitol 172(1–2):23–32, PubMed PMID: 20537798

Feltus DC, Giddings CW, Schneck BL, Monson T, Warshauer D, McEvoy JM (2006) Evidence supporting zoonotic transmission of Cryptosporidium spp. in Wisconsin. J Clin Microbiol 44 (12):4303–4308, PubMed PMID: 17005736. Pubmed Central PMCID: 1698413

Feng Y, Li N, Duan L, Xiao L (2009) Cryptosporidium genotype and subtype distribution in raw wastewater in Shanghai, China: evidence for possible unique *Cryptosporidium hominis* transmission. J Clin Microbiol 47(1):153–157, PubMed PMID: 19005143. Pubmed Central PMCID: 2620847

Feng Y, Yang W, Ryan U, Zhang L, Kvac M, Koudela B et al (2011a) Development of a multilocus sequence tool for typing Cryptosporidium muris and Cryptosporidium andersoni. J Clin Microbiol 49(1):34–41, PubMed PMID: 20980577. Pubmed Central PMCID: 3020410

Feng Y, Lal AA, Li N, Xiao L (2011b) Subtypes of Cryptosporidium spp. in mice and other small mammals. Exp Parasitol 127(1):238–242, PubMed PMID: 20692256

Feng Y, Zhao X, Chen J, Jin W, Zhou X, Li N et al (2011c) Occurrence, source, and human infection potential of cryptosporidium and Giardia spp. in source and tap water in Shanghai, China. Appl Environ Microbiol 77(11):3609–3616

Feng Y, Karna SR, Dearen TK, Singh DK, Adhikari LN, Shrestha A et al (2012) Common occurrence of a unique Cryptosporidium ryanae variant in zebu cattle and water buffaloes in the buffer zone of the Chitwan National Park, Nepal. Vet Parasitol 185(2–4):309–314, PubMed PMID: 21996006

Fiuza VR, Cosendey RI, Frazao-Teixeira E, Santin M, Fayer R, de Oliveira FC (2011) Molecular characterization of Cryptosporidium in Brazilian sheep. Vet Parasitol 175(3–4):360–362, PubMed PMID: 21075526

Fontaine M, Guillot E (2002) Development of a TaqMan quantitative PCR assay specific for Cryptosporidium parvum. FEMS Microbiol Lett 214(1):13–17, PubMed PMID: 12204366

Fontaine M, Guillot E (2003) An immunomagnetic separation-real-time PCR method for quantification of Cryptosporidium parvum in water samples. J Microbiol Methods 54(1):29–36, PubMed PMID: 12732419

Fournet N, Deege MP, Urbanus AT, Nichols G, Rosner BM, Chalmers RM, et al. (2013) Simultaneous increase of Cryptosporidium infections in the Netherlands, the United Kingdom and Germany in late summer season, 2012. Eur Surveill Bull Eur Mal Transm Eur Comm Dis Bull 18(2):1–5. PubMed PMID: 23324424

Frost FJ, Fea E, Gilli G, Biorci F, Muller TM, Craun GF et al (2000a) Serological evidence of Cryptosporidium infections in southern Europe. Eur J Epidemiol 16(4):385–390, PubMed PMID: 10959948

Frost FJ, Muller T, Calderon RL, Craun GF (2000b) A serological survey of college students for antibody to Cryptosporidium before and after the introduction of a new water filtration plant. Epidemiol Infect 125(1):87–92, PubMed PMID: 11057963. Pubmed Central PMCID: 2869573

Frost FJ, Muller T, Craun GF, Calderon RL, Roefer PA (2001) Paired city cryptosporidium serosurvey in the southwest USA. Epidemiol Infect 126(2):301–307, PubMed PMID: 11349981. Pubmed Central PMCID: 2869695

Frost FJ, Kunde TR, Muller TB, Craun GF, Katz LM, Hibbard AJ et al (2003) Serological responses to Cryptosporidium antigens among users of surface- vs. ground-water sources. Epidemiol Infect 131(3):1131–1138, PubMed PMID: 14959781. Pubmed Central PMCID: 2870063

Frost F, Craun G, Mihaly K, Gyorgy B, Calderon R, Muller T (2005) Serological responses to Cryptosporidium antigens among women using riverbank-filtered water, conventionally filtered surface water and groundwater in Hungary. J Water Health 3(1):77–82, PubMed PMID: 15952455

Furtado C, Adak GK, Stuart JM, Wall PG, Evans HS, Casemore DP (1998) Outbreaks of waterborne infectious intestinal disease in England and Wales, 1992–5. Epidemiol Infect 121 (1):109–119, PubMed PMID: 9747762. Pubmed Central PMCID: 2809481

Gatei W, Hart CA, Gilman RH, Das P, Cama V, Xiao L (2006) Development of a multilocus sequence typing tool for *Cryptosporidium hominis*. J Eukaryot Microbiol 53(Suppl 1): S43–S48, PubMed PMID: 17169064

Gatei W, Das P, Dutta P, Sen A, Cama V, Lal AA et al (2007) Multilocus sequence typing and genetic structure of *Cryptosporidium hominis* from children in Kolkata, India. Infect Genet Evol J Mol Epidemiol Evol Genet Infect Dis 7(2):197–205, PubMed PMID: 17010677

Gatei W, Barrett D, Lindo JF, Eldemire-Shearer D, Cama V, Xiao L (2008) Unique Cryptosporidium population in HIV-infected persons, Jamaica. Emerg Infect Dis 14(5):841–843, PubMed PMID: 18439378. Pubmed Central PMCID: 2600223

Geurden T, Levecke B, Caccio SM, Visser A, De Groote G, Casaert S et al (2009) Multilocus genotyping of Cryptosporidium and Giardia in non-outbreak related cases of diarrhoea in human patients in Belgium. Parasitology 136(10):1161–1168, PubMed PMID: 19631012

Gherasim A, Lebbad M, Insulander M, Decraene V, Kling A, Hjertqvist M, et al. (2012) Two geographically separated food-borne outbreaks in Sweden linked by an unusual Cryptosporidium parvum subtype, October 2010. Eur Surveill Bull Eur Mal Transm Eur Comm Dis Bull 17 (46):1–8. PubMed PMID: 23171824

Ghouila A, Terrapon N, Gascuel O, Guerfali FZ, Laouini D, Marechal E et al (2011) EuPathDomains: the divergent domain database for eukaryotic pathogens. Infect Genet Evol 11(4):698–707, PubMed PMID: 20920608

Glaberman S, Sulaiman IM, Bern C, Limor J, Peng MM, Morgan U et al (2001) A multilocus genotypic analysis of Cryptosporidium meleagridis. J Eukaryot Microbiol 48(Suppl-1):19S–22S, PubMed PMID: 11906063

Glaberman S, Moore JE, Lowery CJ, Chalmers RM, Sulaiman I, Elwin K et al (2002) Three drinking-water-associated cryptosporidiosis outbreaks, Northern Ireland. Emerg Infect Dis 8 (6):631–633, PubMed PMID: 12023922. Pubmed Central PMCID: 2738494

Gormley FJ, Little CL, Chalmers RM, Rawal N, Adak GK (2011) Zoonotic cryptosporidiosis from petting farms, England and Wales, 1992–2009. Emerg Infect Dis 17(1):151–152, PubMed PMID: 21192888. Pubmed Central PMCID: 3204639

Gow AG, Gow DJ, Hall EJ, Langton D, Clarke C, Papasouliotis K (2009) Prevalence of potentially pathogenic enteric organisms in clinically healthy kittens in the UK. J Feline Med Surg 11(8): 655–662, PubMed PMID: 19249233

Grinberg A, Lopez-Villalobos N, Pomroy W, Widmer G, Smith H, Tait A (2008a) Host-shaped segregation of the Cryptosporidium parvum multilocus genotype repertoire. Epidemiol Infect 136(2):273–278, PubMed PMID: 17394677. Pubmed Central PMCID: 2870797

Grinberg A, Learmonth J, Kwan E, Pomroy W, Lopez Villalobos N, Gibson I et al (2008b) Genetic diversity and zoonotic potential of Cryptosporidium parvum causing foal diarrhea. J Clin Microbiol 46(7):2396–2398, PubMed PMID: 18508944. Pubmed Central PMCID: 2446936

Grinberg A, Pomroy WE, Squires RA, Scuffham A, Pita A, Kwan E (2011) Retrospective cohort study of an outbreak of cryptosporidiosis caused by a rare Cryptosporidium parvum subgenotype. Epidemiol Infect 139(10):1542–1550, PubMed PMID: 21087535

Gupta S, Maiden MC (2001) Exploring the evolution of diversity in pathogen populations. Trends Microbiol 9(4):181–185, PubMed PMID: 11286883

Hadfield SJ, Chalmers RM (2012) Detection and characterization of Cryptosporidium cuniculus by real-time PCR. Parasitol Res 111(3):1385–1390, PubMed PMID: 22392139

Hadfield SJ, Robinson G, Elwin K, Chalmers RM (2011) Detection and differentiation of Cryptosporidium spp. in human clinical samples by use of real-time PCR. J Clin Microbiol 49 (3):918–924, PubMed PMID: 21177904. Pubmed Central PMCID: 3067739

Halstead FD, Lee AV, Couto-Parada X, Polley SD, Ling C, Jenkins C, et al. (2013) Universal extraction method for gastrointestinal pathogens. J Med Microbiol. 62(Pt 10):1535–1539 PubMed PMID: 23831766

Haque R, Roy S, Siddique A, Mondal U, Rahman SM, Mondal D et al (2007) Multiplex real-time PCR assay for detection of Entamoeba histolytica, Giardia intestinalis, and Cryptosporidium spp. Am J Trop Med Hyg 76(4):713–717, PubMed PMID: 17426176

Harper CM, Cowell NA, Adams BC, Langley AJ, Wohlsen TD (2002) Outbreak of Cryptosporidium linked to drinking unpasteurised milk. Comm Dis Intell Q Rep 26(3):449–450, PubMed PMID: 12416712

Heiges M, Wang H, Robinson E, Aurrecoechea C, Gao X, Kaluskar N et al (2006) CryptoDB: a Cryptosporidium bioinformatics resource update. Nucleic Acids Res 34:D419–D422, PubMed PMID: 16381902. Pubmed Central PMCID: 1347441

Helmy YA, Krucken J, Nockler K, von Samson-Himmelstjerna G, Zessin KH (2013) Molecular epidemiology of Cryptosporidium in livestock animals and humans in the Ismailia province of Egypt. Vet Parasitol 193(1–3):15–24, PubMed PMID: 23305974

Herges GR, Widmer G, Clark ME, Khan E, Giddings CW, Brewer M et al (2012) Evidence that Cryptosporidium parvum populations are panmictic and unstructured in the Upper Midwest of the United States. Appl Environ Microbiol 78(22):8096–8101, PubMed PMID: 22983961. Pubmed Central PMCID: 3485935

Higgins JA, Fayer R, Trout JM, Xiao L, Lal AA, Kerby S et al (2001) Real-time PCR for the detection of Cryptosporidium parvum. J Microbiol Methods 47(3):323–337, PubMed PMID: 11714523

Hijjawi N, Ng J, Yang R, Atoum MF, Ryan U (2010) Identification of rare and novel Cryptosporidium GP60 subtypes in human isolates from Jordan. Exp Parasitol 125(2):161–164, PubMed PMID: 20109456

Hira KG, Mackay MR, Hempstead AD, Ahmed S, Karim MM, O'Connor RM et al (2011) Genetic diversity of Cryptosporidium spp. from Bangladeshi children. J Clin Microbiol 49(6): 2307–2310, PubMed PMID: 21471344. Pubmed Central PMCID: 3122776

Hlavsa MC, Roberts VA, Anderson AR, Hill VR, Kahler AM, Orr M et al (2011) Surveillance for waterborne disease outbreaks and other health events associated with recreational water—United States, 2007–2008. Morb Mortal Wkly Rep Surveill Summ 60(12):1–32, PubMed PMID: 21937976

Hunter PR, Nichols G (2002) Epidemiology and clinical features of Cryptosporidium infection in immunocompromised patients. Clin Microbiol Rev 15(1):145–154, PubMed PMID: 11781272. Pubmed Central PMCID: 118064

Hunter PR, Hughes S, Woodhouse S, Syed Q, Verlander NQ, Chalmers RM et al (2004) Sporadic cryptosporidiosis case–control study with genotyping. Emerg Infect Dis 10(7):1241–1249, PubMed PMID: 15324544: Pubmed Central PMCID: 3323324

Hunter PR, Hadfield SJ, Wilkinson D, Lake IR, Harrison FC, Chalmers RM (2007) Subtypes of Cryptosporidium parvum in humans and disease risk. Emerg Infect Dis 13(1):82–88, PubMed PMID: 17370519. Pubmed Central PMCID: 2725800

Hunter PR, Wilkinson DC, Lake IR, Harrison FC, Syed Q, Hadfield SJ et al (2008) Microsatellite typing of Cryptosporidium parvum in isolates from a waterborne outbreak. J Clin Microbiol 46 (11):3866–3867, PubMed PMID: 18768659. Pubmed Central PMCID: 2576591

Ikarashi M, Fukuda Y, Honma H, Kasai K, Kaneta Y, Nakai Y (2013) First description of heterogeneity in 18S rRNA genes in the haploid genome of Cryptosporidium andersoni Kawatabi type. Vet Parasitol 196:200–224, PubMed PMID: 23369454

Imre K, Luca C, Costache M, Sala C, Morar A, Morariu S et al (2013) Zoonotic Cryptosporidium parvum in Romanian newborn lambs (Ovis aries). Vet Parasitol 191(1–2):119–122, PubMed PMID: 22995338

Insulander M, Silverlas C, Lebbad M, Karlsson L, Mattsson JG, Svenungsson B (2013) Molecular epidemiology and clinical manifestations of human cryptosporidiosis in Sweden. Epidemiol Infect 141(5):1009–1020, PubMed PMID: 22877562

Iqbal J, Khalid N, Hira PR (2011) Cryptosporidiosis in Kuwaiti children: association of clinical characteristics with Cryptosporidium species and subtypes. J Med Microbiol 60(Pt 5):647–652, PubMed PMID: 21233297

Iqbal A, Lim YA, Surin J, Sim BL (2012) High diversity of Cryptosporidium subgenotypes identified in Malaysian HIV/AIDS individuals targeting gp60 gene. PloS One 7(2):e31139, PubMed PMID: 22347442. Pubmed Central PMCID: 3275556

Iseki M (1979) Cryptosporidium felis sp. N. (Protozoa. Eimeriorina) from the domestic cat. Jpn J Parasitol 28:285–307

Jellison KL, Lynch AE, Ziemann JM (2009) Source tracking identifies deer and geese as vectors of human-infectious Cryptosporidium genotypes in an urban/suburban watershed. Environ Sci Technol 43(12):4267–4272, PubMed PMID: 19603633

Jex AR, Gasser RB (2008) Analysis of the genetic diversity within *Cryptosporidium hominis* and *Cryptosporidium parvum* from imported and autochtonous cases of human cryptosporidiosis by mutation scanning. Electrophoresis 29(20):4119–4129, PubMed PMID: 18991263

Jex AR, Gasser RB (2010) Genetic richness and diversity in *Cryptosporidium hominis* and *C. parvum* reveals major knowledge gaps and a need for the application of "next generation" technologies—research review. Biotechnol Adv 28(1):17–26, PubMed PMID: 19699288

Jex AR, Ryan UM, Ng J, Campbell BE, Xiao L, Stevens M et al (2007a) Specific and genotypic identification of Cryptosporidium from a broad range of host species by nonisotopic SSCP analysis of nuclear ribosomal DNA. Electrophoresis 28(16):2818–2825, PubMed PMID: 17702061

Jex AR, Whipp M, Campbell BE, Caccio SM, Stevens M, Hogg G et al (2007b) A practical and cost-effective mutation scanning-based approach for investigating genetic variation in Cryptosporidium. Electrophoresis 28(21):3875–3883, PubMed PMID: 17960838

Jex AR, Pangasa A, Campbell BE, Whipp M, Hogg G, Sinclair MI et al (2008) Classification of Cryptosporidium species from patients with sporadic cryptosporidiosis by use of sequence-based multilocus analysis following mutation scanning. J Clin Microbiol 46(7):2252–2262, PubMed PMID: 18448696. Pubmed Central PMCID: 2446878

Jirku M, Valigurova A, Koudela B, Krizek J, Modry D, Slapeta J (2008) New species of Cryptosporidium Tyzzer, 1907 (Apicomplexa) from amphibian host: morphology, biology and phylogeny. Folia Parasitol 55(2):81–94, PubMed PMID: 18666410

Jothikumar N, da Silva AJ, Moura I, Qvarnstrom Y, Hill VR (2008) Detection and differentiation of *Cryptosporidium hominis* and *Cryptosporidium parvum* by dual TaqMan assays. J Med Microbiol 57(Pt 9):1099–1105, PubMed PMID: 18719179

Jothikumar P, Hill V, Narayanan J (2009) Design of FRET-TaqMan probes for multiplex real-time PCR using an internal positive control. BioTechniques 46(7):519–524, PubMed PMID: 19594451

Karanis P, Kourenti C, Smith H (2007) Waterborne transmission of protozoan parasites: a worldwide review of outbreaks and lessons learnt. J Water Health 5(1):1–38, PubMed PMID: 17402277

Koehler AV, Bradbury RS, Stevens MA, Haydon SR, Jex AR, Gasser RB (2013) Genetic characterization of selected parasites from people with histories of gastrointestinal disorders using a mutation scanning-coupled approach. Electrophoresis 34(12):1720–1728

Kozisek F, Craun GF, Cerovska L, Pumann P, Frost F, Muller T (2008) Serological responses to Cryptosporidium-specific antigens in Czech populations with different water sources. Epidemiol Infect 136(2):279–286, PubMed PMID: 17394676. Pubmed Central PMCID: 2870799

Kramer MH, Herwaldt BL, Craun GF, Calderon RL, Juranek DD (1996) Surveillance for waterborne-disease outbreaks–United States, 1993–1994. MMWR CDC Surveill Summ 45 (1):1–33, PubMed PMID: 8600346

Kvac M, Kvetonova D, Sak B, Ditrich O (2009) Cryptosporidium pig genotype II in immunocompetent man. Emerg Infect Dis 15(6):982–983, PubMed PMID: 19523313. Pubmed Central PMCID: 2727335

Kvac M, Kestranova M, Pinkova M, Kvetonova D, Kalinova J, Wagnerova P et al (2013) Cryptosporidium scrofarum n. sp. (Apicomplexa: Cryptosporidiidae) in domestic pigs (Sus scrofa). Vet Parasitol 191(3–4):218–227

Leav BA, Mackay MR, Anyanwu A, RM OC, Cevallos AM, Kindra G et al (2002) Analysis of sequence diversity at the highly polymorphic Cpgp40/15 locus among Cryptosporidium isolates from human immunodeficiency virus-infected children in South Africa. Infect Immun 70(7):3881–3890, PubMed PMID: 12065532. Pubmed Central PMCID: 128099

Lee MB, Greig JD (2010) A review of gastrointestinal outbreaks in schools: effective infection control interventions. J Sch Health 80(12):588–598, PubMed PMID: 21087255

Lee SH, Levy DA, Craun GF, Beach MJ, Calderon RL (2002) Surveillance for waterborne-disease outbreaks–United States, 1999–2000. Morb Mortal Wkly Rep Surveill Summ 51(8):1–47, PubMed PMID: 12489843

LeJeune JT, Davis MA (2004) Outbreaks of zoonotic enteric disease associated with animal exhibits. J Am Vet Med Assoc 224(9):1440–1445, PubMed PMID: 15124883

Leoni F, Gallimore CI, Green J, McLauchlin J (2003) Molecular epidemiological analysis of Cryptosporidium isolates from humans and animals by using a heteroduplex mobility assay and nucleic acid sequencing based on a small double-stranded RNA element. J Clin Microbiol 41(3):981–992, PubMed PMID: 12624019. Pubmed Central PMCID: 150309

Leoni F, Amar C, Nichols G, Pedraza-Diaz S, McLauchlin J (2006) Genetic analysis of Cryptosporidium from 2414 humans with diarrhoea in England between 1985 and 2000. J Med Microbiol 55(Pt 6):703–707, PubMed PMID: 16687587

Leoni F, Mallon ME, Smith HV, Tait A, McLauchlin J (2007) Multilocus analysis of *Cryptosporidium hominis* and *Cryptosporidium parvum* isolates from sporadic and outbreak-related human cases and C. parvum isolates from sporadic livestock cases in the United Kingdom. J Clin Microbiol 45(10):3286–3294, PubMed PMID: 17687021. Pubmed Central PMCID: 2045344

Levine ND (1980) Some corrections of coccidian (Apicomplexa: Protozoa) nomenclature. J Parasitol 66(5):830–834, PubMed PMID: 7463253

Levine WC, Stephenson WT, Craun GF (1990) Waterborne disease outbreaks, 1986–1988. MMWR CDC Surveill Summ 39(1):1–13, PubMed PMID: 2156147

Levy DA, Bens MS, Craun GF, Calderon RL, Herwaldt BL (1998) Surveillance for waterborne-disease outbreaks–United States, 1995–1996. MMWR CDC Surveill Summ 47(5):1–34, PubMed PMID: 9859954

Lim YA, Iqbal A, Surin J, Sim BL, Jex AR, Nolan MJ et al (2011) First genetic classification of Cryptosporidium and Giardia from HIV/AIDS patients in Malaysia. Infect Genet Evol 11(5): 968–974, PubMed PMID: 21439404

Limor JR, Lal AA, Xiao L (2002) Detection and differentiation of Cryptosporidium parasites that are pathogenic for humans by real-time PCR. J Clin Microbiol 40(7):2335–2338, PubMed PMID: 12089244. Pubmed Central PMCID: 120558

Lindergard G, Nydam DV, Wade SE, Schaaf SL, Mohammed HO (2003) The sensitivity of PCR detection of Cryptosporidium oocysts in fecal samples using two DNA extraction methods. Mol Diagn 7(3–4):147–153, PubMed PMID: 15068384

Lindsay DS, Upton SJ, Owens DS, Morgan UM, Mead JR, Blagburn BL (2000) Cryptosporidium andersoni n. sp. (Apicomplexa: Cryptosporiidae) from cattle, Bos taurus. J Eukaryot Microbiol 47(1):91–95, PubMed PMID: 10651302

Liu C, Vigdorovich V, Kapur V, Abrahamsen MS (1999) A random survey of the Cryptosporidium parvum genome. Infect Immun 67(8):3960–3969, PubMed PMID: 10417162. Pubmed Central PMCID: 96679

Liu A, Wang R, Li Y, Zhang L, Shu J, Zhang W et al (2009) Prevalence and distribution of Cryptosporidium spp. in dairy cattle in Heilongjiang Province, China. Parasitol Res 105(3): 797–802

Liu A, Ji H, Wang E, Liu J, Xiao L, Shen Y et al (2011) Molecular identification and distribution of Cryptosporidium and Giardia duodenalis in raw urban wastewater in Harbin, China. Parasitol Res 109(3):913–918, PubMed PMID: 21461728

Liu J, Gratz J, Amour C, Kibiki G, Becker S, Janaki L et al (2013) A laboratory-developed TaqMan Array Card for simultaneous detection of 19 enteropathogens. J Clin Microbiol 51 (2):472–480, PubMed PMID: 23175269. Pubmed Central PMCID: 3553916

Le Blancq SM, Khramtsov NV, Zamani F, Upton SJ, Wu TW (1997) Ribosomal RNA gene organization in Cryptosporidium parvum. Mol Biochem Parasitol 190(2):463–478. PubMed PMID: 9476794

Lv C, Zhang L, Wang R, Jian F, Zhang S, Ning C et al (2009) Cryptosporidium spp. in wild, laboratory, and pet rodents in china: prevalence and molecular characterization. Appl Environ Microbiol 75(24):7692–7699

MacDonald LM, Sargent K, Armson A, Thompson RC, Reynoldson JA (2002) The development of a real-time quantitative-PCR method for characterisation of a Cryptosporidium parvum in vitro culturing system and assessment of drug efficacy. Mol Biochem Parasitol 121(2): 279–282, PubMed PMID: 12034463

Madrid-Aliste CJ, Dybas JM, Angeletti RH, Weiss LM, Kim K, Simon I et al (2009) EPIC-DB: a proteomics database for studying Apicomplexan organisms. BMC Genom 10:38, PubMed PMID: 19159464. Pubmed Central PMCID: 2652494

Maikai BV, Umoh JU, Lawal IA, Kudi AC, Ejembi CL, Xiao L (2012) Molecular characterizations of Cryptosporidium, Giardia, and Enterocytozoon in humans in Kaduna State, Nigeria. Exp Parasitol 131(4):452–456, PubMed PMID: 22664352

Mallon M, MacLeod A, Wastling J, Smith H, Reilly B, Tait A (2003a) Population structures and the role of genetic exchange in the zoonotic pathogen Cryptosporidium parvum. J Mol Evol 56 (4):407–417, PubMed PMID: 12664161

Mallon ME, MacLeod A, Wastling JM, Smith H, Tait A (2003b) Multilocus genotyping of Cryptosporidium parvum type 2: population genetics and sub-structuring. Infect Genet Evol J Mol Epidemiol Evol Genet Infect Dis 3(3):207–218, PubMed PMID: 14522184

Mary C, Chapey E, Dutoit E, Guyot K, Hasseine L, Jeddi F et al (2013) Multicentric evaluation of a new real-time PCR assay for quantification of Cryptosporidium sp and identification of Cryptosporidium parvum and hominis. J Clin Microbiol 51:2556–2563, PubMed PMID: 23720792

Mattsson JG, Insulander M, Lebbad M, Bjorkman C, Svenungsson B (2008) Molecular typing of Cryptosporidium parvum associated with a diarrhoea outbreak identifies two sources of exposure. Epidemiol Infect 136(8):1147–1152, PubMed PMID: 17961283. Pubmed Central PMCID: 2870910

Mayne DJ, Ressler KA, Smith D, Hockey G, Botham SJ, Ferson MJ (2011) A community outbreak of cryptosporidiosis in Sydney associated with a public swimming facility: a case-control study. Interdiscip Perspect Infect Dis 2011:341065, PubMed PMID: 22194741. Pubmed Central PMCID: 3238377

McAuliffe GN, Anderson TP, Stevens M, Adams J, Coleman R, Mahagamasekera P, Young S, Henderson T, Hofmann M, Jennings LC, Murdoch DR (2013) Systematic application of multiplex PCR enhances the detection of bacteria, parasites, and viruses in stool samples. J Infect 67(2):122–129

McDonald V (2011) Cryptosporidiosis: host immune responses and the prospects for effective immunotherapies. Expert Rev Anti-Infect Ther 9(11):1077–1086, PubMed PMID: 22029525

McLauchlin J, Casemore DP, Moran S, Patel S (1998) The epidemiology of cryptosporidiosis: application of experimental sub-typing and antibody detection systems to the investigation of water-borne outbreaks. Folia Parasitol 45(2):83–92, PubMed PMID: 9684318

McLauchlin J, Pedraza-Diaz S, Amar-Hoetzeneder C, Nichols GL (1999) Genetic characterization of Cryptosporidium strains from 218 patients with diarrhea diagnosed as having sporadic cryptosporidiosis. J Clin Microbiol 37(10):3153–3158, PubMed PMID: 10488169. Pubmed Central PMCID: 85515

McLauchlin J, Amar C, Pedraza-Diaz S, Nichols GL (2000) Molecular epidemiological analysis of Cryptosporidium spp. in the United Kingdom: results of genotyping Cryptosporidium spp. in 1,705 fecal samples from humans and 105 fecal samples from livestock animals. J Clin Microbiol 38(11):3984–3990, PubMed PMID: 11060056. Pubmed Central PMCID: 87529

Mejia R, Vicuna Y, Broncano N, Sandoval C, Vaca M, Chico M et al (2013) A novel, multi-parallel, real-time polymerase chain reaction approach for eight gastrointestinal parasites provides improved diagnostic capabilities to resource-limited at-risk populations. Am J Trop Med Hyg 88:1041–1047, PubMed PMID: 23509117

Millard PS, Gensheimer KF, Addiss DG, Sosin DM, Beckett GA, Houck-Jankoski A et al (1994) An outbreak of cryptosporidiosis from fresh-pressed apple cider. JAMA 272(20):1592–1596, PubMed PMID: 7966869

Miller WA, Miller MA, Gardner IA, Atwill ER, Harris M, Ames J et al (2005) New genotypes and factors associated with Cryptosporidium detection in mussels (Mytilus spp.) along the California coast. Int J Parasitol 35(10):1103–1113

Molloy SF, Smith HV, Kirwan P, Nichols RA, Asaolu SO, Connelly L et al (2010) Identification of a high diversity of Cryptosporidium species genotypes and subtypes in a pediatric population in Nigeria. Am J Trop Med Hyg 82(4):608–613, PubMed PMID: 20348508. Pubmed Central PMCID: 2844578

Morgan UM, Constantine CC, Forbes DA, Thompson RC (1997) Differentiation between human and animal isolates of Cryptosporidium parvum using rDNA sequencing and direct PCR analysis. J Parasitol 83(5):825–830, PubMed PMID: 9379285

Morgan-Ryan UM, Fall A, Ward LA, Hijjawi N, Sulaiman I, Fayer R et al (2002) *Cryptosporidium hominis* n. sp. (Apicomplexa: Cryptosporidiidae) from Homo sapiens. J Eukaryot Microbiol 49 (6):433–440, PubMed PMID: 12503676

Morisset D, Stebih D, Milavec M, Gruden K, Zel J (2013) Quantitative analysis of food and feed samples with droplet digital PCR. PloS One 8(5):e62583, PubMed PMID: 23658750. Pubmed Central PMCID: 3642186

Morrison LJ, Mallon ME, Smith HV, MacLeod A, Xiao L, Tait A (2008) The population structure of the Cryptosporidium parvum population in Scotland: a complex picture. Infect Genet Evol 8 (2):121–129, PubMed PMID: 18077222. Pubmed Central PMCID: 2684618

Morse TD, Nichols RA, Grimason AM, Campbell BM, Tembo KC, Smith HV (2007) Incidence of cryptosporidiosis species in paediatric patients in Malawi. Epidemiol Infect 135(8): 1307–1315, PubMed PMID: 17224087. Pubmed Central PMCID: 2870691

Moss DM, Chappell CL, Okhuysen PC, DuPont HL, Arrowood MJ, Hightower AW et al (1998) The antibody response to 27-, 17-, and 15-kDa Cryptosporidium antigens following experimental infection in humans. J Infect Dis 178(3):827–833, PubMed PMID: 9728553

Mueller-Doblies D, Giles M, Elwin K, Smith RP, Clifton-Hadley FA, Chalmers RM (2008) Distribution of Cryptosporidium species in sheep in the UK. Vet Parasitol 154(3–4): 214–219, PubMed PMID: 18468799

Muthusamy D, Rao SS, Ramani S, Monica B, Banerjee I, Abraham OC et al (2006) Multilocus genotyping of Cryptosporidium sp. isolates from human immunodeficiency virus-infected individuals in South India. J Clin Microbiol 44(2):632–634, PubMed PMID: 16455931. Pubmed Central PMCID: 1392691

Nago TT, Tokashiki YT, Kisanuki K, Nakasone I, Yamane N (2010) Laboratory-based evaluation of loop-mediated isothermal amplification (LAMP) to detect Cryptosporidium oocyst and Giardia lamblia cyst in stool specimens. Rinsho Byori 58(8):765–771, PubMed PMID: 20860168

Nakamura AA, Simoes DC, Antunes RG, da Silva DC, Meireles MV (2009) Molecular characterization of Cryptosporidium spp. from fecal samples of birds kept in captivity in Brazil. Vet Parasitol 166(1–2):47–51, PubMed PMID: 19683397

Nazeer JT, El Sayed KK, von Thien H, El-Sibaei MM, Abdel-Hamid MY, Tawfik RA et al (2013) Use of multiplex real-time PCR for detection of common diarrhea causing protozoan parasites in Egypt. Parasitol Res 112(2):595–601, PubMed PMID: 23114927

Nazemalhosseini-Mojarad E, Haghighi A, Taghipour N, Keshavarz A, Mohebi SR, Zali MR et al (2011) Subtype analysis of Cryptosporidium parvum and *Cryptosporidium hominis* isolates from humans and cattle in Iran. Vet Parasitol 179(1–3):250–252, PubMed PMID: 21376469

Nemejc K, Sak B, Kvetonova D, Hanzal V, Jenikova M, Kvac M (2012) The first report on Cryptosporidium suis and Cryptosporidium pig genotype II in Eurasian wild boars (Sus scrofa) (Czech Republic). Vet Parasitol 184(2–4):122–125, PubMed PMID: 21917378

Network ACN (2010) Laboratory-based surveillance for Cryptosporidium in France, 2006–2009. Eur Surveill 15(33):19642, PubMed PMID: 20739000

Ng J, Pavlasek I, Ryan U (2006) Identification of novel Cryptosporidium genotypes from avian hosts. Appl Environ Microbiol 72(12):7548–7553, PubMed PMID: 17028234. Pubmed Central PMCID: 1694252

Ng J, Eastwood K, Durrheim D, Massey P, Walker B, Armson A et al (2008) Evidence supporting zoonotic transmission of Cryptosporidium in rural New South Wales. Exp Parasitol 119(1): 192–195, PubMed PMID: 18343369

Ng J, MacKenzie B, Ryan U (2010a) Longitudinal multi-locus molecular characterisation of sporadic Australian human clinical cases of cryptosporidiosis from 2005 to 2008. Exp Parasitol 125(4):348–356, PubMed PMID: 20206624

Ng JS, Pingault N, Gibbs R, Koehler A, Ryan U (2010b) Molecular characterisation of Cryptosporidium outbreaks in Western and South Australia. Exp Parasitol 125(4):325–328, PubMed PMID: 20219461

Ng J, Yang R, McCarthy S, Gordon C, Hijjawi N, Ryan U (2011) Molecular characterization of Cryptosporidium and Giardia in pre-weaned calves in Western Australia and New South Wales. Vet Parasitol 176(2–3):145–150, PubMed PMID: 21130578

Ng JS, Eastwood K, Walker B, Durrheim DN, Massey PD, Porigneaux P et al (2012) Evidence of Cryptosporidium transmission between cattle and humans in northern New South Wales. Exp Parasitol 130(4):437–441, PubMed PMID: 22333036

Ngouanesavanh T, Guyot K, Certad G, Le Fichoux Y, Chartier C, Verdier RI et al (2006) Cryptosporidium population genetics: evidence of clonality in isolates from France and Haiti. J Eukaryot Microbiol 53(Suppl 1):S33–S36, PubMed PMID: 17169061

Nichols G (2008) Epidemiology. In: Fayer R, Xiao L (eds) Cryptosporidium and cryptosporidiosis. CRC Press, Boca Raton, pp 79–118

Nichols G (1992) The biology, epidemiology and typing of Cryptosporidium spp. Ph.D. thesis. Surrey University

Nichols GC R, Lake I, Sopwith W, Regan M, Hunter P, Grenfell P, Harrison F, Lane C, Health Protection Agency (2006) Cryptosporidiosis: a report on the surveillance and epidemiology of Cryptosporidium infection in England and Wales. Drinking Water Inspectorate, London, pp 1–148

Nichols GL, McLauchlin J, Samuel D (1991) A technique for typing Cryptosporidium isolates. J Protozool 38(6):237S–240S, PubMed PMID: 1818185

Nichols RA, Campbell BM, Smith HV (2003) Identification of Cryptosporidium spp. oocysts in United Kingdom noncarbonated natural mineral waters and drinking waters by using a modified nested PCR-restriction fragment length polymorphism assay. Appl Environ Microbiol 69(7):4183–4189, PubMed PMID: 12839797. Pubmed Central PMCID: 165191

Nichols RA, Campbell BM, Smith HV (2006a) Molecular fingerprinting of Cryptosporidium oocysts isolated during water monitoring. Appl Environ Microbiol 72(8):5428–5435, PubMed PMID: 16885295. Pubmed Central PMCID: 1538703

Nichols RA, Moore JE, Smith HV (2006b) A rapid method for extracting oocyst DNA from Cryptosporidium-positive human faeces for outbreak investigations. J Microbiol Methods 65 (3):512–524, PubMed PMID: 16290112

Nichols G, Lane C, Asgari N, Verlander NQ, Charlett A (2009) Rainfall and outbreaks of drinking water related disease and in England and Wales. J Water Health 7(1):1–8, PubMed PMID: 18957770

Nichols RA, Connelly L, Sullivan CB, Smith HV (2010) Identification of Cryptosporidium species and genotypes in Scottish raw and drinking waters during a one-year monitoring period. Appl Environ Microbiol 76(17):5977–5986, PubMed PMID: 20639357. Pubmed Central PMCID: 2935041

Nina JM, McDonald V, Dyson DA, Catchpole J, Uni S, Iseki M et al (1992a) Analysis of oocyst wall and sporozoite antigens from three Cryptosporidium species. Infect Immun 60(4): 1509–1513, PubMed PMID: 1548074. Pubmed Central PMCID: 257024

Nina JM, McDonald V, Deer RM, Wright SE, Dyson DA, Chiodini PL et al (1992b) Comparative study of the antigenic composition of oocyst isolates of Cryptosporidium parvum from different hosts. Parasite Immunol 14(2):227–232, PubMed PMID: 1570174

O'Brien E, McInnes L, Ryan U (2008) Cryptosporidium GP60 genotypes from humans and domesticated animals in Australia, North America and Europe. Exp Parasitol 118(1): 118–121, PubMed PMID: 17618622

Ogunkolade BW, Robinson HA, McDonald V, Webster K, Evans DA (1993) Isoenzyme variation within the genus Cryptosporidium. Parasitol Res 79(5):385–388, PubMed PMID: 8415544

Okhuysen PC, Chappell CL, Crabb JH, Sterling CR, DuPont HL (1999) Virulence of three distinct *Cryptosporidium parvum* isolates for healthy adults. J Infect Dis 180(4):1275–1281

Ondriska F, Vrabcova I, Brindakova S, Kvac M, Ditrich O, Boldis V et al (2013) The first reported cases of human cryptosporidiosis caused by *Cryptosporidium hominis* in Slovak Republic. Folia Microbiol 58(1):69–73, PubMed PMID: 22826020

Ong CS, Chow S, Gustafson R, Plohman C, Parker R, Isaac-Renton JL et al (2008) Rare *Cryptosporidium hominis* subtype associated with aquatic center use. Emerg Infect Dis 14 (8):1323–1325, PubMed PMID: 18680673. Pubmed Central PMCID: 2600396

Palenzuela O, Alvarez-Pellitero P, Sitja-Bobadilla A (2010) Molecular characterization of Cryptosporidium molnari reveals a distinct piscine clade. Appl Environ Microbiol 76(22):7646–7649, PubMed PMID: 20870791. Pubmed Central PMCID: 2976190

Pangasa A, Jex AR, Nolan MJ, Campbell BE, Haydon SR, Stevens MA et al (2010) Highly sensitive non-isotopic restriction endonuclease fingerprinting of nucleotide variability in the gp60 gene within Cryptosporidium species, genotypes and subgenotypes infective to humans, and its implications. Electrophoresis 31(10):1637–1647, PubMed PMID: 20419704

Paparini A, Jackson B, Ward S, Young S, Ryan UM (2012) Multiple Cryptosporidium genotypes detected in wild black rats (Rattus rattus) from northern Australia. Exp Parasitol 131(4): 404–412, PubMed PMID: 22659228

Parr JB, Sevilleja JE, Samie A, Alcantara C, Stroup SE, Kohli A et al (2007) Detection and quantification of Cryptosporidium in HCT-8 cells and human fecal specimens using real-time polymerase chain reaction. Am J Trop Med Hyg 76(5):938–942, PubMed PMID: 17488919. Pubmed Central PMCID: 2253489

Patel S, Pedraza-Diaz S, McLauchlin J, Casemore DP (1998) Molecular characterisation of Cryptosporidium parvum from two large suspected waterborne outbreaks. Outbreak control team South and West Devon 1995, incident management team and further epidemiological and microbiological studies subgroup North Thames 1997. Comm Dis Pub Health 1(4):231–233, PubMed PMID: 9854879

Pavlasek I, Ryan U (2008) Cryptosporidium varanii takes precedence over C. saurophilum. Exp Parasitol 118(3):434–437, PubMed PMID: 17945215

Pelayo L, Nunez FA, Rojas L, Wilke H, Furuseth Hansen E, Mulder B et al (2008) Molecular and epidemiological investigations of cryptosporidiosis in Cuban children. Ann Trop Med Parasitol 102(8):659–669, PubMed PMID: 19000383

Peng MM, Xiao L, Freeman AR, Arrowood MJ, Escalante AA, Weltman AC et al (1997) Genetic polymorphism among Cryptosporidium parvum isolates: evidence of two distinct human transmission cycles. Emerg Infect Dis 3(4):567–573, PubMed PMID: 9366611. Pubmed Central PMCID: 2640093

Peng MM, Matos O, Gatei W, Das P, Stantic-Pavlinic M, Bern C et al (2001) A comparison of Cryptosporidium subgenotypes from several geographic regions. J Eukaryot Microbiol 48 (Suppl 1):28S–31S, PubMed PMID: 11906067

Peng MM, Meshnick SR, Cunliffe NA, Thindwa BD, Hart CA, Broadhead RL et al (2003) Molecular epidemiology of cryptosporidiosis in children in Malawi. J Eukaryot Microbiol 50 (Suppl):557–559, PubMed PMID: 14736161

Perz JF, Le Blancq SM (2001) Cryptosporidium parvum infection involving novel genotypes in wildlife from lower New York State. Appl Environ Microbiol 67(3):1154–1162, PubMed PMID: 11229905. Pubmed Central PMCID: 92708

Pieniazek NJ, Bornay-Llinares FJ, Slemenda SB, da Silva AJ, Moura IN, Arrowood MJ et al (1999) New cryptosporidium genotypes in HIV-infected persons. Emerg Infect Dis 5 (3):444–449, PubMed PMID: 10341184. Pubmed Central PMCID: 2640774

Piper MB, Bankier AT, Dear PH (1998) A HAPPY map of Cryptosporidium parvum. Genome Res 8(12):1299–1307, PubMed PMID: 9872984. Pubmed Central PMCID: 310802

Pollock KGJ, Nichols G (2012) Foodborne protozoa and outbreak investigations. In: Robertson LJ, Smith HV (eds) Foodborne protozoan parasites. Nova Science, New York, pp 267–92

Ponka A, Kotilainen H, Rimhanen-Finne R, Hokkanen P, Hanninen ML, Kaarna A, et al (2009) A foodborne outbreak due to Cryptosporidium parvum in Helsinki, November 2008. Eur Surveill 14(28):1–3. PubMed PMID: 19607781

Power ML, Ryan UM (2008) A new species of Cryptosporidium (Apicomplexa: Cryptosporidiidae) from eastern grey kangaroos (Macropus giganteus). J Parasitol 94(5): 1114–1117, PubMed PMID: 18973420

Power ML, Cheung-Kwok-Sang C, Slade M, Williamson S (2009) Cryptosporidium fayeri: diversity within the GP60 locus of isolates from different marsupial hosts. Exp Parasitol 121 (3):219–223, PubMed PMID: 19027006

Power ML, Holley M, Ryan UM, Worden P, Gillings MR (2011) Identification and differentiation of Cryptosporidium species by capillary electrophoresis single-strand conformation polymorphism. FEMS Microbiol Lett 314(1):34–41, PubMed PMID: 21087296

Priest JW, Bern C, Xiao L, Roberts JM, Kwon JP, Lescano AG et al (2006) Longitudinal analysis of cryptosporidium species-specific immunoglobulin G antibody responses in Peruvian children. Clin Vaccine Immunol 13(1):123–131, PubMed PMID: 16426009. Pubmed Central PMCID: 1356630

Putignani L, Menichella D (2010) Global distribution, public health and clinical impact of the protozoan pathogen cryptosporidium. Interdisc Perspect Infect Dis. 1–39 PubMed PMID: 20706669. Pubmed Central PMCID: 2913630

Qi M, Wang R, Ning C, Li X, Zhang L, Jian F et al (2011) Cryptosporidium spp. in pet birds: genetic diversity and potential public health significance. Exp Parasitol 128(4):336–340, PubMed PMID: 21557938

Quilez J, Torres E, Chalmers RM, Hadfield SJ, Del Cacho E, Sanchez-Acedo C (2008) Cryptosporidium genotypes and subtypes in lambs and goat kids in Spain. Appl Environ Microbiol 74 (19):6026–6031, PubMed PMID: 18621872. Pubmed Central PMCID: 2565967

Quiroz ES, Bern C, MacArthur JR, Xiao L, Fletcher M, Arrowood MJ et al (2000) An outbreak of cryptosporidiosis linked to a foodhandler. J Infect Dis 181(2):695–700, PubMed PMID: 10669357

Raskova V, Kvetonova D, Sak B, McEvoy J, Edwinson A, Stenger B et al (2013) Human cryptosporidiosis caused by Cryptosporidium tyzzeri and C. parvum isolates presumably transmitted from wild mice. J Clin Microbiol 51(1):360–362, PubMed PMID: 23100342. Pubmed Central PMCID: 3536213

Reid A, Lymbery A, Ng J, Tweedle S, Ryan U (2010) Identification of novel and zoonotic Cryptosporidium species in marine fish. Vet Parasitol 168(3–4):190–195, PubMed PMID: 20031326

Ren X, Zhao J, Zhang L, Ning C, Jian F, Wang R et al (2012) Cryptosporidium tyzzeri n. sp. (Apicomplexa: Cryptosporidiidae) in domestic mice (Mus musculus). Exp Parasitol 130(3): 274–281

Rengifo-Herrera C, Ortega-Mora LM, Gomez-Bautista M, Garcia-Moreno FT, Garcia-Parraga D, Castro-Urda J et al (2011) Detection and characterization of a Cryptosporidium isolate from a southern elephant seal (Mirounga leonina) from the Antarctic peninsula. Appl Environ Microbiol 77(4):1524–1527, PubMed PMID: 21169427. Pubmed Central PMCID: 3067219

Richter B, Nedorost N, Maderner A, Weissenbock H (2011) Detection of Cryptosporidium species in feces or gastric contents from snakes and lizards as determined by polymerase chain reaction analysis and partial sequencing of the 18S ribosomal RNA gene. J Vet Diagn Invest 23(3): 430–435, PubMed PMID: 21908271

Rider SD Jr, Zhu G (2010) Cryptosporidium: genomic and biochemical features. Exp Parasitol 124 (1):2–9, PubMed PMID: 19187778. Pubmed Central PMCID: 2819285

Risebro HL, Doria MF, Andersson Y, Medema G, Osborn K, Schlosser O et al (2007) Fault tree analysis of the causes of waterborne outbreaks. J Water Health 5(Suppl 1):1–18, PubMed PMID: 17890833

Robertson LJ, Chalmers RM (2013) Foodborne cryptosporidiosis: is there really more in Nordic countries? Trends Parasitol 29(1):3–9, PubMed PMID: 23146217

Robinson G, Chalmers RM (2010) The European rabbit (Oryctolagus cuniculus), a source of zoonotic cryptosporidiosis. Zoonoses Pub Health 57(7–8):e1–e13, PubMed PMID: 20042061

Robinson G, Chalmers RM (2012) Assessment of polymorphic genetic markers for multi-locus typing of *Cryptosporidium parvum* and *Cryptosporidium hominis*. Exp Parasitol 132(2): 200–215, PubMed PMID: 22781277

Robinson G, Wright S, Elwin K, Hadfield SJ, Katzer F, Bartley PM et al (2010) Re-description of Cryptosporidium cuniculus Inman and Takeuchi, 1979 (Apicomplexa: Cryptosporidiidae): morphology, biology and phylogeny. Int J Parasitol 40(13):1539–1548, PubMed PMID: 20600069

Roche JK, Rojo AL, Costa LB, Smeltz R, Manque P, Woehlbier U et al (2013) Intranasal vaccination in mice with an attenuated Salmonella enterica Serovar 908htr A expressing Cp15 of Cryptosporidium: impact of malnutrition with preservation of cytokine secretion. Vaccine 31(6):912–918, PubMed PMID: 23246541. Pubmed Central PMCID: 3563240

Rooney RM, Bartram JK, Cramer EH, Mantha S, Nichols G, Suraj R et al (2004) A review of outbreaks of waterborne disease associated with ships: evidence for risk management. Pub Health Rep 119(4):435–442, PubMed PMID: 15219801. Pubmed Central PMCID: 1497646

Ruecker NJ, Matsune JC, Wilkes G, Lapen DR, Topp E, Edge TA et al (2012) Molecular and phylogenetic approaches for assessing sources of Cryptosporidium contamination in water. Water Res 46(16):5135–5150, PubMed PMID: 22841595

Ruecker NJ, Matsune JC, Lapen DR, Topp E, Edge TA, Neumann NF (2013) The detection of Cryptosporidium and the resolution of mixtures of species and genotypes from water. Infect Genet Evol 15:3–9

Ryan UM, Xiao L, Read C, Sulaiman IM, Monis P, Lal AA et al (2003) A redescription of Cryptosporidium galli Pavlasek, 1999 (Apicomplexa: Cryptosporidiidae) from birds. J Parasitol 89(4):809–813, PubMed PMID: 14533694

Ryan UM, Monis P, Enemark HL, Sulaiman I, Samarasinghe B, Read C et al (2004a) Cryptosporidium suis n. sp. (Apicomplexa: Cryptosporidiidae) in pigs (Sus scrofa). J Parasitol 90(4): 769–773

Ryan U, O'Hara A, Xiao L (2004b) Molecular and biological characterization of a Cryptosporidium molnari-like isolate from a guppy (Poecilia reticulata). Appl Environ Microbiol 70(6): 3761–3765, PubMed PMID: 15184187. Pubmed Central PMCID: 427724

Ryan UM, Power M, Xiao L (2008) Cryptosporidium fayeri n. sp. (Apicomplexa: Cryptosporidiidae) from the Red Kangaroo (Macropus rufus). J Eukaryot Microbiol 55(1): 22–26, PubMed PMID: 18251799

Santin M, Zarlenga DS (2009) A multiplex polymerase chain reaction assay to simultaneously distinguish Cryptosporidium species of veterinary and public health concern in cattle. Vet Parasitol 166(1–2):32–37, PubMed PMID: 19713046

Schuster CJ, Ellis AG, Robertson WJ, Charron DF, Aramini JJ, Marshall BJ et al (2005) Infectious disease outbreaks related to drinking water in Canada, 1974–2001. Can J Pub Health 96 (4):254–258, PubMed PMID: 16625790

Semenza JC, Nichols G (2007) Cryptosporidiosis surveillance and water-borne outbreaks in Europe. Eur Surveill 12(5):E13–E14

Seva Ada P, Funada MR, Richtzenhain L, Guimaraes MB, Souza Sde O, Allegretti L et al (2011) Genotyping of Cryptosporidium spp. from free-living wild birds from Brazil. Vet Parasitol 175 (1–2):27–32, PubMed PMID: 21035268

Shanmugasundram A, Gonzalez-Galarza FF, Wastling JM, Vasieva O, Jones AR (2013) Library of apicomplexan metabolic pathways: a manually curated database for metabolic pathways of apicomplexan parasites. Nucleic Acids Res 41(Database issue):D706–713, PubMed PMID: 23193253. Pubmed Central PMCID: 3531055

Shen Y, Yin J, Yuan Z, Lu W, Xu Y, Xiao L et al (2011) The identification of the Cryptosporidium ubiquitum in pre-weaned Ovines from Aba Tibetan and Qiang autonomous prefecture in China. Biomed Environ Sci 24(3):315–320, PubMed PMID: 21784319

Silverlas C, Mattsson JG, Insulander M, Lebbad M (2012) Zoonotic transmission of Cryptosporidium meleagridis on an organic Swedish farm. Int J Parasitol 42(11):963–967, PubMed PMID: 23022616

Singh P, Foley SL, Nayak R, Kwon YM (2013) Massively parallel sequencing of enriched target amplicons for high-resolution genotyping of Salmonella serovars. Mol Cel Probes 27(2): 80–85, PubMed PMID: 23201627

Slavin D (1955) Cryptosporidium meleagridis (sp. nov.). J Comp Pathol 65(3):262–266

Smith HV, Nichols RA, Mallon M, Macleod A, Tait A, Reilly WJ et al (2005) Natural *Cryptosporidium hominis* infections in Scottish cattle. Vet Rec 156(22):710–711, PubMed PMID: 15923554

Smith A, Reacher M, Smerdon W, Adak GK, Nichols G, Chalmers RM (2006) Outbreaks of waterborne infectious intestinal disease in England and Wales, 1992–2003. Epidemiol Infect 134(6):1141–1149, PubMed PMID: 16690002. Pubmed Central PMCID: 2870523

Smith S, Elliot AJ, Mallaghan C, Modha D, Hippisley-Cox J, Large S et al (2010a) Value of syndromic surveillance in monitoring a focal waterborne outbreak due to an unusual Cryptosporidium genotype in Northamptonshire, United Kingdom, June—July 2008. Eur Surveill 15 (33):19643

Smith RP, Chalmers RM, Mueller-Doblies D, Clifton-Hadley FA, Elwin K, Watkins J et al (2010b) Investigation of farms linked to human patients with cryptosporidiosis in England and Wales. Prev Vet Med 94(1–2):9–17, PubMed PMID: 20096944

Soba B, Logar J (2008) Genetic classification of Cryptosporidium isolates from humans and calves in Slovenia. Parasitology 135(11):1263–1270, PubMed PMID: 18664309

Sopwith W, Osborn K, Chalmers R, Regan M (2005) The changing epidemiology of cryptosporidiosis in North West England. Epidemiol Infect 133(5):785–793, PubMed PMID: 16181496: Pubmed Central PMCID: 2870307

Spain CV, Scarlett JM, Wade SE, McDonough P (2001) Prevalence of enteric zoonotic agents in cats less than 1 year old in central New York State. J Vet Intern Med 15(1):33–38

Stark D, Al-Qassab SE, Barratt JL, Stanley K, Roberts T, Marriott D et al (2011) Evaluation of multiplex tandem real-time PCR for detection of Cryptosporidium spp., Dientamoeba fragilis, Entamoeba histolytica, and Giardia intestinalis in clinical stool samples. J Clin Microbiol 49 (1):257–262

Strong WB, Gut J, Nelson RG (2000) Cloning and sequence analysis of a highly polymorphic Cryptosporidium parvum gene encoding a 60-kilodalton glycoprotein and characterization of its 15- and 45-kilodalton zoite surface antigen products. Infect Immun 68(7):4117–4134, PubMed PMID: 10858229. Pubmed Central PMCID: 101708

Stroup SE, Roy S, McHele J, Maro V, Ntabaguzi S, Siddique A et al (2006) Real-time PCR detection and speciation of Cryptosporidium infection using Scorpion probes. J Med Microbiol 55(Pt 9):1217–1222, PubMed PMID: 16914651

Sulaiman IM, Lal AA, Xiao L (2001) A population genetic study of the Cryptosporidium parvum human genotype parasites. J Eukaryot Microbiol 48(Suppl 1):24S–27S, PubMed PMID: 11906066

Sulaiman IM, Hira PR, Zhou L, Al-Ali FM, Al-Shelahi FA, Shweiki HM et al (2005) Unique endemicity of cryptosporidiosis in children in Kuwait. J Clin Microbiol 43(6):2805–2809, PubMed PMID: 15956401. Pubmed Central PMCID: 1151898

Sweeny JP, Ryan UM, Robertson ID, Yang R, Bell K, Jacobson C (2011) Longitudinal investigation of protozoan parasites in meat lamb farms in southern Western Australia. Prev Vet Med 101 (3–4):192–203, PubMed PMID: 21733584

Taghipour N, Nazemalhosseini-Mojarad E, Haghighi A, Rostami-Nejad M, Romani S, Keshavarz A et al (2011) Molecular epidemiology of cryptosporidiosis in Iranian children, Tehran, Iran. Iran J Parasitol 6(4):41–45, PubMed PMID: 22347312. Pubmed Central PMCID: 3279909

Taniuchi M, Verweij JJ, Noor Z, Sobuz SU, Lieshout L, Petri WA Jr et al (2011) High through put multiplex PCR and probe-based detection with Luminex beads for seven intestinal parasites. Am J Trop Med Hyg 84(2):332–337, PubMed PMID: 21292910. Pubmed Central PMCID: 3029193

Tanriverdi S, Widmer G (2006) Differential evolution of repetitive sequences in *Cryptosporidium parvum* and *Cryptosporidium hominis*. Infect Genet Evol J Mol Epidemiol Evol Genet Infect Dis 6(2):113–122, PubMed PMID: 16503512

Tanriverdi S, Tanyeli A, Baslamisli F, Koksal F, Kilinc Y, Feng X et al (2002) Detection and genotyping of oocysts of Cryptosporidium parvum by real-time PCR and melting curve analysis. J Clin Microbiol 40(9):3237–3244, PubMed PMID: 12202559. Pubmed Central PMCID: 130769

Tanriverdi S, Arslan MO, Akiyoshi DE, Tzipori S, Widmer G (2003) Identification of genotypically mixed Cryptosporidium parvum populations in humans and calves. Mol Biochem Parasitol 130 (1):13–22, PubMed PMID: 14550892

Tanriverdi S, Blain JC, Deng B, Ferdig MT, Widmer G (2007) Genetic crosses in the apicomplexan parasite Cryptosporidium parvum define recombination parameters. Mol Microbiol 63(5): 1432–1439, PubMed PMID: 17302818

Tanriverdi S, Grinberg A, Chalmers RM, Hunter PR, Petrovic Z, Akiyoshi DE et al (2008) Inferences about the global population structures of *Cryptosporidium parvum* and

Cryptosporidium hominis. Appl Environ Microbiol 74(23):7227–7234, PubMed PMID: 18836013. Pubmed Central PMCID: 2592928

Thompson HP, Dooley JS, Kenny J, McCoy M, Lowery CJ, Moore JE et al (2007) Genotypes and subtypes of Cryptosporidium spp. in neonatal calves in Northern Ireland. Parasitol Res 100(3): 619–624, PubMed PMID: 17031699

Tosini F, Drumo R, Elwin K, Chalmers RM, Pozio E, Caccio SM (2010) The CpA135 gene as a marker to identify Cryptosporidium species infecting humans. Parasitol Int 59(4):606–609, PubMed PMID: 20831899

Trotz-Williams LA, Martin DS, Gatei W, Cama V, Peregrine AS, Martin SW et al (2006) Genotype and subtype analyses of Cryptosporidium isolates from dairy calves and humans in Ontario. Parasitol Res 99(4):346–352, PubMed PMID: 16565813

Tyzzer EE (1910) An extracellular Coccidium, Cryptosporidium Muris (Gen. Et Sp. Nov.), of the gastric glands of the common mouse. J Med Res 23(3):487–510, 3. PubMed PMID: 19971982. Pubmed Central PMCID: 2098948

Tyzzer EE (1912) Cryptosporidium parvum (sp. Nov.), a coccidian found in the small intestine of the common mouse. Archiv fur Protistenkunde 26:394–418

Tzipori S, Angus KW, Campbell I, Gray EW (1980) Cryptosporidium: evidence for a single-species genus. Infect Immun 30(3):884–886, PubMed PMID: 7228392. Pubmed Central PMCID: 551396

Verweij JJ, Blange RA, Templeton K, Schinkel J, Brienen EA, van Rooyen MA et al (2004) Simultaneous detection of Entamoeba histolytica, Giardia lamblia, and Cryptosporidium parvum in fecal samples by using multiplex real-time PCR. J Clin Microbiol 42(3): 1220–1223, PubMed PMID: 15004079. Pubmed Central PMCID: 356880

Vetterling JM, Jervis HR, Merrill TG, Sprinz H (1971) Cryptosporidium wrairi sp. n. from the guinea pig Cavia porcellus, with an emendation of the genus. J Protozool 18(2):243–247

Vojdani JD, Beuchat LR, Tauxe RV (2008) Juice-associated outbreaks of human illness in the United States, 1995 through 2005. J Food Prot 71(2):356–364, PubMed PMID: 18326187

Waldron LS, Power ML (2011) Fluorescence analysis detects gp60 subtype diversity in Cryptosporidium infections. Infect Genet Evol 11(6):1388–1395, PubMed PMID: 21609784

Waldron LS, Ferrari BC, Power ML (2009) Glycoprotein 60 diversity in C. hominis and C. parvum causing human cryptosporidiosis in NSW, Australia. Exp Parasitol 122(2):124–127

Waldron LS, Cheung-Kwok-Sang C, Power ML (2010) Wildlife-associated Cryptosporidium fayeri in human, Australia. Emerg Infect Dis 16(12):2006–2007, PubMed PMID: 21122247. Pubmed Central PMCID: 3294593

Waldron LS, Dimeski B, Beggs PJ, Ferrari BC, Power ML (2011a) Molecular epidemiology, spatiotemporal analysis, and ecology of sporadic human cryptosporidiosis in Australia. Appl Environ Microbiol 77(21):7757–7765, PubMed PMID: 21908628. Pubmed Central PMCID: 3209161

Waldron LS, Ferrari BC, Cheung-Kwok-Sang C, Beggs PJ, Stephens N, Power ML (2011b) Molecular epidemiology and spatial distribution of a waterborne cryptosporidiosis outbreak in Australia. Appl Environ Microbiol 77(21):7766–7771, PubMed PMID: 21908623. Pubmed Central PMCID: 3209151

Wang R, Zhang X, Zhu H, Zhang L, Feng Y, Jian F et al (2011) Genetic characterizations of Cryptosporidium spp. and Giardia duodenalis in humans in Henan, China. Exp Parasitol 127 (1):42–45

Wang R, Jian F, Zhang L, Ning C, Liu A, Zhao J et al (2012) Multilocus sequence subtyping and genetic structure of Cryptosporidium muris and Cryptosporidium andersoni. PloS One 7(8): e43782, PubMed PMID: 22937094. Pubmed Central PMCID: 3427161

Wang L, Zhang H, Zhao X, Zhang L, Zhang G, Guo M et al (2013) Zoonotic Cryptosporidium species and Enterocytozoon bieneusi genotypes in HIV-positive patients on antiretroviral therapy. J Clin Microbiol 51(2):557–563, PubMed PMID: 23224097. Pubmed Central PMCID: 3553929

Widmer G (2009) Meta-analysis of a polymorphic surface glycoprotein of the parasitic protozoa *Cryptosporidium parvum* and *Cryptosporidium hominis*. Epidemiol Infect 137(12):1800–1808, PubMed PMID: 19527551. Pubmed Central PMCID: 2783587

Widmer G, Lee Y (2010) Comparison of single- and multilocus genetic diversity in the protozoan parasites *Cryptosporidium parvum* and *C. hominis*. Appl Environ Microbiol 76(19):6639–6644

Widmer G, Sullivan S (2012) Genomics and population biology of Cryptosporidium species. Parasite Immunol 34(2–3):61–71, PubMed PMID: 21595702. Pubmed Central PMCID: 3168714

Widmer G, Lee Y, Hunt P, Martinelli A, Tolkoff M, Bodi K (2012) Comparative genome analysis of two Cryptosporidium parvum isolates with different host range. Infect Genet Evol 12(6): 1213–1221, PubMed PMID: 22522000. Pubmed Central PMCID: 3372781

Wielinga PR, de Vries A, van der Goot TH, Mank T, Mars MH, Kortbeek LM et al (2008) Molecular epidemiology of Cryptosporidium in humans and cattle in The Netherlands. Int J Parasitol 38(7):809–817, PubMed PMID: 18054936

Wong PH, Ong CS (2006) Molecular characterization of the Cryptosporidium cervine genotype. Parasitology 133(Pt 6):693–700, PubMed PMID: 16899138

Wu Z, Nagano I, Boonmars T, Nakada T, Takahashi Y (2003) Intraspecies polymorphism of Cryptosporidium parvum revealed by PCR-restriction fragment length polymorphism (RFLP) and RFLP-single-strand conformational polymorphism analyses. Appl Environ Microbiol 69 (8):4720–4726, PubMed PMID: 12902263. Pubmed Central PMCID: 169079

Xiao L (2010) Molecular epidemiology of cryptosporidiosis: an update. Exp Parasitol 124 (1):80–89, PubMed PMID: 19358845

Xiao LR UM (2008) Molecular epidemiology. In: Fayer R, Xiao L (ed) Cryptosporidium and cryptosporidiosis. CRC Press, London, pp 119–171

Xiao L, Ryan UM (2004) Cryptosporidiosis: an update in molecular epidemiology. Curr Opin Infect Dis 17(5):483–490, PubMed PMID: 15353969

Xiao L, Sulaiman I, Fayer R, Lal AA (1998) Species and strain-specific typing of Cryptosporidium parasites in clinical and environmental samples. Mem Inst Oswaldo Cruz 93(5):687–691, PubMed PMID: 9830539

Xiao L, Limor JR, Li L, Morgan U, Thompson RC, Lal AA (1999) Presence of heterogeneous copies of the small subunit rRNA gene in Cryptosporidium parvum human and marsupial genotypes and Cryptosporidium felis. J Eukaryot Microbiol 46(5):44S–45S, PubMed PMID: 10519242

Xiao L, Singh A, Limor J, Graczyk TK, Gradus S, Lal A (2001a) Molecular characterization of cryptosporidium oocysts in samples of raw surface water and wastewater. Appl Environ Microbiol 67(3):1097–1101, PubMed PMID: 11229897. Pubmed Central PMCID: 92700

Xiao L, Bern C, Limor J, Sulaiman I, Roberts J, Checkley W et al (2001b) Identification of 5 types of Cryptosporidium parasites in children in Lima, Peru. J Infect Dis 183(3):492–497, PubMed PMID: 11133382

Xiao L, Bern C, Arrowood M, Sulaiman I, Zhou L, Kawai V et al (2002) Identification of the cryptosporidium pig genotype in a human patient. J Infect Dis 185(12):1846–1848, PubMed PMID: 12085341

Xiao L, Fayer R, Ryan U, Upton SJ (2004) Cryptosporidium taxonomy: recent advances and implications for public health. Clin Microbiol Rev 17(1):72–97, PubMed PMID: 14726456. Pubmed Central PMCID: 321466

Xiao L, Alderisio KA, Jiang J (2006) Detection of Cryptosporidium oocysts in water: effect of the number of samples and analytic replicates on test results. Appl Environ Microbiol 72 (9):5942–5947, PubMed PMID: 16957214: Pubmed Central PMCID: 1563632

Xiao L, Hlavsa MC, Yoder J, Ewers C, Dearen T, Yang W et al (2009) Subtype analysis of Cryptosporidium specimens from sporadic cases in Colorado, Idaho, New Mexico, and Iowa in 2007: widespread occurrence of one *Cryptosporidium hominis* subtype and case history of an infection with the Cryptosporidium horse genotype. J Clin Microbiol 47(9):3017–3020, PubMed PMID: 19587303. Pubmed Central PMCID: 2738111

Xiao L, Ryan UM, Fayer R, Bowman DD, Zhang L (2012) Cryptosporidium tyzzeri and Cryptosporidium pestis: which name is valid? Exp Parasitol 130(3):308–309, PubMed PMID: 22230707

Xu P, Widmer G, Wang Y, Ozaki LS, Alves JM, Serrano MG et al (2004) The genome of *Cryptosporidium hominis*. Nature 431(7012):1107–1112, PubMed PMID: 15510150

Yang W, Chen P, Villegas EN, Landy RB, Kanetsky C, Cama V et al (2008) Cryptosporidium source tracking in the Potomac River watershed. Appl Environ Microbiol 74(21):6495–6504, PubMed PMID: 18776033. Pubmed Central PMCID: 2576682

Yang R, Fenwick S, Potter A, Ng J, Ryan U (2011) Identification of novel Cryptosporidium genotypes in kangaroos from Western Australia. Vet Parasitol 179(1–3):22–27, PubMed PMID: 21402448

Yoder JS, Blackburn BG, Craun GF, Hill V, Levy DA, Chen N et al (2004) Surveillance for waterborne-disease outbreaks associated with recreational water–United States, 2001–2002. Morb Mortal Wkly Rep Surveill Summ 53(8):1–22, PubMed PMID: 15499306

Yoder JS, Hlavsa MC, Craun GF, Hill V, Roberts V, Yu PA et al (2008) Surveillance for waterborne disease and outbreaks associated with recreational water use and other aquatic facility-associated health events–United States, 2005–2006. Morb Mortal Wkly Rep Surveill Summ 57(9):1–29, PubMed PMID: 18784642

Zankari E, Hasman H, Kaas RS, Seyfarth AM, Agerso Y, Lund O et al (2013) Genotyping using whole-genome sequencing is a realistic alternative to surveillance based on phenotypic antimicrobial susceptibility testing. J Antimicrob Chemother 68(4):771–777, PubMed PMID: 23233485

Zhang W, Shen Y, Wang R, Liu A, Ling H, Li Y et al (2012) Cryptosporidium cuniculus and Giardia duodenalis in rabbits: genetic diversity and possible zoonotic transmission. PloS One 7 (2):e31262, PubMed PMID: 22363600. Pubmed Central PMCID: 3281947

Zhao GH, Ren WX, Gao M, Bian QQ, Hu B, Cong MM et al (2013) Genotyping Cryptosporidium andersoni in cattle in Shaanxi Province, Northwestern China. PloS One 8(4):e60112, PubMed PMID: 23560072. Pubmed Central PMCID: 3613348

Zhou L, Fayer R, Trout JM, Ryan UM, Schaefer FW 3rd, Xiao L (2004) Genotypes of Cryptosporidium species infecting fur-bearing mammals differ from those of species infecting humans. Appl Environ Microbiol 70(12):7574–7577, PubMed PMID: 15574965. Pubmed Central PMCID: 535153

Zintl A, Proctor AF, Read C, Dewaal T, Shanaghy N, Fanning S et al (2009) The prevalence of Cryptosporidium species and subtypes in human faecal samples in Ireland. Epidemiol Infect 137(2):270–277, PubMed PMID: 18474128

Zintl A, Ezzaty-Mirashemi M, Chalmers RM, Elwin K, Mulcahy G, Lucy FE et al (2011) Longitudinal and spatial distribution of GP60 subtypes in human cryptosporidiosis cases in Ireland. Epidemiol Infect 139(12):1945–1955, PubMed PMID: 21281547

Chapter 4
Cryptosporidiosis in Farmed Animals

Lucy J. Robertson, Camilla Björkman, Charlotte Axén, and Ronald Fayer

Abstract Cryptosporidiosis was first identified as a disease of veterinary, rather than human medical, importance, and infection of farmed animals with different species of *Cryptosporidium* continues to be of veterinary clinical concern. This chapter provides insights into *Cryptosporidium* infection in a range of farmed animals – cattle, sheep, goats, pigs, cervids, camelids, rabbits, water buffalo and poultry – presenting not only an updated overview of the infection in these animals, but also information on clinical disease, infection dynamics and zoonotic potential. Although extensive data have been accrued on, for example, *Cryptosporidium parvum* infection in calves, and calf cryptosporidiosis continues to be a major veterinary concern especially in temperate regions, there remains a paucity of data for other farmed animals, despite *Cryptosporidium* infection causing

---

L.J. Robertson (✉)
Parasitology Laboratory, Section for Microbiology, Immunology and Parasitology, Department of Food Safety and Infection Biology, Norwegian School of Veterinary Science, Postbox 8146 Dep, Oslo 0033, Norway
e-mail: lucy.robertson@nvh.no

C. Björkman
Division of Ruminant Medicine and Veterinary Epidemiology, Department of Clinical Sciences, Swedish University of Agricultural Sciences, P.O. Box 7054, Uppsala SE-750 07, Sweden
e-mail: camilla.bjorkman@slu.se

C. Axén
Department of Animal Health and Antimicrobial Strategies, National Veterinary Institute, SE-751 89 Uppsala, Sweden
e-mail: charlotte.axen@sva.se

R. Fayer
Agricultural Research Service, United States Department of Agriculture, Environmental Microbial and Food Safety Laboratory, Building 173 Powder Mill Road, Beltsville, MD 20705, USA
e-mail: ronald.fayer@ars.usda.gov

S.M. Cacciò and G. Widmer (eds.), *Cryptosporidium: parasite and disease*,
DOI 10.1007/978-3-7091-1562-6_4,

significant clinical disease and also, for some species, with the potential for transmission of infection to people, either directly or indirectly.

4.1 Introduction

4.1.1 Species of Cryptosporidium *Relevant to Different Farmed Animals: Overview*

Farmed animals, also commonly referred to as livestock or domesticated animals, are those animals that are reared in an agricultural setting in order to produce various commodities – usually food (meat, organs, eggs, dairy products), and/or hair or wool. In some settings farmed animals are also used to supply labour, and the manure of domesticated animals is often used as fertilizer. Animals were probably first farmed, that is their breeding and living conditions controlled by their human owners, around 7000–8000 BC during the first transitions from hunter-gatherer lifestyles to more settled agricultural living. The physiologies, behaviours, lifecycles of farmed animals generally differ quite substantially from those characteristics of the equivalent wild animals, and this difference impacts the interactions of these farmed animals with their parasites. Farmed animals are exposed to different stresses than wild animals, are kept at different densities, and their lifecycles regulated to such an extent that a parasite-host interaction in a farmed animal may differ significantly from that in a wild animal. Additionally, for infections that are of significant clinical importance, farmers may implement control measures (including treatment or prophylaxis) that alter the infection dynamics. With respect to *Cryptosporidium* infection, for which a satisfactory chemotherapeutic cure or prophylaxis is not yet available, different species infect different species of farmed animal, and may or may not be of clinical relevance.

Table 4.1 provides an overview of the farmed animals included in this chapter, the species of *Cryptosporidium* to which they are susceptible and brief notes on the clinical relevance. Greater details are provided in the appropriate chapter sections. Various categories of animals that are 'farmed', including mink, foxes, guinea pigs etc. are not included, largely because of a lack of information on *Cryptosporidium* in these animals in the domesticated setting. Additionally, farmed fish are not included in this chapter.

4.1.2 Relevance of Cryptosporidiosis in Farmed Animals to Human Infections

While cryptosporidiosis in farmed animals is of veterinary relevance, resulting in clinical morbidity, mortality, and associated production losses, the zoonotic nature

Table 4.1 Overview of farmed animals and major relevant species of *Cryptosporidium*

Farmed animal	Species of *Cryptosporidium*	Clinical notes
Bovines, including cattle (*Bos taurus* and *B. indicus*), banteng (*Bos javanicus*), gayal (*Bos frontalis*), water buffalo (*Bubalus bubalis*), and yaks (*Bos grunniens*)	*C. parvum*[a]	Common in pre-weaned calves – acute onset diarrhoea. Intestinal location
	C. bovis	Common in post-weaned calves – less pathogenic than *C. parvum*
	C. andersoni	Older post-weaned calves, yearlings and adults- some failure to thrive. Infects the gastric glands of the abomasum
	C. ryanae	Common in post-weaned calves
	A range of other species has been reported from cattle and other bovines. These seem to be unusual and are apparently of minor clinical significance	
Small ruminants, including sheep (*Ovis aries*) and goats (*Capra aegagrus hircus*)	*C. parvum*[a]	Relatively common in pre-weaned lambs, associated with diarrhoea
	C. xiaoi	Common in older lambs and sheep, often apparently asymptomatic
	C. ubiquitum[a]	Common in older lambs and sheep, often apparently asymptomatic
Pigs (*Sus scrofa domesticus*)	*C. parvum*[a]	Less common than in bovines and small ruminants; diarrhoea and vomiting
	C. suis	Relatively common, mild symptoms
	C. scrofarum	Relatively common, mild symptoms
Deer (cervids), including red deer (*Cervus elaphus*) fallow deer (*Dama dama*), elk/wapiti (*Cervus canadensis*), white-tailed deer (*Odocoileus virginianus*), and reindeer (*Rangifer tarandus*)	*C. parvum*[a] *C. ubiquitum*[a]	Information on species detected amongst farmed deer is lacking; diarrhoea in young calves, possibly severe, but can also be asymptomatic
Camelids, including dromedaries (*Camelus dromedarius*), llama (*Lama glama*), and alpaca (*Lama pacos*)	*C. parvum*[a]	Relatively little information on species that are infectious to camelids; diarrhoeal disease, particularly in young alpaca (crias)
Rabbits (*Oryctolagus cuniculus*)	*C. cuniculus*[a]	Clinical symptoms in rabbits are apparently mild or lacking
Poultry, including chickens (*Gallus gallus domesticus*), ducks (*Anas platyrhynchos*), turkeys (*Meleagris gallopavo*), geese (*Anser anser domesticus*), ostriches (*Struthio camelus*), pigeons (*Columba livia domestica*) etc.	*C. meleagridis*[a]	Appears to have a wide host range, including farmed poultry (and mammals). Mostly infects the intestines and has been associated with generally mild clinical symptoms
	C. baileyi[a]	A wide avian host range reported, including various farmed poultry species. Detected in many different anatomical sites including digestive tract, respiratory tract, and urinary tract. Has been associated with high morbidity and mortality

(continued)

Table 4.1 (continued)

Farmed animal	Species of *Cryptosporidium*	Clinical notes
	C. galli	May affect a wide range of avian species, including farmed poultry. Infects the proventriculus, and has been associated with acute diarrhoeal disease
	Various host species-associated genotypes have been described from different poultry, including ducks and geese. Their clinical importance and host specificity is generally not known	

[a]Noted for zoonotic potential

of various *Cryptosporidium* species (see Table 4.1) means that public health may also be affected by infections in farmed animals. Infection may be direct, from animal to human, or indirect, via a transmission vehicle. A large number of small outbreaks associated with *C. parvum* in calves and in veterinarians or veterinary students that have been exposed to calf faeces are documented in the literature (e.g. Grinberg et al. 2011; Gait et al. 2008; Robertson et al. 2006). In addition, a number of outbreaks have been documented associated with children visiting 'petting farms' or similar venues, where interaction with young animals such as lambs or calves is encouraged. Less commonly, transmission between animals such as camels or alpacas and their carers has also been reported. Drinking water, and less often food, has been associated with transmission of *Cryptosporidium* infection from animals to human populations, with *C. parvum* from grazing cattle contaminating water supplies particularly implicated. The high densities of farmed animals in water catchment areas mean that implementation of catchment control measures, including preventing defecation into water courses, may have a significant effect on minimising the risk from this potential transmission pathway.

4.2 *Cryptosporidium* Infection in Bovines

There are various species of farmed bovines, with cattle and zebu (*Bos taurus* and *Bos indicus*, respectively) amongst the most important livestock worldwide. Both provide meat, milk and other dairy products, and are also used as draught animals, with an estimated 1.5 billion head globally (3 cattle for every 14 people). A comprehensive overview of *Cryptosporidium* infection in cattle has been published by Santín and Trout (2008) and the information presented here is largely an update built on this solid basis. Domesticated water buffalo (*Bubalus bubalis*) is also important domestic livestock in the bovinae sub-family. Domesticated buffalo consist of swamp buffalo and river buffalo. In 2011, the world population of buffalo was estimated to approximately 195 million animals, of which 97 % were in Asia (FAOSTAT 2012).

4.2.1 *Occurrence (Prevalence)*

4.2.1.1 Cattle

The first report on bovine cryptosporidiosis was published in 1971, when parasites were identified in an 8-month-old heifer with chronic diarrhoea (Panciera et al. 1971). Since then, *Cryptosporidium* infection in cattle has been documented in most countries worldwide. Four major *Cryptosporidium* species infect cattle: *C. parvum*, *C. bovis*, *C. ryanae* and *C. andersoni* (Table 4.2; Fayer et al. 2007; Feng et al. 2007; Santín et al. 2004; Langkjaer et al. 2007). *Cryptosporidium parvum* has a broad host range and apparently has the ability to infect most mammals, including humans and cattle. In contrast, the other three species have almost exclusively been found in cattle. In addition to these four common species, sporadic natural infections with *C. felis*, *C. hominis*, *C. scrofarum*, *C. serpentis*, *C. suis* and *C. suis*-like genotype have been detected in cattle (Bornay-Llinares et al. 1999; Geurden et al. 2006; Langkjaer et al. 2007; Santín et al. 2004; Smith et al. 2005; Chen and Huang 2012). The extent to which these findings reflect true infections or accidental carriage, i.e. ingested oocysts that pass intact through the gastrointestinal tract, remains to be clarified. Cattle have also been experimentally infected with *C. canis*, but natural infection has not been reported (Fayer et al. 2001).

Many reports on *Cryptosporidium* in cattle in different countries and settings have been published over the years, showing that *Cryptosporidium* spp. infections are common worldwide. Both dairy and beef cattle are infected and the prevalence estimates vary considerably among studies. Reported herd level prevalences range from 0 to 100 % (Olson et al. 1997; Chang'a et al. 2011; Maddox-Hyttel et al. 2006; Santín et al. 2004). Infected animals have been reported from all age groups but infection is most common in preweaned calves. When calves up to 2 months of age have been investigated in point prevalence surveys, 5–93 % of the calves shed oocysts (Table 4.2; Maddox-Hyttel et al. 2006; Santín et al. 2004; Uga et al. 2000). Longitudinal studies performed in infected dairy herds showed that all calves in such herds shed oocysts at some time during their first months of life (O'Handley et al. 1999; Santín et al. 2008).

The overall picture is that there is an age-related pattern in the species distribution. *C. parvum* is mostly found in preweaned, monogastric calves up to 2 months of age where it is often the most prevalent species, responsible for more than 80 % of *Cryptosporidium* infections (Brook et al. 2009; Fayer et al. 2007; Plutzer and Karanis 2007; Santín et al. 2004, 2008; Trotz-Williams et al. 2006). In some areas, however, *C. bovis* is the dominating species found in preweaned calves (Budu-Amoako et al. 2012a; Wang et al. 2011a; Silverlås et al. 2010b). The prevalence of *C. parvum* is considerably lower in older calves and young stock, and there are few reports of *C. parvum* infection in adult cows (Fayer et al. 2007; Langkjaer et al. 2007; Silverlås et al. 2010b; Castro-Hermida et al. 2011a; Khan et al. 2010; Ondrácková et al. 2009; Muhid et al. 2011; Budu-Amoako et al. 2012a, b).

Table 4.2 Location, age, and prevalence of *Cryptosporidium* spp. in cattle in some recent studies[a]

Location	Age	No. of animals/farms or locations	Prevalence	Microscopy (M), ELISA (E), Molecular (Mo) or Other (O)	Molecular identification/ species or genotype (% investigated)	Reference
Australia, NSW	Calves	196/20 herds	74 %	Mo	18S rRNA and GP60/ *C. parvum* 59 % *C. bovis* 20 % *C. ryanae* 10 % Mixed infections 10 % Not identified 1 %	Ng et al. 2012
Brazil	≤30 days	196/dairy herds	11 %	Mo	18S rRNA and GP60/ *C. parvum* 33 % *C. bovis* 5 % *C. ryanae* 10 % *C. andersoni* 10 % Not identified 42 %	Meireles et al. 2011
Canada	<2 months 2–6 months >6 months	752/20 dairy herds	17 % 14 % 15 %	M	18S rRNA/ *C. parvum* 5 % *C. bovis* 51 % *C. ryanae* 17 % *C. andersoni* 27 %	Budu-Amoako et al. 2012a
Canada	≤6 months >6 months	739/20 beef herds	18 % 15 %	M	18S rRNA and HSP70/ *C. parvum* 24 % *C. bovis* 20 % *C. ryanae* 7 % *C. andersoni* 49 %	Budu-Amoako et al. 2012b
Czech Republic	20–60 days	750/24 dairy herds	21 %	M	18S rRNA RFLP/ *C. parvum* 86 % *C. bovis* 2 % *C. andersoni* 13 %	Kváč et al. 2011

China	0–>48 months	2,056/14 dairy herds	19 %	M	18s rRNA/ *C. parvum* 48 % *C. bovis* 16 % *C. andersoni* 29 % *C. hominis* 6 % *C. serpentis* 1 %	Chen and Huang 2012
China	0–8 weeks	801/8 herds	21 %	M	18s rRNA and GP60/ *C. parvum* 31 % *C. bovis* 38 % *C. ryanae* 11 % *C. andersoni* 7 % Mixed infections 12 %	Wang et al. 2011b
England and Wales	≤3 months	229 dairy or beef calves/diagnostic lab	45 %	M	18s rRNA/ *C. parvum* 91 % *C. bovis* 2 % Not identified 7 %	Featherstone et al. 2010a
England and Wales	Preweaned Immature Adult	116/11 herds connected with human cryptosporidiosis	81 % 58 % 19 %	M	18s rRNA/ *C. parvum* 77 % *C. bovis* 5 % *C. andersoni* 16 % Not identified 2 %	Smith et al. 2010
Egypt	<6 weeks	96/2 dairy herds	30 %	M	18S rRNA and COWP *C. parvum* 93 % *C. andersoni* 7 %	Amer et al. 2010
Egypt	1 day–3 months >3 months–1 year >1–2 years >2 years	593	30 % 13 % 13 % 5 %	O	18S rRNA and GP60/ *C. parvum* 65 % *C. bovis* 4 % *C. ryanae* 14 % Mixed infections17%	Helmy et al. 2013

(continued)

Table 4.2 (continued)

Location	Age	No. of animals/farms or locations	Prevalence	Microscopy (M), ELISA (E), Molecular (Mo) or Other (O)	Molecular identification/species or genotype (% investigated)	Reference
Hungary	Preweaned calves	79 diarrhoeic/52 herds	49 %	M	18S rRNA and GP60/ *C. parvum* 95 % *C. ryanae* 5 %	Plutzer and Karanis 2007
India	<3 months	461/various	16 %	M	18s sRNA/ *C. parvum* 100 %	Maurya et al. 2013
India	0–2 months 3–12 months >12 months	180/2 dairy herds	20 % 14 % 4 %	M	18S rRNA/ *C. parvum* 29 % *C. bovis* 38 % *C. ryanae* 14 % C. *andersoni* 14 % *C. suis*-like 5 %	Khan et al. 2010
Iran	1–20 weeks	272/15 dairy herds	19 %	M	18S rRNA/ *C. parvum* 73 % *C. bovis* 8 % C. *andersoni* 18 % Atypical isolates 2 %	Keshavarz et al. 2009
Italy	0 day–<12 months	2,024/248 dairy and beef herds	8 %	ELISA, M	COWP and GP60/ *C. parvum* 100 %	Duranti et al. 2009
Japan	3–48 days	80 diarrhoeic/different herds	75 %	Mo	18S rRNA/ *C. parvum* 53 % *C. bovis* 2 % Not identified 45 %	Karanis et al. 2010
Malaysia	1 day–≤4.5 months >4.5–12 months	250/16 herds	31 % 23 %	Mo	18S rRNA and GP60/ *C. parvum* 17 % *C. bovis* 25 % *C. ryanae* 15 %	Muhid et al. 2011

					C. andersoni 20 % Mixed infections 6 % Not identified: 17 %	
Nigeria	2–365 days	194/20 herds	16 %	Mo	18S rRNA/ *C. bovis* 45 % *C. ryanae* 26 % *C. andersoni* 16 % Mixed infections 13 %	Maikai et al. 2011
Northern Ireland	<1 month	779 diarrhoeic/ diagnostic lab	37 %	M	18S rRNA/ *C. parvum* 95 % *C. bovis* 4 % *C. ryanae* 1 %	Thompson et al. 2007
Romania	1–30 days	258 diarrhoeic/9 dairy herds	25 %	M	18S rRNA and GP60/ *C. parvum* 100 %	Imre et al. 2011
Spain	Neonatal Heifers Cows	649/not specified	61 % 15 % 8 %	M	18S rRNA/ *C. parvum* 56 % *C. andersoni* 23 % Not identified 21 %	Castro-Hermida et al. 2011a
Spain	≤21 days	61 diarrhoeic/27 herds	49 %	M	18S rRNA and GP60 *C. parvum* 100 %	Díaz et al. 2010a
Sweden	≤2 months 4–12 months Cows	1,202/50 dairy herds	52 % 29 % 6 %	M	18S rRNA and GP60/ *C. parvum* 14 % *C. bovis* 75 % *C. ryanae* 9 % *C. andersoni* 2 %	Silverlås et al. 2009b; Silverlås et al. 2010b
USA 7 states	5 days–2 months 3–11 months	971/15 dairy herds	50 % 20 %	M	18S rRNA/ *C. parvum* 50 % *C. bovis* 28 % *C. ryanae* 16 % *C. andersoni* 6 %	Santín et al. 2004

(continued)

Table 4.2 (continued)

Location	Age	No. of animals/farms or locations	Prevalence	Microscopy (M), ELISA (E), Molecular (Mo) or Other (O)	Molecular identification/ species or genotype (% investigated)	Reference
USA 7 states	12–24 months	571/14 dairy herds	12 %	Mo	18S rRNA/ *C. parvum* 6 % *C. bovis* 35 % *C. ryanae* 15 % *C. andersoni* 43 % *C. suis* 1 %	Fayer et al. 2006
USA 20 states	6–18 months	819/49 beef herds	20 %	Mo	18S rRNA/ *C. bovis* 23 % *C. ryanae* 9 % *C. andersoni* 68 %	Fayer et al. 2010

[a]See e.g. Santín and Trout (2008) for a review of older studies

In older calves and young stock, *C. bovis* and *C. ryanae* are the most commonly found species (Santín et al. 2008; Fayer et al. 2007; Muhid et al. 2011; Langkjaer et al. 2007; Silverlås et al. 2010b). *C. andersoni* is mainly found in young stock and adult cattle (Enemark et al. 2002; Wade et al. 2000; Fayer et al. 2007; Ralston et al. 2003). Older studies, based on microscopy alone, overestimated the *C. parvum* prevalence in weaned animals because the similarity in oocyst size makes it impossible to differentiate between *C. parvum* (~5.0 × 4.5 μm), *C. bovis* (~4.9 × 4.6 μm) and *C. ryanae* (~3.7 × 3.2 μm).

Molecular studies have revealed different genetic subtypes within the *C. parvum* and *C. hominis* species and DNA sequence analysis of the 60 kDa glycoprotein gene is commonly used to further characterize the isolates. A number of *C. parvum* GP60 subtype families, designated IIa-IIo, have been described. Of these, IIc and IIe are considered anthroponotic, whereas IIa and IId are commonly found in both humans and animals. The other subtype families are uncommon and their zoonotic potential has not been determined. In cattle, *C. parvum* of the IIa subtype family is especially common. In addition to IIa, IId and occasionally III (sometimes named IIj) subtypes are found (Table 4.3; Xiao 2010). Whether there is a difference in pathogenicity between subtypes is currently unknown, as most genotyping studies to date have focused on herds with a history of calf diarrhoea. It has been suggested that herd management strategies affect subtype distribution. Studies from areas with closed herd management (limited animal movements between herds) have shown a high number of subtypes in the calf population, but only one subtype in each herd (Brook et al. 2009; Mišic and Abe 2007; Soba and Logar 2008; Silverlås et al. 2013). It has also been shown that a unique GP60 subtype can persist over time in a closed dairy herd (Björkman and Mattsson 2006). In contrast, only a few subtypes have been identified in areas with more animal movements between herds, but several subtypes could be present in a herd (Brook et al. 2009; Peng et al. 2003; Trotz-Williams et al. 2006).

4.2.1.2 Water Buffalo

The information on the distribution of *Cryptosporidium* infection in water buffalo is rather fragmentary. Reported prevalences vary between 3 % and 38 % (Table 4.4). An association between prevalence of infection and age of the animals, with the highest prevalence in young calves, has been found (Helmy et al. 2013; Maurya et al. 2013; Nasir et al. 2009; Bhat et al. 2012; Díaz de Ramírez et al. 2012).

The first report on *Cryptosporidium* species identification in water buffalo was published in 2005, in which Gómez-Couso et al. (2005) used molecular tools to characterize *Cryptosporidium* oocysts from an asymptomatic neonatal calf in a dairy buffalo farm in Spain. Sequence analysis of a fragment of the oocyst wall protein (COWP) gene revealed that the isolate was closely related to the *Cryptosporidium* 'pig' genotype. A few years later, *C. parvum* was identified in water buffalo calves from Italy (Cacciò et al. 2007) and today *C. ryanae*, *C. bovis* and

Table 4.3 *Cryptosporidium parvum* GP60 subtypes in cattle in some recent studies[a]

Country	Number subtyped	IIa subtypes[b]	IId subtypes[b]	Reference
Australia	36	IIaA16G3R1 (1), IIaA17G4R1 (4), IIaA18G3R1 (29), IIaA20G3R1 (2)		Waldron et al. 2011
Australia	84	IIaA18G3R1 (57), IIaA19G3R1 (11), IIaA17G2R1 (7), IIaA19G2R1 (6), IIaA16G3R1 (2), IIaA20G3R1 (1)		Ng et al. 2012
Brazil	7	IIaA15G2R1 (7)		Meireles et al. 2011
China	67		IIdA19G1 (54)	Wang et al. 2011b
Czech Republic	1	IIaA16G1R1 (1)		Ondráčková et al. 2009
Czech Republic	131	IIaA16G1R1 (56), IIaA15G2R1 (48), IIaA22G2R1 (12), IIaA15G1R1 (12), IIA18G1R1 (3)		Kváč et al. 2011
England and Wales	13	IIaA17G1R1(10), IIaA15G2R1 (3)		Smith et al. 2010
Egypt	24	IIaA15G2R1 (1)	IIdA20G1 (23)	Amer et al. 2010
Egypt	71	IIaA15G1R1 (16)	IIdA20G1 (54), IIdA19G1 (1)	Helmy et al. 2013
Hungary	21	IIaA16G1R1 (15), IIaA17G1R1 (3), IIaA18G1R1 (1)	IIdA19G1(1), IIdA22G1 (1)	Plutzer and Karanis 2007
Iran	25	IIaA15G2R1 (22), IIaA16G3R1 (1)	IIdA15G1 (2)	Nazemalhosseini-Mojarad et al. 2011
Italy	62	IIaA15G2R1 (34), IIaA18G2R1 (10), IIaA17G2R1 (9), IIa A14 (5), IIaA13 (4)		Duranti et al. 2009
Malaysia	8	IIaA17G2R1 (2), IIaA18G3R1 (1)	IIdA15G1 (5)	Muhid et al. 2011
Romania	13	IIaA15G2R1 (8), IIaA16G1R1 (5)		Imre et al. 2011
Spain	27	IIaA15G2R1 (26), IIaA13G1R1 (1)		Díaz et al. 2010a
Sweden	13	IIaA15G1R1 (2), IIaA18G1R1 (2), IIaA21G1R1 (2), IIaA16G1R1 (1)	IIdA20G1 (2), IIdA23G1 (2) IIdA16G1 (1), IIdA22G1 (1)	Silverlås et al. 2010b

Sweden	171	IIaA16G1R1 (58), IIaA17G1R1c (17), IIaA21G1R1 (12), IIaA17G1R1 (6), IIaA18G1R1c (7), IIaA22G1R1 (7), IIaA20G1R1 (5), IIaA15G2R1 (4), IIaA13G1R1 (3), IIaA18G1R1 (3), IIaA14R1 (2), IIaA16G1R1b (2), IIaA18G1R1d (2), IIaA13G1R2 (1), IIaA14G1R1b (1), IIaA17R1 (1), IIaA21G1R1b (1), IIaA23G1R1 (1)	IIdA20G1e (14), IIdA16G1b (4), IIdA22G1 (4), IIdA24G1c (4), IIdA22G1c (3), IIdA23G1 (3), IIdA17G1d (2), IIdA19G1 (2), IIdA26G1b (2)	Silverlås et al. 2013

[a]See Xiao et al. (2010) for a summary of results of older studies
[b]Numbers in parentheses are number of samples with the subtype

Table 4.4 Location, age, and prevalence of *Cryptosporidium* spp. and *C. parvum* subtypes in water buffalo[a]

Location	Symptoms	Host age	Study design	Prevalence % positive (no. positive/no. examined)	Diagnostic technique	Molecular analyses gene/species or subtype[b]	Reference
Egypt	Diarrhoeic and non-diarrhoeic	<3 months	Cross-sectional	Individual 14 % (65/458) Herd 55	Modified Ziehl-Neelsen (mZN)		El-Khodery and Osman 2008
Egypt	Diarrhoeic and non-diarrhoeic	Not stated	Cross-sectional	22 % (16/71)	mZN and Sheather's flotation followed by mZN		Shoukry et al. 2009
Egypt	Not stated	1 week–4 months and adults	Cross-sectional	Calves: 10 % (17/179) Adults: 0 Herds: 71 % (5/7)	Sheather's flotation followed by mZN	18S rRNA/ *C. parvum* (7) *C. ryanae* (10) GP60/ IIdA20G1 (5) IIaA15G1R1 (2)	Amer et al. 2013
Egypt	Not stated for calves, diarrhoeic adults	Calves and adults	Cross-sectional	Overall: 20 %. (43/211) 1 day–3 months: 40 % (34/85) <3 months–1 year: 11 % (6/56) >1 year: 4 % (3/70)	Antibody based copro-antigen test	18S rRNA/ *C. parvum* (30) *C. bovis* (2) *C. ryanae* (3) Mixed infections (10) GP60/ IIdA20G1 (28) IIaA15G1R1 (4)	Helmy et al. 2013
India	Not stated	Not stated	Cross-sectional	17 % (10/60)	Kinyon acid fast stain		Dubey et al. 1992

India	Diarrhoeic and non-diarrhoeic	Not stated	Cross-sectional	25 % (76/305)	mZN and formol-ether sedimentation followed by mZN		Mohanty and Panda 2012
India	Diarrhoeic and non-diarrhoeic	<5 months	Cross-sectional	Overall prevalence: 38 % (62/162) 1-2 months: 49 % (18/37) 2–3 months: 34 % (13/38) 3–4 months: 20 % (76/305) 4–5 months: 6 % (1/17)	mZN		Bhat et al. 2012
India	Diarrhoeic and non-diarrhoeic	<3 months	Cross-sectional	24 % (64/264)	mZN or Sheather's flotation followed by mZN	18S rRNA/ *C. parvum* (16)	Maurya et al. 2013
Nepal	Not stated	2–7 months	Cross-sectional	37 % (30/81)	Water-ethyl sedimentation followed by mZN	18S rRNA/ *C. ryanae* (16) Unique variant	Feng et al. 2012
Pakistan	Diarrhoeic and non-diarrhoeic	1 day–>1 year	Cross sectional	Overall prevalence: 24 % (60/250) 1–30 days: 42 % (29/69) 1–3 months: 26 % (17/66) 4–8 months: 15 % (9/60) 9 months->1 year: 9 % (5/55)	Sheather's flotation followed by mZN		Nasir et al. 2009

(continued)

Table 4.4 (continued)

Location	Symptoms	Host age	Study design	Prevalence % positive (no. positive/no. examined)	Diagnostic technique	Molecular analyses gene/species or subtype[b]	Reference
Philippines	Not stated	1–12 days	Longitudinal	3 % (1/38)	Formalin ether sedimentation followed by Kinyon acid fast stain		Villanueva et al. 2010
Venezuela	Not stated	0–12 weeks	Longitudinal	88 % (22/25)	Formalin-ethylacetate sedimentation and sodium chloride flotation followed by carbol-fuchsin staining		Díaz de Ramírez et al. 2012
Italy	Asymptomatic	Calves	Cross-sectional	14 % (8/57)	ELISA, followed by formol-ether extraction and IFAT	18S rRNA/ *C. parvum*	Cacciò et al. 2007
Italy	Not stated	2–60 days/ 43 farms	Cross-sectional	20 % (35/177) Herds 35 % (15/43)	ELISA		Rinaldi et al. 2007a
Italy	Not stated	1–9 weeks/ 90 farms	Cross-sectional	15 % (51/347) Herd 24 % (22/90)	ELISA		Rinaldi et al. 2007b

[a]A few studies are not included in the table due to inaccessibility of the publications

C. ubiquitum have also been reported (Table 4.4). Most molecular investigations have been done in calves and thus it is not known if the species distribution differs between animals of different age. One of the few studies that include faecal samples from both calves and older animals was done in Egyptian smallholder herds (Helmy et al. 2013). *C. parvum* was most common in calves younger than 3 months but was detected in animals up to 2 years of age. In another study, also from Egypt but from another part of the country, *C. parvum* was only found in calves, whereas none of the sampled cows shed any *Cryptosporidium* oocysts (Amer et al. 2013). When the *C. parvum* isolates were subtyped by sequence analysis of the GP60 gene, subtype families IId and IIa were found, with a majority of IId in both studies. Subtype IId is the dominating subtype family also in cattle in Egypt (Amer et al. 2010; Helmy et al. 2013).

4.2.2 Association of Infection with Clinical Disease

4.2.2.1 Cattle

Cryptosporidium parvum and *C. andersoni* are the two species that have been associated with clinical disease in cattle. *C. parvum* infection is considered a major cause of diarrhoea in young calves (Blanchard 2012; Radostits et al. 2007). Calves are often already infected during the first week of life (Uga et al. 2000) and clinical cryptosporidiosis is mostly seen in calves up to 6 weeks of age. The most prominent finding is pasty to watery diarrhoea, sometimes accompanied by lethargy, inappetence, fever, dehydration and/or poor condition. The calf most often recovers spontaneously within 1–2 weeks, but there is a large variation between individuals in how they respond to, and recover from, infection. In some cases the infection may be fatal (Tzipori et al. 1983; Fayer et al. 1998). A decrease in growth rate may be seen in the weeks after the calves have recovered from the acute phase of the disease (Klein et al. 2008), but no long-term effects on growth and performance have been reported. The pathogenesis of bovine cryptosporidiosis is not fully understood but the clinical signs are attributed to both malabsorption and an increase in fluid secretion in the ileum and proximal portions of the large intestine. For references and a brief overview see O'Handley and Olson (2006). Cryptosporidiosis may be seen in individual calves, but frequently it soon develops among the calves into a herd problem. Concomitant infection with other pathogens, e.g. rotavirus, coronavirus and enteropathogenic *Escherichia coli* (*E. coli* F5+) can worsen the clinical signs and prolong the duration of illness (Blanchard 2012).

A number of studies have reported an association between *C. parvum* infection and diarrhoea in young calves. Many of these were published at the time when all *Cryptosporidium* oocysts of around 3–5 μm in diameter were considered to be *C. parvum*. More recent investigations, applying molecular methods to analyse faecal samples from diarrhoeic calves, corroborate these earlier findings. When samples from young calves with diarrhoea were analysed, *C. parvum* is found to be the dominant species (Quílez et al. 2008b; Imre et al. 2011; Karanis et al. 2010; Soba and Logar 2008; Plutzer and Karanis 2007). Interestingly, this dominance of *C. parvum* in diarrheic calves was also seen in a recent Swedish investigation of diarrheic calves (Silverlås et al. 2013), although *C. bovis* is the predominant species in randomly selected calves in Sweden.

Only one experimental trial has been performed with *C. bovis* (Fayer et al. 2005). Three calves under 1–8 weeks of age were orally inoculated with oocysts. This resulted in subclinical infection in 2 of 3 calves. Both animals had, however, previously been infected with *C. parvum* and cross-protective immunity could not be excluded. Calves with diarrhoea are significantly more likely to be infected with *C. parvum* than with *C. bovis* (Silverlås et al. 2013; Starkey et al. 2006; Kváč et al. 2011). Based on these findings, and based on the fact that *C. bovis* is not common in calves, but is a widespread subclinical infection in older animals in most countries, *C. bovis* is commonly considered to be apathogenic to cattle. However, the pathogenic potential deserves further attention as high numbers of *C. bovis* oocysts in samples from diarrhoeic calves have been reported, even in the absence of *C. parvum* or other diarrhoeal agents (Silverlås et al. 2010a, b, 2013).

Cryptosporidium ryanae was first described as a separate species by Fayer et al. (2008), and until then it was known as *Cryptosporidium* deer-like genotype. An experimental trial was performed in two colostrum-deprived calves 17–18 days old. Both calves started excreting oocysts 11 days after inoculation, but neither of them showed any clinical signs (Fayer et al. 2008). There are several reports of the distribution of *Cryptosporidium* deer-like genotype and *C. ryanae*. Most studies found a predominance of the parasite in older calves and young stock. So far no association with clinical disease has been reported.

In contrast to the other species, *C. andersoni* infects the abomasum. It does not cause diarrhoea, but *C. andersoni* infections have been associated with maldigestion. The infection may cause moderate to severe weight gain impairment in young stock and reduced milk production in cows (Anderson 1998; Esteban and Anderson 1995; Lindsay et al. 2000).

A major obstacle from a disease control perspective is the lack of effective means to control *Cryptosporidium* infection and decrease the level of contamination of the environment with oocysts. Preventive hygiene measures and good management are currently the most important tools to control cryptosporidiosis. Reducing the number of oocysts ingested by neonatal calves may reduce the severity of infection and allow immunity to develop. A common recommendation is to ensure good hygiene in calf facilities and ascertain that all newborn calves ingest an adequate amount of colostrum during their first 24 h of life. Sick calves should be housed in a clean, warm, and dry environment and isolated to prevent spreading of the infection to other calves. Acutely infected animals may need supportive care with fluid and electrolytes, and milk should be given in small quantities several times daily to optimise digestion and minimise weight loss.

Over the years, several substances have been tested for potential anti-cryptosporidial effects with limited success (Santín and Trout 2008). Halofuginone lactate has shown some beneficial effects such as milder clinical signs and reduced oocyst output when used as prophylactic treatment (De Waele et al. 2010; Silverlås et al. 2009a). This drug is approved in Europe to treat calf cryptosporidiosis. However, the safety margin is narrow and the substance is toxic at only twice the therapeutic dose, so careful dosage is necessary. Halofuginone lactate treatment should only be considered in herds with severe diarrhoeal problems strongly associated with *C. parvum*. When treatment is used, it should always be in conjunction with applying measures to reduce environmental contamination and risk of infection.

A recent study investigated if an antibody-biocide fusion consisting of a monoclonal antibody "armed" with membrane-disruptive peptides (biocides) could be used for treatment of cryptosporidiosis in calves (Imboden et al. 2012). Calves 36–48 h of age were challenged once with *C. parvum* oocysts and were simultaneously administered the antibody-biocide fusion mixed with milk replacer. The antibody-biocide fusion treatment was repeated 5–8 times. Control calves were given milk replacer with placebo. Calves receiving the antibody-biocide fusion had a significantly higher health score and shed fewer oocysts than control calves. These results suggest that this concept might be effective in cattle, but further testing is necessary (Imboden et al. 2012).

Vaccination is successfully used to control many infectious diseases in livestock. However, it takes weeks for a protective immune response to develop after a vaccine has been administered, and as calves may be exposed to *Cryptosporidium* oocysts immediately after birth, vaccination of newborn calves is unlikely to be successful in preventing cryptosporidiosis. Thus it has been suggested that the most feasible approach is likely to involve passive immunisation (Innes et al. 2011). Dams are immunised in late gestation and their colostrum is fed to the calves. A recent study investigated antibody responses in calves fed colostrum from heifers vaccinated with a recombinant *C. parvum* oocyst surface protein (rCP15/60). The calves had measurable quantities of the specific antibody in their serum. However, as the calves were not subsequently challenged with oocysts it remains to be seen whether this immunisation scheme can also prevent symptomatic infection and eliminate oocyst shedding (Burton et al. 2011).

4.2.2.2 Water Buffalo

An association between *Cryptosporidium* oocyst shedding and diarrhoea in buffalo calves has been reported from investigations performed in Egypt, India and Venezuela (Mohanty and Panda 2012; Bhat et al. 2012; Maurya et al. 2013; El-Khodery and Osman 2008; Díaz de Ramírez et al. 2012) suggesting the *Cryptosporidium* infection is part of the calf diarrhoea syndrome in water buffalo, as it is in cattle. Species identification was only performed in one of these studies, and *C. parvum* was the only species that was found (Maurya et al. 2013).

4.2.3 *Infection Dynamics: Oocyst Excretion and Transmission*

4.2.3.1 Cattle

Calves begin shedding *C. parvum* oocysts 2–6 days after infection and shedding continues for 1–13 days (Fayer et al. 1998; Tzipori et al. 1983). During the first 2 weeks an infected calf can shed millions of oocysts (Fayer et al. 1998; Uga et al. 2000) resulting in heavy environmental contamination, and efficient dissemination of the parasite within the herd and to the environment. In faecal samples obtained from symptomatic calves naturally infected with *C. parvum* 10^6–10^8 oocysts per gram faeces (OPG) are often seen (Silverlås et al. 2013). In herds with established *C. parvum* infection, most calves are excreting oocysts between 2 and 4 weeks of age (O'Handley et al. 1999; Santín et al. 2008; Uga et al. 2000). When Santín et al. (2008) repeatedly sampled the same 30 calves in a dairy herd from birth to 2 years of age they found *C. parvum* oocysts in faeces of all individuals before they were 3 weeks old, i.e. a cumulative prevalence of 100 %. *C. parvum* oocysts were also found in samples collected from a calf at 16 weeks of age and from another at 6 months of age, indicating that oocysts can be shed intermittently over a long period after the initial infection. Alternatively, these late-shedding individuals might not have developed a fully protective immunity after the first infection, and rather than this being a sign of prolonged infection, they had acquired new infections. In this study, molecular analyses indicated the same sub-genotype at the GP60 locus. However, this does not necessarily indicate prolonged infection, as re-infection with the same genotype in the environment may occur if the immunity is not protective. It has been suggested that an increase in *C. parvum* oocyst shedding may occur in adult cows around calving (so called periparturient rise), but to date there have been few reports to support this. In a recent study, however, dams in a suckler beef herd were found to shed low levels of *C. parvum* oocysts around the time of calving (De Waele et al. 2012).

Only one experimental infection for each of *C. bovis* and *C. ryanae* has been reported so far. Regarding *C. bovis*, one calf shed oocysts from 10 to 28 days after infection and the other only for 1 day (day 12) (Fayer et al. 2005). For *C. ryanae*, oocyst shedding started 11 days after inoculation. Both infected calves excreted oocysts during 15 and 17 consecutive days, respectively (Fayer et al. 2008). Shorter prepatent periods have been seen for both *C. bovis* and *C. ryanae* in natural infections (Silverlås et al. 2010b; Silverlås and Blanco-Penedo 2013). No oocyst excretion rate values were determined from the experimental infections, but in naturally infected calves 300 to 8×10^6 OPG and 100–835,000 OPG have been reported for *C. bovis* and *C. ryanae*, respectively (Silverlås and Blanco-Penedo 2013).

Young stock and adults may also be infected by the larger *C. andersoni* (oocyst size ~7.4×5.5 μm) and may shed oocysts intermittently for many years (Olson et al. 2004; Ralston et al. 2003). A periparturient rise in *C. andersoni* oocyst shedding, seen both as increase in prevalence and in number of oocysts in faeces, has been reported (Ralston et al. 2003).

Several studies have shown that age is associated with *Cryptosporidium* infection and that young calves have the highest risk of being infected (Maddox-Hyttel et al. 2006; Santín et al. 2004; Sturdee et al. 2003; Fayer et al. 2007). This is also the age group that is most often infected with *C. parvum* and suffers from clinical cryptosporidiosis. Thus, from clinical and zoonotic perspective, knowledge on the epidemiology of cryptosporidiosis in young calves is highly valuable. When potential risk factors for *Cryptosporidium* infection in pre-weaned calves have been explored, the results differ between studies. One factor that recurs in several studies is the type of flooring in the calf housing area. In Spain, Castro-Hermida et al. (2002) found that straw on the floor or earth floors in the calf pens increased the risk for infection compared with cement flooring, and in Malaysia calves kept in pens with slatted floors and sand floors had an increased risk compared with those in pens with cement floors (Muhid et al. 2011). A protective effect of cement floors was also reported from the USA (Trotz-Williams et al. 2008). It was suggested that the reason for this protective effect of cement floors is that they facilitate thorough cleaning. This assumption corroborates the finding that a low frequency of cleaning of the calf pens increased the risk for infection (Castro-Hermida et al. 2002). It is also consistent with the finding that the use of an empty period in the calf pen between introductions of calves was associated with a lower risk for infection in Danish dairy herds (Maddox-Hyttel et al. 2006). When cows as a cause of infection were investigated, a higher risk of infection was identified in calves that were housed separately from their dams (Duranti et al. 2009), and a lower risk of infection in dairy calves kept with the cow for more than 6 h after birth (Silverlås et al. 2009b).

In one of the few reports to investigate risk factors for infection with different *Cryptosporidium* species in pre-weaned dairy calves to date (Szonyi et al. 2012), risk of infection with *C. parvum* differed to some extent from that of *C. bovis*. Both *C. parvum* and *C. bovis* were more common in the younger calves, but herd size and hay bedding were associated with an increased risk for *C. parvum* infection, whereas Jersey breed was a risk factor for *C. bovis* infection.

4.2.3.2 Water Buffalo

Experimental *Cryptosporidium* infections in water buffalo have not been reported. However, oocyst shedding dynamics were investigated in naturally infected buffalo calves in a farm located in a tropical dry forest area in Venezuela. Twenty-five calves were sampled from birth to 12 weeks of age. Oocysts were detected from day 5 and 72 % of the calves shed oocysts before they were 30 days (Díaz de Ramírez et al. 2012).

Regarding risk factors for infection, there are some reports of seasonal variations in prevalence (Bhat et al. 2012; El-Khodery and Osman 2008; Maurya et al. 2013; Mohanty and Panda 2012), and (El-Khodery and Osman 2008) identified type of flooring, frequency of cleaning and water source as risk factors for infection in small-scale herds in Egypt.

4.2.4 Zoonotic Transmission

4.2.4.1 Cattle

There are numerous reports of cryptosporidiosis outbreaks in humans after contact with infected calves. These have often involved veterinary students and students at farm schools (see, e.g., Gait et al. 2008; Grinberg et al. 2011; Pohjola et al. 1986; Robertson et al. 2006; Kiang et al. 2006), but also young children have fallen ill after visiting petting zoos or open farms (Gormley et al. 2011; Smith et al. 2004). Contact with cattle has been identified as a risk factor for disease also in case-control studies of sporadic human cryptosporidiosis (Hunter et al. 2004; Robertson et al. 2002; Roy et al. 2004). Altogether, there is plenty of evidence to conclude that *Cryptosporidium* can be transmitted from calves to humans by direct contact or by contaminated equipment. The risk for zoonotic transmission is likely to be highest in herds with *Cryptosporidium* associated calf diarrhoeal problems, where oocyst contamination in the barn can reach high levels and where contact with naïve individuals is most likely to occur. Key measures to prevent visitors becoming infected are to ensure good hygiene in the visitor area, providing suitable hand-washing facilities and ensure that they are used when workers and visitors leave the premises. *C. bovis* infections have recently been detected in a few persons living or working on cattle farms (Khan et al. 2010; Ng et al. 2012). It is not known if these were active infections and the implication of these findings is thus unclear.

As molecular typing methods become more accessible, epidemiological studies can investigate *C. parvum* GP60 subtype distribution in cattle and human populations in different regions. The reports so far indicate that in many areas the subtypes that are most common in cattle are those most often found in humans. For example, *C. parvum* IIaA15G2R1 was the predominant subtype in both bovine and human infections in Slovenia and Portugal (Soba and Logar 2008; Alves et al. 2006). In New South Wales, Australia *C. parvum* IIaA18G3R1 dominated in both calves and people living on cattle farms (Ng et al. 2012), whereas IIaA16G2R1 was the predominant genotype identified in beef cattle and humans in Prince Edward Island, Canada (Budu-Amoako et al. 2012c). Further information is provided in the review by Xiao (2010). That the same subtypes are found in cattle and humans might be taken as an indication of zoonotic transmission. However, it is important to note that even when zoonotic *C. parvum* subtypes are identified in humans, cattle are not necessarily the source of the infection. These zoonotic subtypes can circulate and propagate in the human population in addition to the anthroponotic subtypes. The occasional finding of *C. hominis* (Smith et al. 2005; Chen and Huang 2012) in cattle highlights the fact that cryptosporidiosis may be transmitted not only from cattle to humans, but also from humans to cattle.

Food-related cryptosporidiosis outbreaks have sometimes been associated with cattle. Foodborne transmission was implicated in cases of children who had drunk unpasteurized milk (Harper et al. 2002) or cider, made from apples collected in an orchard where calves from an infected herd had grazed (Millard et al. 1994).

Other outbreaks in which cattle were suspected as the source involved vegetables that had been sprayed with water that could have been contaminated with cattle faeces. Often there was only circumstantial evidence that cattle were the source of contamination, and it was not possible to exclude other potential sources (see e.g. CDC 2011; Robertson and Chalmers 2013).

Outbreaks of cryptosporidiosis associated with drinking water have often been attributed to contamination of water catchments by cattle manure. The evidence implicating cattle has sometimes been substantial (Bridgman et al. 1995; Smith et al. 1989), but for others the evidence was not conclusive. Grazing cattle or slaughterhouse effluent contaminating Lake Michigan were mentioned as two possible sources of *Cryptosporidium* oocysts in the large outbreak in Milwaukee, Wisconsin in 1993 (Mac Kenzie et al 1994), but retrospective analysis of clinical isolates revealed that it was caused by the anthroponotic species *C. hominis* (Sulaiman et al. 1998). This was also the case in the most recent outbreaks in UK (compiled by Chalmers 2012) and a large drinking waterborne outbreak in Sweden in 2010 (Anonymous 2011). Given that pre-weaned calves are the most likely age group to shed *C. parvum* oocysts, any measure to prevent waterborne zoonotic transmission should be directed towards this age group. Protective measures could be to prevent young ruminants from accessing water catchments, and compost or spread calf manure on fields where runoff cannot occur. The manure of older ruminants is generally not a zoonotic concern with respect to *Cryptosporidium*.

4.2.4.2 Water Buffalo

The seemingly common occurrence of *C. parvum* in buffalo calves highlights the potential role of water buffalo in zoonotic transmission. Thus the same precautions to prevent transmission of the parasite to humans, by direct contact or through food or water, are also applicable to water buffalo.

4.3 *Cryptosporidium* spp. in Small Ruminants

Sheep (*Ovis aries*) and goats (*Capra aegagrus hircus*) are important in the global agricultural economy – producing meat, milk and wool – both in developing countries such as India and Iran, and industrialised countries such as Australia and The United Kingdom (de Graaf et al. 1999; Noordeen et al. 2000; Robertson 2009). In 2010, the world stocks were approximately one billion sheep and 910 million goats (FAOSTAT 2013a). Asia has the largest populations of both species, with 42 % and 60 % of the total world populations, followed by Africa (FAOSTAT 2013a). According to a FAO report, over 90 % of the goat population can be found in developing countries (FAO 2012). Sheep and goats tend to be managed differently to cattle, with flocks grazing large enclosures rather than being kept indoors. There have been fewer studies on *Cryptosporidium* infection in sheep than in cattle, and even fewer studies have been performed on goats. Nevertheless, it is known that these protozoans are economically important parasites in both ruminant species (Noordeen et al. 2000; Robertson 2009). Infection and disease was first described in 1974 for sheep (Barker and Carbonell 1974) and in 1981 for goats (Mason et al. 1981). Younger animals are more susceptible to infection than older ones, reflected in high shedding rates and diarrhoeal prevalences in lambs and kids up to 1 month of age, whereas infection in older animals is usually subclinical with lower shedding rates (Vieira et al. 1997).

4.3.1 Occurrence (Prevalence)

As for other animals, ovine and caprine *Cryptosporidium* infection can be found throughout the world. The prevalence varies widely between studies, from 0 % to 77 % in sheep and from 0 % to 100 % in goats. All age groups are susceptible, but infection is more common in lambs and kids than in older animals (Tables 4.5 and 4.6). Study design factors other than age of sampled animals, such as whether only diarrhoeal animals were sampled or not, if a point prevalence study or a longitudinal study was performed and the diagnostic method(s) used, also can affect prevalence data. The effect of using different diagnostic methods is evident in, for example, Giadinis et al. (2012; see also Tables 4.5 and 4.6) and Ryan et al. (2005), where microscopy resulted in lower prevalences than detected by ELISA and PCR, respectively. Prevalence and species distribution for studies conducted on sheep dating back to 2007 are summarised in Table 4.5. Specific data for studies on sheep published before 2007 can be found in "*Cryptosporidium* and Cryptosporidiosis" (Santín and Trout 2008; Tables 18.9 and 18.10). Prevalence rates and species distribution for all identified surveys of goats are summarised in Table 4.6.

Several species and genotypes have been identified in sheep, and the species distribution varies between studies and with age of the animals. *Cryptosporidium parvum*, *C. ubiquitum* (previously *Cryptosporidium* cervine/cervid genotype) and *C. xiaoi* (previously *C. bovis*-like genotype) are the most common species. Sporadic infection with *C. hominis*, *C. suis*, *C. andersoni*, *C. fayeri* (previously marsupial genotype I), *C. scrofarum* (previously pig genotype II), sheep genotype I, and unknown/novel genotypes have been identified (Chalmers et al. 2005; Giles et al. 2009; Karanis et al. 2007; Ryan et al. 2005, 2008; Sweeny et al. 2011b; Wang et al. 2010c). Species distribution differs between studies and between age groups within studies. For instance, *C. parvum* is commonly found in lambs in Italy, Romania, Spain and the UK (Díaz et al. 2010a; Imre et al. 2013; Mueller-Doblies et al. 2008; Paoletti et al. 2009). In other studies, *C. ubiquitum* or *C. xiaoi* is the most common species (Fiuza et al. 2011a; Geurden et al. 2008; Robertson et al. 2010; Sweeny et al. 2011b). For example, Wang et al. (2010c) identified *C. ubiquitum* in 90 % of all analysed samples, and the species dominated in all age groups, whereas Sweeny et al. (2011b) found *C. xiaoi* to be the most common

Table 4.5 Studies on prevalence *Cryptosporidium* infections in sheep published in 2007–2012

Country	Number of animals	Age	Animal prevalence %	Positive herds	*Cryptosporidium* spp. (% of all determined)	Reference
Australia	477	≤8 w	9.3–56.3	5/5	*C. bovis* 36.5	(Yang et al. 2009)
			Overall 24.5		*C. parvum* 46.1	
					C. ubiquitum 8.7	
					Mixed infections 8.7	
Australia[a]	235	2 w–8 m	Age group point prevalences: 18.5–42.6	2/2	*C. andersoni* 1.0[d]	(Sweeny et al. 2011b)
			Cumulative prevalences:		*C. parvum* 5.9	
			Herd A: 81.3		*C. ubiquitum* 9.8	
			Herd B: 71.4		*C. xiaoi* 73.7	
					Sheep genotype I 1.6	
					Mixed infections 8.0	
Australia[a]	96	Ewes, 4 m prp	Herd A, B: 6.3	2/2	*C. xiaoi* 100[d]	(Sweeny et al. 2011b)
		Ewes, 2 w prp	Herd A, B: 8.3			
Australia[a]	200	2–3 m	Herd I: 33.6, II: 31.9	2/2	*C. parvum* 21.0[d]	(Sweeny et al. 2012)
		4–5 m	Herd I: 28.1, II: 23.6		*C. ubiquitum* 44.5	
			Cumulative prevalences:		*C. xiaoi* 31.9	
			Herd I: 40.6		Mixed infections 2.5	
			Herd II: 31.9			
Belgium	137	≤10 w	13.1	4/10	*C. parvum* 10.0	(Geurden et al. 2008)
					C. ubiquitum 90.0	
Brazil	90	2–6 m	2.2		*C. ubiquitum* 100	(Fiuza et al. 2011a)
	35	>12 m	0			
			Overall 1.6			
Brazil	1	<3 m	0	0/9	–	(Sevá et al. 2010)
	10	>3 m	0			
China	378	preweaned	10.8	5/5	*C. andersoni* 4.9	(Wang et al. 2010c)
	585	postweaned	4.3		*C. ubiquitum* 90.2	
	580	ewe, prp	2.1		*C. xiaoi* 4.9	
	158	ewe, pp	2.5			
China	213	6–9 m	14.6	?/8	*C. ubiquitum* 100[e]	(Shen et al. 2011)

(continued)

Table 4.5 (continued)

Country	Number of animals	Age	Animal prevalence %	Positive herds	*Cryptosporidium* spp. (% of all determined)	Reference
Cyprus[c]	39	4–15 d	61.5 (mZN) 76.9 (E)	12/15	ND	(Giadinis et al. 2012)
India	55	<3 m	1.8	–	ND	(Maurya et al. 2013)
Italy[c]	21	–	100	–	*C. parvum* 100	(Drumo et al. 2012)
Italy	149	2 w–3 m	17.4	5/6	*C. parvum* 100	(Paoletti et al. 2009)
Norway[a]	567	5–6 w	14.6	6/6	*C. ubiquitum* 83.3	(Robertson et al. 2010)
	528	6–10 w	24.1		*C. xiaoi* 16.7	
Romania[c]	58	1–7 d	13.8	5/7	*C. parvum* 83.3	(Imre et al. 2013)
	68	8–14 d	16.2		*C. ubiquitum* 8.3	
	49	15–21 d	10.2		*C. xiaoi* 8.3	
Spain[c]	127	<21 d	30.7	17/28	*C. parvum* 63.6	(Díaz et al. 2010a)
					C. ubiquitum 36.4	
Spain	446	Adult	5.3	13/38	ND	(Castro-Hermida et al. 2007)
St Kilda[a, b]		4–$\geq$40 m	9.0–28.6	–	ND	(Craig et al. 2007)
Tunisia	30	<3 m	16.7	3/3	*C. xiaoi*[f]	(Soltane et al. 2007)
	59	>1 y	8.5			
Turkey[c]	151	$\leq$7 d	44.4	26/34	ND	(Sari et al. 2009)
	104	8–14 d	37.5			
	95	15–21 d	40.0			
	50	22 d$\leq$1 m	22.0			
			Overall 38.8			
United Kingdom	260	Mixed	39.6	11/17	*C. parvum* 88.9 *C. xiaoi*[g] 11.1	(Mueller-Doblies et al. 2008; Smith et al. 2010)

d days, *w* weeks, *m* months, *ND* not done, *Prp* prepartus, *pp* post partum, *mZN* modified Ziehl-Neelsen, *E* ELISA

[a]Longitudinal study

[b]Feral population

[c]Only diarrhoeal samples collected

[d]Species distribution for all positive samples independent of age

[e]Only four isolates successfully amplified and sequenced

[f]Reported as similar to *C. bovis*, 99.5–99.7 % identical to *C. xiaoi* isolates in GenBank

[g]Reported as *C. bovis*, 100 % identical to *C. xiaoi* isolates in GenBank

Table 4.6 Published studies on prevalence of *Cryptosporidium* infections in goats

Country	Number of animals	Age	Animal prevalence %	Positive herds	*Cryptosporidium* spp. (% of all determined)	Reference
Belgium	148	≤10 w	9.5	6/10	*C. parvum* (100)	(Geurden et al. 2008)
Brazil	49	<12 m	10.2	2/6	ND	(Bomfim et al. 2005)
	56	>12 m	0			
			Overall 4.8			
China	42	–	35.7	–	*C. xiaoi*[c]	(Karanis et al. 2007)
					Novel genotype[d]	
Cyprus[b]	75	4–15 d	64(mZN) 86.7 (E)	25/28	ND	(Giadinis et al. 2012)
France[a]	40	0-weaning	Age related point prevalences: 0–100	1/1	*C. xiaoi* (94.7)	(Rieux et al. 2013)
			Cumulative 77.5		*C. parvum* (5.3)	
France	200	5–12 m	2.5	–	ND	(Castro-Hermida et al. 2005)
France	879	5–30 d	16.2	32/60 (53.3)	ND	(Delafosse et al. 2006)
India	116	<3 m	3.5	–	*C. parvum* (100)	(Maurya et al. 2013)
Italy[b]	21	–	100	–	*C. parvum* (100)	(Drumo et al. 2012)
Spain[b]	5	<21 d	40	1/1	*C. xiaoi*	(Díaz et al. 2010b)
Spain	116	Adult	7.7	6/20 (30)	ND	(Castro-Hermida et al. 2007)
Spain	134	≤15 d	10.4	4/4	ND	(Sanz Ceballos et al. 2009)
	144	>15 d–2 m	13.4			
	304	>2 m	25.2			
Sri Lanka	558	≤6 m	33.0	23/24	ND	(Noordeen et al. 2000)
	133	6–12 m	30.8			
	329	>12 m	20.1			
			Overall 28.5			
Sri Lanka	72	1–12 m	0–75[b]	1/1	ND	(Noordeen et al. 2001)
Tunisia	184	1–7 y	0	0/5	–	(Soltane et al. 2007)
UK	15	Adult	20	1/4	Negative	(Smith et al. 2010)

d days, *w* weeks, *m* months, *ND* not done, *Prp* prepartus, *pp* post partum, *mZN* modified Ziehl-Neelsen, *E* ELISA
[a]Longitudinal study
[b]Only diarrhoeal samples collected
[c]Only two samples successfully amplified
[d]Clustered with *C. bovis*, *C. ryanae* (previously *Cryptosporidium* deerlike genotype) and *Cryptosporidium* deer genotype

species in lambs (73.1 %) and the only species in ewes (Table 4.5). *Cryptosporidium bovis* has also been reported in sheep (Mueller-Doblies et al. 2008; Soltane et al. 2007; Wang et al. 2010c). Whether *C. bovis* has actually been identified in sheep, or if it is the closely related species *C. xiaoi* is uncertain. For instance, Soltane et al. (2007) reported isolates similar to *C. bovis*, and Mueller-Doblies et al. (2008) reported *C. bovis*, but a BLAST search of the GenBank accession numbers identified these isolates as *C. xiaoi*. For the isolates reported as *C. bovis* by Yang et al. (2009), no GenBank accession numbers are available, so the true identity of those isolates is uncertain. Similarly, no GenBank records are available from the study of Ryan et al. (2005) reporting the "New bovine B genotype" in sheep. Since *C. bovis* was first identified as the bovine B genotype, this could actually be the "*C. bovis*-like genotype", i.e. *C. xiaoi*.

Cryptosporidium andersoni has been identified in a few naturally infected adult sheep (Wang et al. 2010c), but experimental infection in 4-month-old lambs failed (Kváč et al. 2004).

A couple of apparently related surveys from Mexico have been published, but because of lack of clarity in the data, they will not be reviewed in this text. Two studies from Brazil (Sevá et al. 2010) and Mongolia (Burenbaatar et al. 2008) failed to identify *Cryptosporidium* in any of the collected samples, but the number of sampled sheep was small – 11 and 5 animals, respectively.

Because of the small number of studies and isolates analysed, it is hard to draw any conclusion about the species distribution in goats. *Cryptosporidium parvum*, *C. xiaoi* and a novel genotype have been identified in naturally infected goats (Table 4.5). In addition, one report of natural infection with *C. hominis* is also available (Giles et al. 2009). The identification of *C. xiaoi* in a number of samples is in contrast with a failed attempt to infect 36-week-old goats to determine the host range of *C. xiaoi* (Fayer and Santín 2009). Because there is only scant information about cryptosporidiosis in goats, we do not know if an age-related resistance or immunity from a previous *Cryptosporidium* infection could have affected this experiment, as natural infections indicate that *C. xiaoi* is infectious to goats.

Two studies from Mongolia (Burenbaatar et al. 2008) and the United Kingdom (Smith et al. 2010) failed to identify *Cryptosporidium* in any of the collected samples, but the number of sampled goats was small – 16 and 15 animals, respectively.

4.3.2 Association of Infection with Clinical Disease

Cryptosporidiosis has been associated with high morbidity and mortality rates in both lambs and goat kids (Cacciò et al. 2013; Chartier et al. 1995; de Graaf et al. 1999; Giadinis et al. 2007, 2012; Johnson et al. 1999; Munoz et al. 1996; Paraud et al. 2011; Vieira et al. 1997). High mortality has been described both from natural infection and from experimental studies, where infection doses are generally high (Chartier et al. 1995; Giadinis et al. 2007; Paraud et al. 2010). In fact, it has been stated to be one of the most important pathogens associated with diarrhoeal disease and mortality in neonatal lambs and kids (Quílez et al. 2008a).

Anorexia and apathy/depression are common symptoms, accompanied by abdominal pain and pasty to watery, yellow and foul-smelling diarrhoea (de Graaf et al. 1999; Snodgrass et al. 1984). Diarrhoea can last from a few days up to 2 weeks (de Graaf et al. 1999). Faecal consistency is correlated with oocyst excretion (de Graaf et al. 1999; Paraud et al. 2010, 2011), and a longer duration of diarrhoea is potentially associated with infection early in life (Paraud et al. 2010). Body condition score and growth are affected (de Graaf et al. 1999), probably due to both anorexia and the intestinal damage, that can reduce nutrient uptake for weeks (de Graaf et al. 1999; Klein et al. 2008).

Infection in animals older than 1 month is usually subclinical, and even younger animals can be subclinically infected. However, the infection can still affect production, with reduced body condition score (Sweeny et al. 2011a, 2012), reduced growth rate, and reduced carcass weight and dressing percentage at slaughter (Sweeny et al. 2011a).

As discussed above for cattle, before molecular methods were developed *C. parvum* was the only species considered to infect and cause disease in sheep and goats (Chartier et al. 1995; de Graaf et al. 1999; Munoz et al. 1996). *Cryptosporidium parvum* infection has since been associated with diarrhoea in studies using molecular methods (Caccio et al. 2013; Díaz et al. 2010a; Drumo et al. 2012; Imre et al. 2013; Mueller-Doblies et al. 2008). However, *C. xiaoi* has also been associated with mild to severe diarrhoea and mortality (Díaz et al. 2010b; Navarro-i-Martinez et al. 2007; Rieux et al. 2013), and *C. ubiquitum* too has been found in a few diarrhoeal samples from lambs (Díaz et al. 2010a), indicating that *C. parvum* is not the only pathogenic species in small ruminants.

4.3.3 Infection Dynamics: Oocyst Excretion and Transmission

The prepatent period is 3–4 days in goat kids (Paraud et al. 2010) and 2–7 days in lambs (de Graaf et al. 1999). The patent period can last for at least 13 days (Paraud et al. 2010). Shedding peaks a few days to a week into the patent period, and maximum shedding can be as high as 2×10^7 OPG (Rieux et al. 2013). The length of the patent period and shedding intensity are determined by age, immune status and infection dose (de Graaf et al. 1999). A natural age-related resistance to infection seems to be present. In one study, the prepatent period increased and intensity of shedding decreased in lambs with increasing age at infection (Ortega-Mora and Wright 1994). In another study, one naturally infected group of goat kids started shedding at 17 days of age and excretion peaked at a mean of 23×10^4 OPG 5–11 days later (Rieux et al. 2013), whereas another group of animals studied by the same authors started shedding at the age of 10 days, with a mean peak of 3×10^6

OPG 0–4 days later, indicating higher virulence, infection pressure or an age-related higher sensitivity in the latter group.

Factors such as hygienic conditions, milking practices, herd size (population density), season, climatic zone (within a country), and lambing/kidding season are factors that have been associated with prevalence of infection, prevalence of clinical cryptosporidiosis and intensity of oocyst shedding (Alonso-Fresan et al. 2009; Bomfim et al. 2005; Craig et al. 2007; Delafosse et al. 2006; Giadinis et al. 2012; Maurya et al. 2013; Noordeen et al. 2000, 2001). However, factors associated with infection and shedding intensity are also impacted by different management systems and climatic conditions; results from small farms in, for example, India cannot be extrapolated to large-scale farming in, for example, the United Kingdom.

4.3.4 Zoonotic Transmission

The significance of sheep and goats as reservoirs for zoonotic cryptosporidiosis is unclear (Robertson 2009). The first case of suspected zoonotic transmission from sheep was described in 1989 (Casemore 1989), but at that time diagnosis was based solely on microscopy and thus zoonotic transmission cannot be confirmed. However, since the introduction of molecular tools in diagnostics, a number of cases and outbreaks with suspected or confirmed zoonotic transfer from sheep have been described (Cacciò et al. 2013; Gormley et al. 2011). In the UK, where sheep farming is an important industry, a seasonal pattern with spring and autumn peaks of human cryptosporidiosis cases is observed, with the spring peak concurring with the lambing season (Anonymous 2002; Gormley et al. 2011; McLauchlin et al. 2000; Nichols et al. 2006). Lambs in petting zoos seem to be a common infection source (Chalmers et al. 2005; Elwin et al. 2001; Gormley et al. 2011; Pritchard et al. 2007). The incidence of human cryptosporidiosis, especially due to *C. parvum*, dropped significantly during the foot-and-mouth outbreak in the spring and summer 2001 (Hunter et al. 2003) when >6 million livestock animals (~4.9 million sheep; Anonymous 2001) were slaughtered and there were restrictions in animal movements and farm visits, providing further evidence of the importance of zoonotic transmission in this region.

In 2012, an outbreak occurred in Norwegian schoolchildren visiting a farm raising several animal species. *Cryptosporidium parvum* oocysts of an identical and unusual GP60 subtype were identified in faecal samples from six human patients, two lambs and one goat kid. Another human outbreak with the same *C. parvum* subtype had occurred at the same farm 3 years previously, but at that time very few oocysts were detected in animal faecal samples and molecular analyses were not conducted (Lange et al. submitted). Cacciò et al. (2013) described a case where a farmer's son fell ill with cryptosporidiosis, being infected with the same and unusual *C. parvum* subtype that caused high morbidity and mortality in the farm's lambs (Table 4.7).

Studies on *C. parvum* GP60 subtypes have been performed with sheep and goat isolates, and all isolates were found to belong to the IIa and IId families (Table 4.7).

Table 4.7 *Cryptosporidium parvum* GP60 subtypes identified in sheep and goats

Host Country	GP60 subtype	Number of samples	Reference
Sheep			
Australia[a]	IIa	1	(O'Brien et al. 2008)
Italy	IIaA20G2R1	3	(Cacciò et al. 2013)
	IIaA15G2R1	1	
Belgium	IIaA15G2R1	1	(Geurden et al. 2008)
Norway	IIaA19G1R1	2	(Lange et al. submitted)
Romania	IIaA17G1R1	2	(Imre et al. 2013)
	IIaA16G1R1	1	
	IIdA20G1	2	
	IIdA24G1	1	
	IIdA22G2R1	1	
Spain	IIaA15G2R1	3	(Díaz et al. 2010a)
	IIaA16G3R1	7	
Spain	IIaA15G2R1	2	(Quílez et al. 2008a)
	IIaA18G3R1	1	
	IIdA14G1	2	
	IIdA15G1	3	
	IIdA17G1a	44	
	IIdA17G1b	26	
	IIdA18G1	15	
	IIdA19G1	33	
	IIdA21G1	1	
	IIdA22G1	2	
	IIdA24G1	2	
United Kingdom	IIaA15G2R1	3	(Chalmers et al. 2005)
United Kingdom	IIaA15G2R1	1	(Smith et al. 2010)
	IIaA17G1R1	9	
	IIaA17G2R1	1	
Goats			
Belgium	IIaA15G2R1	3	(Geurden et al. 2008)
	IIdA22G1	8	
Norway	IIaA19G1R1	1	(Lange et al. submitted)
Spain	IIdA17G1a	8	(Quílez et al. 2008a)
	IIdA19G1	4	
	IIdA25G1	2	
	IIdA26G1	3	
Italy[a]	IIa	1	(O'Brien et al. 2008)

[a]IIaA12G2R1 or IIaA15G2R1 or IIaA19G4R1 were identified in a sheep, a goat and an alpaca isolate, but it is not reported which subtype was identified in which sample

Several studies using multi-locus genotyping (MLG) have found evidence of specific host associated *C. parvum* populations (Drumo et al. 2012; Mallon et al. 2003a, b; Morrison et al. 2008). In the United Kingdom, sheep MLGs clustered with human and bovine isolates (Mallon et al. 2003a, b), indicating frequent zoonotic transmission, whereas only one of the 34 MLGs identified in sheep/goats in Italy was also identified in human samples (Drumo et al. 2012), indicating a low rate of zoonotic transmission. However, the latter study included very few human isolates.

Zoonotic transmission is commonly observed with *C. parvum*, but *C. ubiquitum* has also been identified in a number of sporadic human cases (Cieloszyk et al. 2012; Elwin et al. 2012; Feltus et al. 2006; Leoni et al. 2006; Ong et al. 2002; Soba et al. 2006; Trotz-Williams et al. 2006). Oocysts of this species have been identified in storm water, wastewater, raw water and drinking water (Jiang et al. 2005; Liu et al. 2011; Nichols et al. 2010; Van Dyke et al. 2012). In Scotland, *C. ubiquitum* was the third most common species in raw water and the most common species identified in drinking water (Nichols et al. 2010). Thus, in areas where *C. ubiquitum* is common in sheep and goats, this species could be a more important cause of zoonotic infection than *C. parvum*. In addition, the relatively common presence of this species in water indicates a potential for waterborne outbreaks.

Natural infection with *C. hominis* has been reported in one goat and two sheep (Giles et al. 2009; Ryan et al. 2005) and in three lambs following experimental infection (Ebeid et al. 2003; Giles et al. 2001), but since animals are not natural hosts for this species, risk of zoonotic transmission with this species should be negligible compared with the risk of human-to-human transmission.

It is important to note that because *Cryptosporidium* infection can be subclinical, the zoonotic potential is not restricted to contact with diarrhoeic *Cryptosporidium*-infected animals (Pritchard et al. 2007).

4.4 *Cryptosporidium* spp. in Pigs

Since domestication around 4900 BC in China, the pig has been an important food source (Moeller and Crespo 2009). Pigs are farmed worldwide, with the global swine inventory estimated at over 800 million in 2002. Because Asian countries are major consumers but do not produce sufficient pigs for their needs, there is a significant international trade in live and slaughtered pigs. China has the world's largest pig population, mostly small herds consisting of only a few animals, and is a net importer of pigs. The United States, European Union, and Canada are major exporters with relatively few but very large production units (Moeller and Crespo 2009). The global trend is for fewer producers responsible for larger numbers of pigs, and more concentration within the swine industry.

4.4.1 Occurrence (Prevalence)

Pigs are the primary host for *C. suis* (Ryan et al. 2004) formerly identified as *Cryptosporidium* pig genotype I and for *C. scrofarum* (Kváč et al. 2013) formerly identified as *Cryptosporidium* pig genotype II. Farm pigs have also been found infected with *C. parvum*, *C. muris*, on one occasion with *C. tyzzeri*, a species common to mice, and with the novel genotype isolate Eire 65.5 (Kváč et al. 2013). Cryptosporidiosis occurs in pigs of all ages in 21 countries on 6 continents (Table 4.8). Before molecular methods were developed *C. parvum* was thought to infect 152 species of mammals and to consist of several genotypes. Consequently some early studies erroneously reported *C. parvum* infection in pigs based on the identification of oocysts in faeces by microscopy. Subsequent use of molecular methods provided the necessary tools to identify and distinguish species.

Overall, prevalence data for locations, herds and age groups vary greatly and are not directly comparable because some data represent pooled samples (some from litters, others from fecal slurry), some data originate from single farms while other data come from multiple farms. Some surveys have studied individual pigs at various ages, or only those pigs with diarrhoea, or simply specimens submitted to diagnostic laboratories from unspecified locations (Table 4.8). Even in comparable populations, such as preweaned pigs in the same country or indifferent countries, data differences are too great to draw any conclusions on prevalence. For example: in Australia- reports of 32.7 % versus 6.0 % prevalence (Johnston et al. 2008 vs. Ryan et al. 2003a); in the Czech Republic- reports of 21.8 % versus 5.7 % prevalence (Kváč et al. 2009a vs. Vitovec et al. 2006); or between Serbia and Spain – reports of 32 versus 0 % prevalence (Mišić et al. 2003 vs. Quílez et al. 1996a). Some studies found significant association between the presence of a particular species and the pigs' age, with *C. suis* prominent in piglets and *C. scrofarum* prominent in weaners (Enemark et al. 2003). In contrast, others found no significant association between species and age or housing conditions (Featherstone et al. 2010b). These prevalence data reflect vast differences in management practices from location to location with too many unknown factors to draw valid conclusions on cause and effect or location within the 49 cited studies in Table 4.8 that reported a prevalence of infection between 0.1 % and 100 %. The only variable repeatedly associated with detection of *Cryptosporidium* is age. Most positive samples were from weaners and growers (Table 4.8). Generally, prevalence increased until pigs were 10 weeks of age, then gradually declined.

4.4.2 Association of Infection with Clinical Disease

The first reports of cryptosporidiosis in pigs found one piglet among 81 herds of nursing piglets with necrotic enteritis, but the significance of this finding was described as unknown (Bergeland 1977) and *Cryptosporidium* was found at necropsy in three pigs without clinical signs (Kennedy et al. 1977).

Although a higher prevalence of diarrhoea was found in *Cryptosporidium*-infected pigs than in uninfected pigs (Hamnes et al. 2007), others found no significant relationship between infection and diarrhoea (Quílez et al. 1996b; Guselle et al. 2003; Maddox-Hyttel et al. 2006; Vitovec et al. 2006; Suárez-Luengas et al. 2007).

Table 4.8 Location, age, and prevalence of *Cryptosporidium* in pigs identified by microscopy and molecular methods

Location	Age	No. of animals/farms or locations	Prevalence (% positive)	Microscopy (M)	Molecular identification/species or genotype (no. or % identified)[a]	Reference
Australia	Not specified	78/diagnostic lab	0	M	No/unknown	O'Donoghue et al. 1987
Australia	3–8 weeks	Samples from 3 herds with clinical disease	3 herds: 2, 5, and 4 positive samples	M	18S rRNA/*C. suis* (8) *C. parvum* (4)	Morgan et al. 1999
Australia	1–8 weeks	648/22 indoor and outdoor herds	6.03 0.5 17.2	M	18S rRNA/*C. suis* *C. scrofarum*	Ryan et al. 2003a
Australia	pre-weaned weaned	156 123/4 piggeries	32.7 10.6	M	18S rRNA *C. suis* (6) *C. scofarum* (32)	Johnson et al. 2008
Brazil São Paulo	Not specified	25	0	M	No/unknown	Sevá et al. 2010
Brazil Rio de Janeiro	Not specified	91/10 piggeries	2.2	Not indicated	18S rRNA *C. scrofarum*	Fiuza et al. 2011b
Canada	5–15 days	1,453 diarrhoeic samples/diagnostic lab	0.5	M	No/unknown	Sanford 1983
Canada	1–30 weeks	3491samples/diagnostic lab	5.3	M	No/unknown	Sanford 1987
Canada	Not specified	100/5 farms	11	M	No/unknown	Olson et al. 1997b
Canada Alberta	Not specified	25–50 samples/1 farm	0	M	No/unknown	Heitman et al. 2002
Canada	21 days–6 months	33/1 farm	100 longitudinal study, cumulative prevalence	M	HSP70/*C. suis* (10)	Guselle et al. 2003

Canada Ontario		122 pooled samples/ 10 farms	55.7	M	18S rRNA/*C. parvum* *C. scrofarum*	Farzan et al. 2011
Canada Prince Edward Island		633/21 herds	26	M	18S rDNA HSP70/*C. suis*, *S. scofarum*, *C. parvum*, *C. tyzzeri*	Budu-Amoako et al. 2012d
China Henan	<1 month 1–2 months 3–6 months >6 months	1,350/14 farms in 10 prefectures	10.2 20.6 0 1.4 Avg 8.2	M	18S rRNA *C. suis* *C. scrofarum*	Wang et al. 2010b
China Chongqing	Boars and sows fatteners, growers, weaners	2,971/14 intensive 29 extensive farms in 13 counties	6.6	M	No/unknown	Lai et al. 2011
China Shanghai	60–180 days of age	2,323/8 farms	34.4	M	18S rRNA *C. suis* (82.6 %) *C. scrofarum* (8.7 %) mixed infection (8.7 %)	Chen et al. 2011
China Yangtze River Delta	Not specified	94/6 farms	14.3–25.0	Not indicated	18S rRNA *C. scrofarum*	Yin et al. 2011
Czech Republic	Sows Pre-weaned weaned	135 3,368 835/8 farms	0 5.7 24.1	M	18S rRNA *C. suis* (394)	Vitovec et al. 2006
Czech Republic	Pre-weaners Starters Pre-growers Sows	119 131 123 40 Total 413/1 farm	21.8 29.0 17.1 2.5 Avg 21.2	M	18S rRNA *C. suis*, *C. muris* *C. scrofarum*	Kváč et al. 2009a
Czech Republic South Bohemia	Abattoir	Finishers sows	123/14 farms	M	18S rRNA GP60 *C. suis*, *C. parvum* *C. scrofarum*	Kváč et al. 2009b

(continued)

Table 4.8 (continued)

Location	Age	No. of animals/farms or locations	Prevalence (% positive)	Microscopy (M)	Molecular identification/species or genotype (no. or % identified)[a]	Reference
Croatia	Not specified	5 pigs each/ 38 production units	65.8 of all units	M	No/unknown	Bilic and Bilkei 2006
Cuba Villa Clara	Suckling weaned	45	6.7	E	No/*C. parvum* based on DAS/ELISA	de la Fe et al. 2013
		45	13.3			
		All diarrhoeic/ 6 piggeries				
Denmark	<1 month	488	6	M	No/unknown	Maddox-Hyttel et al. 2006
	1–4 months	504	71			
	sows	245/50 herds	4			
Denmark	Same as above	1,237/50 herds	Same as above	Not indicated	18S rRNA, HSP70	Langkjaer et al. 2007
					C. suis (50)	
					C. scofarum (133)	
England East Anglia	Suckler	20	20.0	M	*18S rRNA*	Featherstone et al. 2010b
	Weaner	36	38.9		*C. scrofarum* (64.1 %)	
	Grower	93	48.4			
	Finisher	119	36.9		*C parvum* (20.5 %)	
	Adult	39	30.8		*C suis* (15.4 %)/39 samples	
		Total 308/72 farms	Avg 38.6			
		Diagnostic lab				
Germany	1–42 days	287 diarrhoeic samples/	1.4	M	No/unknown	Wieler et al. 2001
		24 farms				
Germany	Not specified	1,427 samples/ diagnostic lab	0.1	M	No/unknown	Epe et al. 2004

Ireland	Not applicable	56 slurry samples/ 33 farms	50 at 1 farm 40.6 at all other farms	M	No/unknown	Xiao et al. 2006
Ireland	Weaners	342/5 units in 4 counties	15.0	Not indicated	18S rRNA *C. suis*, *C. scofarum*, *C. parvum*, *C. muris*	Zintl et al. 2007
	Finishers		7.4			
	Sows		13.3			
	Gilts		Rare			
	Boars		Rare			
			Mean: 11.4			
Italy	Piglets	200 litters (pooled feces from each litter)	0.5	M	No/unknown	Canestri-Trotti et al. 1984
Japan Kanagawa	1–3 months	232	33.2	M	No/unknown	Izumiyama et al. 2001
	6 months	252/8 farms	NS			
Japan Hokkaido	6 months	108/abattoir	0	M	No/unknown	Koyama et al. 2005
Korea	6–8 months	500/location(s) not specified	19.6	M	No/unknown	Rhee et al. 1991
Korea	Not specified	589/4 slaughter locations	62	M	No/unknown	Yu and Seo 2004
Malawi	Not specified	92/2 regions over 3 seasons	17.7–60	M	No/unknown	Banda et al. 2009
Norway	4–33 days	684 litters (each pooled)/100 indoor farms	8.3	M	18S rRNA COWP, Actin/*C. parvum* *C. suis* *C. scrofarum*	Hamnes et al. 2007
Serbia	1–30 days	50	32	M	No/unknown	Mišić et al. 2003
	1–3 months	40	62.5			
	3–6 months	38	44.7			

(continued)

Table 4.8 (continued)

Location	Age	No. of animals/farms or locations	Prevalence (% positive)	Microscopy (M)	Molecular identification/species or genotype (no. or % identified)[a]	Reference
	6–9 months	45	31.1			
	9–12 months	42	23.8			
	>12 months	45	15.5			
		Total 260				
Spain	<2 months	329/farms and abattoirs	3 (all in 1 litter of ten 50-day-old pigs)	M	No/unknown	Villacorta et al. 1991
Spain	1–2 months	36	36	M	No/unknown	Fleta et al. 1995
	2–6 months	11	9			
	mature	10 abattoirs	0			
Spain Aragon	<1 months	49	0	M	No/unknown	Quílez et al. 1996a
	1–2 months	49	59.2			
	2–6 months	312	34.3			
	mature	210	0			
		Total 620/27 farms				
Spain Aragon	<1 months	14	0	M	No/unknown	Quílez et al. 1996b
	1–2 months	16	87.5			
	2–6 months	30	56.7			
	Mature	30	0			
		Total 90				
Spain Zaragosa	Weaned (1–2 months)	75	30.7	M	18S rRNA	Suárez-Luengas et al. 2007
	Fattening (2–6 months)	42	11.9		*C. suis*	
	Sows	25	16.0		*C. scrofarum*	
		Total 142/24 farms	22.5			
Trinidad and Tobago	Not specified	275/locations not specified	19.6	M	No/unknown	Kaminjolo et al. 1993

Trinidad	Not specified	52/3 locations	1.9	M	No/unknown	Adesiyun et al. 2001
USA-California	Not specified	200/stockyard	5	M	No/unknown	Tacal et al. 1987
USA Ohio	7–25 days	44 1	15.9 of litters	M	No/unknown	Xiao et al. 1994
	5–8 weeks	176	8			
	Sows	62/2 farms	0			
USA-California	≤8 months	62	11	M	No/unknown	Atwill et al. 1997
	≥9 months	159/10 feral groups	3			
USA Georgia	Separate farrowing, nursing, finishing, and gestation effluents	10 swine waste lagoons; 12 monthly samples from each	100 (every sample was positive)	M	18S rRNA/*C. suis* and *C. scofarum* (95–100 %); also *C. muris* and *C. parvum*	Jenkins et al. 2010
Vietnam	4–8 weeks	farms	23.5	M	No/unknown	Koudela et al. 1986

[a]The species names *C. suis* and *C. scrofarum* are used throughout. In some places they replaced the former designations *Cryptosporidium* pig genotype I and II that appeared in the original references published before these species were named

Cryptosporidium was detected histologically in the microvillus brush border of 5.3 % of 3,491 pigs from 133 farms examined for routine diagnostic evaluation (Sanford 1987). Most infected pigs were 6–12 weeks old. Organisms were detected in the jejunum, ileum, caecum, and colon, but primarily in microvilli of dome epithelium in the ileum. Twenty six percent of *Cryptosporidium*-infected pigs had diarrhoea but most of these also had other primary agents capable of causing diarrhoea. Similar observations have been made by others. Whereas most infections are asymptomatic or cause only mild, non-specific colitis (Higgins 1999), pigs known to be naturally infected with *C. suis* or *C. scrofarum* have not been found with clinical signs of infection while pigs infected with *C. parvum* or co-infected other enteropathogens such as rotavirus, *Salmonella*, or *Isospora* have had associated diarrhoea and some have died (Enemark et al. 2003; Núñez et al. 2003; Hamnes et al. 2007).

Experimental infections with different species have helped to clarify the relationship of species with clinical disease. Pigs experimentally infected with *C. suis* (Enemark et al. 2003) or *C. scrofarum* (Kváč et al. 2013) showed no clinical signs. The pathogenicity of *C. parvum* isolated from calves was demonstrated in early transmission studies to pigs (Moon and Bemrick 1981; Tzipori et al. 1981b, 1982; Argenzio et al. 1990; Vitovec and Koudela 1992; Pereira et al. 2002). Experimental infection with the avian species, *C. meleagridis*, obtained from a human infection, consistently resulted in oocyst excretion and diarrhoea in pigs, although mucosal changes were milder than those described for *C. parvum* (Akiyoshi et al. 2003). Piglets infected with *C. suis* had mild or no clinical signs despite excreting large numbers of oocysts, in contrast to those infected with *C. parvum* that had diarrhoea for a mean duration of 3.5 days and developed inappetence, depression and vomiting (Enemark et al. 2003).

Developmental stages of *Cryptosporidium* have been observed throughout the intestinal tract. Villous atrophy, villous fusion, crypt hyperplasia, and cellular infiltration of the lamina propria have been observed (Kennedy et al. 1977; Moon and Bemrick 1981; Tzipori et al. 1982, 1994; Sanford 1987; Vitovec and Koudela 1992; Argenzio et al. 1990; Pereira et al. 2002; Enemark et al. 2003; Núñez et al. 2003; Vitovec et al. 2006). Lesions caused by *C. parvum* were the most severe, as were clinical signs associated with that species. Changes in the location of stages have been noted. In the first days of infection more stages were found in the proximal intestine, but later more stages were found in distal locations (Tzipori et al. 1982; Vitovec and Koudela 1992). Extra-intestinal infections also have been reported in pigs. In two naturally infected piglets, the gall bladder was infected (Fleta et al. 1995). In experimentally immunosuppressed piglets, the gall bladder, bile ducts, and pancreatic ducts were found infected (Healey et al. 1997). Infections in the trachea and conjunctiva were detected in experimentally infected normal piglets (Heine et al. 1984).

4.4.3 Infection Dynamics: Oocyst Excretion and Transmission

A survey of faecal slurry from swine finishing operations in Ireland found *C. suis*, *C. scrofarum* and *C. muris* and concluded that *Cryptosporidium* oocysts can persist in treated slurry and potentially contaminate surface water through improper discharge or uncontrolled runoff (Xiao et al. 2006). Hamnes et al. (2007) reported *C. suis* and *C. scrofarum* in faeces of suckling pigs in Norway and reasoned that farrowing operations were sources of these parasites. Additional data on oocyst concentrations, numbers of oocysts excreted, how long oocysts remain infectious under environmental condition, and modes of transmission of *Cryptosporidium* species and genotypes are rare or non-existent. A year-long investigation was conducted at four types of swine operations (finishing, farrowing, nursery and gestation) in Georgia, USA (Jenkins et al. 2010). Mean oocyst concentrations ranged from 11 to 354 oocysts per ml of lagoon effluent; the nursery had the highest concentration of oocysts and the greatest percentage of viable oocysts (24.2 %), *C. suis* and *C. scrofarum* were the dominant species with some *C. muris* and *C. parvum*. Experimental attempts to transmit *C. scrofarum* to adult SCID mice, adult BALB/c mice, Mongolian gerbils, southern multimammate mice, yellow-necked mice, and guinea pigs were unsuccessful, suggesting that rodents are an unlikely source of transmission of this species under natural conditions (Kváč et al. 2013).

4.4.4 Zoonotic Transmission

Cryptosporidium suis was detected by immunofluorescence microscopy and RFLP analysis of PCR products in stools from an HIV patient in Peru (Xiao et al. 2002a). The patient was not severely immunosuppressed and was asymptomatic. He had a dog but reported no contact with other animals or animal faeces, including pigs and pig faeces, so the source and method of transmission are unknown.

4.5 *Cryptosporidium* spp. in Other Farmed Mammals

4.5.1 Cryptosporidium *in Farmed Deer*

Many countries, including Australia, Canada, China, Korea, Norway, Russia, Sweden, UK, USA and Vietnam, have thriving deer farming industries. New Zealand, a country where deer are not native, has the world's largest and most advanced deer farming industry. Although it is difficult to find estimates on the numbers of deer farmed worldwide, more than one million deer were being farmed in New Zealand in 2011 (Sources: Statistics New Zealand and Deer Industry New Zealand http://www.deernz.org/about-deer-industry/nz-deer-industry) – compared with five million dairy cows – and there are over 2,800 deer farmers. More than 90 % of the New Zealand deer industry's products are exported, with approximately half of the export going to Germany and Benelux.

Species of deer which are commercially farmed varies regionally, but the following species are now being farmed in various parts of the world: red deer (*Cervus elaphus*), wapiti or elk (*Cervus canadensis*), fallow deer (*Dama dama*), sika (*Cervus nippon*), rusa deer (*Rusa timorensis*), and reindeer (*Rangifer tarandus*) (FAO http://www.fao.org/docrep/004/X6529E/X6529E02.htm).

4.5.1.1 Occurrence (Prevalence)

Although farmed deer are an important resource in many countries, much of the published information on *Cryptosporidium* in deer refers to studies on wild or free-ranging cervids (e.g. white-tailed deer in USA, roe deer in Spain, caribou in Canada, and moose, red deer, roe deer and reindeer in Norway; Rickard et al. 1999; Castro-Hermida et al. 2011b; Johnson et al 2010; Hamnes et al. 2002). While these studies on free-ranging cervids may give useful information regarding the species or genotype of *Cryptosporidium* that might infect farmed deer (*C. ubiquitum*, *C. parvum*), as farmed deer probably differ quite substantially from their wild counterparts regarding exposures and stresses, extrapolation of prevalence data from wild to farmed deer may give an incorrect picture. Indeed, a study in Poland found that the prevalence of *Cryptosporidium* was significantly higher in wild red deer than farmed red deer (27 % compared with 4.5 %), and mean oocyst concentration was also five times higher in faecal samples from wild red deer (Paziewska et al 2007). However as the sample size was relatively small (52 wild deer, 66 farmed deer) and from only single locations and as age and symptoms were not indicated, it is not possible to determine the reason for these differences.

The few studies on the prevalence of *Cryptosporidium* infection in different farmed or domesticated cervids are summarised in Table 4.9.

4.5.1.2 Association of Infection with Clinical Disease

The lack of surveys for *Cryptosporidium* infection in farmed deer is surprising, given the clear association of infection with clinical disease in farmed cervids. Some of the first published studies on *Cryptosporidium* infection in cervids are case reports of severe (high mortality) outbreaks among farmed red deer calves. In one outbreak in Scotland, UK among 82 artificially reared red deer calves, 56 developed cryptosporidiosis and 20 subsequently died; 80 % of the calves with diarrhoea and 50 % of apparently asymptomatic calves excreted oocysts and post mortem histopathological examination of the intestines demonstrated lesions similar to

Table 4.9 *Cryptosporidium* in different species of deer identified by microscopy

Location	Host species	Symptoms	Host age	Study design	Prevalence (% positive – no. positive/ no. examined)	Diagnostic technique	Molecular analyses	Reference
Ireland	Red deer	Asymptomatic	Adult hinds and calves	Longitudinal – monthly samples	Hinds: 0–63 % according to month (overall prevalence: 39.3 % – 114/290) Calves: 40–100 % according to month (overall prevalence: 60 % – 21/35)	Water-ether sedimentation followed by sucrose flotation and IFAT	Not conducted	Skerret and Holland 2001
Poland	Red deer	Not stated	Not stated	Semi-longitudinal – seasonal sampling	Overall prevalence: 4.5 % – 3/66	Sheather's flotation followed by mZN and IFAT	Unsuccessful	Paziewska et al 2007
China	Red deer, Pere David's deer, Sika deer	Asymptomatic	4 months–10 years	Cross-sectional study at 4 deer farms	Red deer: 0 (0/8) Pere David's deer: 0 (0/33) Sika deer: 2.4 (2/83)	Sheather's flotation followed by light microscopy	1 sample *C. ubiquitum* – PCR and sequence analysis at 4 genes	Wang et al. 2008a

those seen in other species (Tzipori et al. 1981a). Another outbreak among new-born red deer calves in the UK also resulted in high mortality, with calves dying at 24–72 h of age. However, this outbreak was not characterized by diarrhoea, and terminal uraemia was proposed as the symptom leading to death (Simpson 1992). Outbreaks of cryptosporidiosis in red deer calves have also been reported from New Zealand, again with relatively high mortality (10 out of 10 calves dying within a few days of illness onset in one outbreak, and 7 out of 10 calves dying within 3 days of illness in another outbreak) (Orr et al. 1985). Severe subactute enteritis in the small and large intestine were reported in both outbreaks.

Information from other species of farmed deer is more scanty, but a retrospective study of neonatal mortality in farmed elk (Pople et al 2001) identified *Cryptosporidium* infection as one of the most important causes of enteritis leading to death (7 cases out of 11 of infectious enteritis from a total of 111 cases in which 46 had no specific cause of death identified). Among these 7 cases, 4 were associated with an outbreak on a single farm (Pople et al 2001).

Unfortunately, information on the species of *Cryptosporidium* infection associated with clinical disease in deer and *Cryptosporidium* associated with asymptomatic infection is lacking. One outbreak on a Scottish farm occurred when deer were put to graze on a pasture that had previously been grazed by a *Cryptosporidium*-infected herd of cattle (Angus 1988), and therefore it seems probable that this might indicate infection with *C. parvum*; it might be speculated that infection of deer with deer-adapted *C. ubiquitum* is less likely to cause severe symptoms.

4.5.1.3 Infection Dynamics: Oocyst Excretion and Transmission

Published information about infection dynamics in farmed deer is minimal. However, a longitudinal study in asymptomatic farmed red deer in Ireland (Skerrett and Holland 2001) provides some interesting data. Asymptomatic low-level (<10 OPG) oocyst shedding from adult hinds appeared to continue throughout the year, except during the calving season (May - June), when there was a 2-log increase in oocyst excretion rate; the highest level recorded was 67,590 OPG. The authors speculate that this may be related to hormonal or immunological changes, or perhaps alterations in stress levels. The authors note that this preparturient rise in oocyst shedding results in contamination of the environment for the new born calves. However, in this study, although calves became infected, oocyst excretion was low (not exceeding 150 OPG, and usually less), and clinical disease was not observed. Again, the species of *Cryptosporidium* in these infections is unknown.

4.5.1.4 Zoonotic Transmission

Both *C. parvum* and *C. ubiquitum* have zoonotic potential, but there appear to be no documented cases of proven zoonotic transmission from/to deer. Those studies

(USA and Australia) that have investigated deer as sources of contamination in watersheds, have focused on wild deer only and provided contrasting results (Cinque et al 2008; Jellison et al 2009).

4.5.2 **Cryptosporidium *in Farmed Camelids***

Camelids are members of the family *Camelidae*, and include the tribe *Camelini* (including dromedaries and Bactrian camels), and the tribe *Lamini* (llamas, alpacas, vicuñas, and guanacos).

There are two species of camel. Approximately 14 million domesticated one-humped dromedaries (*Camelus dromedarius*) are found in Middle Eastern countries including the Sahel and Horn of Africa, as well as parts of Southern Asia where they provide people with milk, food, and transportation. Nearly two million domesticated two-humped bactrians (*Camelus bactrianus*) are native to the steppes of central Asia the Gobi and Taklamakan Deserts in Mongolia and China.

Alpacas (*Vicugna pacos*) and llamas (*Lama glama*) exist only in the domesticated state and are found worldwide. However, both are native to South America and are raised primarily for fibre production although llamas were once used extensively as work animals. The young of both are called crias.

4.5.2.1 Occurrence (Prevalence)

Camels and dromedaries: In the relatively few studies of dromedaries in Northern Africa and the Middle East, the prevalence of cryptosporidiosis varied greatly (Table 4.10). None of 23 camels from Iraq (Mahdi and Ali 1992) and none of 110 camels on farms in Tunisia were found positive for *Cryptosporidium* (Soltane et al. 2007). *Cryptosporidium* was detected in one of four camel calves in Egypt (Abou-Eisha 1994). However, in an abattoir in Yazd Province in Iran, microscopic examination of 300 faecal specimens detected 61 (20.3 %) positive for *Cryptosporidium* and 12 (12 %) positive abomasal mucosa specimens. At an abattoir in Isfahan Province in central Iran, of 63 adult male and 40 adult female dromedary camels examined, 39 (37.9 %) were *Cryptosporidium*-positive (Razawi et al. 2009). In northwestern Iran, of 170 faecal samples from camels 17 (10 %) were positive for *Cryptosporidium*-like oocysts (Yakhchali and Moradi 2012). The prevalence was significantly higher (20 %) in calves less than a year old.

Oocysts have also been recovered from wild and zoo-housed camels. Faeces from a 3-year-old Bactrian camel in the Wild Animals Rescue Centre of Henan Province in China were found positive for *C. andersoni* (Wang et al. 2008b). Oocysts from a zoo-housed Bactrian camel (Fayer et al. 1991) were infectious for mice (Anderson 1991) and were identified as *C. muris* (Xiao et al. 1999; Morgan et al. 2000); those from camels in the Czech Republic were identified as *C. andersoni*. Other zoo-housed

Table 4.10 *Cryptosporidium* detected in camels

Location	Host	Age	No. infected/ no. examined	Method	Reference
Basrah, Iraq	Camels	Adults	0/24	Microscopy	Mahdi and Ali 1992
Australia	Camel	Not stated	1	Molecular	Morgan et al. 2000
Tunisia	Camels	Not stated	0/110	Microscopy	Soltane et al. 2007
China	Bactrian camel	3-year-old	1	Molecular	Wang et al. 2008b
Isfahan Province, Iran abattoir	Camels	2–14 years	39/103	Microscopy	Razawi et al. 2009
NW Iran	Dromedary camels	Calves and adults	17/170	Microscopy	Yakhchali and Moradi 2012
Yazd Province, Iran abattoir	Camels	<5–>10 years	61/300 faeces 12/100 abomasums	Microscopy	Sazmand et al. 2012

camels have been found to be infected with *Cryptosporidium* (Abou-Eisha 1994; Gomez et al. 2000; Gracenea et al. 2002).

Alpacas and llamas *Cryptosporidium* oocysts have been detected in both these species. However, in California none of 354 llamas from 33 facilities were found positive (Rulofson et al. 2001) nor were 61 alpacas on two farms in Maryland (Trout et al. 2008). Elsewhere in North America, Europe, and Australia small numbers of alpacas, llamas and guanacos have been examined and a few have been found positive for *Cryptosporidium* (Table 4.11). Most examinations were conducted by microscopy, but those that utilized molecular methods identified only *C. parvum* (Morgan et al. 1998; Starkey et al. 2007; O'Brien et al. 2008; Twomey et al. 2008). The exception is a study in which a cria was found infected with *C. ubiquitum* (Gomez-Couso et al. 2012). A national survey of 5,163 1–15-day-old alpacas in 105 Andean herds in Peru, the natural habitat for nearly 80 % of the world's alpacas, found 2 % of the youngest alpacas increasing to 20 % of the oldest alpacas, infected with *Cryptosporidium* spp., with an overall prevalence of 13 % (Lopez-Urbina et al. 2009). More recently in Peru, 4.4 % of 274 alpacas from 12 herds were found positive for *Cryptosporidium* spp. (Gomez-Couso et al. 2012). Herd prevalence was 58.3 % (7/12 herds) for *Cryptosporidium*. The highest prevalence (20 %) was found in the 8-week-old group (Gomez-Couso et al. 2012).

4.5.2.2 Association of Infection with Clinical Disease

Camels and dromedaries Few data are available on the subject of clinical illness associated with cryptosporidiosis in camels. Of 170 faecal samples, 17 camels (10 %) were positive for *Cryptosporidium*-like organisms (Yakchali and Moradi 2012). The prevalence was significantly higher in camel calves (<1 years old) (20 %) than other age groups, in which the diarrhoeic calves had a prevalence of 16 %.

Table 4.11 *Cryptosporidium* in llamas and alpacas

Location	Host	Age	No. infected/ no. examined	Detection method	Reference
Wisconsin, USA	Llama	Cria	1/1	Not stated	Hovda et al.1990
Peru	Alpaca	Not given	1	Molecular	Spano et al. 1997
					Morgan et al. 1998
England	Alpaca	9–30 days old	3/3	Microscopy	Bidewell and Cattell 1998
California, USA	Llama	Crias and Adults	0/354	Microscopy	Rulofson et al 2001
Oregon, USA	Llama and Alpaca	31–210 days old	4/45	Microscopy	Cebra et al 2003
Czech Republic	Alpaca	Not given	1/1	Molecular	Ryan et al. 2003c
Canada	Alpaca	Crias	Multiple (number not given)	Microscopy	Shapiro et al. 2005
Scotland	Alpaca	Not given	Not given	Not stated	Stewart et al. 2005
New York, USA	Alpaca	Crias	8/total not given	Microscopy and Molecular	Starkey et al. 2007
Maryland, USA	Alpaca	10 weeks–10 years	0/61	Molecular	Trout et al. 2008
Ohio	Llama and Alpaca	7–100 days old	15/58	Not stated	Whitehead and Anderson 2008
SW England	Alpaca	Crias; Adults	3/6; 0/24	Molecular	Twomey et al. 2008
		Crias	5/14		
Australia	Alpaca	Not given	1/1	Molecular	O'Brien et al.2008
Washington, USA	Alpaca	8–18 days old	20/20	Microscopy	Waitt et al. 2008
Peru	Alpaca	1–15 days old	24/241	Microscopy	Lopez-Urbina et al. 2009
			666/5,163		
Peru	Alpaca	≥5 weeks	12/274	Microscopy and Molecular	Gomez-Couso et al. 2012
New York and Pennsylvania, USA	Alpaca	Crias	8/110	Microscopy	Burton et al. 2012
		Adults	9/110		

Alpacas and llamas Not all alpacas and llamas infected with *Cryptosporidium* show clinical signs of infection. Of 110 healthy crias and their 110 dams 7 % and 8 %, respectively, were found excreting oocysts (Burton et al. 2012). Oocysts of *C. parvum* were detected in 4 of 14 faecal samples from healthy crias and in one sample from a cria with diarrhoea (Twomey et al. 2008).

Cryptosporidium was observed in a post-operative neonatal llama with diarrhoea, cachexia, dehydration and electrolyte abnormalities (Hovda et al. 1990). During 8 days that intravenous fluids and nutritional support were provided, these signs were not observed.

Of 20 *Cryptosporidium*-infected alpaca crias with diarrhoea, 15 exhibited weight loss and 5 had a poor appetite (Waitt et al. 2008). Most were 8–18 days old when examined. Additional potential gastrointestinal pathogens were found in 7 of these crias. Sixteen crias recovered after supportive therapy that included intravenous rehydration, with partial parenteral administration of nutrients, antimicrobials, oral nutrients, plasma, insulin and other palliative treatments.

Additional reports of diarrhoea associated with cryptosporidiosis have been reported in alpaca and llama crias (Cebra et al. 2003; Shapiro et al. 2005; Whitehead and Anderson 2006; Starkey et al. 2007). Three fatal cases (2 with diarrhoea) of cryptosporidiosis were reported in alpaca crias less than 30 days of age (Bidewell and Cantell 1998). At necropsy, intestinal congestion and distension were noted, oocysts were detected in Ziehl-Neelsen stained smears, and no other significant organisms or toxins were detected.

In South America, llama and alpaca husbandry is a vital economic activity and neonatal diarrhoea syndrome (NDS) is the most common and costly enteric disease in newborn llamas and alpacas (Lopez-Urbina et al. 2009). However, the role of cryptosporidiosis in NDS has not been clearly identified.

4.5.2.3 Zoonotic Transmission

Camels and dromedaries Only rare circumstantial data of zoonoses are available and the link is very tenuous. In Yazd Province in Iran, 24 of 100 people in long-term contact with camels were found infected with *Cryptosporidium* spp. (Sazmand et al. 2012). Infection was higher in winter than summer (16/50 compared with 8/50).

Alpacas and llamas In New York, *Cryptosporidium parvum* infection was identified in 5 crias, 3 of their caretakers were confirmed to have cryptosporidiosis, and three others were suspected to have cryptosporidiosis, suggesting zoonotic transmission (Starkey et al. 2007).

4.5.3 Cryptosporidium *in Farmed Rabbits*

Rabbit farming (cuniculture) for meat, wool, and fur production occurs in a variety of settings around the world, and mostly involves the European (or common) rabbit (*Oryctolagus cuniculus*). Small-scale backyard cuniculture is common in many countries (especially in Africa and South America), but commercial operations on a larger scale are found in Europe (particularly Italy, Spain and France) and Asia (particularly China and Indonesia). In the EU, rabbit meat production was estimated to be around 520,000 tonnes carcass-weight equivalent in 2005 (EFSA-AHAW 2005). In addition, rabbits continue to be bred for biomedical purposes – but this type of rabbit breeding will not be considered further in this chapter. Production and consumption of rabbit meat is relatively low in North America. Different rabbit breeds are used for meat, wool, and fur – with the most commonly used meat breeds being New Zealand, Californian, Florida White and Altex, all having good growth rates and desirable reproductive characteristics.

Much of the information presented in this section is derived from a comprehensive review article from 2010 (Robinson and Chalmers 2010).

4.5.3.1 Occurrence (Prevalence)

The majority of published prevalence information on *Cryptosporidium* in rabbits refers to studies on wild rabbits. Nevertheless, there have been several studies on the occurrence of infection in farmed rabbits and also in laboratory rabbits. The majority of these studies (involving both wild and domestic rabbits) are summarised in Robinson and Chalmers (2010). In Table 4.12, selected prevalence studies (rather than case reports) from farmed rabbits only are summarized, including two recent studies from China. Additionally, a further three studies from China and referenced in Zhang et al. (2012) are not included in Table 4.12 due to inaccessibility of the original publications. Zhang et al. (2012) do not provide any details on these studies and it is not certain that they refer to farmed rabbits. Although some surveys refer to *Cryptosporidium parvum*, all those studies in which genotyping has been used (including from wild rabbits; e.g. Nolan et al. 2010) suggest that the majority of natural infections in rabbits, if not all, are caused by *C. cuniculus*. Nevertheless, experimental infections with other species of *Cryptosporidium* have been established in rabbits, as summarised by Robinson and Chalmers (2010).

4.5.3.2 Association of Infection with Clinical Disease

Although the majority of surveys do not report symptoms associated with cryptosporidiosis in rabbits, experimental infections in preweaned rabbits have been associated with diarrhoea and high mortality (e.g. as reported by Robinson and Chalmers 2010; Mosier et al 1997) and also as described by Pavlásek et al. (1996) in farmed rabbits. However, even asymptomatic infection may result in some pathology, as noted by Inman and Takeuchi (1979), who reported blunted villi, a decrease in villus-crypt ratio, and mild oedema in the lamina propria in an apparently asymptomatic adult rabbit. Thus, even asymptomatic infection may reduce stock productivity.

Table 4.12 *Cryptosporidium* in farmed rabbits

Location	Breed	Symptoms	Host age	Study design	Prevalence (% positive – no. positive/no. examined)	Diagnostic technique	Molecular analyses	Reference
Czech Republic	Broiler rabbits of 6 different breeds or crossbreeds	Variable, including diarrhoea and inappetence. Seven deaths recorded	23–33 days and 82–92 days (all post-weaning)	Longitudinal – with pooled samples	Variable throughout study, but at peak pooled samples from 12/28 cages	Giemsa staining in faeces; post mortem examination of digesta and intestinal scrapings. Histology	Not conducted	Pavlásek et al. 1996
Tunisia	Not stated	Not stated	Not stated	Cross-sectional at 1 farm	Overall prevalence: 0 % – 0/178	Formol-ether sedimentation followed by mZN	Not applicable	Soltane et al 2007
China (Henan Province)	Various, including Standard Rex and New Zealand White	Asymptomatic	5 age groups:	Cross-sectional at 8 farms	Overall prevalence: 3.4 % – –37/1,081	Sheather's flotation followed by modified acid-fast stain	All (36/37successful) *C. cuniculus*[a] – PCR and RFLP and sequence analysis at 18S rRNA gene. 8 samples further analysed at 3 other genes	Shi et al. 2010
			<1 month,		<1 month: 4.1 %–3/73			
			1–3 months,		1–3 months: 10.9 %–27/247			
			4–6 months,		4–6 months: 1.3 %–6/474		GP60 subtyping – 30 of 37 successful; VbA29 (18 samples), VbA35 (4 samples), VbA36 (8 samples)	
			7–12 months,		7–12 months: 0.4 %–1/230			
			>12 months		>12 months: 0 %–0/57			
China (Heilongjiang province	Not stated	Not stated	4–6 months	Cross-sectional at 8 farms	Overall prevalence: 2.4 %–9/378 (positive samples from 4 farms only)	Sheather's flotation followed by bright-field microscopy	All (9/9successful) *C. cuniculus* – PCR and sequence analysis at 18S rRNA gene GP60 subtyping – 9 of 9 successful; VbA21 (6 samples), VbA32 (3 samples)	Zhang et al 2012

[a]Described as rabbit genotype in publication

Although no outbreaks of cryptosporidiosis in rabbit farms have been documented in the literature, acute outbreaks of diarrhoea with high mortality rates are frequently observed in rabbits (Banerjee et al 1987). Although bacterial agents are frequently considered to be the aetiological agent, it seems probable that some may be due to undiagnosed cryptosporidiosis. For example, the parasitological techniques (direct microscopy and flotation) used for investigating epizootic outbreaks of diarrhoea, characterized by a high morbidity and mortality, in different commercial rabbit farms in Mexico (Rodríguez-De Lara et al 2008) may have been insufficient for detecting *Cryptosporidium* infection, particularly if the operators had little experience in diagnosing this infection.

4.5.3.3 Infection Dynamics: Oocyst Excretion and Transmission

Information on the dynamics of *Cryptosporidium* infection in farmed rabbits is mostly lacking, although low oocyst excretion rates were reported in the majority of studies on rabbits in general (not just farmed rabbits). The studies from the Czech Republic provide some data, but, as the animals were not sampled individually, the data are difficult to interpret, and suggest that the source of infection for young rabbits may be low-level excretion of oocysts from mother rabbits at around parturition (Pavlásek et al. 1996).

4.5.3.4 Zoonotic Transmission

C. cuniculus is rarely, but sporadically, identified in human infections. In 3030 *Cryptosporidium*-positive faecal samples submitted for routine typing in UK between 2007 and 2008, 37 (1.2 %) were identified as *C. cuniculus*, with both GP60 Va and Vb subtype families detected (Chalmers et al 2011). However, the greatest evidence for *C. cuniculus* from rabbits having a significant zoonotic potential came from a waterborne outbreak of cryptosporidiosis in England in 2008 affecting 29 people; *C. cuniculus*, subtype VaA18 was identified in eight patients, a water sample from the implicated supply, and from the colon of a carcass of a rabbit (presumably wild) that was found in a tank at the water treatment works (Chalmers et al 2009). Nevertheless, transmission of *Cryptosporidium* to humans from farmed rabbits has not been recorded, and an investigation exploring associations between farm animals and human patients with cryptosporidiosis did not implicate rabbits as a source of infection (Smith et al 2010).

4.6 *Cryptosporidium* spp. in Poultry

The world stock of birds in production in 2011 was estimated to 22×10^9 animals (FAOSTAT 2013b). Approximately 56 % of the world stock was found in Asia, whereas Europe, North America and South America had approximately 10–11 %

each of the population. The largest group was chickens, with 90 % of the total stock. Ducks, turkeys and geese/guinea fowls constituted 6.1 %, 2.1 % and 1.7 % respectively, and other birds (ratites, pigeons etc.) only constituted 0.1 % of the world stock. The main chicken, duck and goose/guinea fowl production is in Asia (54 %, 90 % and 91 % within each group respectively), and most of the turkey production in North America (54 %), followed by Europe (23 %). For other birds, 50 % of the reported production was located in Asia, 41 % in Africa and 9 % in Europe.

Chickens (*Gallus gallus domesticus*) are descendants of the Red jungle fowl (*Gallus gallus*), with some hybridization with the Grey junglefowl (*G. sonneratii*). Broilers are usually kept in intense systems and reach slaughter size at about 6 weeks of age. Organically bred broilers and broilers kept on free range grow a bit more slowly. Laying hens can produce over 300 eggs in their first production year, but after that production declines rapidly.

Domesticated ducks (*Anas platyrhynchos domesticus*) are, except for the Moscovy duck (*Cairina moschata*), descendants of the Mallard (*Anas platyrhynchos*). The majority of domesticated geese (*Anser anser domesticus*) descend from the Greylag goose (*Anser anser*), but the breeds Chinese goose and African goose are derived from the Swan goose (*Anser cygnoides*). Ducks and geese are bred for meat, eggs and down, and ducks, to a lesser degree, also for the production of foie gras.

The domestic turkey (*Meleagris gallopavo*) is a progeny of the wild turkey, which is found in the wild in the United States http://www.turkeyfed.com.au/Turkey_Info.php. Turkeys are bred for meat production. The breed used is the white broad breasted turkey, introduced into commercial production in the 1950s http://bizfil.com/turkey-raising-primer. As with commercial chicken broiler farming, turkey farming is intense. The poults are extremely fast-growing, and reach approximately 6 kg at 10 weeks of age if given proper nutrition http://bizfil.com/turkey-raising-primer/. The United States has the highest consumption of turkey meat per person, and they are also the largest turkey producer, with 7.32 billion pounds of turkey meat produced in 2011 http://www.agmrc.org/commodities__products/livestock/poultry/turkey.

Among ratites, mainly ostriches (*Struthio camelus*) are farmed, but rheas (*Rhea americana*) and emus (*Dromaius novaehollandiae*) are also kept for production. Ratites are bred for meat, egg, and feather and leather production. Farming for feather production began already in the nineteenth century. Partridges, such as the Grey or English partridge (*Perdix perdix*) and red-legged partridge (*Alectoris rufa*), are gallinaceous birds used as game, and have been introduced in different parts of the world for this purpose. Another gallinaceous bird is the helmeted guinea fowl (*Numida meleagris*). They are used for pest control, eating ticks and other insects, and can be kept as an alarm system among other domesticated birds due to their loud and shrieking warning call. The meat is considered a delicacy. The Japanese quail (*Coturnix japonica*) is bred for meat and eggs. Domestic pigeons (*Columba livia domestica*) are the progeny of the world's oldest domesticated bird, the Rock pigeon. Pigeons are bred for meat, sporting competitions, homing, as exhibition birds or pets.

Cryptosporidium infection has been associated with large morbidity and mortality in different bird species (Bezuidenhout et al. 1993; Hoerr et al. 1986; Pages-Mante et al. 2007; Penrith et al. 1994; Ritter et al. 1986; Santos et al. 2005) and can thus be of great economic importance.

4.6.1 Prevalence

Avian cryptosporidiosis was first described in chickens (Tyzzer 1929). The infection was subclinical and situated in the caecum. Invasive stages looked identical to those of *C. parvum*, but no oocyst description was made, and no name was proposed. Today, three valid species have been identified in poultry. In addition, five genotypes have been identified in wild ducks and geese, and five additional genotypes have been described from other birds.

The *Cryptosporidium* oocysts identified by Slavin in 1955 were morphologically similar to *C. parvum*, described in mice in 1912 (Tyzzer 1912), and the infection site was the distal ileum. Slavin identified this bird *Cryptosporidium* as a unique species, *C. meleagridis*. When molecular methods were introduced as a means of species determination, it was verified that *C. parvum* and *C. meleagridis* were indeed different species (Sreter et al. 2000).

A species with a larger oocyst, first identified in chickens, and infecting the intestine, bursa and cloaca, was described and named *C. baileyi* (Current et al. 1986). This species is also involved in respiratory cryptosporidiosis, infecting the epithelium of sinuses, air sacs, nasopharynx, trachea and bronchi (Itakura et al. 1984; Lindsay et al. 1987). Infection of the conjunctiva (Chvala et al. 2006) and urinary tract, including the kidneys has also been shown (Abbassi et al. 1999; Trampel et al. 2000).

A third species, *C. galli*, infecting the proventriculus of chickens, was described by Pavlásek in 1999 and 2001, and re-described in 2003 (Pavlásek 1999, 2001; Ryan et al. 2003b). The species was probably described in finches already in 1990 (Blagburn et al. 1990) and later the name *C. blagburni* was proposed (Morgan et al. 2001). However, molecular analyses have shown that *C. blagburni* is the same species already described as *C. galli*, and thus the latter is considered to be the valid species name.

In addition, isolates referred to as goose genotypes I-IV have been identified in Canada geese and a duck genotype has been described in a Black duck and Canada geese. Of the other genotypes described in birds (avian genotypes I-IV and the Eurasian woodcock genotype), avian genotype II has been detected in ostriches.

Two proposed species are today considered as *nomen nudum* due to lack of sufficient data. *Cryptosporidium tyzzeri* in chickens was described in 1961 (Levine 1961) and later *C. anserinum*, found in the large intestine of geese was described (Proctor and Kemp 1974).

Based on 18S rDNA phylogeny, *C. galli* and the woodcock genotype belong to the clade of gastric cryptosporidia together with *C. andersoni*, *C. muris* and

C. serpentis, whereas *C. meleagridis*, *C. baileyi*, goose genotypes I and II and the duck genotype belong to the intestinal clade (Xiao et al. 2004). *Cryptosporidium meleagridis* is closely related to the group including *C. parvum* and *C. hominis*; *C. baileyi* is closely related to the snake genotype, goose genotypes I, II and the duck genotype cluster together and are closely related to *C. scrofarum*, *C. bovis*, *C. ryanae* and the deer genotype. Goose genotypes III-IV and avian genotypes I-IV were not included in the phylogenetic tree. In another publication, goose genotypes I (goose #1, 2, 3, 6 and 8), II (goose #9) and the duck genotype (goose #5) are closely related, whereas goose genotypes III (goose #3b) and IV (goose #7) are more distant (Jellison et al. 2004). The avian genotypes are more scattered. Avian genotypes I and II belong to the intestinal clade and are closely related to *C. baileyi*. Genotypes III and IV belong to the gastric clade, where genotype III is closely related to the Eurasian woodcock genotype and *C. serpentis*, and genotype IV is closely related to *C. galli* (Ng et al. 2006).

Prevalence data based on fecal examination could be affected by the time from sampling to analysis. This has been observed when oocyst numbers in chicken faeces dropped to approximately one third in samples stored for a week from first to second analysis, and where first analysis was performed on the day after sampling (C. Axén, unpublished data). It is possible that oocysts die and are quickly degraded by detrimental effects (extreme pH) due to the high ammonium content of bird droppings.

4.6.1.1 Chickens

A flock prevalence of 41 % (23/56), with 10–60 % within-flock prevalence, was reported for *C. baileyi* respiratory infection in broilers in the USA (Goodwin et al. 1996). In Morocco, *Cryptosporidium* sp. were found in 14 (37 %) of 38 investigated flocks. Within-flock prevalence ranged from 14 % to 100 %, and the highest prevalence (52 %) was identified in broilers aged 36–45 days, with no infection prior to 25 days of age (Kichou et al. 1996). Diagnosis was based on histopathology.

An overall *Cryptosporidium* prevalence of 10.6 % for layer chickens and 3.4 % for broiler chickens was shown in a study of faecal samples from 2015 birds in China (Wang et al. 2010a). The highest prevalence (24.6 %) was found in 31–60-day old laying chickens, whereas prevalences in broiler chickens never exceeded 5 %. DNA analysis identified *C. baileyi* as the major species, with 92/95 investigated samples, and only 3 samples were positive for *C. meleagridis* (Wang et al. 2010a). In contrast, another recent study identified *C. meleagridis* as the major species in chickens (Baroudi et al. 2013). The overall *Cryptosporidium* prevalence was 34.4 % by histopathology, and the highest prevalence (46.2 %) was identified in 16–30-day-old chickens, which is in line with the results from Kichou et al. (1996) and Wang et al. (2010a). The majority of the birds were infected with *C. meleagridis* only ($n = 25$). *Cryptosporidium baileyi* only was detected in four birds and a mixed *C. meleagridis*/*C. baileyi* infection was found in one bird. However, these chickens had died from diarrhoea, which could affect the outcome regarding *Cryptosporidium* sp.

4.6.1.2 Turkeys

A morbidity of 5–10 % due to sinusitis was reported for a flock where *Cryptosporidium* sp. could be isolated from diseased poults (Glisson et al. 1984). It was stated that macroscopic *post mortem* examination of the infraorbital sinuses of healthy birds was normal compared with those of diseased birds, but it was not clearly stated whether *Cryptosporidium* sp. was also identified in the healthy birds and thus the infection prevalence cannot be estimated. Goodwin et al (1988b) identified invasive *Cryptosporidium* stages in turkey poults form a farm where the poults suffered from self-limiting diarrheal of unknown aetiology, but no prevalence estimation was given. Prevalences of 80 % in 17-day-old poults, 38 % in 24-day-old and 0 % in $\geq$60-day-old poults was found by Woodmansee et al. (1988). Oocysts were identified as *C. meleagridis* based on morphology and infection site. A 35.5 % (17/60) prevalence in diarrhoeic or just unthrifty poults was reported in Iran (Gharagozlou et al. 2006). Prevalence was based on histological examination of intestinal, bursal and cloacal tissues. Examination of faeces revealed that only 29 % of the infected birds shed oocysts. Infection was identified in 1–7-week-old poults, whereas the 43 uninfected poults all were older than 7 weeks. DNA analysis was not performed and oocyst size was not stated in the publication, but based on infection site, host species and symptoms the authors suggested that *C. meleagridis* was the species responsible. A 10.0 %, 10.5 % and 2.5 % pre-slaughter prevalence respectively (age 4–9 weeks) was detected upon faecal examination of three flocks from the same farm (McEvoy and Giddings 2009). One of 59 turkeys was positive at post-slaughter examination (age 14 weeks). Upon DNA analysis, all six positive samples were identified as *C. parvum*. In a recent study, a 43.9 % prevalence of *C. meleagridis* was shown in deceased turkeys, with the highest prevalence (57.9 %) in poults aged 16–30 days (Baroudi et al. 2013).

4.6.1.3 Ducks and Geese

In one study, 73 (57 %) of 128 ducklings and 44 (59 %) of goslings aged 8–35 days were infected with *Cryptosporidium* (Richter et al. 1994). Infection was present in both intestinal and respiratory tract, but oocyst morphology was not described.

In a study on experimental infection with Usutu virus in geese, *Cryptosporidium* developmental stages in tissue samples were an accidental finding. This was further investigated by *in situ*-hybridization, and *Cryptosporidium* infection was detected in 89 % of conjunctival tissue samples and 88 % of bursal tissue samples. DNA analysis revealed presence of *C. baileyi* (Chvala et al. 2006). *C. baileyi* was also identified in two ducks in Rio de Janeiro (Huber et al. 2007).

4.6.1.4 Other Birds

Ratites *Cryptosporidium* infection in ostriches was first described in the early 1990s (Allwright and Wessels 1993; Bezuidenhout et al. 1993; Gajadhar 1993, 1994; Penrith et al. 1994; Penrith and Burger 1993). Infection was first identified in faecal samples from 14 (8.5 %) of 165 ostriches imported from Africa to Canada (Gajadhar 1993). Penrith and Burger (1993) identified invasive stages in a section of the small intestine of a 4-week old chick that has suffered from rectal prolapse, and Allwright and Wessels (1993) identified *Cryptosporidium* in histology sections of the bursa, intestine and pancreatic ducts. In 1994, Gajadhar et al. characterized the isolated oocysts and investigated host specificity. The oocysts were morphologically similar to those of *C. meleagridis*, but attempts to infect suckling mice, chickens, turkeys and quail failed, indicating that this was probably another species. In addition, only faecal samples were investigated, so the infection site was not determined (Gajadhar 1994). As this study was conducted before molecular tools were commonly used for *Cryptosporidium* species determination, the true identity of this isolate will remain unknown.

A low prevalence, with only 2 (0.6 %) of 336 investigated samples from ostriches aged 2 months–5 years being *Cryptosporidium* positive, was found in Greece (Ponce Gordo et al. 2002). Oocysts were of two sizes, 3.8 × 3.8 μm and 5.7 × 4.8 μm, indicating the presence of two different species. In contrast, in a Spanish study a 60 % *Cryptosporidium* prevalence in adult rheas and ostriches was found (Ponce Gordo et al. 2002). The authors reported an oocyst diameter of 3–5 μm, which is similar to the description provided by Gajadhar (1994). Molecular analysis of the isolates was not performed. Oliveira et al. (2008) found 44 % prevalence in 77 ostriches based on microscopy. Oocysts were generally morphologically similar to *C. baileyi* and *Cryptosporidium* avian genotype II (Ryan and Xiao 2008). However, the morphometric variation was so large that the authors suggested that more than one species had been identified (Oliveira et al. 2008), but this was not verified by molecular analysis. An isolate similar to *C. baileyi* in both oocyst morphology and PCR-RFLP banding pattern was described from Brazilian ostriches (Santos et al. 2005). The isolate was characterized as a sister taxon to *C. baileyi* by sequence analysis of the 18S rDNA, HSP70 and actin genes (Meireles et al. 2006), and was named *Cryptosporidium* avian genotype II by another research group (Ng et al. 2006). Experimental infection (oral or intratracheal) with the Brazilian isolate in chickens failed (Meireles et al. 2006). The avian genotype II has also been identified in Vietnam. On a single ostrich farm 110 (23.7 %) of 464 samples were positive for *Cryptosporidium* oocysts. The highest prevalence as well as the highest shedding intensity (35.2 %) was found in 2–3 month-old animals. Of 17 samples used for molecular characterization, all were found to be avian genotype II (Nguyen et al. 2013).

Wang et al (2011b) reported *Cryptosporidium* infection in 53 (11.7 %) of 452 investigated ostrich samples. Prevalence peaked at the age of 4–8 weeks with 16.2 %. No infection was detected in birds younger than 1 week or older than 12 months. Molecular analysis of positive samples identified only *C. baileyi*.

Quails and partridges Enteric cryptosporidiosis in quails, with oocysts similar to *C. meleagridis*, was first described in 1986 (Hoerr et al. 1986; Ritter et al. 1986). Early attempts at experimental infection of quail with *C. baileyi* isolated from chickens failed (Current et al. 1986; Lindsay et al. 1986), but were later successful (Cardozo et al. 2005). Natural infection was first documented in 2001 (Morgan et al. 2001). Since then, natural *C. baileyi* infection has been described in two reports (Murakami et al. 2002; Wang et al. 2012). One large survey of *Cryptosporidium* infection in quails was performed in China (Wang et al. 2012). Out of 1,818 faecal samples, 239 (13.1 %) from 29 (61.7 %) farms were positive. Infection was most common among 72–100-day old quails (23.6 %). DNA analysis revealed *C. baileyi* in 237 samples and *C. meleagridis* in two samples. One case of *Cryptosporidium* infection in partridges was described (Pages-Mante et al. 2007).

Pigeons There are a few reports of cryptosporidiosis in pigeons (Ozkul and Aydin 1994; Qi et al. 2011; Radfar et al. 2012; Rodriguez et al. 1997). Radfar et al. (2012) describe an overall prevalence of 2.9 % in 102 examined adult and nestling birds, with 3.4 % prevalence in adults and 2.3 % prevalence in nestlings. The other articles are case reports (Ozkul and Aydin 1994; Rodriguez et al. 1997) and a study on pet birds in general, where *C. meleagridis* was found in one pigeon (Qi et al. 2011).

4.6.2 *Association of Infection with Clinical Disease*

4.6.2.1 Chickens

Respiratory as well as intestinal and bursal *Cryptosporidium* infections cause disease in chickens, but infection without clinical symptoms has also been observed (Fletcher et al. 1975; Taylor et al. 1994).

In Spain, a 90 % morbidity due to respiratory infection in one flock was caused by *Cryptosporidium* sp. Weekly mortality rates were 0.9–1.5 % (Fernandez et al. 1990). Infection was detected in the trachea and oesophagus. In another flock investigated in the same study, weight loss was the primary symptom, and bursal cryptosporidiosis was diagnosed (Fernandez et al. 1990). Goodwin et al. (1996) found a correlation between *C. baileyi* infection of the trachea and severity of tracheitis symptoms, airsacculitis and condemnation of birds.

In respiratory cryptosporidiosis, co-infection with other pathogens has been identified in a number of studies. *Cryptosporidium* sp. and concurrent adenovirus infection was identified in a large broiler flock with respiratory disease (Dhillon et al. 1981). In a retrospective study on *post mortem* diagnoses of respiratory cryptosporidiosis, it was found that co-infection with virus or bacteria was common (Goodwin et al. 1988a). In another study, *Cryptosporidium* sp. and *Aspergillus* or bacteria were detected in the lungs of four layer chickens that died from pneumonia. *Cryptosporidium* were also found in the ureters and kidneys (Nakamura and Abe 1988). The effect of *Cryptosporidium* infection alone on development of clinical symptoms in these cases cannot be estimated, but there is probably a synergistic effect of co-infections, increasing the severity. Such a synergistic effect of co-infection with infectious bronchitis virus or *Escherichia coli* has been reported (Blagburn et al. 1991).

Respiratory symptoms were reported from chickens that had been experimentally inoculated intra-tracheally with *C. baileyi*, whereas infection was successful but caused no symptoms in orally inoculated chickens (Lindsay et al. 1988).

C. meleagridis infection was associated with diarrhoea and mortality in one study of Algerian chickens (Baroudi et al. 2013). Experimental *C. meleagridis* infection of chickens has been observed to result in the chickens becoming indolent and having soiled feathers. Growth retardation was reported, but compensatory growth occurred after a few weeks (Tumova et al. 2002).

4.6.2.2 Turkeys

Turkey was the first animal species in which clinical cryptosporidiosis was described (Slavin 1955). Infection was associated with diarrhoea at 10–14 days of age, but other parasites (including *Histomonas*, *Trichomonas* and Strongylides) were also detected. Experimental infection (crop inoculation) with *C. meleagridis* produced infection of the ileum, caecum and bursa, but was not associated with clinical symptoms (Bermudez et al. 1988). The isolate used was from symptomatic poults; however, these were simultaneously infected with reovirus (causing enteritis and hepatitis). Co-infection with *Cryptosporidium* and reovirus in turkeys with enteritis and hepatitis, leading to increased mortality, has also been shown in another study (Wages and Ficken 1989). The presence of other pathogens in these studies could indicate a low to moderate primary pathogenicity of *C. meleagridis*. Self-limiting diarrhoea (moderate to severe in character), a slower growth rate and growth deformities were reported from one farm where diseased poults were diagnosed with *Cryptosporidium* infection (Goodwin et al. 1988b). Other pathogens were not excluded, as was also mentioned by the authors.

In dead poults that had suffered from depression and diarrhoea (faeces adhered on the hind part of the body), necropsy revealed lesions in the small intestine. Microscopic investigation identified *Cryptosporidium* sp. in the respiratory tract and kidneys, as well as in the gastrointestinal tract (Tacconi et al. 2001). Diarrhoea, emaciation, lethargy and reduced growth associated with natural *C. meleagridis* infection have been reported from Iran, but the presence of other pathogens was not excluded (Gharagozlou et al. 2006). Of 60 diarrhoeal and/or unthrifty birds, 17 (35.3 %) were identified as *Cryptosporidium* positive by histology, and *C. meleagridis* was reported based on oocyst morphology. Baroudi et al. (2013) identified *C. meleagridis* in 25 (44 %) of 57 examined turkeys that died from diarrhoea, but infection with other pathogens was not investigated.

Respiratory cryptosporidiosis has also been described in turkeys (Ranck and Hoerr 1987; Tarwid et al. 1985). Tarwid et al. (1985) identified *Cryptosporidium* sp. in necropsied birds from two outbreaks of colibacillosis. Colibacillosis is, according to the authors, a secondary disease in turkeys, and *Cryptosporidium* sp. was identified as the primary pathogen. Symptoms were frothy conjunctivitis and increased mortality. Necropsy revealed pathological changes such as pericarditis, peritonitis and air-sacculitis in addition to the conjunctivitis that was observed in live birds.

Thirteen birds with respiratory disease were all positive for *Cryptosporidium* sp. by histology (Ranck and Hoerr 1987). Microscopy of sinus and/or tracheal exudates revealed oval oocysts in some samples, but oocyst size was not described, and both *C. baileyi* and *C. meleagridis* can appear oval (length/width ratios of 1.05–1.79 and 1.00–1.33 respectively (Ryan and Xiao 2008)). Symptoms such as coughing, rattling, sneezing, frothy eyes and swollen sinuses were reported. Other pathogens were present in all but two of the examined birds, and it is unclear whether the infection with *Cryptosporidium* played a primary role in the pathogenesis or not. Studies on cryptosporidiosis in turkeys, including clinical symptoms, are summarised in Table 4.13.

4.6.2.3 Ducks and Geese

Clinical cryptosporidiosis in ducks and geese seems to be less common and milder (see Table 4.14) than in other poultry. Only mild respiratory symptoms resulted from experimental *C. baileyi* infection (both oral and intratracheal inoculation) in ducks (Lindsay et al. 1989). Respiratory and intestinal infection occurred for both infection routes, but symptoms (sneezing, rales, mild dyspnea) were only present in animals infected by the intratracheal route. Mason (1986) described a case of conjunctival cryptosporidiosis. However, since only one of 97 affected ducks was *Cryptosporidium* positive, the author concluded that the parasite was not the cause of the disease. Similarly, no symptoms occurred in geese in which *Cryptosporidium* infection was detected in the conjunctivas and bursas (Chvala et al. 2006); and Richter et al. (1994) noted that enteritis and upper respiratory tract symptoms were equally present in infected and non-infected ducks and geese. Mortality was not increased in the positive flocks (Richter et al. 1994).

4.6.2.4 Other Birds

Ratites *Cryptosporidium* infection in ostrich chicks has been associated with cloacal and phallus prolapse, leading to high mortality (Bezuidenhout et al. 1993; Penrith et al. 1994; Santos et al. 2005). Bezuidenhout et al (1993) found that prolapsed cloacas were heavily infected, whereas Penrith et al. (1994) described heavy infection of both the bursa and cloaca in affected chicks, but healthy chicks were not infected. Santos et al (2005) also identified *Cryptosporidium* infection in the rectum, coprodeum, urodeum and bursa of two dead chicks with cloacal prolapse, both originating from a farm with high mortality rates in 7–30-day-old chicks. However, the authors did not associate the problems with the infection, since changed management practices decreased clinical symptoms and mortality,

Table 4.13 Studies on *Cryptosporidium* infection in turkeys

Country	Age	Symptoms	Location of parasites	Oocyst size	*Cryptosporidium* sp.	Diagnostic method	Reference
United Kingdom	10–14 days	Diarrhea, mortality	Distal jejunum, ileum	4.5 × 4.0 μm	*C. meleagridis*	Feces (smears) histopathology	(Slavin 1955)
United States	7 weeks	Sinusitis, serous conjunctivitis	Infraorbital sinuses	Not stated	*Cryptosporidium* spp.	Histopathology	(Glisson et al. 1984)
Canada	5 weeks	Bronchopneumonia, conjunctivitis, mortality	Trachea	Not stated	*Cryptosporidium* spp.	Histopathology	(Tarwid et al. 1985)
United States	2.5–11 weeks	Coughing, gasping, sneezing, rattling, sinusitis	Turbinates, sinuses, trachea, bronchi	Not stated	*Cryptosporidium* spp.	Exudate (smears) histopathology	(Ranck and Hoerr 1987)
United States[a]	5–26 days	None	Ileum, caecum, bursa	4.9 μm Ø	*C. meleagridis*	Feces (smears + auramine O) histopathology	(Bermudez et al. 1988)
United States	Not stated	Diarrhea, depression, growth retardation, abnormal feathers, misshapen bones	Mid - to distal small intestine	Not stated, developmental stages 2–4 μm Ø	*C. meleagridis*	Pathology, histopathology	(Goodwin et al. 1988b)
United States	25 days	Enteritis, depression, stunted growth, mortality	Ileum, ileocaecal junction	5 μm	*C. meleagridis*	Feces (smears + auramine O) histopathology	(Wages and Ficken 1989)
Hungary[a]	1 weeks	None stated	Mainly small intestine[b]	4.8 × 4.2 μm	*C. meleagridis*	Histopathology, 18S rDNA PCR +	(Sreter et al. 2000)
Italy	30 days	Diarrhea, depression, huddling	Ileum, caecal tonsil, caecum, rectum, bursa, duodenum, proventriculus, kidney, trachea, lung	4.5–5.0 μm Ø	*C. meleagridis*	mucosal scrapings (flotation + mZN) histopathology	(Tacconi et al. 2001)

Iran	1–7 weeks	diarrhea, emaciation, lethargy, growth retardation	Duodenum, jejunum, ileum, caecum, colon, cloaca, bursa	Not stated, developmental stages <5 μm Ø	*C meleagridis*	Feces (flotation + mZN) pathology, histopathology	(Gharagozlou et al. 2006)
United States	4, 9, 18 weeks	None stated	unknown/caecum[c]	Not stated	*C. parvum*	18S rDNA analysis of fecal droppings and post-slaughter caecal content	(McEvoy and Giddings 2009)

[a]Experimental infection
[b]Determined in chickens and mice, oocysts first passaged through turkey poults
[c]One caecum-positive

Table 4.14 Studies on *Cryptosporidium* spp. in ducks and geese

Country	Age	Species	Symptoms	Location of parasites	Oocyst size	*Cryptosporidium* sp.	Diagnostic method	Reference
Ducks								
United States[b]	4 days	Domestic ducks	None or mild respiratory disease	Trachea, Bursa	Not stated	*C. baileyi*	Histology	(Lindsay et al. 1987)
Germany	9–35 days	Peking ducks	Not stated	Bursa, cloaca, intestine, respiratory tract, conjunctiva	Not stated	*Cryptosporidium* spp.	Tissue scrapings + mZN, histology + IFA	(Richter et al. 1994)
Australia	Not stated	Black duck (wild)	Not stated	Not stated	Not stated	*Cryptosporidium* duck genotype	18S rDNA PCR + sequencing	(Morgan et al. (2001)
United States	Unknown	Wild ducks	Unknown	Intestine	Not stated	*Cryptosporidium* spp.	Faecal flotation + IFA, 18S rDNA PCR	(Kuhn et al. 2002)
Brazil	Not stated	Domestic ducks	Not stated	Not done	Not stated	*C. baileyi*	18S rDNA PCR-RFLP and sequencing	(Huber et al. 2007)
Geese								
United States	25 days	Domestic geese	Not stated	Large intestine	Not stated Schizont 3.6 μm Macrogamete 4.5 μm	*Cryptosporidium anserium*, nomen nudum	Histology	(Proctor and Kemp 1974)
Germany	8–35 days	Domestic geese (Danish breed)	Not stated	Bursa, gastro-intestinal and respiratory tract	Not stated	*Cryptosporidium* spp.	Tissue scrapings + mZN, histology + IFA	(Richter et al. 1994)

United States	Unknown	Canada geese	Unknown	Not done	Not done	*C. parvum*[c]	TRAP C2 and β-tubulin PCR + genotyping	(Graczyk et al. 1998)
Austria[a]	16–36 days	Domestic geese	Not stated	Bursa, conjunctiva	Not done	*C. baileyi*	Histology with in-situ hybridization, 18S rDNA PCR + sequencing	(Chvala et al. 2006)
United States	Unknown	Canada geese	Unknown	Not done	Not done	*Cryptosporidium* goose genotypes I, II, III, IV, V[d]	18S rDNA PCR and sequencing	(Jellison et al. 2004)
United States	Unknown	Canada geese	Unknown	Not done	Not done	*Cryptosporidium* goose genotypes I, II, *Cryptosporidium* duck genotype, *C. parvum*[c], *C. hominis*[c]	18S rDNA PCR-RFLP and sequencing	(Zhou et al. 2004)

[a]Experimental infection study for Usutu virus, *Cryptosporidium* accidental finding
[b]Experimental infection
[c]Finding reported as passage of oocysts, not manifest infection
[d]Not named in this publication

although *Cryptosporidium* infection was still present on the farm. Enteritis was indicated by the presence of intestinal invasive stages and rectal prolapse in one chick examined by Penrith and Burger (1993). Because diarrhoea was not reported it is unknown whether the prolapse was caused by intense bowel movements or something else. *Cryptosporidium* infection has also been associated with pancreatic necrosis (Allwright and Wessels 1993).

Quails and partridges *Cryptosporidium* infection has been shown in both diarrhoea and respiratory disease in quails (Guy et al. 1987; Hoerr et al. 1986; Murakami et al. 2002; Ritter et al. 1986). Hoerr et al. (1986) reported high mortality rates from 5 days of age in quails infected with *Cryptosporidium* sp., and with no bacterial or viral pathogens detected. Acute fatal diarrhoea with mortality rates of up to 45 % in 0–17-day-old birds was described by Ritter et al (1986). Reovirus was also detected in necropsied birds, but another study reported that experimental infection with reovirus alone did not produce diarrhoea, whereas infection with *Cryptosporidium* sp., either alone or simultaneously with reovirus, resulted in severe diarrhoea and mortality (Guy et al. 1987). A synergistic effect of co-infection was, however, shown, since oocyst shedding was higher and reovirus infection became systemic and liver necrosis occurred (Guy et al. 1987).

Muramaki et al. (2002) reported a daily mortality rate of 5.7 % in one farm, where birds suffered from upper respiratory tract disease and decreased egg production. Respiratory symptoms were head swelling, nasal discharge and increased lacrimation, and necropsy revealed sinusitis, airsacculitis and egg peritonitis. Co-infection of *Cryptosporidium* sp., *Mycoplasma gallisepticum* and other bacteria was shown. The authors concluded that *M. gallisepticum* was the primary pathogen, but that the mixed infections in conjunction with high ammonia concentrations in the air worsened the symptoms. The role of *Cryptosporidium* infection in respiratory disease in quails thus remains unclear. Wang et al. (2012) reported that no clinical symptoms were seen in 1,818 sampled quails, of which 239 were *Cryptosporidium* positive.

C. meleagridis was the only pathogen identified in an outbreak of diarrhoea and cough in red-legged partridge chicks (Pages-Mante et al. 2007). Morbidity rates were 60–70 % and mortality more than 50 %, indicating high pathogenicity. Invasive stages were identified in both the respiratory and intestinal tract, suggesting that not only *C. bailey* might be associated with respiratory avian cryptosporidiosis.

Pigeons Diarrhoea associated with cryptosporidiosis in pigeons has been described in four birds (Ozkul and Aydin 1994; Rodriguez et al. 1997). Rodriguez et al. (1997) described a 40 % morbidity of yellow watery diarrhoea, weight loss, dehydration and weakness in a farm with 280 pigeons. Mortality was 5 % and necropsy of three birds revealed invasive stages of *Cryptosporidium* in the small intestine, caecum, colon, cloaca, and bursa. No viruses or bacteria could be isolated. Ozkul and Aydin (1994) identified invasive stages in the small intestine of a pigeon that had been depressed and had evidence of diarrhoea in the form of faeces in its hind feathers.

4.6.3 *Infection Dynamics: Oocyst Excretion and Transmission*

Isolates of both *C. baileyi* and *C. meleagridis* derived from one domestic bird species have been successfully transmitted to other domestic birds (Current et al. 1986; Lindsay et al. 1987). *C. galli* has not been experimentally transmitted between different domestic birds, but has been shown in finches as well as chickens (Blagburn et al. 1990; Pavlásek 1999, 2001; Ryan et al. 2003b), and thus has the potential to infect different bird species.

4.6.3.1 Chickens

The prepatent period of *C. baileyi* is approximately 4–8 days (Hornok et al. 1998; Lindsay et al. 1988; Rhee et al. 1991; Tumova et al. 2002). However, in the first report on *C. baileyi* infection in chickens, a prepatent period of up to 24 days was described (Current et al. 1986). Older chicks have a slightly longer prepatent periods than younger ones (Lindsay et al. 1988; Rhee et al. 1991; Taylor et al. 1994; Tumova et al. 2002).

The patent period varies more. At oral inoculation of 2-day-old chicks, a patent period of 26 days was seen, whereas it was 11–15 days in chicks inoculated at 14 days of age, 11–12 days in chicks inoculated at 28 days of age and <7 days in chicks inoculated at 42 days of age (Lindsay et al. 1988). With intratracheal inoculation, the same authors described patent periods of 27, 11–19, 10–11 and <7 days in these age groups (Lindsay et al. 1988). Rhee et al. (1991) and Tumova et al. (2002) observed a mean patent period of approximately 14 days. Oocyst excretion peaked on day 12 and days 11–17 post inoculation, respectively (Rhee et al. 1991; Tumova et al. 2002). Taylor et al. (1994) showed shorter patent periods and lower total oocyst output in older than younger chickens. There was also an effect of infection dose, in that oocyst output was higher and declined more slowly with lower infection doses (Taylor et al. 1994). Similar observations were made for 1- and 9-week old chicks (Sreter et al. 1995). In that study, the mean patent period for 1-week old chicks was 32 days, but one chicken shed oocysts for 151 days.

C. meleagridis was shed in the faeces on days 4–7 post infection in two chickens experimentally infected at 6 weeks of age (Woodmansee et al. 1988). Tumova et al. (2002) infected 7-day-old chicks. Oocysts first appeared 3 days later and the patent period lasted for 16–17 days. Shedding rates were significantly lower than in chicks inoculated with the same number of *C. baileyi* oocysts.

The prepatent and patent period of *C. galli* has been described to be 25 and 6 days respectively (Pavlásek 2001), but was later reported as unknown when *C. galli* was redescribed (Ryan et al. 2003b).

4.6.3.2 Turkeys

The prepatent period of *C. meleagridis* in turkeys inoculated at 7–11 days of age was 2–4 days (Bermudez et al. 1988; Sreter et al. 2000; Woodmansee et al. 1988).

Woodmansee et al. (1988) reported that oocysts were shed for only 4 days; Sreter et al. (2000) found the patent period to be 8–10 days, whereas Bermudez et al. (1988) reported oocyst shedding and invasive stages still being present at day 21 post inoculation. Oocyst shedding rates were moderate (Sreter et al. 2000), and low to moderate (Bermudez et al. 1988).

Experimental infection with *C. baileyi* induced mild infection of the bursa (Current et al. 1986). Lindsay et al. (1987) inoculated turkey poults via the intratracheal, oral and intracloacal route. All three experiments caused infection, but only poults inoculated via the trachea developed symptoms (Lindsay et al. 1987).

4.6.3.3 Ducks and Geese

Lindsay et al (1986) described a prepatent period of 5 days and a possible patent period of 9–10 days in experimentally infected Muscovy ducks, based on investigation of pooled faecal samples (Lindsay et al. 1986). Oocyst morphology was not described. The intestine, bursa and cloaca were positive for invasive stages, but these tissues can be infected by both *C. baileyi* and *C. meleagridis* (Table 4.14).

4.6.3.4 Other Birds

For quails, one study describes a prepatent period of 7 days and a patent period of 21 days for *C. baileyi* (Cardozo et al. 2005). Otherwise, no data are available.

4.6.4 *Zoonotic Transmission*

Only one of the species and genotypes commonly infecting birds – *Cryptosporidium meleagridis* – has, so far, proved to be important in human cryptosporidiosis. This species is the third most common species in human cryptosporidiosis worldwide. In the industrialised world, *C. meleagridis* infection is usually associated with cryptosporidiosis cases in travellers to Asia or Africa (Elwin et al. 2012; Insulander et al. 2013; Leoni et al. 2006), but autochthonous cases have also been described (Elwin et al. 2012; Leoni et al. 2006; Silverlås et al. 2012). Studies on *Cryptosporidium* prevalence and species distribution in humans in South America have identified *C. meleagridis* infection at about the same prevalence as *C. parvum* (Cama et al. 2003, 2007, 2008). Although this is a true zoonotic species, there is only one report in which the bird source has been identified, and in that case, chickens and not turkeys were involved (Silverlås et al. 2012). It is not known whether anthroponotic transmission occurs with this species, but it has been indicated by the fact that not all *C. meleagridis*-infected patients in an epidemiological investigation had had bird or animal contact (Elwin et al. 2012). *C. meleagridis* has the potential to infect other mammalian species as well, and experimental infection of mice, rats, rabbits, pigs and calves has been reported (Akiyoshi et al. 2003; Darabus and Olariu 2003). Due to the wide host range and the close relationship of *C. meleagridis* to *C. parvum* and *C. hominis*, it has been proposed that this species originated as a mammalian *Cryptosporidium* species, and later adapted to birds (Xiao et al. 2002b, 2004).

One study has identified *C. parvum* in turkeys (McEvoy and Giddings 2009), indicating that this species could play a role in zoonotic transmission. However, only one of 59 birds post-slaughter was positive compared to 2.5–10.5 % of the 5–10-week younger poults, which means risk of transmission via contaminated meat should be very small. The higher prevalence in poults should not pose a risk as long as the flocks are closed to the public. The shedding intensity was not reported, but prevalence indicates that infection rather than just intestinal passage was present. Some studies have identified *C. parvum*, *C. hominis* and *C. hominis*-like isolates in Canada geese (Jellison et al. 2004, 2009; Zhou et al. 2004). The authors conclude that these findings are probably not associated with infection and parasite proliferation, but rather transient carriage. Nevertheless, this indicates that domesticated ducks and geese can potentially act as transmission vehicles for these species.

Infection with *C. baileyi* has been identified in one immunodeficient patient. Diagnosis was based on oocyst morphology and biology – experimental infection of mice failed whereas inoculated chickens developed infection of the intestine, bursa and trachea (Ditrich et al. 1991). Since this patient was immunodeficient and no other reports exist, this species should not be considered as a true zoonotic agent.

4.7 Conclusion

Ever since animals were first domesticated, and humans became dependent upon them for the commodities that they supply, particularly food and fibre, the infections that affect the health and productivity of livestock have been a concern. Cryptosporidiosis was first identified as a disease of veterinary significance in the 1950s (in turkeys) and then in the early 1970s in calves, but major interest in cryptosporidiosis only developed with the first report of a human cases later that decade, and the recognition that *Cryptosporidium* infection was also of medical importance. Since then our knowledge on the veterinary significance of *Cryptosporidium* infection has expanded enormously – particularly in the livestock sector most impacted by cryptosporidiosis – young calves. However, as demonstrated in this chapter, it should not be forgotten almost all farmed animals may be pathologically affected by at least one species of *Cryptosporidium*, often causing clinical disease that in some instances may be fatal. For some *Cryptosporidium* species in some farmed animal species, transmission may be anthropozoonotic.

Cryptosporidium is a hugely successful parasite, as demonstrated by its host range and wide geographic distribution, and its control has proved challenging. As long as humans raise and depend on animals, there will be a need to control the transmission of cryptosporidiosis amongst livestock species.

References

Abbassi H, Coudert F, Cherel Y, Dambrine G, Brugere-Picoux J, Naciri M (1999) Renal cryptosporidiosis (*Cryptosporidium baileyi*) in specific-pathogen-free chickens experimentally coinfected with Marek's disease virus. Avian Dis 43:738–744

Abou-Eisha AM (1994) Cryptosporidial infection in man and faro animals in Ismailia Governorate. Vet Med J Giza 42:107–111

Adesiyun AA, Kaminjolo JS, Ngeleka M, Mutani A, Borde G, Harewood W, Harper W (2001) A longitudinal study on enteropathogenic infections of livestock in Trinidad. Rev Soc Bras Med Trop 34:29–35

Akiyoshi DE, Dilo J, Pearson C, Chapman S, Tumwine J, Tzipori S (2003) Characterization of *Cryptosporidium meleagridis* of human origin passaged through different host species. Infect Immun 71:1828–1832

Allwright DM, Wessels J (1993) *Cryptosporidium* species in ostriches. Vet Rec 133:24

Alonso-Fresan MU, Vazquez-Chagoyan JC, Velazquez-Ordonez V, Pescador-Salas N, Saltijeral-Oaxaca J (2009) Sheep management and cryptosporidiosis in central Mexico. Trop Anim Health Prod 41:431–436

Alves M, Xiao L, Antunes F, Matos O (2006) Distribution of *Cryptosporidium* subtypes in humans and domestic and wild ruminants in Portugal. Parasitol Res 99:287–292

Amer S, Honma H, Ikarashi M, Tada C, Fukuda Y, Suyama Y, Nakai Y (2010) *Cryptosporidium* genotypes and subtypes in dairy calves in Egypt. Vet Parasitol 169(3–4):382–386

Amer S, Zidan S, Feng Y, Adamu H, Li N, Xiao L (2013) Identity and public health potential of *Cryptosporidium* spp. in water buffalo calves in Egypt. Vet Parasitol 191(1–2):123–127

Anderson BC (1991) Experimental infection in mice of *Cryptosporidium muris* isolated from a camel. J Protozool 38:16S–17S

Anderson BC (1998) Cryptosporidiosis in bovine and human health. J Dairy Sci 81 (11):3036–3041

Angus KW (1988) Mammalian cryptosporidiosis: a veterinary perspective. In: Angus KW, Blewett DA (eds) Cryptosporidiosis. Proceedings of the first international workshop. The Animal Diseases Research Association, pp 43–55

Anonymous (2001) Foot and Mouth disease online database. Animal health and welfare: FMD data archive. Department for Environment, Food and Rural Affairs (DEFRA), UK

Anonymous (2002) The development of a national collection for oocysts of *Cryptosporidium*. UK Drinking Water Inspectorate (DWI) 170/2/125. Marlow (UK), Foundation for Water Research, 2002

Anonymous (2011) *Cryptosporidium* in Östersund (in Swedish). Swedish Institute for Communicable Disease Control. http://www.smittskyddsinstitutet.se/upload/Publikationer/*Cryptosporidium*-i-Ostersund-2011-15-4.pdf

Argenzio RA, Liacos JA, Levy ML, Meuten DJ, Lecce JG, Powell DW (1990) Villous atrophy crypt hyperplasia cellular infiltration and impaired glucose-Na absorption in enteric cryptosporidiosis of pigs. Gastroenterology 98:1129–1140

Atwill ER, Sweitzer RA, Pereira MG, Gardner IA, Van Vuren D, Boyce WM (1997) Prevalence of and associated risk factors for shedding *Cryptosporidium parvum* oocysts and *Giardia* cysts within feral pig populations in California. Appl Environ Microbiol 63:3946–3949

Banda Z, Nichols RA, Grimason AM, Smith HV (2009) *Cryptosporidium* infection in non-human hosts in Malawi. Onderstepoort J Vet Res 76:363–375

Banerjee AK, Angulo AF, Dhasmana KM, Kong-A-San J (1987) Acute diarrhoeal disease in rabbit: bacteriological diagnosis and efficacy of oral rehydration in combination with loperamide hydrochloride. Lab Anim 21(4):314–317

Barker IK, Carbonell PL (1974) *Cryptosporidium agni* sp.n. from lambs and *Cryptosporidium bovis* sp.n. from a calf with observations on the oocyst. Z Parasitenkd 44:289–298

Baroudi D, Khelef D, Goucem R, Adjou KT, Adamu H, Zhang H, Xiao L (2013) Common occurrence of zoonotic pathogen *Cryptosporidium meleagridis* in broiler chickens and turkeys in Algeria. Vet Parasitol. doi:10.1016/j.vetpar.2013.02.022, In press

Bergeland ME (1977) Necrotic enteritis in nursing piglets. In: 20th annual proceedings of American Association Veterinary Laboratory Diagnosticians, pp 151–158

Bermudez AJ, Ley DH, Levy MG, Ficken MD, Guy JS, Gerig TM (1988) Intestinal and bursal cryptosporidiosis in turkeys following inoculation with *Cryptosporidium* sp. isolated from commercial poults. Avian Dis 32:445–450

Bezuidenhout AJ, Penrith ML, Burger WP (1993) Prolapse of the phallus and cloaca in the ostrich (*Struthio camelus*). J S Afr Vet Assoc 64:156–158

Bhat SA, Juyal PD, Singla LD (2012) Prevalence of cryptosporidiosis in neonatal buffalo calves in Ludhiana district of Punjab, India. Asian J Anim Vet Adv 7(6):512–520

Bidewell CA, Cattell JH (1998) Cryptosporidiosis in young alpacas. Vet Rec 142:287

Bilic HR, Bilkei G (2006) *Balantidium*, *Cryptosporidium* and *Giardia* species infections in indoor and outdoor pig production units in Croatia. Vet Rec 158:61

Björkman C, Mattsson JG (2006) Persistent infection in a dairy herd with an unusual genotype of *Cryptosporidium parvum*. FEMS Microbiol Lett 254:71–74

Blagburn BL, Lindsay DS, Hoerr FJ, Atlas AL, Toivio-Kinnucan M (1990) *Cryptosporidium* sp. infection in the proventriculus of an Australian diamond firetail finch (*Staganoplura bella*: Passeriformes, Estrildidae). Avian Dis 34:1027–1030

Blagburn BL, Lindsay DS, Hoerr FJ, Davis JF, Giambrone JJ (1991) Pathobiology of cryptosporidiosis (*C. baileyi*) in broiler chickens. J Protozool 38:25S–28S

Blanchard PC (2012) Diagnostics of dairy and beef cattle diarrhea. Vet Clin North Am Food Anim Pract 28(3):443–464

Bomfim TC, Huber F, Gomes RS, Alves LL (2005) Natural infection by *Giardia* sp. and *Cryptosporidium* sp. in dairy goats associated with possible risk factors of the studied properties. Vet Parasitol 134:9–13

Bornay-Llinares FJ, da Silva AJ, Moura INS, Myjak P, Pietkiewicz H, Kruminis-Lozowska W, Graczyk TK, Pieniazek NJ (1999) Identification of *Cryptosporidium felis* in a cow by morphologic and molecular methods. Appl Environ Microbiol 65(4):1455–1458

Bridgman SA, Robertson RMP, Syed Q, Speed N, Andrews N, Hunter PR (1995) Outbreak of cryptosporidiosis associated with a disinfected groundwater supply. Epidemiol Infect 115 (3):555–566

Brook EJ, Anthony Hart C, French NP, Christley RM (2009) Molecular epidemiology of *Cryptosporidium* subtypes in cattle in England. Vet J 179(3):378–382

Budu-Amoako E, Greenwood SJ, Dixon BR, Barkema HW, McClure JT (2012a) *Giardia* and *Cryptosporidium* on dairy farms and the role these farms may play in contaminating water sources in Prince Edward Island, Canada. J Vet Intern Med 26(3):668–673

Budu-Amoako E, Greenwood SJ, Dixon BR, Barkema HW, McClure JT (2012b) Occurrence of *Cryptosporidium* and *Giardia* on beef farms and water sources within the vicinity of the farms on Prince Edward Island, Canada. Vet Parasitol 184(1):1–9

Budu-Amoako E, Greenwood SJ, Dixon BR, Sweet L, Ang L, Barkema HW, McClure JT (2012c) Molecular epidemiology of *Cryptosporidium* and *Giardia* in humans on Prince Edward Island, Canada: evidence of zoonotic transmission from cattle. Zoonoses Public Health 59(6):424–433

Budu-Amoako E, Greenwood SJ, Dixon BR, Barkema HW, Hurnik D, Estey C, McClure JT (2012d) Occurrence of *Giardia* and *Cryptosporidium* in pigs on Prince Edward Island, Canada. Vet Parasitol 184:18–24

Burenbaatar B, Bakheit MA, Plutzer J, Suzuki N, Igarashi I, Ongerth J, Karanis P (2008) Prevalence and genotyping of *Cryptosporidium* species from farm animals in Mongolia. Parasitol Res 102:901–905

Burton AJ, Nydam DV, Jones G, Zambriski JA, Linden TC, Cox G, Davis R, Brown A, Bowman DD (2011) Antibody responses following administration of a *Cryptosporidium parvum* rCP15/60 vaccine to pregnant cattle. Vet Parasitol 175(1–2):178–181

Burton AJ, Nydam DV, Mitchell KJ, Bowman DD (2012) Fecal shedding of *Cryptosporidium* oocysts in healthy alpaca crias and their dams. J Am Vet Med Assoc 241:496–498

Cacciò SM, Rinaldi L, Cringoli G, Condoleo R, Pozio E (2007) Molecular identification of *Cryptosporidium parvum* and *Giardia duodenalis* in the Italian water buffalo (*Bubalus bubalis*). Vet Parasitol 150(1–2):146–149

Cacciò SM, Sannella AR, Mariano V, Valentini S, Berti F, Tosini F, Pozio E (2013) A rare *Cryptosporidium parvum* genotype associated with infection of lambs and zoonotic transmission in Italy. Vet Parasitol 191(1–2):128–131

Cama VA, Bern C, Sulaiman IM, Gilman RH, Ticona E, Vivar A, Kawai V, Vargas D, Zhou L, Xiao L (2003) *Cryptosporidium* species and genotypes in HIV-positive patients in Lima Peru. J Eukaryot Microbiol 50(Suppl):531–533

Cama VA, Ross JM, Crawford S, Kawai V, Chavez-Valdez R, Vargas D, Vivar A, Ticona E, Navincopa M, Williamson J, Ortega Y, Gilman RH, Bern C, Xiao L (2007) Differences in clinical manifestations among *Cryptosporidium* species and subtypes in HIV-infected persons. J Infect Dis 196:684–691

Cama VA, Bern C, Roberts J, Cabrera L, Sterling CR, Ortega Y, Gilman RH, Xiao L (2008) *Cryptosporidium* species and subtypes and clinical manifestations in children Peru. Emerg Infect Dis 14:1567–1574

Canestri-Trotti G, Pampiglione S, Visconti S (1984) *Cryptosporidium* e *Isospora suis* nel suino in Italia. Parassitologia 26:299–304

Cardozo SV, Teixeira Filho WL, Lopes CW (2005) Experimental transmission of *Cryptosporidium baileyi* (Apicomplexa: Cryptosporidiidae) isolated of broiler chicken to Japanese quail (*Coturnix japonica*). Rev Bras Parasitol Vet 14:119–124

Casemore DP (1989) Sheep as a source of human cryptosporidiosis. J Infect 19:101–104

Castro-Hermida JA, González-Losada YA, Ares-Mazás E (2002) Prevalence of and risk factors involved in the spread of neonatal bovine cryptosporidiosis in Galicia (NW Spain). Vet Parasitol 106(1):1–10

Castro-Hermida JA, Delafosse A, Pors I, Ares-Mazas E, Chartier C (2005) *Giardia duodenalis* and *Cryptosporidium parvum* infections in adult goats and their implications for neonatal kids. Vet Rec 157:623–627

Castro-Hermida JA, Almeida A, Gonzalez-Warleta M, Correia da Costa JM, Rumbo-Lorenzo C, Mezo M (2007) Occurrence of *Cryptosporidium parvum* and *Giardia duodenalis* in healthy adult domestic ruminants. Parasitol Res 101:1443–1448

Castro-Hermida JA, García-Presedo I, Almeida A, González-Warleta M, Correia Da Costa JM, Mezo M (2011a) *Cryptosporidium* spp. and *Giardia duodenalis* in two areas of Galicia (NW Spain). Sci Total Environ 409(13):2451–2459

Castro-Hermida JA, García-Presedo I, González-Warleta M, Mezo M (2011b) Prevalence of *Cryptosporidium* and *Giardia* in roe deer (*Capreolus capreolus*) and wild boars (*Sus scrofa*) in Galicia (NW Spain). Vet Parasitol 179(1–3):216–219

Cebra CK, Mattson DE, Baker RJ, Sonn RJ, Dearing PL (2003) Potential pathogens in feces from unweaned llamas and alpacas with diarrhea. J Am Vet Med Assoc 223:1806–1808

Centers for Disease Control and Prevention (CDC) (2011) Cryptosporidiosis outbreak at a summer camp–North Carolina, 2009. MMWR Morb Mortal Wkly Rep 60(27):918–922

Chalmers RM (2012) Waterborne outbreaks of cryptosporidiosis. Ann Ist Super Sanita 48(4): 429–446

Chalmers RM, Ferguson C, Cacciò S, Gasser RB, Abs EL-Osta YG, Heijnen L, Xiao L, Elwin K, Hadfield S, Sinclair M, Stevens M (2005) Direct comparison of selected methods for genetic categorisation of *Cryptosporidium parvum* and *Cryptosporidium hominis* species. Int J Parasitol 35:397–410

Chalmers RM, Robinson G, Elwin K, Hadfield SJ, Xiao L, Ryan U, Modha D, Mallaghan C (2009) *Cryptosporidium* sp. rabbit genotype a newly identified human pathogen. Emerg Infect Dis 15(5):829–830

Chalmers RM, Elwin K, Hadfield SJ, Robinson G (2011) Sporadic human cryptosporidiosis caused by *Cryptosporidium cuniculus*, United Kingdom 2007–2008. Emerg Infect Dis 17(3): 536–538

Chang'a JS, Robertson LJ, Mtambo MMA, Mdegela RH, Løken T, Reksen O (2011) Unexpected results from large-scale cryptosporidiosis screening study in calves in Tanzania. Ann Trop Med Parasitol 105(7):515–521

Chartier C, Mallereau M-P, Naciri M (1995) Prophylaxis using paromomycin of natural cryptosporidial infection in neonatal kids. Prev Vet Med 25:357–361

Chen F, Huang K (2012) Prevalence and molecular characterization of *Cryptosporidium* spp. in dairy cattle from farms in China. J Vet Sci 13(1):15–22

Chen Z, Mi R, Yu H, Shi Y, Huang Y, Chen Y, Zhou P, Cai Y, Lin J (2011) Prevalence of *Cryptosporidium* spp. in pigs in Shanghai China. Vet Parasitol 181:113–119

Chvala S, Fragner K, Hackl R, Hess M, Weissenbock H (2006) *Cryptosporidium* infection in domestic geese (*Anser anser f. domestica*) detected by in-situ hybridization. J Comp Pathol 134:211–218

Cieloszyk J, Goni P, Garcia A, Remacha MA, Sanchez E, Clavel A (2012) Two cases of zoonotic cryptosporidiosis in Spain by the unusual species Cryptosporidium ubiquitum and Cryptosporidium felis. Enferm Infecc Microbiol Clin 30:549–551

Cinque K, Stevens MA, Haydon SR, Jex AR, Gasser RB, Campbell BE (2008) Investigating public health impacts of deer in a protected drinking water supply watershed. Water Sci Technol 58(1):127–132

Craig BH, Pilkington JG, Kruuk LE, Pemberton JM (2007) Epidemiology of parasitic protozoan infections in Soay sheep (*Ovis aries* L.) on St Kilda. Parasitology 134:9–21

Current WL, Upton SJ, Haynes TB (1986) The life cycle of *Cryptosporidium baileyi* n. sp. (Apicomplexa Cryptosporidiidae) infecting chickens. J Protozool 33:289–296

Darabus G, Olariu R (2003) The homologous and interspecies transmission of *Cryptosporidium parvum* and *Cryptosporidium meleagridis*. Pol J Vet Sci 6:225–228

de Graaf DC, Vanopdenbosch E, Ortega-Mora LM, Abbassi H, Peeters JE (1999) A review of the importance of cryptosporidiosis in farm animals. Int J Parasitol 29:1269–1287

de la Fé Rodríguez PY, Martin LO, Muñoz EC, Imberechts H, Butaye P, Goddeeris BM, Cox E (2013) Several enteropathogens are circulating in suckling and newly weaned piglets suffering from diarrhea in the province of Villa Clara, Cuba. Trop Anim Health Prod 45(2):435–440

De Waele V, Speybroeck N, Berkvens D, Mulcahy G, Murphy TM (2010) Control of cryptosporidiosis in neonatal calves: use of halofuginone lactate in two different calf rearing systems. Prev Vet Med 96:143–151

De Waele V, Berzano M, Speybroeck N, Berkvens D, Mulcahy GM, Murphy TM (2012) Peri-parturient rise of *Cryptosporidium* oocysts in cows: new insights provided by duplex quantitative real-time PCR. Vet Parasitol 189(2–4):366–368

Delafosse A, Castro-Hermida JA, Baudry C, Ares-Mazas E, Chartier C (2006) Herd-level risk factors for *Cryptosporidium* infection in dairy-goat kids in western France. Prev Vet Med 77:109–121

Dhillon AS, Thacker HL, Dietzel AV, Winterfield RW (1981) Respiratory cryptosporidiosis in broiler chickens. Avian Dis 25:747–751

Díaz de Ramírez A, Jiménez-Garzón JM, Materano-Ocanto PA, Ramírez-Iglesia LN (2012) Dynamic of infections by *Cryptosporidum* spp. and *Giardia* spp. in buffaloes (*Bubalus bubalis*) during the first three months of life. Revista Cientifica de la Facultad de Ciencias Veterinarias de la Universidad del Zulia (Venezuela) 22(6):507–515

Díaz P, Quílez J, Chalmers RM, Panadero R, López C, Sánchez-Acedo C, Morrondo P, Díez-Baños P (2010a) Genotype and subtype analysis of *Cryptosporidium* isolates from calves and lambs in Galicia (NW Spain). Parasitology 137(8):1187–1193

Díaz P, Quilez J, Robinson G, Chalmers RM, Diez-Banos P, Morrondo P (2010b) Identification of *Cryptosporidiumxiaoi* in diarrhoeic goat kids (*Capra hircus*) in Spain. Vet Parasitol 172:132–134

Ditrich O, Palkovic L, Sterba J, Prokopic J, Loudova J, Gibodaa M (1991) The first finding of *Cryptosporidium baileyi* in man. Parasitol Res 77:44–47

Drumo R, Widmer G, Morrison LJ, Tait A, Grelloni V, D'Avino N, Pozio E, Caccio SM (2012) Evidence of host-associated populations of *Cryptosporidium parvum* in Italy. Appl Environ Microbiol 78:3523–3529

Dubey JP, Fayer R, Rao JR (1992) Cryptosporidial oocyst in faeces of water buffalo and zebu calves in India. J Vet Parasitol 6(1):55–56

Duranti A, Cacciò SM, Pozio E, Di Egidio A, De Curtis M, Battisti A, Scaramozzino P (2009) Risk factors associated with *Cryptosporidium parvum* infection in cattle. Zoonoses Publ Health 56(4):176–182

Ebeid M, Mathis A, Pospischil A, Deplazes P (2003) Infectivity of *Cryptosporidium parvum* genotype I in conventionally reared piglets and lambs. Parasitol Res 90:232–235

EFSA-AHAW (European Food Safety Authority – Animal Health and Welfare Panel (2005) The impact of the current housing and husbandry systems on the health and welfare of farmed domestic rabbits. EFSA J 267:1–31, EFSA-Q-2004-023

El-Khodery SA, Osman SA (2008) Cryptosporidiosis in buffalo calves (*Bubalus bubalis*): prevalence and potential risk factors. Trop Anim Health Prod 40(6):419–426

Elwin K, Chalmers RM, Roberts R, Guy EC, Casemore DP (2001) Modification of a rapid method for the identification of gene-specific polymorphisms in *Cryptosporidium parvum* and its application to clinical and epidemiological investigations. Appl Environ Microbiol 67: 5581–5584

Elwin K, Hadfield SJ, Robinson G, Chalmers RM (2012) The epidemiology of sporadic human infections with unusual cryptosporidia detected during routine typing in England and Wales 2000–2008. Epidemiol Infect 140:673–683

Enemark HL, Ahrens P, Lowery CJ, Thamsborg SM, Enemark JMD, Bille-Hansen V, Lind P (2002) *Cryptosporidium andersoni* from a Danish cattle herd: identification and preliminary characterisation. Vet Parasitol 107:37–49

Enemark HL, Ahrens P, Bille-Hansen V, Hoogaard PM, Vigre H, Thamsborg SM, Lind P (2003) *Cryptosporidium parvum*: infectivity and pathogenicity of the 'porcine' genotype. Parasitology 126:107–116

Epe C, Coati N, Schnieder T (2004) Results of parasitological examinations of faecal samples from horses, ruminants, pigs, dogs, cats, hedgehogs and rabbits between 1998 and 2002. Dtsch Tierarztl Wochenschr 111:243–247

Esteban E, Anderson BC (1995) *Cryptosporidium muris*: prevalence, persistency and detrimental effect on milk production in a dry-lot dairy. J Dairy Sci 78(5):1068–1072

FAO (2012). http://www.fao.org/ag/againfo/home/en/news_archive/AGA_in_action/2012_Dairy_Goat_Productivity_in_Asia.html. Accessed 15 Jan 2013

FAOSTAT (2012) Food and Agriculture Organization of the United Nations (FAO) http://faostat.fao.org/site/573/DesktopDefault.aspx?PageID=573#ancor

FAOSTAT (2013a). http://faostat.fao.org/site/573/DesktopDefault.aspx?PageID=573. Accessed 26 Jan 2013

FAOTSTAT (2013b). http://faostat.fao.org/site/569/default.aspx#ancor. Accessed 27 Mar 2013

Farzan A, Parrington L, Coklin T, Cook A, Pintar K, Pollari F, Friendship R, Farber J, Dixon B (2011) Detection and characterization of *Giardia duodenalis* and *Cryptosporidium* spp. on swine farms in Ontario, Canada. Foodborne Pathog Dis 8:1207–1213

Fayer R, Santín M (2009) Cryptosporidium xiaoi n. sp. (Apicomplexa: Cryptosporidiidae) in sheep (Ovis aries). Vet Parasitol 164:192–200

Fayer R, Phillips L, Anderson BC, Bush M (1991) Chronic cryptosporidiosis in a Bactrian camel (*Camelus bactrianus*). J Zoo Wildl Med 22:228–232

Fayer R, Gasbarre L, Pasquali P, Canals A, Almeria S, Zarlenga D (1998) *Cryptosporidium parvum* infection in bovine neonates: dynamic clinical parasitic and immunologic patterns. Int J Parasitol 28:49–56

Fayer R, Trout JM, Xiao L, Morgan UM, Lal AA, Dubey JP (2001) *Cryptosporidium canis* n. sp. from domestic dogs. J Parasitol 87(6):1415–1422

Fayer R, Santín M, Xiao L (2005) *Cryptosporidium bovis* n.sp. (Apicomplexa: Cryptosporidiidae) in cattle (*Bos taurus*). J Parasitol 91:624–629

Fayer R, Santín M, Trout JM, Greiner E (2006) Prevalence of species and genotypes of *Cryptosporidium* found in 1-2-year-old dairy cattle in the eastern United States. Vet Parasitol 135:105–112

Fayer R, Santín M, Trout JM (2007) Prevalence of *Cryptosporidium* species and genotypes in mature dairy cattle on farms in eastern United States compared with younger cattle from the same locations. Vet Parasitol 145:260–266

Fayer R, Santín M, Trout JM (2008) *Cryptosporidium ryanae* n. sp. (Apicomplexa: Cryptosporidiidae) in cattle (*Bos taurus*). Vet Parasitol 156:191–198

Fayer R, Santín M, Dargatz D (2010) Species of *Cryptosporidium* detected in weaned cattle on cow-calf operations in the United States. Vet Parasitol 170:187–192

Featherstone CA, Giles M, Marshall JA, Mawhinney IC, Holliman A, Pritchard GC (2010a) *Cryptosporidium* species in calves submitted for postmortem examination in England and Wales. Vet Rec 167(25):979–980

Featherstone CA, Marshall JA, Giles M, Sayers AR, Pritchard GC (2010b) *Cryptosporidium* species infection in pigs in East Anglia. Vet Rec 166:51–52

Feltus DC, Giddings CW, Schneck BL, Monson T, Warshauer D, McEvoy JM (2006) Evidence supporting zoonotic transmission of *Cryptosporidium* spp. in Wisconsin. J Clin Microbiol 44:4303–4308

Feng Y, Ortega Y, He G, Das P, Xu M, Zhang X, Fayer R, Gatei W, Cama V, Xiao L (2007) Wide geographic distribution of *Cryptosporidium bovis* and the deer-like genotype in bovines. Vet Parasitol 144:1–9

Feng Y, Raj Karna S, Dearen TK, Singh DK, Adhikari LN, Shrestha A, Xiao L (2012) Common occurrence of a unique *Cryptosporidium ryanae* variant in zebu cattle and water buffaloes in the buffer zone of the Chitwan National Park, Nepal. Vet Parasitol 185:309–314

Fernandez A, Quezada M, Gomez MA, Navarro JA, Rodriguez J, Sierra MA (1990) Cryptosporidiosis in chickens from southern Spain. Avian Dis 34:224–227

Fiuza VR, Cosendey RI, Frazao-Teixeira E, Santín M, Fayer R, de Oliveira FC (2011a) Molecular characterization of *Cryptosporidium* in Brazilian sheep. Vet Parasitol 175:360–362

Fiuza VR, Gallo SS, Frazão-Teixeira E, Santín M, Fayer R, Oliveira FC (2011b) *Cryptosporidium* pig genotype II diagnosed in pigs from the state of Rio De Janeiro, Brazil. J Parasitol 97: 146–147

Fleta J, Sánchez-Acedo C, Clavel A, Quílez J (1995) Detection of *Cryptosporidium* oocysts in extra-intestinal tissues of sheep and pigs. Vet Parasitol 59:201–205

Fletcher OJ, Munnell JF, Page RK (1975) Cryptosporidiosis of the bursa of Fabricius of chickens. Avian Dis 19:630–639

Gait R, Soutar RH, Hanson M, Fraser C, Chalmers R (2008) Outbreak of cryptosporidiosis among veterinary students. Vet Rec 162(26):843–845

Gajadhar AA (1993) *Cryptosporidium* species in imported ostriches and consideration of possible implications for birds in Canada. Can Vet J 34:115–116

Gajadhar AA (1994) Host specificity studies and oocyst description of a *Cryptosporidium* sp. isolated from ostriches. Parasitol Res 80:316–319

Geurden T, Goma FY, Siwila J, Phiri IGK, Mwanza AM, Gabriel S, Claerebout E, Vercruysse J (2006) Prevalence and genotyping of *Cryptosporidium* in three cattle husbandry systems in Zambia. Vet Parasitol 138:217–222

Geurden T, Thomas P, Casaert S, Vercruysse J, Claerebout E (2008) Prevalence and molecular characterisation of *Cryptosporidium* and *Giardia* in lambs and goat kids in Belgium. Vet Parasitol 155:142–145

Gharagozlou MJ, Dezfoulian O, Rahbari S, Bokaie S, Jahanzad I, Razavi AN (2006) Intestinal cryptosporidiosis in turkeys in Iran. J Vet Med A Physiol Pathol Clin Med 53:282–285

Giadinis ND, Papadopoulos E, Panousis N, Papazahariadou M, Lafi SQ, Karatzias H (2007) Effect of halofuginone lactate on treatment and prevention of lamb cryptosporidiosis: an extensive field trial. J Vet Pharmacol Ther 30:578–582

Giadinis ND, Symeoudakis S, Papadopoulos E, Lafi SQ, Karatzias H (2012) Comparison of two techniques for diagnosis of cryptosporidiosis in diarrhoeic goat kids and lambs in Cyprus. Trop Anim Health Prod 44:1561–1565

Giles M, Webster KA, Marshall JA, Catchpole J, Goddard TM (2001) Experimental infection of a lamb with *Cryptosporidium parvum* genotype 1. Vet Rec 149:523–525

Giles M, Chalmers R, Pritchard G, Elwin K, Mueller-Doblies D, Clifton-Hadley F (2009) *Cryptosporidium hominis* in a goat and a sheep in the UK. Vet Rec 164:24–25

Glisson JR, Brown TP, Brugh M, Page RK, Kleven SH, Davis RB (1984) Sinusitis in turkeys associated with respiratory cryptosporidiosis. Avian Dis 28:783–790

Gómez MS, Torres J, Gracenea M, Fernandez-Morán J, Gonzalez-Moreno O (2000) Further report on *Cryptosporidium* in Barcelona zoo mammals. Parasitol Res 86:318–323

Gómez-Couso H, Amar CFL, McLauchlin J, Ares-Mazás E (2005) Characterisation of a *Cryptosporidium* isolate from water buffalo (Bubalus bubalis) by sequencing of a fragment of the *Cryptosporidium* oocyst wall protein gene (COWP). Vet Parasitol 131(1–2):139–144

Gómez-Couso H, Ortega-Mora LM, Aguado-Martínez A, Rosadio-Alcántara R, Maturrano-Hernández L, Luna-Espinoza L, Zanabria-Huisa V, Pedraza-Díaz S (2012) Presence and molecular characterisation of *Giardia* and *Cryptosporidium* in alpacas (*Vicugna pacos*) from Peru. Vet Parasitol 187:414–420

Goodwin MA, Latimer KS, Brown J, Steffens WL, Martin W, Resurreccion RS, Smeltzer MA, Dickson TG (1988a) Respiratory cryptosporidiosis in chickens. Poult Sci 67:1684–1693

Goodwin MA, Steffens WL, Russell ID, Brown J (1988b) Diarrhea associated with intestinal cryptosporidiosis in turkeys. Avian Dis 32:63–67

Goodwin MA, Brown J, Resurreccion RS, Smith JA (1996) Respiratory coccidiosis (*Cryptosporidium baileyi*) among northern Georgia broilers in one company. Avian Dis 40:572–575

Gormley FJ, Little CL, Chalmers RM, Rawal N, Adak GK (2011) Zoonotic cryptosporidiosis from petting farms, England and Wales, 1992–2009. Emerg Infect Dis 17:151–152

Gracenea M, Gómez MS, Torres J, Carné E, Fernández-Morán J (2002) Transmission dynamics of *Cryptosporidium* in primates and herbivores at the Barcelona zoo: a long-term study. Vet Parasitol 104:19–26

Graczyk TK, Fayer R, Trout JM, Lewis EJ, Farley CA, Sulaiman I, Lal AA (1998) *Giardia* sp. cysts and infectious *Cryptosporidium parvum* oocysts in the feces of migratory Canada geese (*Branta canadensis*). Appl Environ Microbiol 64:2736–2738

Grinberg A, Pomroy WE, Squires RA, Scuffham A, Pita A, Kwan E (2011) Retrospective cohort study of an outbreak of cryptosporidiosis caused by a rare *Cryptosporidium parvum* subgenotype. Epidemiol Infect 139(10):1542–1550

Guselle NJ, Appelbee AJ, Olson ME (2003) Biology of *Cryptosporidium parvum* in pigs: from weaning to market. Vet Parasitol 113:7–18

Guy JS, Levy MG, Ley DH, Barnes HJ, Gerig TM (1987) Experimental reproduction of enteritis in bobwhite quail (*Colinus virginianus*) with *Cryptosporidium* and reovirus. Avian Dis 31: 713–722

Hamnes IS, Gjerde B, Robertson L, Vikøren T, Handeland K (2002) Prevalence of *Cryptosporidium* and *Giardia* in free-ranging wild cervids in Norway. Vet Parasitol 141(1–2):30–41

Hamnes IS, Gjerde BK, Forberg T, Robertson LJ (2007) Occurrence of *Cryptosporidium* and *Giardia* in suckling piglets in Norway. Vet Parasitol 144:222–233

Harper CM, Cowell NA, Adams BC, Langley AJ, Wohlsen TD (2002) Outbreak of *Cryptosporidium* linked to drinking unpasteurised milk. Commun Dis Intell 26(3):449–450

Healey MC, Yang S, Du C, Liao SF (1997) Bovine fallopian tube epithelial cells, adult C57BL/6 mice, and non-neonatal pigs as models for cryptosporidiosis. J Eukaryot Microbiol 44 (6):64S–65S

Heine J, Moon HW, Woodmansee DB, Pohlenz JF (1984) Experimental tracheal and conjunctival infections with *Cryptosporidium* sp. in pigs. Vet Parasitol 17:17–25

Heitman TL, Frederick LM, Viste JR, Guselle NJ, Morgan UM, Thompson RC, Olson ME (2002) Prevalence of *Giardia* and *Cryptosporidium* and characterization of *Cryptosporidium* spp. isolated from wildlife, human and agricultural sources in the North Saskatchewan River Basin in Alberta, Canada. Can J Microbiol 48:530–541

Helmy YA, Krücken J, Nöckler K, von Samson-Himmelstjerna G, Zessin K-H (2013) Molecular epidemiology of *Cryptosporidium* in livestock animals and humans in the Ismailia province of Egypt. Vet Parasitol 193(1–3):15–24

Higgins RJ (1999) Surveillance for cryptosporidiosis. Pig J 43:88–91

Hoerr FJ, Current WL, Haynes TB (1986) Fatal cryptosporidiosis in quail. Avian Dis 30:421–425

Hornok S, Bitay Z, Szell Z, Varga I (1998) Assessment of maternal immunity to *Cryptosporidium baileyi* in chickens. Vet Parasitol 79:203–212

Hovda LR, McGuirk SM, Lunn DP (1990) Total parenteral nutrition in a neonatal llama. J Am Vet Med Assoc 196:319–322

Huber F, da Silva S, Bomfim TC, Teixeira KR, Bello AR (2007) Genotypic characterization and phylogenetic analysis of *Cryptosporidium* sp. from domestic animals in Brazil. Vet Parasitol 150:65–74

Hunter PR, Chalmers RM, Syed Q, Hughes LS, Woodhouse S, Swift L (2003) Foot and mouth disease and cryptosporidiosis: possible interaction between two emerging infectious diseases. Emerg Infect Dis 9:109–112

Hunter PR, Hughes S, Woodhouse S, Syed Q, Verlander NQ, Chalmers RM, Morgan K, Nichols G, Beeching N, Osborn K (2004) Sporadic cryptosporidiosis case-control study with genotyping. Emerg Infect Dis 10(7):1241–1249

Imboden M, Schaefer DA, Bremel RD, Homan EJ, Riggs MW (2012) Antibody fusions reduce onset of experimental *Cryptosporidium parvum* infection in calves. Vet Parasitol 188 (1–2):41–47

Imre K, Lobo LM, Matos O, Popescu C, Genchi C, Darabus G (2011) Molecular characterisation of *Cryptosporidium* isolates from pre-weaned calves in Romania: is there an actual risk of zoonotic infections? Vet Parasitol 181(2–4):321–324

Imre K, Luca C, Costache M, Sala C, Morar A, Morariu S, Ilie MS, Imre M, Darabus G (2013) Zoonotic *Cryptosporidium parvum* in Romanian newborn lambs (*Ovis aries*). Vet Parasitol 191(1–2):119–122

Inman LR, Takeuchi A (1979) Spontaneous cryptosporidiosis in an adult female rabbit. Vet Pathol 16(1):89–95

Innes EA, Bartley PM, Rocchi M, Benavidas-Silvan J, Burrells A, Hotchkiss E, Chianini F, Canton G, Katzer F (2011) Developing vaccines to control protozoan parasites in ruminants: dead or alive? Vet Parasitol 180(1–2):155–163

Insulander M, Silverlas C, Lebbad M, Karlsson L, Mattsson JG, Svenungsson B (2013) Molecular epidemiology and clinical manifestations of human cryptosporidiosis in Sweden. Epidemiol. Infect 141(5):1009–1020

Itakura C, Goryo M, Umemura T (1984) Cryptosporidial infection in chickens. Avian Pathol 13:487–499

Izumiyama S, Furukawa I, Kuroki T, Yamai S, Sugiyama H, Yagita K, Endo T (2001) Prevalence of *Cryptosporidium parvum* infections in weaned piglets and fattening porkers in Kanagawa Prefecture, Japan. Jpn J Infect Dis 54:23–26

Jellison KL, Distel DL, Hemond HF, Schauer DB (2004) Phylogenetic analysis of the hypervariable region of the 18S rRNA gene of *Cryptosporidium* oocysts in feces of Canada geese (*Branta canadensis*): evidence for five novel genotypes. Appl Environ Microbiol 70:452–458

Jellison KL, Lynch AE, Ziemann JM (2009) Source tracking identifies deer and geese as vectors of human-infectious *Cryptosporidium* genotypes in an urban/suburban watershed. Environ Sci Technol 43(12):4267–4272

Jenkins MB, Liotta JL, Lucio-Forster A, Bowman DD (2010) Concentrations, viability and distribution of *Cryptosporidium* genotypes in lagoons of swine facilities in the Southern Piedmont and in coastal plain watersheds of Georgia. Appl Environ Microbiol 76:5757–5763

Jiang J, Alderisio KA, Xiao L (2005) Distribution of *Cryptosporidium* genotypes in storm event water samples from three watersheds in New York. Appl Environ Microbiol 71:4446–4454

Johnson EH, Muirhead DE, Windsor JJ, King GJ, Al-Busaidy R, Cornelius R (1999) Atypical outbreak of caprine cryptosporidiosis in the Sultanate of Oman. Vet Rec 145:521–524

Johnson J, Buddle R, Reid S, Armson A, Ryan UM (2008) Prevalence of *Cryptosporidium* genotypes in pre and post-weaned pigs in Australia. Exp Parasitol 119:418–421

Johnson D, Harms NJ, Larter NC, Elkin BT, Tabel H, Wei G (2010) Serum biochemistry, serology and parasitology of boreal caribou (*Rangifer tarandus* caribou) in the Northwest Territories, Canada. J Wildl Dis 46(4):1096–1107

Kaminjolo JS, Adesiyun AA, Loregnard R, Kitson-Piggott W (1993) Prevalence of *Cryptosporidium* oocysts in livestock in Trinidad and Tobago. Vet Parasitol 45:209–213

Karanis P, Plutzer J, Halim NA, Igori K, Nagasawa H, Ongerth J, Liqing M (2007) Molecular characterization of *Cryptosporidium* from animal sources in Qinghai province of China. Parasitol Res 101:1575–1580

Karanis P, Eiji T, Palomino L, Boonrod K, Plutzer J, Ongerth J, Igarashi I (2010) First description of *Cryptosporidium bovis* in Japan and diagnosis and genotyping of *Cryptosporidium* spp. in diarrheic pre-weaned calves in Hokkaido. Vet Parasitol 169(3–4):387–390

Kennedy GA, Kreitner GL, Strafuss AC (1977) Cryptosporidiosis in three pigs. J Am Vet Med Assoc 170:348–350

Keshavarz A, Haghighi A, Athari A, Kazemi B, Abadi A, Mojarad EN (2009) Prevalence and molecular characterization of bovine *Cryptosporidium* in Qazvin province, Iran. Vet Parasitol 160(3–4):316–318

Khan SM, Debnath C, Pramanik AK, Xiao L, Nozakid T, Gangulya S (2010) Molecular characterization and assessment of zoonotic transmission of *Cryptosporidium* from dairy cattle in West Bengal, India. Vet Parasitol 171(1–2):41–47

Kiang KM, Scheftel JM, Leano FT, Taylor CM, Belle-Isle PA, Cebelinski EA, Danila R, Smith KE (2006) Recurrent outbreaks of cryptosporidiosis associated with calves among students at an Educational Farm Programme, Minnesota, 2003. Epidemiol Infect 134(4):878–886

Kichou F, Saghir F, El Hamidi M (1996) Natural *Cryptosporidium* sp. infection in broiler chickens in Morocco. Avian Pathol 25:103–111

Klein P, Kleinová T, Volek Z, Šimunek J (2008) Effect of *Cryptosporidium parvum* infection on the absorptive capacity and paracellular permeability of the small intestine in neonatal calves. Vet Parasitol 152(1–2):53–59

Koudela B, Vítovec J, Dao Trong D, Phan Dich L (1986) Preliminary communication on cryptosporidiosis of pigs in Viet-Nam. Folia Parasitol (Praha) 33:301–304

Koyama Y, Satoh M, Maekawa K, Hikosaka K, Nakai Y (2005) Isolation of *Cryptosporidium andersoni* Kawatabi type in a slaughterhouse in the northern island of Japan. Vet Parasitol 130:323–326

Kuhn RC, Rock CM, Oshima KH (2002) Occurrence of *Cryptosporidium* and *Giardia* in wild ducks along the Rio Grande River valley in southern New Mexico. Appl Environ Microbiol 68:161–165

Kvác M, Ditrich O, Kouba M, Sak B, Vitovec J, Kvetonova D (2004) Failed attempt of *Cryptosporidium andersoni* infection in lambs. Folia Parasitol (Praha) 51:373–374

Kváč M, Hanzlíková D, Sak B, Kvetonová D (2009a) Prevalence and age-related infection of Cryptosporidium suis, C. muris and Cryptosporidium pig genotype II in pigs on a farm complex in the Czech Republic. Vet Parasitol 160:319–322

Kváč M, Sak B, Hanzlíková D, Kotilová J, Kvetonová D (2009b) Molecular characterization of *Cryptosporidium* isolates from pigs at slaughterhouses in South Bohemia, Czech Republic. Parasitol Res 104:425–428

Kvác M, Hromadová N, Kvetonová D, Rost M, Sak B (2011) Molecular characterization of *Cryptosporidium* spp. in pre-weaned dairy calves in the Czech Republic: absence of *C. ryanae* and management-associated distribution of *C. andersoni*, *C. bovis* and *C. parvum* subtypes. Vet Parasitol 177(3–4):378–382

Kváč M, Kestřánová M, Pinková M, Květoňová D, Kalinová J, Wagnerová P, Kotková M, Vítovec J, Ditrich O, McEvoy J, Stenger B, Sak B (2013) *Cryptosporidium scrofarum* n. sp. (Apicomplexa: Cryptosporidiidae) in domestic pigs (*Sus scrofa*). Vet Parasitol 191 (3–4):218–227

Lai M, Zhou RQ, Huang HC, Hu SJ (2011) Prevalence and risk factors associated with intestinal parasites in pigs in Chongqing, China. Res Vet Sci 91:121–124

Lange H, Johansen ØH, Vold L, Robertson LJ, Nygard K (Submitted) Second outbreak of infection with a rare *Cryptosporidium parvum* genotype among schoolchildren associated with contact with lambs/goat kids at a holiday farm in Norway

Langkjaer RB, Vigre H, Enemark HL, Maddox-Hyttel C (2007) Molecular and phylogenetic characterization of *Cryptosporidium* and *Giardia* from pigs and cattle in Denmark. Parasitology 134:339–350

Leoni F, Amar C, Nichols G, Pedraza-Diaz S, McLauchlin J (2006) Genetic analysis of *Cryptosporidium* from 2414 humans with diarrhoea in England between 1985 and 2000. J Med Microbiol 55:703–707

Levine ND (1961) Protozoan parasites of domestic animals and of man. Burgess Publishing, Minneapolis

Lindsay DS, Blagburn BL, Sundermann CA (1986) Host specificity of *Cryptosporidium* sp. isolated from chickens. J Parasitol 72:565–568

Lindsay DS, Blagburn BL, Hoerr FJ (1987) Experimentally induced infections in turkeys with *Cryptosporidium baileyi* isolated from chickens. Am J Vet Res 48:104–108

Lindsay DS, Blagburn BL, Sundermann CA, Giambrone JJ (1988) Effect of broiler chicken age on susceptibility to experimentally induced *Cryptosporidium baileyi* infection. Am J Vet Res 49:1412–1414

Lindsay DS, Blagburn BL, Sundermann CA, Hoerr FJ (1989) Experimental infections in domestic ducks with *Cryptosporidium baileyi* isolated from chickens. Avian Dis 33:69–73

Lindsay DS, Upton SJ, Owens DS, Morgan UM, Mead JR, Blagburn BL (2000) *Cryptosporidium andersoni* n. sp. (Apicomplexa: Cryptosporiidae) from cattle, *Bos taurus*. J Eucaryot Microbiol 47(1):91–95

Liu A, Ji H, Wang E, Liu J, Xiao L, Shen Y, Li Y, Zhang W, Ling H (2011) Molecular identification and distribution of *Cryptosporidium* and *Giardia duodenalis* in raw urban wastewater in Harbin, China. Parasitol Res 109:913–918

Lopez-Urbina MT, Gonzalez AE, Gomez-Puerta LA, Romero-Arbizu MA, Oerales-Camacho RA, Rojo-Vasquez FA, Xiao L, Cama V (2009) Prevalence of neonatal cryptosporidiosis in Andean alpacas (*Vicugna pacos*) in Peru. Open Parasitol J 3:9–13

Mac Kenzie WR, Hoxie NJ, Proctor ME, Gradus MS, Blair KA, Peterson DE, Kazmierczak JJ, Addiss DG, Fox KR, Rose JB, Davis JP (1994) A massive outbreak in Milwaukee of *Cryptosporidium* infection transmitted through the public water supply. N Engl J Med 331(3):161–167

Maddox-Hyttel C, Langkjaer RB, Enemark HL, Vigre H (2006) *Cryptosporidium* and *Giardia* in different age groups of Danish cattle and pigs – occurrence and management associated risk factors. Vet Parasitol 141:48–59

Mahdi NK, Ali NH (1992) Cryptosporidiosis among animal handlers and their livestock in Basrah, Iraq. East Afr Med J 79:550–553

Maikai BV, Umoh JU, Kwaga JKP, Lawal IA, Maikai VA, Cama V, Xiao L (2011) Molecular characterization of *Cryptosporidium* spp. in native breeds of cattle in Kaduna State, Nigeria. Vet Parasitol 178(3–4):241–245

Mallon M, MacLeod A, Wastling J, Smith H, Reilly B, Tait A (2003a) Population structures and the role of genetic exchange in the zoonotic pathogen *Cryptosporidium parvum*. J Mol Evol 56:407–417

Mallon ME, MacLeod A, Wastling JM, Smith H, Tait A (2003b) Multilocus genotyping of *Cryptosporidium parvum* Type 2: population genetics and sub-structuring. Infect Genet Evol 3:207–218

Mason RW (1986) Conjunctival cryptosporidiosis in a duck. Avian Dis 30:598–600

Mason RW, Hartley WJ, Tilt L (1981) Intestinal cryptosporidiosis in a kid goat. Aust Vet J 57:386–388

Maurya PS, Rakesh RL, Pradeep B, Kumar S, Kundu K, Garg R, Ram H, Kumar A, Banerjee PS (2013) Prevalence and risk factors associated with *Cryptosporidium* spp. infection in young domestic livestock in India. Trop Anim Health Prod 45:941–946

McEvoy JM, Giddings CW (2009) *Cryptosporidium* in commercially produced turkeys on-farm and postslaughter. Lett Appl Microbiol 48:302–306

McLauchlin J, Amar C, Pedraza-Diaz S, Nichols GL (2000) Molecular epidemiological analysis of *Cryptosporidium* spp. in the United Kingdom: results of genotyping *Cryptosporidium* spp. in 1,705 fecal samples from humans and 105 fecal samples from livestock animals. J Clin Microbiol 38:3984–3990

Meireles MV, Soares RM, dos Santos MM, Gennari SM (2006) Biological studies and molecular characterization of a *Cryptosporidium* isolate from ostriches (*Struthio camelus*). J Parasitol 92(3):623–626

Meireles MV, de Oliveira FP, Teixeira WFP, Coelho WMD, Mendes LCN (2011) Molecular characterization of *Cryptosporidium* spp. in dairy calves from the state of São Paulo, Brazil. Parasitol Res 109(3):949–951

Millard PS, Gensheimer KF, Addiss DG, Sosin DM, Becket GA, Houck-Jankoski A, Hudson A (1994) An outbreak of cryptosporidosis from fresh-pressed apple cider. JAMA 272(20): 1592–1596

Mišic Z, Abe N (2007) Subtype analysis of *Cryptosporidium parvum* isolates from calves on farms around Belgrade, Serbia and Montenegro using the 60 kDa glycoprotein gene sequences. Parasitology 134(3):351–358

Mišic Z, Katic-Radivojevic S, Kulisic Z (2003) *Cryptosporidium* infection in nursing, weaning and post-weaned piglets and sows in the Belgrade district. Acta Vet 53:361–366

Moeller S, Crespo LF (2009) Overview of world swine and pork production. In: Lal R (ed) Agricultural sciences – volume 1. Encyclopedia of life support systems. UNESCO. pp 195–208

Mohanty BN, Panda MR (2012) Prevalence of cryptosporidiosis in buffaloes in and around Bhubaneswar, Odisha. Indian J Field Vet 8(1):55–58

Moon HW, Bemrick WJ (1981) Fecal transmission of calf cryptosporidia between calves and pigs. Vet Pathol 18:248–255

Morgan UM, Sargent KD, Deplazes P, Forbes DA, Spano F, Hertzberg H, Elliot A, Thompson RC (1998) Molecular characterization of *Cryptosporidium* from various hosts. Parasitology 117:31–37

Morgan UM, Buddle JR, Elliott A, Thompson RC (1999) Molecular and biological characterisation of *Cryptosporidium* in pigs. Aust Vet J 77:44–47

Morgan UM, Xiao L, Monis P, Sulaiman I, Pavlasek I, Blagburn B, Olson M, Upton SJ, Khramtsov NV, Lal A, Elliot A, Thompson RC (2000) Molecular and phylogenetic analysis of *Cryptosporidium muris* from various hosts. Parasitology 120:457–464
Morgan UM, Monis PT, Xiao L, Limor J, Sulaiman I, Raidal S, O'Donoghue P, Gasser R, Murray A, Fayer R, Blagburn BL, Lal AA, Thompson RC (2001) Molecular and phylogenetic characterisation of *Cryptosporidium* from birds. Int J Parasitol 31:289–296
Morrison LJ, Mallon ME, Smith HV, MacLeod A, Xiao L, Tait A (2008) The population structure of the *Cryptosporidium parvum* population in Scotland: a complex picture. Infect Genet Evol 8:121–129
Mosier DA, Cimon KY, Kuhls TL, Oberst RD, Simons KR (1997) Experimental cryptosporidiosis in adult and neonatal rabbits. Vet Parasitol 69(3–4):163–169
Mueller-Doblies D, Giles M, Elwin K, Smith RP, Clifton-Hadley FA, Chalmers RM (2008) Distribution of *Cryptosporidium* species in sheep in the UK. Vet Parasitol 154:214–219
Muhid A, Robertson I, Ng J, Ryan U (2011) Prevalence of land management factors contributing to *Cryptosporidium* sp. infection in pre-weaned and post-weaned calves in Johor, Malaysia. Exp Parasitol 127(2):534–538
Munoz M, Alvarez M, Lanza I, Carmenes P (1996) Role of enteric pathogens in the aetiology of neonatal diarrhoea in lambs and goat kids in Spain. Epidemiol Infect 117:203–211
Murakami S, Miyama M, Ogawa A, Shimada J, Nakane T (2002) Occurrence of conjunctivitis, sinusitis and upper region tracheitis in Japanese quail (*Coturnix coturnix japonica*), possibly caused by *Mycoplasma gallisepticum* accompanied by *Cryptosporidium* sp. infection. Avian Pathol 31:363–370
Nakamura K, Abe F (1988) Respiratory (especially pulmonary) and urinary infections of *Cryptosporidium* in layer chickens. Avian Pathol 17:703–711
Nasir A, Avais M, Khan MS, Ahmad N (2009) Prevalence of *Cryptosporidium parvum* infection in Lahore (Pakistan) and its association with diarrhea in dairy calves. Int J Agric Biol 11(2): 221–224
Navarro-i-Martinez L, da Silva AJ, Bornay-Llinares FJ, Moura IN, del Aguila C, Oleaga A, Pieniazek NJ (2007) Detection and molecular characterization of *Cryptosporidium bovis*-like isolate from a newborn lamb in Spain. J Parasitol 93:1536–1538
Nazemalhosseini-Mojarad E, Haghighi A, Taghipour N, Keshavarz A, Mohebi SR, Zali MR, Xiao L (2011) Subtype analysis of *Cryptosporidium parvum* and *Cryptosporidium hominis* isolates from humans and cattle in Iran. Vet Parasitol 179(1–3):250–252
Ng J, Pavlásek I, Ryan U (2006) Identification of novel *Cryptosporidium* genotypes from avian hosts. Appl Environ Microbiol 72:7548–7553
Ng JSY, Eastwood K, Walker B, Durrheim DN, Massey PD, Porigneaux P, Kemp R, McKinnon B, Laurie K, Miller D, Bramley E, Ryan U (2012) Evidence of *Cryptosporidium* transmission between cattle and humans in northern New South Wales. Exp Parasitol 130(4):437–441
Nguyen ST, Fukuda Y, Tada C, Huynh VV, Nguyen DT, Nakai Y (2013) Prevalence and molecular characterization of *Cryptosporidium* in ostriches (*Struthio camelus*) on a farm in central Vietnam. Exp Parasitol 133(1):8–11
Nichols G, Chalmers R, Lake I, Sopwith W, Regan M, Hunter P, Grenfell P, Harrison F, Lane C (2006) Cryptosporidiosis: a report on the surveillance and epidemiology of *Cryptosporidium* infection in England and Wales. Research Contract Number DWI 70/2/201 (849)
Nichols RA, Connelly L, Sullivan CB, Smith HV (2010) Identification of *Cryptosporidium* species and genotypes in Scottish raw and drinking waters during a one-year monitoring period. Appl Environ Microbiol 76:5977–5986
Nolan MJ, Jex AR, Haydon SR, Stevens MA, Gasser RB (2010) Molecular detection of *Cryptosporidium cuniculus* in rabbits in Australia. Infect Genet Evol 10(8):1179–1187
Noordeen F, Rajapakse RP, Faizal AC, Horadagoda NU, Arulkanthan A (2000) Prevalence of *Cryptosporidium* infection in goats in selected locations in three agroclimatic zones of Sri Lanka. Vet Parasitol 93:95–101

Noordeen F, Faizal AC, Rajapakse RP, Horadagoda NU, Arulkanthan A (2001) Excretion of *Cryptosporidium* oocysts by goats in relation to age and season in the dry zone of Sri Lanka. Vet Parasitol 99:79–85

Núñez A, McNeilly F, Perea A, Sánchez-Cordón PJ, Huerta B, Allan G, Carrasco L (2003) Coinfection by *Cryptosporidium parvum* and porcine circovirus type 2 in weaned pigs. J Vet Med B Infect Dis Vet Public Health 50(5):255–258

O'Brien E, McInnes L, Ryan U (2008) *Cryptosporidium* GP60 genotypes from humans and domesticated animals in Australia, North America and Europe. Exp Parasitol 118:118–121

O'Donoghue PJ, Tham VL, de Saram WG, Paull KL, McDermott S (1987) *Cryptosporidium* infection in birds and mammals and attempted cross transmission studies. Vet Parasitol 26:1–11

O'Handley RM, Olson ME (2006) Giardiasis and cryptosporidiosis in ruminants. Vet Clin Food Anim 22:623–643

O'Handley RM, Cockwill C, McAllister TA, Jelinski M, Morck DW, Olson ME (1999) Duration of naturally acquired giardiosis and cryptosporidiosis in dairy calves and their association with diarrhea. J Am Vet Med Assoc 214(3):391–396

Oliveira FC, Ederli NB, Ederli BB, Albuquerque MC, Dos Santos MD (2008) Occurrence of *Cryptosporidium* spp. oocysts (Apicomplexa, Cryptosporidiidae) in ostriches, *Struthio camelus* L., 1758 (Aves, Struthionidae) reared in North and Lowered Coastline regions of the state of Rio de Janeiro, Brazil. Rev Bras Parasitol Vet 17(Suppl 1):322–325

Olson ME, Thorlakson CI, Deselliers L, Morck DW, McAllister TA (1997) *Giardia* and *Cryptosporidium* in Canadian farm animals. Vet Parasitol 68:375–381

Olson ME, O'Handley RM, Ralston BJ, McAllister TA, Thompson RCA (2004) Update on *Cryptosporidium* and *Giardia* infections in cattle. Trends Parasitol 20(4):185–191

Ondrácková Z, Kvác M, Sak B, Kvetonová D, Rost M (2009) Prevalence and molecular characterization of *Cryptosporidium* spp. in dairy cattle in South Bohemia, the Czech Republic. Vet Parasitol 165(1–2):141–144

Ong CS, Eisler DL, Alikhani A, Fung VW, Tomblin J, Bowie W, Isaac-Renton JL (2002) Novel *Cryptosporidium* genotypes in sporadic cryptosporidiosis cases: first report of human infections with a cervine genotype. Emerg Infect Dis 8:263–268

Orr MB, Mackintosh CG, Suttie JM (1985) Cryptosporidiosis in deer calves. New Zeal Vet J 33 (9):151–152

Ortega-Mora LM, Wright SE (1994) Age-related resistance in ovine cryptosporidiosis: patterns of infection and humoral immune response. Infect Immun 62:5003–5009

Ozkul IA, Aydin Y (1994) Small-intestinal cryptosporidiosis in a young pigeon. Avian Pathol 23:369–372

Pages-Mante A, Pages-Bosch M, Majo-Masferrer N, Gomez-Couso H, Ares-Mazas E (2007) An outbreak of disease associated with cryptosporidia on a red-legged partridge (*Alectoris rufa*) game farm. Avian Pathol 36:275–278

Panciera RJ, Thomassen RW, Garner FM (1971) Cryptosporidial infection in a calf. Vet Pathol 8:479–484

Paoletti B, Giangaspero A, Gatti A, Iorio R, Cembalo D, Milillo P, Traversa D (2009) Immunoenzymatic analysis and genetic detection of *Cryptosporidium parvum* in lambs from Italy. Exp Parasitol 122:349–352

Paraud C, Pors I, Chartier C (2010) Evaluation of oral tilmicosin efficacy against severe cryptosporidiosis in neonatal kids under field conditions. Vet Parasitol 170:149–152

Paraud C, Pors I, Journal JP, Besnier P, Reisdorffer L, Chartier C (2011) Control of cryptosporidiosis in neonatal goat kids: efficacy of a product containing activated charcoal and wood vinegar liquid (Obionekk(R)) in field conditions. Vet Parasitol 180:354–357

Pavlásek I (1999) Cryptosporidia: biology diagnosis, host spectrum specificity and the environment. Remedia-Klinicka Mikrobiologie 3:290–301

Pavlásek I (2001) Findings of cryptosporidia in the stomach of chickens and of exotic and wild birds. Veterinarstvi 51:103–108

Pavlásek I, Lávicka M, Tumová E, Skrivan M (1996) Spontaneous *Cryptosporidium* infection in weaned rabbits (Article in Czech). Vet Med (Praha) 41(12):361–366
Paziewska A, Bednarska M, Nieweglowski H, Karbowiak G, Bajer A (2007) Distribution of *Cryptosporidium* and *Giardia* spp. in selected species of protected and game mammals from North-Eastern Poland. Ann Agric Environ Med 14(2):265–270
Peng MM, Wilson ML, Holland RE, Meshnick SR, Lal AA, Xiao L (2003) Genetic diversity of *Cryptosporidium* spp. in cattle in Michigan: implications for understanding the transmission dynamics. Parasitol Res 90:175–180
Penrith ML, Burger WP (1993) A *Cryptosporidium* sp in an ostrich. J S Afr Vet Assoc 64:60–61
Penrith ML, Bezuidenhout AJ, Burger WP, Putterill JF (1994) Evidence for cryptosporidial infection as a cause of prolapse of the phallus and cloaca in ostrich chicks (*Struthio camelus*). Onderstepoort J Vet Res 61:283–289
Pereira SJ, Ramirez NE, Xiao L, Ward LA (2002) Pathogenesis of human and bovine *Cryptosporidium parvum* in gnotobiotic pigs. J Infect Dis 186:715–718
Plutzer J, Karanis P (2007) Genotype and subtype analyses of *Cryptosporidium* isolates from cattle in Hungary. Vet Parasitol 146:357–362
Pohjola S, Oksanen H, Jokipii L, Jokipii AM (1986) Outbreak of cryptosporidiosis among veterinary students. Scand J Infect Dis 18:173–178
Ponce Gordo F, Herrera S, Castro AT, Garcia Duran B, Martinez Diaz RA (2002) Parasites from farmed ostriches (*Struthio camelus*) and rheas (*Rhea americana*) in Europe. Vet Parasitol 107:137–160
Pople NC, Allen AL, Woodbury MR (2001) A retrospective study of neonatal mortality in farmed elk. Can Vet J 42(12):925–928
Pritchard GC, Marshall JA, Giles M, Chalmers RM, Marshall RN (2007) *Cryptosporidium parvum* infection in orphan lambs on a farm open to the public. Vet Rec 161:11–14
Proctor SJ, Kemp RL (1974) *Cryptosporidium anserinum* sp. n. (Sporozoa) in a domestic goose Anser anser L, from Iowa. J Protozool 21:664–666
Qi M, Wang R, Ning C, Li X, Zhang L, Jian F, Sun Y, Xiao L (2011) *Cryptosporidium* spp. in pet birds: genetic diversity and potential public health significance. Exp Parasitol 128:336–340
Quílez J, Ares-Mazás E, Sánchez-Acedo C, del Cacho E, Clavel A, Causapé AC (1996a) Comparison of oocyst shedding and the serum immune response to *Cryptosporidium parvum* in cattle and pigs. Parasitol Res 82:529–534
Quílez J, Sánchez-Acedo C, Clavel A, del Cacho E, López-Bernad F (1996b) Prevalence of *Cryptosporidium* infections in pigs in Aragón, northeastern Spain. Vet Parasitol 67:83–88
Quílez J, Torres E, Chalmers RM, Hadfield SJ, Del Cacho E, Sanchez-Acedo C (2008a) *Cryptosporidium* genotypes and subtypes in lambs and goat kids in Spain. Appl Environ Microbiol 74:6026
Quílez J, Torres E, Chalmers RM, Robinson G, Del Cacho E, Sanchez-Acedo C (2008b) *Cryptosporidium* species and subtype analysis from dairy calves in Spain. Parasitology 135(14): 1613–1620
Radfar MH, Asl EN, Seghinsara HR, Dehaghi MM, Fathi S (2012) Biodiversity and prevalence of parasites of domestic pigeons (*Columba livia domestica*) in a selected semiarid zone of South Khorasan, Iran. Trop Anim Health Prod 44:225–229
Radostits OM, Gay CC, Hinchcliff KW, Constable PD (2007) Diseases of the newborn. In: Veterinary medicine, 10th edn. WB Saunders Company Ltd, London, pp 127–160
Ralston BJ, McAllister TA, Olson ME (2003) Prevalence and infection pattern of naturally acquired giardiasis and cryptosporidiosis in range beef calves and their dams. Vet Parasitol 114(2):113–122
Ranck FM Jr, Hoerr FJ (1987) Cryptosporidia in the respiratory tract of turkeys. Avian Dis 31:389–391
Razawi SM, Oryan A, Bahrami S, Mohammadalipour A, Gowhari M (2009) Prevalence of *Cryptosporidium* infection in camels (*Camelus dromedarius*) in a slaughterhouse in Iran. Trop Biomed 26:267–273

Rhee JK, Seuk Y, Park BK (1991) Isolation and identification of *Cryptosporidium* from various animals in Korea. Korean J Parasitol 29:139–148

Richter D, Wiegand-Tripp G, Burkhardt E, Kaleta EF (1994) Natural infections by *Cryptosporidium* sp. in farm-raised ducks and geese. Avian Pathol 23:277–286

Rickard LG, Siefker C, Boyle CR, Gentz EJ (1999) The prevalence of *Cryptosporidium* and *Giardia* spp. in fecal samples from free-ranging, white-tailed deer (*Odocoileus virginianus*) in the southeastern United States. J Vet Diagn Invest 11(1):65–72

Rieux A, Paraud C, Pors I, Chartier C (2013) Molecular characterization of *Cryptosporidium* spp. in pre-weaned kids in a dairy goat farm in western France. Vet Parasitol 192:268–272

Rinaldi L, Condoleo RU, Condoleo R, Saralli G, Bruni G, Cringoli G (2007a) *Cryptosporidium* and *Giardia* in water buffaloes (*Bubalus bubalis*) of the Italian Mediterranean bred. Vet Res Commun 31(S1):253–255

Rinaldi L, Musella V, Condoleo R, Saralli G, Veneziano V, Bruni G, Condoleo RU, Cringoli G (2007b) *Giardia* and *Cryptosporidium* in water buffaloes (*Bubalus bubalis*). Parasitol Res 100(5):1113–1118

Ritter GD, Ley DH, Levy M, Guy J, Barnes HJ (1986) Intestinal cryptosporidiosis and reovirus isolation from bobwhite quail (*Colinus virginianus*) with enteritis. Avian Dis 30:603–608

Robertson LJ (2009) *Giardia* and *Cryptosporidium* infections in sheep and goats: a review of the potential for transmission to humans via environmental contamination. Epidemiol Infect 137:913–921

Robertson LJ, Chalmers RM (2013) Foodborne cryptosporidiosis: is there really more in Nordic countries? Trends Parasitol 29(1):3–9

Robertson B, Sinclair MI, Forbes AB, Veitch M, Kirk M, Cunliffe D, Willis J, Fairley CK (2002) Case-control studies of sporadic cryptosporidiosis in Melbourne and Adelaide, Australia. Epidemiol Infect 128(3):419–431

Robertson L, Gjerde B, Forberg T, Haugejorden G, Kielland C (2006) A small outbreak of human cryptosporidiosis associated with calves at a dairy farm in Norway. Scand J Infect Dis 23(9): 810–813

Robertson LJ, Gjerde BK, Furuseth Hansen E (2010) The zoonotic potential of *Giardia* and *Cryptosporidium* in Norwegian sheep: a longitudinal investigation of 6 flocks of lambs. Vet Parasitol 171:140–145

Robinson G, Chalmers RM (2010) The European rabbit (*Oryctolagus cuniculus*), a source of zoonotic cryptosporidiosis. Zoonoses Publ Health 57(7–8):e1–e13

Rodriguez F, Oros J, Rodriguez JL, Gonzalez J, Castro P, Fernandez A (1997) Intestinal cryptosporidiosis in pigeons (*Columba livia*). Avian Dis 41:748–750

Rodríguez-De Lara R, Cedillo-Peláez C, Constantino-Casas F, Fallas-López M, Cobos-Peralta MA, Gutiérrez-Olvera C, Juárez-Acevedo M, Miranda-Romero LA (2008) Studies on the evolution pathology, and immunity of commercial fattening rabbits affected with epizootic outbreaks of diarrhoeas in Mexico: a case report. Res Vet Sci 84(2):257–268

Roy SL, DeLong SM, Stenzel SA, Shiferaw B, Roberts JM, Khalakdina A, Marcus R, Segler SD, Shah DD, Thomas S, Vugia DJ, Zansky SM, Dietz V, Beach MJ (2004) Risk factors for sporadic cryptosporidiosis among immunocompetent persons in the United States from 1999 to 2001. J Clin Microbiol 42(7):2944–2951

Rulofson FC, Atwill ER, Holmberg CA (2001) Fecal shedding of Giardia duodenalis, *Cryptosporidium parvum*, salmonella organisms and Escherichia coli O157:H7 from llamas in California. Am J Vet Res 62:637–642

Ryan UM, Xiao L (2008) Birds. In: Fayer R, Xiao L (eds) Cryptosporidium and cryptosporidiosis. CRC Press, Boca Raton

Ryan UM, Samarasinghe B, Read C, Buddle JR, Robertson ID, Thompson RC (2003a) Identification of a novel *Cryptosporidium* genotype in pigs. Appl Environ Microbiol 69:3970–3974

Ryan UM, Xiao L, Read C, Sulaiman IM, Monis P, Lal AA, Fayer R, Pavlásek I (2003b) A redescription of *Cryptosporidium galli* Pavlasek, 1999 (Apicomplexa: Cryptosporidiidae) from birds. J Parasitol 89:809–813

Ryan U, Xiao L, Read C, Zhou L, Lal AA, Pavlasek I (2003c) Identification of novel *Cryptosporidium* genotypes from the Czech Republic. Appl Environ Microbiol 69(7):4302–4307

Ryan UM, Monis P, Enemark HL, Sulaiman I, Samarasinghe B, Read C, Buddle R, Robertson I, Zhou L, Thompson RCA, Xiao L (2004) *Cryptosporidium suis* n. sp. (Apicomplexa: Cryptosporidiidae) in pigs (*Sus scrofa*). J Parasitol 90:769–773

Ryan UM, Bath C, Robertson I, Read C, Elliot A, McInnes L, Traub R, Besier B (2005) Sheep may not be an important zoonotic reservoir for *Cryptosporidium* and *Giardia* parasites. Appl Environ Microbiol 71:4992–4997

Ryan UM, Power M, Xiao L (2008) *Cryptosporidium fayeri* n. sp. (Apicomplexa: Cryptosporidiidae) from the Red Kangaroo (*Macropus rufus*). J Eukaryot Microbiol 55:22–26

Sanford E (1983) Porcine neonatal coccidiosis: clinical, pathological, epidemiological and diagnostic features. Calif Vet 37:26–27

Sanford E (1987) Enteric cryptosporidial infection in pigs: 184 cases (1981–1985). Am Vet Med Assoc 190:695–698

Santín M, Trout JM (2008) Livestock. In: Fayer R, Xiao L (eds) *Cryptosporidium* and cryptosporidiosis. CRC Press, Florida, pp 451–483

Santín M, Trout JM, Xiao L, Zhou L, Greiner E, Fayer R (2004) Prevalence and age-related variation of *Cryptosporidium* species and genotypes in dairy calves. Vet Parasitol 122:103–117

Santín M, Trout JM, Fayer R (2008) A longitudinal study of cryptosporidiosis in dairy cattle from birth, to 2 years of age. Vet Parasitol 155:15–23

Santos MMAB, Peiró JR, Meireles MV (2005) *Cryptosporidium* infection in ostriches (*Struthio camelus*) in Brazil: clinical, morphological and molecular studies. Braz J Poult Sci 7:109–113

Sanz Ceballos L, Illescas Gomez P, Sanz Sampelayo MR, Gil Extremera F, Rodriguez Osorio M (2009) Prevalence of *Cryptosporidium* infection in goats maintained under semi-extensive feeding conditions in the southeast of Spain. Parasite 16:315–318

Sari B, Arslan MO, Gicik Y, Kara M, Tasci GT (2009) The prevalence of *Cryptosporidium* species in diarrhoeic lambs in Kars province and potential risk factors. Trop Anim Health Prod 41:819–826

Sazmand A, Rasooli A, Nouri M, Hamidinejat H, Hekmatimoghaddam S (2012) Prevalence of *Cryptosporidium* spp. in camels and involved people in Yazd Province, Iran. Iran J Parasitol 7:80–84

Sevá AP, Funada MR, Souza Sde O, Nava A, Richtzenhain LJ, Soares RM (2010) Occurrence and molecular characterization of *Cryptosporidium* spp. isolated from domestic animals in a rural area surrounding Atlantic dry forest fragments in Teodoro Sampaio municipality, State of São Paulo, Brazil. Rev Bras Parasitol Vet 19:249–253

Shapiro JL, Watson P, McEwen B, Carman S (2005) Highlights of camelid diagnoses from necropsy submissions to the Animal Health Laboratory, University of Guelph from 1998 to 2004. Can Vet J 46:317–318

Shen Y, Yin J, Yuan Z, Lu W, Xu Y, Xiao L, Cao J (2011) The identification of the *Cryptosporidium ubiquitum* in pre-weaned ovines from Aba Tibetan and Qiang autonomous prefecture in China. Biomed Environ Sci 24:315–320

Shoukry NM, Dawoud HA, Haridy FM (2009) Studies on zoonotic cryptosporidiosisparvum in Ismailia governorate, Egypt. J Egypt Soc Parasitol 39(2):479–488

Silverlås C, Blanco-Penedo I (2013) *Cryptosporidium* spp. in calves and cows from organic and conventional dairy herds. Epidemiol Infect 141(03):529–539

Silverlås C, Björkman C, Egenvall A (2009a) Systematic review and meta-analyses of the effects of halofuginone against calf cryptosporidiosis. Prev Vet Med 91(2–4):73–84

Silverlås C, Emanuelson U, de Verdier K, Björkman C (2009b) Prevalence and associated management factors of *Cryptosporidium* shedding in 50 Swedish dairy herds. Prev Vet Med 90:242–253

Silverlås C, de Verdier K, Emanuelson U, Mattsson JG, Björkman C (2010a) *Cryptosporidium* infection in herds with and without calf diarrhoeal problems. Parasitol Res 107(6):435–444

Silverlås C, Näslund K, Björkman C, Mattsson JG (2010b) Molecular characterisation of *Cryptosporidium* isolates from Swedish dairy cattle in relation to age, diarrhoea and region. Vet Parasitol 169:289–295

Silverlås C, Mattsson JG, Insulander M, Lebbad M (2012) Zoonotic transmission of *Cryptosporidium meleagridis* on an organic Swedish farm. Int J Parasitol 42:963–967

Silverlås C, Bosaeus-Reineck H, Näslund K, Björkman C (2013) Is there a need for improved *Cryptosporidium* diagnostics in Swedish calves? Int J Parasitol 43:155–161

Simpson VR (1992) Cryptosporidiosis in newborn red deer (Cervus elaphus). Vet Rec 130(6): 116–118

Skerrett HE, Holland CV (2001) Asymptomatic shedding of *Cryptosporidium* oocysts by red deer. Vet Parasitol 94(4):239–246

Slavin D (1955) *Cryptosporidium meleagridis* (sp. nov.). J Comp Pathol 65:262–266

Smith HV, Patterson WJ, Hardie R, Greene LA, Benton C, Tulloch W, Gilmour RA, Girdwood RW, Sharp JC, Forbes GI (1989) An outbreak of waterborne cryptosporidiosis caused by post-treatment contamination. Epidemiol Infect 103(3):703–715

Smith KE, Stenzel SA, Bender JB, Wagstrom E, Soderlund D, Leano FT, Taylor CM, Belle-Isle PA, Danila R (2004) Outbreaks of enteric infections caused by multiple pathogens associated with calves at a farm day camp. Pediatr Infect Dis J 23:1064–1104

Smith HV, Nichols RAB, Mallon M, MacLeod A, Tait A, Reilly WJ, Browning LM, Gray D, Reid SWJ, Wastling JM (2005) Natural *Cryptosporidium hominis* infections in Scottish cattle. Vet Rec 156:710–711

Smith RP, Chalmers RM, Mueller-Doblies D, Clifton-Hadley FA, Elwin K, Watkins J, Paiba GA, Hadfield SJ, Giles M (2010) Investigation of farms linked to human patients with cryptosporidiosis in England and Wales. Prev Vet Med 94:9–17

Snodgrass DR, Angus KW, Gray EW (1984) Experimental cryptosporidiosis in germfree lambs. J Comp Pathol 94:141–152

Soba B, Logar J (2008) Genetic classification of *Cryptosporidium* isolates from humans and calves in Slovenia. Parasitology 135(11):1263–1270

Soba B, Petrovec M, Mioc V, Logar J (2006) Molecular characterisation of *Cryptosporidium* isolates from humans in Slovenia. Clin Microbiol Infect 12:918–921

Soltane R, Guyot K, Dei-Cas E, Ayadi A (2007) Prevalence of *Cryptosporidium* spp. (Eucoccidiorida: Cryptosporiidae) in seven species of farm animals in Tunisia. Parasite 14: 335–338

Spano F, Putignani L, McLauchlin J, Casemore DP, Crisanti A (1997) PCR-RFLP analysis of the *Cryptosporidium* oocyst wall protein (COWP) gene discriminates between *C. wrairi* and *C. parvum*, and between *C. parvum* isolates of human and animal origin. FEMS Microbiol Lett 150:209–217

Sreter T, Varga I, Bekesi L (1995) Age-dependent resistance to *Cryptosporidium baileyi* infection in chickens. J Parasitol 81:827–829

Sreter T, Kovacs G, da Silva AJ, Pieniazek NJ, Szell Z, Dobos-Kovacs M, Marialigeti K, Varga I (2000) Morphologic host specificity and molecular characterization of a Hungarian *Cryptosporidium meleagridis* isolate. Appl Environ Microbiol 66:735–738

Starkey SR, Kimber KR, Wade SE, Schaaf SL, White ME, Mohammed HO (2006) Risk factors associated with *Cryptosporidium* infection on dairy farms in a New York State watershed. J Dairy Sci 89:4229–4236

Starkey SR, Johnson AL, Ziegler PE, Mohammed HO (2007) An outbreak of cryptosporidiosis among alpaca crias and their human caregivers. J Am Vet Med Assoc 231:1562–1267

Stewart WC, Pollock KG, Browning LM, Young D, Smith-Palmer A, Reilly WJ (2005) Survey of zoonoses recorded in Scotland between 1993 and 2002. Vet Rec 157:697–702

Sturdee AP, Bodley-Tickell AT, Archer A, Chalmers RM (2003) Long-term study of *Cryptosporidium* prevalence on a lowland farm in the United Kingdom. Vet Parasitol 116:97–113

Suárez-Luengas L, Clavel A, Quílez J, Goñi-Cepero MP, Torres E, Sánchez-Acedo C, del Cacho E (2007) Molecular characterization of *Cryptosporidium* isolates from pigs in Zaragoza, northeastern Spain. Vet Parasitol 148:231–235

Sulaiman IM, Xiao L, Yang C, Escalante L, Moore A, Beard CB, Arrowood MJ, Lal AA (1998) Differentiating human from animal isolates of *Cryptosporidium parvum*. Emerg Infect Dis 4:681–685

Sweeny JP, Ryan UM, Robertson ID, Jacobson C (2011a) *Cryptosporidium* and *Giardia* associated with reduced lamb carcase productivity. Vet Parasitol 182:127–139

Sweeny JP, Ryan UM, Robertson ID, Yang R, Bell K, Jacobson C (2011b) Longitudinal investigation of protozoan parasites in meat lamb farms in southern Western Australia. Prev Vet Med 101:192–203

Sweeny JP, Robertson ID, Ryan UM, Jacobson C, Woodgate RG (2012) Impacts of naturally acquired protozoa and strongylid nematode infections on growth and faecal attributes in lambs. Vet Parasitol 184:298–308

Szonyi B, Chang YF, Wade SE, Mohammed HO (2012) Evaluation of factors associated with the risk of infection with *Cryptosporidium parvum* in dairy calves. Am J Vet Res 73(1):76–85

Tacal JV, Sobieh M, El-Ahraf A (1987) *Cryptosporidium* in market pigs in southern California USA. Vet Rec 120:615–616

Tacconi G, Pedini AV, Gargiulo AM, Coletti M, Piergili-Fioretti D (2001) Retrospective ultramicroscopic investigation on naturally cryptosporidial-infected commercial turkey poults. Avian Dis 45:688–695

Tarwid JN, Cawthorn RJ, Riddell C (1985) Cryptosporidiosis in the respiratory tract of turkeys in Saskatchewan. Avian Dis 29:528–532

Taylor MA, Catchpole J, Norton CC, Green JA (1994) Variations in oocyst output associated with *Cryptosporidium baileyi* infections in chickens. Vet Parasitol 53:7–14

Thompson H, Dooley JG, Kenny J, McCoy M, Lowery C, Moore J, Xiao L (2007) Genotypes and subtypes of *Cryptosporidium* spp. in neonatal calves in Northern Ireland. Parasitol Res 100(3):619–624

Trampel DW, Pepper TM, Blagburn BL (2000) Urinary tract cryptosporidiosis in commercial laying hens. Avian Dis 44:479–484

Trotz-Williams LA, Martin DS, Gatei W, Cama V, Peregrine AS, Martin SW, Nydam DV, Jamieson F, Xiao L (2006) Genotype and subtype analyses of *Cryptosporidium* isolates from dairy calves and humans in Ontario. Parasitol Res 99:346–352

Trotz-Williams LA, Martin SW, Leslie KE, Duffield T, Nydam DV, Peregrine AS (2008) Association between management practices and within-herd prevalence of *Cryptosporidium parvum* shedding on dairy farms in southern Ontario. Prev Vet Med 83(1):11–23

Trout JM, Santín M, Fayer R (2008) Detection of assemblage A *Giardia duodenalis* and *eimeria* spp. In alpacas on two Maryland farms. Vet Parasitol 153:203–208

Tumova E, Skrivan M, Marounek M, Pavlásek I, Ledvinka Z (2002) Performance and oocyst shedding in broiler chickens orally infected with *Cryptosporidium baileyi* and *Cryptosporidium meleagridis*. Avian Dis 46:203–207

Twomey DF, Barlow AM, Bell S, Chalmers RM, Elwin K, Giles M, Higgins RJ, Robinson G, Stringer RM (2008) Cryptosporidiosis in two alpaca (Lama pacos) holdings in the South-West of England. Vet J 175:419–422

Tyzzer EE (1912) *Cryptosporidium parvum* (sp. nov.), a coccidium found in the small intestine of the common mouse. Arch Protistenkd 26:394–412

Tyzzer EE (1929) Coccidiosis in gallinaceous birds. Am J Hyg 10:269–383

Tzipori S, Angus KW, Campbell I, Sherwood D (1981a) Diarrhea in young red deer associated with infection with *Cryptosporidium*. J Infect Dis 144(2):170–175

Tzipori S, McCartney E, Lawson GH, Rowland AC, Campbell I (1981b) Experimental infection of piglets with *Cryptosporidium*. Res Vet Sci 31:358–368

Tzipori S, Smith M, Makin T, Halpin C (1982) Enterocolitis in piglets caused by *Cryptosporidium* purified from calf faeces. Vet Parasitol 11:121–126

Tzipori S, Smith M, Halpin C, Angus KW, Sherwood D, Campbell I (1983) Experimental cryptosporidiosis in calves: clinical manifestations and pathological findings. Vet Rec 112: 116–120

Uga S, Matsuo J, Kono E, Kimura K, Inoue M, Rai SK, Ono K (2000) Prevalence of *Cryptosporidium parvum* infection and pattern of oocyst shedding in calves in Japan. Vet Parasitol 94: 27–32

Van Dyke MI, Ong CS, Prystajecky NA, Isaac-Renton JL, Huck PM (2012) Identifying host sources, human health risk and indicators of *Cryptosporidium* and *Giardia* in a Canadian watershed influenced by urban and rural activities. J Water Health 10:311–323

Vieira LS, Silva MB, Tolentino AC, Lima JD, Silva AC (1997) Outbreak of cryptosporidiosis in dairy goats in Brazil. Vet Rec 140:427–428

Villacorta I, Ares-Mazas E, Lorenzo MJ (1991) *Cryptosporidium parvum* in cattle, sheep and pigs in Galicia (N.W. Spain). Vet Parasitol 38:249–252

Villanueva MA, Domingo CYJ, Abes NS, Minala CN (2010) Incidence an risk factors of *Cryptosporidium* sp. infection in water buffaloes confined in a communal management system in the Philippines. Internet J Vet Med 8(1):5. doi:10.5580/1227

Vítovec J, Koudela B (1992) Pathogenesis of intestinal cryptosporidiosis in conventional and gnotobiotic piglets. Vet Parasitol 43:25–36

Vítovec J, Hamadejová K, Landová L, Kvác M, Kvetonová D, Sak B (2006) Prevalence and pathogenicity of *Cryptosporidium suis* in pre- and post-weaned pigs. J Vet Med B Infect Dis Vet Publ Health 53:239–243

Wade SE, Mohammed HO, Schaaf SL (2000) Prevalence of Giardia sp. *Cryptosporidium parvum* and *Cryptosporidium muris* (*C. andersoni*) in 109 dairy herds, in five counties of southeastern New York. Vet Parasitol 93:1–11

Wages DP, Ficken MD (1989) Cryptosporidiosis and turkey viral hepatitis in turkeys. Avian Dis 33:191–194

Waitt LH, Cebra CK, Firshman AM, McKenzie EC, Schlipf JW Jr (2008) Cryptosporidiosis in 20 alpaca crias. J Am Vet Med Assoc 233:294–298

Waldron LS, Dimeski B, Beggs PJ, Ferrari BC, Power ML (2011) Molecular epidemiology spatiotemporal analysis and ecology of sporadic human cryptosporidiosis in Australia. Appl Environ Microbiol 77(21):7757–7765

Wang R, Wang J, Sun M, Dang H, Feng Y, Ning C, Jian F, Zhang L, Xiao L (2008a) Molecular characterization of the *Cryptosporidium* cervine genotype from a sika deer (*Cervus nippon* Temminck) in Zhengzhou, China and literature review. Parasitol Res 103(4):865–869

Wang R, Zhang L, Ning C, Feng Y, Jian F, Xiao L, Lu B, Ai W, Dong H (2008b) Multilocus phylogenetic analysis of *Cryptosporidium andersoni* (Apicomplexa) isolated from a bactrian camel (*Camelus bactrianus*) in China. Parasitol Res 102:915–920

Wang R, Jian F, Sun Y, Hu Q, Zhu J, Wang F, Ning C, Zhang L, Xiao L (2010a) Large-scale survey of *Cryptosporidium* spp. in chickens and Pekin ducks (*Anas platyrhynchos*) in Henan, China: prevalence and molecular characterization. Avian Pathol 39:447–451

Wang R, Qiu S, Jian F, Zhang S, Shen Y, Zhang L, Ning C, Cao J, Qi M, Xiao L (2010b) Prevalence and molecular identification of *Cryptosporidium* spp. in pigs in Henan, China. Parasitol Res 107:1489–1494

Wang Y, Feng Y, Cui B, Jian F, Ning C, Wang R, Zhang L, Xiao L (2010c) Cervine genotype is the major *Cryptosporidium* genotype in sheep in China. Parasitol Res 106:341–347

Wang R, Qi M, Jingjing Z, Sun D, Ning C, Zhao J, Zhang L, Xiao L (2011a) *Prevalence of Cryptosporidium baileyi* in ostriches (*Struthio camelus*) in Zhengzhou, China. Vet Parasitol 175(1–2):151–154

Wang R, Wang H, Sun Y, Zhang L, Jian F, Qi M, Ning C, Xiao L (2011b) Characteristics of *Cryptosporidium* transmission in preweaned dairy cattle in Henan, China. J Clin Microbiol 49(3):1077–1082

Wang R, Wang F, Zhao J, Qi M, Ning C, Zhang L, Xiao L (2012) *Cryptosporidium* spp. in quails (*Coturnix coturnix japonica*) in Henan, China: molecular characterization and public health significance. Vet Parasitol 187:534–537

Whitehead CE, Anderson DE (2006) Neonatal diarrhea in llamas and alpacas. Small Ruminant Res 61:207–215

Wieler LH, Ilieff A, Herbst W, Bauer C, Vieler E, Bauerfeind R, Failing K, Klös H, Wengert D, Baljer G, Zahner H (2001) Prevalence of enteropathogens in suckling and weaned piglets with diarrhoea in southern Germany. J Vet Med B Infect Dis Vet Publ Health 48:151–159

Woodmansee DB, Pavlásek I, Pohlenz JF, Moon HW (1988) Subclinical cryptosporidiosis of turkeys in Iowa. J Parasitol 74:898–900

Xiao L (2010) Molecular epidemiology of cryptosporidiosis: an update. Exp Parasitol 124(1): 80–89

Xiao L, Herd RP, Bowman GL (1994) Prevalence of *Cryptosporidium* and *Giardia* infections on two Ohio pig farms with different management systems. Vet Parasitol 52:331–336

Xiao L, Escalante L, Yang C, Sulaiman I, Escalante AA, Montali RJ, Fayer R, Lal AA (1999) Phylogenetic analysis of *Cryptosporidium* parasites based on the small-subunit r-RNA gene locus. Appl Environ Microbiol 65:1578–1583

Xiao L, Bern C, Arrowood M, Sulaiman I, Zhou L, Kawai V, Vivar A, Lal AA, Gilman RH (2002a) Identification of the *Cryptosporidium* pig genotype in a human patient. J Infect Dis 185:1846–1848

Xiao L, Sulaiman IM, Ryan UM, Zhou L, Atwill ER, Tischler ML, Zhang X, Fayer R, Lal AA (2002b) Host adaptation and host-parasite co-evolution in *Cryptosporidium*: implications for taxonomy and public health. Int J Parasitol 32:1773–1785

Xiao L, Fayer R, Ryan U, Upton SJ (2004) *Cryptosporidium* taxonomy: recent advances and implications for public health. Clin Microbiol Rev 17:72–97

Xiao L, Moore JE, Ukoh U, Gatei W, Lowery CJ, Murphy TM, Dooley JS, Millar BC, Rooney PJ, Rao JR (2006) Prevalence and identity of *Cryptosporidium* spp. in pig slurry. Appl Environ Microbiol 72:4461–4463

Yakhchali M, Moradi T (2012) Prevalence of *Cryptosporidium*-like infection in one-humped camels (*Camelus dromedarius*) of northwestern Iran. Parasite 19:71–75

Yang R, Jacobson C, Gordon C, Ryan U (2009) Prevalence and molecular characterisation of *Cryptosporidium* and *Giardia* species in pre-weaned sheep in Australia. Vet Parasitol 161: 19–24

Yin J, Shen Y, Yuan Z, Lu W, Xu Y, Cao J (2011) Prevalence of the *Cryptosporidium* pig genotype II in pigs from the Yangtze River Delta, China. PLoS One 6:e20738

Yu JR, Seo M (2004) Infection status of pigs with *Cryptosporidium parvum*. Korean J Parasitol 42:45–47

Zhang W, Shen Y, Wang R, Liu A, Ling H, Li Y, Cao J, Zhang X, Shu J, Zhang L (2012) *Cryptosporidium cuniculus* and *Giardia duodenalis* in rabbits: genetic diversity and possible zoonotic transmission. PLoS One 7(2):e31262. doi:10.1371/journal.pone.0031262

Zhou L, Kassa H, Tischler ML, Xiao L (2004) Host-adapted *Cryptosporidium* spp. in Canada geese (*Branta canadensis*). Appl Environ Microbiol 70:4211–4215

Zintl A, Neville D, Maguire D, Fanning S, Mulcahy G, Smith HV, De Waal T (2007) Prevalence of *Cryptosporidium* species in intensively farmed pigs in Ireland. Parasitology 134:1575–1582

Chapter 5
Cryptosporidiosis in Other Vertebrates

Martin Kváč, John McEvoy, Brianna Stenger, and Mark Clark

Abstract *Cryptosporidium* has adapted to a broad range of hosts in all major vertebrate classes, and the species associated with humans and livestock represent a small fraction of the diversity in the genus. This review focuses on *Cryptosporidium* and cryptosporidiosis in terrestrial vertebrates other than humans and livestock. As the known host range of *Cryptosporidium* continues to expand, major orders of amphibians (Anura), reptiles (Squamata and Testudines), avians (17 out of 26 orders), and mammals (18 out of 29 orders) are now represented. The greatest *Cryptosporidium* diversity appears to be in mammals, which may be an Artifact of undersampling in other classes, but more likely reflects a different mechanism of *Cryptosporidium* diversification in mammals relative to other classes.

M. Kváč (✉)
Institute of Parasitology, Biology Centre of the Academy of Sciences of the Czech Republic, Branišovská 31, 370 05 České Budějovice, Czech Republic

University of South Bohemia in České Budějovice, Studentská 13, 370 05 České Budějovice, Czech Republic
e-mail: kvac@paru.cas.cz; kvac@centrum.cz

J. McEvoy
Department of Veterinary and Microbiological Sciences, North Dakota State University Department 7690, Fargo, ND 58108-6050, USA
e-mail: john.mcevoy@ndsu.edu

B. Stenger
Department of Veterinary and Microbiological Sciences, North Dakota State University Department 7690, Fargo, ND 58108-6050, USA

Department of Biological Sciences, North Dakota State University Department 2715, Fargo, ND 58108-6050, USA
e-mail: brianna.l.schneck@my.ndsu.edu

M. Clark
Department of Biological Sciences, North Dakota State University Department 2715, Fargo, ND 58108-6050, USA
e-mail: m.e.clark@ndsu.edu

S.M. Cacciò and G. Widmer (eds.), *Cryptosporidium: parasite and disease*,
DOI 10.1007/978-3-7091-1562-6_5,

5.1 Introduction

The emergence of *Cryptosporidium* as a serious human pathogen in the 1980s spurred research efforts to understand the biology and ecology of this enigmatic parasite. While early studies were hindered by a lack of tools to type isolates, the widespread use of genotyping during the last 15 years has begun to uncover the enormous diversity that exists in the genus *Cryptosporidium*.

The major focus of *Cryptosporidium* research over the past 30 years has been the control of cryptosporidiosis in humans and livestock, and the primary motivation to study other vertebrates has been to understand the ecology of human cryptosporidiosis. Wildlife-associated cryptosporidia account for a relatively small but significant proportion of human cryptosporidiosis cases (Feltus et al. 2006; Robinson et al. 2008; Elwin et al. 2011), the recent emergence of *C. cuniculus* as a human pathogen illustrates the connection between wildlife and public health. Rabbit-adapted *C. cuniculus* was not on the public health radar until the misadventures of a rabbit at a UK water treatment facility in 2008 resulted in a waterborne outbreak of cryptosporidiosis (Chalmers et al. 2009). Genetically, *C. cuniculus* is remarkably similar to the major human pathogen *C. hominis*, and the two cannot be differentiated by some routine molecular diagnostics (Robinson et al. 2010). However, prompted by the waterborne outbreak, samples from 3,030 sporadic cryptosporidiosis cases in the UK were reexamined using enhanced diagnostics, and *C. cuniculus* was detected in 1.2 % of samples (Elwin et al. 2012a). Apart from the obvious benefit of knowing the species, and hence the likely source of the *Cryptosporidium* causing human disease, these studies raise interesting and fundamental questions about the biology of *Cryptosporidium*; such as, how does *Cryptosporidium* diversify and how does diversification affect host specificity and disease potential? These questions can be addressed by examining *Cryptosporidium* diversity across the range of vertebrate hosts, not just humans and livestock.

This chapter addresses the occurrence of *Cryptosporidium* in vertebrate hosts other than fish (Chap. 1), humans (Chaps. 2 and 3), and livestock (Chap. 4). Information is organized by vertebrate class (Amphibia, Reptilia, Aves, and Mammalia), and by order within each class. Background information is provided on each order, including orders that do not currently contain identified *Cryptosporidium* hosts. It is hoped that this approach will facilitate the identification of knowledge gaps and will stimulate ideas for future research.

5.2 *Cryptosporidium* and Cryptosporidiosis of Amphibians

To date, *Cryptosporidium* has been detected in only one of the three orders of amphibians: the Anura. It has not been detected in the Caudata, an order that contains more than 550 species in ten families and approximately 30 genera of salamanders, or the Gymnophonia, which are snake-like amphibians with about

180 species in ten families and 35 genera. The Caudata require freshwater for reproduction and they consume invertebrates. Gymnophonia are tropical in distribution, have larvae that may only be partially aquatic, and typically inhabit loose soil layers.

5.2.1 *Anura*

Anura is the largest order of Amphibians, with over 6,000 species of frogs and toads in 33 families and over 65 genera. Frogs and toads require at least some water for reproduction, but they occupy a wide variety of terrestrial and freshwater habitats. There are no marine species, and only a few inhabit brackish waters. All Anurans consume invertebrate prey, lacking teeth or grinding organs in their digestive tract.

Cryptosporidium fragile, the only amphibian-adapted species described to date, was isolated from the doubtful toad (*Duttaphrynus melanostictus*) originating in Malaysia (Jirků et al. 2008). Oocysts of *C. fragile* measure 6.2 μm (5.5–7.0 μm) × 5.5 μm (5.0–6.5 μm) and are readily lysed in hypertonic solutions. *Cryptosporidium fragile* infects the gastric epithelium and clusters with gastric cryptosporidia in molecular phylogenies (Table 5.1 and Fig. 5.1). Other reports of *Cryptosporidium* and cryptosporidiosis in reptiles have been rare. Cryptosporidiosis was diagnosed in a captive South African clawed frog (*Xenopus laevis*) that was euthanized after becoming grossly emaciated from illness (Green et al. 2003). Endogenous *Cryptosporidium* stages were identified in the gastric mucosa, and oocysts were isolated from water in the infected frog's enclosure. *Cryptosporidium* oocysts also have been reported in a captive Bell's horned frog (*Ceratophrys ornata*) (Crawshaw and Mehren 1987).

5.3 *Cryptosporidium* and Cryptosporidiosis of Reptiles

Cryptosporidium has been detected in two of the four orders of reptiles: Squamata and Testudines. It has not yet been reported in Crocodylia, a small order of 25 species in three families and eight genera, or the Rhynchocephalia, which is represented by two living species: the Brother's Island tuatara (*Sphenodon guntheri*) and the northern tuatara (*Sphenodon punctatus*). Crocodylians, which include crocodiles and alligators, have a two-chambered stomach that is unique among reptiles. The first chamber contains stones and functions much like the avian gizzard, and the second chamber is extremely acidic to facilitate almost complete digestion of prey. Tuataras, which are found only in New Zealand, are similar to the Squamates, but have different dentition and skull characteristics. They feed primarily on invertebrates, but occasionally eat small vertebrates or eggs.

Table 5.1 GenBank accession numbers of representative *Cryptosporidium* small subunit rRNA gene sequences from amphibians and reptiles. A phylogeny constructed from sequences in this table is presented in Fig. 5.1

Class	Order	Common names/groups	Host species (scientific name) and GenBank accession numbers
Amphibia	Anura	Toad	**Doubtful toad** (*Duttaphrynus melanostictus*) [EU162751]
Reptilia	Squamata	Lizards	**Central bearded dragon** (*Pogona vitticeps*) [AY382169, AY382170]
			Desert monitor (*Varanus griseus*) [AF112573]
			Leopoard gecko (*Eublepharis macularius*) [AY120915, AY282714, EU553556]
			Savannah monitor (*Varanus exanthematicus*) [AF093500, AY282713]
			Schneiders skink (*Eumeces schneideri*) [AY282715]
			Veiled chameleon (*Chamaeleo calyptratus*) [EU553587]
		Snakes	**Australian taipan** (*Oxyuranus scutellatus*) [AF108866]
			Ball phyton (*Python regius*) [EU553589, EU553590]
			Boa constrictor (*Boa constrictor*) [AY268584]
			Corn snake (*Pantherophis guttatus guttatus*) [EF50204, AF151376]
			Japanese grass snake (*Rhabdophis tigrinus*) [AB222185]
			New Guinea boa snake (*Candoia aspera*) [AY120913]
	Testudines	Tortoises and turtles	**Marginated tortoise** (*Testudo marginata*) [EF547155]
			Pancake tortoise (*Malacochersus tornieri*) [GQ504270]
			Russian tortoise (*Agrionemys horsfieldii*) [GQ504268]
			Indian star tortoise (*Geochelone elegans*) [AY120914]

5.3.1 Squamata

The order Squamata – snakes and lizards – is the largest and most diverse reptilian order, with more than 9,000 species in 50 families. They occupy diverse habitats, though most species inhabit drier terrestrial areas. All are predators that consume a wide variety of prey. Many produce venom or toxins to assist in immobilization of prey, which are swallowed without mastication.

Two *Cryptosporidium* species have been described in the Squamata: *Cryptosporidium serpentis*, which infects the gastric epithelium, and *C. varanii* (syn. *C. saurophilum*), which is an intestinal species (Table 5.1 and Fig. 5.1).

Cryptosporidium serpentis, which was described by Brownstein et al. (1977) and later named by Levine (1980), can cause chronic, insidious, and often fatal hypertrophic gastritis in both immature and mature ophidian hosts (Brownstein et al. 1977; Cranfield and Graczyk 1994; Kimbell et al. 1999). Clinical signs include

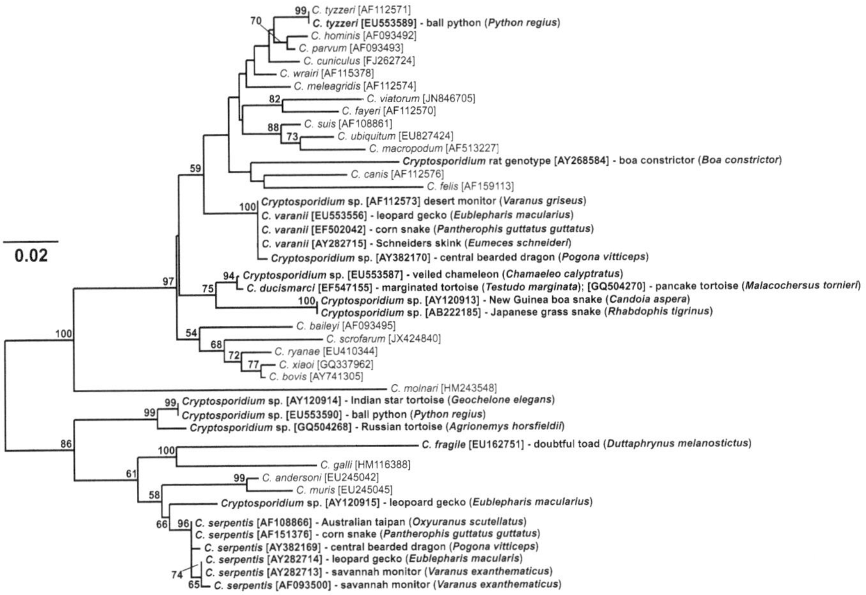

Fig. 5.1 A SSU rDNA-based maximum likelihood (GTRCAT model) tree of *Cryptosporidium* spp. sequences showing clades of cryptosporidia infecting amphibians and reptiles. *Cryptosporidium* sp. detected in amphibians and reptiles are in bold. The percentage of replicate trees in which the associated taxa clustered together in the bootstrap test (1,000 replicates) is shown at nodes. Bootstrap values <50 are not shown. Interrupted branches have been shortened fivefold

mid-abdominal swelling and postprandial regurgitation, but not all cases are symptomatic (Brownstein et al. 1977; Cranfield and Graczyk 1994).

Although it is primarily a parasite of snakes (Table 5.2), *C. serpentis* also causes asymptomatic infections in a number of lizard species (Table 5.3). Unsuccessful attempts to experimentally infect Balb/c mice (Fayer et al. 1995), Pekin ducklings (*Anas platyrhynchos*) (Graczyk et al. 1998b), African clawed frogs (*Xenopus laevis*), and wood frogs (*Rana sylvatica*) (Graczyk et al. 1998c) suggest that *C. serpentis* is restricted to reptiles. However, strict reptile-specificity is brought into question by a recent report that *C. serpentis* was isolated from cattle in China, and was infectious for immunosuppressed and non-immunosuppressed Balb/c mice under experimental conditions (Chen and Qiu 2012).

Snakes are not susceptible to experimental infection with cryptosporidia from homothermic animals, including *C. andersoni* (calves), *C. baileyi* (chickens), *C. meleagridis* (turkeys), *C. muris* (mice), *C. muris*-like (Bactrian camels), and *C. wrairi* (guinea pigs), but they are susceptible to other reptilian cryptosporidia (Graczyk and Cranfield 1998). Reports of *C. tyzzeri* (previously mouse genotype I) in a black rat snake (*Elaphe obsoleta obsoleta*), boa constrictor (*Boa constrictor ortoni*), California king snake (*Lampropeltis getulus californiae*), corn snake (*Pantherophis guttatus guttatus*), emerald tree boa (*Corallus caninus*), fox snake

Table 5.2 *Cryptosporidium* spp. identified in snakes

Family	Host species (scientific name)	*Cryptosporidium* taxa	Reference
Boidae	**Amazon tree boa** (*Corallus hortulanus*)	*C. serpentis*	(Morgan et al. 1999c; Xiao et al. 2004)
	Boa constrictor (*Boa constrictor*)	*C. muris*, *C. serpentis*, *C. tyzzeri*, *C. varanii*, rat genotype I	(Xiao et al. 2004; Pedraza-Diaz et al. 2009; Rinaldi et al. 2012)
	Brazilian rainbow boa (*Epicrates cenchria*)	*C. serpentis*	(Graczyk and Cranfield 2000; Sevá Ada et al. 2011b)
	Colombian rainbow (*Epicrates cenchria maurus*)	*C. serpentis*	(Ryan et al. 2003a)
	Emerald tree boa (*Corallus caninus*)	*C. serpentis*, *C. tyzzeri*	(Xiao et al. 2004)
	Green anaconda (*Eunectes murinus*)	*C. serpentis*	(Sevá Ada et al. 2011b)
	Kenyan or East African sand boa (*Eryx colubrinus*)	*Cryptosporidium* sp.	(Pedraza-Diaz et al. 2009)
	Madagascar tree boa (*Sanzinia madagascariensis*)	*C. serpentis*	(Levine 1980)
	New Guinean viper boa (Candoia aspera)	*C. serpentis*, *C. varanii*, *Cryptosporidium* sp.	(Graczyk and Cranfield 2000; Xiao et al. 2004)
	Rosy boa (*Lichanura trivirgata*)	*C. serpentis*	(Ryan et al. 2003a)
Colubridae	**Black rat snake** (*Pantherophis obsoleta obsoleta*)	*C. muris*, *C. serpentis*, *C. tyzzeri*, *C. varanii*, *Cryptosporidium* sp.	(Cranfield and Graczyk 1994; Morgan et al. 1999c; Xiao et al. 2004)
	Brazos water snake (*Nerodia harteri harteri*)	*Cryptosporidium* sp.	(Upton et al. 1989)
	Bull snake (*Pituophis melanoleucus melanoleucus*)	*C. serpentis*, *C. varanii*	(Morgan et al. 1999c; Graczyk and Cranfield 2000; Xiao et al. 2004)
	California king snake (*Lampropeltis getulus californiae*)	*C. serpentis*, *C. tyzzeri*	(Xiao et al. 2004; Pedraza-Diaz et al. 2009)
	Common garter snake (*Thamnophis sirtalis*)	*Cryptosporidium* sp.	(Brower and Cranfield 2001)
	Corn snake (*Pantherophis guttatus guttatus*)	*C. muris*, *C. serpentis*, *C. tyzzeri*, *C. varanii*	(Levine 1980; Cranfield and Graczyk 1994; Kimbell et al. 1999; Morgan et al. 1999c; Graczyk and

(continued)

Table 5.2 (continued)

Family	Host species (scientific name)	*Cryptosporidium* taxa	Reference
			Cranfield 2000; Xiao et al. 2004; Plutzer and Karanis 2007; Pedraza-Diaz et al. 2009; Richter et al. 2011; Sevá Ada et al. 2011b; Rinaldi et al. 2012)
	Diamondback water snake (*Nerodia rhombifer rhombifera*)	*Cryptosporidium* sp.	(Upton et al. 1989)
	Eastern indigo snake (*Drymarchon corais couperi*)	*Cryptosporidium* sp.	(Cerveny et al. 2012)
	Fox snake (*Pantherophis vulpina gloydi*)	*C. serpentis*, *C. tyzzeri*	(Xiao et al. 2004)
	Jalisco milk snake (*Lampropeltis triangulum arcifera*)	*Cryptosporidium* sp.	(Upton et al. 1989)
	Louisiana pine snake (*Pituophis ruthveni*)	*C. varanii*	(Xiao et al. 2004)
	Milk snake (*Lampropeltis triangulum*)	*C. serpentis*, *C. tyzzeri*, *C. varanii*, *Cryptosporidium* sp.	(Graczyk and Cranfield 2000; Xiao et al. 2004; Pedraza-Diaz et al. 2009; Sevá Ada et al. 2011b)
	Mexican black kingsnake (*Lampropeltis getulus nigritus*)	*C. serpentis*	(Pedraza-Diaz et al. 2009)
	Mexican kingsnake (*Lampropeltis mexicana*)	*C. serpentis*	(Sevá Ada et al. 2011b)
	Pine snake (*Pituophis melanoleucus*)	*C. serpentis*, *C. varanii*, *Cryptosporidium* sp.	(Xiao et al. 2004)
	Prairie king snake (*Lampropeltis calligaster*)	*C. tyzzeri*	(Xiao et al. 2004)
	Rough green snake (*Opheodrys aestivus*)	*Cryptosporidium* sp.	(Brower and Cranfield 2001)
	Ruthven's kingsnake (*Lampropeltis ruthveni*)	*C. serpentis*	(Graczyk and Cranfield 2000)
	Spotted leaf-nosed snake (*Phyllorhynchus decurtatus*)	*C. serpentis*	(Graczyk and Cranfield 2000)

(continued)

Table 5.2 (continued)

Family	Host species (scientific name)	*Cryptosporidium* taxa	Reference
	Texas rat snake (*Elaphe obsolete lindheimeri*)	*Cryptosporidium* sp.	(Upton et al. 1989)
	Trans-Pecos rat snake (*Bogertophis subocularis*)	*C. serpentis*	(Levine 1980)
	Tricolor hognose snake (*Lystrophis semicinctus*)	*Cryptosporidium* sp.	(Cerveny et al. 2012)
	Western fox snake (*Pantherophis vulpina vulpina*)	*Cryptosporidium* sp.	(Upton et al. 1989)
	Yellow rat snake (*Elaphe obsoleta quadrivittata*)	*C. serpentis*, *C. muris*	(Cranfield and Graczyk 1994; Ryan et al. 2003a)
Elapidae	**Cape coral snake** (*Aspidelaps lubricus lubricus*)	*Cryptosporidium* sp.	(Cerveny et al. 2012)
	Common death adder (*Acanthophis antarticus*)	*C. serpentis*	(Morgan et al. 1999c)
	Eastern/mainland tiger snake (*Notechis scutatus*)	*C. serpentis*	(Morgan et al. 1999c)
	Mulga/king brown snake (*Pseudechis australis*)	*Cryptosporidium* sp.	(Morgan et al. 1999c)
	Red-bellied black snake (*Pseudechis porphyriacus*)	*C. tyzzeri*	(Morgan et al. 1999c)
	Taipan (*Oxyuranus scutellatus*)	*C. serpentis*	(Morgan et al. 1999c; Xiao et al. 2004)
Pythonidae	**Ball python** (*Python regius*)	*C. ducismarci*, *C. muris*, *C. serpentis*, *C. tyzzeri*, tortoise genotype, *Cryptosporidium* sp.	(Xiao et al. 2004; Pedraza-Diaz et al. 2009; Sevá Ada et al. 2011b)
	Burmese python (*Python mollurus*)	*C. serpentis*	(Xiao et al. 2004)
	Boelen's python (*Morelia boeleni*)	*C. serpentis*	(Xiao et al. 2004)
	Green python (*Chondropython viridis*)	*C. serpentis*, *C. varanii*, *Cryptosporidium* sp.	(Xiao et al. 2004)

(continued)

Table 5.2 (continued)

Family	Host species (scientific name)	*Cryptosporidium* taxa	Reference
	Indian rock python (*Python molurus*)	*C. serpentis*	(Rinaldi et al. 2012)
	Woma python (*Aspidites ramsayi*)	*C. tyzzeri*, *Cryptosporidium* sp.	(Morgan et al. 1999c; Cerveny et al. 2012)
Viperidae	**Bornmueller's viper** (*Vipera bornmuelleri*)	*C. serpentis*	(Xiao et al. 2004)
	Eastern diamondback rattlesnake (*Crotalus adamanteus*)	*C. serpentis*	(Graczyk and Cranfield 2000)
	Jararaca (*Bothropoides jararaca*)	*C. serpentis*	(Sevá Ada et al. 2011b)
	Jararacussu (*Bothrops jararacussu*)	*C. serpentis*	(Sevá Ada et al. 2011b)
	Mountain viper (*Vipera wagneri*)	*C. serpentis*	(Xiao et al. 2004)
	Nikolski viper (*Vipera nikolski*)	*C. serpentis*	(Ryan et al. 2003a)
	Northwestern tropical rattlesnake (*Crotalus durissus culminatus*)	*Cryptosporidium* sp.	(Upton et al. 1989)
	Pit viper (*Crotalus viridis viridis*)	*Cryptosporidium* sp.	(Cerveny et al. 2012)
	Timber rattlesnake (*Crotalus horridus*)	*C. serpentis*	(Levine 1980; Heuschele et al. 1986)
	Tropical rattlesnake (*Caudisona durissa*)	*C. serpentis*	(Sevá Ada et al. 2011b)

(*Elaphe vulpina gloydi*), milk snake (*Lampropeltis triangulum*), mangrove monitor (*Varanus indicus*), and prairie king snake (*Lamproletis calligaster*); *C. muris* in a corn snake (*Pantherophis guttatus guttatus*); and rat genotype I in a boa constrictor probably represent passive transmission as a consequence of feeding on infected rodents (Xiao et al. 2004).

Koudela and Modrý (1998) identified and named the species *C. saurophilum* from five lizard species: the desert monitor (*Varanus griseus*), emerald monitor (*Varanus prasinus*), leopard gecko (*Eublepharis macularius*), Schneider's skink (*Eumeces schneideri*), and skink (*Mabuya perrotetii*). Pavlásek and Ryan (2008) subsequently showed that *C. saurophilum* was indistinguishable from *C. varanii*, a species described 4 years earlier in an emerald monitor (Pavlásek et al. 1995), and proposed that *C. varanii* should take precedence as the species name. *Cryptosporidium varanii* (syn. *saurophilum*) infects the intestine and cloaca of a number of lizard species (Table 5.3), causing weight loss and abdominal swelling (Koudela and Modrý 1998). In contrast to *C. serpentis* in snakes, the disease is observed in

Table 5.3 *Cryptosporidium* spp. identified in lizards

Family	Host species (scientific name)	*Cryptosporidium* taxa	Reference
Agamidae	**Bearded dragon** (*Pogona vitticeps*)	*C. serpentis*, *C. varanii*	(Xiao et al. 2004)
	Damara rock agama (*Agama planiceps*)	*Cryptosporidium* sp.	(Upton et al. 1989)
	Frilled lizard (*Chlamydosaurus kingui*)	*C. serpentis*, *Cryptosporidium* sp.	(Pedraza-Diaz et al. 2009; Rinaldi et al. 2012)
	Ground agama (*Agama aculeata*)	*Cryptosporidium* sp.	(Upton et al. 1989)
Chamaeleonidae	**Giant Madagascar chameleon** (*Chamaeleo oustaleti*)	*C. serpentis*	(Pedraza-Diaz et al. 2009)
	Mountain chameleon (*Chamaeleo montium*)	*C. serpentis*, *C. varanii*	(Xiao et al. 2004)
	Veiled chameleon (*Chamaeleo calyptratus*)	*C. ducismarci*, *C. varanii*, *Cryptosporidium* sp.	(Koudela and Modrý 1998; Pedraza-Diaz et al. 2009; Rinaldi et al. 2012)
Gekkonidae	**African fat-tailed gecko** (*Hemiteconyx caudicintus*)	*C. varanii*	(Pedraza-Diaz et al. 2009)
	Gargoyle gecko (*Rhodocodactylus auriculatus*)	*C. serpentis*, *C. varanii*	(Xiao et al. 2004)
	Gecko (Gekkoninae sp.)	*C. parvum*, *C. varanii*	(Xiao et al. 2004)
	Giant ground gecko (*Chondrodactylus angulifer*)	*Cryptosporidium* sp.	(Upton et al. 1989)
	Leopard gecko (*Eublepharis macularius*)	*C. parvum*, *C. serpentis*, *C. varanii*, *Cryptosporidium* sp.	(Koudela and Modrý 1998; Xiao et al. 2004; Pedraza-Diaz et al. 2009; Richter et al. 2011; Rinaldi et al. 2012)
	Madagascar giant day gecko (*Phelsuma madagascariensis grandis*)	*Cryptosporidium* sp.	(Upton et al. 1989)
	Mediterranean house gecko (*Hemidactylus turcicus turcicus*)	*Cryptosporidium* sp.	(Upton et al. 1989)
Gerrhosauridae	**Plated lizard** (*Gerrhosaurus* sp.)	*C. varanii*	(Xiao et al. 2004)

(continued)

Table 5.3 (continued)

Family	Host species (scientific name)	*Cryptosporidium* taxa	Reference
Iguanidae	**Green iguana** (*Iguana iguana*)	*C. varanii*, *C. parvum*, avian genotype V	(Xiao et al. 2004: Kik et al. 2011)
Lacertidae	**European green lizard** (*Lacerta viridis*)	*C. varanii*	(Koudela and Modrý 1998)
Scincidae	**Ocellated skink** (*Chalcides ocellatus*)	*C. varanii*	(Koudela and Modrý 1998)
	Schneider's skink (*Eumeces schneideri*)	*C. varanii*	(Koudela and Modrý 1998)
	Skink (*Mabuya perrotetii*)	*C. serpentis*, *C. varanii*	(Koudela and Modrý 1998; Xiao et al. 2004)
Varanidae	**Crocodile monitor** (*Varanus salvadori*)	*C. muris*	(Ryan et al. 2003a)
	Desert monitor (*Varanus griseus*)	*C. serpentis*, *C. varanii*	(Koudela and Modrý 1998) (Ryan et al. 2003a; Xiao et al. 2004)
	Emerald monitor (*Varanus prasinus*)	*C. varanii*, *C. serpentis*	(Pavlásek et al. 1995; Koudela and Modrý 1998)
	Mangrove monitor (*Varanus indicus*)	*C. tyzzeri*	(Xiao et al. 2004)
	Monitor sp. (*Varanus* sp.)	*C. varanii*, *C. parvum*	(Xiao et al. 2004)
	Nile monitor (*Varanus niloticus*)	*C. serpentis*	(Xiao et al. 2004)
	Savannah monitor (*Varanus exanthematicus*)	*C. serpentis*	(Morgan et al. 1999c; Xiao et al. 2004)

juveniles but not adults (Koudela and Modrý 1998). Oocysts are smaller than *C. serpentis* at 5.0 μm (4.4–5.6 μm) × 4.7 μm (4.2–5.2 μm) with a shape index of 1.09 (1.04–1.12) (Koudela and Modrý 1998). Under experimental conditions, *C. varanii* was not infectious for snakes, birds, or mice (Koudela and Modrý 1998); however, natural *C. varanii* infections have been detected in at least nine snake species from three families (Table 5.2).

Other *Cryptosporidium* taxa identified in Squamata include a genotype sharing 99 % identity with *C. serpentis* from a leopard gecko (Richter et al. 2011), avian genotype V from a green iguana (*Iguana iguana*) (Kik et al. 2011), *C. ducismarci* and the tortoise genotype from a ball python (Pedraza-Diaz et al. 2009), *C. ducismarci* from a veiled chameleon (Pedraza-Diaz et al. 2009), and a genotype from a New Guinea boa (*Candoia aspera*) and Japanese grass snake (*Rhabdophis tigrinus*) sharing 98 % similarity with *C. ducismarci* and avian genotype II (Xiao et al. 2002; Kuroki et al. 2008).

5.3.2 Testudines

The order Testudines is comprised of 250 species of turtles in 14 families and more than 20 genera. All species possess a bony and cartilaginous carapace. Turtles are found in a great variety of habitats, including dry upland regions as well as marine and fresh waters. Diet is similarly varied, with mostly omnivorous species, but also some herbivorous and carnivorous species.

Cryptosporidium ducismarci (previously known as *Cryptosporidium* sp. ex *Testudo marginata* CrIT-20) has been proposed as a new *Cryptosporidium* species (Traversa et al. 2008; Traversa 2010). This species, which infects the intestinal epithelium of *Testudo marginata*, also has been detected in an asymptomatic veiled chameleon and ball phyton in Spain (Pedraza-Diaz et al. 2009) and in the pancake tortoise (*Malacochersus tornieri*) and Russian tortoise (*Testudo horsfieldi*) in the US (Griffin et al. 2010). Biological data, including descriptions of pathology, and oocyst morphometry data are lacking for this species description.

The *Cryptosporidium* tortoise genotype has been identified in an asymptomatic Indian star tortoise (*Geochelone elegans*) at the Lisbon Zoo (Alves et al. 2005). The SSU rRNA sequence of the isolate identified by Alves et al. (2005) is identical to a sequence previously reported by Xiao et al. (2002) in an Indian star tortoise at the St. Louis Zoo (GenBank accession no. AY120914), and a sequence from an environmental isolate in Maryland, USA (Yang et al. 2008). In a study of pet reptiles, Pedraza-Diaz et al. (2009) detected *Cryptosporidium* sp. in 17.6 % of tortoise samples and identified the *Cryptosporidium* tortoise genotype in a Hermann's tortoise (*Testudo hermanni*) and a ball python (*Python regius*).

5.4 *Cryptosporidium* and Cryptosporidiosis of Birds

GenBank accession numbers and phylogenetic relationships among representative *Cryptosporidium* small subunit rRNA sequences from birds are presented in Table 5.4 and Fig. 5.2. Three avian-associated *Cryptosporidium* species are recognized: *Cryptosporidium meleagridis*, *Cryptosporidium baileyi*, and *Cryptosporidium galli*. These species infect a broad range of birds; however, they differ in host range and site of infection. *Cryptosporidium meleagridis* also causes disease in humans (McLauchlin et al. 2000; Alves et al. 2003; Cama et al. 2003).

Cryptosporidium meleagridis is primarily a parasite of intestinal epithelial cells of birds, particularly turkeys (Slavin 1955; Sreter and Varga 2000). *Cryptosporidium baileyi* inhabits the respiratory tract, bursa of Fabricius, and cloaca of the domestic chicken and a broad range of other birds (Current et al. 1986; Sreter and Varga 2000). Endogenous stages of *C. galli* are localized on glandular epithelial cells of the proventriculus (Ryan et al. 2003b).

The three avian species can be distinguished based on oocyst morphology. Oocysts of *C. meleagridis* are the smallest of the three, measuring

Table 5.4 GenBank accession numbers of representative *Cryptosporidium* small subunit rRNA gene sequences from avians. A phylogeny constructed from sequences in this table is presented in Fig. 5.2

Order	Common names/groups	Host species (scientific name) and GenBank accession numbers
Anseriformes	Waterfowl: geese and ducks	**Black duck** (species unspecified) [AF316630]
		Goose (species unspecified) [AY120912, EF641009, FJ607874, FJ607886, FJ607887, FJ607896, FJ607898, FJ607910, FJ607918]
		Canada goose (*Branta canadensis*) [AY324635, AY324637–AY324639, AY324641, AY324643, AY504512–AY504517]
		Goose (*Anser anser f. domestica*) [FJ984564]
		Mallard (*Anas platyrhynchos*) [GU082388]
Charadriiformes	Gulls and woodcock	**Eurasian woodcock** (*Scolopax rusticola*) [AY273769]
		Kelp gull (*Larus dominicanus*) [GQ355891]
Columbiformes	Pigeons and doves	**Fan-tailed pigeon** (*Columba livia*) [HM116382]
		Pigeon (species unspecified) [EU032319–EU032324]
		Rufous turtle dove (*Streptopelia orientalis*) [HM116384]
Falconiformes	Black vulture	**Black vulture** (*Coragyps atratus*) [GQ227474]
Galliformes	Fowl: chickens, turkey, quail, Indian peafowl	**Chicken** (*Gallus gallus domesticus*) [AF093495, JX548291, JX548292, JX548299, GQ227476, AY168847, AY168848]
		Cockatiel (*Nymphicus hollandicus*) [GQ227477]
		Indian peafowl (*Pavo cristatus*) [GQ227478]
		Quail (*Coturnix coturnix*) [JQ217141, JQ217142]
		Quail (species unspecified) [AF316631]
		Turkey (*Meleagris gallopavo*) [AF112574]
Gruiformes	Coot	**Eurasian coot** (*Fulica atra*) [FJ984565]
Passeriformes	Perching birds: finches, canaries, sparrows, nightingale, waxwings, corvids	**Atlantic canary** (*Serinus canaria*) [GU074388, GU074389]
		Aurora finch (*Pytilia phoencoptera*) [AF316627–AF316629]
		Blackbilled magpie (*Pica pica*) [HM116380]

(continued)

Table 5.4 (continued)

Order	Common names/groups	Host species (scientific name) and GenBank accession numbers
		Bohemian waxwing (*Bombycilla garrulus*) [HM116383, HM116388]
		Canary (*Serinus canaria*) [GQ227479, EU543269]
		Common myna (*Acridotheres tristis*) [HM116374]
		Crested lark (*Galerida cristata*) [HM116379]
		Gold finch (*Carduelis carduelis*) [AY168846]
		Gouldian finch (*Erythrura gouldiae*) [AF316623–AF316625, HM116377]
		Java sparrow (*Padda oryzivora*) [GU074384]
		Lesser seed finch (*Oryzoborus angolensis*) [EU543270]
		Plum-headed finch (*Neochmia modesta*) [AF316626]
		Red-billed Leiothrix (*Leiothrix lutea*) [HM116375]
		Red-billed blue magpie (*Urocissa erythrorhyncha*) [HM116386]
		Silver eared leiothrix (*Leiothrix argentauris*) [HM116387]
		Society finch (*Lonchura striata domestica*) [GU074390]
		White Java sparrow (*Padda oryzivora*) [HM116376]
		Zebra finch (*Taeniopygia guttata*) [HM116378]
Psittaciformes	Parrots, cockatiels, lovebirds	**Cockatiel** (*Nymphicus hollandicus*) [AB471645–AB471648, EU543268, GQ227477, GQ227481, GU074385–GU074387, HM116381, HM116385]
		Indian rose parakeet (*Psittacula krameri*) [AF180339]
		Peach-faced lovebird (*Agapornis roseicollis*) [GQ227480]
Struthioniformes	Ostrich	**Ostrich** (*Struthio camelus*) [DQ002931]

4.5–6.0 μm × 4.2–5.3 μm with a shape index (length/width) of 1.0–1.3. *Cryptosporidium baileyi* oocysts measure 6.0–7.5 μm × 4.8–5.7 μm with a shape index of 1.1–1.8 (Lindsay et al. 1989), and oocysts of *C. galli* measure 8.2 × 6.3 μm with a shape index of 1.3 (Ryan et al. 2003b). Relatively little is known about the biology of other cryptosporidia infecting birds.

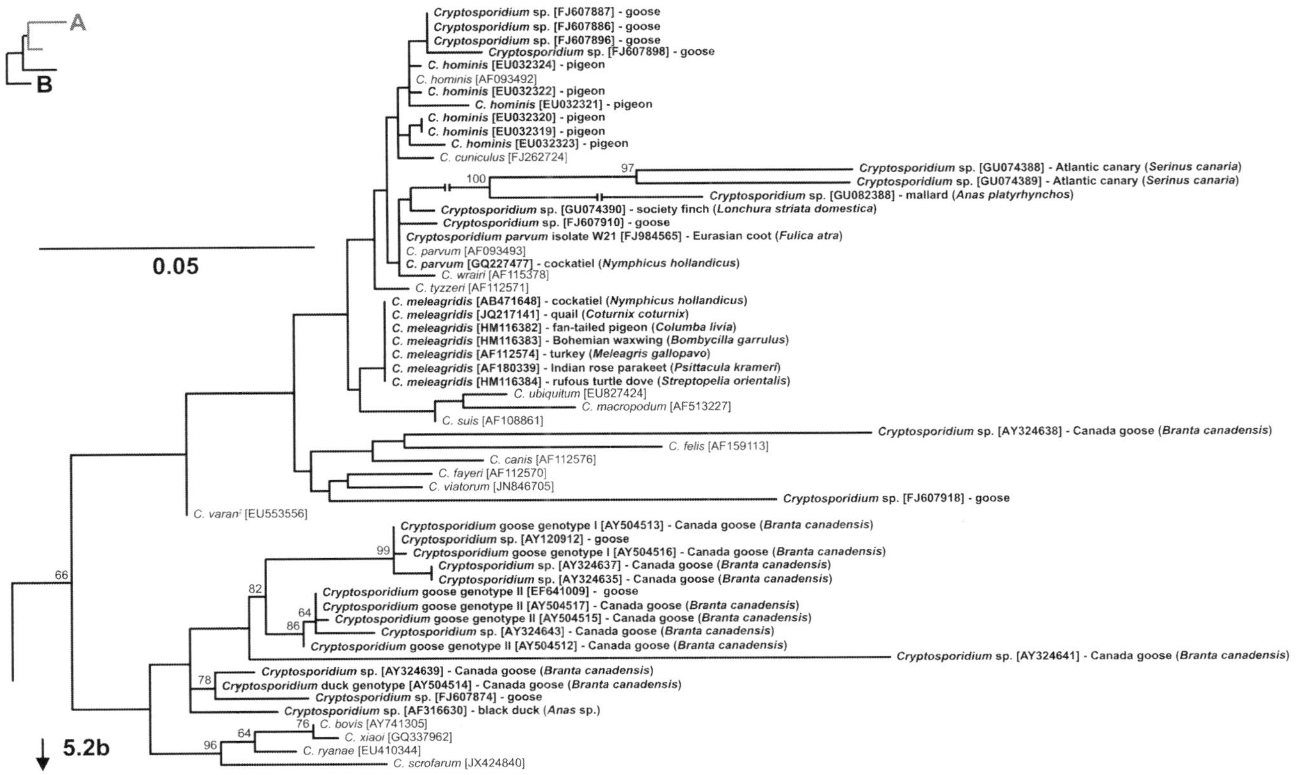

Fig. 5.2 (continued)

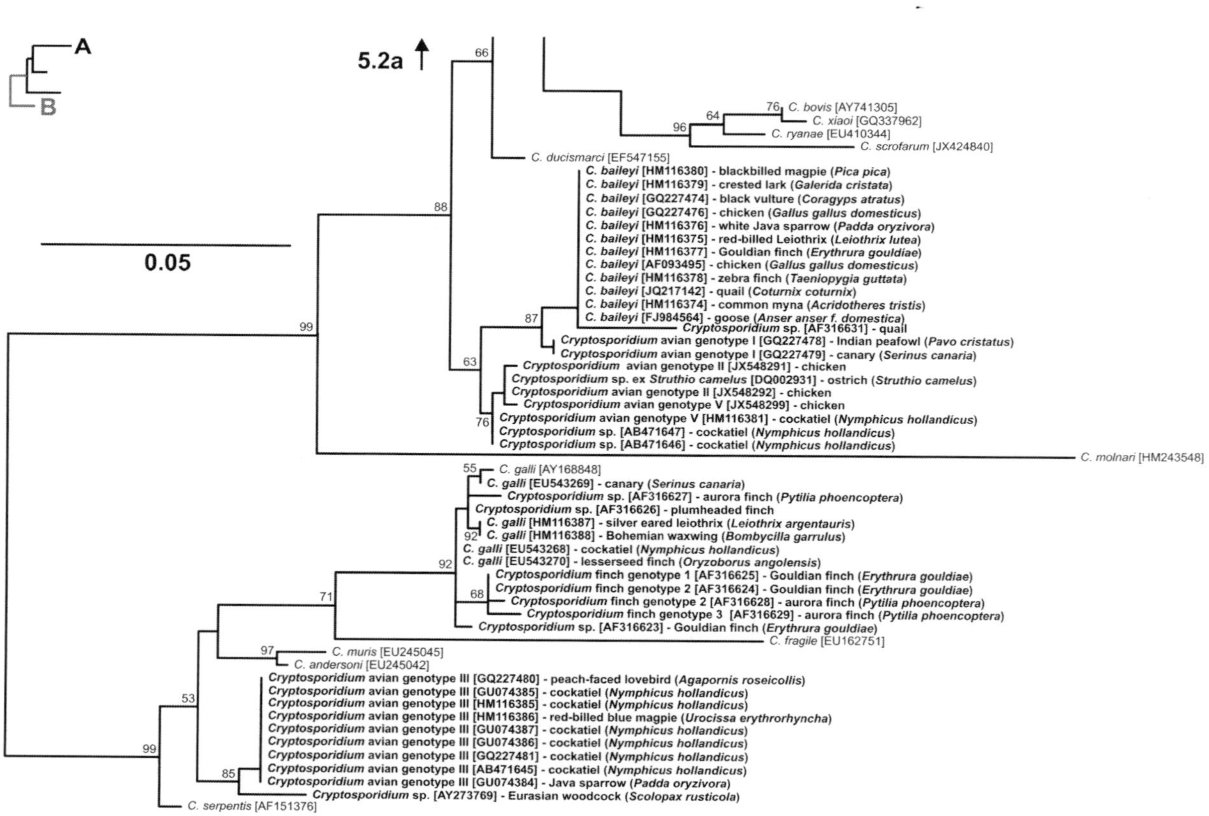

Fig. 5.2 A SSU rDNA-based maximum likelihood (GTRCAT model) tree of *Cryptosporidium* spp. sequences showing clades of cryptosporidia infecting avians. *Cryptosporidium* sp. detected in avians are in bold. The percentage of replicate trees in which the associated taxa clustered together in the bootstrap test (1,000 replicates) is shown at nodes. Bootstrap values <50 are not shown. Interrupted branches have been shortened fivefold

Cryptosporidium has been reported in 17 out of 26 avian orders. Members of the following orders have not yet been identified as hosts:

Tinamiformes: This is a very old avian lineage comprising one family, nine genera, and 47 species. Members are ominvorous, feeding on a variety of seeds and invertebrates. Many species occur in forests or open grasslands, but they generally prefer dryland habitats.

Gaviiformes: This order comprises one family, one genus, and five species of aquatic birds that primarily inhabit North America and Eurasia. They feed on fish, amphibians, or aquatic invertebrates and inhabit both freshwater and marine environments during their annual cycle.

Podicipediformes: This order comprises 22 species in one family and six genera of aquatic birds. Molecular phylogenies indicate that they are closely related to loons and flamingoes. They feed on fish, amphibians, and aquatic invertebrates.

Procellariiformes: This order comprises four families, 26 genera, and about 112 species of seabirds. All inhabit offshore marine environments and most return to land only for nesting. Although, most feed exclusively on marine invertebrates or fish, a couple of species scavenge carrion on islands.

Opisthocomiformes: This order comprises one species, genus, and family of tropical birds that inhabit swamps and mangroves of South America. They feed on vegetable matter (seeds, nuts, leaves, and fruits), and have well-developed ceca that facilitates fermentation in the digestive tract.

Cuculiformes: This is a medium sized order comprising six families, 30 genera, and 143 species. Members inhabit Africa, Asia, Australasia, Eurasia, and the Americas (South, Central, and North).

Apodiformes: This order comprises two families containing approximatley 24 genera and 100 species of small birds. They mostly inhabit the northern hemisphere and can be found in a variety of habitats. They feed aerially on insects.

Trochiliformes: These small, nectar-feeding birds include hermits and hummingbirds. There are approximately 330 species in one family with about 100 genera. The majority of species occur in tropical and subtropical Central and South America.

Trogoniformes: Trogons comprise approximately 39 species in one family with six to eight genera. They inhabit tropical forests, where they feed mainly on insects or fruit.

5.4.1 Struthioniformes

The order Struithioniformes (including cassowaries, emus, kiwis, ostriches, and rheas) is comprised of some of the oldest avian species. There are about 12 species in five families and six genera. All are flightless, forage on insects or vegetation, and lay large eggs with thick shells. Ceca are often well developed in these species.

Cryptosporidium infections in ostriches can be subclinical (Gajadhar 1993), or characterized by a prolapsed phallus/cloaca (Bezuidenhout et al. 1993; Santos et al. 2005), and edema/necrosis of the pancreas (Jardine and Verwoerd 1997). In Brazil, ostriches with a prolapsed cloaca shed oocysts of *Cryptosporidium* avian genotype II measuring 6.0 μm (5.0–6.5 μm) × 4.8 μm (4.2–5.3 μm) with a shape index of 1.31 (Santos et al. 2005; Meireles et al. 2006). *Cryptosporidium* avian genotype II is not infectious for 1-day-old chicks under experimental conditions (Meireles et al. 2006), although it has been detected in a naturaly infected chicken (GenBank accession no. JX548291; unpublished). *Cryptosporidium* avian genotype II was detected in 23 % of farmed ostriches (*Struthio camelus*) in Vietnam (Nguyen et al. 2013). The highest prevalence and oocyst shedding intensity was observed in birds aged 61–90 days; younger and older birds had a lower prevalence and shed fewer oocysts (Nguyen et al. 2013). Oocysts from subclinical *Cryptosporidium* infections in ostriches being imported into Canada measured 4.6 μm (3.9–6.1 μm) × 4.0 μm (3.3–5.0 μm) with a shape index of 1.15 (range 1.0–1.4) (Gajadhar 1993), which is similar to oocysts of *C. meleagridis*. Oocysts were not infectious for suckling mice, chickens, turkeys, or quail (Gajadhar 1993, 1994). *Cryptosporidium baileyi* oocysts were identified in an ostrich from the Czech Republic (Ryan et al. 2003a). *Cryptosporidium* sp. was detected in 60 % of farmed ostriches and rheas originating from Belgium, France, Netherlands Portugal, and Spain (Gordo et al. 2002).

5.4.2 *Anseriformes*

Waterfowl, including ducks, geese and swans, form the order Anseriformes, with about 162 species in three families and 52 genera. These birds are omnivorous, feeding on combinations of vegetation and invertebrates, with a few groups feeding on small aquatic vertebrates.

Three species and five named genotypes of *Cryptosporidium* have been identified in ducks, geese, mergansers, and swans (Table 5.5). In addition, five unnamed genotypes have been reported in Canada geese (*Branta canadensis*) (Jellison et al. 2004), however, sequences have not been deposited in GenBank, and it is not clear if they differ from subsequently named genotypes. Canada geese have also been shown to host the human pathogens *C. hominis* and *C. parvum*, and the rodent-specific muskrat genotype I (Graczyk et al. 1998d; Zhou et al. 2004b; Jellison et al. 2009). It is hypothesized that birds serve as mechanical vectors for mammalian cryptosporidia, and this is supported by data showing that *C. parvum* does not infect Canada geese or Pekin ducks (*Anas platyrhynchos*) under experimental conditions but oocysts remain viable following passage (Graczyk et al. 1996b, 1997).

Cryptosporidium has been reported to cause varying degrees of respiratory distress in ducks and geese (O'Donoghue et al. 1987). *Cryptosporidium baileyi* has been identified in the bursa of Fabricius and cloaca of ducks with mild symptoms (Lindsay et al. 1989), and the bursa of Fabricius and conjunctiva of domestic geese with no clinical symptoms (Chvala et al. 2006).

Table 5.5 *Cryptosporidium* spp. identified in Anseriformes species

Host species (scientific name)	*Cryptosporidium* taxa	References
American widgeon (*Anas americana*)	*Cryptosporidium* sp.	(Kuhn et al. 2002)
Black swan (*Cygnus atratus*)	*Cryptosporidium* sp.	(Rohela et al. 2005)
Blue-winged teal (*Anas discors*)	*Cryptosporidium* sp.	(Kuhn et al. 2002)
Canada goose (*Branta canadensis*)	*C. hominis*, *C. hominis*-like, *C. parvum*, muskrat genotype I, duck genotype, goose genotype I, goose genotype II, 5 unnamed genotypes	(Graczyk et al. 1998d; Jellison et al. 2004, 2009; Zhou et al. 2004b)
Common merganser (*Mergus merganser*)	*Cryptosporidium* sp.	(Kuhn et al. 2002)
Domestic goose (*Anser anser f. domestica*)	*C. baileyi*	(Chvala et al. 2006)
Green-winged teal (*Anas cercca carolinensis*)	*Cryptosporidium* sp.	(Kuhn et al. 2002)
Hooded merganser (*Lophodytes cucullatus*)	*Cryptosporidium* sp.	(Kuhn et al. 2002)
Mallard (*Anas platyrhynchos*)	*Cryptosporidium* sp.	(O'Donoghue et al. 1987; Kuhn et al. 2002)
Swan goose (*Anser cygnoides*)	*Cryptosporidium* sp.	(Rohela et al. 2005)

5.4.3 Galliformes

There are approximately 290 species in five families and 80 genera of gallinaceous birds. These terrestrial species feed mostly on seeds and small insects. This group has been extensively domesticated for agricultural purposes, and many species have been introduced to areas well outside of their historic range.

A number of gallinaceous birds including the domestic chicken (*Gallus gallus domesticus*), grouse (*Tetrastes bonasia rupestris*), and capercaillie (*Tetrao urogallus*), are natural hosts of *C. galli* (Ryan et al. 2003b; Ng et al. 2006).

Non-genotyped cryptosporidia have been detected in the common quail (*Coturnix coturnix*), common peafowl (*Pavo cristatus*), great argus (*Argusianus argus*), great currasow (*Crax rubra*), red-legged partridge (*Alectoris rufa*), and ring-necked pheasant (*Phasianus colchicus*) (O'Donoghue et al. 1987; Rohela et al. 2005; Lim et al. 2007). Infections in the common quail and ring-necked pheasant have been associated with clinical signs of respiratory disease, and oocysts recovered from these hosts were infectious for chickens (O'Donoghue et al. 1987).

Cryptosporidium meleagridis caused an outbreak of respiratory and diarrheal cryptosporidiosis with greater than 50 % mortality in red-legged partridges (*Alectoris rufa*) on a game farm in Spain (Pages-Mante et al. 2007).

5.4.4 Phoenicopteriformes

Phoenicopteriformes is a small order of five species in one family and three genera of flamingoes. These long-legged wading birds use a specialized bill to filter-feed invertebrates or algae from shallow waters. The intestine is relatively long compared to other avian species.

In the only report from this order, *C. galli* was identified in a Cuban flamingo (*Phoenicopterus ruber*) from the Czech Republic (Ng et al. 2006).

5.4.5 Sphenisciformes

Sphenisciformes is comprised of 17 penguin species in one family and six genera. Penguins are flightless birds of the southern oceans, feeding on fish and krill in offshore areas. Very little is known about cryptosporidiosis in this group of birds.

Cryptosporidium oocysts have been detected in 6.6 % (11/167) of stools from the Adelie penguin (*Pygoscelis adeliae*) from the Antarctic territory (Fredes et al. 2007), and in 32.8 % (21/64) of gentoo penguins (*Pygoscelis papua*) from Ardley Island, King George Island, South Shetland Islands, and the Antarctic Specially Protected Area no. 150 (Fredes et al. 2008).

5.4.6 Pelecaniformes

Pelecaniformes includes waterbirds comprising approximately 65 species in eight families and ten genera. These birds feed mostly on aquatic vertebrates. A gular pouch is usually present, and prey is often swallowed whole. Almost all inhabit marine waters for part of their annual cycle.

There have been just two reports of *Cryptosporidium* from the order Pelecaniformes. *Cryptosporidium baileyi* and a non-genotyped *Cryptosporidium* sp. were reported in a cormorant and great cormorant (*Phalacrocorax carbo*), respectively (Jellison et al. 2004; Plutzer and Tomor 2009).

5.4.7 *Ciconiiformes*

There are approximately 116 ciconiiform species in three families and 39 genera. All are long-legged wading birds, feeding in aquatic environments on a variety of invertebrates and vertebrates.

Cryptosporidium has been reported in a marabou stork (*Leptoptilos crumeniferus*) from a zoo in Malaysia (Rohela et al. 2005) and 12.5 % (3/24) of white storks (*Ciconia ciconia*) from Poland (Majewska et al. 2009). The species infecting white storks was identified as *C. parvum* using fluorescence in situ hybridization.

5.4.8 *Falconiformes*

There are approximately 304 species in three to five families and 83 genera of falconiform birds. These birds feed on vertebrate prey or carrion. Most nest in trees or cliffs and are terrestrial in their habits.

Cryptosporidium baileyi and *C. parvum* have been identified in Falconiforms, and both are associated with disease. *Cryptosporidium parvum* caused anorexia, mild respiratory difficulties, and mild bilateral ocular discharge in a 3-month old gyrfalcon (*Falco rusticolus*), and *C. baileyi* caused mild bilateral conjunctivities and sinusitis in a 13-year-old gyrfalcon × Saker falcon hybrid (*Falco rusticolus* × *Falco cherrug*) (Barbon and Forbes 2007). Both raptors recovered following treatment with paromomycin. *Cryptosporidium baileyi* similarly caused an upper respiratory tract infection in three mixed-bred falcons (*Falco rusticolus* × *Falco cherrug*) (van Zeeland et al. 2008). One of the birds had epiglottal swelling and laryngeal stridor, and two had nasal discharge and sneezing. Oocyst shedding was not detected in any of the birds. *Cryptosporidium baileyi* also was identified in a Saker falcon (*Falco cherrug*) with inflammation of the middle ears, conjunctivae, third eyelids, choanae, larynx, trachea, salivary glands of the tongue, syrinx, and turbinates (Bougiouklis et al. 2012), and in a captive black vulture (*Coragyps atratus*) (Nakamura et al. 2009).

5.4.9 *Charadriiformes*

Charadriiformes is a diverse order of shorebirds, with about 367 species distributed among 17 families and 88 genera. Most species feed on invertebrate prey, and utilize marine habitats during part of their annual cycle.

Cryptosporidium sp. has been reported in black-headed gulls (*Chroicocephalus ridibundus*) (Pavlásek 1993; Smith et al. 1993; Ryan et al. 2003a), herring gulls (*Larus argentatus*) (Smith et al. 1993; Bogomolni et al. 2008), and kelp gulls

(*Larus dominicanus*) (GenBank accession no. GQ355891; unpublished). Smith et al. (1993) detected *Cryptosporidium* oocysts measuring 4.7 μm (4.4–5.2 μm) × 5.2 μm (4.7–5.8 μm) in feces from gulls trapped at two refuse sites and other locations in Scotland, and found no difference in *Cryptosporidium* prevalence between the black-headed gull and the herring gull. *Cryptosporidium baileyi* was identified as a cause of significant morbidity and mortality in 28–100 % of black-headed gulls in the Czech Republic (Pavlásek 1993). *Cryptosporidium baileyi* from black-headed gulls infected 4-day-old chickens (*Gallus gallus f. domestica*), causing 40 % mortality (Pavlásek 1993). The intestinal locations of developmental stages are similar to those in the black-headed gull (Pavlásek 1993). *Cryptosporidium baileyi* was reported in a black-headed gull, and a novel *Cryptosporidium* genotype (GenBank accession no. AY273769) was identified in a wild-caught Eurasian woodcock (*Scolopax rusticola*) from the Czech Republic (Ryan et al. 2003a).

5.4.10 Gruiformes

Gruiformes contains 212 species distributed among 11 families and 61 genera. Most species inhabit marshland areas, and feed on invertebrates, seeds, or other vegetation. Ceca and hindgut fermentation are common in this group.

Cryptosporidium baileyi has been identified in a crane from the Czech Republic (Ng et al. 2006). In Korea, a 4-month old white naped crane (*Grus vipio*) that died from a disseminated *Eimeria* infection also had an incidental *Cryptosporidium* infection of the cloaca (Kim et al. 2005). In Hungary, Plutzer and Tomor (2009) identified *C. parvum* in a wild Eurasian coot (*Fulica atra*) (GenBank accession no. FJ984565).

5.4.11 Columbiformes

Pigeons and doves make up the order Columbiformes. Although there are 308 species worldwide, only one or two families containing 42 genera are recognized. These birds occupy a diverse range of terrestrial habitats, feeding almost exclusively on seeds and grains.

There have been relatively few reports of *Cryptosporidium* in pigeons (Ozkul and Aydin 1994; Rodriguez et al. 1997; Abreu-Acosta et al. 2009; Qi et al. 2011; Radfar et al. 2012). *Cryptosporidium hominis* has been reported in rock pigeons (*Columba livia*) on Tenerife, one of the Canary Islands, suggesting that pigeons can play a role in the transmission of human-pathogenic cryptosporidia (Abreu-Acosta et al. 2009). In a study of *Cryptosporidium* in birds at pet stores in China, Qi et al. (2011) detected *C. meleagridis* in a fan-tailed pigeon and a rufous turtle dove (*Streptopelia orientalis*).

5.4.12 *Psittaciformes*

Birds in the order Psittaciformes order, which contains 364 species from 85 genera in the family Psittacidae, inhabit trees in dense, tropical or subtropical forests, and feed almost exclusively on fruits and seeds. Although there have been few studies of psittaciforms in their natural environment, captive psittaciforms host a number of *Cryptosporidium* species and genotypes (Table 5.6). The greatest diversity of taxa has been detected in the cockatiel (*Nymphicus hollandicus*), which hosts four species (*C. baileyi*, *C. galli*, *C. meleagridis*, and *C. parvum*) and three genotypes (avian genotypes II, III, and V) (Abe and Iseki 2004; Ng et al. 2006; Antunes et al. 2008; Nakamura et al. 2009; Abe and Makino 2010; Qi et al. 2011; Gomes et al. 2012). The galah (*Eolophus roseicapilla*), which is in the same subfamily as the cockatiel, also hosts avian genotype II (Ng et al. 2006).

Although most cases of *Cryptosporidium* infection in psittaciform birds have been asymptomatic, there have been reports of clinical disease. Latimer et al. (1992) diagnosed cryptosporidiosis in four cockatoos with psittacine beak-and-feather disease. *Cryptosporidium* infection was confined to the bursa of Fabricius in three of the birds, and was more widespread in the intestine of the fourth. All birds had intermittent diarrhea. Makino et al. (2010) detected avian genotype II in 35 % (13/37) of infected peach-faced lovebirds (*Agapomis roseicollis*). All birds had symptoms of infection including weight loss and chronic vomiting.

5.4.13 *Strigiformes*

The owls make up a small order of about 180 species in two families and 29 genera. All species are predators of small vertebrates, principally rodents. Most species are nocturnal and inhabit forested areas, though a few nest underground in burrows.

In a study of 12 adult owls held in captivity in south Brazil, da Silva et al. (2009) isolated *Cryptosporidium* oocysts measuring 5–6 μm × 4–5 μm from a barn owl (*Tyto alba*), great horned owl (*Bubo virginianus*), and striped owl (*Phinoptynx clamator*). Owls showed no clincal signs of disease.

Molina-Lopez et al. (2010) diagnosed ocular and respiratory cryptosporidiosis in 16 wild fledgling scops owls (*Otus scops*) up to 2 months after they were admitted to a wildlife rehabilitation center in Catalonia, northern Spain; the owls were born in the wild and were healthy when they arrived at the center. Blepharoedema, conjunctival hyperaemia, and mucopurulent ocular discharge were diagnosed unilaterally in 75 % (12/16) of the birds and bilaterally in 25 % (4/16). Five owls (31 %) developed diffuse epithelial corneal edema, one exhibited mild anterior exudative uveitis, and another developed rhinitis. *Cryptosporidium baileyi*, measuring 6.5–7.0 μm × 5.0–5.5 μm, was identified in samples from two birds that were

Table 5.6 *Cryptosporidium* spp. identified in Psittaciformes species

Host species (scientific name)	*Cryptosporidium* taxa	References
Alexandrine parakeet (*Psittacula eupatria*)	Avian genotype II, *Cryptosporidium* sp.	(Ng et al. 2006; Papini et al. 2012)
Blue-fronted Amazon (*Amazona aestiva*)	*Cryptosporidium* sp.	(Papini et al. 2012)
Budgerigar (*Melopsittacus undulatus*)	*Cryptosporidium* sp.	(Nakamura et al. 2009)
Cockatiel (*Nymphicus hollandicus*)	*C. baileyi*, *C. galli*, *C. meleagridis*, *C. parvum*, avian genotype II, avian genotype III, avian genotype V	(Abe and Iseki 2004; Ng et al. 2006; Antunes et al. 2008; Nakamura et al. 2009; Abe and Makino 2010; Qi et al. 2011; Gomes et al. 2012)
Eastern rosella (*Piatycercus eximius*)	*Cryptosporidium* sp.	(Papini et al. 2012)
Eclectus parrot (*Eclectus roratus*)	Avian genotype II	(Ng et al. 2006)
Galah (*Eolophus roseicapilla*)	Avian genotype II	(Ng et al. 2006)
Goffin's cockatoo (*Cacatua goffini*)	*Cryptosporidium* sp.	(Nakamura et al. 2009)
Indian ring-necked parrot (*Psittacula krameri*)	*C. meleagridis*	(Morgan et al. 2000b)
Major Mitchell's cockatoo (*Lophochroa leadbeateri*)	Avian genotype II	(Ng et al. 2006)
Peach faced lovebirds (*Agapomis roseicollis*)	Avian genotype III	(Makino et al. 2010)
Princess parrot (*Polytelis alexandrae*)	Avian genotype II	(Ng et al. 2006)
Red-bellied macaw (*Orthopsittaca manilata*)	*Cryptosporidium* sp.	(Nakamura et al. 2009)
Red-crowned Amazon (*Amazona viridigenalis*)	*C. baileyi*	(Ryan et al. 2003a)
Rose-ringed parakeet (*Psittacula krameri*)	*C. meleagridis*	(Ryan et al. 2003a)

(continued)

Table 5.6 (continued)

Host species (scientific name)	*Cryptosporidium* taxa	References
Salmon-crested cockatoo (*Cacatua moluccensis*)	*Cryptosporidium* sp.	(Rohela et al. 2005)
Sun conure (*Aratinga solstitialis*)	Avian genotype II	(Ng et al. 2006)
Turquoise parrot (*Neophema pulchella*)	*C. galli*	(Ng et al. 2006)
White-eyed para-keet (*Aratinga leucophthalma*)	Avian genotype II	(Sevá Ada et al. 2011a)

euthanized due to the severity of their disease. The remaining owls recovered following a 15-day treatment with azithromycin.

5.4.14 Caprimulgiformes

This small order of insectivorous birds has about 118 species in five families and 22 genera. Most feed nocturnally, on the wing. These species generally occur near upland forests, and almost all species nest on the ground.

Cryptosporidium muris has been detected in the tawny frogmouth (*Podargus strigoides*); although, the bird may have been a mechanical vector (Ryan et al. 2003a; Ng et al. 2006).

5.4.15 Piciformes

The piciform birds include woodpeckers and toucans, with approximately 398 species in five to nine families and eight genera. Most species occur in forests of the Americas, and almost all nest in cavities. The woodpeckers feed mostly on insects, but toucans are fruigivorous and the honeyguides have the unique ability to eat beeswax.

Cryptosporidium infection has been reported in a channel-billed toucan (*Rhamphastus vitellinus*), chestnut-eared aracari (*Pteroglossus castanotis*), and Toco toucan (*Ramphastos toco*) (Ryan et al. 2003a; Nakamura et al. 2009). The channel-billed toucan was infected with *C. baileyi* (Ryan et al. 2003a).

5.4.16 Coraciiformes

Hornbills, kingfishers, and rollers make up the order Coraciiformes, which has about 209 species in 11 families and 51 genera that are found mostly in Eurasia. These birds are generally omnivorous, feeding on a variety of invertebrates, small vertebrates, or fruits and seeds. The coraciiform birds typically inhabit forests, but ground dwelling species occur in sub-Saharan Africa. Most nest in cavities or burrows.

To date, only a few species from the order Coraciiforme have been identified as hosts for *Cryptosporidium*. *Cryptosporidium galli* was identified in a captive rhinocerous hornbill (*Buceros rhinoceros*) in the Czech Republic (Ng et al. 2006). In Malaysia, *Cryptosporidium* sp. was detected in a wrinkled hornbill (*Aceros corrugatus*) and a wreathed hornbill (*Aceros undulates*) at the Kuala Lumpur National Zoo and Zoo Negara, respectively (Rohela et al. 2005; Lim et al. 2007).

5.4.17 Passeriformes

Passeriformes is the largest avian order, with over 5,700 species in approximately 96 families and more than 1,200 genera. All species have feet for perching, and they occupy a wide range of habitats from dense forests to open grasslands. Fruits, seeds and invertebrates are the dominant foods in their diet.

Passerines are relatively frequent hosts of *Cryptosporidium* sp., some of which are human pathogens (Table 5.7). Yet, despite having more than 50 % of the avian diversity, passerines host only four species (*C. baileyi*, *C. galli*, *C. meleagridis*, and *C. parvum*) and three genotypes (avian genotypes I, III, and IV) of *Cryptosporidium* (Table 5.6).

Australian passerines infected with avian genotypes I, III, and IV showed no clinical signs; whereas, birds from the Czech Republic, which were primarily infected with *C. baileyi* and *C. galli*, had diarrhea and anorexia (Ng et al. 2006).

5.5 *Cryptosporidium* and Cryptosporidiosis of Mammals

Oocyst size and gastrointestinal localization of *Cryptosporidium* species reported in wild mammals are shown in Table 5.8. In contrast to cryptosporidia of domestic animals, oocyst morphometry, infection site, and course of infection are unknown for many *Cryptosporidium* genotypes infecting wild mammals, including bear, beaver, brushtail possum II, chipmunk II and III, *C. bovis*-like, *C. hominis*-monkey, *C. muris*-like, *C. ryanae*-variant, *C. suis*-like, deer, deer mouse I–IV, elephant seal, fox, giant panda, guinea pig, hamster, hedgehog, horse, kangaroo I, mink, mouse II,

Table 5.7 *Cryptosporidium* spp. identified in Passeriformes species

Family	Host species (scientific name)	*Cryptosporidium* taxa	Reference
Alaudidae	**Crested lark** (*Galerida cristata*)	*C. baileyi*	(Qi et al. 2011)
Bombycillidae	**Bohemian waxwing** (*Bombycilla garrulous*)	*C. galli*, *C. meleagridis*	(Qi et al. 2011; Sevá Ada et al. 2011a)
Cardinalidae	**Green-winged saltator** (*Saltator similis*)	*C. galli*	(Sevá Ada et al. 2011a)
Corvidae	**Black-billed magpie** (*Pica pica*)	*C. baileyi*	(Qi et al. 2011)
	Hooded crow (*Corvus cornix*)	*Cryptosporidium* sp.	(Plutzer and Tomor 2009)
	Red-billed blue magpie (Urocissa erythrorhyncha)	Avian genotype III	(Qi et al. 2011)
Estrildidae	**Aurora finch** (*Pytilia hypogrammica*)	*Cryptosporidium* sp.	(Morgan et al. 2001)
	Australian diamond firetail finch (*Stagonoplura bella*)	*Cryptosporidium* sp.[a]	(Blagburn et al. 1990)
	Bengalese finch (*Lonchura striata domestica*)	*C. parvum*[b]	(Gomes et al. 2012)
	Black-throated finch (*Poephila cincta*)	*Cryptosporidium* sp.	(Gardiner and Imes 1984)
	Bronze mannikin finch (*Lonchura cucullata*)	*Cryptosporidium* sp.	(Lindsay et al. 1991)
	Chestnut finch (*Lonchura castaneothorax*)	*C. galli*	(Ng et al. 2006)
	Diamond firetail finch (*Stagonopleura guttata*)	*Cryptosporidium* sp.	(Lindsay et al. 1991)
	Gouldian finch (*Erythrura gouldiae*)	*C. baileyi*, *C. galli*	(Morgan et al. 2001; Qi et al. 2011)
	Java sparrow (*Padda oryzivora*)	*C. baileyi*, avian genotype III	(Qi et al. 2011; Gomes et al. 2012)
	Painted firetail finch (*Emblema pictum*)	*C. galli*	(Ng et al. 2006)
	Parson's finch (*Poephila cincta*)	*C. galli*	(Ng et al. 2006)
	Plum-headed finch (*Neochmia modesta*)	*Cryptosporidium* sp.	(Morgan et al. 2001)
	Red-face aurora finch (*Pytilia hypogrammica*)	*Cryptosporidium* sp.	(Morgan et al. 2001)
	Zebra finch (*Taeniopygia guttata*)	*C. galli*, *C. baileyi*	(Ng et al. 2006; Qi et al. 2011)
Fringillidae	**Canary** (*Serinus canaria*)	*C. galli*, avian genotype I	(Ng et al. 2006; Nakamura et al. 2009)
	Goldfinch (*Carduelis tristis*)	*C. galli*	(Sevá Ada et al. 2011a)
	Pine grosbeak (*Pinicola enucleator*)	*C. galli*	(Ryan et al. 2003b)
	White lored euphonia (*Euphonia chrysopasta*)	*Cryptosporidium* sp.	(Lindsay et al. 1991)

(continued)

Table 5.7 (continued)

Family	Host species (scientific name)	*Cryptosporidium* taxa	Reference
Hirundinidae	**Cliff swallow** (*Petrochelidon pyrrhonota*)	*Cryptosporidium* sp.	(Ley et al. 2012)
Icteridae	**Chopi blackbird** (*Gnorimopsar chopi*)	*Cryptosporidium* sp.	(Nakamura et al. 2009)
	Crested oropendola (*Psarocolius decumanus*)	*C. baileyi*	(Ryan et al. 2003b)
	Red-rumped cocique (*Cacicus haemorrhous*)	*C. baileyi*	(Ryan et al. 2003b)
Leiothrichidae	**Red-billed leiothrix** (*Leiothrix lutea*)	*C. baileyi*	(Qi et al. 2011)
	Silver-eared mesia (*Leiothrix argentauris*)	*C. galli*	(Qi et al. 2011)
Ploceidae	**Eastern golden-backed weaver** (*Ploceus jacksoni*)	*C. baileyi*	(Ng et al. 2006)
Pycnonotidae	**Gray-bellied bulbul** (*Pycnonotus cyaniventris*)	*C. baileyi*	(Ng et al. 2006)
Sturnidae	**Common myna** (*Acridotheres tristis*)	*C. baileyi*	(Qi et al. 2011)
Thraupidae	**Double-collared seedeater** (*Sporophila caerulescens*)	*Cryptosporidium* sp.	(Nakamura et al. 2009)
	Lesser seed finch (*Oryzoborus* sp.)	*C. galli*	(Antunes et al. 2008; Nakamura et al. 2009)
	Red-cowled cardinal (*Paroaria dominicana*)	*C. galli*	(Ng et al. 2006)
	Saffron finch (*Sicalis flaveola*)	*C. baileyi*, *C. galli*	(Nakamura et al. 2009; Sevá Ada et al. 2011a)
	Slate-coloured seedeater (*Sporophila schistacea*)	*C. galli*	(Sevá Ada et al. 2011a)
Turdidae	**Rufous-bellied thrush** (*Turdus rufiventris*)	*C. galli*, *Cryptosporidium* sp.	(Sevá Ada et al. 2011a)
Zosteropidae	**Japaneese white eye** (*Zosterops japonicus*)	Avian genotype IV	(Ng et al. 2006)

[a]Associated with clinical disease

[b]*C. parvum* was identified using molecular tools; however, oocysts from the sample measured 7.2 μm × 5.8 μm, which suggests that another species was also present

muskrat I and II, opposuum II, raccon, rat I–IV, Sbey/Sbld A, Sbey B, Sbey/Sbld/Sltl C, Sbld D, seal I and II, shrew, skunk, vole, W12, W18, and many other unnamed genotypes (Fig. 5.3).

Table 5.9 shows GenBank accession numbers for representative SSU rRNA gene sequences from mammal-associated cryptosporidia. *Cryptosporidium* has been identified in the three major groups of mammals: the egg-laying mammals (Protheria), marsupials (Metatheria), and placental mammals (Eutheria).

Table 5.8 Oocyst morphology and infection site of *Cryptosporidium* taxa detected in wild mammals

Cryptosporidium taxa	Oocysts size (μm)	Infection site	Reference
C. andersoni	6.0–8.1 × 5.0–6.5	Abomasum	(Lindsay et al. 2000)
C. baileyi	6.0–7.5 × 4.8–5.7	Brusa of Fabricius, cloaca	(Lindsay et al. 1989)
C. canis	3.68–5.88 × 3.68–5.88	Small intestine	(Fayer et al. 2001)
C. cuniculus	5.55–6.40 × 5.02–5.92	Small intestine	(Robinson et al. 2010)
C. fayeri	4.5–5.1 × 3.8–5.0	ND	(Ryan et al. 2008)
C. felis	4.6 × 4.0 (3.2–5.1 × 3.0–4.0)	Small intestine	(Iseki 1979)
C. macropodum	4.5–6.0 × 5.0–6.0	ND	(Power and Ryan 2008)
C. muris	7.5–9.8 × 4.6–6.3	Stomach	(Upton and Current 1985)
C. parvum	5.2–5.7 × 4.7–5.3	Small intestine	(Vítovec et al. 2006)
C. scrofarum	4.81–5.96 × 4.23–5.29	Small intestine	(Kváč et al. 2013a)
C. suis	6.0–6.8 × 5.3–5.7	Large intestine	(Vítovec et al. 2006)
C. tyzzeri	4.64 ± 0.05 × 4.19 ± 0.06	Small intestine	(Ren et al. 2012)
C. ubiquitum	4.71–5.32 × 4.33–4.98	Small intestine	(Fayer et al. 2010)
C. wrairi	5.4 × 4.6 (4.8–5.6 × 4.0–5.0)	Small intestine	(Tilley et al. 1991)
Brushtail genotype I	3.92 ± 0.25 × 4.12 ± 0.34	ND	(Hill et al. 2008)
Chipmunk genotype I	5.3–6.6 × 4.7–5.9	ND	(Kváč et al. 2008a)
Ferret genotype	4.9–6.0 × 4.7–5.6	ND	(Kváč et al. 2008a)

ND not determined

5.6 Egg-Laying Mammals

Monotremata, which is the only order of egg-laying mammals, contains two families, three genera, and five species. All lack teeth as adults, forage on insects or other invertebrates, and are native to Australia and New Guinea.

The only report of *Cryptosporidium* from this order has been in a short-beaked echidna (*Tachyglossus aculeatus*), a species that inhabits Australia and New Guinea (O'Donoghue 1995).

5.7 Marsupials

Marsupials comprise seven orders of about 334 species primarily inhabiting Australia and surrounding islands, and South America. *Cryptosporidium* has thus far been detected in species from the orders Dasyuromorphia, Peramelemorphia, and Diprotodontia in Australia, and Didelphimorphia in South America. Members

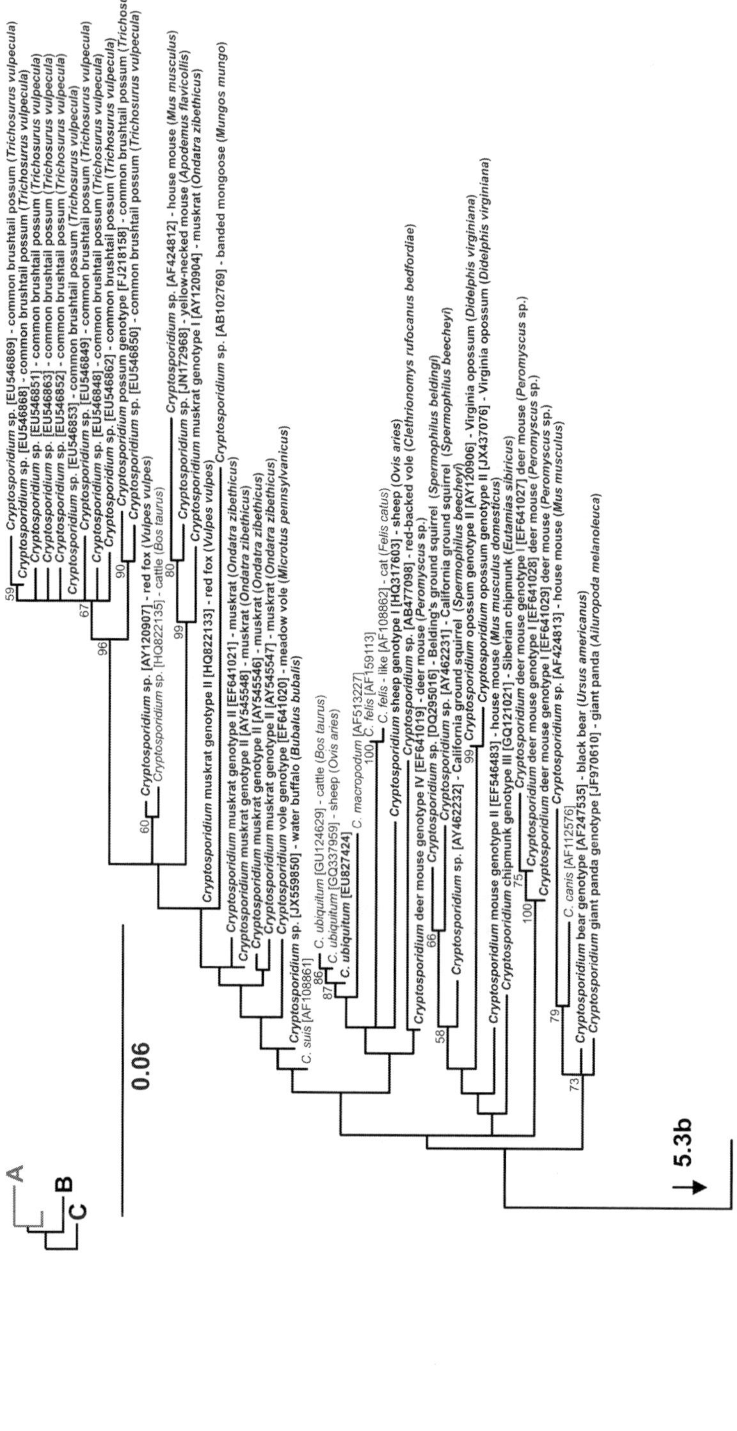

Fig. 5.3 (continued)

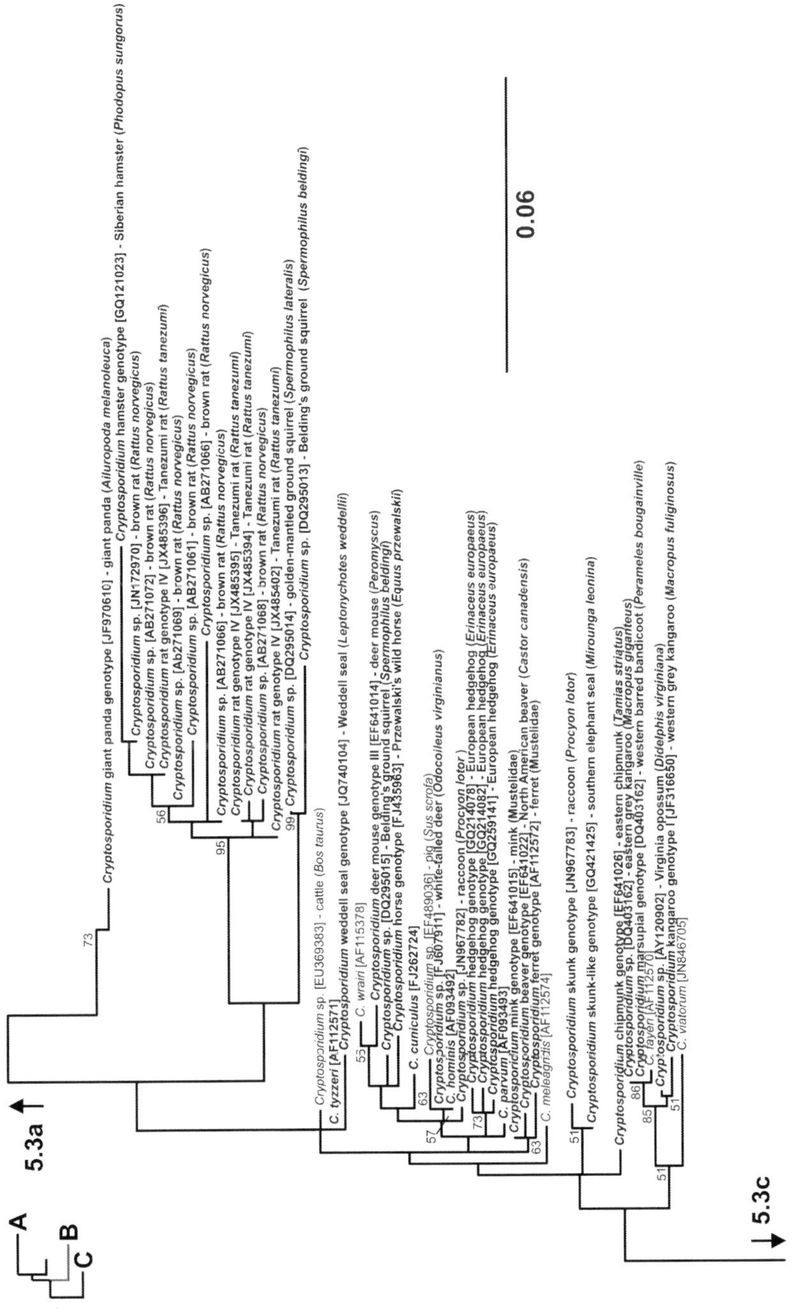

Fig. 5.3 (continued)

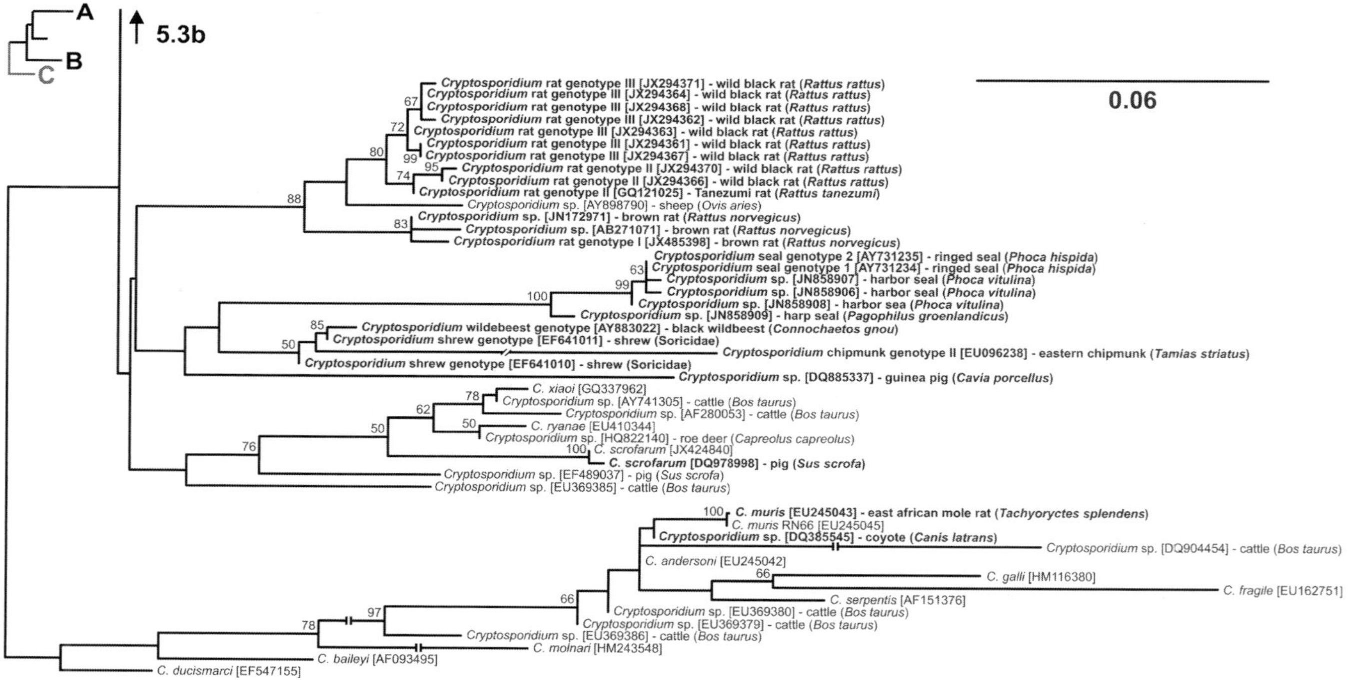

Fig. 5.3 A SSU rDNA-based maximum likelihood (GTRCAT model) tree of *Cryptosporidium* spp. sequences showing clades of cryptosporidia infecting mammals. *Cryptosporidium* sp. detected in mammals are in bold. The percentage of replicate trees in which the associated taxa clustered together in the bootstrap test (1,000 replicates) is shown at nodes. Bootstrap values <50 are not shown. Interrupted branches have been shortened fivefold

Table 5.9 Select GenBank accession numbers for *Cryptosporidium* small subunit rRNA gene sequences from mammals. Hosts are separated by taxonomic order and common name groupings

Infraclass	Order	Common names/groups	Host species (scientific name) and GenBank accession numbers
Placental mammals	Artiodactyla	Ungulates: sheep, cows, pigs, deer, alpine ibex, Black wildebeest, water buffalo	**Alpine ibex** (*Capra ibex*) [EF613340]
			Black wildbeest (*Connochaetos gnou*) [AY883022]
			Deer [AY120910, FJ607911, FJ607928]
			Roe deer (*Capreolus capreolus*) [HQ822140]
			Sika deer (*Cervus nippon*) [DQ898159]
			Water buffalo (*Bubalus bubalis*) [JX559850]
			Wild boar (*Sus scrofa*) [U96770]
	Carnivora	Bears	**Black bear** (*Ursus americanus*) [AF247535]
			Giant panda (*Ailuropoda melanoleuca*) [JF970610, JN790957]
		Canines: dog, foxes, coyote	**Coyote** (*Canis latrans*) [DQ385545]
			Dog (*Canis lupus*) [AF112576]
			Red fox (*Vulpes vulpes*) [AY120907, HQ822133]
			Raccoon dog (*Nyctereutes procyonoides viverrinus*) [AB104730]
		Cats	**Cat** (*Felis catus*) [AF108862, AF112575, AF159113]
		Mongoose	**Banded mongoose** (Mungos mungo) [AB102769]
		Skunk, raccoons, otter, mink, ferret	**Ferret** (*Mustela*) [AF112572]
			Mink (*Mustela vison*) [EF428186, EF428187, EF428189, EF641015]
			Raccoon (*Procyon lotor*) [AY120903,FJ607943, GQ426097, JN967782–JN967784]
			River otter (*Lontra canadensis*) [DQ288166]

(continued)

Table 5.9 (continued)

Infraclass	Order	Common names/groups	Host species (scientific name) and GenBank accession numbers
		Seals	**Harbor seal** (*Phoca vitulina*) [JN858906–JN858908]
			Harp seal (*Pagophilus groenlandicus*) [JN858909]
			Hooded seal (*Cystophora cristata*) [JN858905]
			Southern elephant seal (*Mirounga leonina*) [GQ421425, JQ740100–JQ740102]
			Ringed seal (*Phoca hispida*) [AY731234, AY731235]
			Weddell seal (*Leptonychotes weddellii*) [JQ740103, JQ740104]
	Hyracoidea	Hyrax	**Rock hyrax** (*Procavia capensis*) [AF161579]
	Erinaceomorpha	Hedgehogs	**European hedgehog** (*Erinaceus europaeus*) [GQ214078, GQ214082, [GQ259141]
	Soricomorpha	Shrews and moles	**Shrew** (Soricidae) [EF641010, EF641011]
	Lagomorpha	Rabbits	**European rabbit** (*Oryctolagus cuniculus*) [AY120901, AY273771, FJ262725, GQ865536, HQ397716]
	Perissodactyla	Horse	**Prezewalski's wild horse** (*Equus przewalskii*) [FJ435963]
	Non-human Primates	Lemur	**Black-and-white colobus** (*Colobus guereza*) [F342450]
			Coquerel's sifaka (*Propithecus verreauxi coquereli*) [AF442484]
			Gray lagur (*Semnopithecus entellus thersites*) [EF446673 EF446678]
			Mountain gorilla (*Gorilla gorilla gorilla*) [JQ837801]
			Olive baboon (*Papio anubis*) [JF681172–JF681174]

(continued)

Table 5.9 (continued)

Infraclass	Order	Common names/groups	Host species (scientific name) and GenBank accession numbers
			Purple-faced lagur (*Trachypithecus vetulus philbricki*) [EF446679]
			Red colobus (*Procolobus rufomitratus*) [JF342488–JF342495]
			Rhesus macaque (*Macaca mulatta*) [HM234173, JX000568–JX000570]
			Toque macaque (*Macaca sinica sinica*) [EF446672; EF446674–EF446677]
	Rodentia	Mice and rats	**Brown rat** (*Rattus norvegicus*) [AB271061, AB271062, AB271064, AB271066, AB271068, AB271069, AB271071–AB271073, JX485398, EU245045, JN172970, JN172971]
			Deer mouse (*Peromyscus* sp.) [AY120905, EF641014, EF641019, EF641027–EF641030]
			Mouse (species unspecified) [AF108863, AF112571]
			House mouse (*Mus musculus*) [AF424812, AF424813]
			House mouse (*Mus domesticus*) [EF546483]
			Tanezumi rat (*Rattus tanezumi*) [GQ121025, JX485394–JX485396, JX485402]
			Wild black rat (*Rattus rattus*) [JX29435–JX294371]
			Yellow-necked mouse (*Apodemus flavicolis*) [JN172968]
		Squirrels: chipmunks, tree and ground squirrels	**Belding's ground squirrel** (*Spermophilus beldingi*) [DQ295013, DQ295015, DQ295016]
			California ground squirrel (*Spermophilus beecheyi*) [AY462231–AY462233, DQ295012]

(continued)

Table 5.9 (continued)

Infraclass	Order	Common names/groups	Host species (scientific name) and GenBank accession numbers
			Golden-mantled ground squirrel (*Spermophilus lateralis*) [DQ295014]
			Eastern gray squirrel (*Sciurus carolinensis*) [EU096237]
			Eastern chipmunk (*Tamias striatus*) [EU096238, EF641026]
			Red squirrel (*Sciurus vulgaris*) [EU250844, EU250845]
			Siberian chipmunk (*Eutamias sibiricus*) [GQ121021]
		Muskrats, voles, hamster	**Boreal red-backed vole** (*Myodes gapperi*) [EF641012, EF641013, or EF641016][a]
			Meadow vole (*Microtus pennsylvanicus*) [EF641020]
			Muskrat (*Ondatra zibethicus*) [AY120904, AY545546–AY545548, EF641021][b]
			Red-backed vole (*Clethrionomys rufocanus bedfordiae*) [AB477098]
			Siberian hamster (*Phodopus sungorus*) [GQ121023]
			Brazilian porcupine (*Coendou prehensiles*) [HM209375]
		Guinea pigs, porcupine	**Brazilian porcupine** (*Coendou prehensiles*) [HM209375]
			Guinea pig (*Cavia porcellus*) [AF115378, DQ885337]
		Beaver	**North American beaver** (*Castor canadensis*) [EF641022]
Marsupials	Didelphimorphia	American opossums	**Opossum** (*Didelphis virginiana*) [AY120902, AY120906, JX437075–JX437079]

(continued)

Table 5.9 (continued)

Infraclass	Order	Common names/groups	Host species (scientific name) and GenBank accession numbers
	Diprotodontia	Kangaroos, possums, koala	**Eastern grey kangaroo** (*Macropus giganteus*) [AF513227, AY237630, DQ403162]
			Common brushtail possum (*Trichosurus vulpecula*) [EU546848–EU546853, EU546862, EU546863, EU546868 EU546869, FJ218158]
			Red Kangaroo (*Macropus rufus*) [AF108860, AF112570]
			Western grey kangaroo (*Macropus fuliginosus*) [JF316650, JF316651]
	Peramelemorphia	Bandicoot	**Western barred bandicoot** (*Perameles bougainville*) [DQ403162]

[a]It is not clear from the GenBank entries which of these three sequences was isolated from the boreal red-backed vole
[b]It is not clear from the GenBank entries which of these three sequences was isolated from the muskrat

of the Notoryctemporphia, which contains two species of marsupial moles (*Notoryctes typhlops* and *Notoryctes caurinus*) that are native to Australia; the Microbiotheria, which contains a single species (*Dromiciops gliroides*) that is native to the southwestern part of South America; and the Paucituberculata, which contains six species of shrew opossum (*Rhyncholestes raphanurus*, *Lestoros inca*, *Caenolestes caniventer*, *Caenolestes condorensis*, *Caenolestes convelatus*, and *Caenolestes fuliginosus*) that are native to South America, have yet to be identified as hosts for *Cryptosporidium*.

5.7.1 Dasyuromorphia

This order of carnivorous species includes the Tasmanian Devil (*Sarcophilus harrisii*), numbat (*Myrmecobius fasciatus*), and the shrew-like antechinus species.

To date, only the brown antechinus (*Antechinus staurtii*) has been identified as a host for *Cryptosporidium* (Barker et al. 1978).

5.7.2 *Peramelemorphia*

This order of omnivorous marsupials, which includes bandicoots and bilbies, is found in Australia and New Guinea.

Cryptosporidium sp. and *C. fayeri* have been reported in a southern brown bandicoot (*Isoodon obesulus*) and western-barred bandicoot (*Peremeles bougainville*), respectively (O'Donoghue 1995; Power 2010). *Cryptosporidium muris* was detected in greater bilbies (*Macrotis lagotis*) at a captive breeding colony in Australia. Although some animals cleared the infection within 2 months, others remained infected for 6 months (Warren et al. 2003). Mice trapped in the pens of infected bilbies were positive for *C. muris* and were considered to be the likely source of the infection.

5.7.3 *Diprotodontia*

Diprotodontia, a large and diverse order of herbivorous marsupials that are native to Australia and surrounding islands, hosts a number of *Cryptosporidium* species and genotypes. *Cryptosporidium fayeri*, which was first isolated from a koala (*Phascolarctos cincereus*) (Morgan et al. 1997), has been reported in the red kangaroo (*Macropus rufus*), eastern grey kangaroo (*Macropus giganteus*), and yellow-footed rock wallaby (*Petrogale xanthopus*) (Ryan et al. 2008). *Cryptosporidium macropodum* (previously marsupial genotype II) has been reported in the red kangaroo, eastern grey kangaroo, western grey kangaroo, and swamp wallaby (*Wallabia bicolor*). Neither *C. fayeri* nor *C. macropodum* are known to cause clinical disease in diprodonts or any other marsupial. The *Cryptosporidium* kangaroo genotype I, which was identified in the western grey kangaroo (Yang et al. 2011), clusters with the opossum genotype in a neighbour-joining phylogeny of SSU rRNA sequences (Fig. 5.3b). Brushtail possum genotypes I and II have been identified in the brushtail possum (*Trichasuris vulpecula*) (Power et al. 2003; Hill et al. 2008). Non-genotyped *Cryptosporidium* sp. have been reported in the red-necked wallaby (*Macropus rufogriseus*), pademelon (*Thylogale billardierii*), and koala (Jakob 1992; O'Donoghue 1995).

5.7.4 *Didelphimorphia*

The didelphimorphs (opossums) are native to the American continent.

The *C. fayeri*-opossum genotype (previously opossum genotype I; this genotype is 99 % similar to *C. fayeri* at the SSU rRNA locus), *C. fayeri* (identified by RFLP analysis), and opossum genotype II have been reported in Virginia opossums (*Didelphis virginiana*) from California and New York, USA (Xiao et al. 2002;

Ziegler et al. 2007b). *Cryptosporidium* also has been detected in the white-eared opossum (*Didelphis albiventris*) (Zanette et al. 2008).

5.8 Placental Mammals

Of the 21 orders of placental mammals, 13 contain species that have been identified as hosts for *Cryptosporidium*. The following eight orders have yet to be identified as hosts:

Cingulata and Pilosa: These two orders – totalling 19 species in five families and 18 genera – constitute the superorder Xenarthra. Representatives of these orders include armadillos (Cingulata), anteaters (Cingulata), and sloths (Pilosa). Many species are now extinct. The habitats and diets within the group are quite varied. Most are omnivores, but many are insectivores or herbivores.

Macroscelididae, Afrosordida, and Tublidentata: These are three of the six orders in the superorder Afrotheria. They include tenrecs (Afrosoricida), elephant shrews (Macroscelidia), and aardvarks (Tubulidentata).

Pholidota: This order compromises about eight species in one family and one genus. They are insectivores, inhabiting Africa and southern areas of Asia.

Dermoptera: These comprise a small family of two genera and two arboreal species of gliding mammals, colugos, in Southeast Asia.

Scandentia: These are native of Indonesia and comprise two families, five genera, and 20 species. These small omnivores, commonly referred to as treeshrews, are found in densely forested habitats.

5.8.1 Hyracoidea

Hyracoidea (hyraxes) are in the superorder Afrotheria. They have a digestive system that is similar in function to that of ruminants.

A single species, *C. muris*, has been identified in the rock hyrax (*Procavia capensis*) (Graczyk et al. 1996a; Xiao et al. 1999a).

5.8.2 Sirenia

The order Sirenia, which contains aquatic herbivores such as the dugong and manatee, is most closely related to elephants in the order Proboscidea.

Reports of *Cryptosporidium* in the sirenians have been rare. Hill et al. (1997) described a case of intestinal cryptosporidiosis in a dugnong (*Dugong dugon*) from Queensland, Australia. Three dugongs died as a consequence of the disease and a fourth, the subject of the study, was humanely euthanized (Hill et al. 1997). Sequence analysis of SSU rRNA and acetyl CoA synthethase genes amplified from preserved tissue specimens revealed that the dugong was infected with *C. hominis* (Morgan et al. 2000a).

5.8.3 Proboscidea

There have been few studies of *Cryptosporidium* in the herbivorous elephants that comprise the order Proboscidea.

Cryptosporidium sp. has been detected in African elephants (*Loxodonta africana*) at Kruger National Park, South Africa (6/144 positive) (Abu Samra et al. 2011) and the Barcelona Zoo (Gracenea et al. 2002).

5.8.4 Erinaceomorpha

Hedgehogs (Erinceomorpha) contain 24 species in one family (Erinceidae) and ten genera. Some recent phylogenies include this group in the order Soricomorpha (moles and shrews); hedgehogs are similar to shrews in many ways, including an omnivorous diet that may include carrion.

Cryptosporidium can cause clinical and sometimes fatal infections in hedgehogs. Graczyk et al. (1998a) reported fatal cryptosporidiosis in juvenile African hedgehogs (*Ateletrix albiventris*) housed at the Baltinore Zoo. *Cryptosporidium* developmental stages were detected in the ileum, jejunum, and colon, and moderate to severe villous atrophy was detected in the ileum and jejunum. Meredith and Milne (2009) reported on a case of cryptosporidiosis in an adult European hedgehog (*Erinaceus europaenus*) with hemorrhagic diarrhoea. Similar to the juvenile African hedgehogs, moderate to severe villous atrophy was detected in the ileum and jejunum. Sturdee et al. (1999) detected *Cryptosporidium* oocysts in feces of a free-living European hedgehog from the UK with an estimated infection intensity of 3,000 oocysts per gram. *Cryptosporidium* spp. was detected in 30.0 % (56/188) of European hedgehogs in Germany (Dyachenko et al. 2010). *Cryptosporidium parvum* from gp60 subtype families IIa and IIc, and *Cryptosporidium* hedgehog genotype from gp60 subtype family XIIa (previously VIIa) were identified in positive samples. Although *Cryptosporidium* positive hedgehogs had diarrhea, any association between a *Cryptosporidium* sp. and clinical signs was not determined.

5.8.5 *Soricomorpha*

The Soricomorpha (moles and shrews) represent about 420 species in four families and 44 genera. These are small omnivorous mammals, though invertebrates (especially insects) form a large component of the diet. Many are nocturnal or fossorial. Shrews and moles occupy a variety of habitats, and often live in close association with water.

Cryptosporidium has been reported in the greater white-toothed shrew (*Crocidura russula*), common shrew (*Sorex araneus*), masked shrew (*Sorex cinereus*), northern short-tailed shrew (*Blarina brevicauda*), pygmy shrew (*Sorex minutus*), and Brewer's mole (*Parascalops brewer*) (Siński 1993; Sturdee et al. 1999; Torres et al. 2000; Ziegler et al. 2007a). The shrew genotype (also known as W5) has been identified in 2/5 northern short-tailed shrews in New York (Feng et al. 2007). It is not known to what extent *Cryptosporidium* infection of shrews is associated with clinical disease.

5.8.6 *Cetacea*

Cetaceans include approximately 90 species in 11 families and 40 genera of dolphins and whales. There are two main groups. The Mysticeti (baleen whales) have baleen rather than teeth for filtering invertebrates from water or bottom sediments. The Odontoceti (toothed whales) have teeth and generally are piscivorous. Most species inhabit marine waters, but a few occur in coastal rivers. Cetaceans are most closely related to the hippopotamus.

Cryptosporidium has been detected in fecal samples from 5.1 % (2/39) of bowhead whales (*Balaena mysticetus*) and 24.5 % (12/49) of North Atlantic right whales (*Eubalaena glacialis*) (Hughes-Hanks et al. 2005). Isolates were not genotyped.

5.8.7 *Artiodactyla*

The Artiodactyl mammals include about 240 species in 10 families and 89 genera. These even-toed ungulates include three main groups that differ in their digestive system: Suiformes (pigs) have non-ruminating stomachs, Tylopoda (camels) have three-chambered ruminating stomachs, and Ruminantia (deer, antelope, cows, and hippos) have four-chambered ruminating stomachs. They occupy a wide variety of habitats, but many utilize grasslands or savannahs.

5.8.7.1 Pigs

Most research has focused on the domestic pig, and there is relatively limited data regarding *Cryptosporidium* and cryptosporidiosis in wild pigs (described variously as wild boars and feral pigs). Although domestic pigs can host several *Cryptosporidium* taxa, including *C. felis*, *C. hominis*, *C. meleagridis*, *C. muris*, *C. parvum*, *C. scrofarum*, *C. suis*, *C. tyzzeri*, *Cryptosporidium* sp. Eire w65.5, *Cryptosporidium* rat genotype, and *C. suis*-like (Morgan et al. 1999a; Ebeid et al. 2003; Chen and Huang 2007; Kváč et al. 2009a, c), only *C. parvum*, *C. suis*, and *C. scrofarum* have been detected in wild pigs (Table 5.10) (Atwill et al. 1997; Němejc et al. 2012; Garcia-Presedo et al. 2013). The occurrence of non-adapted cryptosporidia in domestic pigs, and their absence from wild pigs, may be a consequence of farming operations that place pigs in close proximity to humans and cattle (e.g. *C. parvum*), cats (*C. felis*), and rodents (*C. tyzzeri*, *C. muris*, and rat genotype).

Similar to domestic pigs, it appears that shedding of *Cryptosporidium* oocysts by wild pigs is associated with age and population density. Atwill et al. (1997) reported a 4.2-fold greater likelihood of oocyst shedding in animals younger than 8 months compared to older animals. Also, wild pigs from sites with $\geq$2.0 wild pigs/km^2 had approximately a tenfold greater likelihood of shedding *Cryptosporidium* oocysts compared to animals from sites with $\leq$1.9 feral pigs/km^2.

There is no association between diarrhea in wild pigs and the presence of cryptosporidia, and infection intensity is generally less than 2,000 oocyts per gram of feces (Castro-Hermida et al. 2011a; Němejc et al. 2012).

5.8.7.2 Camels and Llamas

Camels are susceptible to infection with two closely related gastric cryptosporidia, *C. muris* and *C. andersoni*, which prior to 2001 were considered a single species, *C. muris* (Lindsay et al. 2000). A key difference between these species is that only *C. muris* is infectious for neonatal mice under experimental conditions (Lindsay et al. 2000). Therefore, when Anderson (1991) isolated *C. muris*-like oocysts from a camel and demonstrated their infectivity for 2–20-day-old mice, it is probable that the isolate was *C. muris*, and not *C. andersoni*. Molecular studies have subsequently shown that *C. muris* from a Bactrian camel clusters with *C. muris* from mice, a hamster, and a rock hyrax in phylogenies constructed from SSU rRNA, Internal transcribed spacer region 1, and HSP-70 sequences (Xiao et al. 1999a; Morgan et al. 2000c).

A *C. andersoni* isolate with oocysts measuring 7.0–7.2 μm × 5.1–5.2 μm (shape index: 1.37–1.39) from a 3-year-old Bactrian camel did not infect immunosuppressed or immunocompetent calves, immunosuppressed or immunocompetent Kun-ming mice, or severe combined immunodeficiency mice (Wang et al. 2007).

Table 5.10 *Cryptosporidium* spp. identified in wild pigs

Cryptosporidium taxa	Country	Prevalence	Reference
Cryptosporidium sp.	Spain	11.5 % (20/175)	(Castro-Hermida et al. 2011a)
	Spain	ND	(Gómez et al. 2000)
	USA	ND	(Pereira et al. 1998)
C. parvum[a]	Spain	11.1 % (3/27)	(Garcia-Presedo et al. 2013)
	USA	5.4 % (12/221)	(Atwill et al. 1997)
C. scrofarum	Austria	2.3 % (1/44)[b] 13.6 % (6/44)[c]	(Němejc et al. 2013)
	Czech Republic	0 % (0/193)[b] 13.0 % (25/193)[c]	(Němejc et al. 2012)
	Czech Republic	1.7 % (4/231)[b] 11.3 % (26/231)[c]	(Němejc et al. 2013)
	Poland	2.3 % (3/129)[b] 7.8 % (10/129)[c]	(Němejc et al. 2013)
	Slovak Republic	0 % (0/56)[b] 1.8 % (1/56)[c]	(Němejc et al. 2013)
	Spain	70.4 % (19/27)[c]	(Garcia-Presedo et al. 2013)
C. suis	Austria	2.3 % (1/44)[b] 11.4 % (5/44)[c]	(Němejc et al. 2013)
	Czech Republic	0 % (0/193)[b] 13.0 % (25/193)[c]	(Němejc et al. 2013)
	Czech Republic	0.9 % (2/231)[b] 10.8 % (25/231)[c]	(Němejc et al. 2013)
	Poland	0 % (0/129)[b] 0.8 % (1/129)[c]	(Němejc et al. 2013)
	Slovak Republic	0 % (0/56)[b] 3.6 % (2/56)[c]	(Němejc et al. 2013)
	Spain	18.5 % (5/27)[c]	(Garcia-Presedo et al. 2013)

[a]Although, the sequence obtained from the feral pig isolate was confirmed as *C. parvum* by the appropriate location of and 100 % sequence homology with two internal probes and three internal primers, in retrospect is impossible to determinate the species and genotypes
[b]Determined using microscopy and
[c]Determined using PCR
ND not determined

Kváč et al. (2008b) infected two out of three lambs and three out of three goat kids with *C. muris* isolate CB03, which originated from a naturally infected Bactrian camel. *Cryptosporidium muris* CB03 was also infectious for Balb/c mice and southern multimammate mice (*Mastomys coucha*). Developmental stages were detected in the *plicae spirales curvature major* in the abomasum of infected lambs and kids. Gastric glands were slightly dilated, and covered with cuboidal and metaplased cells. Infiltration or congestion in the *lamina propria* was not detected, and there was no diarrhea.

Oocysts isolated from rectal samples of camels at a slaughterhouse in Iran had a diameter of 4.56 ± 0.65 μm (range: 4.20–5.70 μm), which is consistent with the morphology of intestinal cryptosporidia (Razawi et al. 2009). However, the isolates were not genotyped so this could not be confirmed. Ryan et al. (2003a) identified *C. parvum* in an alpaca (*Lama pacos*), and Gómez et al. (2000) detected *Cryptosporidium* sp. in a guanaco (*Lama guanicoe*).

A study of dromedary camels (*Camelus dromedaries*) in Iran showed a high *Cryptosporidium* prevalence in animals younger than 1 year (20 %) relative to adult camels (6.5 %) (Yakhchali and Moradi 2012). Another study in Iran showed a relatively high prevalence of *Cryptosporidium* sp. (20.3 %; 61/300) in asymptomatic camels (Sazmand et al. 2012).

5.8.7.3 Giraffe and Okapi

Cryptosporidium muris and *Cryptosporidium* sp. have been detected in giraffes (*Giraffa camelopardalis*) from zoos in the Czech Republic and Spain, respectively (Gómez et al. 1996; Kodádková et al. 2010). The *C. muris* isolate was identical to isolates from the rock hyrax and Bactrian camel at the SSU rRNA locus, and was not infectious for Balb/c mice under experimental conditions (Kodádková et al. 2010).

5.8.7.4 Deer and Moose

Cryptosporidium parvum, *C. ubiquitum*, and the *Cryptosporidium* deer genotype have been detected at varying rates in deer and moose worldwide. *Cryptosporidium parvum* and *C. ubiquitum* have been reported in 12.5 % (4/32) of swamp deer (*Cervus duvauceli*) from Nepal (Feng et al. 2012), in a roe deer from England (Robinson et al. 2011), in a red deer (*Cervus elaphus*) from the Czech Republic (Hajdušek et al. 2004), and in a sika deer (*Cervus nippon*) from China (Wang et al. 2008a). The *Cryptosporidium* deer genotype, which is closely related to *C. ryanae*, has been reported in white-tailed deer (*Odoileus virginianus*) from the US (Xiao et al. 2002). *Cryptosporidium* sp. has been detected in 1.3 % (2/149) and 6 % (3/49) of caribou (*Rangifer tarandus*) from Canada and Alaska, respectively; 8.3 % (2/24), 8.8 % (4/35), and 5.0 % (72/360) of white-tailed deer from the US; 7.9 % (3/38) of black-tailed deer (*Odocoileus hemionus columbianus*) from the US; 25 % (10/40) of tule elk (*Cervus canadensis* ssp. nannodes) from the US; 1.3 % (3/224), 9.1 % (2/22), and 6.2 % (18/291) of roe deer (*Capreolus capreolus*) from Spain, Poland, and Norway, respectively; 14.4 % (17/118), 0.3 % (1/289), 41.5 % (135/325), and 100 % (2/2) of red deer from Poland, Norway, Ireland, and the US, respectively; and 3.3 % (15/455) of moose (*Alces alces*) from Norway (Simpson 1992; Fayer et al. 1996; Deng and Cliver 1999; Rickard et al. 1999; Skerrett and Holland 2001; Siefker et al. 2002; Hamnes et al. 2006; Ziegler et al. 2007b; Johnson et al. 2010; Castro-Hermida et al. 2011b). Other deer that have been identified as hosts of *Cryptosporidium* sp., include the fallow deer (*Dama dama*), axis deer (*Axis axis*), barasingha deer (*Cervus duvauceli*), Eld's deer, muntjac deer, sambar (*Rusa unicolor*), Thorold's deer (*Cervus albirostris*), and Père David's deer (*Elaphurus davidianus*) (Heuschele et al. 1986; Sturdee et al. 1999).

5.8.7.5 Bovids

Cryptosporidium has thus far been identified in wild bovid species representing eight of the ten bovid subfamilies (Table 5.11). It has not yet been identified in the Tibetan antelope (*Pantholops hodgsonii*) or grey rhebok (*Pelea capreolus*), which are the sole representatives of the Pantholopinae and Peleinae subfamilies, respectively. Most studies of wild bovid species have been limited to captive animals outside their native habitat.

In a study of cryptosporidiosis in wild bovids, *Cryptosporidium* developmental stages were identified in the small intestine, cecum, spiral colon, and colon (Van Winkle 1985). Four mountain gazelles (*Gazella cuvieri*) at the Munich Zoo with cryptosporidiosis had anorexia and weight loss but no diarrhea (Pospischil et al. 1987). *Cryptosporidium* developmental stages were detected in the abomasum only. The mucosa was hyperplastic and diffusely infiltrated with small lymphocytes, and the mucosal glands were elongated and hypercellular.

Hippopotamus

Cryptosporidium sp. has been detected in a pygmy hippopotamus (*Choeropsis liberiensis*) at the Barcelona Zoo (Gómez et al. 2000).

5.9 Chiroptera

Bats represent a diverse order of flying mammals with over 1,100 species in 18 families and 202 genera. The group is split into the Megachiropterans, which are large, fruit or nectar-eating bats that lack certain ear structures, and the Microchiropterans, which are small, omnivorous or insectivorous species that utilize echolocation and possess associated ear structures. Most (but not all) bats hibernate, and are generally associated with forest habitats.

Despite representing about 20 % of the mammalian fauna, few bat species have been identified as hosts for *Cryptosporidium*. Cryptosporidiosis was diagnosed in a big brown bat (*Eptesicus fuscus*) from Oregon, USA (Dubey et al. 1998). In New York, *Cryptosporidium* sp. and *C. parvum* were identified in a small brown bat (*Myotis lucifugus*) and big brown bat, respectively (Ziegler et al. 2007b). *Cryptosporidium tyzzeri* was detected in a large-footed bat (*Myotus adversus*) from New South Wales, Australia (Morgan et al. 1999b).

Table 5.11 *Cryptosporidium* in wild bovids grouped by subfammily

Subfamily	Host species (scientific name)	*Cryptosporidium* taxa	Reference
Aepycerotinae	**Impala** (*Aepyceros melampus*)	*C. ubiquitum*, *Cryptosporidium* sp.	(Abu Samra et al. 2011; 2013) (Heuschele et al. 1986)
Alcelaphinae	**Black wildebeest** (*Connochaetes gnou*)	Similar to shrew genotype (W5), *Cryptosporidium* sp.	(Mtambo et al. 1997; Alves et al. 2005)
	Blesbok (*Damaliscus dorcas philipsi*)	*C. ubiquitum*	(Ryan et al. 2003a)
	Blue wildebeest (*Connochaetes taurinus*)[a]	*C. parvum*	(Gómez et al. 1996; Morgan et al. 1999b)
Antilopinae	**Addra gazelle** (*Gazella dama ruficollis*)[a]	*Cryptosporidium* sp.	(Heuschele et al. 1986)
	Blackbuck (*Antilope cervicapra*)	*Cryptosporidium* sp.	(Van Winkle 1985; Heuschele et al. 1986)
	Dorcas gazella (*Gazella dorcas neglecta*)[a]	*Cryptosporidium* sp.	(Gómez et al. 1996)
	Mountain gazelle (*Gazella cuvieri*)[a]	*Cryptosporidium* sp.	(Pospischil et al. 1987)
	Persian gazelle (Gazella subgutturosa)[a]	*Cryptosporidium* sp.	(Heuschele et al. 1986)
	Slender-horned gazelle (*Gazella leptoceros*)[a]	*C. parvum*, *Cryptosporidium* sp.	(Heuschele et al. 1986; Geurden et al. 2009)
	Springbok (*Antidorcas marsupialis*)[a]	*Cryptosporidium* sp.	(Heuschele et al. 1986)
	Thomson's gazelle (*Eudorcus thomsonii*)	*Cryptosporidium* sp.	(Canestri-Trotti 1989)
Bovinae	**African buffalo** (*Syncerus caffer*)	*C. ubiquuitum*, *C. bovis*, *Cryptosporidium* sp.	(Gómez et al. 1996; Mtambo et al. 1997; Abu Samra et al. 2011; 2013)
	Bison, American (*Bison bison*)[a]	*C. tyzzeri*, *Cryptosporidium* sp.	(Alves et al. 2005; Geurden et al. 2009)
	Bison, European (*Bison bonasus*)[a]	*C. andersoni*, *Cryptosporidium* sp.	(Paziewska et al. 2007) (Ryan et al. 2003a)
	Bongo antelope (*Tragelaphus eurycerus*)[a]	*C. parvum*, *Cryptosporidium* sp.	(Geurden et al. 2009) (Gómez et al. 2000)

(continued)

Table 5.11 (continued)

Subfamily	Host species (scientific name)	*Cryptosporidium* taxa	Reference
	Eland (*Taurotragus oryx*)[a]	*C. parvum*, *Cryptosporidium* sp.	(Heuschele et al. 1986; Geurden et al. 2009)
	Lowland anoa (*Bubalus depressicornis*)[a]	*Cryptosporidium* sp.	(Gómez et al. 2000)
	Nilgai (*Boselaphus tragocamelus*)[a]	*Cryptosporidium* sp.	(Heuschele et al. 1986)
	Nyala (*Tragelaphus angasi*)[a]	*C. ubiquitum*	(Ryan et al. 2003a)
	Water buffalo (*Bubalus bubalis*)	*C. ryanae* variant	(Feng et al. 2012)
	Yak (*Bos mutus*)[a]	*C. parvum*, *Cryptosporidium* sp.	(Geurden et al. 2009) (Karanis et al. 2007)
	Zebu (*Bos primigenius indicus*)	*C. ryanae* variant	(Feng et al. 2012)
Caprinae	**Alpine ibex** (*Capra ibex*)[a]	*C. ubiquitum*	(Karanis et al. 2007)
	Angora goat (*Capra hircus*)	*Cryptosporidium* sp.	(Mason et al. 1981)
	Armenian mouflon (*Ovis orientalis gmelini*)[a]	*Cryptosporidium* sp.	(Heuschele et al. 1986)
	Barbary sheep (*Ammotragus lervia*)[a]	*C. tyzzeri*	(Karanis et al. 2007)
	Mouflon sheep (*Ovis musimon*)[a]	*C. ubiquitum*, *Cryptosporidium* sp.	(Gómez et al. 2000; Ryan et al. 2003a)
	Takin (*Budorcas taxicolor*)[a]	*C. tyzzeri*	(Karanis et al. 2007)
	Turkomen markhor (*Capra falconeri*)[a]	*Cryptosporidium* sp.	(Heuschele et al. 1986)
	Urial (*Ovis orientalis*)	*Cryptosporidium* sp.	(Ducatelle et al. 1983)
Hippotraginae	**Addax** (*Addax nasomaculatus*)	*Cryptosporidium* sp.	(Van Winkle 1985) (Heuschele et al. 1986)
	Fringe-eared oryx (*Oryx beisa callotis*)	*Cryptosporidium* sp.	(Van Winkle 1985)
	Sable antelope (*Hippotragus niger*)[a]	*C. parvum*, *Cryptosporidium* sp.	(Hajdušek et al. 2004) (Van Winkle 1985; Heuschele et al. 1986)
	Scimitar-horned oryx (*Oryx dammah*)	*Cryptosporidium* sp.	(Van Winkle 1985) (Heuschele et al. 1986)

(continued)

Table 5.11 (continued)

Subfamily	Host species (scientific name)	*Cryptosporidium* taxa	Reference
	White antelope (*Addax nasomaculatus*)	*Cryptosporidium* sp.	(Van Winkle 1985)
Reduncinae	**Nile lechwe** (*Kobus megaceros*)[a]	*Cryptosporidium* sp.	(Heuschele et al. 1986)
	Waterbuck (*Kobus ellipsiprymnus*)	*Cryptosporidium* sp.	(Gómez et al. 1996)

[a]Wild bovids under captive condition

5.10 Perissodactyla

The Perissodactyl mammals include about 16 species in three families and six genera. These are odd-toed ungulates that include horses, tapirs and rhinoceroses. The digestive system is much simpler than that of the ruminants, with cellulose broken down by fermentation in the latter portion of the gut. They occupy mainly grassland or savannah habitat.

5.10.1 Equids

Equids include the zebra, ass, kiang, and wild horse. *Cryptosporidium* was detected in 28 % (7/25) of zebras (*Equus zebra*) at Mikumi National Park in Tanzania (Mtambo et al. 1997). *Cryptosporidium* horse genotype was identified in a Przewalski's wild horse foal at the Prague Zoo (Ryan et al. 2003a). Although, no other cryptosporidia have been reported in wild Equids, domestic horses have been identified as hosts for *C. parvum*, *Cryptosporidium* horse genotype, and *Cryptosporidium* hedgehog genotype (Laatamna et al. 2013).

5.10.2 Rhinoceros and Tapirs

Cryptosporidium sp. has been reported in a southern white rhinoceros (*Ceratotherium simum simum*), rhinoceros (*Rhinoceros unicornis*), and a South American tapir (*Tapirus terrestris*) at the Barcelona Zoo (Wang and Liew 1990; Gómez et al. 1996, 2000).

5.11 Carnivora

5.11.1 Fin-Footed Mammals (Pinnipedia)

Cryptosporidium muris, which was isolated from a ringed seal (*Phoca hispida*) in Northern Quebec, Canada (Santín et al. 2005), remains the only species identified in Pinnipeds to date. *Cryptosporidium* sp. has been reported in the ringed seal (*Phoca hispida*), harbor seal (*Phoca vitulina*), harp seal (*Pagophilus groenlandicus*), Weddell seal (*Leptonychotes weddellii*), California sea lion (*Zalophus californianus*), grey seal (*Halichoerus grypus*), and southern elephant seal (*Mirounga leonina*) (Deng et al. 2000; Hughes-Hanks et al. 2005; Santín et al. 2005; Bogomolni et al. 2008; Rengifo-Herrera et al. 2011, 2013; Bass et al. 2012). In a neighbor-joining phylogeny, SSU rRNA gene sequences from the ringed seal (seal genotype I, GenBank accession no. AY731234 and seal genotype II, GenBank accession no. AY731235), harbor seal (GenBank accession nos. JN858906–JN858908), harp seal (GenBank accession no. JN858909), and hooded seal (GenBank accession no. JN858905) form a separate clade that is most closely related to a group containing sequences from the shrew (Soricidae), wildebeest (*Connochaetes gnou*), and eastern chipmunk (*Tamias striatus*) (Fig. 5.3c). Isolates from southern elephant seals (*Mirounga leonina*) on the west coast of the Antarctic Peninsula shared 99 % identity with a skunk genotype and *C. tyzzeri* at the SSU locus and are named southern elephant seal genotype (Rengifo-Herrera et al. 2011, 2013). Isolates from the Weddell seal are most closely related to the ferret genotype and *C. parvum* (99 % identity at the SSU locus). *Cryptosporidium* sp. detected in California sea lions (*Zalophus californianus*) from the northern California coastal area share 98 % identity with *C. parvum* at the COWP locus (Deng et al. 2000).

5.11.2 Domestic Dogs

There have been few molecular studies of *Cryptosporidium* in dogs (Table 5.12). *Cryptosporidium canis* is the most frequently identified species in dogs (Abe et al. 2002) and it appears to be relatively host adapted. Other species found in dogs include *C. parvum* (Hajdušek et al. 2004), *C. muris* (Ellis et al. 2010), and *C. meleagridis* (Hajdušek et al. 2004). *Cryptosporidium muris* caused chronic gastritis in an 18-month male, mixed-breed dog (Ellis et al. 2010). It appeared that the dog had a concurrent infection with *Helicobacter*.

In a study of parasites infecting sled dogs in Poland, Bajer et al. (2011) found a higher prevalence of *Cryptosporidium* in dogs infected concurrently with *Giardia* (35.5 %) than in dogs without *Giardia* (2.7 %). In contrast, *Cryptosporidium* prevalence was three times lower in dogs infected with nematodes. Other studies have shown a higher prevalence of *Cryptosporidium* in dogs with diarrhea (Mirzaei

Table 5.12 Prevalence of *Cryptosporidium* in dogs from different countries

Country	Prevalence	Cryptosporidium taxa	Reference
Australia	10.9 % (54/493)	*Cryptosporidium* sp.	(Bugg et al. 1999)
Argentina	2.2 % (48/2,193)	*Cryptosporidium* sp.	(Fontanarrosa et al. 2006)
	1.0 % (1/100)	*Cryptosporidium* sp.	(Soriano et al. 2010)
Brazil	–	*C. canis*	(Thomaz et al. 2007)
	9.6 % (43/450)	*Cryptosporidium* sp.	(Lallo and Bondan 2006)
	4.9 % (4/81)	*Cryptosporidium* sp.	(Mandarino-Pereira et al. 2010)
Canada	74.3 % (52/70)	*Cryptosporidium* sp.	(Shukla et al. 2006)
	3.2 % (5/155)	*Cryptosporidium* sp.	(Himsworth et al. 2010)
Costa-rica	1.7 % (1/58)	*Cryptosporidium* sp.	(Scorza et al. 2011)
Czech Republic	ND	*C. meleagridis*, *C. parvum*	(Hajdušek et al. 2004)
Iran	2.0 % (11/548)	*Cryptosporidium* sp.	(Mirzaei 2012)
Iran	5.2 % (4/77)	*Cryptosporidium* sp.	(Beiromvand et al. 2013)
Japan	3.9 % (3/77)	*C. canis*	(Yoshiuchi et al. 2010)
Korea	9.7 % (25/257)	*Cryptosporidium* sp.	(Kim et al. 1998)
Netherlands	8.6 % (13/152)	*Cryptosporidium* sp.	(Overgaauw et al. 2009)
Poland	13 % (14/108)	*Cryptosporidium* sp.	(Bajer et al. 2011)
Spain	5.1 % (4/79)	*Cryptosporidium* sp.	(Dado et al. 2012)
	6.3 % (32/505)	*Cryptosporidium* sp.	(Gracenea et al. 2009)
USA	10.2 % (5/49)	*Cryptosporidium* sp.	(Jafri et al. 1993)
	17 % (17/100)	*Cryptosporidium* sp.	(Juett et al. 1996)
	3.8 % (5/130)	*Cryptosporidium* sp.	(Hackett and Lappin 2003)
	2.0 % (4/200)	*Cryptosporidium* sp.	(el-Ahraf et al. 1991)
	2.3 % (3/129)	*C. canis*	(Wang et al. 2012)
	ND	*C. muris*	(Ellis et al. 2010)
	12.0 % (6/50) with diarrhea; 2.0 % (1/50) without diarrhea	*Cryptosporidium* sp.	(Tupler et al. 2012)
	2.5 % (3/120)	*Cryptosporidium* sp.	(McKenzie et al. 2010)

ND not determined

2012; Tupler et al. 2012), and dogs less than 1-year-old (Mirzaei 2012). It is not known if the association with diarrhea is dependent on the *Cryptosporidium* species/genotype causing the infection.

Cryptosporidium canis does occasionally cause human disease (Leoni et al. 2006; Hijjawi et al. 2010), and a UK study found that dogs are the most likely companion animals to shed *Cryptosporidium* (46/139) (Smith et al. 2009).

5.11.3 Raccoon Dogs

Cryptosporidium parvum has been identified in a raccoon dog (*Nyctereutes procyonoides viverrinus*) at a zoo in Osaka, Japan (Matsubayashi et al. 2005).

5.11.4 Foxes

Cryptosporidium has been reported in the two genera of true foxes: *Vulpes* and *Urocyon*. Current (1989) identified *Cryptosporidium* sp. in the gray fox (*Urocyon cinereoargenteus*), which is native to southern parts of North America and northern parts of South America. *Cryptosporidium canis* fox genotype ($n = 4$), *C. canis* ($n = 1$), and muskrat genotype I ($n = 1$) were identified in 6/76 *Cryptosporidium* positive samples from unidentified fox species in Maryland, USA (Zhou et al. 2004a). *Cryptosporidium* was detected in 38.7 % (24/62) and 8.1 % (10/124) of red foxes (*Vulpes vulpes*) from the Slovak Republic and Ireland, respectively (Nagano et al. 2007; Ravaszová et al. 2012). Two of the isolates from the study in Ireland were identified as *C. parvum* by sequence analysis of the SSU rRNA and gp60 loci. In contrast to the relatively high prevalence reported in the Slovak Republic, *Cryptosporidium* sp. was detected in only 2.2 % (6/269) and 8.7 % (2/23) of red foxes from Norway and England, respectively (Sturdee et al. 1999; Hamnes et al. 2007).

5.11.5 Wolves

Two large Canadian studies were consistent in showing a relatively low prevalence of *Cryptosporidium* in gray wolves. The prevalence was 1.7 % in a study of 1,558 fecal samples from wolves in British Columbia (Bryan et al. 2012), and 1.2 % in a study of 601 fecal samples from wolves in Manitoba (Stronen et al. 2011). In contrast, 54.9 % of fecal samples (28/51) from wolves in northeast Poland were positive for *Cryptosporidium* (Kloch et al. 2005). A subsequent study in the same region of Poland reported a prevalence of 35.7 %, and *C. parvum* was identified by sequence analysis of the COWP gene (GenBank accession number: AF266273) (Paziewska et al. 2007).

5.11.6 Coyotes

Cryptosporidium has been detected in 27 % (6/22) and 26.3 % (5/19) of fecal samples from coyotes (*Canis latrans*) in northeastern Pennsylvania (Trout et al. 2006) and New York (Ziegler et al. 2007a), respectively. One isolate from the Pennsylvania study shared 99.7 % sequence identity with *C. muris* at the SSU rRNA locus, and the remaining five isolates were identical to the *C. canis* coyote genotype (Trout et al. 2006). In a study of coyotes in southern Alberta and Saskatchewan, Canada, *Cryptosporidium* was not found during winter, and the prevalence was 17.4 % during summer. SSU rRNA sequences from two of the isolates were identified as the *C. canis* coyote genotype (Thompson et al. 2009).

5.11.7 Bears

Cryptosporidium has been detected in three of the five genera in the family Ursidae (*Ailuropoda*, *Helarctus*, and *Ursus*). There have been no reports in the sloth bear (*Melursus ursinus*) or spectacled bear (*Tremarctos ornatus*), which are the only extant representatives of their genera.

Cryptosporidium has been reported in two captive Malayan sun bears (*Helarctos malayanus*) at a zoological park in Taiwan (Wang and Liew 1990).

Cryptosporidium has been detected in black bears (*Ursus americanus*), brown bears (*Ursus arctos*), and polar bears (*Ursus maritimus*) in the genus *Ursus* (Siam et al. 1994; Duncan et al. 1999; Xiao et al. 2000; Ravaszová et al. 2012). Xiao et al. (2000) identified the *Cryptosporidium* bear genotype in a black bear and showed that it is most closely related to *C. canis* at the SSU rRNA and HSP-70 loci.

Karanis et al. (2007) detected *C. tyzzeri* (previously mouse genotype I) in a lesser panda (*Ailurus fulgens*). An isolate from a giant panda (*Ailuropoda melanoleuca*) was named the *Cryptosporidium* giant panda genotype based on sequences of the SSU rRNA, actin, COWP, and HSP-70 genes. Although the study has not been published, the sequences are available in GenBank under the accession numbers JF970610, JN969985, JN588570, and JN588571.

5.11.8 Raccoons

A number of studies have identified *Cryptosporidium* in raccoons (*Procyon lotor*) (Snyder 1988; Perz and Le Blancq 2001; Zhou et al. 2004a; Ziegler et al. 2007a;

Chavez et al. 2012). Zhou et al. (2004a) identified the skunk genotype in 3.9 % (2/51) of raccoon samples collected in the Chesapeake Bay area of Maryland, USA. *Cryptosporidium* was detected in 11 out of 44 fecal samples from raccoons in Colorado, USA. Five out of six of the isolates that were genotyped were identified as the skunk genotype, and one was identified as *C. parvum* (Chavez et al. 2012).

Cryptosporidium also causes clinical disease in raccoons. A juvenile raccoon, estimated to be between 6 and 12-months-old was found in a moribund condition in Fort Collins, Colorado. The animal was emaciated and dehydrated, and had diarrhea and a mucoid oculonasal discharge. *Cryptosporidium* sp. was identified on intestinal villi (Martin and Zeidner 1992).

5.11.9 Ferrets

Rehg et al. (1988) reported cases of subclinical cryptosporidiosis in 40 % of ferrets at an animal research facility in Tennessee. Subsequent studies identified the *Cryptosporidium* ferret genotype in the black-footed ferret (*Mustela nigripes*) and domestic ferret (*Mustela putorius furo*) in the USA and Japan (Xiao et al. 1999b; Sulaiman et al. 2000; Abe and Iseki 2003). The ferret genotype is genetically closely related to *C. parvum*, sharing 99 %, 98 %, 98 %, and 97 % identity at the SSU rRNA, actin, COWP, and HSP-70 genes, respectively. Ferret genotype isolates from the USA (Sulaiman et al. 2000) and Japan (Abe and Iseki 2003) have identical sequences, suggesting that this genotype is geographically conserved.

Cryptosporidium also causes clinical disease in ferrets. Gómez-Villamandos et al. (1995) reported an outbreak of fatal cryptosporidiosis in captive 11–12 month-old non-pregnant female ferrets (*Mustela putorius furo*) kept on a goat farm. Animals died 48–72 h after the onset of signs, which included anorexia, depression, and diarrhea. *Cryptosporidium* developmental stages were identified in feces and tissue, but the species/genotype causing the infection was not identified.

5.11.10 Otters

Cryptosporidium has been detected in the European or wild otter (*Lutra lutra*), river otter (*Lontra canadensis*), and sea otter (*Enhydra lutris nereis*) (Feng et al. 2007; Gaydos et al. 2007; Méndez-Hermida et al. 2007; Oates et al. 2012). In Spain, *Cryptosporidium* oocysts were detected in 17 out of 437 fecal samples from wild otters (Méndez-Hermida et al. 2007). In the USA, nine river otters from the Puget Sound Georgia Basin were positive for the *Cryptosporidium* mink genotype (Gaydos et al. 2007), and a river otter in New York was positive for the *Cryptosporidium* skunk genotype (Feng et al. 2007). *Cryptosporidium* sp. was detected in a single sea otter in California (Oates et al. 2012).

5.11.11 Mink and Martens

Rademacher et al. (1999) isolated *Cryptosporidium* oocysts measuring 3–5 μm from captive beech martens (*Martes foina*) with episodes of diarrhea. *Cryptosporidium* oocysts (4.5–5.5 μm) were also detected in 24.2 % (8/33) of fecal samples from captive American minks (*Mustela vison*) in Spain. The animals did not show clinical signs associated with cryptosporidiosis. Three isolates had COWP sequences that differed from the ferret genotype by a single nucleotide polymorphism (Gómez-Couso et al. 2007). In China, eight *Cryptosporidium* isolates were obtained from 469 fecal samples of American minks (*Mustela vison*) originated from a farm in Hebei Province in China. Six of the eight *Cryptosporidium*-positive samples contained a novel genotype, the *Cryptosporidium* mink genotype, that was most closely related to the *Cryptosporidium* ferret genotype (Wang et al. 2008b). In New York, one of four mink (*Mustela vison*) in a watershed study was positive for *Cryptoporidium* mink genotype, and an ermine (*Mustela erminea*) had a mixed infection with *Cryptosporidium* shrew genotype (also known as W5 genotype) and W18 genotype (Feng et al. 2007). Also in New York, Ziegler et al. (2007a) detected *Cryptosporidium* sp. in one of three ermine (*Mustela erminea*) and in one of 58 American mink (*Musela vison*).

5.11.12 Skunk

Cryptosporidium has been detected in the striped skunk (*Mephitis mephitis*) in New York, USA (Perz and Le Blancq 2001; Ziegler et al. 2007a), and the *Cryptosporidium* skunk genotype has been described from this host (Xiao et al. 2002; Feng et al. 2011). *Cryptosporidium* prevalence in the striped skunk in New York was 14 % (12/86) (Ziegler et al. 2007a).

5.11.13 Badger

Cryptosporidium has been reported in 15.4 % (4/26) of European badgers (*Meles meles*) in England (Sturdee et al. 1999). The *Cryptosporidium* species/genotype infecting badgers is not yet known.

Table 5.13 Prevalence of *Cryptosporidium* in cats from different countries

Country	Prevalence	*Cryptosporidium* taxa	Reference
Argentina	2.2 % (48/2,193)	*Cryptosporidium* sp.	(Fontanarrosa et al. 2006)
Canada	7.1 % (5/70)	*Cryptosporidium* sp.	(Shukla et al. 2006)
Germany	5.3 % (1/19)	*Cryptosporidium* sp.	(Sotiriadou et al. 2013)
Japan	12.7 % (7/55)	*C. felis*	(Yoshiuchi et al. 2010)
Korea	9.7 % (25/257)	*Cryptosporidium* sp.	(Kim et al. 1998)
Netherlands	8.6 %	*Cryptosporidium* sp.	(Overgaauw et al. 2009)
Spain	6.3 % (32/505)	*Cryptosporidium* sp.	(Gracenea et al. 2009)
UK	5.9 % (3/51)	*Cryptosporidium* sp.	(Smith et al. 2009)
USA	5.3 % (11/206)	*Cryptosporidium* sp.	(Hill et al. 2000)
	4.7 % (16/344)	*Cryptosporidium* sp.	(Mekaru et al. 2007)
	3.8 % (10/263)	*Cryptosporidium* sp.	(Spain et al. 2001)
	6.4 % (11/173)	*Cryptosporidium* sp.	(Nutter et al. 2004)
	12.0 % (30/250) – IFA	*Cryptosporidium* sp.	(Ballweber et al. 2009)
	4.8 % (12/250) – PCR	*C. felis*	

5.11.14 Banded Mongoose

The banded mongoose (*Mungos mungo*) has been identified as a host of a *Cryptosporidium* genotype that is most closely related to the bear genotype at the SSU rRNA and HSP-70 loci (Abe et al. 2004).

5.11.15 Domestic Cat

Two *Cryptosporidium* species, *C. felis* and *C. muris*, have been reported in domestic cats (*Felis catus*). *Cryptosporidium felis* was first identified in cats in Japan (Iseki 1979) and has since been reported in cats worldwide. *Cryptosporidium muris* has been reported in cats less frequently (Santín et al. 2006; Pavlásek and Ryan 2007; FitzGerald et al. 2011). FitzGerald et al. (2011) reported a *C. felis*/*C. muris* coinfection in a cat with chronic diarrhea. *Cryptosporidium muris* was identified within the glands of the gastric mucosa. Interestingly, the cat appeared to be persistently infected with *C. muris*, but not *C. felis*. *Cryptosporidium felis* has been identified as an infrequent cause of human cryptosporidiosis (Pedraza-Diaz et al. 2001; Caccio et al. 2002; Caccio 2005; Leoni et al. 2006). The prevalence of *Cryptosporidium* sp. in cats from different countries is shown in Table 5.13.

5.11.16 Wild Cats

Cryptosporidium tyzzeri has been identified in black leopards (*Pantera pardus*) (Karanis et al. 2007). *Cryptosporidium* sp. also has been reported in bobcats (*Lynx rufus*) in California (Carver et al. 2012) and New York, USA (Ziegler et al. 2007a).

5.12 Lagomorpha

Lagomorpha (rabbits, hares and pikas) includes about 87 species in two families and 12 genera. The phylogeny of the lagomorphs is not well understood. They have four upper incisors, and all are herbivorous. All produce both a hard fecal pellet and a soft, grease-like pellet derived from the cecum. The soft pellet is reingested, and contains a much higher concentration of vitamins and minerals than the uningested hard pellet. Lagomorphs occupy both forested and grassland areas.

Rabbits are the major host of *C. cuniculus* (previously *Cryptosporidium* rabbit genotype). *Cryptosporidium cuniculus* is closely related to *C. hominis*, sharing 99.5, 100, 99.9, and 99.5 % nucleotide sequence identity at the SSU rRNA, COWP, actin, and HSP-70 loci, respectively (Robinson et al. 2010). It emerged as a human pathogen in 2009 when it was identified as the cause of a waterborne cryptosporidiosis outbreak in Northhamptonshire, England (Robinson et al. 2010). A subsequent study identified *C. cuniculus* in 1.2 % of 3,030 samples from human cryptosporidiosis cases in the UK during the period 2007–2008 (Chalmers et al. 2011). In a comprehensive review of the literature on *Cryptosporidium* in rabbits, which will not be duplicated here, Robinson and Chalmers (2010) noted that in studies employing genotyping tools, *C. cuniculus* was the only species/genotype detected. More recently, *C. cuniculus* was the only species detected in rabbits in Australia (Nolan et al. 2010, 2013) and China (Shi et al. 2010; Zhang et al. 2012), and prevalence was highest in younger (1–3-month-old) rabbits (Shi et al. 2010).

5.13 Rodentia

Rodents comprise about 40 % of the mammalian diversity, with over 2,200 species in 31 families and 481 genera. Five suborders are typically recognized: Myomorpha (mice, rats, gerbils, and relatives), Sciuromorpha (tree squirrles and relatives), Castorimorpha (beavers, gophers, and relatives), Anumaluromorpha (scaly-tailed squirrels, and springhares), and Hystricomorpha (gundis, capybaras, and relatives). With such diversity, rodents occupy a wide range of habitats, and generalizations regarding their natural history are difficult. All rodents have two upper and two lower incisors that grow continuously (and are self-sharpening) but lack canines. Most species are herbivorous or omnivorous.

5.13.1 Muridae

Old world mice and rats comprise the subfamily Murinae in the largest mammal family, the Muridae. Gerbils, which also are members of the family Muridae, have yet to be identified as a natural host for *Cryptosporidium*, although they are susceptible to experimental infections (Baishanbo et al. 2005; Kváč et al. 2009d).

Four *Cryptosporidium* species (*C. tyzzeri*, *C. muris*, *C. parvum*, and *C. scrofarum*) and six genotypes (*C. suis*-like genotype, rat genotypes I–IV and mouse genotype II) have been reported as natural infections in several species of *Mus*, *Apodemus*, and *Rattus* in Australia, China, the Czech Republic, Germany, Kenya, New Zealand, the Philippines, Poland, Portugal, Spain, the UK, and the US (Table 5.13). The *C. suis*-like genotype and *C. scrofarum* were reported in rats in the Philippines; *C. suis*-like genotype was found exclusively in Asian house rats, and *C. scrofarum* was additionally identified in brown rats (Ng-Hublin et al. 2013). The *C. suis*-like genotype, which is 99.7 % similar to *C. suis* at the SSU rRNA locus, has additionally been reported in cattle and humans (Ong et al. 2002; Langkjær et al. 2007; Robinson et al. 2011). *Cryptosporidium scrofarum* is a pig-adapted species that also has been reported in cattle and humans (Kváč et al. 2009b; Ng et al. 2011). *Cryptosporidium muris*, a rodent adapted species, has been detected the domestic mouse (*Mus musculus domesticus*) in the UK (Chalmers et al. 1997), the Algerian mouse (*Mus spretus*) in Spain (Torres et al. 2000), and the East African mole rat (*Tachyoryctes splendens*) in Kenya (Kváč et al. 2008b). *Cryptosporidium* mouse genotype II has been identified in 11 domestic mice in Australia (Foo et al. 2007). Rat genotypes I–IV have been reported in various rat species in Australia, China, and the Philippines (Table 5.14). Although *C. tyzzeri* (previously mouse genotype I) has been reported in the yellow-necked mouse, voles, snakes, and rats (Morgan et al. 1998, 1999b; Bajer et al. 2003; Xiao and Ryan 2004; Karanis et al. 2007), the house mouse is considered the major host for this species (Ren et al. 2012).

The fate of the house mouse (*Mus musculus*) has been intimately connected to humans since the establishment of a commensal relationship at the dawn of civilization. Two house mouse subspecies, *Mus musculus musculus* and *Mus musculus domesticus*, diverged approximately 0.5 million years ago in the Middle East (Geraldes et al. 2008; Duvaux et al. 2011; Auffray and Britton-Davidian 2012; Bonhomme and Searle 2012). *Mus m. musculus* spread to northern Eurasia and migrated westward through Europe. *Mus m. domesticus* expanded westward through Asia Minor to southern and western Europe, northern Africa, and the New World (Boursot et al. 1993; Guénet and Bonhomme 2003; Rajabi-Maham et al. 2008; Duvaux et al. 2011; Auffray and Britton-Davidian 2012; Bonhomme and Searle 2012; Cucchi et al. 2012). About 6,000 years ago, westward migrating

Table 5.14 *Cryptosporidium* spp. identified in species of old world mice and rats

Host (scientific name)	*Cryptosporidium* taxa	Reference
Algerian mouse (*Mus spretus*)	*C. muris*, *C. parvum*	(Torres et al. 2000)
Asian house rat (*Rattus tanezumi*)	*C. scrofarum*, *C. tyzzeri*, *C. suis*-like, rat genotypes II–IV	(Lv et al. 2009; Ng-Hublin et al. 2013)
Black rat (*Rattus rattus*)	*C. parvum*, rat genotypes II–III, *Cryptosporidium* sp.	(Miyaji et al. 1989; Yamura et al. 1990; Webster and Macdonald 1995; Chilvers et al. 1998; Torres et al. 2000; Paparini et al. 2012)
Brown rat (*Rattus norvegicus*)	*C. tyzzeri*, *C. scrofarum*, *C. muris*, rat genotypes I–IV, *Cryptosporidium* sp.	(Iseki 1986; Miyaji et al. 1989; Yamura et al. 1990; Quy et al. 1999; Kimura et al. 2007; Lv et al. 2009; Ng-Hublin et al. 2013)
House mouse (*Mus musculus*)	*C. tyzzeri*, *C. muris*, mouse genotype II, *Cryptosporidium* sp.	(Klesius et al. 1986; Chalmers et al. 1997; Chilvers et al. 1998; Xiao et al. 1999b; Foo et al. 2007; Ziegler et al. 2007a; 2007b; Lv et al. 2009; Kváč et al. 2013b)
Japanese field mouse (*Apodemus speciosus*)	*Cryptosporidium* sp.	(Nakai et al. 2004; Hikosaka and Nakai 2005)
Striped field mouse (*Apodemus agrarius*)	*Cryptosporidium* sp.	(Sińksi et al. 1998)
East African mole rat (*Tachyoryctes splendens*)	*C. muris*	Kváč et al. 2008b
Wood mouse (*Apodemus sylvaticus*)	*C. parvum*, *C. muris*, *Cryptosporidium* sp.	(Chalmers et al. 1997; Torres et al. 2000; Hajdušek et al. 2004)
Yellow-necked mouse (*Apodemus flavicollis*)	*C. tyzzeri*, *C. parvum*, *Cryptosporidium* sp.	(Sińksi 1993; Torres et al. 2000; Bajer et al. 2002; 2003; Bednarska et al. 2003)

M. m. musculus and eastward-migrating *M. m. domesticus* reestablished contact in central Europe, and formed a stable, narrow (approximately 20 km wide) hybrid zone that stretches 2,500 km from Norway to the Black Sea (Macholán et al. 2003; Jones et al. 2011; Ďureje et al. 2012). This hybrid zone affords a rare opportunity to study speciation through the interactions between two subspecies that have been separated for about 500,000 years.

The hybrid zone was recently used to test the hypothesis that the house mouse-adapted species *C. tyzzeri* is coevolving with subspecies of *M. musculus*. *Cryptosporidium tyzzeri* isolates from naturally infected *M. m. musculus* and *M. m. domesticus* in the hybrid zone differed genetically, morphometrically, and biologically (Kváč et al. 2013b), which supports the coevolution hypothesis.

5.13.2 Cricetidae

Cricetidae, which includes muskrats, deer mice, voles, cotton rats, and hamsters, are the second largest family of mammals with greater than 681 species in 130 genera. To date, members of this family have been reported as hosts for four *Cryptosporidium* species (*C. muris*, *C. parvum*, *C. andersoni*, and *C. ubiquitum*) and ten genotypes (vole genotype, muskrat genotypes I and II, W12, hamster genotype, chipmunk genotype I, and deer mouse genotypes I–IV) (Table 5.15).

Within the Cricetidae, *C. muris* and *C. andersoni* have been detected only in hamsters (Ryan et al. 2003a; Lv et al. 2009). *Cryptosporidium andersoni*, a ruminant adapted species, was detected in 5.9 % (8/136) of pet hamsters from three different species in China (Lv et al. 2009), suggesting that hamsters may be a significant host for *C. andersoni*, at least in that country. *Cryptosporidium muris* and *C. parvum* are the only *Cryptosporidium* species or genotypes known to infect members of both the Muridae and Cricetidae family. These species also are found in the family Sciuridae (see below).

5.13.3 Sciuridae

Five species (*C. andersoni*, *C. baileyi*, *C. muris*, *C. parvum*, and *C. ubiquitum*) and ten genotypes have been detected in the family Sciuridae, which includes squirrels and chipmunks (Table 5.16). In addition to *C. parvum* and *C. muris*, which are found in Muridae, Cricetidae, and Sciuridae, two species (*C. andersoni* and *C. ubiquitum*) and two genotypes (chipmunk genotype I and deer mouse genotype III) are common to the Cricetidae and Sciuridae (Fig. 5.4).

5.13.4 Hystricomorpha

Hystricomorphs include the guinea pig, capybara, and chinchilla, which were the first rodents to colonize South America about 40 million years ago, probably from Africa, and they remained the only rodents in South America until the arrival of murids about five million years ago (Antoine et al. 2012).

Table 5.17 shows the *Cryptosporidium* sp. that have been reported in Hystricomorphs. *Cryptosporidium wrairi* is found only in guinea pigs (Vetterling et al. 1971; Lv et al. 2009), and the prevalence of infection can be as high as 85 % in pet guinea pigs (Lv et al. 2009). Guinea pigs also host a guinea pig genotype

Table 5.15 *Cryptosporidium* spp. identified in the Cricetidae

Host (scientific name)	*Cryptosporidium* taxa	Reference
Bank vole (*Myodes glareolus*)	*Cryptosporidium* sp.	(Siński 1993; Laakkonen et al. 1994; Chalmers et al. 1997; Bull et al. 1998; Siński et al. 1998; Torres et al. 2000; Bajer et al. 2002; Bednarska et al. 2007)
Campbell hamster (*Phodopus campbelli*)	*C. andersoni*, *C. muris*, *C. parvum*	(Lv et al. 2009)
Common vole (*Microtus arvalis*)	*Cryptosporidium* sp.	(Siński et al. 1998; Bednarska et al. 2007)
Cotton rat (*Sigmodon hispidus*)	*Cryptosporidium* sp.	(Elangbam et al. 1993)
Deer mouse or white-footed mice (*Peromyscus* spp.)	*C. parvum*, *C. ubiquitum*, muskrat genotype II (W16), chipmunk genotype I (W17), deer mouse genotypes I–IV	(Perz and Le Blancq 2001; (Xiao et al. 2002; Feng et al. 2007; Ziegler et al. 2007a, b)
Golden hamster (*Mesocricetus auratus*)	*C. andersoni*, *C. muris*, *C. parvum*	(Ryan et al. 2003a; Lv et al. 2009)
Meadow vole (*Microtus pennsylvanicus*)	Vole genotype (W15), muskrat genotype II (W16)	(Feng et al. 2007; Ziegler et al. 2007a; 2007b)
Muskrat (*Ondatra zibethicus*)	Muskrat genotype I (W7), muskrat genotype II (W16)	(Siński et al. 1998; Perz and Le Blancq 2001; Xiao et al. 2002; Zhou et al. 2004a; Feng et al. 2007; Ziegler et al. 2007b)
Siberian hamster (*Phodopus sungorus*)	*C. andersoni*, *C. muris*, *C. parvum*, hamster genotype	(Lv et al. 2009)
Southern red-backed vole (*Myodes gapperi*)	*C. parvum*, vole cluster, muskrat genotype I (W7), Muskrat genotype II (W16), W12	(Feng et al. 2007; Ziegler et al. 2007a, b)

(GenBank Accession nos. DQ885337 and DQ885338) that is only 93 % similar to *C. wrairi* at the SSU rRNA locus. *Cryptosporidium parvum* has been identified in capybara (*Hydrochoerus hydrochaeris*) in Brazil (Meireles et al. 2007); however, it is likely that the infections resulted from exposure of capybara to water polluted by anthroponotic activities.

Table 5.16 *Cryptosporidium* spp. identified in the Sciuridae – chipmunks and squirrels

Host (scientific name)	*Cryptosporidium* taxa	Reference
American red squirrel (*Tamiasciurus hudsonicus*)	*C. ubiquitum*, chipmunk genotype I	(Ziegler et al. 2007a, b)
Bobak marmot (*Marmota bobac*)	*C. andersoni*	(Ryan et al. 2003a)
Belding's ground squirrel (*Spermophilus beldingi*)	Sbey/Sbld A, Sbey/Sbld/Sltl C, Sbld D	(Pereira et al. 2010)
California ground squirrels (*Spermophilus beecheyi*)	Sbey/Sbld A, Sbey B, Sbey/Sbld/Sltl C	(Atwill et al. 2001; 2004)
Eastern chipmunk (*Tamias striatus*)	*C. andersoni*, *C. baileyi*, *C. ubiquitum*, chipmunk genotypes I–II	(Perz and Le Blancq 2001; Feng et al. 2007; Ziegler et al. 2007a, 2007b)
Eastern grey squirrel (*Sciurus carolinensis*)	*C. baileyi*, *C. muris*, *C. parvum*, *C. ubiquitum*, chipmunk genotype I, deer mouse genotype III, skunk genotype	(Sundberg et al. 1982; Feng et al. 2007; Ziegler et al. 2007a, 2007b)
Eurasian red squirrel (*Sciurus vulgaris*)	*C. ubiquitum*, ferret genotype, chipmunk genotype I (W17)	(Bertolino et al. 2003; Feng et al. 2007; Kváč et al. 2008a; Lv et al. 2009)
Fox squirrel (*Sciurus niger*)	*C. ubiquitum*, *Cryptosporidium* sp.	(Current 1989) (Stenger et al. unpublished)
Golden-mantled ground squirrel (*Spermophilus lateralis*)	Sbey/Sbld/Sltl C	(Pereira et al. 2010)
Siberian chipmunk (*Tamias sibiricus*)	*C. muris*, *C. parvum*, chipmunk genotype III, Ferret genotype	(Matsui et al. 2000; Lv et al. 2009)
Southern flying squirrel (*Glaucomys volans*)	*Cryptosporidium* sp.	(Current 1989)
Thirteen-lined ground squirrel (*Ictidomys tridecemlineatus*)	*Cryptosporidium* sp.	(Current 1989)
Woodchuck or Groundhog (*Marmota monax*)	*C. ubiquitum*	(Feng et al. 2007; Ziegler et al. 2007b)

5.13.5 *Beaver and Gophers*

Cryptosporidium has been detected in the American and European beavers and the plains pocket gopher (Table 5.18). Combining the data from three studies in the eastern US (Fayer et al. 2006; Feng et al. 2007; Ziegler et al. 2007b), the prevalence of *Cryptosporidium* in American beavers (*Castor Canadensis*) was 4.0 % (7/176).

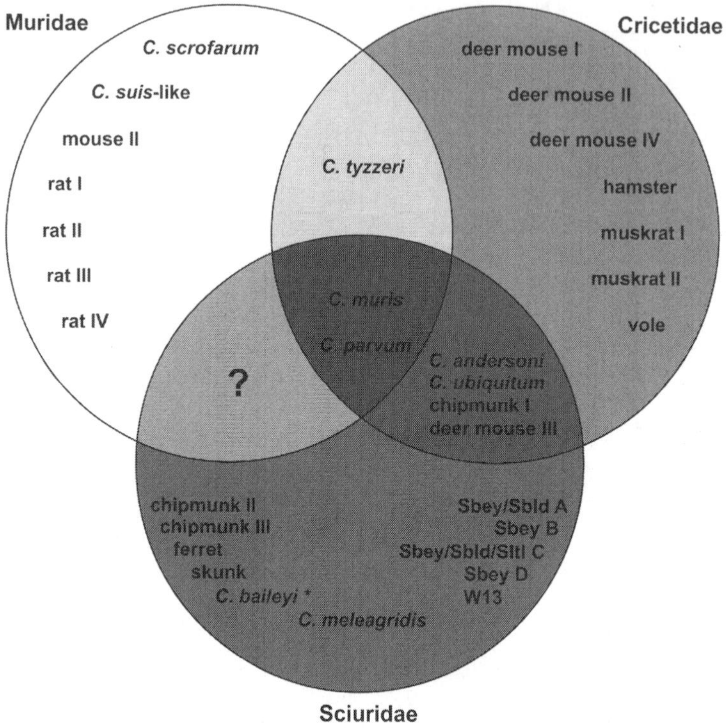

Fig. 5.4 Venn diagram showing group specific and overlapping *Cryptosporidium* taxa from the muridae, cricetidae, and scuridae. **C. baileyi* is generally associated with avian hosts

In Europe, *Cryptosporidium* was detected in 19.2 % (10/52) of European beavers (*Castor fiber*) in northeastern Poland (Paziewska et al. 2007). *Cryptosporidium ubiquitum* and a novel beaver genotype were identified in beavers in New York. The beaver genotype clusters with mink and ferret genotypes in a neighbor-joining phylogeny of SSU rRNA sequences, (Fig. 5.3b) and is most similar to a ferret-like genotype, sharing 99.6 % identity at the SSU rRNA locus.

5.14 Non-Human Primates

There are about 375 species of non-human primates in 15 families and 68 genera. All are omnivorous, and occupy both forest and open grassland habitats. There is potential for *Cryptosporidium* transmission between humans and endangered non-human primates in areas where the two live in close proximity (Nizeyi et al. 1999, 2002; Graczyk et al. 2001). Yet, knowledge of *Cryptosporidium* and cryptosporidiosis in our closest relatives remains relatively poor, and few studies

Table 5.17 *Cryptosporidium* spp. identified in hystricomorphs

Host (scientific name)	*Cryptosporidium* taxa	Reference
Capybara (*Hydrochoerus hydrochaeris*)	*C. parvum*	(Meireles et al. 2007)
Chinchilla (*Chinchilla laniger*)	*Cryptosporidium* sp.	(Yamini and Raju 1986)
Guinea pig (*Cavia porcellus*)	*C. wrairi*, guinea pig genotype	(Vetterling et al. 1971; Xiao et al. 1999b; Huber et al. 2007; Lv et al. 2009)
Indian crested porcupine (*Hystrix indica*)	*Cryptosporidium* sp.	(Fayer et al. 2000)
North American porcupine (*Erethizon dorsatum*)	*Cryptosporidium* sp.	(Ziegler et al. 2007a, 2007b)
Nutria/coypu (*Myocastor coypus*)	*Cryptosporidium* sp.	(Ryan et al. 2003a)

Table 5.18 *Cryptosporidium* spp. identified in the beaver and gophers

Host (scientific name)	*Cryptosporidium* taxa	Reference
American beaver (*Castor canadensis*)	*C. ubiquitum*, beaver genotype	(Fayer et al. 2006; Feng et al. 2007; Ziegler et al. 2007b)
European beaver (*Castor fiber*)	*Cryptosporidium* sp.	(Siński et al. 1998; Paziewska et al. 2007)
Plains pocket gopher (*Geomys bursarius*)	*Cryptosporidium* sp.	(Current 1989)

have genotyped isolates (Table 5.19). Among those that have, the potential for zoonotic transmission is clearly evident in the species detected.

Cryptosporidium parvum has been identified in the mountain gorilla (*Gorilla gorilla beringei*) from Uganda; toque macaque (*Macaca sinica sinica*), gray lagur (*Semnopithecus entellus thersites*), and purple-faced lagur (*Trachypithecus vetulus philbricki*) from Sri Lanka; and rhesus macaque (*Macaca mulatta*) from China (Graczyk et al. 2001; Ekanayake et al. 2007; Ye et al. 2012). *Cryptosporidium hominis* has been reported in rhesus macaques from the US and China and olive baboons (*Papio anubis*) from Kenya (Feng et al. 2011; Li et al. 2011; Ye et al. 2012). Isolates from the red-tailed guenon, red colobus, and black-and-white colobus were indistinguishable from *C. hominis*, *C. parvum* and *C. cuniculus* at the COWP locus (Salyer et al. 2012). Other *Cryptosporidium* taxa detected in primates include *C. felis* in rhesus macaques at a park in China (Ye et al. 2012) and a *Cryptosporidium* sp. closely related to *C. suis* in a Coquerel's sifaka (*Propithecus coquereli*) in the US (da Silva et al. 2003).

The clinical signs of cryptosporidiosis in infant non-human primates resemble those seen in human infants. Cryptosporidiosis was diagnosed in 81/157 infant primates, predominantly pigtailed macaques (*Macaca nemestrina*), housed in a nursery unit at the Washington Regional Primate Research Center. The mean age at onset of oocyst shedding was 38 ± 25 days, and animals shed oocysts for

Table 5.19 *Cryptosporidium* spp. identified in non-human primates

Host (scientific name)	Country	*Cryptosporidium* taxa	Reference
Captive – zoo			
Black-and-white ruffed lemur (*Lemur variegatus*)	Spain	*Cryptosporidium* sp.	(Gómez et al. 2000)
Black-capped squirrel monkey (*Saimiri sciureus boliviensis*)	Spain	*Cryptosporidium* sp.	(Gómez et al. 2000)
Black lemur (*Eulemur macaco*)	Spain	*Cryptosporidium* sp.	(Gómez et al. 2000; Gracenea et al. 2002)
Bonnet macaque (*Macaca radiata*)	Malaysia	*Cryptosporidium* sp.	(Lim et al. 2008)
Brown spider monkey (*Ateles belzebuth hybridus*)	Spain	*Cryptosporidium* sp.	(Gómez et al. 2000)
Brown lemur (*Lemur macacomayottensis*)	Spain	*Cryptosporidium* sp.	(Gómez et al. 1992)
Campbell's mona (*Cercopithecus campbelli*)	Spain	*Cryptosporidium* sp.	(Gómez et al. 1992)
Collared mangabey (*Cercocebus torquatus lunulatus*)	Spain	*Cryptosporidium* sp.	(Gracenea et al. 2002)
Common marmoset (*Calithrix jacchus*)	USA	*Cryptosporidium* sp.	(Kalishman et al. 1996)
Cotton-top tamarin (*Saguinus oedipus*)	USA	*Cryptosporidium* sp.	(Heuschele et al. 1986)
Drill (*Mandrillus leucophaeus*)	Spain	*Cryptosporidium* sp.	(Gómez et al. 2000)
Gray langur (*Semnopithecus*)	Malaysia	*Cryptosporidium* sp.	(Lim et al. 2008)
Lesser slow loris (*Nycticebus pygmaeus*)	Spain	*Cryptosporidium* sp.	(Gómez et al. 1992)
Mangabey (*Cercocebus albigena*)	Spain	*Cryptosporidium* sp.	(Gómez et al. 1992, 2000; Gracenea et al. 2002)
Marimonda spider monkey (*Ateles belzebuth*)	Spain	*Cryptosporidium* sp.	(Gómez et al. 1992)
Olive baboon (*Papio anubis*)	Italy	*Cryptosporidium* sp.	(Fagiolini et al. 2010)
Patas monkey (*Erythrocebus patas*)	Spain	*Cryptosporidium* sp.	(Gómez et al. 1992)
Pig-tailed macaque (*Macaca leonina*)	Malaysia	*Cryptosporidium* sp.	(Lim et al. 2008)
Red ruffed lemur (*Vaecia variegate rubra*)	USA	*Cryptosporidium* sp.	(Heuschele et al. 1986)
Ring-tailed lemur (*Lemur catta*)	Spain Italy	*Cryptosporidium* sp.	(Gómez et al. 2000) (Fagiolini et al. 2010)
Siamang (*Hylobates syndactylus*)	Spain	*Cryptosporidium* sp.	(Gómez et al. 2000; Gracenea et al. 2002)
Southern talapoin monkey (*Miopithecus talapoin*)	Spain	*Cryptosporidium* sp.	(Gómez et al. 2000)
Stump-tailed macaque (*Macaca arctoides*)	Malaysia	*Cryptosporidium* sp.	(Lim et al. 2008)

(continued)

Table 5.19 (continued)

Host (scientific name)	Country	*Cryptosporidium* taxa	Reference
Talapoin monkey (*Cercopithecus talapoin*)	Spain	*Cryptosporidium* sp.	(Gómez et al. 1992)
Tibetan Macaque (*Macaca thibetana*)	Spain	*Cryptosporidium* sp.	(Gómez et al. 2000)
Velvet monkey (*Cercopithecus aethiops*)	Spain	*Cryptosporidium* sp.	(Gómez et al. 1992)
Western gorilla (*Gorilla gorilla gorilla*)	Spain	*Cryptosporidium* sp.	(Gómez et al. 2000; van Zijll Langhout et al. 2010)
White-collared monkey (*Cercocebus torquatus*)	Spain	*Cryptosporidium* sp.	(Gómez et al. 1992)
White-crowned mangabeys (*Cercocebus torquatus lunulatus*)	Spain	*Cryptosporidium* sp.	(Gómez et al. 2000)
White-faced saki (*Pithecia pithecia*)	Spain	*Cryptosporidium* sp.	(Gómez et al. 2000)
Captive – non-zoo			
Crab-eating macaque (*Macaca fascicularis*)	USA	*C. muris*-like	(Dubey et al. 2002)
Coquerel's sifaka (*Propithecus coquereli*)	USA	*Cryptosporidium* sp.	(Charles-Smith et al. 2010) (da Silva et al. 2003)
Rhesus macaque (*Macaca mulatta*)	USA	*Cryptosporidium* sp.	(Yanai et al. 2000)
	USA	*Cryptosporidium* sp.	(Osborn et al. 1984; Feng et al. 2011; Ye et al. 2012)
	Not reported	*C. hominis* monkey genotype	Feng et al. 2011;
Southern pig-tailed macaque (*Macaca nemestrina*)	USA	*Cryptosporidium* sp.	(Miller et al. 1990a)
Free living			
Black-and-white colobus (*Colobus guereza*)	Uganda	*Cryptosporidium* sp.	(Salzer et al. 2007; Salyer et al. 2012)
Gray lagur (*Semnopithecus entellus thersites*)	Sri Lanka	*C. parvum*, *Cryptosporidium* sp.	(Ekanayake et al. 2006)
Green monkey (*Cercopithecus aethiops*)	Ethiopia	*Cryptosporidium* sp.	(Legesse and Erko 2004)
Mountain gorilla (*Gorilla beringei beringei*)	Uganda	*Cryptosporidium* sp.	(Nizeyi et al. 1999; Sleeman et al. 2000; Graczyk et al. 2001)
	Rwanda	*C. parvum*	Graczyk et al. 2001)
		Cryptosporidium sp.	Sleeman et al. 2000,

(continued)

Table 5.19 (continued)

Host (scientific name)	Country	*Cryptosporidium* taxa	Reference
Olive baboon (*Papio anubis*)	Kenya	*Cryptosporidium* sp.	(Muriuki et al. 1997, 1998; Hope et al. 2004; Legesse and Erko 2004; Li et al. 2011)
		C. hominis	Li et al. 2011)
	Uganda	*Cryptosporidium* sp.	Hope et al. 2004
	Ethiopia	*Cryptosporidium* sp.	Legesse and Erko 2004
Purple-faced lagur (*Trachypithecus vetulus*)	Sri Lanka	*Cryptosporidium* sp.	(Ekanayake et al. 2006)
		C. parvum	(Ekanayake et al. 2007)
Rhesus macaque (*Macaca mulatta*)	China	*C. felis*, *C. hominis*, *C. parvum*	Ye et al. 2012
Red colobus (*Procolobus tephrosceles*)	Uganda	*Cryptosporidium* sp.	(Salzer et al. 2007, Salyer et al. 2012)
Red-tailed guenon (*Cercopithecus ascanius*)	Uganda	*Cryptosporidium* sp.	(Salzer et al. 2007) (Salyer et al. 2012)
Savanna chimpamzee (*Pan troglodytes schweinfurthii*)	Tanzania	*Cryptosporidium* sp.	(Gonzalez-Moreno et al. 2013) (Ekanayake et al. 2006)
Toque macaque (*Macaca sinica*)	Sri Lanka	*Cryptosporidium* sp.	Ekanayake et al. 2007)
Vervet monkey (*Chlorocebus pygerythrus*)	Kenya	*Cryptosporidium* sp.	(Muriuki et al. 1997, 1998)
Western gorilla (*Gorilla gorilla gorilla*)	Gabon	*Cryptosporidium* sp.	(van Zijll Langhout et al. 2010)

7–78 days. All but one of the animals had clinical symptoms of enteric infection, which included severe diarrhea and dehydration. The outbreak was confined to the nursery and no cases were detected among juvenile or adult animals (Miller et al. 1990a). A follow-up study (Miller et al. 1990b) characterized the course of experimental infections in pigtailed macaques infected with 10 or 2×10^5 oocysts from naturally infected macaques (the species/genotype of the isolate was not reported). Infected animals suffered clinical enteritis, watery stools, lethargy, and loss of appetite. Animals infected with 10 and 2×10^5 oocysts began shedding oocysts on day 8 and day 7, respectively, and there was no difference in the duration of intense oocyst shedding or clinical symptoms between the treatments.

Cryptosporidiosis in immunocompromised non-human primates resembles cryptosporidiosis in immunocompromised humans with respect to extraintestinal involvement. Kaup et al. (1994) diagnosed biliary and pancreatic cryptosporidiosis in several rhesus monkeys following an experimental simian immunodeficiency virus (SIV) infection. Yanai et al. (2000) reported moderate to severe bronchopneumonia with cryptosporidiosis in macaques experimentally infected with SIV. Endogenous stages were detected in the trachea, lungs, bile ducts, pancreas, and intestine. Conjunctival infections were detected in six SIV-infected, immunodeficient rhesus monkeys (Baskin 1996). Although conjunctival cryptosporidiosis has been reported in birds, there have been no reported cases in humans. Singh et al. (2011) examined the time during an SIV infection when a self-limiting *Cryptosporidium parvum* infection becomes persistent due to mucosal immune defects. One group of SIV-infected macaques was challenged with *C. parvum* during acute SIV infection and the second group was challenged during the chronic infection phase. Interestingly, persistent cryptosporidiosis developed during acute SIV infection.

5.15 Diversity of *Cryptosporidium* in Different Vertebrate Classes

The diversity of mammalian cryptosporidia appears to reflect mammal diversity. We compared *Cryptosporidium* diversity in the order Rodentia – the most diverse order of mammals – to diversity in other mammalian orders. Figure 5.5 presents 68 mammalian *Cryptosporidium* taxa (species and genotypes) in a Venn diagram with two sets: 'Rodentia' and 'other mammals' Among the taxa that are members of both sets (the intersection), *C. andersoni*, *C. scrofarum*, and the ferret genotype primarily infect non-rodent hosts and are therefore included with 'other mammals'. Similarly, *C. tyzzeri* and *C. muris* are primarily parasites of rodents and are included with the Rodentia set. *Cryptosporidium baileyi* was excluded from the analysis because it primarily infects avian hosts. Forty percent (27/67) of mammalian *Cryptosporidium* taxa are associated with the order Rodentia, which reflects the diversity of Rodentia within mammals (Rodentia contains 40 % of mammalian species). Three percent (2/67) of *Cryptosporidium* taxa (*C. parvum* and *C. ubiquitum*) can be considered generalists, and 57 % (38/67) of taxa are associated with non-rodent mammals. These data suggest that cryptosporidia diverged in close association with mammal species.

In contrast to mammals, it appears that the diversity of avian cryptosporidia does not reflect avian diversity. The order Passeriformes comprises about 50 % of avian species; species diversity in the Passeriforme order is comparable to that of the entire mammal class. Figure 5.6 presents 17 *Cryptosporidium* taxa from avians in a Venn diagram with two sets: 'Passeriformes' and 'other avians'. Muskrat genotype I, *C. hominis*, *C. hominis*-like, *C. muris*, and *C. parvum* were excluded from the analsysis because they are not considered avian-associated taxa. Seventeen percent (2/12) of avian *Cryptosporidium* taxa are exclusively found in passerines (avian genotypes I and IV), 50 %

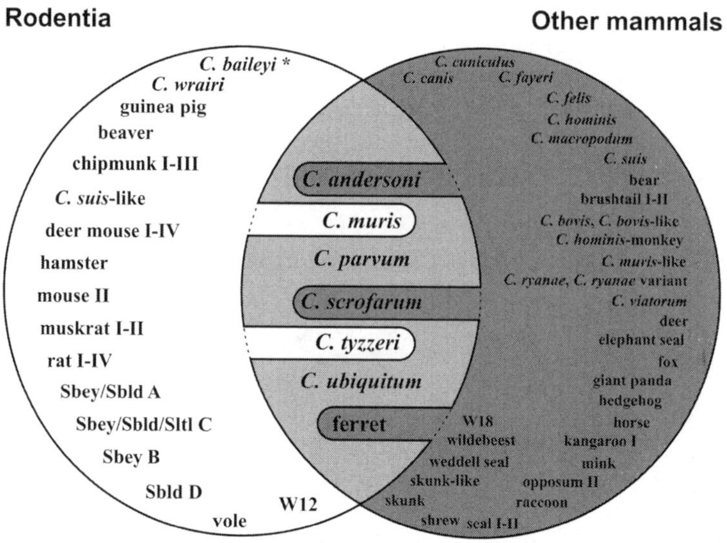

Fig. 5.5 Venn diagram showing group-specific and overlapping *Cryptosporidium* taxa in rodent and non-rodent mammal groups. **C. baileyi* is generally associated with avian hosts

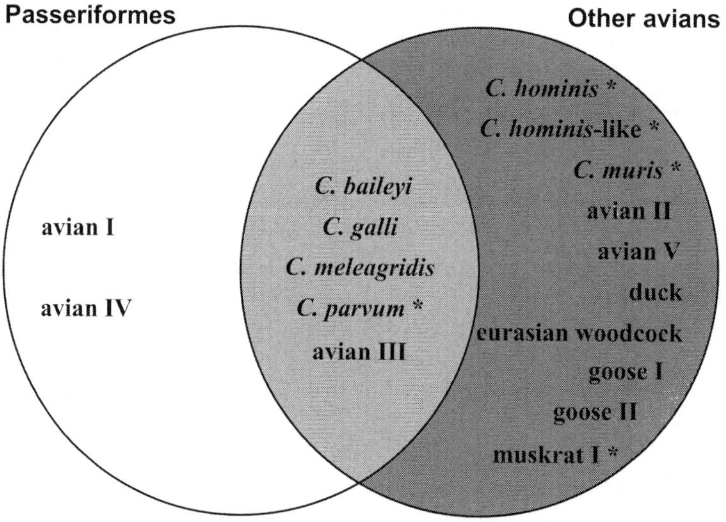

Fig. 5.6 Venn diagram showing group-specific and overlapping *Cryptosporidium* taxa from passerine and non-passerine bird groups

(6/12) are found only in non-passerine orders, and 33 % (4/12; *C. baileyi*, *C. meleagridis*, *C. galli*, and avian genotype III) infect multiple avian orders and can be considered generalists. From these data, it can be hypothesized that the mechanisms of *Cryptosporidium* diversification in avians and mammals are fundamentally different.

The diversity of *Cryptosporidium* in reptiles and amphibians appears to be low. Despite a comparable faunal diversity with birds, and considerably greater faunal diversity than mammals, relatively few *Cryptosporidium* taxa have been reported in reptiles or amphibians.

5.16 Gaps in Knowledge and Suggestions for Future Work

- *Cryptosporidium* remains poorly characterized in many vertebrate orders. For example, Chiroptera (bats), an order that comprises 20 % of mammal species, has not been well studied. If *Cryptosporidium* diversity reflects bat diversity, then we should expect to find many bat-adapted *Cryptosporidium* taxa.
- *Cryptosporidium viatorum* is a recently described species that has thus far been identified only in cases of human cryptosporidiosis in the UK ($n = 10$) and Sweden ($n = 2$); all cases were associated with travel to countries in the Indian subcontinent, Africa, and South America (Elwin et al. 2012b; Insulander et al. 2013). Studies are needed to determine if there is a major animal host for this species. Greater sampling in general, particularly of animals in their natural habitat, will help to help to identify emerging sources of human disease.
- Relatively little is known about the general biology of many of the *Cryptosporidium* genotypes identified. Experimental evidence of infection characteristics and host specificity will enhance understanding of the biological diversity and will clarify taxonomy.

References

Abe N, Iseki M (2003) Identification of genotypes of *Cryptosporidium parvum* isolates from ferrets in Japan. Parasitol Res 89:422–424

Abe N, Iseki M (2004) Identification of *Cryptosporidium* isolates from cockatiels by direct sequencing of the PCR-amplified small subunit ribosomal RNA gene. Parasitol Res 92:523–526

Abe N, Makino I (2010) Multilocus genotypic analysis of *Cryptosporidium* isolates from cockatiels, Japan. Parasitol Res 106:1491–1497

Abe N, Sawano Y, Yamada K, Kimata I, Iseki M (2002) *Cryptosporidium* infection in dogs in Osaka, Japan. Vet Parasitol 108:185–193

Abe N, Takami K, Kimata I, Iseki M (2004) Molecular characterization of a *Cryptosporidium* isolate from a banded mongoose *Mungos mungo*. J Parasitol 90:167–171

Abreu-Acosta N, Foronda-Rodriguez P, Lopez M, Valladares B (2009) Occurrence of *Cryptosporidium hominis* in pigeons (*Columba livia*). Acta Parasitiol 54:1–5

Abu Samra N, Jori F, Samie A, Thompson P (2011) The prevalence of *Cryptosporidium* spp. oocysts in wild mammals in the Kruger National Park, South Africa. Vet Parasitol 175:155–159

Abu Samra N, Jori F, Xiao L, Rikhotso O, Thompson PN (2013) Molecular characterization of *Cryptosporidium* species at the wildlife/livestock interface of the Kruger National Park, South Africa. Comp Immunol Microbiol Infect Dis 36:295–302

Alves M, Xiao L, Sulaiman I, Lal AA, Matos O, Antunes F (2003) Subgenotype analysis of *Cryptosporidium* isolates from humans, cattle, and zoo ruminants in Portugal. J Clin Microbiol 41:2744–2747

Alves M, Xiao L, Lemos V, Zhou L, Cama V, da Cunha MB et al (2005) Occurrence and molecular characterization of *Cryptosporidium* spp. in mammals and reptiles at the Lisbon Zoo. Parasitol Res 97:108–112

Anderson BC (1991) Experimental infection in mice of *Cryptosporidium muris* isolated from a camel. J Protozool 38:16S–17S

Antoine PO, Marivaux L, Croft DA, Billet G, Ganerod M, Jaramillo C et al (2012) Middle Eocene rodents from Peruvian Amazonia reveal the pattern and timing of caviomorph origins and biogeography. Proc Biol Sci 279:1319–1326

Antunes RG, Simões DC, Nakamura AA, Meireles MV (2008) Natural infection with *Cryptosporidium galli* in canaries (*Serinus canaria*), in a cockatiel (*Nymphicus hollandicus*), and in lesser seed-finches (*Oryzoborus angolensis*) from Brazil. Avian Dis Dig 3:e40–e42

Atwill ER, Sweitzer RA, Pereira MG, Gardner IA, Van VD, Boyce WM (1997) Prevalence of and associated risk factors for shedding *Cryptosporidium parvum* oocysts and *Giardia* cysts within feral pig populations in California. Appl Environ Microbiol 63:3946–3949

Atwill ER, Camargo SM, Phillips R, Alonso LH, Tate KW, Jensen WA et al (2001) Quantitative shedding of two genotypes of *Cryptosporidium parvum* in California ground squirrels (*Spermophilus beecheyi*). Appl Environ Microbiol 67:2840–2843

Atwill ER, Phillips R, Pereira MD, Li X, McCowan B (2004) Seasonal shedding of multiple *Cryptosporidium* genotypes in California ground squirrels (*Spermophilus beecheyi*). Appl Environ Microbiol 70:6748–6752

Auffray JC, Britton-Davidian J (2012) The house mouse and its relatives: systematics and taxonomy. In: Macholán M, Baird SJE, Munclinger P, Piálek J (eds) In evolution of the house mouse. Cambridge University Press, Cambridge, pp 1–34

Baishanbo A, Gargala G, Delaunay A, Francois A, Ballet JJ, Favennec L (2005) Infectivity of *Cryptosporidium hominis* and *Cryptosporidium parvum* genotype 2 isolates in immunosuppressed Mongolian gerbils. Infect Immun 73:5252–5255

Bajer A, Bednarska M, Pawełczyk A, Behnke JM, Gilbert FS, Sinski E (2002) Prevalence and abundance of Cryptosporidium parvum and giardia spp. in wild rural rodents from the Mazury Lake District region of Poland. Parasitology 125:21–34

Bajer A, Cacciò S, Bednarska M, Behnke JM, Pieniazek NJ, Sinski E (2003) Preliminary molecular characterization of *Cryptosporidium parvum* isolates of wildlife rodents from Poland. J Parasitol 89:1053–1055

Bajer A, Bednarska M, Rodo A (2011) Risk factors and control of intestinal parasite infections in sled dogs in Poland. Vet Parasitol 175:343–350

Ballweber LR, Panuska C, Huston CL, Vasilopulos R, Pharr GT, Mackin A (2009) Prevalence of and risk factors associated with shedding of *Cryptosporidium felis* in domestic cats of Mississippi and Alabama. Vet Parasitol 160:306–310

Barbon AR, Forbes N (2007) Use of paromomycin in the treatment of a *Cryptosporidium* infection in two falcons. Falco 30:22–24

Barker IK, Beveridge I, Bradley AJ, Lee AK (1978) Observations on spontaneous stress-related mortality among males of Dasyurid marsupial *Antechinus-Stuartii* Macleay. Aust J Zool 26:435–447

Baskin GB (1996) Cryptosporidiosis of the conjunctiva in SIV-infected rhesus monkeys. J Parasitol 82:630–632

Bass AL, Wallace CC, Yund PO, Ford TE (2012) Detection of *Cryptosporidium* sp. in two new seal species, *Phoca vitulina* and *Cystophora cristata*, and a novel *Cryptosporidium* genotype in a third seal species, *Pagophilus groenlandicus*, from the Gulf of Maine. J Parasitol 98:316–322

Bednarska M, Bajer A, Kulis K, Sinski E (2003) Biological characterisation of *Cryptosporidium parvum* isolates of wildlife rodents in Poland. Ann Agric Environ Med 10:163–169

Bednarska M, Bajer A, Sinski E, Girouard AS, Tamang L, Graczyk TK (2007) Fluorescent in situ hybridization as a tool to retrospectively identify *Cryptosporidium parvum* and *Giardia lamblia* in samples from terrestrial mammalian wildlife. Parasitol Res 100:455–460

Beiromvand M, Akhlaghi L, Fattahi Massom SH, Meamar AR, Motevalian A, Oormazdi H et al (2013) Prevalence of zoonotic intestinal parasites in domestic and stray dogs in a rural area of Iran. Prev Vet Med 109:162–167

Bertolino S, Wauters LA, De Bruyn L, Canestri-Trotti G (2003) Prevalence of coccidia parasites (Protozoa) in red squirrels (*Sciurus vulgaris*): effects of host phenotype and environmental factors. Oecologia 137:286–295

Bezuidenhout AJ, Penrith ML, Burger WP (1993) Prolapse of the phallus and cloaca in the ostrich (*Struthio camelus*). J S Afr Vet Assoc 64:156–158

Blagburn BL, Lindsay DS, Hoerr FJ, Atlas AL, Toivio-Kinnucan M (1990) *Cryptosporidium* sp. infection in the proventriculus of an Australian diamond firetail finch (*Staganoplura bella*: Passeriformes, Estrildidae). Avian Dis 34:1027–1030

Bogomolni AL, Gast RJ, Ellis JC, Dennett M, Pugliares KR, Lentell BJ et al (2008) Victims or vectors: a survey of marine vertebrate zoonoses from coastal waters of the Northwest Atlantic. Dis Aquat Organ 81:13–38

Bonhomme F, Searle JB (2012) House mouse phylogeography. In: Macholán M, Baird SJE, Munclinger P, Piálek J (eds) In evolution of the house mouse. Cambridge University Press, Cambridge, pp 278–296

Bougiouklis PA, Weissenböck H, Wells A, Miller WA, Palmieri C, Shivaprasad HL (2012) Otitis media associated with *Cryptosporidium baileyi* in a Saker Falcon (*Falco cherrug*). J Comp Pathol 148:419–423

Boursot P, Auffray JC, Brittondavidian J, Bonhomme F (1993) The evolution of house mice. Annu Rev Ecol Syst 24:119–152

Brower AI, Cranfield MR (2001) *Cryptosporidium* sp.-associated enteritis without gastritis in rough green snakes (*Opheodrys aestivus*) and a common garter snake (*Thamnophis sirtalis*). J Zoo Wildl Med 32:101–105

Brownstein DG, Strandberg JD, Montali RJ, Bush M, Fortner J (1977) *Cryptosporidium* in snakes with hypertrophic gastritis. Vet Pathol 14:606–617

Bryan HM, Darimont CT, Hill JE, Paquet PC, Thompson RC, Wagner B et al (2012) Seasonal and biogeographical patterns of gastrointestinal parasites in large carnivores: wolves in a coastal archipelago. Parasitology 139:781–790

Bugg RJ, Robertson ID, Elliot AD, Thompson RC (1999) Gastrointestinal parasites of urban dogs in Perth, Western Australia. Vet J 157:295–301

Bull S, Chalmers R, Sturdee AP, Curry A, Kennaugh J (1998) Cross-reaction of an anti-*Cryptosporidium* monoclonal antibody with sporocysts of *Monocystis* species. Vet Parasitol 77:195–197

Caccio SM (2005) Molecular epidemiology of human cryptosporidiosis. Parassitologia 47:185–192

Caccio S, Pinter E, Fantini R, Mezzaroma I, Pozio E (2002) Human infection with *Cryptosporidium felis*: case report and literature review. Emerg Infect Dis 8:85–86

Cama VA, Bern C, Sulaiman IM, Gilman RH, Ticona E, Vivar A et al (2003) *Cryptosporidium* species and genotypes in HIV-positive patients in Lima, Peru. J Eukaryot Microbiol 50 (Suppl):531–533

Canestri-Trotti G (1989) Studies on *Cryptosporidium* sp. In: Angus KW, Blewett DA (eds) *Cryptosporidiosis*. Proceedings of the 1st international workshop, 7–8 Sept 1988. Moredun Research Institute, Edinburgh. The Animal Diseases Research Association, Edinburgh. p 118

Carver S, Scorza AV, Bevins SN, Riley SP, Crooks KR, Vandewoude S et al (2012) Zoonotic parasites of bobcats around human landscapes. J Clin Microbiol 50:3080–3083

Castro-Hermida JA, García-Presedo I, Almeida A, González-Warleta M, Correia Da Costa JM, Mezo M (2011a) *Cryptosporidium* spp. and *Giardia* duodenalis in two areas of Galicia (NW Spain). Sci Total Environ 409:2451–2459

Castro-Hermida JA, García-Presedo I, González-Warleta M, Mezo M (2011b) Prevalence of *Cryptosporidium* and *Giardia* in roe deer (*Capreolus capreolus*) and wild boars (*Sus scrofa*) in Galicia (NW, Spain). Vet Parasitol 179:216–219

Cerveny SN, Garner MM, D'Agostino JJ, Sekscienski SR, Payton ME, Davis MR (2012) Evaluation of gastroscopic biopsy for diagnosis of *Cryptosporidium* sp. infection in snakes. J Zoo Wildl Med 43:864–871

Chalmers RM, Sturdee AP, Bull SA, Miller A, Wright SE (1997) The prevalence of *Cryptosporidium parvum* and *C. muris* in *Mus domesticus*, *Apodemus sylvaticus* and *Clethrionomys glareolus* in an agricultural system. Parasitol Res 83:478–482

Chalmers RM, Robinson G, Elwin K, Hadfield SJ, Xiao L, Ryan U et al (2009) *Cryptosporidium* sp. rabbit genotype, a newly identified human pathogen. Emerg Infect Dis 15:829–830

Chalmers RM, Elwin K, Hadfield SJ, Robinson G (2011) Sporadic human cryptosporidiosis caused by *Cryptosporidium cuniculus*, United Kingdom, 2007–2008. Emerg Infect Dis 17:536–538

Charles-Smith LE, Cowen P, Schopler R (2010) Environmental and physiological factors contributing to outbreaks of *Cryptosporidium* in Coquerel's sifaka (*Propithecus coquereli*) at the Duke Lemur Center: 1999–2007. J Zoo Wildl Med 41:438–444

Chavez DJ, LeVan IK, Miller MW, Ballweber LR (2012) *Baylisascaris procyonis* in raccoons (*Procyon lotor*) from eastern Colorado, an area of undefined prevalence. Vet Parasitol 185:330–334

Chen F, Huang K (2007) Prevalence and phylogenetic analysis of *Cryptosporidium* in pigs in Eastern China. Zoonoses Public Health 54:393–400

Chen F, Qiu H (2012) Identification and characterization of a Chinese isolate of *Cryptosporidium serpentis* from dairy cattle. Parasitol Res 111:1785–1791

Chilvers BL, Cowan PE, Waddington DC, Kelly PJ, Brown TJ (1998) The prevalence of infection of *Giardia* spp. and *Cryptosporidium* spp. in wild animals on farmland, Southeastern North Island, New Zealand. Int J Environ Health Res 8:59–64

Chvala S, Fragner K, Hackl R, Hess M, Weissenbock H (2006) *Cryptosporidium* infection in domestic geese (*Anser anser f. domestica*) detected by in-situ hybridization. J Comp Pathol 134:211–218

Cranfield MR, Graczyk TK (1994) Experimental infection of elaphid snakes with *Cryptosporidium serpentis* (Apicomplexa: Cryptosporidiidae). J Parasitol 80:823–826

Crawshaw GJ, Mehren KG (1987) Cryptosporidiosis in zoo and wild animals. In: Ippen R, Schroder HD (ed) Erkrankungen der Zootiere, Verhandlungsbericht des 29 International Symposium uber die Erkrankungen der Zootiere von 20, pp 353–362

Cucchi T, Auffray JC, Vigne JD (2012) History of house mouse synanthropy and dispersal in the near East and Europe: a zooarchaeological insight. In: Macholán M, Baird SJE, Munclinger P, Piálek J (eds) In evolution of the house mouse. Cambridge University Press, Cambridge, pp 65–93

Current WL (1989) *Cryptosporidium* spp. In: Genta RM, Walzer PD (eds) *Parasitic* infections in the compromised host. Marcel Dekker, New York, pp 281–341

Current WL, Upton SJ, Haynes TB (1986) The life cycle of *Cryptosporidium baileyi* n. sp. (Apicomplexa, Cryptosporidiidae) infecting chickens. J Protozool 33:289–296

da Silva AJ, Cacciò S, Williams C, Won KY, Nace EK, Whittier C et al (2003) Molecular and morphologic characterization of a *Cryptosporidium* genotype identified in lemurs. Vet Parasitol 111:297–307 da Silva AS, Zanette RA, Lara VM, Gressler LT, Carregaro AB, Santurio JM et al (2009) Gastrointestinal parasites of owls (Strigiformes) kept in captivity in the Southern region of Brazil. Parasitol Res 104:485–487

Dado D, Izquierdo F, Vera O, Montoya A, Mateo M, Fenoy S et al (2012) Detection of zoonotic intestinal parasites in public parks of Spain. Potential epidemiological role of microsporidia. Zoonoses Public Health 59:23–28

Deng MQ, Cliver DO (1999) Improved immunofluorescence assay for detection of *Giardia* and *Cryptosporidium* from asymptomatic adult cervine animals. Parasitol Res 85:733–736

Deng MQ, Peterson RP, Cliver DO (2000) First findings of *Cryptosporidium* and *Giardia* in California sea lions (*Zalophus californianus*). J Parasitol 86:490–494

Dubey JP, Hamir AN, Sonn RJ, Topper MJ (1998) Cryptosporidiosis in a bat (*Eptesicus fuscus*). J Parasitol 84:622–623

Dubey JP, Markovits JE, Killary KA (2002) *Cryptosporidium muris*-like infection in stomach of cynomolgus monkeys (*Macaca fascicularis*). Vet Pathol 39:363–371

Ducatelle R, Maenhout D, Charlier G, Miry C, Coussement W (1983) Cryptosporidiosis in goats and in mouflon sheep. Vlaams Diergen Tijds 52:7–17

Duncan RB, Caudell D, Lindsay DS, Moll HD (1999) Cryptosporidiosis in a black bear in Virginia. J Wildl Dis 35:381–383

Ďureje L, Macholán M, Baird SJE, Piálek J (2012) The mouse hybrid zone in Central Europe: from morphology to molecules. Folia Zool 61:308–318

Duvaux L, Belkhir K, Boulesteix M, Boursot P (2011) Isolation and gene flow: inferring the speciation history of European house mice. Mol Ecol 20:5248–5264

Dyachenko V, Kuhnert Y, Schmaeschke R, Etzold M, Pantchev N, Daugschies A (2010) Occurrence and molecular characterization of *Cryptosporidium* spp. genotypes in European hedgehogs (*Erinaceus europaeus L.*) in Germany. Parasitology 137:205–216

Ebeid M, Mathis A, Pospischil A, Deplazes P (2003) Infectivity of *Cryptosporidium parvum* genotype I in conventionally reared piglets and lambs. Parasitol Res 90:232–235

Ekanayake DK, Arulkanthan A, Horadagoda NU, Sanjeevani GK, Kieft R, Gunatilake S et al (2006) Prevalence of *Cryptosporidium* and other enteric parasites among wild non-human primates in Polonnaruwa, Sri Lanka. Am J Trop Med Hyg 74:322–329

Ekanayake DK, Welch DM, Kieft R, Hajduk S, Dittus WP (2007) Transmission dynamics of *Cryptosporidium* infection in a natural population of non-human primates at Polonnaruwa, Sri Lanka. Am J Trop Med Hyg 77:818–822

el-Ahraf A, Tacal JV Jr, Sobih M, Amin M, Lawrence W, Wilcke BW (1991) Prevalence of cryptosporidiosis in dogs and human beings in San Bernardino County, California. J Am Vet Med Assoc 198:631–634

Elangbam CS, Qualls CW, Ewing SA, Lochmiller RL (1993) Cryptosporidiosis in a cotton rat (*Sigmodon hispidus*). J Wildl Dis 29:161–164

Ellis AE, Brown CA, Miller DL (2010) Diagnostic exercise: chronic vomiting in a dog. Vet Pathol 47:991–993

Elwin K, Hadfield SJ, Robinson G, and Chalmers RM (2011) The epidemiology of sporadic human infections with unusual cryptosporidia detected during routine typing in England and Wales, 2000–2008. Epidemiol Infect 140:673–683

Elwin K, Hadfield SJ, Robinson G, Chalmers RM (2012a) The epidemiology of sporadic human infections with unusual cryptosporidia detected during routine typing in England and Wales, 2000–2008. Epidemiol Infect 140:673–683

Elwin K, Hadfield SJ, Robinson G, Crouch ND, Chalmers RM (2012b) *Cryptosporidium viatorum* n. sp. (Apicomplexa: Cryptosporidiidae) among travellers returning to great Britain from the Indian subcontinent, 2007–2011. Int J Parasitol 42:675–682

Fagiolini M, Lia RP, Laricchiuta P, Cavicchio P, Mannella R, Cafarchia C et al (2010) Gastrointestinal parasites in mammals of two Italian zoological gardens. J Zoo Wildl Med 41:662–670

Fayer R, Graczyk TK, Cranfield MR (1995) Multiple heterogenous isolates of *Cryptosporidium serpentis* from captive snakes are not transmissible to neonatal BALB/c mice (*Mus musculus*). J Parasitol 81:482–484

Fayer R, Fischer JR, Sewell CT, Kavanaugh DM, Osborn DA (1996) Spontaneous cryptosporidiosis in captive white-tailed deer (*Odocoileus virginianus*). J Wildl Dis 32:619–622

Fayer R, Morgan U, Upton SJ (2000) Epidemiology of *Cryptosporidium*: transmission, detection and identification. Int J Parasitol 30:1305–1322

Fayer R, Trout JM, Xiao L, Morgan UM, Lai AA, Dubey JP (2001) *Cryptosporidium canis* n. sp. from domestic dogs. J Parasitol 87:1415–1422

Fayer R, Santín M, Trout JM, DeStefano S, Koenen K, Kaur T (2006) Prevalence of microsporidia, *Cryptosporidium* spp., and *Giardia* spp. in beavers (*Castor canadensis*) in Massachusetts. J Zoo Wildl Med 37:492–497

Fayer R, Santín M, Macarisin D (2010) *Cryptosporidium ubiquitum* n. sp. in animals and humans. Vet Parasitol 172:23–32

Feltus DC, Giddings CW, Schneck BL, Monson T, Warshauer D, McEvoy JM (2006) Evidence supporting zoonotic transmission of *Cryptosporidium* spp. in Wisconsin. J Clin Microbiol 44:4303–4308

Feng Y, Alderisio KA, Yang W, Blancero LA, Kuhne WG, Nadareski CA et al (2007) *Cryptosporidium* genotypes in wildlife from a New York watershed. Appl Environ Microbiol 73:6475–6483

Feng Y, Lal AA, Li N, Xiao L (2011) Subtypes of *Cryptosporidium* spp. in mice and other small mammals. Exp Parasitol 127:238–242

Feng Y, Karna SR, Dearen TK, Singh DK, Adhikari LN, Shrestha A et al (2012) Common occurrence of a unique *Cryptosporidium ryanae* variant in zebu cattle and water buffaloes in the buffer zone of the Chitwan National Park, Nepal. Vet Parasitol 185:309–314

FitzGerald L, Bennett M, Ng J, Nicholls P, James F, Elliot A et al (2011) Morphological and molecular characterisation of a mixed *Cryptosporidium muris*/*Cryptosporidium felis* infection in a cat. Vet Parasitol 175:160–164

Fontanarrosa MF, Vezzani D, Basabe J, Eiras DF (2006) An epidemiological study of gastrointestinal parasites of dogs from Southern Greater Buenos Aires (Argentina): age, gender, breed, mixed infections, and seasonal and spatial patterns. Vet Parasitol 136:283–295

Foo C, Farrell J, Boxell A, Robertson I, Ryan UM (2007) Novel *Cryptosporidium* genotype in wild Australian mice (*Mus domesticus*). Appl Environ Microbiol 73:7693–7696

Fredes F, Raffo E, Muñoz P (2007) Short note: first report of *Cryptosporidium* spp. oocysts in stool of Adélie penguin from the Antarctic using acid-fast stain. Antarct Sci 19:437–438

Fredes F, Diaz A, Raffo E, Munoz P (2008) *Cryptosporidium* spp. oocysts detected using acid-fast stain in faeces of gentoo penguins (*Pygoscelis papua*) in Antarctica. Antarct Sci 20:495–496

Gajadhar AA (1993) *Cryptosporidium* species in imported ostriches and consideration of possible implications for birds in Canada. Can Vet J 34:115–116

Gajadhar AA (1994) Host specificity studies and oocyst description of a *Cryptosporidium* sp. isolated from ostriches. Parasitol Res 80:316–319

Garcia-Presedo I, Pedraza-Diaz S, Gonzalez-Warleta M, Mezo M, Gomez-Bautista M, Ortega-Mora LM et al (2013) Presence of *Cryptosporidium scrofarum*, *C. suis* and *C. parvum* subtypes IIaA16G2R1 and IIaA13G1R1 in Eurasian wild boars (*Sus scrofa*). Vet Parasitol 196:497–502

Gardiner CH, Imes GD (1984) *Cryptosporidium* sp. in the kidneys of a black-throated finch. J Am Vet Med Assoc 185:1401–1402

Gaydos JK, Miller WA, Gilardi KV, Melli A, Schwantje H, Engelstoft C et al (2007) *Cryptosporidium* and *Giardia* in marine-foraging river otters (*Lontra canadensis*) from the Puget Sound Georgia Basin ecosystem. J Parasitol 93:198–202

Geraldes A, Basset P, Gibson B, Smith KL, Harr B, Yu HT et al (2008) Inferring the history of speciation in house mice from autosomal, X-linked, Y-linked and mitochondrial genes. Mol Ecol 17:5349–5363

Geurden T, Goossens E, Levecke B, Vercammen F, Vercruysse J, Claerebout E (2009) Occurrence and molecular characterization of *Cryptosporidium* and *Giardia* in captive wild ruminants in Belgium. J Zoo Wildl Med 40:126–130
Gomes RS, Huber F, da Silva S, do Bomfim TC (2012) *Cryptosporidium* spp. parasitize exotic birds that are commercialized in markets, commercial aviaries, and pet shops. Parasitol Res 110:1363–1370
Gómez MS, Gracenea M, Gosalbez P, Feliu C, Enseñat C, Hidalgo R (1992) Detection of oocysts of *Cryptosporidium* in several species of monkeys and in one prosimian species at the Barcelona Zoo. Parasitol Res 78:619–620
Gómez MS, Vila T, Feliu C, Montoliu I, Gracenea M, Fernandez J (1996) A survey for *Cryptosporidium* spp. in mammals at the Barcelona Zoo. Int J Parasitol 26:1331–1333
Gómez MS, Torres J, Gracenea M, Fernandez-Moran J, Gonzalez-Moreno O (2000) Further report on *Cryptosporidium* in Barcelona zoo mammals. Parasitol Res 86:318–323
Gómez-Couso H, Méndez-Hermida F, Ares-Mazas E (2007) First report of *Cryptosporidium parvum* 'ferret' genotype in American mink (*Mustela vison* Shreber 1777). Parasitol Res 100:877–879
Gómez-Villamandos JC, Carrasco L, Mozos E, Hervás J (1995) Fatal Cryptosporidiosis in Ferrets (*Mustela putorius furo*): a morphopathologic study. J Zoo Wildl Med 26:539–544
Gonzalez-Moreno O, Hernandez-Aguilar RA, Piel AK, Stewart FA, Gracenea M, Moore J (2013) Prevalence and climatic associated factors of *Cryptosporidium* sp. infections in savanna chimpanzees from Ugalla, Western Tanzania. Parasitol Res 112:393–399
Gordo F, Herrera S, Castro A, Duran B, Diaz R (2002) Parasites from farmed ostriches (*Struthio camelus*) and rheas (*Rhea americana*) in Europe. Vet Parasitol 107:137–160
Gracenea M, Gómez MS, Torres J, Carné E, Fernández-Morán J (2002) Transmission dynamics of *Cryptosporidium* in primates and herbivores at the Barcelona zoo: a long-term study. Vet Parasitol 104:19–26
Gracenea M, Gómez MS, Torres J (2009) Prevalence of intestinal parasites in shelter dogs and cats in the metropolitan area of Barcelona (Spain). Acta Parasitol 54:73–77
Graczyk TK, Cranfield MR (1998) Experimental transmission of *Cryptosporidium* oocyst isolates from mammals, birds and reptiles to captive snakes. Vet Res 29:187–195
Graczyk TK, Cranfield MR (2000) *Cryptosporidium serpentis* oocysts and microsporidian spores in feces of captive snakes. J Parasitol 86:413–414
Graczyk TK, Cranfield MR, Fayer R (1996a) Evaluation of commercial enzyme immunoassay (EIA) and immunofluorescent antibody (FA) test kits for detection of *Cryptosporidium* oocysts of species other than *Cryptosporidium parvum*. Am J Trop Med Hyg 54:274–279
Graczyk TK, Cranfield MR, Fayer R, Anderson MS (1996b) Viability and infectivity of *Cryptosporidium parvum* oocysts are retained upon intestinal passage through a refractory avian host. Appl Environ Microbiol 62:3234–3237
Graczyk TK, Cranfield MR, Fayer R, Trout J, Goodale HJ (1997) Infectivity of *Cryptosporidium parvum* oocysts is retained upon intestinal passage through a migratory water-fowl species (Canada goose, *Branta canadensis*). Trop Med Int Health 2:341–347
Graczyk TK, Cranfield MR, Dunning C, Strandberg JD (1998a) Fatal cryptosporidiosis in a juvenile captive African Hedgehog (*Ateletrix albiventris*). J Parasitol 84:178–180
Graczyk TK, Cranfield MR, Fayer R (1998b) Oocysts of *Cryptosporidium* from snakes are not infectious to ducklings but retain viability after intestinal passage through a refractory host. Vet Parasitol 77:33–40
Graczyk TK, Cranfield MR, Geitner ME (1998c) Multiple *Cryptosporidium serpentis* oocyst isolates from captive snakes are not transmissible to amphibians. J Parasitol 84:1298–1300
Graczyk TK, Fayer R, Trout JM, Lewis EJ, Farley CA, Sulaiman I et al (1998d) *Giardia* sp. cysts and infectious *Cryptosporidium parvum* oocysts in the feces of migratory Canada geese (*Branta canadensis*). Appl Environ Microbiol 64:2736–2738

Graczyk TK, DaSilva AJ, Cranfield MR, Nizeyi JB, Kalema GR, Pieniazek NJ (2001) *Cryptosporidium parvum* genotype 2 infections in free-ranging mountain gorillas (*Gorilla gorilla beringei*) of the Bwindi Impenetrable National Park, Uganda. Parasitol Res 87:368–370

Green SL, Bouley DM, Josling CA, Fayer R (2003) Cryptosporidiosis associated with emaciation and proliferative gastritis in a laboratory-reared South African clawed frog (*Xenopus laevis*). Comp Med 53:81–84

Griffin C, Reavill DR, Stacy BA, Childress AL, Wellehan JF Jr (2010) Cryptosporidiosis caused by two distinct species in Russian tortoises and a pancake tortoise. Vet Parasitol 170:14–19

Guénet JL, Bonhomme F (2003) Wild mice: an ever-increasing contribution to a popular mammalian model. Trends Genet 19:24–31

Hackett T, Lappin MR (2003) Prevalence of enteric pathogens in dogs of North-Central Colorado. J Am Anim Hosp Assoc 39:52–56

Hajdušek O, Ditrich O, Šlapeta J (2004) Molecular identification of *Cryptosporidium* spp. in animal and human hosts from the Czech Republic. Vet Parasitol 122:183–192

Hamnes IS, Gjerde B, Robertson L, Vikoren T, Handeland K (2006) Prevalence of Cryptosporidium and Giardia in free-ranging wild cervids in Norway. Vet Parasitol 141:30–41

Hamnes IS, Gjerde BK, Forberg T, Robertson LJ (2007) Occurrence of *Giardia* and *Cryptosporidium* in Norwegian red foxes (*Vulpes vulpes*). Vet Parasitol 143:347–353

Heuschele WP, Oosterhuis J, Janssen D, Robinson PT, Ensley PK, Meier JE et al (1986) Cryptosporidial infections in captive wild animals. J Wildl Dis 22:493–496

Hijjawi N, Ng J, Yang R, Atoum MF, Ryan U (2010) Identification of rare and novel *Cryptosporidium* GP60 subtypes in human isolates from Jordan. Exp Parasitol 125:161–164

Hikosaka K, Nakai Y (2005) A novel genotype of *Cryptosporidium muris* from large Japanese field mice, *Apodemus speciosus*. Parasitol Res 97:373–379

Hill BD, Fraser IR, Prior HC (1997) *Cryptosporidium* infection in a dugong (*Dugong dugon*). Aust Vet J 75:670–671

Hill SL, Cheney JM, Taton-Allen GF, Reif JS, Bruns C, Lappin MR (2000) Prevalence of enteric zoonotic organisms in cats. J Am Vet Med Assoc 216:687–692

Hill NJ, Deane EM, Power ML (2008) Prevalence and genetic characterization of *Cryptosporidium* isolates from common brushtail possums (*Trichosurus vulpecula*) adapted to urban settings. Appl Environ Microbiol 74:5549–5555

Himsworth CG, Skinner S, Chaban B, Jenkins E, Wagner BA, Harms NJ et al (2010) Multiple zoonotic pathogens identified in canine feces collected from a remote Canadian indigenous community. Am J Trop Med Hyg 83:338–341

Hope K, Goldsmith ML, Graczyk T (2004) Parasitic health of olive baboons in Bwindi Impenetrable National Park, Uganda. Vet Parasitol 122:165–170

Huber F, da Silva S, Bomfim TC, Teixeira KR, Bello AR (2007) Genotypic characterization and phylogenetic analysis of *Cryptosporidium* sp. from domestic animals in Brazil. Vet Parasitol 150:65–74

Hughes-Hanks JM, Rickard LG, Panuska C, Saucier JR, O'Hara TM, Dehn L et al (2005) Prevalence of *Cryptosporidium* spp. and *Giardia* spp. in five marine mammal species. J Parasitol 91:1225–1228

Insulander M, Silverlås C, Lebbad M, Karlsson L, Mattsson JG, Svenungsson B (2013) Molecular epidemiology and clinical manifestations of human cryptosporidiosis in Sweden. Epidemiol Infect 141:1009–1020

Iseki M (1979) *Cryptosporidium felis* sp. n. (protozoa: Eimeriorina) from the domestic cat. Jpn J Parasitol 28:285–307

Iseki M (1986) Two species of *Cryptosporidium* naturally infecting house rats. Jpn J Parasitol 35:251–256

Jafri HS, Reedy T, Moorhead AR, Dickerson JW, Schantz PM, Bryan RT (1993) Detection of pathogenic protozoa in fecal specimens from urban dwelling dogs. Am J Trop Med Hyg 49: S269

Jakob W (1992) Cryptosporidien- und andere kokzidienoozysten bei zoound wildtierren im nach ziehl-neelsen gefarbten kotausstrich. In: Ippen R (ed) Erkrankungen der Zootiere. Verhandlungsbericht des 34. Internationalen Symposium uber die Erkrankungen der Zootiere vom 27 Mai bis 31 Mai, 1992, Santander, Spain, vol 34, Akademie-Verlag, Berlin, pp 291–299

Jardine JE, Verwoerd DJ (1997) Pancreatic cryptosporidiosis in ostriches. Avian Pathol 26:665–670

Jellison KL, Distel DL, Hemond HF, Schauer DB (2004) Phylogenetic analysis of the hypervariable region of the 18S rRNA gene of *Cryptosporidium* oocysts in feces of Canada geese (*Branta canadensis*): evidence for five novel genotypes. Appl Environ Microbiol 70:452–458

Jellison KL, Lynch AE, Ziemann JM (2009) Source tracking identifies deer and geese as vectors of human-infectious *Cryptosporidium* genotypes in an urban/suburban watershed. Environ Sci Technol 43:4267–4272

Jirků M, Valigurová A, Koudela B, Křížek J, Modrý D, Šlapeta J (2008) New species of *Cryptosporidium* Tyzzer, 1907 (Apicomplexa) from amphibian host: morphology, biology and phylogeny. Folia Parasitol (Praha) 55:81–94

Johnson D, Harms NJ, Larter NC, Elkin BT, Tabel H, Wei G (2010) Serum biochemistry, serology, and parasitology of boreal caribou (*Rangifer tarandus caribou*) in the Northwest Territories, Canada. J Wildl Dis 46:1096–1107

Jones EP, Jensen JK, Magnussen E, Gregersen N, Hansen HS, Searle JB (2011) A molecular characterization of the charismatic Faroe house mouse. Biol J Linn Soc 102:471–482

Juett BW, Otero RB, Bischoff WH (1996) *Cryptosporidium* in the domestic dog population of Central Kentucky. Trans Ky Acad Sci 57:18–21

Kalishman J, Paul-Murphy J, Scheffler J, Thomson JA (1996) Survey of *Cryptosporidium* and *Giardia* spp. in a captive population of common marmosets. Lab Anim Sci 46:116–119

Karanis P, Plutzer J, Halim NA, Igori K, Nagasawa H, Ongerth J et al (2007) Molecular characterization of *Cryptosporidium* from animal sources in Qinghai province of China. Parasitol Res 101:1575–1580

Kaup FJ, Kuhn EM, Makoschey B, Hunsmann G (1994) Cryptosporidiosis of liver and pancreas in rhesus monkeys with experimental SIV infection. J Med Primatol 23:304–308

Kik MJ, van Asten AJ, Lenstra JA, Kirpensteijn J (2011) Cloaca prolapse and cystitis in green iguana (Iguana iguana) caused by a novel *Cryptosporidium* species. Vet Parasitol 175:165–167

Kim JT, Wee SH, Lee CG (1998) Detection of *Cryptosporidium* oocysts in canine fecal samples by immunofluorescence assay. Korean J Parasitol 36:147–149

Kim Y, Howerth EW, Shin NS, Kwon SW, Terrell SP, Kim DY (2005) Disseminated visceral coccidiosis and cloacal cryptosporidiosis in a Japanese white-naped crane (*Grus vipio*). J Parasitol 91:199–201

Kimbell LM 3rd, Miller DL, Chavez W, Altman N (1999) Molecular analysis of the 18S rRNA gene of *Cryptosporidium serpentis* in a wild-caught corn snake (*Elaphe guttata guttata*) and a five-species restriction fragment length polymorphism-based assay that can additionally discern *C. parvum* from *C. wrairi*. Appl Environ Microbiol 65:5345–5349

Kimura A, Edagawa A, Okada K, Takimoto A, Yonesho S, Karanis P (2007) Detection and genotyping of *Cryptosporidium* from brown rats (*Rattus norvegicus*) captured in an urban area of Japan. Parasitol Res 100:1417–1420

Klesius PH, Haynes TB, Malo LK (1986) Infectivity of *Cryptosporidium* sp. isolated from wild mice for calves and mice. J Am Vet Med Assoc 189:192–193

Kloch A, Bednarska M, Bajer A (2005) Intestinal macro—and microparasites of wolves (*Canis lupus* L.) from North-Eastern Poland recovered by coprological study. Ann Agric Environ Med 12:237–245

Kodádková A, Kváč M, Ditrich O, Sak B, Xiao L (2010) *Cryptosporidium muris* in a reticulated giraffe (*Giraffa camelopardalis reticulata*). J Parasitol 96:211–212

Koudela B, Modrý D (1998) New species of *Cryptosporidium* (Apicomplexa : Cryptosporidiidae) from lizards. Folia Parasitol (Praha) 45:93–100

Kuhn RC, Rock CM, Oshima KH (2002) Occurrence of *Cryptosporidium* and *Giardia* in wild ducks along the Rio Grande River valley in Southern New Mexico. Appl Environ Microbiol 68:161–165

Kuroki T, Izumiyama S, Yagita K, Une Y, Hayashidani H, Kuro-o M et al (2008) Occurrence of *Cryptosporidium* sp. in snakes in Japan. Parasitol Res 103:801–805

Kváč M, Hofmannová L, Bertolino S, Wauters L, Tosi G, Modrý D (2008a) Natural infection with two genotypes of *Cryptosporidium* in red squirrels (*Sciurus vulgaris*) in Italy. Folia Parasitol (Praha) 55:95–99

Kváč M, Sak B, Květoňová D, Ditrich O, Hofmannová L, Modrý D et al (2008b) Infectivity, pathogenicity, and genetic characteristics of mammalian gastric *Cryptosporidium* spp. in domestic ruminants. Vet Parasitol 153:363–367

Kváč M, Hanzlíková D, Sak B, Květoňová D (2009a) Prevalence and age-related infection of *Cryptosporidium suis*, C. muris and *Cryptosporidium* pig genotype II in pigs on a farm complex in the Czech Republic. Vet Parasitol 160:319–322

Kváč M, Květoňová D, Sak B, Ditrich O (2009b) *Cryptosporidium* pig genotype II in immunocompetent man. Emerg Infect Dis 15:982–983

Kváč M, Sak B, Hanzlíková D, Kotilová J, Květoňová D (2009c) Molecular characterization of *Cryptosporidium* isolates from pigs at slaughterhouses in South Bohemia, Czech Republic. Parasitol Res 104:425–428

Kváč M, Sak B, Kvetonová D, Secor WE (2009d) Infectivity of gastric and intestinal *Cryptosporidium* species in immunocompetent Mongolian gerbils (*Meriones unguiculatus*). Vet Parasitol 163:33–38

Kváč M, Kestřánová M, Pinková M, Květoňová D, Kalinová J, Wagnerová P et al (2013a) *Cryptosporidium scrofarum* n. sp. (Apicomplexa: Cryptosporidiidae) in domestic pigs (*Sus scrofa*). Vet Parasitol 191:218–227

Kváč M, McEvoy J, Loudová M, Stenger B, Sak B, Květoňová D et al (2013b) Coevolution of *Cryptosporidium tyzzeri* and the house mouse (*Mus musculus*). Int J Parasitol 43:805–817

Laakkonen J, Soveri T, Henttonen H (1994) Prevalence of *Cryptosporidium* sp. in peak density *Microtus agrestis*, *Microtus oeconomus* and *Clethrionomys glareolus* populations. J Wildl Dis 30:110–111

Laatamna AE, Wagnerová P, Sak B, Květoňová D, Aissi M, Rost M et al (2013) Equine cryptosporidial infection associated with *Cryptosporidium* hedgehog genotype in Algeria. Vet Parasitol 197:350–353

Lallo MA, Bondan EF (2006) Prevalence of *Cryptosporidium* sp. in institutionalized dogs in the city of Sao Paulo, Brazil. Rev Saude Publica 40:120–125

Langkjær RB, Vigre H, Enemark HL, Maddox-Hyttel C (2007) Molecular and phylogenetic characterization of *Cryptosporidium* and *Giardia* from pigs and cattle in Denmark. Parasitology 134:339–350

Latimer KS, Steffens WL 3rd, Rakich PM, Ritchie BW, Niagro FD, Kircher IM et al (1992) Cryptosporidiosis in four cockatoos with psittacine beak and feather disease. J Am Vet Med Assoc 200:707–710

Legesse M, Erko B (2004) Zoonotic intestinal parasites in *Papio anubis* (baboon) and *Cercopithecus aethiops* (vervet) from four localities in Ethiopia. Acta Trop 90:231–236

Leoni F, Amar C, Nichols G, Pedraza-Diaz S, McLauchlin J (2006) Genetic analysis of *Cryptosporidium* from 2414 humans with diarrhoea in England between 1985 and 2000. J Med Microbiol 55:703–707

Levine ND (1980) Some corrections of coccidian (Apicomplexa: Protozoa) nomenclature. J Parasitol 66:830–834

Ley DH, Moresco A, Frasca S Jr (2012) Conjunctivitis, rhinitis, and sinusitis in cliff swallows (*Petrochelidon pyrrhonota*) found in association with *Mycoplasma sturni* infection and cryptosporidiosis. Avian Pathol 41:395–401

Li W, Kiulia NM, Mwenda JM, Nyachieo A, Taylor MB, Zhang X et al (2011) *Cyclospora papionis*, *Cryptosporidium hominis*, and human-pathogenic *Enterocytozoon bieneusi* in captive baboons in Kenya. J Clin Microbiol 49:4326–4329

Lim YAL, Rohela M, Shukri MM (2007) Cryptosporidiosis among birds and bird handlers at Zoo Negara, Malaysia. Southeast Asian J Trop Med Public Health 38:20–26

Lim YA, Ngui R, Shukri J, Rohela M, Mat Naim HR (2008) Intestinal parasites in various animals at a zoo in Malaysia. Vet Parasitol 157:154–159

Lindsay DS, Blagburn BL, Sundermann CA, Hoerr FJ (1989) Experimental infections in domestic ducks with *Cryptosporidium baileyi* isolated from chickens. Avian Dis 33:69–73

Lindsay DS, Blagburn BL, Hoerr FJ, Smith PC (1991) Cryptosporidiosis in zoo and pet birds. J Protozool 38:180S–181S

Lindsay DS, Upton SJ, Owens DS, Morgan UM, Mead JR, Blagburn BL (2000) *Cryptosporidium andersoni* n. sp. (Apicomplexa: Cryptosporiidae) from cattle, *Bos taurus*. J Eukaryot Microbiol 47:91–95

Lv C, Zhang L, Wang R, Jian F, Zhang S, Ning C et al (2009) *Cryptosporidium* spp. in wild, laboratory, and pet rodents in China: prevalence and molecular characterization. Appl Environ Microbiol 75:7692–7699

Macholán M, Kryštufek B, Vohralík V (2003) The location of the *Mus musculus/M. domesticus* hybrid zone in the Balkans: clues from morphology. Acta Theriol (Warsz) 48:177–188

Majewska AC, Graczyk TK, Slodkowicz-Kowalska A, Tamang L, Jedrzejewski S, Zduniak P et al (2009) The role of free-ranging, captive, and domestic birds of Western Poland in environmental contamination with *Cryptosporidium parvum* oocysts and *Giardia lamblia* cysts. Parasitol Res 104:1093–1099

Makino I, Abe N, Reavill DR (2010) *Cryptosporidium* avian genotype III as a possible causative agent of chronic vomiting in peach-faced lovebirds (*Agapornis roseicollis*). Avian Dis 54:1102–1107

Mandarino-Pereira A, de Souza FS, Lopes CW, Pereira MJ (2010) Prevalence of parasites in soil and dog feces according to diagnostic tests. Vet Parasitol 170:176–181

Martin HD, Zeidner NS (1992) Concomitant cryptosporidia, coronavirus and parvovirus infection in a raccoon (*Procyon lotor*). J Wildl Dis 28:113–115

Mason RW, Hartley WJ, Tilt L (1981) Intestinal cryptosporidiosis in a kid goat. Aust Vet J 57:386–388

Matsubayashi M, Takami K, Kimata I, Nakanishi T, Tani H, Sasai K et al (2005) Survey of *Cryptosporidium* spp. and *Giardia* spp. infections in various animals at a zoo in Japan. J Zoo Wildl Med 36:331–335

Matsui T, Fujino T, Kajima J, Tsuji M (2000) Infectivity to experimental rodents of *Cryptosporidium parvum* oocysts from Siberian chipmunks (*Tamias sibiricus*) originated in the People's Republic of China. J Vet Med Sci 62:487–489

McKenzie E, Riehl J, Banse H, Kass PH, Nelson S Jr, Marks SL (2010) Prevalence of diarrhea and enteropathogens in racing sled dogs. J Vet Intern Med 24:97–103

McLauchlin J, Amar C, Pedraza-Diaz S, Nichols GL (2000) Molecular epidemiological analysis of *Cryptosporidium* spp. in the United Kingdom: results of genotyping *Cryptosporidium* spp. in 1,705 fecal samples from humans and 105 fecal samples from livestock animals. J Clin Microbiol 38:3984–3990

Meireles MV, Soares RM, dos Santos MM, Gennari SM (2006) Biological studies and molecular characterization of a *Cryptosporidium* isolate from ostriches (*Struthio camelus*). J Parasitol 92:623–626

Meireles MV, Soares RM, Bonello F, Gennari SM (2007) Natural infection with zoonotic subtype of *Cryptosporidium parvum* in Capybara (*Hydrochoerus hydrochaeris*) from Brazil. Vet Parasitol 147:166–170

Mekaru SR, Marks SL, Felley AJ, Chouicha N, Kass PH (2007) Comparison of direct immunofluorescence, immunoassays, and fecal flotation for detection of *Cryptosporidium* spp. and

Giardia spp. in naturally exposed cats in 4 Northern California animal shelters. J Vet Intern Med 21:959–965
Méndez-Hermida F, Gómez-Couso H, Romero-Suances R, Ares-Mazas E (2007) *Cryptosporidium* and *Giardia* in wild otters (*Lutra lutra*). Vet Parasitol 144:153–156
Meredith AL, Milne EM (2009) Cryptosporidial infection in a captive European hedgehog (*Erinaceus europaeus*). J Zoo Wildl Med 40:809–811
Miller RA, Bronsdon MA, Kuller L, Morton WR (1990a) Clinical and parasitologic aspects of cryptosporidiosis in nonhuman primates. Lab Anim Sci 40:42–46
Miller RA, Bronsdon MA, Morton WR (1990b) Experimental cryptosporidiosis in a primate model. J Infect Dis 161:312–315
Mirzaei M (2012) Epidemiological survey of *Cryptosporidium* spp. in companion and stray dogs in Kerman, Iran. Vet Ital 48:291–296
Miyaji S, Tanikawa T, Shikata J (1989) Prevalence of *Cryptosporidium* in *Rattus rattus* and *R. norvegicus* in Japan. Jpn J Parasitol 38:368–372
Molina-Lopez RA, Ramis A, Martin-Vazquez S, Gomez-Couso H, Ares-Mazas E, Caccio SM et al (2010) *Cryptosporidium baileyi* infection associated with an outbreak of ocular and respiratory disease in otus owls (*Otus scops*) in a rehabilitation centre. Avian Pathol 39:171–176
Morgan UM, Constantine CC, Forbes DA, Thompson RC (1997) Differentiation between human and animal isolates of *Cryptosporidium parvum* using rDNA sequencing and direct PCR analysis. J Parasitol 83:825–830
Morgan UM, Sargent KD, Deplazes P, Forbes DA, Spano F, Hertzberg H et al (1998) Molecular characterization of *Cryptosporidium* from various hosts. Parasitology 117:31–37
Morgan UM, Buddle JR, Armson A, Elliot A, Thompson RC (1999a) Molecular and biological characterisation of *Cryptosporidium* in pigs. Aust Vet J 77:44–47
Morgan UM, Sturdee AP, Singleton G, Gomez MS, Gracenea M, Torres J et al (1999b) The *Cryptosporidium* "mouse" genotype is conserved across geographic areas. J Clin Microbiol 37:1302–1305
Morgan UM, Xiao L, Fayer R, Graczyk TK, Lal AA, Deplazes P et al (1999c) Phylogenetic analysis of *Cryptosporidium* isolates from captive reptiles using 18S rDNA sequence data and random amplified polymorphic DNA analysis. J Parasitol 85:525–530
Morgan UM, Xiao L, Hill BD, O'Donoghue P, Limor J, Lal A et al (2000a) Detection of the *Cryptosporidium parvum* "human" genotype in a dugong (*Dugong dugon*). J Parasitol 86:1352–1354
Morgan UM, Xiao L, Limor J, Gelis S, Raidal SR, Fayer R et al (2000b) *Cryptosporidium meleagridis* in an Indian ring-necked parrot (*Psittacula krameri*). Aust Vet J 78:182–183
Morgan UM, Xiao L, Monis P, Sulaiman I, Pavlásek I, Blagburn B et al (2000c) Molecular and phylogenetic analysis of *Cryptosporidium muris* from various hosts. Parasitology 120:457–464
Morgan UM, Monis PT, Xiao L, Limor J, Sulaiman I, Raidal S et al (2001) Molecular and phylogenetic characterisation of *Cryptosporidium* from birds. Int J Parasitol 31:289–296
Mtambo MM, Sebatwale JB, Kambarage DM, Muhairwa AP, Maeda GE, Kusiluka LJ et al (1997) Prevalence of *Cryptosporidium* spp. oocysts in cattle and wildlife in Morogoro region Tanzania. Prev Vet Med 31:185–190
Muriuki SM, Farah IO, Kagwiria RM, Chai DC, Njamunge G, Suleman M et al (1997) The presence of *Cryptosporidium* oocysts in stools of clinically diarrhoeic and normal nonhuman primates in Kenya. Vet Parasitol 72:141–147
Muriuki SM, Murugu RK, Munene E, Karere GM, Chai DC (1998) Some gastro-intestinal parasites of zoonotic (public health) importance commonly observed in old world non-human primates in Kenya. Acta Trop 71:73–82
Nagano Y, Finn MB, Lowery CJ, Murphy T, Moriarty J, Power E et al (2007) Occurrence of *Cryptosporidium parvum* and bacterial pathogens in faecal material in the red fox (*Vulpes vulpes*) population. Vet Res Commun 31:559–564

Nakai Y, Hikosaka K, Sato M, Sasaki T, Kaneta Y, Okazaki N (2004) Detection of *Cryptosporidium muris* type oocysts from beef cattle in a farm and from domestic and wild animals in and around the farm. J Vet Med Sci 66:983–984

Nakamura AA, Simoes DC, Antunes RG, da Silva DC, Meireles MV (2009) Molecular characterization of *Cryptosporidium* spp. from fecal samples of birds kept in captivity in Brazil. Vet Parasitol 166:47–51

Němejc K, Sak B, Květoňová D, Hanzal V, Jeníková M, Kváč M (2012) The first report on *Cryptosporidium suis* and *Cryptosporidium* pig genotype II in Eurasian wild boars (*Sus scrofa*) (Czech Republic). Vet Parasitol 184:122–125

Němejc K, Sak B, Květoňová D, Kernerová N, Rost M, Cama VA et al (2013) Occurrence of *Cryptosporidium suis* and *Cryptosporidium scrofarum* on commercial swine farms in the Czech Republic and its associations with age and husbandry practices. Parasitol Res 112:1143–1154

Ng J, Pavlásek I, Ryan U (2006) Identification of novel *Cryptosporidium* genotypes from avian hosts. Appl Environ Microbiol 72:7548–7553

Ng J, Yang R, McCarthy S, Gordon C, Hijjawi N, Ryan U (2011) Molecular characterization of *Cryptosporidium* and *Giardia* in pre-weaned calves in Western Australia and New South Wales. Vet Parasitol 176:145–150

Ng-Hublin JS, Singleton GR, Ryan U (2013) Molecular characterization of *Cryptosporidium* spp. from wild rats and mice from rural communities in the Philippines. Infect Genet Evol 16:5–12

Nguyen ST, Fukuda Y, Tada C, Huynh VV, Nguyen DT, Nakai Y (2013) Prevalence and molecular characterization of *Cryptosporidium* in ostriches (*Struthio camelus*) on a farm in Central Vietnam. Exp Parasitol 133:8–11

Nizeyi JB, Mwebe R, Nanteza A, Cranfield MR, Kalema GR, Graczyk TK (1999) *Cryptosporidium* sp. and *Giardia* sp. infections in mountain gorillas (*Gorilla gorilla beringei*) of the Bwindi Impenetrable National Park, Uganda. J Parasitol 85:1084–1088

Nizeyi JB, Sebunya D, Dasilva AJ, Cranfield MR, Pieniazek NJ, Graczyk TK (2002) Cryptosporidiosis in people sharing habitats with free-ranging mountain gorillas (*Gorilla gorilla beringei*), Uganda. Am J Trop Med Hyg 66:442–444

Nolan MJ, Jex AR, Haydon SR, Stevens MA, Gasser RB (2010) Molecular detection of *Cryptosporidium cuniculus* in rabbits in Australia. Infect Genet Evol 10:1179–1187

Nolan MJ, Jex AR, Koehler AV, Haydon SR, Stevens MA, Gasser RB (2013) Molecular-based investigation of *Cryptosporidium* and *Giardia* from animals in water catchments in Southeastern Australia. Water Res 47:1726–1740

Nutter FB, Dubey JP, Levine JF, Breitschwerdt EB, Ford RB, Stoskopf MK (2004) Seroprevalences of antibodies against *Bartonella henselae* and *Toxoplasma gondii* and fecal shedding of *Cryptosporidium* spp., *Giardia* spp., and *Toxocara cati* in feral and pet domestic cats. J Am Vet Med Assoc 225:1394–1398

O'Donoghue PJ (1995) *Cryptosporidium* and cryptosporidiosis in man and animals. Int J Parasitol 25:139–195

O'Donoghue PJ, Tham VL, de Saram WG, Paull KL, McDermott S (1987) *Cryptosporidium* infections in birds and mammals and attempted cross-transmission studies. Vet Parasitol 26:1–11

Oates SC, Miller MA, Hardin D, Conrad PA, Melli A, Jessup DA et al (2012) Prevalence, environmental loading, and molecular characterization of *Cryptosporidium* and *Giardia* isolates from domestic and wild animals along the Central California Coast. Appl Environ Microbiol 78:8762–8772

Ong CS, Eisler DL, Alikhani A, Fung VW, Tomblin J, Bowie WR et al (2002) Novel *Cryptosporidium* genotypes in sporadic cryptosporidiosis cases: first report of human infections with a cervine genotype. Emerg Infect Dis 8:263–268

Osborn KG, Prahalada S, Lowenstine LJ, Gardner MB, Maul DH, Henrickson RV (1984) The pathology of an epizootic of acquired immunodeficiency in rhesus macaques. Am J Pathol 114:94–103

Overgaauw PA, van Zutphen L, Hoek D, Yaya FO, Roelfsema J, Pinelli E et al (2009) Zoonotic parasites in fecal samples and fur from dogs and cats in The Netherlands. Vet Parasitol 163:115–122

Ozkul IA, Aydin Y (1994) Small-intestinal cryptosporidiosis in a young pigeon. Avian Pathol 23:369–372

Pages-Mante A, Pages-Bosch M, Majo-Masferrer N, Gomez-Couso H, Ares-Mazas E (2007) An outbreak of disease associated with cryptosporidia on a red-legged partridge (*Alectoris rufa*) game farm. Avian Pathol 36:275–278

Paparini A, Jackson B, Ward S, Young S, Ryan UM (2012) Multiple *Cryptosporidium* genotypes detected in wild black rats (*Rattus rattus*) from Northern Australia. Exp Parasitol 131:404–412

Papini R, Girivetto M, Marangi M, Mancianti F, Giangaspero A (2012) Endoparasite infections in pet and zoo birds in Italy. ScientificWorldJournal 2012:253127

Pavlásek I (1993) The black-headed gull (*Larus ridibundus* L.), a new host for *Cryptosporidium baileyi* (Apicomplexa: Cryptosporidiidae). Vet Med (Praha) 38:629–638

Pavlásek I, Ryan U (2007) The first finding of a natural infection of *Cryptosporidium muris* in a cat. Vet Parasitol 144:349–352

Pavlásek I, Ryan U (2008) *Cryptosporidium varanii* takes precedence over *C. saurophilum*. Exp Parasitol 118:434–437

Pavlásek I, Lávisková M, Horák P, Král J, Král B (1995) *Cryptosporidium varanii* n. sp. (Apicomplexa: Cryptosporidiidae) in emerald monitor (*Varanus prasinus* Schlegal, 1893) in captivity in Prague zoo. Gazella (Zoo Praha) 22:99–108

Paziewska A, Bednarska M, Nieweglowski H, Karbowiak G, Bajer A (2007) Distribution of *Cryptosporidium* and *Giardia* spp. in selected species of protected and game mammals from North-Eastern Poland. Ann Agric Environ Med 14:265–270

Pedraza-Diaz S, Amar C, Iversen AM, Stanley PJ, McLauchlin J (2001) Unusual *Cryptosporidium* species recovered from human faeces: first description of *Cryptosporidium felis* and *Cryptosporidium* 'dog type' from patients in England. J Med Microbiol 50:293–296

Pedraza-Diaz S, Ortega-Mora LM, Carrion BA, Navarro V, Gomez-Bautista M (2009) Molecular characterisation of *Cryptosporidium* isolates from pet reptiles. Vet Parasitol 160:204–210

Pereira M, Atwill ER, Crawford MR, Lefebvre RB (1998) DNA sequence similarity between California isolates of *Cryptosporidium parvum*. Appl Environ Microbiol 64:1584–1586

Pereira MG, Li X, McCowan B, Phillips RL, Atwill ER (2010) Multiple unique *Cryptosporidium* isolates from three species of ground squirrels (*Spermophilus beecheyi*, *S. beldingi*, and *S. lateralis*) in California. Appl Environ Microbiol 76:8269–8276

Perz JF, Le Blancq SM (2001) *Cryptosporidium parvum* infection involving novel genotypes in wildlife from lower New York State. Appl Environ Microbiol 67:1154–1162

Plutzer J, Karanis P (2007) Molecular identification of a *Cryptosporidium saurophilum* from corn snake (*Elaphe guttata guttata*). Parasitol Res 101:1141–1145

Plutzer J, Tomor B (2009) The role of aquatic birds in the environmental dissemination of human pathogenic *Giardia duodenalis* cysts and *Cryptosporidium* oocysts in Hungary. Parasitol Int 58:227–231

Pospischil A, Stiglmair-Herb MT, von Hegel G, Wiesner H (1987) Abomasal cryptosporidiosis in mountain gazelles. Vet Rec 121:379–380

Power ML (2010) Biology of *Cryptosporidium* from marsupial hosts. Exp Parasitol 124:40–44

Power ML, Ryan UM (2008) A new species of *Cryptosporidium* (Apicomplexa: Cryptosporidiidae) from eastern grey kangaroos (*Macropus giganteus*). J Parasitol 94:1114–1117

Power ML, Shanker SR, Sangster NC, Veal DA (2003) Evaluation of a combined immunomagnetic separation/flow cytometry technique for epidemiological investigations of *Cryptosporidium* in domestic and Australian native animals. Vet Parasitol 112:21–31

Qi M, Wang R, Ning C, Li X, Zhang L, Jian F et al (2011) *Cryptosporidium* spp. in pet birds: genetic diversity and potential public health significance. Exp Parasitol 128:336–340

Quy RJ, Cowan DP, Haynes PJ, Sturdee AP, Chalmers RM, Bodley-Tickell AT et al (1999) The Norway rat as a reservoir host of *Cryptosporidium parvum*. J Wildl Dis 35:660–670

Rademacher U, Jakob W, Bockhardt I (1999) *Cryptosporidium* infection in beech martens (*Martes foina*). J Zoo Wildl Med 30:421–422

Radfar MH, Asl EN, Seghinsara HR, Dehaghi MM, Fathi S (2012) Biodiversity and prevalence of parasites of domestic pigeons (*Columba livia domestica*) in a selected semiarid zone of South Khorasan, Iran. Trop Anim Health Prod 44:225–229

Rajabi-Maham H, Orth A, Bonhomme F (2008) Phylogeography and postglacial expansion of *Mus musculus domesticus* inferred from mitochondrial DNA coalescent, from Iran to Europe. Mol Ecol 17:627–641

Ravaszová P, Halánová M, Goldová M, Valenčáková A, Malčeková B, Hurníková Z et al (2012) Occurrence of *Cryptosporidium* spp. in red foxes and brown bear in the Slovak Republic. Parasitol Res 110:469–471

Razawi SM, Oryan A, Bahrami S, Mohammadalipour A, Gowhari M (2009) Prevalence of *Cryptosporidium* infection in camels (*Camelus dromedarius*) in a slaughterhouse in Iran. Trop Biomed 26:267–273

Rehg JE, Gigliotti F, Stokes DC (1988) Cryptosporidiosis in ferrets. Lab Anim Sci 38:155–158

Ren X, Zhao J, Zhang L, Ning C, Jian F, Wang R et al (2012) *Cryptosporidium tyzzeri* n. sp. (Apicomplexa: Cryptosporidiidae) in domestic mice (*Mus musculus*). Exp Parasitol 130:274–281

Rengifo-Herrera C, Ortega-Mora LM, Gomez-Bautista M, Garcia-Moreno FT, Garcia-Parraga D, Castro-Urda J et al (2011) Detection and characterization of a *Cryptosporidium* isolate from a southern elephant seal (*Mirounga leonina*) from the Antarctic Peninsula. Appl Environ Microbiol 77:1524–1527

Rengifo-Herrera C, Ortega-Mora LM, Gomez-Bautista M, Garcia-Pena FJ, Garcia-Parraga D, Pedraza-Diaz S (2013) Detection of a novel genotype of *Cryptosporidium* in Antarctic pinnipeds. Vet Parasitol 191:112–118

Richter B, Nedorost N, Maderner A, Weissenbock H (2011) Detection of *Cryptosporidium* species in feces or gastric contents from snakes and lizards as determined by polymerase chain reaction analysis and partial sequencing of the 18S ribosomal RNA gene. J Vet Diagn Invest 23:430–435

Rickard LG, Siefker C, Boyle CR, Gentz EJ (1999) The prevalence of *Cryptosporidium* and *Giardia* spp. in fecal samples from free-ranging white-tailed deer (*Odocoileus virginianus*) in the Southeastern United States. J Vet Diagn Invest 11:65–72

Rinaldi L, Capasso M, Mihalca AD, Cirillo R, Cringoli G, Caccio S (2012) Prevalence and molecular identification of *Cryptosporidium* isolates from pet lizards and snakes in Italy. Parasite 19:437–440

Robinson G, Chalmers RM (2010) The European rabbit (*Oryctolagus cuniculus*), a source of zoonotic cryptosporidiosis. Zoonoses Public Health 57:e1–e13

Robinson G, Elwin K, Chalmers RM (2008) Unusual Cryptosporidium genotypes in human cases of diarrhea. Emerg Infect Dis 14:1800–1802

Robinson G, Wright S, Elwin K, Hadfield SJ, Katzer F, Bartley PM et al (2010) Re-description of *Cryptosporidium cuniculus* Inman and Takeuchi, 1979 (Apicomplexa: Cryptosporidiidae): morphology, biology and phylogeny. Int J Parasitol 40:1539–1548

Robinson G, Chalmers RM, Stapleton C, Palmer SR, Watkins J, Francis C et al (2011) A whole water catchment approach to investigating the origin and distribution of *Cryptosporidium* species. J Appl Microbiol 111:717–730

Rodriguez F, Oros J, Rodriguez JL, Gonzalez J, Castro P, Fernandez A (1997) Intestinal cryptosporidiosis in pigeons (*Columba livia*). Avian Dis 41:748–750

Rohela M, Lim YA, Jamaiah I, Khadijah PY, Laang ST, Nazri MH et al (2005) Occurrence of *Cryptosporidium* oocysts in Wrinkled Hornbill and other birds in the Kuala Lumpur National Zoo. Southeast Asian J Trop Med Public Health 36(Suppl 4):34–40

Ryan U, Xiao L, Read C, Zhou L, Lal AA, Pavlásek I (2003a) Identification of novel *Cryptosporidium* genotypes from the Czech Republic. Appl Environ Microbiol 69:4302–4307
Ryan UM, Xiao L, Read C, Sulaiman IM, Monis P, Lal AA et al (2003b) A redescription of *Cryptosporidium galli* Pavlásek, 1999 (Apicomplexa: Cryptosporidiidae) from birds. J Parasitol 89:809–813
Ryan UM, Power M, Xiao L (2008) *Cryptosporidium fayeri* n. sp. (Apicomplexa: Cryptosporidiidae) from the Red Kangaroo (*Macropus rufus*). J Eukaryot Microbiol 55:22–26
Salyer SJ, Gillespie TR, Rwego IB, Chapman CA, Goldberg TL (2012) Epidemiology and molecular relationships of *Cryptosporidium* spp. in people, primates, and livestock from Western Uganda. PLoS Negl Trop Dis 6:e1597
Salzer JS, Rwego IB, Goldberg TL, Kuhlenschmidt MS, Gillespie TR (2007) *Giardia* sp. and *Cryptosporidium* sp. infections in primates in fragmented and undisturbed forest in Western Uganda. J Parasitol 93:439–440
Santín M, Dixon BR, Fayert R (2005) Genetic characterization of *Cryptosporidium* isolates from ringed seals (*Phoca hispida*) in Northern Quebec, Canada. J Parasitol 91:712–716
Santín M, Trout JM, Vecino JA, Dubey JP, Fayer R (2006) *Cryptosporidium*, *Giardia* and *Enterocytozoon bieneusi* in cats from Bogota (Colombia) and genotyping of isolates. Vet Parasitol 141:334–339
Santos M, Peiró J, Meireles M (2005) *Cryptosporidium* infection in ostriches (*Struthio camelus*) in Brazil: clinical, morphological and molecular studies. Rev Bras Cienc Avic 7:113–117
Sazmand A, Rasooli A, Nouri M, Hamidinejat H, Hekmatimoghaddam S (2012) Prevalence of *Cryptosporidium* spp. in camels and involved people in Yazd Province, Iran, Iran. J Parasitol 7:80–84
Scorza AV, Duncan C, Miles L, Lappin MR (2011) Prevalence of selected zoonotic and vector-borne agents in dogs and cats in Costa Rica. Vet Parasitol 183:178–183
Sevá Ada P, Funada MR, Richtzenhain L, Guimarães MB, SeO S, Allegretti L et al (2011a) Genotyping of *Cryptosporidium* spp. from free-living wild birds from Brazil. Vet Parasitol 175:27–32
Sevá Ada P, Sercundes MK, Martins J, de Souza SO, da Cruz JB, Lisboa CS et al (2011b) Occurrence and molecular diagnosis of *Cryptosporidium serpentis* in captive snakes in Sao Paulo, Brazil. J Zoo Wildl Med 42:326–329
Shi K, Jian F, Lv C, Ning C, Zhang L, Ren X et al (2010) Prevalence, genetic characteristics, and zoonotic potential of *Cryptosporidium* species causing infections in farm rabbits in China. J Clin Microbiol 48:3263–3266
Shukla R, Giraldo P, Kraliz A, Finnigan M, Sanchez AL (2006) *Cryptosporidium* spp. and other zoonotic enteric parasites in a sample of domestic dogs and cats in the Niagara region of Ontario. Can Vet J 47:1179–1184
Siam MA, Salem GH, Ghoneim NH, Michael SA, El-Refay MAH (1994) Public health importance of enteric parasitosis in captive carnivora. Assiut Vet Med J 32:132–140
Siefker C, Rickard LG, Pharr GT, Simmons JS, O'Hara TM (2002) Molecular characterization of *Cryptosporidium* sp. isolated from Northern Alaskan caribou (*Rangifer tarandus*). J Parasitol 88:213–216
Simpson VR (1992) Cryptosporidiosis in newborn red deer (*Cervus elaphus*). Vet Rec 130:116–118
Singh I, Carville A, Tzipori S (2011) Cryptosporidiosis in rhesus macaques challenged during acute and chronic phases of SIV infection. AIDS Res Hum Retroviruses 27:989–997
Siński E (1993) Cryptosporidiosis in Poland: clinical, epidemiologic and parasitologic aspects. Folia Parasitol (Praha) 40:297–300
Siński E, Bednarska M, Bajer A (1998) The role of wild rodents in ecology of cryptosporidiosis in Poland. Folia Parasitol (Praha) 45:173–174
Skerrett HE, Holland CV (2001) Asymptomatic shedding of *Cryptosporidium* oocysts by red deer hinds and calves. Vet Parasitol 94:239–246
Slavin D (1955) *Cryptosporidium meleagridis* (sp. nov.). J Comp Pathol 65:262–266

Sleeman JM, Meader LL, Mudakikwa AB, Foster JW, Patton S (2000) Gastrointestinal parasites of mountain gorillas (*Gorilla gorilla beringei*) in the Parc National des Volcans, Rwanda. J Zoo Wildl Med 31:322–328

Smith HV, Brown J, Coulson JC, Morris GP, Girdwood RW (1993) Occurrence of oocysts of *Cryptosporidium* sp. in *Larus* spp. gulls. Epidemiol Infect 110:135–143

Smith RP, Chalmers RM, Elwin K, Clifton-Hadley FA, Mueller-Doblies D, Watkins J et al (2009) Investigation of the role of companion animals in the zoonotic transmission of cryptosporidiosis. Zoonoses Public Health 56:24–33

Snyder DE (1988) Indirect immunofluorescent detection of oocysts of *Cryptosporidium parvum* in the feces of naturally infected raccoons (*Procyon lotor*). J Parasitol 74:1050–1052

Soriano SV, Pierangeli NB, Roccia I, Bergagna HF, Lazzarini LE, Celescinco A et al (2010) A wide diversity of zoonotic intestinal parasites infects urban and rural dogs in Neuquen, Patagonia, Argentina. Vet Parasitol 167:81–85

Sotiriadou I, Pantchev N, Gassmann D, Karanis P (2013) Molecular identification of *Giardia* and *Cryptosporidium* from dogs and cats. Parasite 20:8

Spain CV, Scarlett JM, Wade SE, McDonough P (2001) Prevalence of enteric zoonotic agents in cats less than 1 year old in Central New York State. J Vet Intern Med 15:33–38

Sreter T, Varga I (2000) Cryptosporidiosis in birds–a review. Vet Parasitol 87:261–279

Stronen AV, Sallows T, Forbes GJ, Wagner B, Paquet PC (2011) Diseases and parasites in wolves of the riding mountain National Park region, Manitoba, Canada. J Wildl Dis 47:222–227

Sturdee AP, Chalmers RM, Bull SA (1999) Detection of *Cryptosporidium* oocysts in wild mammals of mainland Britain. Vet Parasitol 80:273–280

Sulaiman IM, Morgan UM, Thompson RC, Lal AA, Xiao L (2000) Phylogenetic relationships of *Cryptosporidium* parasites based on the 70-kDa heat shock protein (HSP70) gene. Appl Environ Microbiol 66:2385–2391

Sundberg JP, Hill D, Ryan MJ (1982) Cryptosporidiosis in a gray squirrel. J Am Vet Med Assoc 181:1420–1422

Thomaz A, Meireles MV, Soares RM, Pena HF, Gennari SM (2007) Molecular identification of *Cryptosporidium* spp. from fecal samples of felines, canines and bovines in the state of Sao Paulo, Brazil. Vet Parasitol 150:291–296

Thompson RC, Colwell DD, Shury T, Appelbee AJ, Read C, Njiru Z et al (2009) The molecular epidemiology of *Cryptosporidium* and *Giardia* infections in coyotes from Alberta, Canada, and observations on some cohabiting parasites. Vet Parasitol 159:167–170

Tilley M, Upton SJ, Chrisp CE (1991) A comparative study on the biology of *Cryptosporidium* sp. from guinea pigs and *Cryptosporidium parvum* (Apicomplexa). Can J Microbiol 37:949–952

Torres J, Gracenea M, Gomez MS, Arrizabalaga A, Gonzalez-Moreno O (2000) The occurrence of *Cryptosporidium parvum* and *C. muris* in wild rodents and insectivores in Spain. Vet Parasitol 92:253–260

Traversa D (2010) Evidence for a new species of *Cryptosporidium* infecting tortoises: *Cryptosporidium ducismarci*. Parasit Vectors 3:21

Traversa D, Iorio R, Otranto D, Modrý D, Šlapeta J (2008) *Cryptosporidium* from tortoises: genetic characterisation, phylogeny and zoonotic implications. Mol Cell Probes 22:122–128

Trout JM, Santín M, Fayer R (2006) *Giardia* and *Cryptosporidium* species and genotypes in coyotes (*Canis latrans*). J Zoo Wildl Med 37:141–144

Tupler T, Levy JK, Sabshin SJ, Tucker SJ, Greiner EC, Leutenegger CM (2012) Enteropathogens identified in dogs entering a Florida animal shelter with normal feces or diarrhea. J Am Vet Med Assoc 241:338–343

Upton SJ, Current WL (1985) The species of *Cryptosporidium* (Apicomplexa: Cryptosporidiidae) infecting mammals. J Parasitol 71:625–629

Upton SJ, McAllister CT, Freed PS, Barnard SM (1989) *Cryptosporidium* spp. in wild and captive reptiles. J Wildl Dis 25:20–30

Van Winkle TJ (1985) Cryptosporidiosis in young artiodactyls. J Am Vet Med Assoc 187:1170–1172

van Zeeland YR, Schoemaker NJ, Kik MJ, van der Giessend JW (2008) Upper respiratory tract infection caused by *Cryptosporidium baileyi* in three mixed-bred falcons (*Falco rusticolus* x *Falco cherrug*). Avian Dis 52:357–363

van Zijll Langhout M, Reed P, Fox M (2010) Validation of multiple diagnostic techniques to detect *Cryptosporidium* sp. and *Giardia* sp. in free-ranging western lowland gorillas (*Gorilla gorilla gorilla*) and observations on the prevalence of these protozoan infections in two populations in Gabon. J Zoo Wildl Med 41:210–217

Vetterling JM, Jervis HR, Merrill TG, Sprinz H (1971) *Cryptosporidium wrairi* sp. n. from the guinea pig *Cavia porcellus*, with an emendation of the genus. J Protozool 18:243–247

Vítovec J, Hamadejová K, Landová L, Kváč M, Květoňová D, Sak B (2006) Prevalence and pathogenicity of *Cryptosporidium suis* in pre—and post-weaned pigs. J Vet Med B 53:239–243

Wang JS, Liew CT (1990) Prevalence of *Cryptosporidium* spp. in birds in Taiwan. Taiwan J Vet Med Anim Husb 56:45–57

Wang R, Zhang L, Ning C, Feng Y, Jian F, Xiao L et al (2007) Multilocus phylogenetic analysis of *Cryptosporidium andersoni* (Apicomplexa) isolated from a Bactrian camel (*Camelus bactrianus*) in China. Parasitol Res 102:915–920

Wang R, Wang J, Sun M, Dang H, Feng Y, Ning C et al (2008a) Molecular characterization of the *Cryptosporidium* cervine genotype from a sika deer (*Cervus nippon Temminck*) in Zhengzhou, China and literature review. Parasitol Res 103:865–869

Wang R, Zhang L, Feng Y, Ning C, Jian F, Xiao L et al (2008b) Molecular characterization of a new genotype of *Cryptosporidium* from American minks (*Mustela vison*) in China. Vet Parasitol 154:162–166

Wang A, Ruch-Gallie R, Scorza V, Lin P, Lappin MR (2012) Prevalence of *Giardi*a and *Cryptosporidium* species in dog park attending dogs compared to non-dog park attending dogs in one region of Colorado. Vet Parasitol 184:335–340

Warren KS, Swan RA, Morgan-Ryan UM, Friend JA, Elliot A (2003) *Cryptosporidium muris* infection in bilbies (*Macrotis lagotis*). Aust Vet J 81:739–741

Webster JP, Macdonald DW (1995) Cryptosporidiosis reservoir in wild brown rats (*Rattus norvegicus*) in the UK. Epidemiol Infect 115:207–209

Xiao L, Ryan UM (2004) Cryptosporidiosis: an update in molecular epidemiology. Curr Opin Infect Dis 17:483–490

Xiao L, Escalante L, Yang C, Sulaiman I, Escalante AA, Montali RJ et al (1999a) Phylogenetic analysis of *Cryptosporidium* parasites based on the small-subunit rRNA gene locus. Appl Environ Microbiol 65:1578–1583

Xiao L, Morgan UM, Limor J, Escalante A, Arrowood M, Shulaw W et al (1999b) Genetic diversity within *Cryptosporidium parvum* and related *Cryptosporidium* species. Appl Environ Microbiol 65:3386–3391

Xiao L, Limor JR, Sulaiman IM, Duncan RB, Lal AA (2000) Molecular characterization of a *Cryptosporidium* isolate from a black bear. J Parasitol 86:1166–1170

Xiao L, Sulaiman IM, Ryan UM, Zhou L, Atwill ER, Tischler ML et al (2002) Host adaptation and host-parasite co-evolution in *Cryptosporidium*: implications for taxonomy and public health. Int J Parasitol 32:1773–1785

Xiao L, Ryan UM, Graczyk TK, Limor J, Li L, Kombert M et al (2004) Genetic diversity of *Cryptosporidium* spp. in captive reptiles. Appl Environ Microbiol 70:891–899

Yakhchali M, Moradi T (2012) Prevalence of *Cryptosporidium*-like infection in one-humped camels (*Camelus dromedarius*) of Northwestern Iran. Parasite 19:71–75

Yamini B, Raju NR (1986) Gastroenteritis associated with a Cryptosporidium sp. in a chinchilla. J Am Vet Med Assoc 189:1158–1159

Yamura H, Shirasaka R, Asahi H, Koyama T, Motoki M, Ito H (1990) Prevalence of *Cryptosporidium* infection among house rats, *Rattus rattus* and *R. norvegicus*, in Tokyo, Japan and experimental cryptosporidiosis in roof rats. Jpn J Parasitol 39:439–444

Yanai T, Chalifoux LV, Mansfield KG, Lackner AA, Simon MA (2000) Pulmonary cryptosporidiosis in simian immunodeficiency virus-infected rhesus macaques. Vet Pathol 37:472–475

Yang W, Chen P, Villegas EN, Landy RB, Kanetsky C, Cama V et al (2008) *Cryptosporidium* source tracking in the Potomac River watershed. Appl Environ Microbiol 74:6495–6504

Yang R, Fenwick S, Potter A, Ng J, Ryan U (2011) Identification of novel *Cryptosporidium* genotypes in kangaroos from Western Australia. Vet Parasitol 179:22–27

Ye J, Xiao L, Ma J, Guo M, Liu L, Feng Y (2012) Anthroponotic enteric parasites in monkeys in public park, China. Emerg Infect Dis 18:1640–1643

Yoshiuchi R, Matsubayashi M, Kimata I, Furuya M, Tani H, Sasai K (2010) Survey and molecular characterization of *Cryptosporidium* and *Giardia* spp. in owned companion animal, dogs and cats, in Japan. Vet Parasitol 174:313–316

Zanette RA, da Silva AS, Lunardi F, Santurio JM, Monteiro SG (2008) Occurrence of gastrointestinal protozoa in *Didelphis albiventris* (opossum) in the central region of Rio Grande do Sul state. Parasitol Int 57:217–218

Zhang W, Shen Y, Wang R, Liu A, Ling H, Li Y et al (2012) *Cryptosporidium cuniculus* and *Giardia duodenalis* in rabbits: genetic diversity and possible zoonotic transmission. PLoS One 7:e31262

Zhou L, Fayer R, Trout JM, Ryan UM, Schaefer FW III, Xiao L (2004a) Genotypes of *Cryptosporidium* species infecting fur-bearing mammals differ from those of species infecting humans. Appl Environ Microbiol 70:7574–7577

Zhou L, Kassa H, Tischler ML, Xiao L (2004b) Host-adapted *Cryptosporidium* spp. in Canada geese (*Branta canadensis*). Appl Environ Microbiol 70:4211–4215

Ziegler PE, Wade SE, Schaaf SL, Chang YF, Mohammed HO (2007a) *Cryptosporidium* spp. from small mammals in the New York City watershed. J Wildl Dis 43:586–596

Ziegler PE, Wade SE, Schaaf SL, Stern DA, Nadareski CA, Mohammed HO (2007b) Prevalence of *Cryptosporidium* species in wildlife populations within a watershed landscape in Southeastern New York State. Vet Parasitol 147:176–184

Part II
Molecular Biology

Chapter 6
Cryptosporidium: Current State of Genomics and Systems Biological Research

Aaron R. Jex and Robin B. Gasser

Abstract Recent years have seen an unprecedented expansion in our knowledge of *Cryptosporidium* and cryptosporidiosis through the emergence of the genomics and systems biological age. High-quality draft genome sequences are now published for *C. parvum* and *C. hominis*, and the draft assembly of *C. muris* has been made publicly available. These genome sequences reveal a highly stream-lined parasite with limited metabolic and biosynthetic pathways and a heavy reliance on the host-cell. Bottlenecks in these pathways may be exploited for new drugs, which remain stubbornly illusive for these parasites. As more genomic information becomes available, fundamental research into gene regulation, genomic evolution and genome-wide variation becomes possible. This research will provide new insights into the transmission dynamics of these parasites and markers associated with host-specificity, virulence and pathogenicity, and will allow the identification of novel loci for use as molecular-diagnostic markers and genes under heavy immunoselection, potentially providing a basis for vaccine development. With the accelerating reduction in costs associated with 'omic' research, improved accessibility to analytical tools and in vitro culture of *Cryptospordium*, this field has tremendous potential to shape our understanding of their biology in the coming years.

A.R. Jex (✉)
The University of Melbourne, Werribee, Victoria, Australia
e-mail: ajex@unimelb.edu.au

R.B. Gasser
The University of Melbourne, Parkville, Victoria 3010, Australia
e-mail: robinbg@unimelb.edu.au

S.M. Cacciò and G. Widmer (eds.), *Cryptosporidium: parasite and disease*,
DOI 10.1007/978-3-7091-1562-6_6,

6.1 Genomics of *Cryptosporidium*

Research of *Cryptosporidium* in the genomics age is landmarked by the publication of the complete genome sequence of *C. parvum* Iowa strain (Abrahamsen et al. 2004). This genome was sequenced to ~13-fold coverage using a shotgun Sanger sequencing approach (Table 6.1). Briefly, this process involved isolating total genomic DNA from purified *C. parvum* oocysts, randomly shearing the DNA (2–5 kb) and constructing plasmid libraries, from which 120,000 clones were sequenced. In addition, to facilitate scaffolding and the resolution of repeat regions, the authors sequenced the genome to ~0.5-fold depth using data generated from large-insert (~15 kb) λ-phage libraries constructed from genomic DNA of the same Iowa isolate. The assembly of the sequence data from both libraries was guided by a physical map, which had been generated previously for this *C. parvum* isolate using ~200 oligonucleotide primers to screen haploid amounts (i.e., a "HAPPY" map) of *C. parvum* DNA by the polymerase chain reaction (PCR) for marker regions spaced ~50 kp apart (Piper et al. 1998; Abrahamsen et al. 2004) and a draft sequence for chromosome 6 (Bankier et al. 2003). The combination of the long sequence reads, large-insert libraries and the HAPPY map, combined with selective PCR-based sequencing for the purpose of gap closure, allowed the assembly of the *C. parvum* genome into its eight chromosomes, leaving five physical gaps. The final assembly revealed an AT-rich (70 %) and highly compact genome of ~9.1 Mb, which is relatively small in comparison with related apicomplexans, such as *Eimeria tenella* (~60 Mb) (Shirley 1994, 2000) and *Plasmodium falciparum* (~23 Mb) (Gardner et al. 2002a) and, unlike many apicomplexans, includes no mitochondrial or apicoplast genomes.

The annotation of the *C. parvum* genome was achieved primarily through the prediction of open reading frames (ORFs) of at least 67 amino acids in length from the assembled genome sequence. Although gene prediction methods based on the use of hidden Markov and machine learning models for the analysis of expressed sequence tag (EST) data, and, more recently, deep RNA-Sequencing data (Wang et al. 2009) are considered standard approaches for current genomic sequencing projects (Jex et al. 2011a; Young et al. 2012), extensive EST data are not yet available for *Cryptosporidium* species and are challenging to generate from the endogenous developmental stages. Noting this, the *C. parvum* genome appears to have few intronic sequences and, thus, the straight-forward prediction of ORFs was sufficient to allow the identification of 3,952 (3,807 protein-coding) genes in *C. parvum*. These genes were fewer in number than predicted to be encoded by the genomes of related apicomplexans, such species of *Plasmodium* (~5,300 genes) (Gardner et al. 2002b), *Toxoplasma* (~7,700) (Radke et al. 2005) and *Theileria* (~4,000) (Gardner et al. 2005), which, coupled to a lack of introns and limited non-coding sequence (~25 % of the entire genome), explains the smaller size of the *C. parvum* genome (see also Bankier et al. 2003).

Table 6.1 Comparison of characteristics and metrics of each *Cryptosporidium* genome sequenced and available in *CryptoDB* (www.cryptodb.org)

Feature	*C. parvum* Iowa II	*C. hominis* TU502	*C. parvum* TU114	*C. muris* RN66
Project accession	AAEE01000000	AAEL000000	not available	PRJNA19553
Assembly size (Mb)	9.1	8.7	not available	9.2
Assembly metrics (Contigs: N50)	8: 1.10 Mb	1,422: 14.5 kb	not available	45: 0.72 Mb
Number of genes	3,952	3,994	~3,952	3,986
GC richness	30.3	31.7	30.3	27.8
% identity (to *C. parvum* Iowa II)	–	~97	~99.8	not available
Human infective	Yes	Yes	Yes	No
Human specific	No	Yes	Yes	No

The completion of the *C. parvum* genome was rapidly followed by the publication of the genome of *C. hominis* TU502 strain (Xu et al. 2004). As for *C. parvum*, the *C. hominis* genome was sequenced using a shotgun Sanger method to a depth of ~12-fold from a plasmid insert library. Unlike for *C. parvum*, a physical map was not available for *C. hominis*, although the authors constructed large (~7–8-fold coverage), bacterial artificial chromosome (BAC) libraries for this species and used the *C. parvum* HAPPY map to guide assembly. Using this approach, the *C. hominis* genome is currently represented by ~1,400 contigs and ~300 scaffolds, with <250 gaps. The gene content for the *C. hominis* genome appears to be similar to that of *C. parvum*, with the genome predicted to encode 3,994 genes. Xu et al. (2004) estimated that 5–20 % of the *C. hominis* genes represent introns, with the higher end of this range differing markedly from that predicted for *C. parvum*, emphasizing the need for increased transcriptomic research of these *Cryptosporidium* species to validate their genome annotation.

An alignment of the *C. parvum* and *C. hominis* genomes showed the extent of similarity between the two species (Xu et al. 2004). Upon pairwise comparison, Xu et al. (2004) found that the genomes diverged in sequence by no more than 3–5 %, with no evidence of large insertion, deletion or recombination events. Their respective gene sets appear to be almost identical, with genes of *C. parvum* found to have homologues in *C. hominis*, and the <50 genes predicted for *C. hominis* and apparently absent from *C. parvum* relating to known gaps in the genome of *C. hominis*, suggesting that both parasites have the same or similar biological pathways, with the differences in their phenotypic behaviour (e.g., host specificity, infectivity and symptomology; see Chaps. 4 and 9) possibly relating to transcriptomic differences and/or mutations (Xu et al. 2004).

6.2 The Inferred *Cryptosporidium* Proteome

The compact genomes of *C. parvum* and *C. hominis* seem to reflect the biology of these parasites and have a reduced complement of genes associated with both anabolic and catabolic phases of metabolism. Energy generation (i.e., ATP) appears to be dependent exclusively upon the degradation of simple sugars via anaerobic glycolysis, with no evidence of a mitochondrial genome or many of the nuclear genes associated with the Krebs cycle or electron transport chain (Abrahamsen et al. 2004; Xu et al. 2004). In addition, there is no evidence for the presence of genes associated with energy production via the digestion of fatty acids/proteins or the urea and nitrogen cycles (i.e., for amino acid synthesis) or the shikimate pathway (Xu et al. 2004). The 'minimalistic' metabolome of these organisms substantially reduces their ability to synthesize essential building blocks (e.g., some nucleotides and amino acids), likely leading to a major dependency on the host cell. Supporting this hypothesis is the identification of a variety of amino acid transporter genes, which are proposed to be involved in amino acid salvaging, and a variety of enzymes necessary for the conversion of amino acids and nucleotides (e.g., pyrimidines to purines and purines to pyrimidines) (Abrahamsen et al. 2004; Striepen and Kissinger 2004; Xu et al. 2004).

In addition to the core metabolome predicted from the genomes of *C. parvum* and *C. hominis*, these parasites encode a variety of proteins involved in a wide array of biological processes. Not surprisingly, many of the fundamental biological processes of the cell (e.g., DNA replication, transcription and translation) appear to be linked to standard eukaryotic pathways. A notable exception to this, as highlighted by Abrahamsen et al. (2004) for *C. parvum*, but also consistent with *C. hominis* (see Xu et al. 2004), is a diminished spliceosomal pathway with a reduced U6 spliceosomal complex, and a lack of identifiable homologues to DICER and argonaute, suggesting an altered role for small interferring RNAs in gene regulation in these microorganisms.

Of major interest among the structural proteins encoded by the *Cryptosporidium* genomes are those involved in the structure and function of the apical complex and/or adorning the cell-surface. It appears that most of the known proteins associated with the apical complex in species of *Plasmodium* have homologues in *Cryptosporidium*, suggesting that the overall molecular biology of this crucial organelle is conserved across many, if not all apicomplexans. Among the most studied groups of cell-surface proteins encoded by an apicomplexan genome are the variable surface proteins, which provide *Plasmodium* with its infamous capacity to evade host immune responses (Reeder and Brown 1996). Abrahamsen et al. (2004) explored the *C. parvum* genome in-depth for these sequences but did not find them. These authors did note several groups of molecules with strong SER- and THR-rich glycosylation signals and transmembrane domains, consistent with a diverse panel of cell-surface glycoproteins, which appear to have undergone an expansion via duplication since the evolutionary emergence of *Cryptosporidium*. Many of these molecules were actively transcribed during in vitro development (Abrahamsen et al. 2004), and it would seem likely that at least some of them play important roles in cell-cell adhesion and other mechanisms associated with host invasion and infection, as has been shown to be the case for the 15 and 40 kilodalton (kDa) glycoprotein (G15 and GP40), both of which are encoded by the 60 kDa glycoprotein (*gp60*) gene (Strong and Nelson 2000).

6.3 The Druggable Genome of *Cryptosporidium*

Despite substantial research into the molecular biology of *Cryptosporidium* in the past decades, there are still no highly effective drugs or vaccines for intervention, and few examples of compounds that show or appear to show efficacy against members of the genus (Jex et al. 2010). A major goal of genomic research of these parasites is to identify novel drug targets to address the serious lack of chemotherapeutic treatment options. In sequencing the *C. parvum* genome, Abrahamsen et al. (2004) highlighted that the 'minimalist' metabolic/biosynthetic pathways of this microorganism present numerous, novel avenues for drug target and drug discovery. Most notable among these pathways, given the limited biosynthetic capabilites of *Cryptosporidium* species, is the observation that the parasites have a single enzyme, 5′ inosine monophosphate dehydrogenase (IMPDH), for the conversion of adenosine monophosphate to guanosine monophosphate. Mycophenolic acid is a known inhibitor of IMPDH (Senda et al. 1995), but attempts to inhibit *Cryptosporidium*-encoded IMPDH using this compound class have not been successful. The bacterial origin of the enzyme in these parasites was highlighted as a possible confounding issue (Umejiego et al. 2004). However, more recently, using in silico docking experiments and high-throughput screening, several compounds (Compounds "A", "F", "G" and "H": Umejiego et al. 2008) with specific efficacy against *Cryptosporidium* in cell-culture superior to the performance of paramomycin were identified. Abrahamsen et al. (2004) also predicted the potential of a wide variety of transporter proteins as targets, such as those used by the parasite to salvage nucleotides and essential amino acids from the host, as well as a mitochondrial-associated alternative oxidase, which catalyzes the reduction of oxygen to water without the production of free radicals and/or oxidizing reagents (e.g., superoxide/H_2O_2). These authors (Abrahamsen et al. 2004) highlighted the potential of salicylhydroxamic acid to inhibit this latter enzyme.

In a recent review (Jex et al. 2011b), we conducted an in-depth exploration of the *Cryptosporidium* genome for potential drug targets. We found, for example, that 17 of 31 high priority 'druggable' molecules proposed for *P. falciparum* (see Crowther et al. 2010) had close homologues in *Cryptosporidium* species. Given the recent focus on repurposing anti-malarial drugs via the Medicines for Malaria Venture (www.mmv.org), there are clear prospects for screening these "malaria-box" compounds for efficacy against *Cryptosporidium*. In addition to assessing these homologues, we conducted a global prediction of druggability on the complete gene set of *C. parvum and C. hominis* (see Jex et al. 2011b) by assessing their homology to essential genes in *Saccharomyces cerevisiae* (yeast). Approximately 33 % of the annotated *Cryptosporidium* genes have an homologue in yeast, and >500 of them are associated with a lethal phenotype; 313 of these 'essential' genes have homologous sequences in the BRENDA/CHEMBL databases (Schomburg et al. 2002; Gaulton et al. 2012). Most notable among these are 60 protein-synthesizing GTPases, for which >50 inhibitors (primarily stemming from anti-cancer research) are contained within these database, including a variety of aminoglycosides, such as dihydrostreptomycin, hygromycin, neomycin and viomycin (Jex et al. 2011b). Interestingly, one of the few available compounds with some efficacy against *Cryptosporidium* is an aminoglycoside (i.e., paromomycin) (Rossignol 2010), although it is not known to have an inhibitory effect on GTPases. Aminoglycides (often named with the suffix *-mycin*) have been widely used in humans as topical antibiotics, but are known to be toxic to mammalian cells (Martinez-Salgado et al. 2007; Rizzi and Hirose 2007; Guthrie 2008). Recent research shows that this toxicity can be reduced through careful administration (Murakami et al. 2008; Pannu and Nadim 2008) and/or modified formulations (Tugcu et al. 2006; Jeyanthi and Subramanian 2009; Nagai and Takano 2010), suggesting that the aminoglycosides might be re-considered as anti-cryptosporidial compounds.

6.4 Additional *Cryptosporidium* Genomes

An interesting aspect of the biology of *Cryptosporidium*, and of critical importance in the control of cryptosporidiosis, is the variable host-specificity of the species of this genus. For example, epidemiological studies of these parasites in the past decades, coupled to molecular systematics in more recent times, have provide a wealth of data to indicate that human infections are largely caused by *C. parvum* and *C. hominis*, with most other species of *Cryptosporidium* rarely being detected in otherwise healthy humans and displaying varying levels of adaptation to other hosts (Xiao et al. 2004). Of the *Cryptosporidium* species known to infect humans, *C. hominis* is thought to be human-specific, whereas *C. parvum* is transmitted by either anthroponotic or zoonotic routes (Xiao et al. 2004). However, *C. hominis* has been reported in naturally occurring infections in animals, such as cattle in Scotland (Smith et al. 2005) and, has been detected (although infections were not confirmed) in the faeces of cattle in New Zealand (Abeywardena et al. 2012) and a dugong in Australia (Morgan et al. 2000). In addition, although cattle, in particular, have been identified as a major source of *C. parvum* for transmission to humans, population genetic and molecular epidemiological data suggest that human-specific lineages might yet exist within this well-studied species (Mallon et al. 2003; Jex and Gasser 2010).

Several aspects might explain the variable host-specificity or adaption of these species: (a) the differing gastrointestinal conditions (e.g., pH, and body temperature) between and among vertebrate groups known to be infected by at least one species

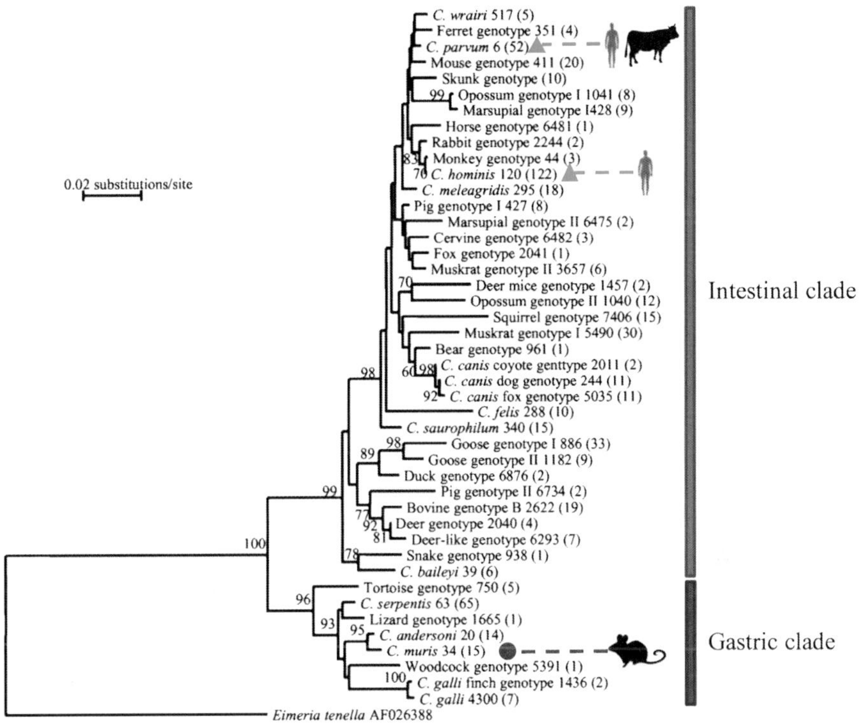

Fig. 6.1 Current molecular phylogeny (based on 18S ribosomal DNA sequence) of *Cryptosporidium*, highlighting the species for which complete genomic data are presently available (phylogram modified from Xiao et al. 2004). Genomes that are both sequenced and published are represented by *green triangles*. Genomes that have been sequenced but not yet published in the peer-reviewed literature are represented by a *red circle*. Human/animal images represent that common hosts for each species for which complete genomic data are presently available

of *Cryptosporidium*; (b) differences between and among the *Cryptosporidium* species or their hosts in the cell-surface molecules involved in host-cell binding and/or invasion by the activated sporozoite; and (c) the differing innate or acquired immunological responses of the host against *Cryptosporidium*. Understanding whether these factors influence the host range of different species will be of critical importance in the effective control of cryptosporidiosis.

In pursuit of this understanding, the recent sequencing of the genomes of *C. parvum* TU114 (Widmer et al. 2012) and *C. muris* RN66 (NCBI bioproject: PRJNA19553) genomes represents a major advance, with the former appearing to be a human-adapted/specific lineage of *C. parvum* (classified subgenotype IIcA5G3R2 at gp60 (Jex and Gasser 2010)) and the latter being representative of a gastric-colonizing species not known to infect humans (Fig. 6.1: Xiao et al. 2004).

To faciltate comparisons among *C. parvum* TU114 (IIcA5G3R2), *C. parvum* IOWA (IIaA15G2R1) and *C. hominis* TU502 (IaA21), two *C. parvum* TU114 samples were selected and their genomes sequenced using Illumina technology to a mean depth of ~20–40-fold, revealing, upon alignment to the *C. parvum* IOWA sequence, 12,748 single nucleotide polymorphisms (SNPs) (i.e. ~1.4 per kilobase) common to both TU114 isolates. The authors (Widmer et al. 2012) hypothesized that the human-infective phenotype of both *C. parvum* TU114 and *C. hominis* was a likely consequence of the same positive selective force, and explored their data for genomic regions in the TU114 samples that had diverged from *C. parvum* IOWA and converged with *C. hominis*. These analyses identified several genomic regions of *C. parvum* TU114 which were significantly more similar to *C. hominis* than to *C. parvum* IOWA, despite the two *C. parvum* genome sequences being tenfold more similar to each other than either was to *C. hominis* overall. Three exonic regions in the genome, corresponding to the genes cgd1_650, cgd3_3370 and cgd6_5260, contained regions that were significantly more similar to orthologues in *C. hominis* than those in the IOWA strain of *C. parvum* (Widmer et al. 2012). This pattern was shown to be consistent for five other *C. parvum* IIc isolates in conventional PCR-based sequencing study of these gene regions (Widmer et al. 2012). Although their specific function is not known, gene regions cgd1_650, cgd3_3370 are predicted to encode signal peptide domains but no transmembrane domains, suggesting that their products are secreted by the parasite. At present, no other information on the function of these genes is available in the literature. Clearly, this aspect warrants detailed exploration. gene cgd6_5260 appears to encode an ABC transporter (ATpase with 2 AAA domains). AAA + ATPases have a wide range of cellular functions, including membrane fusion, proteolysis and DNA replication (IPR003593). In the current context, membrane fusion is noteworthy, as it might suggest cell-cell interactions between the parasite and the host; however, cg6_5260 also appears to encode a nucleotide-binding domain, suggesting that its role more likely relates to DNA replication, which is obviously more difficult to reconcile with host-specificity.

The sequencing of the *C. muris* genome will also provide new insights into aspects of *Cryptosporidium* biology and host-parasite relationships. *C. muris* is infective primarily to rodents and is located in the stomach of the host, whereas the other sequenced *Cryptosporidium* species are pathogens of the small intestine. Currently, although the *C. muris* genome has not been published in the peer-reviewed literature, the complete, annotated genome assembly is available through the GenBank database (accessible via www.ncbi.nlm.nih.gov) and CryptoDB (www.cryptodb.org). The current assembly is ~9.2 Mb and consists of 84 scaffolded sequences, with 50 % of the genome (i.e., N50) represented by contigs of at least 720 kb. The *C. muris* genome is predicted to encode ~3,900 protein-coding genes, with 3,511 and 3,359 of them predicted to have identifiable orthologues in *C. parvum* IOWA and *C. hominis*, respectively, and 3,167 genes being shared among all three species (Fig. 6.2: www.CryptoDB.org). Characterizing the functional similarities and differences associated with these overlapping gene-sets as well as further comparative analyses of the *C. muris*, *C. parvum* and *C. hominis*

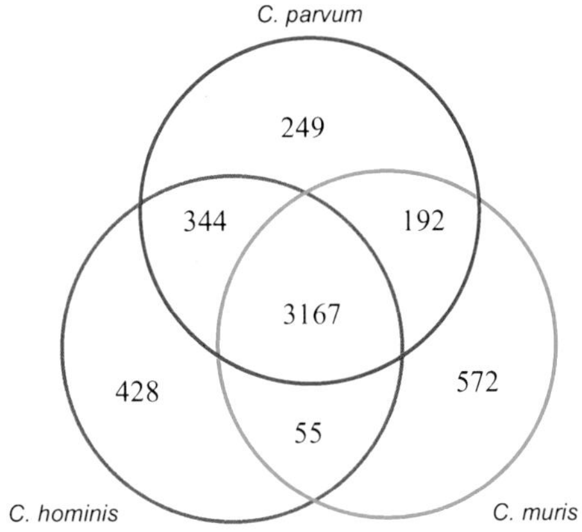

Fig. 6.2 Venn diagram of homology between and among the coding gene sets of *Cryptosporidium parvum*, *C. hominis* and *C. muris* based on available genomic data (Source: www.cryptodb.org)

genomes both at the structural level (e.g., overall gene composition and organisation, synteny and the identification of large insertion/deletion or translocation events) and the characterization of genetic variation (i.e., SNPs) across the genome should be considered a high priority and will likely yield significant insights into the genetics and evolution of members of the genus. For example, an alignment of the genomes of *C. muris*, *C. parvum* IOWA, *C. parvum* TU114 and *C. hominis* provides scope to assess regions of the genome under neutral versus purifying selective pressure at a much deeper level than has been possible previously for this genus. On a genome-by-genome comparison, this information might be useful for assessing phenotypic differences, such as host-specificity and site-selection, among the species. Considering all four genomes, these analyses should provide insight into the genomic 'dark matter' of each species. For example, at present ~75 % of the *Cryptosporidium* genome is annotated as protein-coding. However, as no RNA-Sequencing data are available for *Cryptosporidium* species, there is no direct support for many of these predicted genes. Given that nearly two thirds of the currently annotated *Cryptosporidium* genes represent 'hypothetical proteins', for which no functional data are available, more support for the gene-sets is needed. In addition, very little is known about functional, non-coding regions in *Cryptosporidium* genomes sequenced to date. Studies of the accumulation of mutations in a variety of eukayrotic species (e.g., *Homo sapiens*, *Caenorhabditis elegans* and *Saccharomyces cerevisiae*) have shown that at least 4–8 % of the non-exonic regions of the genomes of these organisms are under purifying selection (Siepel et al. 2005; Gerstein et al. 2010; Ponting and Hardison 2011). In these eukaryotes, many of the non-coding, but evolutionarily constrained regions have important regulatory functions, including non-coding RNAs, transcriptional signal sites, and promotor and/or enhancer regions. Defining regions in *Cryptosporidium* genomes undergoing purifying selection will assist in identifying such regulatory elements are present in these parasites. Indeed, in sequencing the *C. parvum* TU114 genome, some authors (Widmer et al. 2012) found evidence to suggest that at least some non-coding regions of the genome appears to be under purifying selection. Including the genome of the distantly related *C. muris* in these analyses might improve comparisons and provide greater confidence in identifying regions likely to be enriched for regulatory elements (Siepel et al. 2005). In a recent study, Mullapudi et al. (2007) used pattern recognition software to identify conserved motifs in the regions up-stream of the annotated *C. parvum* genes. They proposed these motifs as putative *cis*-regulatory elements (CREs) and found evidence, based on real-time PCR, for the co-transcription of numerous genes sharing the same 'CRE' conserved motif. This research represents one of the first explorations of gene regulation in *Cryptosporidium*. A subsequent study using a similar approach has also predicted conserved CREs associated with the co-transcription of heat shock proteins (HSPs) during *C. parvum* excystation (Cohn et al. 2010). Given the compact nature of *Cryptosporidium* genomes, their 'regulome' may represent a close-approximation of the minimum regulatory network required for eukaryotic cells to function, and might have far reaching biological implications for understanding the eukaryotic cell.

6.5 Systems Biology Research of *Cryptosporidium*

Referred to initially as 'next-generation sequencing' and, more recently, as 'high-throughput sequencing' or 'massively parallel sequencing', advanced technologies, such as the 454 Roche (Margulies et al. 2005) and Illumina (Bentley et al. 2008), have vastly reduced the cost and time associated with conducting genomic research. New developments in large-scale and benchtop sequencing are expected to further accelerate the reduction in costs of such research and drive innovation across all disciplines of biology and biomedicine (Niedringhaus et al. 2011).

6.5.1 Deep Exploration of the Systems Biology of C. parvum *and* C. hominis

The innovation associated with the expansion of genomic, transcriptomic and bioinformatic technologies will have profound implications for *Cryptosporidium* research. Initial applications of these technologies will likely include refinements of existing genome sequences for *C. parvum* and *C. hominis*. For example, although the *C. parvum* genome has been assembled, the *C. hominis* genome is presently represented by several hundred scaffolds of >1,400 individual sequence contigs. Finishing these two genomes will allow the gene sets to be confirmed and genetic and genomic differences between these two important, human-infective parasites to be re-assessed, and will also provide a basis to further explore regulatory pathways associated with non-coding RNAs and other regions of intergenic DNA (Siepel et al. 2005; Gerstein et al. 2010; Ponting and Hardison 2011). Furthermore, although the alignment of genome sequences currently provides little indication of significant structural differences between *C. parvum* and *C. hominis*, the absence of a physcial map for the latter species prevents a thorough comparison. Rearrangements may have relevance, not only in comprehensively determining the differences between the species and their coding gene sets, but also in establishing rates of inter- and intra-specific recombination, which might have implications for the emergence and spread of drug resistance, following the future introduction and administration of effective and safe anti-cryptosporidial compounds on a large scale. Although the in vitro cultivation of *Cryptosporidium* species is presently challenging, recent advances in culture methodologies (reviewed by Jex et al. 2010) might pave the way for the the construction of genetic crossing experiments, as a means of assessing genotypic causes of phenotypic traits, such as virulence, host-specificity and infectivity.

In addition to characterizing and finishing the genome sequence of these species, there is a clear need to support gene annotations. Specifically, most of the protein-coding genes currently predicted for *Cryptosporidium* species are not supported by transcriptomic or expression data. Previously, this lack of information has related to the challenges associated with isolating enough parasite material for molecular study. However, the advent of high-throughput sequencing technologies provides opportunities to enhance our understanding of the genomics and genetics of *Cryptosporidium*. Traditional obstacles to the transcriptomic research of species of *Cryptosporidium* seem to be related primarily to limitations associated with the specific isolation of RNA from the endogenous stages of the life cycle and the relative transcriptional 'dormancy' of the oocyst stage. In vitro culturing of *Cryptosporidium* has, until recent times, been a major obstacle, and, despite advances in these technologies (Hijjawi et al. 2010), the transcriptome might be quite different from that be the host. Laser capture and microdissection technologies, in the hands of skilled microscopists, might provide a means to overcome the difficulty in exploring the transcriptome of specific endogenous stages when coupled with single-cell RNA-Sequencing technologies (Tang et al. 2010). Such information could pave the way for testing of current gene sets for *Cryptosporidium*, the assessment of intron-exon boundaries, definition of promoter and transcription start sites, and the characterization of stage-specific changes in gene transcription, identification of alternative splicing of transcripts associated with multi-exon genes or multi-gene operons (Wang et al. 2009). Clearly, such new developments will have major implications for understanding *Cryptosporidium* molecular biology by allowing the exploraton of key genes and their roles in major life-cycle phases (e.g., infection, reproduction or cellular destruction). In addition, characterizing constitutively transcribed and/or stage-specific genes should also assist significantly toward prioritizing novel drug candidates.

Further expansions of transcriptomic research of *Cryptosporidium* may seek to explore regulatory (i.e., non-coding) RNAs. Although there is little evidence that the genomes of *Cryptosporidium* species have genes encoding argonaute and DICER (Abrahamsen et al. 2004; Xu et al. 2004), commonly involved in RNA silencing pathways, other non-coding RNAs, such as long non-coding, small nuclear and/or *piwi*-RNAs, might play key regulatory roles that remain unexplored in these species to date.

In addition to characterizing genes and differential (including developmentally regulated) transcription, increased insights into the proteome of *Cryptosporidium* is highly desirable, particularly considering that approximately two-thirds of all known genes of *Cryptosporidium* are considered to encode 'hypothetical proteins' (Abrahamsen et al. 2004; Xu et al. 2004). To date, requirements for sufficient amounts of parasite material and the costs associated with proteomic analyses have limited, detailed expression profiling of *Cryptosporidium* species. However, although in vitro cultivation might assist, there have been massive technological enhancements in proteomic technologies, such that low cost and highly sensitive analytical analyses are now feasible (Cox and Mann 2011).

Culturing methods also have major implications for studying other areas of systems biology, particularly in reconstructing the *Cryptosporidium* metabolome. Genomic sequence data indicate that metabolomes of different species of *Cryptosporidium* are 'minimalistic', with a heavy reliance on simple sugars, anaerobic glycolysis and the harvesting of essential building blocks from the host through a range of small molecule salvage pathways. However, as indicated, ~66 % of the protein-coding genes currently annotated for *C. parvum* and *C. hominis* encode hypothetical proteins. Given the large evolutionary distances between *Cryptosporidium* and eukaryotic model organisms for which functional genomic data are available, it is probable that some of these genes may relate to known metabolic pathways, but are too divergent in sequence to be identified using currently available bioinformatic tools. Considering the extent with which the genome of these parasites appears to have been shaped through the horizontal acquistion of genes from bacteria (Abrahamsen et al. 2004; Xu et al. 2004), one cannot discount the possiblity that these *Cryptosporidium* species have novel, highly divergent (i.e., relative to other eukaryotes) biosynthetic pathways to account for their metabolic demands that cannot presently be identified among the 'orphan' proteins encoded in the genome. By coupling advanced in vitro cultivation methods for *Cryptosporidium* to new advances in metabolomic mapping (i.e., highly sensitive gas chromatography - mass spectrophotometry, nuclear magnetic resonance and matrix-assisted laser desorption/ionisation—time of flight (MALDI-TOF) mass spectrophotometric technologies), it will soon be possible to directly explore metabolism in these parasites, as has been done in recent years for species of *Trypanosoma* and *Leishmania* (Creek et al. 2012). Such applications have significant potential to improve our knowledge of *Cryptosporidium* species and assist in finding ways of disrupting metabolic pathways as a new means of intervention. Even without the use of culturing methods, recent advances in molecular and biochemical technologies (Fritzsch et al. 2012) should allow detailed investigations of gene function in *Cryptosporidium* and are likely to have substantial impact in the coming years.

6.5.2 *Exploring the* Cryptosporidium variome

In addition to harnessing new molecular technologies to improve the annotation and our understanding of the biology of *Cryptosporidium*, such technologies, particularly DNA sequencing technologies, may be applied to assess genomic variation within and among species. For example, at present, a major molecular marker for the exploration of genetic variation within, between and among *Cryptosporidium* populations is the *gp60* gene (Jex and Gasser 2010). This highly polymorphic glycoprotein allows the clustering of *Cryptosporidium* populations into genotypes, and microsatellite variability in the form of a serine/threonine rich glycosylation region, allowing a further subdivision of *Cryptosporidium* genotypes into subgenotypes (or subtypes) (Strong et al. 2000). Recently, we reviewed the global richness (= the number of sequence types) and diversity (= the relative proportional distribution and/or domination of these sequence types) of all *gp60* sequence types reported for *C. parvum* and *C. hominis* and the frequency with which they occur among the hosts for which such data are available (Jex and Gasser 2010). This review showed that, of the ~150 subgenotypes (representing 11 *C. parvum* and 6 *C. hominis* genotypes) for which data were available at that time (i.e., 2010), the majority (85 %) of them had been reported from humans. Notably, despite this high richness of subgenotypes, six specific subgenotypes (namely *C. hominis* IbA10G2, *C. homins* IbA9G3, *C. hominis* IeA11G3T3, *C. parvum* IIcA5G3R2, *C. parvum* IIaA15G2R1 and *C. parvum* IIaA18G3R1) were associated with the ~70 % of infections in humans from >40 countries (for which information was available) (Jex and Gasser 2010). Indeed, evidence demonstrates clearly *C. hominis* subgenotype IbA10G2 to be the cause of the most infamous outbreak of cryptosporidiosis in Milwaukee in 1992 (Zhou et al. 2003).

This finding might have profound implications for the treatment and control of cryptosporidiosis, but the extent to which *gp60* sequence is representative of population-level variation in *Cryptosporidium* is unclear. A recent publication by Tanriverdi et al. (2008) explored population structuring in *C. parvum* and *C. hominis* using numerous isolates from a broad geographical range, assessed at several genetic loci, and found evidence of some population structuring but also substantial 'genetic mixing'. This information would appear to suggest that the dominance of a few *gp60* subgenotypes infecting humans might relate to functional constraints on the genes themselves. Given the role of protein GP15/40 in binding to the host cell membrane, some forms of this molecule (expressed by subgenotypes) are likely to bind effectively to host-derived surface proteins, enabling subsequent cellular invasion. Understanding whether human cryptospordiosis is largely dominated by six subgenotypes of *C. parvum* and *C. hominis* and the molecular mechanisms linked to infectivity may be a key to the control of human cryptosporidiosis, with implications for the development of new intervention approaches. Antibody responses against proteins GP15 and G40 have been detected (by immunoblot) in the sera from humans following cryptosporidiosis (Strong et al. 2000). Interestingly, a recent study (Preidis et al. 2007) showed that recombinant *C. hominis* (but not *C. parvum*)

GP15 stimulated the production of interferon-gamma in whole blood samples taken from donors that were seropositive for *Cryptosporidium*. As noted by the authors of that study (Preidis et al. 2007), interferon-gamma is known to play a key role in the host immune response to *Cryptosporidium* infection (Jex et al. 2010). Although this research hints to recombinant *C. hominis* GP15 as a potential vaccine, the absence of a similar response against *C. parvum* GP15 (Preidis et al. 2007) suggests species- or genotype-specific immunogenic responses. Understanding, on a genomic level, the richness and diversity of genetic variants of *Cryptosporidium* infecting humans might have implications for vaccine development against various species of *Cryptosporidium*.

6.6 Concluding Remarks

The year 2007 marked the centenary of the discovery of *Cryptosporidium*. The past century has seen immense changes in our understanding of these organisms, with most advances emerging in the last three decades. That this time-scale corresponds to the advent of PCR and the dawn of the molecular biological age is no co-incidence. The application of PCR technology has revolutionized our understanding of *Cryptosporidium*, from systematics to transmission dynamics, population genetics, diagnostics and molecular/cellular biology. Massively parallel genomic sequencing and high-throughput technologies are just emerging. These platforms bring an analytical and exploratory power beyond anything that we have seen in biology to date, and mark a new era in the biological and biomedical sciences. In the field of *Cryptosporidium* research, the application of these emerging technologies has only just begun and, yet, the promise of this research captures the imagination.

Clearly, whole-genome sequencing of *Cryptosporidium* from humans and other animals should be considered a priority as is the case for an ongoing effort to sequence additional *Cryptosporidium* species. In previous years, such research would have been considered unaffordable and far too time consuming. However, given the capacity of modern sequencing devices, this is no longer the case. Benchtop sequencing platforms, such as the Illumina MiSeq, coupled to sample multiplexing technologies, now provide sufficient data to sequence complete *Cryptosporidium* genomes to a depth of $\geq$100-fold for less than ~\$100 per sample. New developments in DNA library construction further facilitate this research, with the sequencing of whole genomes from nanograms (or less) of genomic DNA being now possible (Ling et al. 2009; Grindberg et al. 2011; Bankevich et al. 2012). Such research can be further enabled through the use of whole genome amplification (WGA) technologies, which allow the synthesis of micrograms of DNA from nanogram or even pictogram amounts of starting template. These methods, also coupled to other 'omic' technologies, allow us to explore the biology of cryptosporidia from humans and other animals to depths never before achieveable.

Nonetheless, significant challenges remain, not least the accessibility of phenetically and genetically well-defined *Cryptosporidium* isolates ("strains") to underpin research efforts. Limitations associated with in vitro cultivation as well as the lack of a known function for most proteins encoded in *Cryptosporidium* genomes represent significant challenges. Despite the complexity of the technologies and the enormity of 'systems biology', as always, addressing these challenges will require persistance, significant technical skill, and the imagination and innovation to solve many unforseen obstacles. After 100 years of research and advances in this field, perhaps Edmund Tyzzer would be pleased to see that some things never change.

Acknowledgements Grant support from the Australian Research Council, Melbourne Water Corporation and the Australian National Health and Medical Research Council (NHMRC) is gratefully acknowledged.

References

Abeywardena H, Jex AR, Nolan MJ, Haydon SR, Stevens MA, McAnulty RW et al (2012) Genetic characterisation of *Cryptosporidium* and *Giardia* from dairy calves: discovery of species/genotypes consistent with those found in humans. Infect Genet Evol 12:1984–1993

Abrahamsen MS, Templeton TJ, Enomoto S, Abrahante JE, Zhu G, Lancto CA et al (2004) Complete genome sequence of the apicomplexan, *Cryptosporidium parvum*. Science 304:441–445

Bankevich A, Nurk S, Antipov D, Gurevich AA, Dvorkin M, Kulikov AS et al (2012) SPAdes: a new genome assembly algorithm and its applications to single-cell sequencing. J Comput Biol 19:455–477

Bankier AT, Spriggs HF, Fartmann B, Konfortov BA, Madera M, Vogel C et al (2003) Integrated mapping, chromosomal sequencing and sequence analysis of *Cryptosporidium parvum*. Genome Res 13:1787–1799

Bentley DR, Balasubramanian S, Swerdlow HP, Smith GP, Milton J, Brown CG et al (2008) Accurate whole human genome sequencing using reversible terminator chemistry. Nature 456:53–59

Cohn B, Manque P, Lara AM, Serrano M, Sheth N, Buck G (2010) Putative cis-regulatory elements associated with heat shock genes activated during excystation of *Cryptosporidium parvum*. PLoS One 5:e9512

Cox J, Mann M (2011) Quantitative, high-resolution proteomics for data-driven systems biology. Annu Rev Biochem 80:273–299

Creek DJ, Anderson J, McConville MJ, Barrett MP (2012) Metabolomic analysis of trypanosomatid protozoa. Mol Biochem Parasitol 181:73–84

Crowther GJ, Shanmugam D, Carmona SJ, Doyle MA, Hertz-Fowler C, Berriman M et al (2010) Identification of attractive drug targets in neglected-disease pathogens using an in silico approach. PLoS Negl Trop Dis 4:e804

Fritzsch FS, Dusny C, Frick O, Schmid A (2012) Single-cell analysis in biotechnology, systems biology, and biocatalysis. Annu Rev Chem Biomol Eng 3:129–155

Gardner MJ, Bishop R, Shah T, de Villiers EP, Carlton JM, Hall N et al (2005) Genome sequence of *Theileria parva*, a bovine pathogen that transforms lymphocytes. Science 309:134–137

Gardner MJ, Hall N, Fung E, White O, Berriman M, Hyman RW et al (2002) Genome sequence of the human malaria parasite *Plasmodium falciparum*. Nature 419:498–511

Gaulton A, Bellis LJ, Bento AP, Chambers J, Davies M, Hersey A et al (2012) ChEMBL: a large-scale bioactivity database for drug discovery. Nucleic Acids Res 40:D1100–1107

Gerstein MB, Lu ZJ, Van Nostrand EL, Cheng C, Arshinoff BI, Liu T et al (2010) Integrative analysis of the *Caenorhabditis elegans* genome by the modENCODE project. Science 330:1775–1787

Grindberg RV, Ishoey T, Brinza D, Esquenazi E, Coates RC, Liu WT et al (2011) Single cell genome amplification accelerates identification of the apratoxin biosynthetic pathway from a complex microbial assemblage. PLoS ONE 6:e18565

Guthrie OW (2008) Aminoglycoside induced ototoxicity. Toxicology 249:91–96

Hijjawi N, Estcourt A, Yang R, Monis P, Ryan U (2010) Complete development and multiplication of *Cryptosporidium hominis* in cell-free culture. Vet Parasitol 169:29–36

Jex AR, Chalmers R, Smith HV, Widmer G, McDonald V, Gasser RB (2010) Cryptospordiosis. In: Palmer S, Torgerson P, Soulsby EJ (eds) Zoonoses: Biology, Clinical Practice, and Public Health Control. Oxford University Press, Oxford

Jex AR, Gasser RB (2010) Genetic richness and diversity in *Cryptosporidium hominis* and *C. parvum* reveals major knowledge gaps and a need for the application of "next generation" technologies – research review. Biotechnol Adv 26:17–26

Jex AR, Liu S, Li B, Young ND, Hall RS, Li Y et al (2011a) Ascaris suum draft genome. Nature 479:529–533

Jex AR, Smith HV, Nolan MJ, Campbell BE, Young ND, Cantacessi C et al (2011b) Cryptic parasite revealed improved prospects for treatment and control of human cryptosporidiosis through advanced technologies. Adv Parasitol 77:141–173

Jeyanthi T, Subramanian P (2009) Nephroprotective effect of *Withania somnifera*: a dose-dependent study. Ren Fail 31:814–821

Ling J, Zhuang G, Tazon-Vega B, Zhang C, Cao B, Rosenwaks Z et al (2009) Evaluation of genome coverage and fidelity of multiple displacement amplification from single cells by SNP array. Mol Hum Reprod 15:739–747

Mallon ME, MacLeod A, Wastling JM, Smith H, Tait A (2003) Multilocus genotyping of *Cryptosporidium parvum* Type 2: population genetics and sub-structuring. Infect Genet Evol 3:207–218

Margulies M, Egholm M, Altman WE, Attiya S, Bader JS, Bemben LA et al (2005) Genome sequencing in microfabricated high-density picolitre reactors. Nature 437:376–380

Martinez-Salgado C, Lopez-Hernandez FJ, Lopez-Novoa JM (2007) Glomerular nephrotoxicity of aminoglycosides. Toxicol Appl Pharmacol 223:86–98

Morgan UM, Xiao L, Hill BD, O'Donoghue P, Limor J, Lal A et al (2000) Detection of the *Cryptosporidium parvum* "human" genotype in a dugong (*Dugong dugon*). J Parasitol 86:1352–1354

Mullapudi N, Lancto CA, Abrahamsen MS, Kissinger JC (2007) Identification of putative *cis*-regulatory elements in *Cryptosporidium parvum* by de novo pattern finding. BMC Genomics 8:13

Murakami S, Nagai J, Fujii K, Yumoto R, Takano M (2008) Influences of dosage regimen and co-administration of low-molecular weight proteins and basic peptides on renal accumulation of arbekacin in mice. J Antimicrob Chemother 61:658–664

Nagai J, Takano M (2010) Molecular-targeted approaches to reduce renal accumulation of nephrotoxic drugs. Expert Opin Drug Metab Toxicol 6:1125–1138

Niedringhaus TP, Milanova D, Kerby MB, Snyder MP, Barron AE (2011) Landscape of next-generation sequencing technologies. Anal Chem 83:4327–4341

Pannu N, Nadim MK (2008) An overview of drug-induced acute kidney injury. Crit Care Med 36: S216–223

Piper MB, Bankier AT, Dear PH (1998) A HAPPY map of *Cryptosporidium parvum*. Genome Res 8:1299–1307

Ponting CP, Hardison RC (2011) What fraction of the human genome is functional? Genome Res 21:1769–1776

Preidis GA, Wang HC, Lewis DE, Castellanos-Gonzalez A, Rogers KA, Graviss EA et al (2007) Seropositive human subjects produce interferon-gamma after stimulation with recombinant *Cryptosporidium hominis* gp15. Am J Trop Med Hyg 77:583–585

Radke JR, Behnke MS, Mackey AJ, Radke JB, Roos DS, White MW (2005) The transcriptome of Toxoplasma gondii. BMC Biol 3:26

Reeder JC, Brown GV (1996) Antigenic variation and immune evasion in *Plasmodium falciparum* malaria. Immunol Cell Biol 74:546–554

Rizzi MD, Hirose K (2007) Aminoglycoside ototoxicity. Curr Opin Otolaryngol Head Neck Surg 15:352–357

Rossignol JF (2010) *Cryptosporidium* and *Giardia*: treatment options and prospects for new drugs. Exp Parasitol 124:45–53

Schomburg I, Chang A, Schomburg D (2002) BRENDA, enzyme data and metabolic information. Nucleic Acids Res 30:47–49

Senda M, DeLustro B, Eugui E, Natsumeda Y (1995) Mycophenolic acid, an inhibitor of IMP dehydrogenase that is also an immunosuppressive agent, suppresses the cytokine-induced nitric oxide production in mouse and rat vascular endothelial cells. Transplantation 60:1143–1148

Shirley M (1994) The genome of *Eimeria tenella*: further studies on its molecular organisation. Parasitol Res 80:366–373

Shirley MW (2000) The genome of *Eimeria* spp., with special reference to *Eimeria tenella*–a coccidium from the chicken. Int J Parasitol 30:485–493

Siepel A, Bejerano G, Pedersen JS, Hinrichs AS, Hou M, Rosenbloom K et al (2005) Evolutionarily conserved elements in vertebrate, insect, worm, and yeast genomes. Genome Res 15:1034–1050

Smith HV, Nichols RA, Mallon M, Macleod A, Tait A, Reilly WJ et al (2005) Natural *Cryptosporidium hominis* infections in Scottish cattle. Vet Rec 156:710–711

Striepen B, Kissinger JC (2004) Genomics meets transgenics in search of the elusive *Cryptosporidium* drug target. Trends Parasitol 20:355–358

Strong WB, Gut J, Nelson RG (2000) Cloning and sequence analysis of a highly polymorphic *Cryptosporidium parvum* gene encoding a 60-kilodalton glycoprotein and characterization of its 15- and 45-kilodalton zoite surface antigen products. Infect Immun 68:4117–4134

Strong WB, Nelson RG (2000) Gene discovery in *Cryptosporidium parvum*: expressed sequence tags and genome survey sequences. Contrib Microbiol 6:92–115

Tang F, Barbacioru C, Nordman E, Li B, Xu N, Bashkirov VI et al (2010) RNA-Seq analysis to capture the transcriptome landscape of a single cell. Nat Protoc 5:516–535

Tanriverdi S, Grinberg A, Chalmers RM, Hunter PR, Petrovic Z, Akiyoshi DE et al (2008) Inferences about the global population structures of Cryptosporidium parvum and *Cryptosporidium hominis*. Appl Environ Microbiol 74:7227–7234

Tugcu V, Ozbek E, Tasci AI, Kemahli E, Somay A, Bas M et al (2006) Selective nuclear factor kappa-B inhibitors, pyrolidium dithiocarbamate and sulfasalazine, prevent the nephrotoxicity induced by gentamicin. BJU Int 98:680–686

Umejiego NN, Gollapalli D, Sharling L, Volftsun A, Lu J, Benjamin NN et al (2008) Targeting a prokaryotic protein in a eukaryotic pathogen: identification of lead compounds against cryptosporidiosis. Chem Biol 15:70–77

Umejiego NN, Li C, Riera T, Hedstrom L, Striepen B (2004) *Cryptosporidium parvum* IMP dehydrogenase: identification of functional, structural, and dynamic properties that can be exploited for drug design. J Biol Chem 279:40320–40327

Wang Z, Gerstein M, Snyder M (2009) RNA-Seq: a revolutionary tool for transcriptomics. Nat Rev Genet 10:57–63

Widmer G, Lee Y, Hunt P, Martinelli A, Tolkoff M, Bodi K (2012) Comparative genome analysis of two *Cryptosporidium parvum* isolates with different host range. Infect Genet Evol 12:1213–1221

Xiao L, Fayer R, Ryan U, Upton SJ (2004) *Cryptosporidium* taxonomy: recent advances and implications for public health. Clin Microbiol Rev 17:72–97

Xu P, Widmer G, Wang Y, Ozaki LS, Alves JM, Serrano MG et al (2004) The genome of *Cryptosporidium hominis*. Nature 431:1107–1112

Young ND, Jex AR, Li B, Liu S, Yang L, Xiong Z et al (2012) Whole-genome sequence of Schistosoma haematobium. Nat Genet 44:221–225

Zhou L, Singh A, Jiang J, Xiao L (2003) Molecular surveillance of *Cryptosporidium* spp. in raw wastewater in Milwaukee: implications for understanding outbreak occurrence and transmission dynamics. J Clin Microbiol 41:5254–5257

Chapter 7
From Genome to Proteome: Transcriptional and Proteomic Analysis of *Cryptosporidium* Parasites

Jonathan M. Wastling and Nadine P. Randle

Abstract The ability to generate large-scale transcriptome and proteome datasets has changed the landscape in parasite biology. These data enable an integrated, whole organism approach to understanding how parasites function and interact with their hosts. The difficulty in propagating *Cryptosporidium* means that work has focused on the oocyst and sporozoite stages of the parasite. Transcriptional studies using expressed sequence tags (ESTs), microarrays and quantitative real-time PCR (qRT-PCR) have given a valuable insight into how both the parasite and host cells respond to infection. In addition, global proteomics analyses have characterised the expressed proteomes of oocysts and sporozoites of *C. parvum*. Currently, *Cryptosporidium* research is lagging behind some other pathogens in terms of global 'omics' analyses. However, there have been significant technological and bioinformatics advances in transcriptome and proteome analyses in recent years, which are set to continue. Exploiting these technologies and capitalising on the resulting "systems-biology" data mean that exciting times are ahead in the field of *Cryptosporidium* biology.

7.1 Introduction

The availability of the genome sequences for *Cryptosporidium parvum* (Abrahamsen et al. 2004) and *C. hominis* (Xu et al. 2004) has greatly advanced our knowledge of the biology of these parasites, but is only part of the story of trying to understand the evolutionary success of *Cryptosporidium*. Although genomes provide information concerning the potential repertoire of genes in *Cryptosporidium*, they provide no information on the actual expression of these genes, or on the translation of genes into functional proteins. In order to investigate the most

J.M. Wastling (✉) • N.P. Randle
Department of Infection Biology, Institute of Infection and Global Health, University of Liverpool, Brownlow Hill, Liverpool L3 5RF, UK
e-mail: J.Wastling@liverpool.ac.uk; Nadine.Randle@liverpool.ac.uk

S.M. Cacciò and G. Widmer (eds.), *Cryptosporidium: parasite and disease*,
DOI 10.1007/978-3-7091-1562-6_7, © Springer-Verlag Wien 2014

interesting biological problems, such as how *Cryptosporidium* adapts to its unique biological niche, how it controls development through its life-cycle and how it interacts with its host environment, it is necessary to determine the expression of both genes and proteins. Analysis of the mRNA transcriptome and protein expression by proteomics has the potential to yield a wealth of "systems-biology" expression data to help answer some of these essential questions.

The difficulty in propagating in *Cryptosporidium* spp. in vitro has hampered gene and protein expression studies. While it appears that it might be possible to maintain *C. parvum* in cell culture for a period of time (Arrowood 2002; Current and Haynes 1984; Hijjawi et al. 2001, 2002; Upton et al. 1994), insufficient material can be produced for the later life-cycle stages to perform comprehensive transcriptomic and proteomic analyses. The oocyst and sporozoites are currently the only life-cycle stages of *C. parvum* that are readily accessible for transcript and proteome analyses, thus, such studies have been restricted to these stages.

This chapter describes the recent advances in the transcriptomic and proteomic analyses of *Cryptosporidium* and how these data can be used to further the understanding of its biology.

7.2 Global mRNA Transcriptional Analysis

The transcriptome describes all the RNA transcripts present in a cell and is highly dynamic, reflecting all the genes that are being actively expressed at a given point in time. Examining the expression profiles of mRNA transcripts in a population of cells gives a good understanding of the functional processes occurring. Gene expression profiles vary depending on the cell type, developmental stage of the organism and environmental conditions, thus unlike the relatively static genome, the transcriptome is highly dynamic and responsive to biological events. Two main approaches can be taken in order to characterise and quantify the transcriptome of a cell or organism: hybridisation- or sequence-based methods. A brief overview of the main techniques is given here, but there are numerous previously published general reviews of these to which reference should be made (Holt and Jones 2008; Mutz et al. 2013; Wang et al. 2009).

Microarrays are the most widely used hybridisation-based technique for simultaneously measuring gene expression levels for thousands of genes, identifying SNPs and for comparative genome analysis (Gobert et al. 2005). They consist of glass or silicon slides with thousands of DNA probes attached in an ordered array; these probes correspond to mRNA sequences from the target species and can be oligonucleotides, cDNA or PCR products. Hybridisation to complementary DNA (cDNA) arrays is more specific than to oligonucleotide arrays, but the latter are considerably cheaper. When performing a microarray experiment, cDNA is usually labelled with a fluorescent dye, such as Cy3 or Cy5, prior to being hybridised to the microarray. Samples can be labelled with different dyes and hybridised to the same array for comparative studies. When the sample binds to the probe, it emits a signal that can be detected at a specific wavelength and is proportional to the amount of target sequence that has bound.

Reverse-transcription PCR (RT-PCR) is used to detect mRNA expression, but is relatively non-quantitative. However, this technique can be readily adapted to obtain relative or absolute gene expression levels using quantitative real-time PCR (qRT-PCR). By incorporating fluorescent dyes or probes, such as SYBR Green or TaqMan, into the PCR products as they are amplified, the amount of DNA present can be measured in real-time. This technique is widely used for quantifying gene expression levels.

There have been significant technological advances in recent years, with numerous high-throughput "next-generation" sequencing platforms now available. This technology has revolutionised the extent and complexity of transcriptome sequence data that can be obtained for an organism. RNA-Seq is highly sensitive, with very low background noise and a large dynamic range that can quantify transcripts of very low to very high abundance (Mortazavi et al. 2008). Briefly, a complementary DNA (cDNA) library is generated from RNA by amplification from adaptors attached to either end of the RNA strands. Single-end or paired-end sequences are generate by sequencing from either one or both adaptors, respectively. The cDNA fragments are sequenced in a high-throughput manner to produce reads of 30–1,000 nucleotides (nt), depending on the sequencing platform used. These reads can either be aligned (mapped) to an existing genome sequence or assembled *de novo*; the resulting transcriptional map yields data about both the structure and expression level of each gene (inferred from the number of reads for a given transcript). The SOLiD sequencing platform is based on ligation and generates short reads of up to 50 nt. The Illumina and 454 platforms both take a "sequencing by synthesis" approach; the Illumina sequencers produce reads of 50–300 nt while the 454 produces much longer reads of up to 1,000 nt. The short read lengths produced by SOLiD and Illumina are ideally suited to analysing transcriptomes of good quality, well annotated genomes and can be useful in identifying intron-exon boundaries and sequence variations (Wang et al. 2009). When there is either an incomplete, or no reference genome available the longer reads produced by 454 sequencing are more appropriate as they maximise the possibility of reads overlapping.

A limiting factor with microarrays and PCR product-based transcriptome analyses is the need for good quality genome sequences. Many parasite genomes are incomplete which can present major challenges when designing probes and primers for these types of studies (Chen et al. 2004). The RNA-Seq approach to transcriptome analysis offers considerable advantages over microarrays: it requires less mRNA (a considerable advantage when working with microorganisms like *Cryptosporidium* which are difficult to propagate in vitro), has a much larger dynamic range for accurately detecting and quantifying transcript expression levels, has very low background noise and is highly reproducible. Correctly mapping the mRNA sequence reads to the appropriate genome and/or assembling sequences *de novo* can still present a challenge, although the use of paired end sequencing has made this task substantially easier (Holt and Jones 2008; Roach et al. 1995).

7.3 Proteomics Technologies

Proteins are arguably the most relevant functional component in a biological system; however, they are in a constant state of flux and activation, presenting a significant analytical challenge in comparison to mRNA. The development of mass spectrometry-based proteomics as a routine technology has transformed the way in which protein expression can be analysed. In common with high-throughput transcriptome sequencing, characteristic signatures (peptide fragmentation patterns in the case of proteins) can be matched by bioinformatics to specific genes. This means that the physical data from a mass spectrometer can be de-convoluted to provide high-throughput detection and analysis platforms for proteins, providing a corresponding genome sequence is available. Mass spectrometry (MS) analysis of proteins involves the ionization, separation and detection of molecular ions based on their mass-to-charge ratio (m/z). The resulting spectra are then searched against a corresponding virtual database of peptide mass data. Briefly, the virtual database uses previously established gene models, or translates the genome sequence in all six frames and the protein sequences are then digested in silico to generate every possible peptide sequence. The m/z for each ion detected by the mass spectrometer is then queried against the virtual database to find a corresponding peptide from an annotated protein with the same m/z. A typical proteomics workflow is shown in Fig. 7.1.

In order to reduce the complexity of a sample for MS analysis, proteins can be separated by gel electrophoresis or chromatography before digesting with an enzyme, such as trypsin, to produce peptides. Alternatively, peptides can be separated after enzymatic digestion using methods such as liquid chromatography. For the former, one- and two-dimensional gel electrophoresis (1-DE and 2-DE) have been widely employed to separate proteins but these techniques are most suited to analysing abundant hydrophilic proteins. In 1-DE proteins are separated by their mass so bands frequently contain multiple proteins. Using 2-DE, proteins are first separated according to their isoelectric point and then by mass, producing a highly resolved protein map in which each spot only contains one protein. The 2-DE approach reveals something of the complexity of the proteome, as proteins may appear in multiple spots due to post-translational modifications and alternative splicing.

Multidimensional protein identification technology (MudPIT) uses liquid chromatography (LC) to separate peptides for identification by MS. This can be done by excising bands from a 1-DE gel, digesting them and then performing the separation step or by digesting the proteins in solution and performing two rounds of LC followed by MS analysis. Advances in mass spectrometry instrumentation and protein separation technology have increased the resolution and accuracy of m/z data acquisition. This, when coupled with the use of more sophisticated bioinformatics tools, has led to more reliable and statistically sound protein identifications with much reduced false discovery rates (FDR).

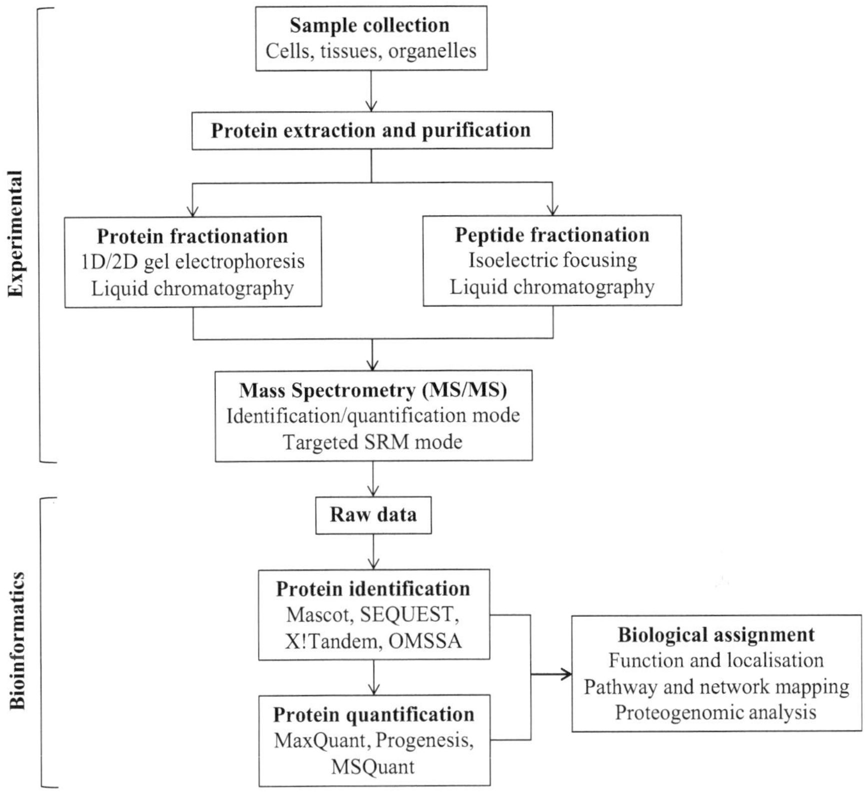

Fig. 7.1 Schematic representing a typical proteomics workflow. The identification and quantification of proteins involves an experimental and bioinformatics stage. In the experimental stage samples are collected, protein is extracted and purified, samples are separated and digested (in either protein or peptide space) and mass spectrometry analysis is performed. The bioinformatics analysis uses the raw spectra data to interrogate protein identification databases and specialised quantification programmes calculate the relative and absolute protein abundances. Biological relevance is then assigned using pathway mapping tools and predictions of the protein function and localisation

7.4 The Transcriptome of *C. parvum* Oocysts and Sporozoites

Large-scale transcriptional analysis from *Cryptosporidium* parasites began with sequenced cDNA libraries of oocysts/sporozoites which generated expressed sequence tag (EST) data (Strong and Nelson 2000). There are currently 98,058 EST transcripts from *C. parvum* and *C. muris* on CryptoDB (version 5.0) (http://cryptodb.org/cryptodb/; Heiges et al. 2006), which provide evidence for 2,386 genes. One of the early studies to profile the genes expressed by *C. parvum* sporozoites sequenced cDNA libraries and generated 567 ESTs (Strong and Nelson

2000). To complement the EST data, and to expand the study to other life-cycle stages, the authors also analysed 1,507 genome survey sequences (GSS) that were generated from randomly sheared genomic DNA. This was performed prior to the release of the genome, but BLAST analyses found that 32 % of the ESTs showed similarity to existing sequences in publicly available databases, leaving a high frequency of sequences with no known homology or function. More recently over 1,000 ESTs were sequenced from oocysts undergoing excystation, this study has shown that transcripts encoding for structural proteins are highly up-regulated during the early stages of the excystation process (Jenkins et al. 2011). In addition, a high percentage of hypothetical proteins were identified as well as a large number of ESTs involved in cell signalling, proliferation and metabolism. In concordance with proteomic evidence, heat shock proteins (HSP) and ribosomal proteins were also abundant (Snelling et al. 2007). In a study to assess the quality of the current gene models, 5′-EST data generated for *C. parvum*, *Toxoplasma gondii* and a variety of *Plasmodium* species were compared with existing genome sequences. In contrast to other *Apicomplexa*, gene annotations for *C. parvum* were found to be highly reliable, undoubtedly due to the small size of the genome and low frequency of introns (Wakaguri et al. 2009).

Recently, a *C. parvum* microarray (CpArray15K) containing oligonucleotide probes for 3,805 genes has been developed (Zhang et al. 2012). The array, coupled with qRT-PCR, was used to compare gene expression profiles between untreated and UV irradiated oocysts to identify genes involved in the response to environmental stress. In the untreated oocysts, genes associated with protein synthesis (transcription, translation and ribosome biogenesis) were highly expressed. From the high expression levels of ubiquitin and proteasome, the authors suggest that the oocyst may degrade and reuse amino acids. Genes involved in DNA repair, intracellular trafficking and cytoskeleton rearrangement were up-regulated in response to UV irradiation, as were members of the T-complex protein 1 (TCP-1) family, and some thioredoxin-related genes which are known to be associated with responses to stress.

Remarkably, at the time of press, no RNASeq data for *Cryptosporidium* has been published, despite the significant advances in sequencing technologies. However, the first comprehensive in vitro analysis of the *C. parvum* transcriptome has recently been completed (Mauzy et al. 2012). The transcript abundances of 3,302 genes (representing 87 % of the genome) were determined by qRT-PCR. This monumental experiment was designed as a temporal study, in which samples of infected human epithelial cells were collected at seven time points over a 72 h infection period. For each of the genes analysed, transcripts were detected for at least one time point. Although expression levels varied throughout over time, two thirds of the transcripts were detected at every time point. Gene expression patterns were analysed by looking at relative fold changes that were normalised to 18S rRNA expression levels. Distinct expression profiles were associated with different developmental stages, as evidenced by the lack of congruence between metabolic, ribosomal and proteasome protein expression, and 18 s rRNA levels. The number of distinct transcripts detected increased from 6 h post infection, with a third of the genes being most highly expressed at either 48 or 72 h. When looking at the complete dataset, cluster analysis revealed nine distinct clusters; genes within each cluster did not map to the same chromosomal region indicating that gene expression is independent of chromosome location. From the cluster analysis, expression of transcription related genes peaked at 2 h, proteins involved in translation dominated at 6 h while transcripts related to structural genes, such as myosin and tubulin, were most prevalent at 12 h post infection. An increase in transcript abundance, coinciding with the sexual reproduction of the parasites, was observed from 36 to 72 h. These data suggest that *C. parvum* has a more complex system of transcription regulation than previously thought (Mauzy et al. 2012; Templeton et al. 2004).

7.5 Modulation of the Host Cell Transcriptome by *Cryptosporidium*

Transcriptional studies have been used to determine host-response, as well as parasite gene expression patterns in cell monolayers, including several microarray studies that characterised the host cell (human) responses to infection with *C. parvum* (Castellanos-Gonzalez et al. 2008; Deng et al. 2004; Liu et al. 2009; Yang et al. 2009, 2010). These studies have shown that significant changes occur in the host cell in response to infection with *C. parvum*, particularly in genes associated with the inflammatory response, apoptosis, cell proliferation and the cytoskeleton. The composition of the host cell membrane is affected by the presence of the parasite (Yang et al. 2009). One study showed that heat shock proteins and pro-inflammatory cytokines were up-regulated in response to *C. parvum* infection (Deng et al. 2004). Interestingly, the differential expression of actin related genes was also observed with those for actin binding proteins being down-regulated while genes for host actin and tubulin were up-regulated. This corresponds well with previous findings of host cell actin polymerisation and cytoskeleton remodelling in response to *C. parvum* infection (Chap. 10) (Chen et al. 2001; Elliott and Clark 2000; Elliott et al. 2001). Apoptosis of host cells was observed to be differentially regulated throughout the infection process with anti-apoptotic genes being more highly expressed during the first 12 h of infection with *C. parvum* and no apoptosis being observed in culture; from 24 h post infection, pro-apoptotic genes were up-regulated (Liu et al. 2009). Genes related to apoptosis have been found to be differentially regulated in the host in other studies (Deng et al. 2004; Yang et al. 2009), consistent with previous evidence of host cell apoptosis in *C. parvum* infected cell monolayers (Griffiths et al. 1994; Widmer et al. 2000). Coupling biopsy infection with microarray analysis revealed that the osteoprotegerin (OPG) gene was up-regulated in response to infection with *C. parvum* and *C. hominis* and that OPG expression was greater in response to *C. hominis*. These findings were confirmed in HCT-8 cells and OPG was also shown to be expressed in response to an in vivo infection of a volunteer with *C. meleagridis*. Thus, a role for OPG in modulating the host cell response to *Cryptosporidium* infection has been proposed (Castellanos-Gonzalez et al. 2008).

A difficulty faced by studies to analyze the in vitro expression of both parasite and host genes during *Cryptosporidium* cell invasion is the relatively low multiplicity of infection that can be achieved using current cell culture techniques. Since the majority of cells remain uninfected in culture, signals from infected cells may appear relatively weak as compared to the majority of uninfected cells. In order to overcome this problem, Yang et al. (2010) used fluorescence-activated cell sorting (FACS) to separate infected and uninfected cells from a single monolayer to compare host cell responses using a human genome microarray. At 24 h post infection, 417 genes were up-regulated and 361 were found to be down-regulated. In sorted cells, glycoproteins predominated in the up-regulated genes of infected cells (Yang et al. 2010); lectin-binding studies confirmed that the sugars on these glycoproteins were mainly N-linked oligosaccharides. This differs from studies using unsorted cells where the majority of the up-regulated genes were related to apoptosis and cell proliferation (Deng et al. 2004; Liu et al. 2009; Yang et al. 2009).

7.6 The Proteome of *C. parvum*

In contrast to other apicomplexan parasites, relatively few proteomic analyses have been performed for *Cryptosporidium* to date. Early studies of *Cryptosporidium* by mass spectrometry were limited to identifying characteristic spectral patterns that could be used to distinguish between species. Prior to the release of the *C. parvum* genome, MALDI-TOF MS was used to identify spectral peaks that could identify *Cryptosporidium* at the genus level and others that could be used to distinguish between *C. parvum* and *C. muris* oocysts (Magnuson et al. 2000). Further work also revealed spectral peaks that were specific to the oocyst and sporozoites (Glassmeyer et al. 2007).

Subsequently, facilitated by the availability of the genome sequence, large-scale global analyses of the expressed proteome of *C. parvum* oocysts and sporozoites have been performed (Sanderson et al. 2008; Snelling et al. 2007). In the first proteomic analysis of *C. parvum*, a combination of LC-MS/MS and iTRAQ isobaric labelling were used to identify around 200 proteins, representing about 6 % of the predicted proteome, from excysted and non-excysted oocysts (Snelling et al. 2007). For both stages of the parasite, the soluble fraction was labelled with iTRAQ reagents for quantitative analysis on a QSTAR QqTOF mass spectrometer and the insoluble fraction was subject to qualitative MudPIT analysis. By using iTRAQ to label the samples, Snelling and co-workers were able to perform a relative quantification of the proteins identified between the excysted and non-excysted parasites. In the soluble fractions, a total of 26 proteins were found to have significantly higher expression levels post-excystation, relative to unexcysted oocysts. These proteins include a variety of heat shock proteins,

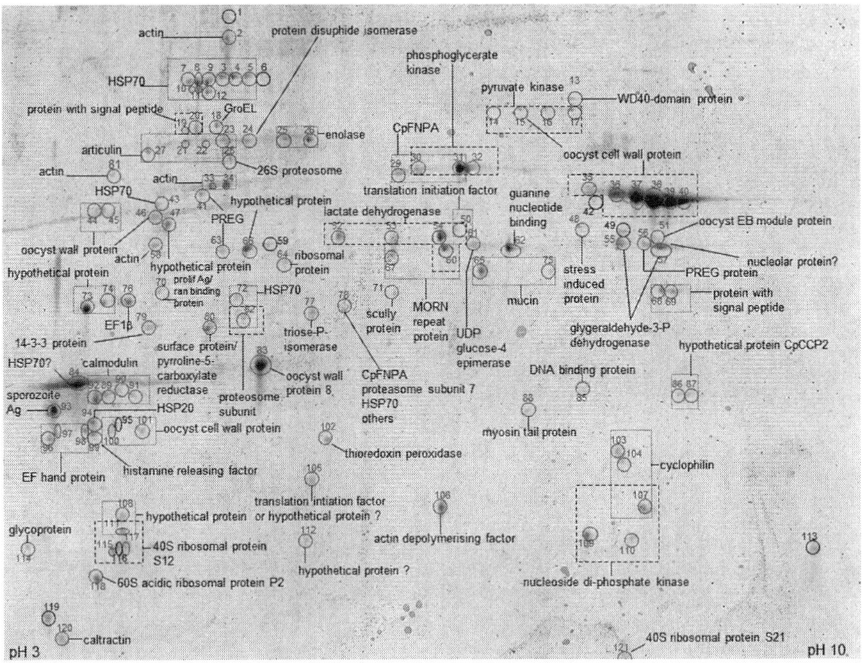

Fig. 7.2 2-DE analysis of *Cryptosporidium parvum* sporozoites. Proteins were resolved using a broad non linear pH3-10 gradient followed by separation on a 12.5 % denaturing poly acrylamide gel before staining with Coomassie Colloidal (Sanderson et al. 2008)

ribosomal proteins and metabolic enzymes as well as three apicomplexan-specific and five *Cryptosporidium*-specific proteins. Several of these proteins contained signal peptides, were predicted to be secreted and were hypothesised to play a role in host cell invasion by the parasite.

In the second, and most comprehensive of these studies, 1,237 non-redundant proteins were identified from excysted oocysts and sporozoites, equivalent to 30 % of the predicted proteome (Sanderson et al. 2008). To maximise the coverage of the proteome, and increase the confidence in the proteins identified, three complementary approaches were used: 1-DE LC-MS/MS, 2-DE LC-MS/MS and MudPIT. Proteins excised from 1-DE and 2-DE gels were digested with trypsin and analysed by either MALDI-ToF or LTQ ion trap mass spectrometry; the resulting spectra were used to interrogate a *C. parvum* genome database in Mascot® (Perkins et al. 1999) (Fig. 7.2). In the MudPIT experiment, spectra from lysed soluble and insoluble fractions were searched using SEQUEST® (Eng et al. 1994). A total of 642, 282 and 1,154 non-redundant proteins were identified from the 1-DE, 2-DE and MudPIT analyses, respectively. Approximately 20 % of the proteins identified were done so based on a single peptide hit from one platform and thus would not have been identified if a single approach had been taken.

Analysis of protein function revealed structural information for 1,165 of the non-redundant proteins identified, with 213 (18 %) predicted to contain transmembrane domains (about 50 % of these proteins have no known function). The MudPIT dataset contained a greater proportion of hydrophobic proteins than the other two platforms. Putative functional categories were assigned to proteins using a combination of gene ontology (GO) classifications and homology and literature searches. However, no information was available for approximately 40 % of the identified proteins, which were designated as unclassified. Key findings include peptide evidence for 18 of the mitochondrial and iron-sulphur metabolism genes predicted in the *C. parvum* and *C. hominis* genomes, consistent with the theory that a functional relic mitochondrion is present in *Cryptosporidium* (Putignani et al. 2004). Peptide evidence was also detected for 14 of the predicted micronemal proteins, 12 putative rhoptry (ROP) proteins, a few dense granule proteins and a large number of surface proteins.

7.7 Sub-proteome Analysis of *C. parvum* Sporozoites

Despite huge advances in the sensitivity of mass-spectrometry as a protein discovery tool, global proteomics experiments are still not able to cover the complete depth of the expressed proteome in a single experiment. One way to increase the depth of coverage is to simplify the proteome by fractionation prior to analysis, creating "sub-proteomes" that contain fewer proteins. For example, sub-proteomes for apical invasion related organelles have been characterised for a range of apicomplexan parasites including the rhoptries of *Toxoplasma gondii*, *Eimeria tenella* and several *Plasmodium* species (Bradley et al. 2005; Bromley et al. 2003; Sam-Yellowe et al. 2004); the micronemes of *E. tenella* (Bromley et al. 2003) and excreted/secreted antigens of *T. gondii* (Zhou et al. 2005). In *Cryptosporidium*, there have been relatively few equivalent studies, largely due to the difficulties of obtaining sufficient amounts of parasite material with which to perform organelle fractionation by techniques such as density gradient centrifugation.

In one study to investigate how sporozoites are tethered to the oocyst wall, mass spectrometry analysis was employed to identify the key components of the oocyst wall and affinity purified glycoproteins from the sporozoites (Chatterjee et al. 2010). Excysted and sonicated oocysts were centrifuged through sucrose to isolate the walls, while the glycoprotein fractions were prepared using Sepharose columns. The authors found that the *Cryptosporidium* oocyst wall proteins (COWPs) constitute about 75 % of the proteins identified by mass spectrometry in the oocyst walls. The dominant oocyst wall protein in *C. parvum* is COWP1, with more than 50 % of peptides identified by mass spectrometry belonging to this protein. COWP8 and COWP6 are also relatively abundant while COWP2, COWP3 and COWP4 were only detected at very low levels and COWP5, COWP7 and COWP9 were not detected. These results confirmed the previous findings of Sanderson et al. (2008). Another protein detected at low levels associated with the oocyst wall was gp40/15, a mucin-like glycoprotein that localises to the sporozoites surface and inner surface of the oocyst wall (Chatterjee et al. 2010; O'Connor et al. 2007). The authors also identified five possible oocyst wall proteins (POWPs) that were detected at low levels and are unique to *C. parvum*. When analysing the affinity purified glycoprotein components of the oocyst, gp40/15, gp900 and six other glycoproteins (CpMPA1 to CpMPA6) were identified, all are unique to *C. parvum*.

7.8 Publicly Accessible Transcriptomic and Proteomic Databases for *Cryptosporidium*

CryptoDB (http://cryptodb.org/cryptodb/) is an integrated database of genomic and functional genomic data from representative strains of *C. parvum*, *C. hominis* and *C. muris* (Heiges et al. 2006). It contains the most recently annotated genome sequences for each species alongside experimental data from the research community. CryptoDB is a part of EuPathDB (Aurrecoechea et al. 2007), a freely available and accessible resource providing easy access to a range of "-omics" data for *Cryptosporidium* and other protozoan parasites. Transcriptomic and proteomics data from a variety of studies, primarily on *C. parvum*, have been fully integrated into CryptoDB. By mapping the transcript and peptide data generated from these global expression studies to the genome, proteins that were not predicted by the original gene models have been identified and the existence and expression of hypothetical proteins have been verified. These expression data have provided empirical evidence that has been instrumental in improving the annotation of the genome (Sanderson et al. 2008).

Currently, CryptoDB (version 5.0) contains 98,058 ESTs, providing transcript evidence for 1,806 *C. parvum* (70 560 ESTs) and 580 *C. muris* RN66 (27,498 ESTs) genes along with RT-PCR temporal expression data for *C. parvum* (Mauzy et al. 2012). In addition, proteomics data from five independent studies have been integrated into CryptoDB (Sanderson et al. 2008; Snelling et al. 2007); at the moment this data is limited to the IOWA II strain of *C. parvum*. Expression data for 1,320 proteins, representing approximately a third of the predicted proteome, has been uploaded to the database for the oocyst and sporozoite stages of the parasite. This is a reasonable coverage (for both transcript and protein evidence) given that these life-cycle stages are unlikely to express stage specific proteins from the sexual stages and merozoites, although lower than for some other Coccidia, such as *Toxoplasma gondii* (Xia et al. 2008).

Genes of interest can be searched for by a variety of methods such as, by gene identifier (e.g. cgd6_2090) or by text search (e.g. oocyst wall protein), as illustrated in Fig. 7.3. Batch searches, such as "identify genes based on EST evidence", can also be performed to search the database for experimental data. An overview of the

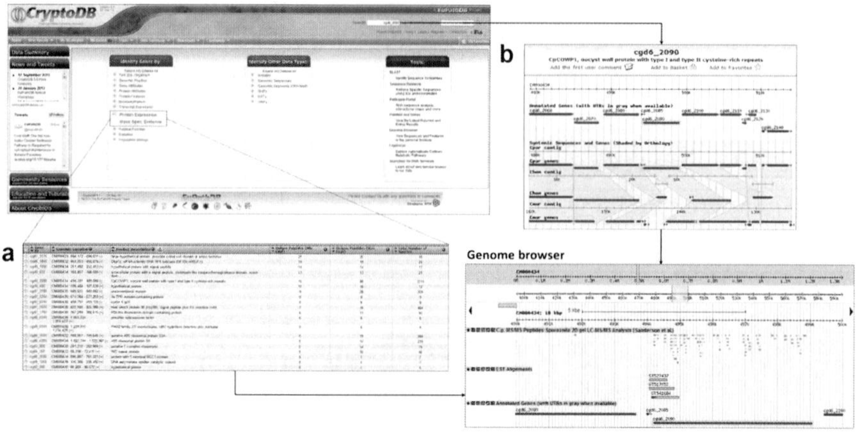

Fig. 7.3 Visualisation and mining of proteomic and transcriptomic data on CryptoDB (http://cryptodb.org/cryptodb/). In this example data for COWP1 (cgd6_2090) has been mined by (**a**) using tools such as 'Identify genes based on Mass Spec Evidence' and (**b**) gene ID or text searching. The resulting peptide, EST, sequence and gene model data can then be viewed in Genome Browser

data available for individual genes is displayed, including the location of the gene in the genome, any orthologs to it in other *Cryptosporidium* species along with any transcript and/or peptide evidence for the gene product and functional information. Data can also be visualised in GBrowse (Stein et al. 2002) for more in-depth data mining.

An alternative proteomics database is EPICDB, which contains protein predictions and mass spectrometry data for *T. gondii* and *C. parvum* (Madrid-Aliste et al. 2009). EPICDB also includes functional annotations, ESTs, open reading frames (ORFs) and experimental data. The curated Library of Apicomplexan Metabolic Pathways (LAMP) provides core metabolic pathways for *C. parvum*, *C. hominis* and *C. muris*, along with other apicomplexan parasites (Shanmugasundram et al. 2013). This database combines proteomic, genomic and transcriptomic data from EuPathDB with evidence from the literature to reconstruct metabolic pathways and networks.

7.9 Conclusions

The analysis of gene and protein expression in pathogens and their hosts has been transformed by the advent of high-throughput transcriptional analysis and proteomics. Both techniques have been applied to the study of *Cryptosporidium*, although there remains much to do. The depth of proteomics coverage of *Cryptosporidium* is still relatively poor compared to other *Apicomplexa* such as *Toxoplasma* and

Plasmodium, and almost nothing is known about gene expression in life-stages other than the oocyst and sporozoite. Nonetheless, the application of both transcriptional and proteomics analysis techniques to *Cryptosporidium* research has done much to help build a platform on which more targeted biological experiments can now be performed. However, both types of data, valuable as they are, need to be seen in context. The increasing abundance of simultaneously collected transcriptomics and proteomics data from both pathogens and their hosts reveals that the relationship between protein and transcript measurements is not always as faithful as might be supposed (Wastling et al. 2009). Since proteins are the functional molecules in a biological system, this adds complexity to making functional interpretations from steady state mRNA expression data alone. Moreover, the abundance of a protein is not always the most important feature, with alternative splicing and posttranslational modifications (PTMs), such as phosphorylation, also influencing its activity. Finally, proteins do not act in isolation and protein-protein interactions play an important role in the functioning of a cell/organism. All of this means that we are some way off from developing a truly comprehensive "systems" view of *Cryptosporidium* biology until we are able to provide greater coverage and complexity in transcriptional and proteomic studies. We anticipate these datasets will be acquired and made available to the research community in the near future, providing significant advances in our knowledge and understanding of this fascinating parasite.

References

Abrahamsen MS, Templeton TJ, Enomoto S et al (2004) Complete genome sequence of the apicomplexan, Cryptosporidium parvum. Science 304:441–445

Arrowood MJ (2002) In vitro cultivation of cryptosporidium species. Clin Microbiol Rev 15:390–400

Aurrecoechea C, Heiges M, Wang H et al (2007) ApiDB: integrated resources for the apicomplexan bioinformatics resource center. Nucleic Acids Res 35:D427–D430

Bradley PJ, Ward C, Cheng SJ et al (2005) Proteomic analysis of rhoptry organelles reveals many novel constituents for host-parasite interactions in Toxoplasma gondii. J Biol Chem 280:34245–34258

Bromley E, Leeds N, Clark J et al (2003) Defining the protein repertoire of microneme secretory organelles in the apicomplexan parasite Eimeria tenella. Proteomics 3:1553–1561

Castellanos-Gonzalez A, Yancey LS, Wang HC et al (2008) Cryptosporidium infection of human intestinal epithelial cells increases expression of osteoprotegerin: a novel mechanism for evasion of host defenses. J Infect Dis 197:916–923

Chatterjee A, Banerjee S, Steffen M et al (2010) Evidence for mucin-like glycoproteins that tether sporozoites of Cryptosporidium parvum to the inner surface of the oocyst wall. Eukaryot Cell 9:84–96

Chen XM, Levine SA, Splinter PL et al (2001) Cryptosporidium parvum activates nuclear factor kappaB in biliary epithelia preventing epithelial cell apoptosis. Gastroenterology 120:1774–1783

Chen YA, McKillen DJ, Wu S et al (2004) Optimal cDNA microarray design using expressed sequence tags for organisms with limited genomic information. BMC Bioinformatics 5:191

Current WL, Haynes TB (1984) Complete development of Cryptosporidium in cell culture. Science 224:603–605
Deng M, Lancto CA, Abrahamsen MS (2004) Cryptosporidium parvum regulation of human epithelial cell gene expression. Int J Parasitol 34:73–82
Elliott DA, Clark DP (2000) Cryptosporidium parvum induces host cell actin accumulation at the host-parasite interface. Infect Immun 68:2315–2322
Elliott DA, Coleman DJ, Lane MA et al (2001) Cryptosporidium parvum infection requires host cell actin polymerization. Infect Immun 69:5940–5942
Eng JK, McCormack AL, Yates JR (1994) An approach to correlate tandem mass spectral data of peptides with amino acid sequences in a protein database. J Am Soc Mass Spectrom 5:976–989
Glassmeyer ST, Ware MW, Schaefer FW 3rd et al (2007) An improved method for the analysis of Cryptosporidium parvum oocysts by matrix-assisted laser desorption/ionization time of flight mass spectrometry. J Eukaryot Microbiol 54:479–481
Gobert GN, Moertel LP, McManus DP (2005) Microarrays: new tools to unravel parasite transcriptomes. Parasitology 131:439–448
Griffiths JK, Moore R, Dooley S et al (1994) Cryptosporidium parvum infection of Caco-2 cell monolayers induces an apical monolayer defect, selectively increases transmonolayer permeability, and causes epithelial cell death. Infect Immun 62:4506–4514
Heiges M, Wang H, Robinson E et al (2006) CryptoDB: a Cryptosporidium bioinformatics resource update. Nucleic Acids Res 34:D419–D422
Hijjawi NS, Meloni BP, Morgan UM et al (2001) Complete development and long-term maintenance of Cryptosporidium parvum human and cattle genotypes in cell culture. Int J Parasitol 31:1048–1055
Hijjawi NS, Meloni BP, Ryan UM et al (2002) Successful in vitro cultivation of Cryptosporidium andersoni: evidence for the existence of novel extracellular stages in the life cycle and implications for the classification of Cryptosporidium. Int J Parasitol 32:1719–1726
Holt RA, Jones SJ (2008) The new paradigm of flow cell sequencing. Genome Res 18:839–846
Jenkins MC, O'Brien C, Miska K et al (2011) Gene expression during excystation of Cryptosporidium parvum oocysts. Parasitol Res 109:509–513
Liu J, Deng M, Lancto CA et al (2009) Biphasic modulation of apoptotic pathways in Cryptosporidium parvum-infected human intestinal epithelial cells. Infect Immun 77:837–849
Madrid-Aliste CJ, Dybas JM, Angeletti RH et al (2009) EPIC-DB: a proteomics database for studying Apicomplexan organisms. BMC Genomics 10:38
Magnuson ML, Owens JH, Kelty CA (2000) Characterization of Cryptosporidium parvum by matrix-assisted laser desorption ionization-time of flight mass spectrometry. Appl Environ Microbiol 66:4720–4724
Mauzy MJ, Enomoto S, Lancto CA et al (2012) The Cryptosporidium parvum transcriptome during in vitro development. PLoS One 7:e31715
Mortazavi A, Williams BA, McCue K et al (2008) Mapping and quantifying mammalian transcriptomes by RNA-Seq. Nat Methods 5:621–628
Mutz KO, Heilkenbrinker A, Lonne M et al (2013) Transcriptome analysis using next-generation sequencing. Curr Opin Biotechnol 24:22–30
O'Connor RM, Wanyiri JW, Cevallos AM et al (2007) Cryptosporidium parvum glycoprotein gp40 localizes to the sporozoite surface by association with gp15. Mol Biochem Parasitol 156:80–83
Perkins DN, Pappin DJC, Creasy DM et al (1999) Probability-based protein identification by searching sequence databases using mass spectrometry data. Electrophoresis 20:3551–3567
Putignani L, Tait A, Smith HV et al (2004) Characterization of a mitochondrion-like organelle in Cryptosporidium parvum. Parasitology 129:1–18
Roach JC, Boysen C, Wang K et al (1995) Pairwise end sequencing: a unified approach to genomic mapping and sequencing. Genomics 26:345–353
Sam-Yellowe TY, Florens L, Wang T et al (2004) Proteome analysis of rhoptry-enriched fractions isolated from Plasmodium merozoites. J Proteome Res 3:995–1001

Sanderson SJ, Xia D, Prieto H et al (2008) Determining the protein repertoire of Cryptosporidium parvum sporozoites. Proteomics 8:1398–1414

Shanmugasundram A, Gonzalez-Galarza FF, Wastling JM et al (2013) Library of Apicomplexan metabolic pathways: a manually curated database for metabolic pathways of apicomplexan parasites. Nucleic Acids Res 41:D706–D713

Snelling WJ, Lin Q, Moore JE et al (2007) Proteomics analysis and protein expression during sporozoite excystation of Cryptosporidium parvum (Coccidia, Apicomplexa). Mol Cell Proteomics 6:346–355

Stein LD, Mungall C, Shu S et al (2002) The generic genome browser: a building block for a model organism system database. Genome Res 12:1599–1610

Strong WB, Nelson RG (2000) Preliminary profile of the Cryptosporidium parvum genome: an expressed sequence tag and genome survey sequence analysis. Mol Biochem Parasitol 107:1–32

Templeton TJ, Iyer LM, Anantharaman V et al (2004) Comparative analysis of apicomplexa and genomic diversity in eukaryotes. Genome Res 14:1686–1695

Upton SJ, Tilley M, Brillhart DB (1994) Comparative development of Cryptosporidium parvum (Apicomplexa) in 11 continuous host cell lines. FEMS Microbiol Lett 118:233–236

Wakaguri H, Suzuki Y, Sasaki M et al (2009) Inconsistencies of genome annotations in apicomplexan parasites revealed by 5'-end-one-pass and full-length sequences of oligo-capped cDNAs. BMC Genomics 10:312

Wang Z, Gerstein M, Snyder M (2009) RNA-Seq: a revolutionary tool for transcriptomics. Nat Rev Genet 10:57–63

Wastling JM, Xia D, Sohal A et al (2009) Proteomes and transcriptomes of the Apicomplexa–where's the message? Int J Parasitol 39:135–143

Widmer G, Corey EA, Stein B et al (2000) Host cell apoptosis impairs Cryptosporidium parvum development in vitro. J Parasitol 86:922–928

Xia D, Sanderson SJ, Jones AR et al (2008) The proteome of Toxoplasma gondii: integration with the genome provides novel insights into gene expression and annotation. Genome Biol 9:R116

Xu P, Widmer G, Wang Y et al (2004) The genome of Cryptosporidium hominis. Nature 431:1107–1112

Yang YL, Serrano MG, Sheoran AS et al (2009) Over-expression and localization of a host protein on the membrane of Cryptosporidium parvum infected epithelial cells. Mol Biochem Parasitol 168:95–101

Yang YL, Buck GA, Widmer G (2010) Cell sorting-assisted microarray profiling of host cell response to Cryptosporidium parvum infection. Infect Immun 78:1040–1048

Zhang H, Guo F, Zhou H et al (2012) Transcriptome analysis reveals unique metabolic features in the Cryptosporidium parvum Oocysts associated with environmental survival and stresses. BMC Genomics 13:647

Zhou XW, Kafsack BF, Cole RN et al (2005) The opportunistic pathogen Toxoplasma gondii deploys a diverse legion of invasion and survival proteins. J Biol Chem 280:34233–34244

Chapter 8
Cryptosporidium Metabolism

Guan Zhu and Fengguang Guo

Abstract Rather than the presence of unique metabolic pathways, it is the absence of many pathways that characterizes the metabolism of *Cryptosporidium*. In fact, this genus of parasites has lost its ability of synthesizing de novo virtually all nutrients such as amino acids, nucleotides and fatty acids, thus relying on a large number of transporters to scavenge nutrients from the host. Members of this genus lack an apicoplast and associated pathways that are present in other apicomplexans. They lack cytochrome-based respiration, and rely mainly on glycolysis for energy production. Core metabolic pathways are highly streamlined, and redundancy is rare. These features make *Cryptosporidium* different from other apicomplexans. This chapter summarizes these features based on the analysis of genome sequences and published biochemical data in the context of drug targets and drug development.

8.1 Introduction

The genus *Cryptosporidium* comprises many species with different host specificities. *C. parvum* and *C. hominis* are the major human pathogens. The genomes of *C. parvum* and *C. hominis* have been reported (Abrahamsen et al. 2004; Xu et al. 2004), that of *C. muris* has been sequenced but remains to be published, and many more are being sequenced by an NIH-funded "Comparative Genomics of *Cryptosporidium* Species" project (http://www.genome.gov/26525388). The three sequenced genomes are accessible at GenBank and EuPathDB (http://www.EuPathDB.org). The availability of these genome sequences has greatly contributed to our understanding of parasite biology, and provided new opportunities to study *Cryptosporidium* metabolism. Currently, our

---

G. Zhu (✉) • F. Guo
Department of Veterinary Pathobiology, College of Veterinary Medicine & Biomedical Sciences, Texas A&M University, College Station, TX 77843-4467, USA
e-mail: gzhu@cvm.tamu.edu

S.M. Cacciò and G. Widmer (eds.), *Cryptosporidium: parasite and disease*,
DOI 10.1007/978-3-7091-1562-6_8, © Springer-Verlag Wien 2014

knowledge on *Cryptosporidium* metabolism is mainly derived from genome annotations and analyses. A growing but limited number of enzymes have been investigated for their functions, biochemical features and potentials as therapeutic targets. However, systematic studies at the pathway level are still limited. Two reviews on *Cryptosporidium* biochemistry and metabolism were published as a book chapter in 2008 and a journal article in 2010 (Rider and Zhu 2010; Zhu 2008), These publications were mainly based on the *C. parvum* and *C. hominis* genomes and biochemical data available then. In this chapter, we will highlight the major metabolic features in *Cryptosporidium* with the addition of the newly sequenced *C. muris* genome and up-to-date biochemical data.

8.2 General Features of *Cryptosporidium* Metabolism

Phylogenetic and phylogenomic analyses have consistently placed *Cryptosporidium* at the base of the Apicomplexa with a closer relationship to gregarines than coccidia and haematozoa (Barta and Thompson 2006; Templeton et al. 2010; Zhu et al. 2000a). This view differs from the conventional taxonomy that considers *Cryptosporidium* as a sister group to the intestinal coccidia. The divergence of cryptosporidia from coccidia is also supported at the genomic and metabolic level. *Cryptosporidium* species have highly compact genomes (i.e., ~10 Mb), which are 3–5 times smaller than those of other apicomplexans and feature short intergenic sequences and very few introns (Abrahamsen et al. 2004; Xu et al. 2004). More importantly, this genus of parasites has lost the ability to synthesize virtually any nutrients de novo, such as amino acids, nucleotides and fatty acids. *Cryptosporidium* also lacks an apicoplast and mitochondrial genomes and associated metabolic pathways that are present in coccidia and haematozoa (Abrahamsen et al. 2004; Xu et al. 2004; Zhu et al. 2000b). Members of the genus possess a mitochondrial remnant that lacks the cytochrome-based respiratory chain, and retains only limited functions such as the assembly of ion-sulfur clusters (Lei et al. 2010; Kang et al. 2008; Keithly et al. 2005; Slapeta and Keithly 2004; Roberts et al. 2004; Riordan et al. 2003). Therefore, *Cryptosporidium* metabolism is extremely simplified. This feature differentiates these species from the evolutionarily more closely related gregarines that are able to synthesize many nutrients (e.g., Templeton et al. 2010).

Because *Cryptosporidium* cannot synthesize most nutrients de novo, it relies on a large family of transporters to scavenge nutrients such as amino acids (~11 transporters), sugars (~9), and nucleotides (at least one) (Abrahamsen et al. 2004; Xu et al. 2004). About 24 putative ATP-binding cassette (ABC) transporters can be identified in the *C. parvum* genome, but their substrates remain to be determined (Benitez et al. 2007; Li and Mun 2005; Bonafonte et al. 2004; Zapata et al. 2002; Perkins et al. 1999). ABC transporters are a large family of proteins with various substrate preferences including ions, sugars, amino acids, peptides, lipids, sterols and drugs. Therefore, many of them could be utilized by the parasite to scavenge nutrients. The parasite also possesses at least seven P-type ATPases (P-ATPases) that are mostly involved in cation transport. Among them, a putative Ca^{2+}-ATPase and a heavy metal ATPase with binding specificity for reduced copper [Cu(I)] have been reported (LaGier et al. 2002, 2001; Zhu and Keithly 1997). One of the CpATPases belongs to the phospholipid transporter family and may be involved in lipid transport or membrane remodeling.

8.3 Carbohydrate and Energy Metabolisms

Carbohydrates are a source of energy and serve as building blocks for various biomolecules. *Cryptosporidium* is able to synthesize amylopectin as an energy storage polysaccharide. This biosynthetic capability is supported by earlier biochemical analyses and the presence of two glycogen branching enzymes (Harris et al. 2004; Zhang et al. 2012). The parasite can use polysaccharides, disaccharides or hexoses (e.g., glucose) to produce pyruvate and acetyl-CoA via the glycolytic pathway (Fig. 8.1). However, it employs two pyrophosphate-dependent phosphofructokinase (PPi-PFK) isoforms, rather than an ATP-PFK, to minimize the consumption of ATP. It also uses a unique pyruvate: $NADP^+$ oxidoreductase comprised of pyruvate:ferredoxin oxidoreductase (PFO) and P450 reductase domains, rather than a pyruvate dehydrogenase complex to convert pyruvate into acetyl-CoA (Abrahamsen et al. 2004; Ctrnacta et al. 2006; Rotte et al. 2001). It has been speculated that $NADP^+$ oxidoreductase is associated with the anti-cryptosporidial action of nitazoxanide (NTZ) that is currently the only drug approved by the FDA to treat cryptosporidial infections in immune-competent patients (Coombs and Muller 2002). Unlike classic inhibitor-enzyme interactions, NTZ may not act on $NADP^+$ oxidoreductase. Rather, it is converted to a biotoxic free radical molecule by the PFO domain, similar to the reductive activation of 5-nitroimidazole metronidazole by PFO in the anaerobic protists *Trichomonas* and *Giardia* (Leitsch et al. 2011; Crossnoe et al. 2002; Yarlett et al. 1986). To better understand the enzymes in the carbohydrate pathway, protein crystals have been obtained for *C. parvum* LDH, glyceraldehyde 3-phosphate dehydrogenase (G3PDH), and pyruvate kinase (Nguyen et al. 2011; Cook et al. 2009; Senkovich et al. 2005). The crystal structures of pyruvate kinase and triosephosphate isomerase are resolved, and the structure of the active site of pyruvate kinase displays no obvious difference compared to the human homologue (Nguyen et al. 2011; Cook et al. 2009).

There are at least two types of cryptosporidial mitochondrial remnants: the "intestinal-type" *C. parvum* and *C. hominis* lack both the tricarboxylic acid (TCA) cycle and the cytochrome-based respiratory chain (Abrahamsen et al. 2004; Xu et al. 2004); whereas the "gastric-type" *C. muris* similarly lacks the respiratory chain but, based on the *C. muris* genome annotation, retains a complete set of enzymes for the TCA cycle and a type II NADH dehydrogenase (unpublished). The absence of an electron transport chain in all three species indicates that their mitochondria are unlikely to be a major source of energy,

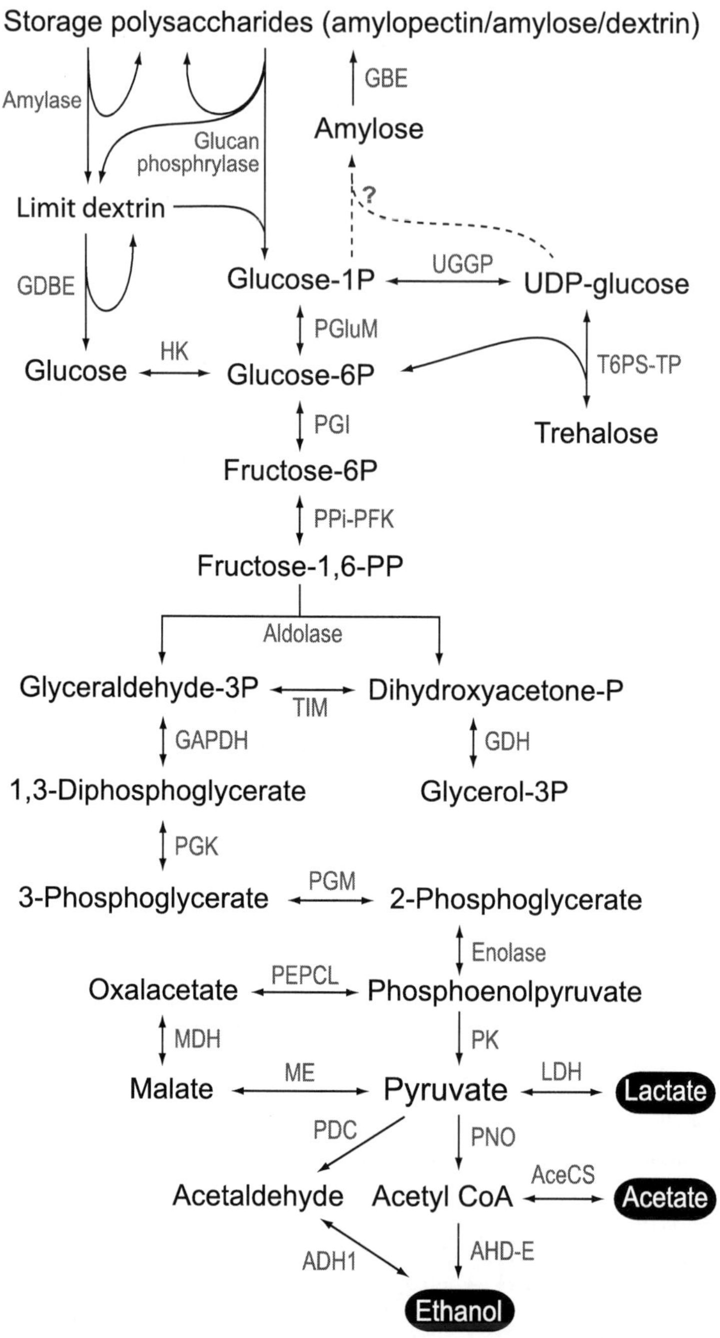

Fig. 8.1 Illustration of the carbohydrate metabolic pathway in *Cryptosporidium*, in which the core components are glycolytic and fermentative enzymes. Abbreviations: *AceCS* acetyl-CoA synthetase (also known as acetate-CoA ligase), *ADH1* alcohol dehydrogenase 1 (monofunctional), *ADH-E* type E alcohol dehydrogenase (bifunctional), *GDH* glycerol phosphate dehydrogenase, *GAPDH* glyceraldehyde phosphate dehydrogenase, *GBE* glycogen branching enzyme, *GDBE* glycogen

although it is possible that a certain level of electron potential may be generated via the type II NADH dehydrogenase in *C. muris*. Without respiration, the TCA cycle is possibly used by *C. muris* to supply intermediate metabolites. Additionally, all three genomes encode a plant-type alternative oxidase, which may be involved in detoxification of oxygen and/or generating certain electron potential.

Cryptosporidium does not carry a mitochondrial genome and the machinery for the replication, transcription and translation of organellar genomes. In addition to the enzymes discussed above, their nuclear genomes also encode a number of other proteins with mitochondrial-targeting signals, which include several translocases of outer and inner membranes, heat-shock proteins (HSPs), solute carriers, nucleotide anti-porters, ferredoxin and ferredoxin reductase and a small set of enzymes involved in ion-sulfur [Fe-S] cluster assembly (Abrahamsen et al. 2004; Lei et al. 2010; Kang et al. 2008; Mogi and Kita 2010; LaGier et al. 2003). It is believed that the [Fe-S] cluster assembly is one of the core functions retained in the mitochondrial remnant.

Collectively, we may conclude that *Cryptosporidium* relies mainly on glycolysis to produce energy. This notion is also supported by the presence of fermentative enzymes for producing three organic end products to avoid the accumulation of pyruvate and acetyl-CoA at the end of glycolysis (Fig. 8.1). These include lactate produced by lactate dehydrogenase (LDH), acetic acid by acetyl-CoA synthetase (AceCS; also known as acetate-CoA ligase, AceCL) and alcohol by a type-E bifunctional alcohol dehydrogenase (ADH-E) from acetyl-CoA or monofunctional ADH coupled with pyruvate decarboxylase from pyruvate (Zhang et al. 2012). Among them, LDH and ADH are bacterial-type enzymes. *Cryptosporidium* LDH originated from malate dehydrogenase (MDH) by a relatively recent gene duplication event which occurred after this genus separated from other apicomplexans. In fact, all apicomplexan MDH and LDH are bacterial-type, derived from an α-proteobacterial MDH (Zhu and Keithly 2002; Madern et al. 2004). Interestingly, a recent microarray-based transcriptome analysis also revealed that LDH has the highest level of expression among all genes in *C. parvum* oocysts, suggesting that the parasite mainly depends on LDH to keep the glycolytic pathway unobstructed in the external environment (Zhang et al. 2012).

Cryptosporidium possesses a plant-type pathway for synthesizing trehalose, which is accomplished by UDP-glucose/galactose pyrophosphorylase (UGGP) and a bifunctional enzyme fusion containing trehalose-6P synthase and trehalose phosphatase (T6PS-TP) (Yu et al. 2010). The presence of trehalose in *C. parvum* oocysts has been confirmed biochemically (Yu et al. 2010). However, the genome

Fig. 8.1 (continued) debranching enzyme, *HK* hexokinase, *LDH* lactate dehydrogenase, *MDH* malate dehydrogenase, *ME* malic-enzyme, *PDC* pyruvate decarboxylase, *PEPCL* phosphoenolpyruvate carboxylase, *PGI* phosphoglucose isomerase, *PGluM* phosphoglucose mutase, *PGK* phosphoglycerate kinase, *PGM* phosphoglycerate mutase, *PK* pyruvate kinase, *PNO* pyruvate: $NADP^+$ oxidoreductase, *PPi-PFK* pyrophosphate-dependent phosphofructokinase, *T6PS-TP* trehalose-6-phosphate synthase-trehalose phosphatase, *TIM* triosephosphate isomerase, *UGGP* UDP-galactose/glucose pyrophosphorylase lacks trehalase, suggesting that the parasite is unlikely to reuse trehalose as a carbon source, unless using reversed reactions by UGGP and T6PS-TP. This differs from the intestinal coccidian *Eimeria* that has a trehalase, but lacks UGGP and T6PS-TP. Instead, *Eimeria* possesses a mannitol cycle that is absent in most other apicomplexans, including *Cryptosporidium* (Coombs and Muller 2002; Schmatz 1989). Trehalose and mannitol are known to function as anti-desiccants, antioxidants or protein-stabilizing agents in microorganisms, plants and some invertebrates, thus likely playing an important role in protecting the parasite against environmental stress.

The glycolytic pathway also provides GDP-mannose derived from fructose-6P or mannose-6P for N-glycan biosynthesis. *Cryptosporidium* appears to have a complete set of enzymes for synthesizing N-glycans in the lumen of the endoplasmic reticulum (ER) (e.g., various asparagine-linked glycosylation [ALG] transferases and an oligosaccharidyl-lipid flippase RFT1). N-glycan synthesis is also connected to the GPI (glycosyl-phosphatidyl-inositol) anchor synthesis. Like other apicomplexans and protists, these parasites lack enzymes to make more complex N-glycans in the Golgi apparatus that are common in fungi and plants. The *C. parvum* genome encodes ~30 mucin-like proteins, many of which are (or are predicted to be) membrane or secretory proteins based on the presence of signal peptides (e.g., Cevallos et al. 2000a, b; Barnes et al. 1998; Chatterjee et al. 2010). Most of the mucins contain both N- and O-glycosylation sites, and at least four enzymes involved in mucin-type O-glycosylation have been identified in the *C. parvum* genome, including UDP-N-acetyl-D-galactosamine-polypetide N-acetylgalactosaminyl transferases (Wanyiri and Ward 2006). The involvement of mucins in parasite attachment to, and invasion of, host cells is being actively investigated. Binding of some of the mucin-like proteins by antibodies can block or reduce infection in vitro and/or in vivo, suggesting that mucins may be targets for developing immunotherapeutics (Wanyiri and Ward 2006).

8.4 Amino Acid Metabolism

Cryptosporidium cannot synthesize any amino acids de novo, which differs from other apicomplexans that possess complete pathways to make at least some amino acids. Instead, it retains only enzymes to interconvert certain amino acids coupled with other metabolic pathways. These include: glutamine synthetase for recycling glutamate produced by GMP synthetase back to glutamine; serine hydroxymethyl transferase for converting glycine to serine within the folate cycle; asparagine synthetase for producing asparagine from aspartate (which may be required to recycle ammonia released by AMP-deaminase); S-adenosylmethionine (SAM or AdoMet) synthetase to catalyze the formation of SAM to serve as an important methyl donor for transmethylation; and S-adenosylhomocysteine (SAH) synthase (SAHS; also known as SAH hydrolase, SAHH) for converting SAH derived from SAM after transmethylation to homocysteine and adenosine.

Among these enzymes, the general molecular and biochemical features of *C. parvum* SAHH (CpSAHH) has been characterized. Its inhibitors D-eritadenine and 9-(S)-(2, 3-dihydroxypropyl)adenine [(S)-DHPA] display efficacy at low micromolar levels against growth of *C. parvum* in vitro (Ctrnacta et al. 2007, 2010).

8.5 Nucleotide Metabolism

Most apicomplexans scavenge purines, but are capable of synthesizing pyrimidines de novo. However, *Cryptosporidium* lacks synthetic pathways for both purines and pyrimidines. The highly simplified purine salvage pathway starts with the uptake of adenosine by a nucleoside transporter. Adenosine is converted to AMP by adenosine kinase, IMP by AMP deaminase, XMP by IMP dehydrogenase (IMPDH), and GMP by GMP synthase (Fig. 8.2) (Striepen and Kissinger 2004). IMPDH genes in *C. parvum* and *C. hominis* were acquired from ε-proteobacteria, and differ from the eukaryotic type IMPDHs found in humans, animals and other apicomplexans (Striepen et al. 2002). However, an IMPDH gene has yet to be identified in the current version of the *C. muris* genome. Based on the essential role of this enzyme, *C. muris* needs IMPDH unless it directly scavenges GMP from the host. The *C. parvum* and *C. hominis* IMPDH genes are located at the end of chromosome 6. As this region is not represented in the sequenced *C. muris* genome, it is likely that a *C. muris* IMPDH gene is located in an unsequenced region.

Because *Cryptosporidium* lacks hypoxanthine-xanthine-guanine phosphoribosyl transferase to serve as an alternative purine salvaging pathway, blocking this AMP-GMP pathway can effectively kill the parasite. The bacterial-type IMPDH that is highly divergent from humans and animals has been considered an attractive drug target. Its protein structure has been determined, and a number of potent inhibitors have been identified and are being evaluated for drug development (Johnson et al. 2013; Gorla et al. 2012; Sharling et al. 2010; Umejiego et al. 2004, 2008).

For pyrimidine salvaging, *Cryptosporidium* may utilize uracil, uridine and cystidine by converting them to UMP and CMP by uridine kinase (UK) and a bifunctional enzyme with UK fused to uracil phosphoribosyltransferase (UK-UPRT) (Fig. 8.2). Thymidine can also be used and converted to dTMP by a bacterial-type thymidine kinase (TK), which is subsequently converted to dUMP by thymidylate synthase (TS) coupled with the folate cycle, and to dCMP by dCMP deaminase. dCMP, CMP and UMP can be further converted to dCDP/dCTP, CDP/CTP and UDP/UTP by a multi-functional UMP kinase and UDP kinase. The three pyrimidine nucleotide pathways are interconnected by a ribonucleoside-diphosphate reductase enzyme. Therefore, inhibition of a single pathway may be insufficient to block the pyrimidine synthesis, unless interconvertion is restricted by the presence of rate-limiting enzymes and/or if the supply of any single source of precursors is limited. Although the pyrimidine salvaging pathways appear to be redundant, a recent study has shown that

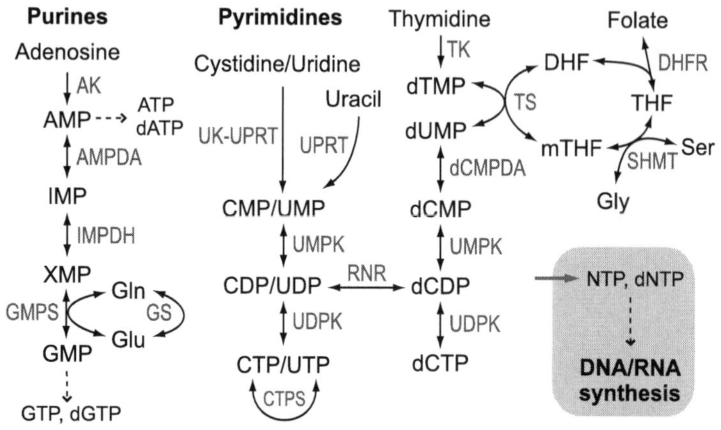

Fig. 8.2 Streamlined purine and pyrimidine salvaging pathways and folate cycle in *Cryptosporidium*. Abbreviations: *AK* adenosine kinase, *AMPDA* AMP deaminase, *CTPS* CTP synthase, *dCMPDA* dCMP deaminase, *DHFR* dihydrofolate reductase, *GMPS* GMP synthase (glutamine-hydrolyzing), *IMPDH* IMP dehydrogenase, *RNR* ribonucleoside-diphosphate reductase, *SHMT* serine hydroxymethyl transferase, *TK* thymidine kinase, *TS* thymidylate synthase, *UDPK* UDP/CDP kinase, *UK* uridine kinase, *UMPK* UMP/CMP kinase, *UPRT* uracil phosphoribosyltransferase

TK-mediated pro-drug activation may be utilized as an effective strategy for treating cryptosporidiosis (Sun et al. 2010).

In the folate cycle, dihydrofolate reductase and thymidylate synthase (DHFR-TS) are fused into a bifunctional enzyme in apicomplexans and some other protists (Vasquez et al. 1996). The linker between the DHFR and TS domains in *Cryptosporidium* is unique as it contains an 11-residue α-helix with extensive interactions with the opposite DHFR-TS monomer of the homodimeric enzyme (O'Neil et al. 2003). The active site of *C. parvum* DHFR contains unique residues that are analogous to the point mutations associated with antifolate resistance in other DHFRs, suggesting CpDHFR may be intrinsically resistant to some antifolate inhibitors (Vasquez et al. 1996). However, several novel CpDHFR inhibitors have been identified using a yeast complementation system and structure-based virtual screens, but their anti-cryptosporidial activity in vitro or in vivo remains to be determined (Senkovich et al. 2009; Martucci et al. 2009; Bolstad et al. 2008; Popov et al. 2006; Anderson 2005; Lau et al. 2001; Brophy et al. 2000).

8.6 Lipid Metabolism

Fatty acids are a source of energy in many organisms and major components of all biomembranes. However, *Cryptosporidium* is unable to use fatty acids as an energy source due to the absence of the β-oxidation pathway. It also lacks an apicoplast and its associated pathways such as isoprenoid synthesis and the Type II fatty acid synthase (FAS) system. Therefore, the parasite cannot synthesize fatty acids (Zhu 2004). However, *Cryptosporidium* possesses a 25-kb intronless Type I FAS gene that resembles bacterial polyketide synthase (PKS) and predicts a ~900 kDa megasynthase comprised of at least 21 enzymatic domains (Zhu et al. 2000c). The basic biochemical features have been studied using recombinant proteins. Its N-terminal loading unit containing an acyl-ligase (AL) and an acyl-carrier protein (ACP) has a substrate preference towards long chain fatty acids (LCFAs), indicating that CpFAS1 functions as a fatty acid "elongase" rather than synthesizing fatty acids (Zhu et al. 2004). This notion is further supported by functional analysis of its C-terminal reductase domain that is only active with very long chain (VLC) fatty acyl-CoAs (i.e., >C20:0) (Zhu et al. 2010). The reductase domain-catalyzed reductive reaction may release final products as fatty acyl aldehydes or fatty acyl alcohol, which differs from classic Type I FAS (in humans and animals) and Type II FAS (in prokaryotes, plants and plastid-containing apicomplexans) that use thioesterase to release acyl chains as fatty acids by hydrolysis (Zhu 2004). Between the loading unit and the reductase domain are three internal acyl elongation modules, each consisting of a complete set of 6 enzymes: (1) ACP for carrying acyl chains; (2) acyl-transferase (AT) for loading malonyl-CoA and transferring an acyl-chain from the previous module to ACP; (3) ketoacyl-ACP synthase (KS) for condensing a two-carbon (C2) unit from malonyl-CoA into the acyl-chain by a carboxylation reaction; (4) ketoacyl-ACP reductase for the reduction of a keto group; (5) hydroxyacyl-ACP dehydrase for the dehydration of a hydroxyl group; and (6) enoyl-ACP reductase for the reduction of double bonds (Zhu 2004; Zhu et al. 2000c, 2004, 2010). Therefore, at least three C2 units can be added into the acyl precursors (e.g., C16:0 palmitic acid) to form very long fatty acyl chains (e.g., C22:0) that are released as fatty acyl aldehydes or alcohol. Because aldehydes are biotoxic, fatty acyl alcohols are likely the final products that can be produced by two series of reductive reactions.

In parallel to Type I FAS, the *Cryptosporidium* genome also encodes a giant 45-kb intronless PKS, which represents the first PKS discovered in a protist (Zhu et al. 2002). Molecular and biochemical analysis reveals that CpPKS1 is similarly structured as CpFAS1, but contains seven internal acyl elongation modules that lack one or more of the five enzymatic domains. Therefore, elongated acyl chains will contain keto groups, hydroxyl groups and/or double bonds, which are characteristic of polyketides. The CpPKS1 loading unit also displays substrate preference towards LCFAs, suggesting the final product(s) may contain 30 or more carbons (Fritzler and Zhu 2007).

The ACP domains in all types of FAS and PKS systems require a post-translational modification by phosphopantetheinyl transferase (PPT) to add a prosthetic phosphopantetheine to a serine residue to become a functional holo-ACP. There are two types of PPT with different substrate preferences: SFP-type for activating Type I ACPs and ACPS-type for Type II ACPs. In fact, the types of PPT present in various apicomplexans match well with the types of FAS systems. For example, *Cryptosporidium* possesses only Type I FAS and SFP-PPT,

Plasmodium has only Type II FAS and ACPS-PPT, while *Toxoplasma* contains both types I and II FAS and both SFP-PPT and ACPS-PPT. The activation of CpFAS1-ACP domains by SFP-PPT has been biochemically characterized and demonstrated (Cai et al. 2005).

In addition to type I FAS and PKS, *Cryptosporidium* possesses another set of enzymes capable of elongating fatty acyl chains, in which the substrates are fatty acyl-CoA thioesters, rather than acyl-ACP. The hallmark enzyme is a long chain acyl elongase (LCE). The biochemical features of CpLCE1 have been characterized using recombinant protein expressed in HEK-293 T cells as its expression in bacterial systems was found to be difficult (Fritzler et al. 2007). CpLCE1 is a membrane protein localized on the surface of sporozoites and the parasitophorous vacuole membrane (PVM). Localization on the surface contrasts with CpFAS1 and CpPKS1 that are mainly cytosolic. CpLCE1 is able to add a single C2 unit to LCFAs with substrate preference towards C14:0 myristoyl-CoA and C16:0 palmitoyl-CoA (Fritzler et al. 2007).

Long chain fatty acyl-CoA synthetase (ACS; aka fatty acid-CoA ligase, ACL) is another family of important enzymes in lipid metabolism. It catalyzes the first reaction in all fatty acid metabolisms by activating free fatty acids to form fatty acyl-CoA thioesters except for the Type I and II FAS/PKS systems. In fact, AL domains in Type I FAS/PKS systems share similar molecular and biochemical properties with ACS, and can also catalyze the formation of fatty acyl-CoA (Fritzler and Zhu 2007). *Cryptosporidium* possesses three ACS genes, which are under investigation in our laboratory. Our data have shown that CpACS enzymes prefer LCFAs as substrates and their inhibitors could inhibit the growth of *C. parvum*, suggesting that ACS can be explored as a novel therapeutic target in the parasite (unpublished observation).

Cryptosporidium also possesses a long-type fatty acyl-CoA binding protein (ACBP) that is responsible for restraining the movement of fatty acyl-CoA and/or forming an acyl-CoA pool in cells. This function is important, as free fatty acyl-CoA may be harmful to cellular membranes due to its "detergent effect" if it is not restrained or immediately routed into other metabolic pathways. CpACBP1 is also a membrane protein localized to PVM. It prefers binding to LC and VLC fatty acyl-CoA thioesters (Zeng et al. 2006). More recently, a fluorescence assay was developed for CpACBP1 and used to screen 1,040 known drugs, from which 28 drugs displayed inhibitory effects on CpACBP1 at sub-micromolar concentrations. Among them, four drugs (i.e., broxyquinoline, cloxyquin, cloxacillin sodium and sodium dehydrocholate) displayed efficacies against the growth of *C. parvum* with ID_{50} values at low micromolar levels. This observation raises hopes for potential repurposing of known drugs to treat cryptosporidiosis (Fritzler and Zhu 2012).

Two distinct oxysterol binding protein (OSBP)-related proteins (ORPs) have also been identified in *C. parvum*, designated as CpORP1 and CpORP2 (Zeng and Zhu 2006). The short-type CpOPR1 contains only a ligand binding domain, while the long-type CpORP2 contains Pleckstrin homology and ligand-binding domains. Lipid–protein overlay assays have revealed that CpORP1 and CpORP2 could specifically bind to phosphatidic acid, various phosphatidylinositol phosphates

(PIPs), and sulfatide, but not to other types of lipids with simple heads. However, cholesterol was not a ligand for these two proteins. Like CpLCE1 and ACBP, CpORP1 also localized to the PVM, while CpORP2 localized only in intracellular merozoites (Zeng and Zhu 2006).

Due to the incapability of synthesizing fatty acids de novo, *Cryptosporidium* needs to scavenge fatty acids/lipids from the host. However, it is unclear how the parasite scavenges lipids as no specific fatty acid or lipid transporters have been identified or experimentally validated. A recent study has provided strong evidence that *C. parvum* is able to scavenge cholesterol from host cells and from the intestinal lumen (Ehrenman et al. 2013). Among lipoproteins, LDL is an important source of cholesterol, and *C. parvum* can obtain cholesterol that is incorporated into micelles and internalized into enterocytes by the NPC1L1 transporter. Pharmacological blockage of NPC1L1 function by ezetimibe or moderate down-regulation of NPC1L1 expression decreases parasite infectivity (Ehrenman et al. 2013).

The presence of acyl-CoA binding protein, long-chain acyl elongase, acyl-CoA synthetase and OSBP-related protein in PVM indicates that this unique membrane structure is involved in lipid metabolism including transport, activation and/or elongation of fatty acids in *Cryptosporidium*. Some ACS proteins in bacteria and yeast are also known to function as fatty acid transporters (Black and DiRusso 2007; DiRusso and Black 1999). Based on the most recent data, we have formulated a working hypothesis that fatty acids may be directly transported into the parasite via an undefined pathway(s) as free fatty acids, or by a PVM-specific ACS coupled with the formation of fatty acyl-CoA (Fig. 8.3). Free fatty acids may be elongated by the Type I FAS or PKS, or activated by ACS within the parasite to form fatty acyl-CoA. All fatty acyl-CoA thioesters may be immediately routed into subsequent metabolic pathways (such synthesis of complex lipids and biomembranes, or protein palmitoylation), undergo chain elongation by LCE, or bound to ACBP to form an acyl-CoA pool before entering subsequent pathways. *Cryptosporidium* may also scavenge phospholipids as implied by the presence of a type IV P-ATPase (cgd7_1760) with predicted substrate affinity towards phospholipids. However, it remains to be determined if this transporter is truly involved in lipid scavenging or is simply responsible for intracellular trafficking of phospholipids in parasite cells.

Cryptosporidium has a small set of enzymes involved in synthesis of complex lipids. In glycerolipid synthesis, it only retains enzymes for conversions between 1, 2-diacyl-sn-glyccrol-3-phosphate, 1,2 diacyl-sn-glycerol and triacylglycerol by phosphatidate phosphatase, diacylglycerol kinase, and diacylglycerol acyltransferase 1, indicating that the parasite may store fatty acids in the form of triacylglycerol. Phosphatidyl-ethanolamine may be synthesized from diacyl-sn-glycerol and CDP-ethanolamine (produced from phospho-ethanolamine) by ethanolamine-phosphotransferase (ETHPT), or via the diacyl-sn-glycerol-3-phosphate to CDP-diacyl-glycerol to phosphatidyl-L-serine to phosphatidyl-entholamine pathway by CDP-diacylglycerol synthase (CDS), phosphatidylserine synthase and phosphatidylserine decarboxylase (PSDC). However, a phosphatidylserine synthase gene has not be identified in the *Cryptosporidium* genomes.

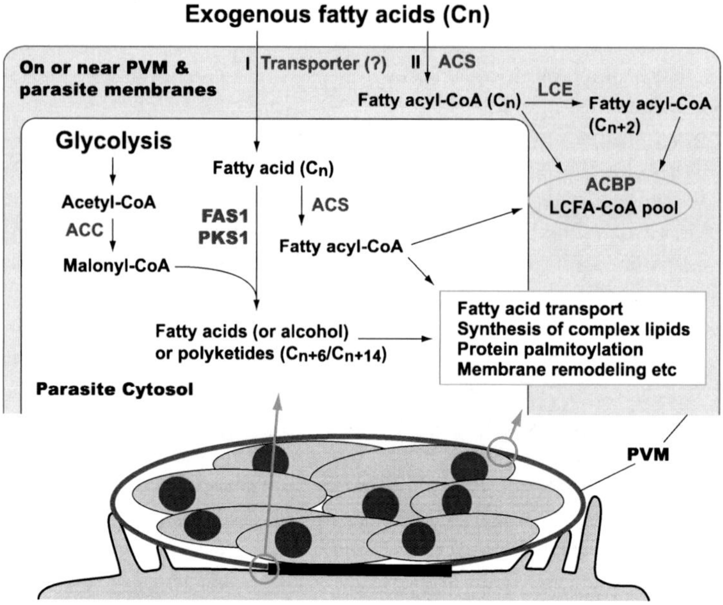

Fig. 8.3 Working hypothesis on the fatty acid metabolism associated with the parasitophorous vacuole membrane (PVM) in *Cryptosporidium*. Abbreviations: *ACBP* fatty acyl-CoA binding protein, *ACC* acetyl-CoA carboxylase, *ACS* Fatty acyl-CoA synthetase, *FAS1* Type I FAS, *LCE* long chain fatty acyl-elongase, *LCFA-CoA* long chain fatty acyl-CoA, *PKS1* Type I PKS

Additionally, enzymes involved in synthesizing lipoproteins and glycolipids are present in the *Cryptosporidium* genomes, which include up to nine DHHC family palmitoyl transferases for post-translational S-palmitoylation of proteins, and PIG-A, PIG-C, PIG-H, PIG-P, PIG-M and PIG-U involved in GPI anchor synthesis.

8.7 Stress-Related Pathways

Cryptosporidium must face various external and internal environmental stresses, including UV irradiation, temperature changes and dehydration in the natural environment, as well as hazardous chemicals, free radical molecules, drugs and host immune responses at various life cycle stages. The parasite genome encodes several classes of proteins that are important in handling these stresses, including heat shock proteins (HSPs) and various anti-oxidant molecules. HSPs are well known stress proteins that are generally up-regulated as part of the stress responses by participating in protein folding, maintaining proper protein conformation, and monitoring cellular proteins. The expression of the *CpHSP70* gene has been found to be highly up-regulated in response to chlorine-based oxidants and heat treatment (Bajszar and Dekonenko 2010).

Cryptosporidium possesses a number of anti-oxidant molecules, including superoxide dismutase (SOD), glutathione S-transferase, glutathione peroxidase, thioredoxin reductase, and a number of thioredoxin related proteins (e.g., Kang et al. 2008; Zhang et al. 2012; Yoon et al. 2012). In *C. parvum* oocysts, all three glutaredoxin-associated genes and five out of 13 thioredoxin-associated genes are expressed at various levels. Upon UV treatment, three putative t-complex protein 1 (TCP-1) subunits and several thioredoxin-associated genes are up-regulated in oocysts, whereas various HSP/DNAj family members, SOD and glutaredoxin-related genes are not up-regulated or are even down-regulated, suggesting that different stress proteins play different roles in responses to different stresses (Zhang et al. 2012).

DNA damage may occur more frequently in the natural environment than during DNA replication in the host cell due to the exposure of oocysts to UV irradiation. *Cryptosporidium* possesses a machinery for DNA excision repair. Several genes encoding excision repair enzymes were found to be up-regulated upon UV treatment, including the replication protein large subunit 1B (CpRPA1B) (Zhang et al. 2012; Rider and Zhu 2008; Rochelle et al. 2005). There are two types of RPA1 proteins in *Cryptosporidium* (i.e., RPA1A and RPA1B) which are involved in DNA replication, repair and recombination (Rider and Zhu 2008; Zhu et al. 1999; Millership and Zhu 2002). Several studies have indicated that RPA1A is mainly responsible for general DNA replication in the parasite, whereas RPA1B may play a role in DNA recombination and repair (Rider and Zhu 2008; Rider et al. 2005).

Trehalose synthesis is another important anti-stress pathway as described in Sect. 8.3 (Yu et al. 2010). Despite *Cryptosporidium* lacking amino acid synthetic pathways, a single standalone bacterial-type tryptophan synthase β-subunit gene is present in the *C. parvum* and *C. hominis* genomes. It is known that tryptophan starvation is one of the innate immunity's strategies in humans and animals to kill cells infected with certain pathogens including *Toxoplasma* and probably *Cryptosporidium* by activating the tryptophan degradation pathway (MacKenzie et al. 2007; Habara-Ohkubo et al. 1993). It is possible that *Cryptosporidium* may use tryptophan synthase β-subunit to synthesize tryptophan from indole present in the gut to evade tryptophan depletion.

8.8 Conclusions

Cryptosporidium is extremely well adapted to a parasitic life style. These parasites rely on the host to supply virtually all nutrients for their highly streamlined metabolic pathways. The insensitivity of *Cryptosporidium* to many anti-apicomplexan drugs is explained by the absence of the drug targets which are common in other apicomplexans. Examples of such pathways are de novo biosynthetic pathways and the cytochrome-based respiratory chain. Evolutionary divergence from other apicomplexans (e.g., bacterial type IMPDH and DHFR-TS with unusual amino acids at the active site) was also observed and may be relevant to drug design. The availability of whole genome sequences provides opportunities not only to better understand the unique metabolic features of these parasites, but also to identify key enzymes and study their biochemical features of interest to drug development. Indeed, a number of potential drug targets have been proposed and/or are currently pursued by various laboratories, including protein kinases and enzymes in the carbohydrate, energy, nucleotide, and fatty acid metabolic pathways.

Currently, research on *Cryptosporidium* is still hampered by the lack of genetic tools and by technical difficulties in manipulating of the parasite in the laboratory. Therefore, recombinant protein-based biochemical analysis is still the best approach to study the metabolism and characterize potential drug targets. On the other hand, although gene knockout or knockdown tools are not available to validate drug targets, we can still effectively predict potential drug targets in the streamlined core metabolic pathways that are essential to the parasite. Drug development against cryptosporidial infection has been progressing slowly, but promising new data are being reported (see also Chap. 11). Research will be accelerated by new knowledge generated by biochemical analysis, target-based high-throughput screening of drugs, and structure-based analysis of protein-inhibitor interactions.

Acknowledgements We thank Dr. J. M. Fritzler at Weber State University and Dr. S. D. Rider at Wright State University for their critical reading of the manuscript. Studies derived from the author's laboratory have been mainly supported by grants from the National Institute of Allergy and Infectious Diseases, National Institutes of Health.

References

Abrahamsen MS, Templeton TJ, Enomoto S, Abrahante JE, Zhu G, Lancto CA et al (2004) Complete genome sequence of the apicomplexan, *Cryptosporidium parvum*. Science 304 (5669):441–445. doi:10.1126/science.1094786

Anderson AC (2005) Targeting DHFR in parasitic protozoa. Drug Discov Today 10(2):121–128. doi:10.1016/S1359-6446(04)03308-2

Bajszar G, Dekonenko A (2010) Stress-induced Hsp70 gene expression and inactivation of *Cryptosporidium parvum* oocysts by chlorine-based oxidants. Appl Environ Microbiol 76 (6):1732–1739. doi:10.1128/AEM.02353-09

Barnes DA, Bonnin A, Huang JX, Gousset L, Wu J, Gut J et al (1998) A novel multi-domain mucin-like glycoprotein of *Cryptosporidium parvum* mediates invasion. Mol Biochem Parasitol 96(1–2):93–110

Barta JR, Thompson RC (2006) What is *Cryptosporidium*? Reappraising its biology and phylogenetic affinities. Trends Parasitol 22(10):463–468. doi:10.1016/j.pt.2006.08.001

Benitez AJ, McNair N, Mead J (2007) Modulation of gene expression of three *Cryptosporidium parvum* ATP-binding cassette transporters in response to drug treatment. Parasitol Res 101 (6):1611–1616. doi:10.1007/s00436-007-0701-x

Black PN, DiRusso CC (2007) Vectorial acylation: linking fatty acid transport and activation to metabolic trafficking. Novartis Found Symp 286:127–138, discussion 38–41, 62–3, 96–203

Bolstad DB, Bolstad ES, Frey KM, Wright DL, Anderson AC (2008) Structure-based approach to the development of potent and selective inhibitors of dihydrofolate reductase from *Cryptosporidium*. J Med Chem 51(21):6839–6852. doi:10.1021/jm8009124

Bonafonte MT, Romagnoli PA, McNair N, Shaw AP, Scanlon M, Leitch GJ et al (2004) *Cryptosporidium parvum*: effect of multi-drug reversing agents on the expression and function of ATP-binding cassette transporters. Exp Parasitol 106(3–4):126–134. doi:10.1016/j.exppara.2004.03.012

Brophy VH, Vasquez J, Nelson RG, Forney JR, Rosowsky A, Sibley CH (2000) Identification of *Cryptosporidium parvum* dihydrofolate reductase inhibitors by complementation in *Saccharomyces cerevisiae*. Antimicrob Agents Chemother 44(4):1019–1028

Cai X, Herschap D, Zhu G (2005) Functional characterization of an evolutionarily distinct phosphopantetheinyl transferase in the apicomplexan *Cryptosporidium parvum*. Eukaryot Cell 4(7):1211–1220. doi:10.1128/EC.4.7.1211-1220.2005

Cevallos AM, Bhat N, Verdon R, Hamer DH, Stein B, Tzipori S et al (2000a) Mediation of *Cryptosporidium parvum* infection in vitro by mucin-like glycoproteins defined by a neutralizing monoclonal antibody. Infect Immun 68(9):5167–5175

Cevallos AM, Zhang X, Waldor MK, Jaison S, Zhou X, Tzipori S et al (2000b) Molecular cloning and expression of a gene encoding *Cryptosporidium parvum* glycoproteins gp40 and gp15. Infect Immun 68(7):4108–4116

Chatterjee A, Banerjee S, Steffen M, O'Connor RM, Ward HD, Robbins PW et al (2010) Evidence for mucin-like glycoproteins that tether sporozoites of *Cryptosporidium parvum* to the inner surface of the oocyst wall. Eukaryot Cell 9(1):84–96. doi:10.1128/EC.00288-09

Cook WJ, Senkovich O, Chattopadhyay D (2009) An unexpected phosphate binding site in glyceraldehyde 3-phosphate dehydrogenase: crystal structures of apo, holo and ternary complex of *Cryptosporidium parvum* enzyme. BMC Struct Biol 9:9. doi:10.1186/1472-6807-9-9

Coombs GH, Muller S (2002) Recent advances in the search for new anti-coccidial drugs. Int J Parasitol 32(5):497–508

Crossnoe CR, Germanas JP, LeMagueres P, Mustata G, Krause KL (2002) The crystal structure of *Trichomonas vaginalis* ferredoxin provides insight into metronidazole activation. J Mol Biol 318(2):503–518. doi:10.1016/S0022-2836(02)00051-7

Ctrnacta V, Ault JG, Stejskal F, Keithly JS (2006) Localization of pyruvate: $NADP^+$ oxidoreductase in sporozoites of *Cryptosporidium parvum*. J Eukaryot Microbiol 53(4):225–231. doi:10.1111/j.1550-7408.2006.00099.x

Ctrnacta V, Stejskal F, Keithly JS, Hrdy I (2007) Characterization of S-adenosylhomocysteine hydrolase from *Cryptosporidium parvum*. FEMS Microbiol Lett 273(1):87–95. doi:10.1111/j.1574-6968.2007.00795.x

Ctrnacta V, Fritzler JM, Surinova M, Hrdy I, Zhu G, Stejskal F (2010) Efficacy of S-adenosylhomocysteine hydrolase inhibitors, D-eritadenine and (S)-DHPA, against the growth of *Cryptosporidium parvum* in vitro. Exp Parasitol 126(2):113–116. doi:10.1016/j.exppara.2010.04.007

DiRusso CC, Black PN (1999) Long-chain fatty acid transport in bacteria and yeast. Paradigms for defining the mechanism underlying this protein-mediated process. Mol Cell Biochem 192 (1–2):41–52

Ehrenman K, Wanyiri JW, Bhat N, Ward HD, Coppens I (2013). *Cryptosporidium parvum* scavenges LDL-derived cholesterol and micellar cholesterol internalized into enterocytes. Cell Microbiol 15(7):1182–1197. doi:10.1111/cmi.12107

Fritzler JM, Zhu G (2007) Functional characterization of the acyl-[acyl carrier protein] ligase in the *Cryptosporidium parvum* giant polyketide synthase. Int J Parasitol 37(3–4):307–316. doi:10.1016/j.ijpara.2006.10.014

Fritzler JM, Zhu G (2012) Novel anti-*Cryptosporidium* activity of known drugs identified by high-throughput screening against parasite fatty acyl-CoA binding protein (ACBP). J Antimicrob Chemother 67(3):609–617. doi:10.1093/jac/dkr516

Fritzler JM, Millership JJ, Zhu G (2007) *Cryptosporidium parvum* long-chain fatty acid elongase. Eukaryot Cell 6(11):2018–2028. doi:10.1128/EC.00210-07

Gorla SK, Kavitha M, Zhang M, Liu X, Sharling L, Gollapalli DR et al (2012) Selective and potent urea inhibitors of *Cryptosporidium parvum* inosine 5′-monophosphate dehydrogenase. J Med Chem 55(17):7759–7771. doi:10.1021/jm3007917

Habara-Ohkubo A, Shirahata T, Takikawa O, Yoshida R (1993) Establishment of an antitoxoplasma state by stable expression of mouse indoleamine 2,3-dioxygenase. Infect Immun 61(5):1810–1813

Harris JR, Adrian M, Petry F (2004) Amylopectin: a major component of the residual body in *Cryptosporidium parvum* oocysts. Parasitology 128(Pt 3):269–282

Johnson CR, Gorla SK, Kavitha M, Zhang M, Liu X, Striepen B et al (2013) Phthalazinone inhibitors of inosine-5′-monophosphate dehydrogenase from *Cryptosporidium parvum*. Bioorg Med Chem Lett 23(4):1004–1007. doi:10.1016/j.bmcl.2012.12.037

Kang JM, Cheun HI, Kim J, Moon SU, Park SJ, Kim TS et al (2008) Identification and characterization of a mitochondrial iron-superoxide dismutase of *Cryptosporidium parvum*. Parasitol Res 103(4):787–795. doi:10.1007/s00436-008-1041-1

Keithly JS, Langreth SG, Buttle KF, Mannella CA (2005) Electron tomographic and ultrastructural analysis of the *Cryptosporidium parvum* relict mitochondrion, its associated membranes, and organelles. J Eukaryot Microbiol 52(2):132–140. doi:10.1111/j.1550-7408.2005.04-3317.x

LaGier MJ, Zhu G, Keithly JS (2001) Characterization of a heavy metal ATPase from the apicomplexan *Cryptosporidium parvum*. Gene 266(1–2):25–34

LaGier MJ, Keithly JS, Zhu G (2002) Characterisation of a novel transporter from *Cryptosporidium parvum*. Int J Parasitol 32(7):877–887

LaGier MJ, Tachezy J, Stejskal F, Kutisova K, Keithly JS (2003) Mitochondrial-type iron-sulfur cluster biosynthesis genes (IscS and IscU) in the apicomplexan *Cryptosporidium parvum*. Microbiology 149(Pt 12):3519–3530

Lau H, Ferlan JT, Brophy VH, Rosowsky A, Sibley CH (2001) Efficacies of lipophilic inhibitors of dihydrofolate reductase against parasitic protozoa. Antimicrob Agents Chemother 45 (1):187–195. doi:10.1128/AAC.45.1.187-195.2001

Lei C, Rider SD Jr, Wang C, Zhang H, Tan X, Zhu G (2010) The apicomplexan *Cryptosporidium parvum* possesses a single mitochondrial-type ferredoxin and ferredoxin: $NADP^+$ reductase system. Protein Sci 19(11):2073–2084. doi:10.1002/pro.487

Leitsch D, Burgess AG, Dunn LA, Krauer KG, Tan K, Duchene M et al (2011) Pyruvate: ferredoxin oxidoreductase and thioredoxin reductase are involved in 5-nitroimidazole activation while flavin metabolism is linked to 5-nitroimidazole resistance in *Giardia lamblia*. J Antimicrob Chemother 66(8):1756–1765. doi:10.1093/jac/dkr192

Li LC, Mun YF (2005) Partial characterization of genes encoding the ATP-binding cassette proteins of *Cryptosporidium parvum*. Trop Biomed 22(2):115–122

MacKenzie CR, Heseler K, Muller A, Daubener W (2007) Role of indoleamine 2,3-dioxygenase in antimicrobial defence and immuno-regulation: tryptophan depletion versus production of toxic kynurenines. Curr Drug Metab 8(3):237–244

Madern D, Cai X, Abrahamsen MS, Zhu G (2004) Evolution of *Cryptosporidium parvum* lactate dehydrogenase from malate dehydrogenase by a very recent event of gene duplication. Mol Biol Evol 21(3):489–497. doi:10.1093/molbev/msh042

Martucci WE, Udier-Blagovic M, Atreya C, Babatunde O, Vargo MA, Jorgensen WL et al (2009) Novel non-active site inhibitor of *Cryptosporidium hominis* TS-DHFR identified by a virtual screen. Bioorg Med Chem Lett 19(2):418–423. doi:10.1016/j.bmcl.2008.11.054

Millership JJ, Zhu G (2002) Heterogeneous expression and functional analysis of two distinct replication protein A large subunits from *Cryptosporidium parvum*. Int J Parasitol 32 (12):1477–1485

Mogi T, Kita K (2010) Diversity in mitochondrial metabolic pathways in parasitic protists *Plasmodium* and *Cryptosporidium*. Parasitol Int 59(3):305–312. doi:10.1016/j.parint.2010. 04.005

Nguyen TN, Abendroth J, Leibly DJ, Le KP, Guo W, Kelley A et al (2011) Structure of triosephosphate isomerase from *Cryptosporidium parvum*. Acta Crystallogr Sect F Struct Biol Cryst Commun 67(Pt 9):1095–1099. doi:10.1107/S1744309111019178

O'Neil RH, Lilien RH, Donald BR, Stroud RM, Anderson AC (2003) Phylogenetic classification of protozoa based on the structure of the linker domain in the bifunctional enzyme, dihydrofolate reductase-thymidylate synthase. J Biol Chem 278(52):52980–52987. doi:10.1074/jbc.M310328200

Perkins ME, Riojas YA, Wu TW, Le Blancq SM (1999) CpABC, a *Cryptosporidium parvum* ATP-binding cassette protein at the host-parasite boundary in intracellular stages. Proc Natl Acad Sci U S A 96(10):5734–5739

Popov VM, Chan DC, Fillingham YA, Atom Yee W, Wright DL, Anderson AC (2006) Analysis of complexes of inhibitors with *Cryptosporidium hominis* DHFR leads to a new trimethoprim derivative. Bioorg Med Chem Lett 16(16):4366–4370. doi:10.1016/j.bmcl.2006.05.047

Rider SD Jr, Zhu G (2008) Differential expression of the two distinct replication protein A subunits from *Cryptosporidium parvum*. J Cell Biochem 104(6):2207–2216. doi:10.1002/jcb.21784

Rider SD Jr, Zhu G (2010) *Cryptosporidium*: genomic and biochemical features. Exp Parasitol 124 (1):2–9. doi:10.1016/j.exppara.2008.12.014

Rider SD Jr, Cai X, Sullivan WJ Jr, Smith AT, Radke J, White M et al (2005) The protozoan parasite *Cryptosporidium parvum* possesses two functionally and evolutionarily divergent replication protein A large subunits. J Biol Chem 280(36):31460–31469. doi:10.1074/jbc.M504466200

Riordan CE, Ault JG, Langreth SG, Keithly JS (2003) *Cryptosporidium parvum* Cpn60 targets a relict organelle. Curr Genet 44(3):138–147. doi:10.1007/s00294-003-0432-1

Roberts CW, Roberts F, Henriquez FL, Akiyoshi D, Samuel BU, Richards TA et al (2004) Evidence for mitochondrial-derived alternative oxidase in the apicomplexan parasite *Cryptosporidium parvum*: a potential anti-microbial agent target. Int J Parasitol 34(3):297–308. doi:10.1016/j.ijpara.2003.11.002

Rochelle PA, Upton SJ, Montelone BA, Woods K (2005) The response of *Cryptosporidium parvum* to UV light. Trends Parasitol 21(2):81–87. doi:10.1016/j.pt.2004.11.009

Rotte C, Stejskal F, Zhu G, Keithly JS, Martin W (2001) Pyruvate : $NADP^+$ oxidoreductase from the mitochondrion of *Euglena gracilis* and from the apicomplexan *Cryptosporidium parvum*: a biochemical relic linking pyruvate metabolism in mitochondriate and amitochondriate protists. Mol Biol Evol 18(5):710–720

Schmatz DM (1989) The mannitol cycle–a new metabolic pathway in the Coccidia. Parasitol Today 5(7):205–208

Senkovich O, Speed H, Grigorian A, Bradley K, Ramarao CS, Lane B et al (2005) Crystallization of three key glycolytic enzymes of the opportunistic pathogen *Cryptosporidium parvum*. Biochim Biophys Acta 1750(2):166–172. doi:10.1016/j.bbapap.2005.04.009

Senkovich O, Schormann N, Chattopadhyay D (2009) Structures of dihydrofolate reductase-thymidylate synthase of *Trypanosoma cruzi* in the folate-free state and in complex with two antifolate drugs, trimetrexate and methotrexate. Acta Crystallogr D Biol Crystallogr 65 (Pt 7):704–716. doi:10.1107/S090744490901230X

Sharling L, Liu X, Gollapalli DR, Maurya SK, Hedstrom L, Striepen B (2010) A screening pipeline for antiparasitic agents targeting *Cryptosporidium* inosine monophosphate dehydrogenase. PLoS Negl Trop Dis 4(8):e794. doi:10.1371/journal.pntd.0000794

Slapeta J, Keithly JS (2004) *Cryptosporidium parvum* mitochondrial-type HSP70 targets homologous and heterologous mitochondria. Eukaryot Cell 3(2):483–494

Striepen B, Kissinger JC (2004) Genomics meets transgenics in search of the elusive *Cryptosporidium* drug target. Trends Parasitol 20(8):355–358. doi:10.1016/j.pt.2004.06.003

Striepen B, White MW, Li C, Guerini MN, Malik SB, Logsdon JM Jr et al (2002) Genetic complementation in apicomplexan parasites. Proc Natl Acad Sci U S A 99(9):6304–6309. doi:10.1073/pnas.092525699

Sun XE, Sharling L, Muthalagi M, Mudeppa DG, Pankiewicz KW, Felczak K et al (2010) Prodrug activation by *Cryptosporidium* thymidine kinase. J Biol Chem 285(21):15916–15922. doi:10.1074/jbc.M110.101543

Templeton TJ, Enomoto S, Chen WJ, Huang CG, Lancto CA, Abrahamsen MS et al (2010) A genome-sequence survey for *Ascogregarina taiwanensis* supports evolutionary affiliation but metabolic diversity between a Gregarine and *Cryptosporidium*. Mol Biol Evol 27(2):235–248. doi:10.1093/molbev/msp226

Umejiego NN, Li C, Riera T, Hedstrom L, Striepen B (2004) *Cryptosporidium parvum* IMP dehydrogenase: identification of functional, structural, and dynamic properties that can be exploited for drug design. J Biol Chem 279(39):40320–40327. doi:10.1074/jbc.M407121200

Umejiego NN, Gollapalli D, Sharling L, Volftsun A, Lu J, Benjamin NN et al (2008) Targeting a prokaryotic protein in a eukaryotic pathogen: identification of lead compounds against cryptosporidiosis. Chem Biol 15(1):70–77. doi:10.1016/j.chembiol.2007.12.010

Vasquez JR, Gooze L, Kim K, Gut J, Petersen C, Nelson RG (1996) Potential antifolate resistance determinants and genotypic variation in the bifunctional dihydrofolate reductase-thymidylate synthase gene from human and bovine isolates of *Cryptosporidium parvum*. Mol Biochem Parasitol 79(2):153–165

Wanyiri J, Ward H (2006) Molecular basis of *Cryptosporidium*-host cell interactions: recent advances and future prospects. Future Microbiol 1(2):201–208. doi:10.2217/17460913.1.2.201

Xu P, Widmer G, Wang Y, Ozaki LS, Alves JM, Serrano MG et al (2004) The genome of *Cryptosporidium hominis*. Nature 431(7012):1107–1112. doi:10.1038/nature02977

Yarlett N, Yarlett NC, Lloyd D (1986) Ferredoxin-dependent reduction of nitroimidazole derivatives in drug-resistant and susceptible strains of *Trichomonas vaginalis*. Biochem Pharmacol 35(10):1703–1708

Yoon S, Park WY, Yu JR (2012) Recombinant thioredoxin peroxidase from *Cryptosporidium parvum* has more powerful antioxidant activity than that from *Cryptosporidium muris*. Exp Parasitol 131(3):333–338. doi:10.1016/j.exppara.2012.04.018

Yu Y, Zhang H, Zhu G (2010) Plant-type trehalose synthetic pathway in *Cryptosporidium* and some other apicomplexans. PLoS One 5(9):e12593. doi:10.1371/journal.pone.0012593

Zapata F, Perkins ME, Riojas YA, Wu TW, Le Blancq SM (2002) The *Cryptosporidium parvum* ABC protein family. Mol Biochem Parasitol 120(1):157–161

Zeng B, Zhu G (2006) Two distinct oxysterol binding protein-related proteins in the parasitic protist *Cryptosporidium parvum* (Apicomplexa). Biochem Biophys Res Commun 346(2):591–599. doi:10.1016/j.bbrc.2006.05.165

Zeng B, Cai X, Zhu G (2006) Functional characterization of a fatty acyl-CoA-binding protein (ACBP) from the apicomplexan *Cryptosporidium parvum*. Microbiology 152(Pt 8):2355–2363. doi:10.1099/mic.0.28944-0

Zhang H, Guo F, Zhou H, Zhu G (2012) Transcriptome analysis reveals unique metabolic features in the *Cryptosporidium parvum* oocysts associated with environmental survival and stresses. BMC Genomics 13:647. doi:10.1186/1471-2164-13-647

Zhu G (2004) Current progress in the fatty acid metabolism in *Cryptosporidium parvum*. J Eukaryot Microbiol 51(4):381–388

Zhu G (2008) Biochemistry. In: Fayer R, Xiao L (eds) *Cryptosporidium* and cryptosporidiosis, 2nd edn. CRC Press, Boca Raton, pp 57–77

Zhu G, Keithly JS (1997) Molecular analysis of a P-type ATPase from *Cryptosporidium parvum*. Mol Biochem Parasitol 90(1):307–316

Zhu G, Keithly JS (2002) Alpha-proteobacterial relationship of apicomplexan lactate and malate dehydrogenases. J Eukaryot Microbiol 49(3):255–261

Zhu G, Marchewka MJ, Keithly JS (1999) *Cryptosporidium parvum* possesses a short-type replication protein A large subunit that differs from its host. FEMS Microbiol Lett 176(2):367–372

Zhu G, Keithly JS, Philippe H (2000a) What is the phylogenetic position of *Cryptosporidium*? Int J Syst Evol Microbiol 50(Pt 4):1673–1681

Zhu G, Marchewka MJ, Keithly JS (2000b) *Cryptosporidium parvum* appears to lack a plastid genome. Microbiology 146(Pt 2):315–321

Zhu G, Marchewka MJ, Woods KM, Upton SJ, Keithly JS (2000c) Molecular analysis of a Type I fatty acid synthase in *Cryptosporidium parvum*. Mol Biochem Parasitol 105(2):253–260

Zhu G, LaGier MJ, Stejskal F, Millership JJ, Cai X, Keithly JS (2002) *Cryptosporidium parvum*: the first protist known to encode a putative polyketide synthase. Gene 298(1):79–89

Zhu G, Li Y, Cai X, Millership JJ, Marchewka MJ, Keithly JS (2004) Expression and functional characterization of a giant Type I fatty acid synthase (CpFAS1) gene from *Cryptosporidium parvum*. Mol Biochem Parasitol 134(1):127–135

Zhu G, Shi X, Cai X (2010) The reductase domain in a Type I fatty acid synthase from the apicomplexan *Cryptosporidium parvum*: restricted substrate preference towards very long chain fatty acyl thioesters. BMC Biochem 11:46. doi:10.1186/1471-2091-11-46

Part III
Host-parasite Interaction

Chapter 9
Human Cryptosporidiosis: A Clinical Perspective

Henry Shikani and Louis M. Weiss

Abstract Cryptosporidiosis is a diarrheal illness that in humans is most commonly due to infection by *Cryptosporidium parvum* or *C. hominis*. It has a world-wide distribution and is associated with both human to human and zoonotic infections, depending on the *Cryptosporidium* species causing the infection. Cryptosporidiosis can occur in both immune competent individuals, where it causes a self limiting diarrheal illness, or in patients with immune deficiency where the diarrhea can be extensive and result in death. The severity of illness is dependent upon factors such as age, environment, co-existent diseases, and host immune status. Infection typically involves the gastrointestinal cells that line the epithelial surface of the small and large intestines, but, depending on the species of *Cryptosporidium*, can also involve other epithelial surfaces such as the respiratory tract, particularly in immune compromised hosts. This chapter reviews the available data on the pathogenesis of cryptosporidiosis in humans, the host response to this infection, the clinical complications associated with illness, and the therapeutic and preventative strategies for management of this infection.

9.1 Introduction

Cryptosporidiosis is a common cause of diarrheal illness due to infection with species of the Apicomplexan protozoan parasite *Cryptosporidium*. Although *Cryptosporidium* was described in mice in 1907, it was not until 1976 that it was first

H. Shikani
Department of Pathology, Albert Einstein College of Medicine, 1300 Morris Park Avenue, Bronx, NY 10461, USA

L.M. Weiss (✉)
Department of Pathology, Division of Parasitology, Department of Medicine, Division of Infectious Diseases, Albert Einstein College of Medicine, 1300 Morris Park Avenue, Bronx, NY 10461, USA
e-mail: louis.weiss@einstein.yu.edu

S.M. Cacciò and G. Widmer (eds.), *Cryptosporidium: parasite and disease*,
DOI 10.1007/978-3-7091-1562-6_9,

reported to have an association with diarrhea in humans, in a healthy child and an adult with immune suppression. To date, about 20 *Cryptosporidium* species have been identified (White 2010). While *Cryptosporidium parvum* and *Cryptosporidium hominis* are the most common causative agents of human disease, a number of other species can also infect humans including *Cryptosporidium canis*, *C. meleagridis*, *C. felis*. *C. suis*, *C. baylei*, *C. muris*, and *C. andersoni* (Cama et al. 2007, 2008; Gatei et al. 2006; Leoni et al. 2006; Muthusamy et al. 2006; Nichols 2008; Pedraza-Diaz et al. 2001; Fayer 2008; Tumwine et al. 2003) Infection characteristics and clinical features associated with cryptosporidiosis are shown in Tables 9.1 and 9.2 (Chalmers and Davies 2010; Guerrant 2008). These organisms usually cause diarrhea. Spontaneous recovery is the rule, and there is no universally effective specific therapeutic agent. Cryptosporidiosis is commonly identified as an opportunistic infection in immune compromised patients, such as those with acquired immunodeficiency syndrome (AIDS) (Bushen and Guerrant 2006; Chen et al. 2002; Kosek et al. 2001). It is also, however, a cause of diarrhea in immune competent individuals in both developed and developing countries, with a significant effect on public health (Bushen and Guerrant 2006; Chen et al. 2002; Kosek et al. 2001). Cryptosporidiosis-associated diarrheal syndromes can be stratified into three types, which are dependent on the specific population affected: (1) a self-limited diarrhea most often seen in immune competent individuals; (2) a persistent diarrhea which most commonly afflicts children in developing countries and can be associated with nutritional and growth effects in afflicted children; and (3) a chronic diarrhea syndrome most often seen in immune compromised individuals (Bushen and Guerrant 2006; Chen et al. 2002; Kosek et al. 2001). Many of the studies in the literature were done prior to the recognition of *C. hominis* as a distinct species from *C. parvum* and, therefore, whenever *C. parvum* is mentioned, in older literature, it is likely that these reports include cases due to both organisms. *Cryptosporidium hominis* is found primarily in humans and is responsible for most water borne outbreaks of this parasite, including the outbreak in 1993 in Milwaukee that involved 403,000 people (Chalmers and Davies 2010; MacKenzie et al. 1995).

9.2 Clinical Disease

Following *Cryptosporidium spp.* infection, humans are asymptomatic for typically 1 week during which the parasite multiplies in its target tissue, the epithelial lining of the intestine (Jokipii and Jokipii 1986; Mac Kenzie et al. 1994; Okhuysen et al. 1999, 2002). The incubation period between infection and presentation of symptoms can range, however, from 1 day to 1 month, depending on strain/species of parasite and the immune status of the host (Okhuysen et al. 1999, 2002; Pereira et al. 2002) (Table 9.1). Cryptosporidiosis typically presents as diarrhea, however, as mentioned above, the severity of disease and exact diarrheal syndrome can vary depending on the immune status and age of affected individuals (White 2010). Three distinct susceptible populations have been described: (1) immune competent

Table 9.1 Infection characteristics and clinical features associated with various *Cryptosporidium* isolates. Adapted from Chalmers and Davies (2010)

Study isolate	*C. parvum* IOWA[a]	*C. parvum* IOWA[b]	*C. parvum* TAMU[b]	*C. parvum* UCP[b]	*C. parvum*[c]	*C. hominis* TU502[d]
Source of isolate	Calf[a]	Calf[b]	Foal[b]	Calf[b]	Moredun Deer[c]	Human[d]
Infectious dose ID50	132[a] (clinical definition with microbiological confirmation)	87[b,e]; 74.5[b,f]	9[b,e]; 125[b,f]	1,042[b,e]; 2788[b,f]	300[c,e]; 375[c,f]	10[d,e]; 83[d,f]
Clinical attack rate (percent of subjects developing symptoms)	100 %[a,g]; 88 %[a,h]; 20 %[a,i]	52 %[b,g]	86 %[b,g]	59 %[b,g]	69 %[c,g]	62 %[d,g]; 75 %[d,j]; 71 %[d,k]; 60 %[d,i]; 40 %[d,l]
Asymptomatic shedding of oocysts	Yes[a]	Yes[b]	Yes[b]	Yes[b]	No[c]	No[d]
Prepatent period (mean days)	6[a,g]; 9[a,m]; 10[a,n]	7.7[b,g]	4[b,g]	7[b,g]	5[c,g]	
Incubation period: mean; median (days)	9[a]; 6.5[a]	9[b]; 7[b]	5[b]; 5[b]	11[b]; 6[b]	Median: 3[c]	5.4[d]; 4[d]
Patent period: mean duration of shedding (days)	12[a,g]; 10[a,m]; 2[a,n]	8.4[b,g]	3.4[b,g]	3.3[b,g]	5[c,g]	
Duration of diarrhea: mean (hours)	74[a]	64.2[b]	94.5[b]	81.6[b]	Median: 169[c]	1,373[d]
Severity (mean number of unformed stools)	12.7[a]	7[b]	9[b]	8[b]	19[c]	
Relapse rate (%)		18[b]	58[b]	46[b]		

[a]DuPont et al. (1995)
[b]Okhuysen et al. (1999)
[c]Okhuysen et al. (2002)
[d]Chappell et al. (2006)
[e]Clinical = symptoms reported
[f]Microbiological = oocysts detected in stools
[g]Greater than 1,000 oocysts
[h]Greater than 300 oocysts
[i]30 oocysts
[j]500 oocysts
[k]100 oocysts
[l]10 oocysts
[m]300–500 oocysts
[n]30–100 oocysts

Table 9.2 Clinical features of healthy adults infected with *C. parvum* or *C. hominis* Oocysts. Adapted from Warren and Guerrant (2008)

Clinical features	*C. parvum*: 30–1,000,000 oocysts/dose (29 adults)	*C. hominis*: 10–500 oocysts/ dose (21 adults)
Infection rate[a]	18 (62 %)	9 (43 %)
Enteric symptoms rate[b]	11 (38 %)	14 (67 %)
Diarrhea attack rate[c]	7 (24 %)[d]	13 (62 %)[e]
Incubation period (mean days)	9	5.4
Duration of illness (mean hours)	74	137.3
Maximum number of unformed stools per day (mean)	6.4	3.2
Number of unformed stools per illness (mean)	12.7	8.9
Total stool weight per episode of diarrhea (mean, kg)	1.23	1.08

[a]Infection = excretion of oocysts in stools more than 36 h after ingestion of oocysts
[b]Rate of development of fever, nausea, vomiting, abdominal pain/cramps, tenesmus, gas-related intestinal symptoms, fecal urgency or fecal incontinence
[c]Diarrhea in *C. parvum* study: passage of three unformed stools in 8 h or passage of more than three unformed stools in 24 h in addition to the presence of at least one enteric symptom; Diarrhea in *C. hominis* study: passage of $\geq$ 200 g of unformed stools per day or $\geq$ 3 unformed stools in 8 h or $\geq$ unformed stools in 24 h
[d]All volunteers excreted oocysts
[e]6/13 volunteers excreted detectable oocysts

individuals in developed countries including tourists travelling in developing countries, (2) young children in developing countries whose immune status is not yet fully functional, and (3) immune compromised individuals (typically those with AIDS) (White 2010). Tables 9.3 and 9.4 describe the clinical characteristics of individuals in these patient populations (Chalmers and Davies 2010; Guerrant 2008).

9.2.1 Immune Competent Individuals

Infection of immune competent hosts in developed countries usually occurs because of water-related outbreaks, travel or animal/human contact (Hunter et al. 2004b; Roy et al. 2004). Children and the elderly are most frequently affected, and disease normally presents as a watery diarrhea that lasts for 1–2 weeks, but occasionally can persist for a month (Gambhir et al. 2003, Guerrant 2008; Mac Kenzie et al. 1994; Naumova et al. 2003; Neill et al. 1996). Oocysts are shed for a mean period of 7 days (range 1-15 days) after symptoms have ceased, although exceptionally for up to 2 months (Chalmers and Davies 2010; Jokipii and Jokipii 1986). Associated symptoms include abdominal cramps, nausea, vomiting and fever (Mac Kenzie et al. 1994). Recurrence of gastrointestinal symptoms can occur, e.g. in the Milwaukee outbreak this was reported by 30 % of patients

Table 9.3 Clinical characteristics of cryptosporidiosis in different patient groups. Adapted from Chalmers and Davies (2010)

	Immune-competent:	**Immune-compromised:**	**Immune-competent:**
	Laboratory-diagnosed patients seeking medical care in United Kingdom	AIDS, haematological malignancy, primary T-cell deficiency	Children in prospective studies in developing countries
Length/ severity of illness	Self-limiting (mean: 13 days, median: 11 days)[a]	Severe/chronic/intractable; increased morbidity/mortality	Acute or chronic disease/carriage
Clinical features	**Gastrointestinal:**	**Gastrointestinal**[c,d]:	**Gastrointestinal**[e]:
	Diarrhea (98 %)[b]: watery (81 %)[b]; loose (17 %)[b]; bloody (11 %)[a]; relapsing (40 %)[a]	Diarrhea: transient; relapsing; chronic; cholera-like	Watery diarrhea (96 %)
	Abdominal pain (60–96 %)[b,a]	**Abdominal pain**	**Vomiting (57 %)**
	Vomiting (49–65 %)[b,a]	**Vomiting**	**Fever (37 %)**
	Fever (36–59 %)[b,a]	**Fever**	**Malnutrition**[f,g,h,i]
	Nausea (35 %)[b]	**Nausea**	**Failure to thrive**[f,k]
		Severe weight loss	**Weight loss**[f]
		Biliary tract involvement[d]:	**Impaired cognitive function**[j]
		Cholangitis; pancreatitis; sclerosing cholangitis;	
		Liver cirrhosis	
		Respiratory involvement[d]:	
		cough; sinusitis	

[a]Hunter et al. (2004a, b)
[b]Palmer and Biffin (1990)
[c]Manabe et al. (1998)
[d]Hunter and Nichols (2002)
[e]Khan et al. (2004)
[f]Molbak et al. (1997)
[g]Sallon et al. (1988)
[h]Sarabia-Arce et al. (1990)
[I]Lima et al. (1992)
[j]Guerrant et al. (1999)
[k]Checkley et al. (1997)

(Chalmers and Davies 2010; MacKenzie et al. 1995) and in an English study of sporadic cases by 40 % of patients (Chalmers and Davies 2010; Hunter et al. 2004a). The clinical presentation of infection may vary due to the infecting *Cryptosporidium* species. Oocysts are shed for longer and the number of oocysts detected in stools is higher in *C. hominis* infections than *C. parvum* (Bushen et al. 2007; Cama et al. 2008; Chalmers and Davies 2010; McLauchlin et al. 1999; Xiao et al. 2001). Cryptosporidiosis-associated disease in immune

Table 9.4 Clinical features of human Cryptosporidiosis. Adapted from Warren and Guerrant (2008)

Characteristics	Immune competent individuals	Immune compromised individuals	Other populations
Susceptible population	Healthy adults/ volunteers	AIDS patients, solid-organ transplant recipients, malignant individuals, individuals suffering from malnutrition, individuals undergoing hemodialysis, individuals with primary immunodeficiency diseases	Children (primarily those younger than 2 years of age), elderly adults
Site of infection	Intestine	Intestine or extraintestine	Intestine (primarily)
Enteric presentation	Asymptomatic or acute	Asymptomatic, transient, chronic or fulminant	Asymptomatic, acute or persistent
Signs and symptoms	Diarrhea, abdominal discomfort, fatigue	Diarrhea, weight loss, abdominal pain, nausea, vomiting, fever	Diarrhea, dehydration, malnutrition, weight loss
Duration of illness	4–14 days	14 days to several months	Outbreak: 5–17 days; Greater than 14 days to several weeks in children
Clinical outcome	Self-limiting	Increased morbidity and mortality	Long-term developmental impact on malnourished children; increased transmission and hospitalization in elderly

competent individuals is generally mild and can often be asymptomatic (White 2010), reflecting the importance of a functional immune system in controlling infection. This is reflected in reports of outbreaks and studies that have demonstrated that *C. parvum* infected immune competent hosts can have no symptoms (Chappell et al. 2006; Cicirello et al. 1997; DuPont et al. 1995; Mac Kenzie et al. 1994; Okhuysen et al. 1998, 1999, 2002; Pozio et al. 1997), but shed oocysts. Thus, as with many infections, the prevalence of individuals with symptomatic infection underestimates the extent of an infection in the community.

9.2.2 *Young Children in Developing Countries*

In developing countries, cryptosporidiosis-associated diarrhea most frequently occurs in children younger than 5 years (White 2010). In developing countries in Asia, Africa and Latin America, 5–10 % of diarrheal disease in children is due to infection with *Cryptosporidium spp.* (White 2010). Symptoms include watery diarrhea, abdominal cramps and, less frequently, fever and cough (White 2010).

Almost half of infected children develop a persistent diarrhea that lasts over 2 weeks (Newman et al. 1994; Sodemann et al. 1999), which increases the likelihood of additional episodes of diarrhea and death in these children (Agnew et al. 1998; Amadi et al. 2001; Lima et al. 2000; Molbak et al. 1993). Infection with *Cryptosporidium* is one of the major causes of persistent diarrhea in developing countries and can also manifest as chronic diarrhea and lead to malnutrition (Amadi et al. 2002; Behera et al. 2008). Clinical presentation varies dependent on *Cryptosporidium spp.*, and perhaps the specific strain, with *C. hominis* being associated with the most severe disease syndromes in children and *C. meleagridis* with milder disease (Ajjampur et al. 2007; Bushen et al. 2007; Cama et al. 2008; Hunter et al. 2004a). Children in a birth cohort in Brazil with *C. hominis* infection had increased fecal lactoferrin and delayed growth compared with those with *C. parvum* (Bushen et al. 2007; Chalmers and Davies 2010). Children in Lima, Peru with *C. hominis* had diarrhea, nausea, vomiting and general malaise while only diarrhea was reported in those with *C. parvum*, *C. meleagridis*, *C. canis* or *C. felis* infection (Cama et al. 2008; Chalmers and Davies 2010). At a subtype level (GP60 gene subtyping) *C. hominis* subtypes Ia, Ib, Id and Ie were associated with diarrhea and Ib was also associated with nausea, vomiting and general malaise (Cama et al. 2008; Chalmers and Davies 2010).

Cryptosporidiosis is strongly associated with malnutrition in children in developing countries (Hunter and Nichols 2002). Infection causes malnutrition, and disease is significantly more severe, and can be fatal, in malnourished individuals (Amadi et al. 2001; Behera et al. 2008; Hunter and Nichols 2002; Javier Enriquez et al. 1997; Macfarlane and Horner-Bryce 1987; Newman et al. 1999; Sallon et al. 1988, 1994; Sarabia-Arce et al. 1990; Tumwine et al. 2003). A study conducted in the West Indies found that 15 out of 77 fecal specimens from malnourished children were positive for *Cryptosporidium* (Hunter and Nichols 2002; Macfarlane and Horner-Bryce 1987). Of the 15 cases, all experienced fever and diarrhea, most demonstrated dehydration and vomiting and two died due to the infection (Hunter and Nichols 2002; Macfarlane and Horner-Bryce 1987). A study in Israel reported that 30 out of 221 malnourished children with diarrhea also had *Cryptosporidium* infection (Hunter and Nichols 2002; Sallon et al. 1988). In a study performed in India, 50 children with diarrhea were examined (Hunter and Nichols 2002; Jaggi et al. 1994). Seven of those children were infected with *Cryptosporidium*, six of which were also malnourished (Hunter and Nichols 2002; Jaggi et al. 1994). A report from Gabon found the *Cryptosporidium* carriage rate in malnourished children (31.8 %) to be twice that of children with proper nourishment (16.8 %) (Duong et al. 1995; Hunter and Nichols 2002). In a Tanzanian study, 7 out of 55 children with acute diarrhea were malnourished and infected with *Cryptosporidium* (Cegielski et al. 1999; Hunter and Nichols 2002).

Malnourished children infected with *Cryptosporidium* before the age of one commonly demonstrate significant and unrecoverable weight loss and frequently experience stunted growth (Agnew et al. 1998; Checkley et al. 1998; Molbak et al. 1997). A study of infected infants in Guinea-Bissau found that, on average, boys with cryptosporidiosis lost 392 g and girls lost 294 g of body weight (Hunter and Nichols 2002; Molbak et al. 1997). A study conducted in Bangkok, Thailand found that diarrhea was more common in children infected with *Cryptosporidium* compared to another intestinal parasite, *Giardia lamblia* (Hunter and Nichols 2002; Janoff et al. 1990). *Cryptosporidium*-infected children also lost more weight relative to those with *G. lamblia* (Hunter and Nichols 2002; Janoff et al. 1990). These reports are consistent with the idea that cryptosporidiosis is a precursor to malnutrition, however this cause-effect notion remains controversial (Hunter and Nichols 2002). The available studies also demonstrate that malnourished children infected with *Cryptosporidium* experience a more serious and potentially fatal disease characterized by a significant loss in weight (Hunter and Nichols 2002).

9.2.3 Immune Compromised Individuals

Cryptosporidiosis in immune compromised hosts often presents as a complication of HIV-1 infection, usually when the patient has clinically defined AIDS with a CD4+ T cell count under 200 cells/mm^3 (White 2010). The disease variability seen in the presentation of cryptosporidiosis in various immune compromised patients is remarkable and currently unexplainable. Oddly, a large portion of HIV/*Cryptosporidium* co-infected individuals, even with CD4+ T cells under 200 cells/mm^3, demonstrate mild or no symptoms, while others experience chronic watery diarrhea, significant weight loss and severe malabsorption (White 2010). This variability might occur because of co-infections with other pathogens that are either exacerbating or preventing symptoms of cryptosporidiosis (Hashmey et al. 1997; Lumadue et al. 1998). With the emergence of effective combination antiretroviral therapy (cART), the prevalence of cryptosporidiosis in the HIV-1 infected population in the areas with widespread use of cART has dramatically decreased (Kim et al. 1998; Le Moing et al. 1998). Therapy with cART helps maintain adequate levels of CD4$^+$ T cells necessary to control the infection (Blanshard et al. 1992; Hashmey et al. 1997; Lumadue et al. 1998; Manabe et al. 1998).

Blanshard et al. examined 128 HIV positive patients with cryptosporidiosis and divided them into four clinical presentations: (1) asymptomatic infection, (2) transient infection, (3) chronic infection, and (4) fulminant infection (Blanshard et al. 1992; Farthing 2000). Infected, asymptomatic cases demonstrated normal bowel patterns, with less than three stools per day (Blanshard et al. 1992; Farthing 2000). This was the least common presentation, in this study, occurring in 3.9 % of patients (Blanshard et al. 1992; Farthing 2000). This study, however, did not specifically sample all HIV positive patients, but was based on samples submitted to a diagnostic laboratory, and thus likely underestimated the asymptomatic carriers of this organism. Transiently infected individuals represented 28.7 % of the patients and were defined as those who experienced diarrhea for under 2 months (Blanshard et al. 1992; Farthing 2000). At the 2-month time point, all symptoms ceased and the parasite could no longer be found in fecal samples (Blanshard et al. 1992; Farthing 2000). Chronic infection was the most common disease outcome (Blanshard

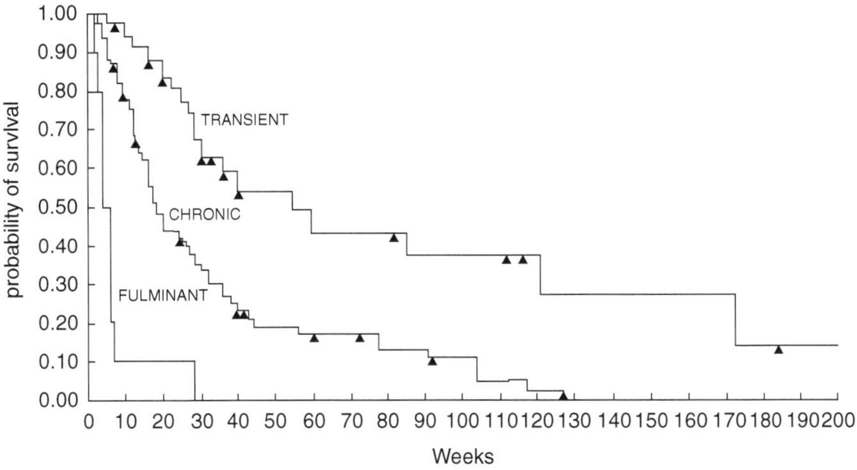

Fig. 9.1 Kaplan-Meier plots of the differential survivals of patients with transient, chronic and fulminant cryptosporidiosis (Reprinted with permission from Blanshard et al. (1992))

et al. 1992; Farthing 2000) comprising 59.6 % of these patients. In these chronically infected cases, diarrhea and *Cryptosporidium spp.* shedding in stools lasted over 2 months (Blanshard et al. 1992; Farthing 2000). Lastly, 7.8 % of patients experienced fulminant infection (Blanshard et al. 1992; Farthing 2000). These individuals experienced the most aggressive clinical outcome, passing at least two liters of watery diarrhea per day, i.e., a cholera-like diarrheal syndrome (Blanshard et al. 1992; Farthing 2000). Asymptomatic or transiently infected cases typically exhibited higher $CD4^+$ T cell counts than the other two groups (Blanshard et al. 1992; Farthing 2000). All patients with fulminant infection had $CD4^+$ T cell counts lower than 50 cells/mm^3 (Blanshard et al. 1992; Farthing 2000), demonstrating the importance of T cells and the cellular immune response in controlling this disease. Survival (Fig. 9.1) and weight were also strongly associated with clinical presentation, as asymptomatic and transiently infected individuals survived much longer (median: 36 weeks) and exhibited a higher weight (median: 64.7 kg) than chronic (median: 20 weeks; 59.3 kg) and fulminant (median: 5 weeks; 55.2 kg) cases (Blanshard et al. 1992; Farthing 2000).

Chronic and fulminant patterns of disease are frequent outcomes in individuals in developing areas of the world (Farthing 2000). Sub-Saharan African AIDS patients co-infected with *Cryptosporidium* commonly demonstrate the chronic diarrhea wasting syndrome known as 'slim disease' (Farthing 2000). This illness has a major effect on morbidity and mortality, particular in areas with limited use of cART.

9.2.4 Additional Gastrointestinal Infections Due to Cryptosporidiosis

The severity of cryptosporidiosis is associated with the ability of the parasite to spread within the human gastrointestinal tract. The ability of the organism to infect various organs of the gastrointestinal tract varies by host and likely is highly dependent on the host immune status. When infection is controlled and asymptomatic, *Cryptosporidium spp.* infection is usually restricted to the colon (Clayton et al. 1994; Farthing 2000). In circumstances where disease becomes more severe, *Cryptosporidium* colonization occurs throughout the gastrointestinal tract, including such areas as the pharynx, esophagus, stomach, small bowel, small intestine and rectum (Berk et al. 1984; Farthing 2000; Godwin 1991; Kazlow et al. 1986; Kelly et al. 1998). Due to a defective immune response, it is well known that *Cryptosporidium spp.* infection in severely immune compromised individuals, such as AIDS patients, can involve the biliary system epithelium (Clavel et al. 1996; Farthing 2000; Hashmey et al. 1997; Lopez-Velez et al. 1995; Meynard et al. 1996; Vakil et al. 1996). *Cryptosporidium spp.* have also been reported in the duodenal papilla and the pancreatic duct, and infection has been associated with the production of overt pancreatitis (Farthing 2000; Gross et al. 1986). Additionally, ampullar infection can lead to severe constriction and obstruction of the bile duct, increasing susceptibility to cholangitis (Farthing 2000; Hasan et al. 1991). Gallbladder and biliary system infection is common in HIV-positive patients with cryptosporidiosis, occurring in 10–26 % of cases (Farthing 2000; Hasan et al. 1991; McGowan et al. 1993; Vakil et al. 1996).

AIDS-related sclerosing cholangitis is a serious clinical complication that manifests in patients with AIDS and is strongly associated with *C. parvum* intestinal infection (Farthing 2000). Symptoms include abdominal pain, fever and jaundice. Patients with this syndrome demonstrate increased serum alkaline phosphatase and significant anatomical damage to the biliary system on various imaging studies (Farthing 2000). Cryptosporidiosis patients with biliary disease have lower $CD4^+$ T cell counts and a significantly higher rate of mortality than individuals without biliary symptoms (Farthing 2000; Vakil et al. 1996). Infection of the biliary tract with *C. parvum* has been associated with rapid onset of death (Hashmey et al. 1997; Vakil et al. 1996). Among 24 patients with AIDS and biliary involvement and symptoms from the large waterborne outbreak of cryptosporidiosis in Milwaukee, it was reported that biliary symptoms was a strong indicator of the prognosis since 83 % of patients with symptoms died within the following year compared to only 48 % of those without these symptoms (Hunter and Nichols 2002; Vakil et al. 1996). This may have reflected the advanced state AIDS in patients with biliary tract involvement. Lastly, widespread squamous metaplasia of the epithelial lining of the bile duct has been reported in the context of chronic *Cryptosporidium* infection (Farthing 2000; Kline et al. 1993).

9.2.5 *Extraintestinal Infection*

Cryptosporidiosis is known to spread as far as the respiratory tract (Farthing 2000). In addition, this may be more common with other species of *Cryptosporidium*, e.g. *Cryptosporidium meleagridis* is known to be associated with respiratory disease in birds, *Cryptosporidium spp.* have been found in humans in sputum, in bronchial aspirates, and in the bronchial epithelium (Blanshard et al. 1992; Clavel et al. 1996; Farthing 2000; Kocoshis et al. 1984; Travis et al. 1990). Symptoms of respiratory infection include cough, dyspnea, fever and thoracic pain with pneumonia being described as the most severe clinical outcome (Farthing 2000). Respiratory disease frequently results from co-infection with *Cryptosporidium spp.* and other pathogens, namely *Mycobacterium tuberculosis* or *Pneumocystis carinii*, however single infections with this parasite have also been observed (Clavel et al. 1996; Farthing 2000). Parasitic sinusitis has also been found in individuals co-infected with HIV and *Cryptosporidium spp.* (Dunand et al. 1997; Hunter and Nichols 2002). These cases typically presented fever, chills, and local tenderness and discharge (Dunand et al. 1997; Hunter and Nichols 2002). Diseases such as reactive arthritis and Reiter's syndrome have also been seen in patients with cryptosporidiosis.

9.3 Additional Complications of Cryptosporidiosis

9.3.1 *Pneumatosis Cystoides Intestinales*

A number of case reports on AIDS patients with cryptosporidiosis have described additional consequences of infection and in particular, pneumatosis cystoides intestinales has been described (Collins et al. 1992; Hunter and Nichols 2002; Samson and Brown 1996; Sidhu et al. 1994). Patients with pneumatosis cystoides intestinales are defined by the presence of rupture-prone, gas-containing cysts in their intestinal wall that are associated with the production of pneumoretroperitonea (air in the retroperitoneum) and pneumomediastina (air in the mediastinum) (Hunter and Nichols 2002).

9.3.2 *Esophageal Damage and Appendicitis*

A case was reported of a 2-year-old child with cryptosporidiosis who experienced damage to the esophagus, causing vomiting and dysphagia (Hunter and Nichols 2002; Kazlow et al. 1986). Cryptosporidiosis has also been associated with appendicitis (Hunter and Nichols 2002; Oberhuber et al. 1991). In a review of the literature, 16 cases of gastric cryptosporidiosis were identified (Hunter and Nichols 2002; Ventura et al. 1997). In an endoscopic study of 24 patients with AIDS and chronic diarrhea due to cryptosporidiosis, 16 (67 %) had parasites in the gastric epithelium (Hunter and Nichols 2002; Rossi et al. 1998). The majority of these patients had no specific symptoms related to this gastric colonization. A significant complication of gastric involvement by *Cryptosporidium spp.* is antral narrowing and gastric outlet obstruction (Cersosimo et al. 1992; Garone et al. 1986; Hunter and Nichols 2002; Iribarren et al. 1997; Moon et al. 1999). This obstruction causes nausea, vomiting and a reduction in nutrient intake.

9.4 Risk Factors for Cryptosporidiosis

Susceptibility to cryptosporidiosis depends on several factors, including environmental conditions, host immune status, age, geographic location and contact with infected humans/animals (Table 9.5). The prevalence of cryptosporidiosis ranges from as low as 1 % in more westernized regions of the world, such as North America and Europe, to as high as 30 % in tropical areas, especially in developing countries (Farthing 2000). A tropical climate is ideal for survival of the *Cryptosporidium* oocyst. The average prevalence of diarrheal cases due to infection with the parasite is almost threefold higher in developing (6.1 %) versus developed (2.1 %) countries (Farthing 2000; Adal and Guerrant 1995).

Cryptosporidiosis is a very common cause of waterborne disease and has been associated with drinking water and swimming pool contact. Cryptosporidiosis is responsible for one of the largest waterborne outbreaks ever described (Hunter and Nichols 2002; Mac Kenzie et al. 1994). The infectious dose of *Cryptosporidium* in human hosts is thought to be low (Blagburn and Current 1983; Guerrant 2008). In one study involving infection of immune competent volunteers with the Iowa isolate of *C. parvum* (Abrahamsen et al. 2004), a 50 % infectious dose (ID_{50}) of 132 oocysts was estimated (DuPont et al. 1995; Guerrant 2008). There is significant unpredictability in infectivity, however, among different strains of *C. parvum*, as demonstrated by distinct ID_{50}s of individual isolates in various host species (Guerrant 2008; Okhuysen et al. 1999). In a study where immune competent volunteers were infected with *C. hominis*, the ID_{50} ranged from 10 to 83 oocysts (Chappell et al. 2006; Guerrant 2008). Susceptibility to re-infection and illness following primary infection was investigated by re-challenging volunteers (DuPont et al. 1995) with the homologous isolate 1 year later (Okhuysen et al. 1999). This study demonstrated that infection still occurred and that the onset and duration was similar to the primary infection, although the degree of diarrhea and number of oocysts shed was decreased.

Table 9.5 Risk factors for human Cryptosporidiosis

	Immune competent hosts	Immune compromised hosts
Developed countries	Age (young most susceptible)	Recreational water (i.e. swimming pools, lakes, rivers, beaches)
	Recreational water (i.e. swimming pools, lakes, rivers, beaches)	Poorly sanitized sources of water
	Poorly sanitized sources of water	Primary immunodeficiency diseases (e.-g. common variable immunodeficiency)
	Travel	Agents that cause immunodeficiency (HIV/AIDS)
	Hospitals	Transplantation
	Animal reservoirs	Immune suppressant medications
Developing countries	Age (young most susceptible)	Agents that cause immunodeficiency
	Lack of proper breast-feeding	(HIV/AIDS)
	Poorly sanitized sources of water	Poorly sanitized sources of water
	Rain	Rain
	Animal reservoirs	Animal reservoirs

9.4.1 Immunodeficiency Diseases

As noted previously, patients with immune deficiency, particularly those with AIDS, are prone to developing severe cryptosporidiosis. Several other immune defects have been associated with cryptosporidiosis. Another immunodeficiency complication associated with cryptosporidiosis is severe combined immunodeficiency syndrome (Hunter and Nichols 2002). One case report described how, despite receiving both extensive therapy as well as a thymus transplantation, an infected individual with this syndrome died 5 months after being diagnosed with cryptosporidiosis (Hunter and Nichols 2002; Kocoshis et al. 1984). Furthermore, this patient exhibited a widespread infection with extraintestinal spread as *Cryptosporidium* parasites were identified in the small intestine, pancreatic duct and bronchioles (Hunter and Nichols 2002; Kocoshis et al. 1984). In another case of cryptosporidiosis, deficiencies in selective immunoglobulin A and *Saccharomyces* opsonin were discovered (Hunter and Nichols 2002; Jacyna et al. 1990). Several patients with X-linked hyper-immunoglobulin M syndrome have simultaneously demonstrated chronic diarrhea and liver damage due to cryptosporidiosis (Hayward et al. 1997; Hunter and Nichols 2002; Levy et al. 1997). This is a highly fatal immunodeficiency syndrome, characterized by a mutation in the gene for CD40 ligand and significant impairment of the T lymphocytes (Hunter and Nichols 2002). Mouse models have suggested that the CD40/CD40 ligand system is critical in the immune response to this pathogen. Lastly, Gomez Morales et al. described a child with chronic cryptosporidosis and diarrhea with a marked loss in weight who demonstrated deficiencies in interferon-γ production (IFN-γ) (Gomez Morales et al. 1996; Hunter and Nichols 2002).

9.4.2 Cancer

Several studies have looked at associations of cancer and cryptosporidiosis, however, a consistent association has not been demonstrated. A study performed in Turkey on 106 fecal specimens from patients with diarrhea and cancer found that 17.0 % were positive for parasite oocysts (Hunter and Nichols 2002; Tanyuksel et al. 1995). In a similar study conducted in India, only 1.3 % of diarrhea/cancer cases had *Cryptosporidium* in their stools (Hunter and Nichols 2002; Sreedharan et al. 1996), suggesting a possible country-specific susceptibility to infection in this patient population. In two separate studies performed on children with cancer and diarrhea (one in New South Wales and the other in Malaysia), 0 % and 2 % of patients, respectively, were positive for *Cryptosporidium* (Burgner et al. 1999; Hunter and Nichols 2002; Menon et al. 1999), further suggesting a possible country-specific susceptibility.

9.4.3 Leukemia/Lymphoma

The relationship of cryptosporidiosis and leukemia has not been definitely determined. In one study, a child with acute lymphocytic leukemia and severe diarrhea recovered (Hunter and Nichols 2002; Stine et al. 1985), while, in a different study, an infected child with acute lymphoblastic leukemia had a relapse (Hunter and Nichols 2002; Lewis et al. 1985). A study on six children with leukemia or lymphoma found that four cases survived with chemotherapy while persistently infected patients died (Foot et al. 1990; Hunter and Nichols 2002).

9.4.4 Bone Marrow and Solid-Organ Transplantation

Severe cryptosporidiosis with diarrhea can occur in patients undergoing bone marrow transplantation (Gentile et al. 1991; Hunter and Nichols 2002; Manivel et al. 1985). Pulmonary cryptosporidiosis, sometimes fatal, has also been described in bone marrow transplant cases (Hunter and Nichols 2002; Kibbler et al. 1987; Nachbaur et al. 1997). Additionally, one bone marrow transplantation facility experienced an outbreak of cryptosporidiosis (Hunter and Nichols 2002; Martino et al. 1988). Cryptosporidiosis has been observed in children undergoing liver transplantations (Campos et al. 2000; Gerber et al. 2000; Hunter and Nichols 2002; Vajro et al. 1991). A Belgian study found that three out of 461 children experienced a diffuse *Cryptosporidium*-associated cholangitis post-transplant (Campos et al. 2000; Hunter and Nichols 2002). In a Pittsburgh study, it was discovered that, of 1,160 patients receiving abdominal organ transplants, four children had cryptosporidiosis and resolved their disease (Gerber et al. 2000;

Hunter and Nichols 2002). Three of the four children were undergoing liver transplants while the fourth was a small bowel transplant patient (Gerber et al. 2000; Hunter and Nichols 2002). The rate of cryptosporidiosis in patients undergoing renal transplantation has also been explored in multiple studies (Hunter and Nichols 2002). In two separate studies conducted in Brazil and Turkey, renal transplantation was associated with an increase in infection with *Cryptosporidium spp*. (Chieffi et al. 1998; Hunter and Nichols 2002; Ok et al. 1997).

9.4.5 Diabetes

An association between cryptosporidiosis and diabetes has been proposed. In separate reports, three diabetes patients also experienced chronic diarrhea due to *Cryptosporidium* infection (Chan et al. 1989; Hunter and Nichols 2002; Trevino-Perez et al. 1995). The significance of this observation is not clear as the majority of individuals with diabetes do not demonstrate simultaneous evidence of cryptosporidiosis and in large studies of cryptosporidiosis, diabetes has not emerged as a risk factor (Hunter and Nichols 2002).

9.5 Host Immune Response

9.5.1 CD4$^+$ and CD8$^+$ T Cells

As with all diseases, an effective host response to infection with *Cryptosporidium* is critical to controlling the pathology and clinical outcome associated with the parasite. Arguably, the most important cell type in this response are CD4$^+$ T cells, as evidenced by the fact that *Cryptosporidium* presents as one of the main opportunistic infections in patients with low cell counts due to HIV (Blanshard et al. 1992; Flanigan et al. 1992; Hashmey et al. 1997; Pantenburg et al. 2008). The importance of this T cell subtype has also been seen in in vivo mouse studies. Mice that lack functional CD4$^+$ T cells demonstrate chronic *Cryptosporidium* infection. However, mice given an infusion of these cells demonstrate significant improvement (Aguirre et al. 1994; Chen et al. 1993; Culshaw et al. 1997; McDonald et al. 1992, 1994; Perryman et al. 1994; Ungar et al. 1991). The role of the effector CD8$^+$ T cell subtype in controlling *Cryptosporidium* infection is not as clear (White 2010). While these cells localize to sites of infection (Abrahamsen 1998; Wyatt et al. 1997) and are known to produce IFN-γ, a crucial cytokine for combating infection, it is not known whether a lack of CD8$^+$ T cells exacerbates disease (Kirkpatrick et al. 2008; Pantenburg et al. 2008; Preidis et al. 2007). Furthermore, CD8$^+$ T cells do not appear to be important for preventing cryptosporidiosis in mice

(Aguirre et al. 1994; McDonald et al. 1994), making it more challenging to characterize their role during infection.

9.5.2 IFN-γ

The importance of IFN-γ in controlling cryptosporidiosis is well established. Mice that lack this pro-inflammatory cytokine develop a persistent infection described as worse than in mice without $CD4^+$ T cells (Mead and You 1998; Theodos et al. 1997; Tzipori et al. 1995). Moreover, studies in mice also demonstrate that interleukin (IL)-12, a cytokine which contributes to the production of IFN-γ, is also critical for regulating *Cryptosporidium* infection (Campbell et al. 2002; Urban et al. 1996). As has been shown in vitro with *C. parvum*, IFN-γ may help clear the parasite by activating intestinal epithelial cells (Pollok et al. 2001). While evidence points strongly towards IFN-γ playing a key role in controlling *Cryptosporidium* infection, this cytokine does not always help regulate the disease. Infected individuals with low levels of IFN-γ have been shown to control this infection (Gomez Morales et al. 1999; Kirkpatrick et al. 2002). Additionally, murine studies demonstrate that functional IFN-γ is critical to the survival of C57BL/6, but not BALB/c mice, suggesting that other immune factors must also be working to reverse or prevent disease (Gomez Morales et al. 1999; Kirkpatrick et al. 2002; Mead and You 1998). A number of other cytokines have been proposed to function in limiting infection, including the pro-inflammatory cytokine, tumor necrosis factor-alpha (TNF-α), which has been shown to clear the parasite in vivo and in vitro (Lacroix et al. 2001; Pollok et al. 2001; Wyatt et al. 1997).

9.5.3 Antibody

While the role of cellular immunity, specifically $CD4^+$ T cells, in responding to *Cryptosporidium* infection is clear, the function of the humoral arm of the immune response remains unknown (Pantenburg et al. 2008). Human studies have demonstrated that individuals infected with *C. parvum* exhibit high antibody levels (Benhamou et al. 1995; Cozon et al. 1994; Dann et al. 2000), but the reasons for this are not clear. Murine studies have shown that the disease is not more severe in infected mice with non-functional B cells, however, treatment of infected mice with antibody to *Cryptosporidium* has been shown to facilitate parasite clearance (Arrowood et al. 1989; Chen et al. 2003; Jenkins et al. 1999; Perryman et al. 1999; Riggs 2002; Sagodira et al. 1999; Taghi-Kilani et al. 1990).

9.6 Diagnosis

Cryptosporidium infection is typically diagnosed by stool examination (White 2010). A number of approaches have been attempted to increase diagnostic sensitivity, i.e. concentration methods involving formalin buffer, flotation methods and the use of immunomagnetic beads aimed at isolating the organism (Pereira et al. 2002; Smith 2008; Webster et al. 1996). Various staining methods have also been tested for diagnosis, including acid-fast staining, fluorescence and immunofluorescence with antibodies that target the oocyst (White 2010). Because of its high sensitivity compared to other assays, immunofluorescence is now considered the gold standard for detecting *Cryptosporidium* in stool samples (Alles et al. 1995; Balatbat et al. 1996; Dagan et al. 1995; Garcia and Shimizu 1997; Ignatius et al. 1997; Johnston et al. 2003; Kehl et al. 1995). Other approaches that focus on detecting *Cryptosporidium*-specific antigens, such as enzyme-linked immunosorbent assay (ELISA) and immunochromatography, have been generally described as similar in sensitivity to other techniques, but more specific (Cryptosporidiosis 2009; Dagan et al. 1995; Garcia and Shimizu 1997, 2000; Garcia et al. 2003; Johnston et al. 2003; Smith 2008). Lastly, PCR tests for detection of *Cryptosporidium* DNA have been used and are considered more sensitive than stool examination (Ajjampur et al. 2008; Amar et al. 2007; Kaushik et al. 2008; Nair et al. 2008; ten Hove et al. 2007). Identification of *Cryptosporidium* species, in clinical specimens, is most commonly done using molecular techniques, such as PCR.

9.7 Pathology

Cryptosporidium invasion of host cells is restricted to the luminal border of the epithelial lining of the gastrointestinal tract (Fig. 9.2). Severe infection leads to displacement of the microvillous border and loss of the surface epithelium, causing changes in the villous architecture, with villous atrophy, blunting and crypt cell hyperplasia, and mononuclear cell infiltration in the lamina propria. The cause of diarrhea has yet to be elucidated but osmotic, inflammatory and secretory processes have been suggested. It is likely, that pathogenesis is due to several factors including the effect of the parasite and the substances it produces on the epithelial layer and the immunological and inflammatory responses of the host, leading to impaired intestinal absorption and enhanced secretion (Farthing 2000).

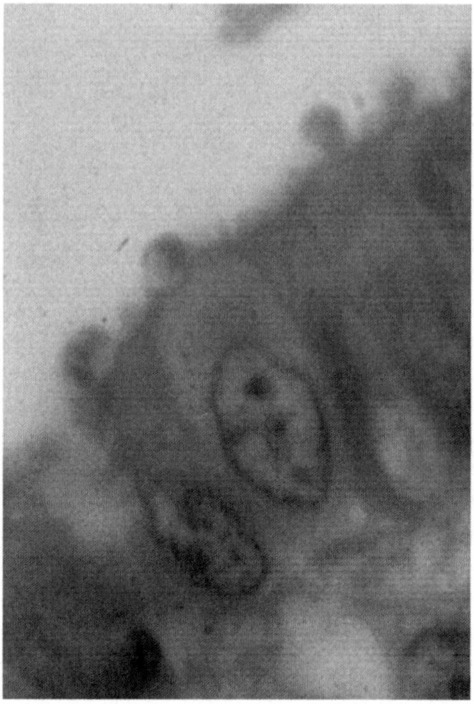

Fig. 9.2 Intestinal biopsy demonstrating *Cryptosporidium parvum* at the surface of the intestinal epithelial cells. Hematoxylin and eosin stain, 100X oil objective

9.8 Pathogenesis

The extent of pathology associated with *Cryptosporidium* infection largely depends on the individual's immune status. These parasites typically infect gastrointestinal epithelial cells. In immune competent cases, organisms normally concentrate in the distal small intestines and proximal colon, however immunodeficiency can cause the parasite to spread to other areas in the body, such as the gut and respiratory tract (Godwin 1991; Greenberg et al. 1996; Kelly et al. 1998). Additionally, the infectious dose can also contribute to pathogenesis as a large dose of oocysts has the potential to lead to a disseminated infection, causing a dysregulated involvement of such immune cells as lymphocytes and neutrophils (Clayton et al. 1994; Genta et al. 1993; Goodgame et al. 1995; Greenberg et al. 1996; Lumadue et al. 1998).

A balanced immune response is critical to controlling any pathogenic infection, including that of *Cryptosporidium*. While the importance of pro-inflammatory cytokines such as IFN-γ in regulating cryptosporidiosis is well documented, their dysregulated activity may also contribute to some of the characteristic features of disease, including increased permeability in the intestines (Roche et al. 2000). Anti-inflammatory cytokines such as transforming growth factor beta (TGF-β) have been shown in vitro to prevent *C. parvum*-mediated epithelial barrier breakdown (Roche et al. 2000), demonstrating the necessity of a regulated immune response.

The major clinical feature associated with cryptosporidiosis is diarrhea (White 2010). Physiologically, diarrhea is believed to manifest because of several factors, including the malabsorption of sodium, the secretion of chloride and the increased permeability of the intestinal epithelial layer (White 2010). These consequences likely result, in part, from the cascade of nuclear factor kappa-light-chain-enhancer of activated B cells (NF-κB)-mediated events that occur intracellularly following *Cryptosporidium* infection of epithelial cells in the intestine. Infection induces activation of the multi-functional transcription factor NF-κB (Chen et al. 2001; McCole et al. 2000). NF-κB subsequently activates a variety of anti-apoptotic factors as well as pro-inflammatory cytokines and chemokines, including TNF-α, IL-1β and IL-8 (Alcantara et al. 2003; Kandil et al. 1994; Kirkpatrick et al. 2002, 2006; Lacroix et al. 2001; Pantenburg et al. 2008; Robinson et al. 2001; Seydel et al. 1998; Wyatt et al. 1997). Inflammation causes a dysregulated influx of immune and inflammatory cells to infection sites, which probably contributes to enhanced epithelial permeability as well as other aspects of disease (White 2010).

Other factors which have been implicated in the pathogenesis of cryptosporidiosis include the TNF-α-mediated increased production of prostaglandins (Kandil et al. 1994). This increased production is believed to contribute to sodium malabsorption and chloride secretion in porcine and bovine cryposporidiosis (Argenzio et al. 1996; Cole et al. 2003; Gookin et al. 2004; Kandil et al. 1994; Laurent et al. 1998). Evidence thus far suggests that prostaglandins do not play a role, however, in human cryptosporidiosis (Okhuysen et al. 2001; Robinson et al. 2001; Sharpstone et al. 1996; Snijders et al. 1995).

The specific neuropeptide substance P has also been associated with cryptosporidiosis (Garza et al. 2008; Hernandez et al. 2007; Robinson et al. 2003, 2008). It has been shown that levels of this neuropeptide correlate with severity of disease both in infected immune competent and immune compromised individuals (Robinson et al. 2003). Both in non-human primate and murine models of infection, it was demonstrated that therapeutically targeting substance P itself or its receptor significantly ameliorates disease (Garza et al. 2008; Hernandez et al. 2007; Robinson et al. 2008; Sonea et al. 2002). Substance P is thought to contribute to both chloride secretion and enhanced intestinal permeability (Garza et al. 2008; Hernandez et al. 2007).

Cell death in the intestine, both by apoptosis and necrosis, is a predominant feature of cryptosporidiosis. While infected epithelial cells are programmed initially to survive, uninfected adjacent cells undergo apoptosis (Chen et al. 1999, 2001; McCole et al. 2000; Motta et al. 2002). Interactions between FAS ligand and its receptor are thought to facilitate cell death (Chen et al. 1999, 2001; McCole et al. 2000; Motta et al. 2002). Moreover, once *Cryptosporidium* organisms complete their intracellular cycle, they induce necrosis of infected cells (Chen et al. 1998; Elliott and Clark 2003; Griffiths et al. 1994). All of these events together likely contribute to the decreased intestinal surface area, increased permeability and malabsorption that occur during disease (White 2010).

9.8.1 Malabsorption

Disease caused by *Cryptosporidium* typically increases in severity when the organism spreads to the proximal small intestine (Clayton et al. 1994; Farthing 2000). Infection of both humans and animals results in significant anatomical disturbances to the crypt and villous architecture of the small intestine, with complete atrophy and hyperplasia occurring in certain circumstances (Argenzio et al. 1990; Clayton et al. 1994; Farthing et al. 1996, Farthing 2000; Genta et al. 1993; Phillips et al. 1992). These alterations are associated with an uncontrolled influx of inflammatory cells to the lamina propria and epithelium, leading to both a decrease in surface area and a decline in the enzymatic activity of disaccharidases and aminopeptidases in the microvillous membrane (Farthing 2000). The reduced activity of the above enzymes is likely related, in part, to the damage to the microvilli which takes place during infection (Bird and Smith 1980; Farthing 2000). This microvillous damage is most severe at points where *Cryptosporidium* organisms are attaching to epithelial cells (Bird and Smith 1980; Farthing 2000). Furthermore, it has been demonstrated that intracellular organelles undergo significant structural change during infection, e.g. swelling of mitochondria. These alterations in structure correlate with an impairment in microvillous function, including a reduction in both the absorption of monosaccharides and the co-transport of sodium and glucose (Argenzio et al. 1990; Farthing 2000; Modigliani et al. 1985). In cases of more severe disease, i.e. AIDS patients with cryptosporidiosis, poor absorption of vitamin B_{12} and fat has also been described (Farthing 2000; Goodgame et al. 1995; Modigliani et al. 1985).

9.8.2 Dysregulated Intestinal Secretion

Characterized by aggressive bouts of watery diarrhea, the clinical presentation of fulminant cryptosporidiosis has been compared to that caused by enterotoxin-producing organisms (i.e., cholera) (Farthing 2000). This comparison evokes the possibility that cryptosporidiosis is a diarrheal disease which results from both defective absorption and increased secretion of important electrolytes (Farthing 2000). Studies on human stool samples attempting to confirm this "secretory" characterization have thus far been controversial, however, enterotoxin-like activity has been found in fecal specimens from *C. parvum*-infected calves (Farthing 2000; Guarino et al. 1994, 1995, 1997; Kelly et al. 1996). It was also discovered that individuals with cryptosporidiosis have increased intestinal levels of the potent secretagogue, hydroxytryptamine (5-HT) (Farthing 2000; Sears and Guerrant 1994). The secretory diarrhea which results from infection with *Cryptosporidium* may therefore result from a combination of enterotoxin-like activity and 5-HT (Farthing 2000).

9.9 Clinical Management of Cryptosporidiosis

Effective management of *Cryptosporidium* infection can minimize the severity of this diarrheal disease (White 2010). Diarrhea can result in extreme dehydration, therefore management begins with replacing lost fluids and electrolytes, whether through oral rehydration or intravenous approaches. Rehydration and repletion of electrolyte losses by either the oral or intravenous (IV) route are important. Severe diarrhea can exceed 10 L/day among patients with AIDS, often requiring intensive support. Oral rehydration should be pursued aggressively with oral rehydration solutions. Patients should be monitored closely for signs and symptoms of volume depletion, electrolyte and weight loss, and malnutrition. Total parenteral nutrition might be indicated in certain patients. Unfortunately, cryptosporidiosis causes a loss in enzymes such as lactase, therefore rehydration regimens should be lactose-free (Phillips et al. 1992). In addition, in the recovery stage following infection, lactase deficiency can persist, contributing to diarrhea. Therefore, after treatment it may be necessary to restrict milk products and/or use lactase until the brush border is restored. To date, there are limited options for the direct treatment of infection, as a large number of antiparasitic agents have no effect especially in immune compromised hosts. The cornerstone of treatment for cryptosporidiosis is, therefore, the restoration of host immune function, e.g. cART in patients with AIDS. Case reports had suggested that oral therapy with anti-Cryptosporidium immunoglobulin preparations (bovine colostrum) was effective; however, a clinical trial of 20 g/day for treatment did not demonstrate any benefit (Fries et al. 1994). Patients with biliary tract involvement may require endoscopic retrograde choledocoduodenoscopy for diagnosis and may benefit from sphincterotomy and/or stenting (Hashmey et al. 1997).

9.9.1 Restoration of the Immune System

Cryptosporidiosis is a disease which frequently presents in immune compromised individuals, such as patients with AIDS. Any therapeutic agent, therefore, that either restores a functional immune system or targets the pathogen responsible for damaging the immune response could ameliorate clinical outcome (White 2010). In AIDS patients with cryptosporidiosis, antiretroviral therapy (cART) has been shown to significantly improve diarrheal disease (Carr et al. 1998; Foudraine et al. 1998; Grube et al. 1997; Maggi et al. 2000; Miao et al. 2000; Okhuysen et al. 2001). Interestingly, the antiretroviral medications that function by inhibiting HIV proteases have demonstrated specific activity against *Cryptosporidium* species both in vivo and in vitro (Hommer et al. 2003; Mele et al. 2003). Immune reconstitution inflammatory syndrome (IRIS) has not been described in association with treatment of cryptosporidiosis.

9.9.2 Antimotility Agents

The intestinal epithelial cell layer breakdown which occurs during cryptosporidiosis results in an increase in intestinal transit (Brantley et al. 2003; Sharpstone et al. 1999) that contributes to malabsorption (Brantley et al. 2003; Sharpstone et al. 1999). Therapeutics that regulate motility in the intestine are, therefore, critical for regulating disease severity (White 2010). Opiates are typically the primary antimotility drugs used for treatment (White 2010). Tincture of opium may be more effective than loperamide. Octreotide in case reports was also effective, however, a clinical trial did not demonstrate benefit for severe diarrhea due to cryptosporidiosis (Cello et al. 1991).

9.9.3 Antiparasitic Agents

There have been a large number of studies aimed at developing a satisfactory therapy for cryptosporidiosis, particularly in patients with AIDS. In vitro screening of over 200 agents failed to identify a highly active compound (White 2010). Recently halofuginone lactate (Halocur™) was identified and developed as a commercial veterinary drug for the prevention and treatment of this disease in cattle. To date, there have been no antiparasitic agents found which have clinical activity for the treatment of cryptosporidiosis in immune compromised patients. Multiple agents have been investigated in small randomized controlled clinical trials of HIV-infected adults, including nitazoxanide, paromomycin, spiramycin, bovine hyperimmune colostrum, and bovine dialyzable leukocyte extract. No pharmacologic or immunologic therapy directed specifically against *Cryptosporidium* has been shown to be consistently effective when used without cART (Cabada and White 2010). Agents which have variable activity, but are used clinically, include macrolides, e.g. clarithromycin and azithromycin, and the aminoglycoside paromomycin (Griffiths 1998; Hashmey et al. 1997; Hunter and Nichols 2002; Tzipori 1998). Nitazoxanide has been approved by the Food and Drug Administration (FDA) for treatment of *C. parvum* in immune competent children and adults. Because of the clinical significance of cryptosporidiosis, a trial of nitazoxanide or other anti-parasitic drugs in conjunction with cART, but never instead of cART, may be considered in patients with AIDS. Combination therapy with multiple antiparasitic agents has not been studied for the treatment of cryptosporidiosis and merits further investigation.

9.9.4 Nitazoxanide

To date, the only licensed treatment for cryptosporidiosis is the thiazole-containing therapeutic, nitazoxanide with in vivo activity against a broad range of helminths, bacteria, and protozoa (Shirley et al. 2012). It has been approved by the Food and Drug Administration (FDA) for treatment of *C. parvum* in children and adults (Chap. 11). Nitazoxanide has demonstrated efficacy in reducing the severity and duration of disease in immune competent individuals infected with the parasite (Shirley et al. 2012). Regardless of dosage and length of treatment, nitazoxanide has not proven to be highly effective in immune compromised patients, specifically those with AIDS (Amadi et al. 2009; Shirley et al. 2012). In patients with AIDS and cryptosporidiosis it has been administered at 500 to 1,000 mg twice daily in adults with some evidence of response when used on a compassionate expanded use program in patients with advanced AIDS (Rossignol 2006, Rossignol et al. 1998, 2001). Response depended on immune status as 70 % of patients with CD4+ counts > 50 cells/mm^3 responded compared to 20 % of those with lower CD4+ cell counts. HIV-infected adults with cryptosporidiosis with $CD4^+$ >50 cells/μL treated with 500 to 1,000 mg twice daily of nitazoxanide for 14 days experienced substantially higher rates of parasitological cure and resolution of diarrhea than persons receiving placebo treatment (Rossignol et al. 1998); however, this was not confirmed in two randomized trials in children (Amadi et al. 2002, 2009). This is probably due to key deficiencies in the host immune system, which are necessary for the thiazole to be effective. Ideal treatment for AIDS cases with cryptosporidiosis would therefore consist of cART to target the virus and antiparasitic agents to target the parasite (Shirley et al. 2012).

Adverse events associated with nitazoxanide are limited and typically mild, and no important drug-drug interactions have been reported. Nitazoxanide is not teratogenic in animals but human data on use in pregnancy are not available.

9.9.5 Paromomycin

Paromomycin is a nonabsorbable aminoglycoside that is used for the treatment of giradiasis and intestinal amebiasis. In animal models, it has shown efficacy in the treatment of cryptosporidiosis (Tzipori et al. 1994) A meta-analysis of 11 published human studies of paromomycin noted an overall response rate of 67 %; however, relapses were common, with a long-term success rate of only 33 % (Hashmey et al. 1997) The dose in adults is 25 to 35 mg/kg/day in 2 to 4 divided doses. Two randomized trials and a meta-analysis of these trials comparing paromomycin with placebo among patients with AIDS and cryptosporidiosis demonstrated a limited effectiveness of this drug; however, this analysis is limited by the small sample size and methodological issues (Cabada and White 2010; Hewitt et al. 2000; White et al. 1994). There appeared to be an improved response rate if paromomycin was combined with cART in AIDS patients (Maggi et al. 2000, 2001). Some clinicians believed that paromomycin response correlated with the infection being primarily present in the colon rather than small bowel. In AIDS patients, paromomycin could be used instead of nitazoxanide, but needs to be combined with cART.

9.9.6 *Macrolides*

Macrolides including spiramycin, azithromycin, roxithromycin and clarithromycin have shown some activity for cryptosporidiosis, but this has been variable. Spiramycin in children at a dose of 100 mg/kg/day was associated with a more rapid resolution of cryptosporidiosis in an initial trial; however a second trial demonstrated no efficacy (Wittenberg et al. 1989). Intravenous spiramycin was associated with therapeutic benefit; however, this was associated with significant drug induced side effects (Stockdale et al. 2008). Azithromycin has efficacy in vitro and in animal models (Stockdale et al. 2008) and in many case reports in humans (Hicks et al. 1996; Kadappu et al. 2002; Nachbaur et al. 1997). In placebo controlled trials of azithromycin (900 mg/day) there was no effect seen (Blanshard et al. 1997; Stockdale et al. 2008; White 2010). Clarithromycin was active in animal and in vitro studies (Fayer and Ellis 1993) as well as in some case reports from children in Egypt with cryptosporidiosis. Data from patients on this drug for MAI prophylaxis also suggested a protective effect. Roxithromycin demonstrated activity in two uncontrolled trials (Sprinz et al. 1998; Uip et al. 1998). Overall, in immune compromised patients with AIDS, macrolides could be used instead of nitazoxanide, but would need to be combined with cART.

9.9.7 *Preventing Infection*

Immune suppressed patients should be counseled regarding the routes of transmission of *Cryptosporidium spp*., e.g. direct contact with infected adults, diaper-aged children and infected animals; coming into contact with contaminated water during recreational activities; drinking contaminated water; and eating contaminated food. Handwashing and careful hygiene can reduce the risk for diarrhea (Huang and Zhou 2007). Patients with HIV-infection should be advised to wash their hands after any potential contacts that could result in transmission of cryptosporidiosis. Immune suppressed individuals should avoid unprotected sex practices that could lead to direct (e.g., oral-anal) or indirect (e.g., penile-anal) contact with feces. HIV-infected patients (especially if their $CD4^+$ is < 200 cells/μL), should avoid direct contact with diarrhea or stool. Immune suppressed patients should not, in general, consume raw shellfish as oocysts can survive in oysters for >2 months and have been found in oysters taken from commercial oyster beds (Fayer et al. 1999;

Freire-Santos et al. 2000; Gomez-Couso et al. 2004; Graczyk and Schwab 2000). It should be appreciated, however, that there is no direct evidence, in humans, of transmission by this route.

Outbreaks of cryptosporidiosis have been linked to drinking water from municipal water supplies (Berkelman 1994; Bouzid et al. 2008; Leclerc et al. 2002). During outbreaks or in other situations that warrant a community advisory to boil water, boiling water for at least 3 min will eliminate the risk for cryptosporidiosis. Freezing does not kill this organism, so ice made from contaminated water is not safe for consumption. Immune suppressed patients can consider the routine use of submicron point of use water filters or bottled water as this can also reduce the risk for infection from municipal or well water sources.

As chronic cryptosporidiosis occurs primarily in persons with advanced immune deficiency, the initiation of cART in AIDS patients will prevent the majority of these infections. Rifabutin and clarithromycin when taken for *Mycobacterium avium-intracellulare* prophylaxis appear to provide protection against cryptosporidiosis; however, data are insufficient to recommend this as routine chemoprophylaxis for cryptosporidiosis (Fichtenbaum et al. 2000; Holmberg et al. 1998).

9.10 Conclusion

Cryptosporidiosis is a worldwide public health concern which can affect a variety of individuals irrespective of their immune status. Employing means to prevent both contamination of food and water is crucial for limiting the spread of this easily transmissible disease. Research needs to continue to focus on the development of effective therapeutic agents and preventative strategies, including vaccination. Significant work remains to identify optimal treatment regimens. In the meantime, public health measure such as consistent hand hygiene and safe food and water practices are critical for preventing infection, particularly in patients with a compromised immune system.

Acknowledgment We thank Dr. Herbert B. Tanowitz for his critical reading of this manuscript and his editorial advice. This was supported by NIH R01AI93220 (LMW) and T32 NS007098-31 (HJS).

References

Abrahamsen MS (1998) Bovine T cell responses to Cryptosporidium parvum infection. Int J Parasitol 28(7):1083–1088

Abrahamsen MS, Templeton TJ, Enomoto S, Abrahante JE, Zhu G, Lancto CA, Deng M, Liu C, Widmer G, Tzipori S, Buck GA, Xu P, Bankier AT, Dear PH, Konfortov BA, Spriggs HF, Iyer L, Anantharaman V, Aravind L, Kapur V (2004) Complete genome sequence of the apicomplexan, Cryptosporidium parvum. Science 304(5669):441–445. doi:10.1126/science.1094786

Adal KA SC, Guerrant RL (1995) *Cryptosporodium* and related species. In: Infections of the gastrointestinal tract. Raven Press, New York

Agnew DG, Lima AA, Newman RD, Wuhib T, Moore RD, Guerrant RL, Sears CL (1998) Cryptosporidiosis in northeastern Brazilian children: association with increased diarrhea morbidity. J Infect Dis 177(3):754–760

Aguirre SA, Mason PH, Perryman LE (1994) Susceptibility of major histocompatibility complex (MHC) class I- and MHC class II-deficient mice to Cryptosporidium parvum infection. Infect Immun 62(2):697–699

Ajjampur SS, Gladstone BP, Selvapandian D, Muliyil JP, Ward H, Kang G (2007) Molecular and spatial epidemiology of cryptosporidiosis in children in a semiurban community in South India. J Clin Microbiol 45(3):915–920. doi:10.1128/JCM.01590–06

Ajjampur SS, Rajendran P, Ramani S, Banerjee I, Monica B, Sankaran P, Rosario V, Arumugam R, Sarkar R, Ward H, Kang G (2008) Closing the diarrhoea diagnostic gap in Indian children by the application of molecular techniques. J Med Microbiol 57 (Pt 11):1364–1368. doi:10.1099/jmm.0.2008/003319-0

Alcantara CS, Yang CH, Steiner TS, Barrett LJ, Lima AA, Chappell CL, Okhuysen PC, White AC Jr, Guerrant RL (2003) Interleukin-8, tumor necrosis factor-alpha, and lactoferrin in immunocompetent hosts with experimental and Brazilian children with acquired cryptosporidiosis. Am J Trop Med Hyg 68(3):325–328

Alles AJ, Waldron MA, Sierra LS, Mattia AR (1995) Prospective comparison of direct immunofluorescence and conventional staining methods for detection of Giardia and Cryptosporidium spp. in human fecal specimens. J Clin Microbiol 33(6):1632–1634

Amadi B, Kelly P, Mwiya M, Mulwazi E, Sianongo S, Changwe F, Thomson M, Hachungula J, Watuka A, Walker-Smith J, Chintu C (2001) Intestinal and systemic infection, HIV, and mortality in Zambian children with persistent diarrhea and malnutrition. J Pediatr Gastroenterol Nutr 32(5):550–554

Amadi B, Mwiya M, Musuku J, Watuka A, Sianongo S, Ayoub A, Kelly P (2002) Effect of nitazoxanide on morbidity and mortality in Zambian children with cryptosporidiosis: a randomised controlled trial. Lancet 360(9343):1375–1380. doi:10.1016/S0140-6736(02)11401-2

Amadi B, Mwiya M, Sianongo S, Payne L, Watuka A, Katubulushi M, Kelly P (2009) High dose prolonged treatment with nitazoxanide is not effective for cryptosporidiosis in HIV positive Zambian children: a randomised controlled trial. BMC Infect Dis 9:195. doi:10.1186/1471-2334-9-195

Amar CF, East CL, Gray J, Iturriza-Gomara M, Maclure EA, McLauchlin J (2007) Detection by PCR of eight groups of enteric pathogens in 4,627 faecal samples: re-examination of the English case–control infectious intestinal disease study (1993–1996). Eur J Clin Microbiol Infect Dis 26(5):311–323. doi:10.1007/s10096-007-0290-8

Argenzio RA, Liacos JA, Levy ML, Meuten DJ, Lecce JG, Powell DW (1990) Villous atrophy, crypt hyperplasia, cellular infiltration, and impaired glucose-Na absorption in enteric cryptosporidiosis of pigs. Gastroenterology 98(5 Pt 1):1129–1140

Argenzio RA, Armstrong M, Rhoads JM (1996) Role of the enteric nervous system in piglet cryptosporidiosis. J Pharmacol Exp Ther 279(3):1109–1115

Arrowood MJ, Mead JR, Mahrt JL, Sterling CR (1989) Effects of immune colostrum and orally administered antisporozoite monoclonal antibodies on the outcome of Cryptosporidium parvum infections in neonatal mice. Infect Immun 57(8):2283–2288

Balatbat AB, Jordan GW, Tang YJ, Silva J Jr (1996) Detection of Cryptosporidium parvum DNA in human feces by nested PCR. J Clin Microbiol 34(7):1769–1772

Behera B, Mirdha BR, Makharia GK, Bhatnagar S, Dattagupta S, Samantaray JC (2008) Parasites in patients with malabsorption syndrome: a clinical study in children and adults. Dig Dis Sci 53 (3):672–679. doi:10.1007/s10620-007-9927-9

Benhamou Y, Kapel N, Hoang C, Matta H, Meillet D, Magne D, Raphael M, Gentilini M, Opolon P, Gobert JG (1995) Inefficacy of intestinal secretory immune response to Cryptosporidium in acquired immunodeficiency syndrome. Gastroenterology 108(3):627–635

Berk RN, Wall SD, McArdle CB, McCutchan JA, Clemett AR, Rosenblum JS, Premkumer A, Megibow AJ (1984) Cryptosporidiosis of the stomach and small intestine in patients with AIDS. AJR Am J Roentgenol 143(3):549–554. doi:10.2214/ajr.143.3.549

Berkelman RL (1994) Emerging infectious diseases in the United States, 1993. J Infect Dis 170 (2):272–277

Bird RG, Smith MD (1980) Cryptosporidiosis in man: parasite life cycle and fine structural pathology. J Pathol 132(3):217–233. doi:10.1002/path.1711320304

Blagburn BL, Current WL (1983) Accidental infection of a researcher with human Cryptosporidium. J Infect Dis 148(4):772–773

Blanshard C, Jackson AM, Shanson DC, Francis N, Gazzard BG (1992) Cryptosporidiosis in HIV-seropositive patients. Q J Med 85(307–308):813–823

Blanshard C, Shanson DC, Gazzard BG (1997) Pilot studies of azithromycin, letrazuril and paromomycin in the treatment of cryptosporidiosis. Int J STD AIDS 8(2):124–129

Bouzid M, Steverding D, Tyler KM (2008) Detection and surveillance of waterborne protozoan parasites. Curr Opin Biotechnol 19(3):302–306. doi:10.1016/j.copbio.2008.05.002

Brantley RK, Williams KR, Silva TM, Sistrom M, Thielman NM, Ward H, Lima AA, Guerrant RL (2003) AIDS-associated diarrhea and wasting in Northeast Brazil is associated with subtherapeutic plasma levels of antiretroviral medications and with both bovine and human subtypes of Cryptosporidium parvum. Braz J Infect Dis 7(1):16–22

Burgner D, Pikos N, Eagles G, McCarthy A, Stevens M (1999) Epidemiology of Cryptosporidium parvum in symptomatic paediatric oncology patients. J Paediatr Child Health 35(3):300–302

Bushen OY, Lima AA, Guerrant RL (2006) Cryptosporidiosis. In: Tropical infectious diseases. Principle, pathogens, and practice. Churchill Livingstone, Philadelphia, pp 1003–1014

Bushen OY, Kohli A, Pinkerton RC, Dupnik K, Newman RD, Sears CL, Fayer R, Lima AA, Guerrant RL (2007) Heavy cryptosporidial infections in children in northeast Brazil: comparison of Cryptosporidium hominis and Cryptosporidium parvum. Trans R Soc Trop Med Hyg 101(4):378–384. doi:10.1016/j.trstmh.2006.06.005

Cabada MM, White AC Jr (2010) Treatment of cryptosporidiosis: do we know what we think we know? Curr Opin Infect Dis 23(5):494–499. doi:10.1097/QCO.0b013e32833de052

Cama VA, Ross JM, Crawford S, Kawai V, Chavez-Valdez R, Vargas D, Vivar A, Ticona E, Navincopa M, Williamson J, Ortega Y, Gilman RH, Bern C, Xiao L (2007) Differences in clinical manifestations among Cryptosporidium species and subtypes in HIV-infected persons. J Infect Dis 196(5):684–691. doi:10.1086/519842

Cama VA, Bern C, Roberts J, Cabrera L, Sterling CR, Ortega Y, Gilman RH, Xiao L (2008) Cryptosporidium species and subtypes and clinical manifestations in children, Peru. Emerg Infect Dis 14(10):1567–1574. doi:10.3201/eid1410.071273

Campbell LD, Stewart JN, Mead JR (2002) Susceptibility to Cryptosporidium parvum infections in cytokine- and chemokine-receptor knockout mice. J Parasitol 88(5):1014–1016. doi:10.1645/0022-3395(2002)088[1014:STCPII]2.0.CO;2

Campos M, Jouzdani E, Sempoux C, Buts JP, Reding R, Otte JB, Sokal EM (2000) Sclerosing cholangitis associated to cryptosporidiosis in liver-transplanted children. Eur J Pediatr 159 (1–2):113–115

Carr A, Marriott D, Field A, Vasak E, Cooper DA (1998) Treatment of HIV-1-associated microsporidiosis and cryptosporidiosis with combination antiretroviral therapy. Lancet 351 (9098):256–261. doi:10.1016/S0140-6736(97)07529-6

Cegielski JP, Ortega YR, McKee S, Madden JF, Gaido L, Schwartz DA, Manji K, Jorgensen AF, Miller SE, Pulipaka UP, Msengi AE, Mwakyusa DH, Sterling CR, Reller LB (1999) Cryptosporidium, enterocytozoon, and cyclospora infections in pediatric and adult patients with diarrhea in Tanzania. Clin Infect Dis 28(2):314–321. doi:10.1086/515131

Cello JP, Grendell JH, Basuk P, Simon D, Weiss L, Wittner M, Rood RP, Wilcox CM, Forsmark CE, Read AE et al (1991) Effect of octreotide on refractory AIDS-associated diarrhea. A prospective, multicenter clinical trial. Ann Intern Med 115(9):705–710

Cersosimo E, Wilkowske CJ, Rosenblatt JE, Ludwig J (1992) Isolated antral narrowing associated with gastrointestinal cryptosporidiosis in acquired immunodeficiency syndrome. Mayo Clin Proc 67(6):553–556

Chalmers RM, Davies AP (2010) Minireview: clinical cryptosporidiosis. Exp Parasitol 124 (1):138–146. doi:10.1016/j.exppara.2009.02.003

Chan AW, MacFarlane IA, Rhodes JM (1989) Cryptosporidiosis as a cause of chronic diarrhoea in a patient with insulin-dependent diabetes mellitus. J Infect 19(3):293

Chappell CL, Okhuysen PC, Langer-Curry R, Widmer G, Akiyoshi DE, Tanriverdi S, Tzipori S (2006) Cryptosporidium hominis: experimental challenge of healthy adults. Am J Trop Med Hyg 75(5):851–857

Checkley W, Gilman RH, Epstein LD, Suarez M, Diaz JF, Cabrera L, Black RE, Sterling CR (1997) Asymptomatic and symptomatic cryptosporidiosis: their acute effect on weight gain in Peruvian children. Am J Epidemiol 145(2):156–163

Checkley W, Epstein LD, Gilman RH, Black RE, Cabrera L, Sterling CR (1998) Effects of Cryptosporidium parvum infection in Peruvian children: growth faltering and subsequent catch-up growth. Am J Epidemiol 148(5):497–506

Chen W, Harp JA, Harmsen AG (1993) Requirements for CD4+ cells and gamma interferon in resolution of established Cryptosporidium parvum infection in mice. Infect Immun 61 (9):3928–3932

Chen XM, Levine SA, Tietz P, Krueger E, McNiven MA, Jefferson DM, Mahle M, LaRusso NF (1998) Cryptosporidium parvum is cytopathic for cultured human biliary epithelia via an apoptotic mechanism. Hepatology 28(4):906–913. doi:10.1002/hep.510280402

Chen XM, Gores GJ, Paya CV, LaRusso NF (1999) Cryptosporidium parvum induces apoptosis in biliary epithelia by a Fas/Fas ligand-dependent mechanism. Am J Physiol 277(3 Pt 1): G599–G608

Chen XM, Levine SA, Splinter PL, Tietz PS, Ganong AL, Jobin C, Gores GJ, Paya CV, LaRusso NF (2001) Cryptosporidium parvum activates nuclear factor kappaB in biliary epithelia preventing epithelial cell apoptosis. Gastroenterology 120(7):1774–1783

Chen XM, Keithly JS, Paya CV, LaRusso NF (2002) Cryptosporidiosis. N Engl J Med 346 (22):1723–1731. doi:10.1056/NEJMra013170

Chen W, Harp JA, Harmsen AG (2003) Cryptosporidium parvum infection in gene-targeted B cell-deficient mice. J Parasitol 89(2):391–393. doi:10.1645/0022-3395(2003)089[0391: CPIIGB]2.0.CO;2

Chieffi PP, Sens YA, Paschoalotti MA, Miorin LA, Silva HG, Jabur P (1998) Infection by Cryptosporidium parvum in renal patients submitted to renal transplant or hemodialysis. Rev Soc Bras Med Trop 31(4):333–337

Cicirello HG, Kehl KS, Addiss DG, Chusid MJ, Glass RI, Davis JP, Havens PL (1997) Cryptosporidiosis in children during a massive waterborne outbreak in Milwaukee, Wisconsin: clinical, laboratory and epidemiologic findings. Epidemiol Infect 119(1):53–60

Clavel A, Arnal AC, Sanchez EC, Cuesta J, Letona S, Amiguet JA, Castillo FJ, Varea M, Gomez-Lus R (1996) Respiratory cryptosporidiosis: case series and review of the literature. Infection 24(5):341–346

Clayton F, Heller T, Kotler DP (1994) Variation in the enteric distribution of cryptosporidia in acquired immunodeficiency syndrome. Am J Clin Pathol 102(4):420–425

Cole J, Blikslager A, Hunt E, Gookin J, Argenzio R (2003) Cyclooxygenase blockade and exogenous glutamine enhance sodium absorption in infected bovine ileum. Am J Physiol Gastrointest Liver Physiol 284(3):G516–G524. doi:10.1152/ajpgi.00172.2002

Collins CD, Blanshard C, Cramp M, Gazzard B, Gleeson JA (1992) Case report: pneumatosis intestinalis occurring in association with cryptosporidiosis and HIV infection. Clin Radiol 46 (6):410–411

Cozon G, Biron F, Jeannin M, Cannella D, Revillard JP (1994) Secretory IgA antibodies to Cryptosporidium parvum in AIDS patients with chronic cryptosporidiosis. J Infect Dis 169 (3):696–699

Cryptosporidiosis. http://www.dpd.cdc.gov/dpdx/HTML/Cryptosporidiosis.htm. Accessed 15 Jan 2009

Culshaw RJ, Bancroft GJ, McDonald V (1997) Gut intraepithelial lymphocytes induce immunity against Cryptosporidium infection through a mechanism involving gamma interferon production. Infect Immun 65(8):3074–3079

Dagan R, Fraser D, El-On J, Kassis I, Deckelbaum R, Turner S (1995) Evaluation of an enzyme immunoassay for the detection of Cryptosporidium spp. in stool specimens from infants and young children in field studies. Am J Trop Med Hyg 52(2):134–138

Dann SM, Okhuysen PC, Salameh BM, DuPont HL, Chappell CL (2000) Fecal antibodies to Cryptosporidium parvum in healthy volunteers. Infect Immun 68(9):5068–5074

Dunand VA, Hammer SM, Rossi R, Poulin M, Albrecht MA, Doweiko JP, DeGirolami PC, Coakley E, Piessens E, Wanke CA (1997) Parasitic sinusitis and otitis in patients infected with human immunodeficiency virus: report of five cases and review. Clin Infect Dis 25 (2):267–272

Duong TH, Dufillot D, Koko J, Nze-Eyo'o R, Thuilliez V, Richard-Lenoble D, Kombila M (1995) Digestive cryptosporidiosis in young children in an urban area in Gabon. Sante 5(3):185–188

DuPont HL, Chappell CL, Sterling CR, Okhuysen PC, Rose JB, Jakubowski W (1995) The infectivity of Cryptosporidium parvum in healthy volunteers. N Engl J Med 332 (13):855–859. doi:10.1056/NEJM199503303321304

Elliott DA, Clark DP (2003) Host cell fate on Cryptosporidium parvum egress from MDCK cells. Infect Immun 71(9):5422–5426

Farthing MJG (2000) Clinical aspects of human Cryptosporidiosis. In: Cryptosporidiosis and microsporidiosis (F. Petry Editor), Contributions to Microbiology, Karger, Basel, Switzerland

Farthing MJ, Kelly MP, Veitch AM (1996) Recently recognised microbial enteropathies and HIV infection. J Antimicrob Chemother 37(Suppl B):61–70

Fayer R, Ellis W (1993) Glycoside antibiotics alone and combined with tetracyclines for prophylaxis of experimental cryptosporidiosis in neonatal BALB/c mice. J Parasitol 79(4):553–558

Fayer R, Lewis EJ, Trout JM, Graczyk TK, Jenkins MC, Higgins J, Xiao L, Lal AA (1999) Cryptosporidium parvum in oysters from commercial harvesting sites in the Chesapeake Bay. Emerg Infect Dis 5(5):706–710. doi:10.3201/eid0505.990513

Fayer R (2008) General biology. In: Cryptosporidium and cryptosporidiosis, 2nd edn. CRC Press, Boca Raton, pp 1–42

Fichtenbaum CJ, Zackin R, Feinberg J, Benson C, Griffiths JK (2000) Rifabutin but not clarithromycin prevents cryptosporidiosis in persons with advanced HIV infection. AIDS 14 (18):2889–2893

Flanigan T, Whalen C, Turner J, Soave R, Toerner J, Havlir D, Kotler D (1992) Cryptosporidium infection and CD4 counts. Ann Intern Med 116(10):840–842

Foot AB, Oakhill A, Mott MG (1990) Cryptosporidiosis and acute leukaemia. Arch Dis Child 65 (2):236–237

Foudraine NA, Weverling GJ, van Gool T, Roos MT, de Wolf F, Koopmans PP, van den Broek PJ, Meenhorst PL, van Leeuwen R, Lange JM, Reiss P (1998) Improvement of chronic diarrhoea in patients with advanced HIV-1 infection during potent antiretroviral therapy. AIDS 12 (1):35–41

Freire-Santos F, Oteiza-Lopez AM, Vergara-Castiblanco CA, Ares-Mazas E, Alvarez-Suarez E, Garcia-Martin O (2000) Detection of Cryptosporidium oocysts in bivalve molluscs destined for human consumption. J Parasitol 86(4):853–854. doi:10.1645/0022-3395(2000)086[0853:DOCOIB]2.0.CO;2

Fries L, Hillman K, Crabb J, et al (1994) Clinical and microbiologic effects of bovine anti-Cryptosporidium immunoglobulin (BACI) on cryptosporidial diarrhea in AIDS (abstract

M31). In: Interscience conference on antimicrobial agents and chemotherapy, American Society for Microbiology, Orlando
Gambhir IS, Jaiswal JP, Nath G (2003) Significance of Cryptosporidium as an aetiology of acute infectious diarrhoea in elderly Indians. Trop Med Int Health 8(5):415–419
Garcia LS, Shimizu RY (1997) Evaluation of nine immunoassay kits (enzyme immunoassay and direct fluorescence) for detection of Giardia lamblia and Cryptosporidium parvum in human fecal specimens. J Clin Microbiol 35(6):1526–1529
Garcia LS, Shimizu RY (2000) Detection of Giardia lamblia and Cryptosporidium parvum antigens in human fecal specimens using the ColorPAC combination rapid solid-phase qualitative immunochromatographic assay. J Clinical Microbiol 38(3):1267–1268
Garcia LS, Shimizu RY, Novak S, Carroll M, Chan F (2003) Commercial assay for detection of Giardia lamblia and Cryptosporidium parvum antigens in human fecal specimens by rapid solid-phase qualitative immunochromatography. J Clin Microbiol 41(1):209–212
Garone MA, Winston BJ, Lewis JH (1986) Cryptosporidiosis of the stomach. Am J Gastroenterol 81(6):465–470
Garza A, Lackner A, Aye P, D'Souza M, Martin P Jr, Borda J, Tweardy DJ, Weinstock J, Griffiths J, Robinson P (2008) Substance P receptor antagonist reverses intestinal pathophysiological alterations occurring in a novel ex-vivo model of Cryptosporidium parvum infection of intestinal tissues derived from SIV-infected macaques. J Med Primatol 37(3):109–115. doi:10.1111/j.1600-0684.2007.00251.x
Gatei W, Wamae CN, Mbae C, Waruru A, Mulinge E, Waithera T, Gatika SM, Kamwati SK, Revathi G, Hart CA (2006) Cryptosporidiosis: prevalence, genotype analysis, and symptoms associated with infections in children in Kenya. Am J Trop Med Hyg 75(1):78–82
Genta RM, Chappell CL, White AC Jr, Kimball KT, Goodgame RW (1993) Duodenal morphology and intensity of infection in AIDS-related intestinal cryptosporidiosis. Gastroenterology 105 (6):1769–1775
Gentile G, Venditti M, Micozzi A, Caprioli A, Donelli G, Tirindelli C, Meloni G, Arcese W, Martino P (1991) Cryptosporidiosis in patients with hematologic malignancies. Rev Infect Dis 13(5):842–846
Gerber DA, Green M, Jaffe R, Greenberg D, Mazariegos G, Reyes J (2000) Cryptosporidial infections after solid organ transplantation in children. Pediatr Transplant 4(1):50–55
Godwin TA (1991) Cryptosporidiosis in the acquired immunodeficiency syndrome: a study of 15 autopsy cases. Hum Pathol 22(12):1215–1224
Gomez Morales MA, Ausiello CM, Guarino A, Urbani F, Spagnuolo MI, Pignata C, Pozio E (1996) Severe, protracted intestinal cryptosporidiosis associated with interferon gamma deficiency: pediatric case report. Clin Infect Dis 22(5):848–850
Gomez Morales MA, La Rosa G, Ludovisi A, Onori AM, Pozio E (1999) Cytokine profile induced by Cryptosporidium antigen in peripheral blood mononuclear cells from immunocompetent and immunosuppressed persons with cryptosporidiosis. J Infect Dis 179(4):967–973. doi:10.1086/314665
Gomez-Couso H, Freire-Santos F, Amar CF, Grant KA, Williamson K, Ares-Mazas ME, McLauchlin J (2004) Detection of Cryptosporidium and Giardia in molluscan shellfish by multiplexed nested-PCR. Int J Food Microbiol 91(3):279–288. doi:10.1016/j.ijfoodmicro.2003.07.003
Goodgame RW, Kimball K, Ou CN, White AC Jr, Genta RM, Lifschitz CH, Chappell CL (1995) Intestinal function and injury in acquired immunodeficiency syndrome-related cryptosporidiosis. Gastroenterology 108(4):1075–1082
Gookin JL, Duckett LL, Armstrong MU, Stauffer SH, Finnegan CP, Murtaugh MP, Argenzio RA (2004) Nitric oxide synthase stimulates prostaglandin synthesis and barrier function in C. - parvum-infected porcine ileum. Am J Physiol Gastrointes Liver Physiol 287(3):G571–G581. doi:10.1152/ajpgi.00413.2003
Graczyk TK, Schwab KJ (2000) Foodborne infections vectored by molluscan shellfish. Curr Gastroenterol Rep 2(4):305–309

Greenberg PD, Koch J, Cello JP (1996) Diagnosis of Cryptosporidium parvum in patients with severe diarrhea and AIDS. Dig Dis Sci 41(11):2286–2290

Griffiths JK (1998) Human cryptosporidiosis: epidemiology, transmission, clinical disease, treatment, and diagnosis. Adv Parasitol 40:37–85

Griffiths JK, Moore R, Dooley S, Keusch GT, Tzipori S (1994) Cryptosporidium parvum infection of Caco-2 cell monolayers induces an apical monolayer defect, selectively increases transmonolayer permeability, and causes epithelial cell death. Infect Immun 62 (10):4506–4514

Gross TL, Wheat J, Bartlett M, O'Connor KW (1986) AIDS and multiple system involvement with cryptosporidium. Am J Gastroenterol 81(6):456–458

Grube H, Ramratnam B, Ley C, Flanigan TP (1997) Resolution of AIDS associated cryptosporidiosis after treatment with indinavir. Am J Gastroenterol 92(4):726

Guarino A, Canani RB, Pozio E, Terracciano L, Albano F, Mazzeo M (1994) Enterotoxic effect of stool supernatant of Cryptosporidium-infected calves on human jejunum. Gastroenterology 106(1):28–34

Guarino A, Canani RB, Casola A, Pozio E, Russo R, Bruzzese E, Fontana M, Rubino A (1995) Human intestinal cryptosporidiosis: secretory diarrhea and enterotoxic activity in Caco-2 cells. J Infect Dis 171(4):976–983

Guarino A, Castaldo A, Russo S, Spagnuolo MI, Canani RB, Tarallo L, DiBenedetto L, Rubino A (1997) Enteric cryptosporidiosis in pediatric HIV infection. J Pediatr Gastroenterol Nutr 25 (2):182–187

Guerrant DI, Moore SR, Lima AA, Patrick PD, Schorling JB, Guerrant RL (1999) Association of early childhood diarrhea and cryptosporidiosis with impaired physical fitness and cognitive function four-seven years later in a poor urban community in northeast Brazil. Am J Trop Med Hyg 61(5):707–713

Hasan FA, Jeffers LJ, Dickinson G, Otrakji CL, Greer PJ, Reddy KR, Schiff ER (1991) Hepatobiliary cryptosporidiosis and cytomegalovirus infection mimicking metastatic cancer to the liver. Gastroenterology 100(6):1743–1748

Hashmey R, Smith NH, Cron S, Graviss EA, Chappell CL, White AC Jr (1997) Cryptosporidiosis in Houston, Texas. A report of 95 cases. Medicine 76(2):118–139

Hayward AR, Levy J, Facchetti F, Notarangelo L, Ochs HD, Etzioni A, Bonnefoy JY, Cosyns M, Weinberg A (1997) Cholangiopathy and tumors of the pancreas, liver, and biliary tree in boys with X-linked immunodeficiency with hyper-IgM. J Immunol 158(2):977–983

Hernandez J, Lackner A, Aye P, Mukherjee K, Tweardy DJ, Mastrangelo MA, Weinstock J, Griffiths J, D'Souza M, Dixit S, Robinson P (2007) Substance P is responsible for physiological alterations such as increased chloride ion secretion and glucose malabsorption in cryptosporidiosis. Infect Immun 75(3):1137–1143. doi:10.1128/IAI.01738-05

Hewitt RG, Yiannoutsos CT, Higgs ES, Carey JT, Geiseler PJ, Soave R, Rosenberg R, Vazquez GJ, Wheat LJ, Fass RJ, Antoninievic Z, Walawander AL, Flanigan TP, Bender JF (2000) Paromomycin: no more effective than placebo for treatment of cryptosporidiosis in patients with advanced human immunodeficiency virus infection. AIDS Clinical Trial Group. Clin Infect Dis 31(4):1084–1092. doi:10.1086/318155

Hicks P, Zwiener RJ, Squires J, Savell V (1996) Azithromycin therapy for Cryptosporidium parvum infection in four children infected with human immunodeficiency virus. J Pediatr 129(2):297–300

Holmberg SD, Moorman AC, Von Bargen JC, Palella FJ, Loveless MO, Ward DJ, Navin TR (1998) Possible effectiveness of clarithromycin and rifabutin for cryptosporidiosis chemoprophylaxis in HIV disease. HIV Outpatient Study (HOPS) Investigators. JAMA 279(5):384–386

Hommer V, Eichholz J, Petry F (2003) Effect of antiretroviral protease inhibitors alone, and in combination with paromomycin, on the excystation, invasion and in vitro development of Cryptosporidium parvum. J Antimicrob Chemother 52(3):359–364. doi:10.1093/jac/dkg357

Huang DB, Zhou J (2007) Effect of intensive handwashing in the prevention of diarrhoeal illness among patients with AIDS: a randomized controlled study. J Med Microbiol 56(Pt 5):659–663. doi:10.1099/jmm.0.46867-0

Hunter PR, Nichols G (2002) Epidemiology and clinical features of Cryptosporidium infection in immunocompromised patients. Clin Microbiol Rev 15(1):145–154

Hunter PR, Hughes S, Woodhouse S, Raj N, Syed Q, Chalmers RM, Verlander NQ, Goodacre J (2004a) Health sequelae of human cryptosporidiosis in immunocompetent patients. Clin Infect Dis 39(4):504–510. doi:10.1086/422649

Hunter PR, Hughes S, Woodhouse S, Syed Q, Verlander NQ, Chalmers RM, Morgan K, Nichols G, Beeching N, Osborn K (2004b) Sporadic cryptosporidiosis case–control study with genotyping. Emerg Infect Dis 10(7):1241–1249. doi:10.3201/eid1007.030582

Ignatius R, Eisenblatter M, Regnath T, Mansmann U, Futh U, Hahn H, Wagner J (1997) Efficacy of different methods for detection of low Cryptosporidium parvum oocyst numbers or antigen concentrations in stool specimens. Eur J Clin Microbiol Infect Dis 16(10):732–736

Iribarren JA, Castiella A, Lobo C, Lopez P, Von Wichmann MA, Arrizabalaga J, Rodriguez-Arrondo FJ, Alzate LF (1997) AIDS-associated cryptosporidiosis with antral narrowing. A new case. J Clin Gastroenterol 25(4):693–694

Jacyna MR, Parkin J, Goldin R, Baron JH (1990) Protracted enteric cryptosporidial infection in selective immunoglobulin A and saccharomyces opsonin deficiencies. Gut 31(6):714–716

Jaggi N, Rajeshwari S, Mittal SK, Mathur MD, Baveja UK (1994) Assessment of the immune and nutritional status of the host in childhood diarrhoea due to cryptosporidium. J Commun Dis 26 (4):181–185

Janoff EN, Mead PS, Mead JR, Echeverria P, Bodhidatta L, Bhaibulaya M, Sterling CR, Taylor DN (1990) Endemic Cryptosporidium and Giardia lamblia infections in a Thai orphanage. Am J Trop Med Hyg 43(3):248–256

Javier Enriquez F, Avila CR, Ignacio Santos J, Tanaka-Kido J, Vallejo O, Sterling CR (1997) Cryptosporidium infections in Mexican children: clinical, nutritional, enteropathogenic, and diagnostic evaluations. Am J Trop Med Hyg 56(3):254–257

Jenkins MC, O'Brien C, Trout J, Guidry A, Fayer R (1999) Hyperimmune bovine colostrum specific for recombinant Cryptosporidium parvum antigen confers partial protection against cryptosporidiosis in immunosuppressed adult mice. Vaccine 17(19):2453–2460

Johnston SP, Ballard MM, Beach MJ, Causer L, Wilkins PP (2003) Evaluation of three commercial assays for detection of Giardia and Cryptosporidium organisms in fecal specimens. J Clin Microbiol 41(2):623–626

Jokipii L, Jokipii AM (1986) Timing of symptoms and oocyst excretion in human cryptosporidiosis. N Engl J Med 315(26):1643–1647. doi:10.1056/NEJM198612253152604

Kadappu KK, Nagaraja MV, Rao PV, Shastry BA (2002) Azithromycin as treatment for cryptosporidiosis in human immunodeficiency virus disease. J Postgrad Med 48(3):179–181

Kandil HM, Berschneider HM, Argenzio RA (1994) Tumour necrosis factor alpha changes porcine intestinal ion transport through a paracrine mechanism involving prostaglandins. Gut 35(7):934–940

Kaushik K, Khurana S, Wanchu A, Malla N (2008) Evaluation of staining techniques, antigen detection and nested PCR for the diagnosis of cryptosporidiosis in HIV seropositive and seronegative patients. Acta Trop 107(1):1–7. doi:10.1016/j.actatropica.2008.02.007

Kazlow PG, Shah K, Benkov KJ, Dische R, LeLeiko NS (1986) Esophageal cryptosporidiosis in a child with acquired immune deficiency syndrome. Gastroenterology 91(5):1301–1303

Kehl KS, Cicirello H, Havens PL (1995) Comparison of four different methods for detection of Cryptosporidium species. J Clin Microbiol 33(2):416–418

Kelly P, Thillainayagam AV, Smithson J, Hunt JB, Forbes A, Gazzard BG, Farthing MJ (1996) Jejunal water and electrolyte transport in human cryptosporidiosis. Dig Dis Sci 41 (10):2095–2099

Kelly P, Makumbi FA, Carnaby S, Simjee AE, Farthing MJ (1998) Variable distribution of Cryptosporidium parvum in the intestine of AIDS patients revealed by polymerase chain reaction. Eur J Gastroenterol Hepatol 10(10):855–858
Khan WA, Rogers KA, Karim MM, Ahmed S, Hibberd PL, Calderwood SB, Ryan ET, Ward HD (2004) Cryptosporidiosis among Bangladeshi children with diarrhea: a prospective, matched, case–control study of clinical features, epidemiology and systemic antibody responses. Am J Trop Med Hyg 71(4):412–419
Kibbler CC, Smith A, Hamilton-Dutoit SJ, Milburn H, Pattinson JK, Prentice HG (1987) Pulmonary cryptosporidiosis occurring in a bone marrow transplant patient. Scand J Infect Dis 19 (5):581–584
Kim LS, Hadley WK, Stansell J, Cello JP, Koch J (1998) Declining prevalence of cryptosporidiosis in San Francisco. Clin Infect Dis 27(3):655–656
Kirkpatrick BD, Daniels MM, Jean SS, Pape JW, Karp C, Littenberg B, Fitzgerald DW, Lederman HM, Nataro JP, Sears CL (2002) Cryptosporidiosis stimulates an inflammatory intestinal response in malnourished Haitian children. J Infect Dis 186(1):94–101. doi:10.1086/341296
Kirkpatrick BD, Noel F, Rouzier PD, Powell JL, Pape JW, Bois G, Alston WK, Larsson CJ, Tenney K, Ventrone C, Powden C, Sreenivasan M, Sears CL (2006) Childhood cryptosporidiosis is associated with a persistent systemic inflammatory response. Clin Infect Dis 43 (5):604–608. doi:10.1086/506565
Kirkpatrick BD, Haque R, Duggal P, Mondal D, Larsson C, Peterson K, Akter J, Lockhart L, Khan S, Petri WA Jr (2008) Association between Cryptosporidium infection and human leukocyte antigen class I and class II alleles. J Infect Dis 197(3):474–478. doi:10.1086/525284
Kline TJ, Delas Morenas T, O'Brien M, Smith BF, Afdhal NH (1993) Squamous metaplasia of extrahepatic biliary system in an AIDS patient with cryptosporidia and cholangitis. Dig Dis Sci 38(5):960–962
Kocoshis SA, Cibull ML, Davis TE, Hinton JT, Seip M, Banwell JG (1984) Intestinal and pulmonary cryptosporidiosis in an infant with severe combined immune deficiency. J Pediatr Gastroenterol Nutr 3(1):149–157
Kosek M, Alcantara C, Lima AA, Guerrant RL (2001) Cryptosporidiosis: an update. Lancet Infect Dis 1(4):262–269. doi:10.1016/S1473-3099(01)00121-9
Lacroix S, Mancassola R, Naciri M, Laurent F (2001) Cryptosporidium parvum-specific mucosal immune response in C57BL/6 neonatal and gamma interferon-deficient mice: role of tumor necrosis factor alpha in protection. Infect Immun 69(3):1635–1642. doi:10.1128/IAI.69.3.1635-1642.2001
Laurent F, Kagnoff MF, Savidge TC, Naciri M, Eckmann L (1998) Human intestinal epithelial cells respond to Cryptosporidium parvum infection with increased prostaglandin H synthase 2 expression and prostaglandin E2 and F2alpha production. Infect Immun 66(4):1787–1790
Le Moing V, Bissuel F, Costagliola D, Eid Z, Chapuis F, Molina JM, Salmon-Ceron D, Brasseur P, Leport C (1998) Decreased prevalence of intestinal cryptosporidiosis in HIV-infected patients concomitant to the widespread use of protease inhibitors. AIDS 12(11):1395–1397
Leclerc H, Schwartzbrod L, Dei-Cas E (2002) Microbial agents associated with waterborne diseases. Crit Rev Microbiol 28(4):371–409. doi:10.1080/1040-840291046768
Leoni F, Amar C, Nichols G, Pedraza-Diaz S, McLauchlin J (2006) Genetic analysis of Cryptosporidium from 2414 humans with diarrhoea in England between 1985 and 2000. J Med Microbiol 55(Pt 6):703–707. doi:10.1099/jmm.0.46251-0
Levy J, Espanol-Boren T, Thomas C, Fischer A, Tovo P, Bordigoni P, Resnick I, Fasth A, Baer M, Gomez L, Sanders EA, Tabone MD, Plantaz D, Etzioni A, Monafo V, Abinun M, Hammarstrom L, Abrahamsen T, Jones A, Finn A, Klemola T, DeVries E, Sanal O, Peitsch MC, Notarangelo LD (1997) Clinical spectrum of X-linked hyper-IgM syndrome. J Pediatr 131 (1 Pt 1):47–54
Lewis IJ, Hart CA, Baxby D (1985) Diarrhoea due to Cryptosporidium in acute lymphoblastic leukemia. Arch Dis Child 60(1):60–62

Lima AA, Fang G, Schorling JB, De Albuquerque L, McAuliffe JF, Mota S, Leite R, Guerrant RL (1992) Persistent diarrhea in Northeast Brazil: etiologies and interactions with malnutrition. Acta Paediatr 81(Suppl 381):39–44

Lima AA, Moore SR, Barboza MS Jr, Soares AM, Schleupner MA, Newman RD, Sears CL, Nataro JP, Fedorko DP, Wuhib T, Schorling JB, Guerrant RL (2000) Persistent diarrhea signals a critical period of increased diarrhea burdens and nutritional shortfalls: a prospective cohort study among children in northeastern Brazil. J Infect Dis 181(5):1643–1651. doi:10.1086/315423

Lopez-Velez R, Tarazona R, Garcia Camacho A, Gomez-Mampaso E, Guerrero A, Moreira V, Villanueva R (1995) Intestinal and extraintestinal cryptosporidiosis in AIDS patients. Eur J Clin Microbiol Infect Dis 14(8):677–681

Lumadue JA, Manabe YC, Moore RD, Belitsos PC, Sears CL, Clark DP (1998) A clinicopathologic analysis of AIDS-related cryptosporidiosis. AIDS 12(18):2459–2466

Mac Kenzie WR, Hoxie NJ, Proctor ME, Gradus MS, Blair KA, Peterson DE, Kazmierczak JJ, Addiss DG, Fox KR, Rose JB et al (1994) A massive outbreak in Milwaukee of cryptosporidium infection transmitted through the public water supply. N Engl J Med 331(3):161–167. doi:10.1056/NEJM199407213310304

Macfarlane DE, Horner-Bryce J (1987) Cryptosporidiosis in well-nourished and malnourished children. Acta Paediatr Scand 76(3):474–477

MacKenzie WR, Schell WL, Blair KA, Addiss DG, Peterson DE, Hoxie NJ, Kazmierczak JJ, Davis JP (1995) Massive outbreak of waterborne cryptosporidium infection in Milwaukee, Wisconsin: recurrence of illness and risk of secondary transmission. Clin Infect Dis 21 (1):57–62

Maggi P, Larocca AM, Quarto M, Serio G, Brandonisio O, Angarano G, Pastore G (2000) Effect of antiretroviral therapy on cryptosporidiosis and microsporidiosis in patients infected with human immunodeficiency virus type 1. Eur J Clin Microbiol Infect Dis 19(3):213–217

Maggi P, Larocca AM, Ladisa N, Carbonara S, Brandonisio O, Angarano G, Pastore G (2001) Opportunistic parasitic infections of the intestinal tract in the era of highly active antiretroviral therapy: is the CD4(+) count so important? Clin Infect Dis 33(9):1609–1611. doi:10.1086/323017

Manabe YC, Clark DP, Moore RD, Lumadue JA, Dahlman HR, Belitsos PC, Chaisson RE, Sears CL (1998) Cryptosporidiosis in patients with AIDS: correlates of disease and survival. Clin Infect Dis 27(3):536–542

Manivel C, Filipovich A, Snover DC (1985) Cryptosporidiosis as a cause of diarrhea following bone marrow transplantation. Dis Colon Rectum 28(10):741–742

Martino P, Gentile G, Caprioli A, Baldassarri L, Donelli G, Arcese W, Fenu S, Micozzi A, Venditti M, Mandelli F (1988) Hospital-acquired cryptosporidiosis in a bone marrow transplantation unit. J Infect Dis 158(3):647–648

McCole DF, Eckmann L, Laurent F, Kagnoff MF (2000) Intestinal epithelial cell apoptosis following Cryptosporidium parvum infection. Infect Immun 68(3):1710–1713

McDonald V, Deer R, Uni S, Iseki M, Bancroft GJ (1992) Immune responses to Cryptosporidium muris and Cryptosporidium parvum in adult immunocompetent or immunocompromised (nude and SCID) mice. Infect Immun 60(8):3325–3331

McDonald V, Robinson HA, Kelly JP, Bancroft GJ (1994) Cryptosporidium muris in adult mice: adoptive transfer of immunity and protective roles of CD4 versus CD8 cells. Infect Immun 62 (6):2289–2294

McGowan I, Hawkins AS, Weller IV (1993) The natural history of cryptosporidial diarrhoea in HIV-infected patients. AIDS 7(3):349–354

McLauchlin J, Pedraza-Diaz S, Amar-Hoetzeneder C, Nichols GL (1999) Genetic characterization of Cryptosporidium strains from 218 patients with diarrhea diagnosed as having sporadic cryptosporidiosis. J Clin Microbiol 37(10):3153–3158

Mead JR, You X (1998) Susceptibility differences to Cryptosporidium parvum infection in two strains of gamma interferon knockout mice. J Parasitol 84(5):1045–1048

Mele R, Gomez Morales MA, Tosini F, Pozio E (2003) Indinavir reduces Cryptosporidium parvum infection in both in vitro and in vivo models. Int J Parasitol 33(7):757–764
Menon BS, Abdullah MS, Mahamud F, Singh B (1999) Intestinal parasites in Malaysian children with cancer. J Trop Pediatr 45(4):241–242
Meynard JL, Meyohas MC, Binet D, Chouaid C, Frottier J (1996) Pulmonary cryptosporidiosis in the acquired immunodeficiency syndrome. Infection 24(4):328–331
Miao YM, Awad-El-Kariem FM, Franzen C, Ellis DS, Muller A, Counihan HM, Hayes PJ, Gazzard BG (2000) Eradication of cryptosporidia and microsporidia following successful antiretroviral therapy. J Acquir Immune Defic Syndr 25(2):124–129
Modigliani R, Bories C, Le Charpentier Y, Salmeron M, Messing B, Galian A, Rambaud JC, Lavergne A, Cochand-Priollet B, Desportes I (1985) Diarrhoea and malabsorption in acquired immune deficiency syndrome: a study of four cases with special emphasis on opportunistic protozoan infestations. Gut 26(2):179–187
Molbak K, Hojlyng N, Gottschau A, Sa JC, Ingholt L, da Silva AP, Aaby P (1993) Cryptosporidiosis in infancy and childhood mortality in Guinea Bissau, west Africa. BMJ 307 (6901):417–420
Molbak K, Andersen M, Aaby P, Hojlyng N, Jakobsen M, Sodemann M, da Silva AP (1997) Cryptosporidium infection in infancy as a cause of malnutrition: a community study from Guinea-Bissau, west Africa. Am J Clin Nutr 65(1):149–152
Moon A, Spivak W, Brandt LJ (1999) Cryptosporidium-induced gastric obstruction in a child with congenital HIV infection: case report and review of the literature. J Pediatr Gastroenterol Nutr 28(1):108–111
Motta I, Gissot M, Kanellopoulos JM, Ojcius DM (2002) Absence of weight loss during Cryptosporidium infection in susceptible mice deficient in Fas-mediated apoptosis. Microbes Infect 4 (8):821–827
Muthusamy D, Rao SS, Ramani S, Monica B, Banerjee I, Abraham OC, Mathai DC, Primrose B, Muliyil J, Wanke CA, Ward HD, Kang G (2006) Multilocus genotyping of Cryptosporidium sp. isolates from human immunodeficiency virus-infected individuals in South India. J Clin Microbiol 44(2):632–634. doi:10.1128/JCM.44.2.632-634.2006
Nachbaur D, Kropshofer G, Feichtinger H, Allerberger F, Niederwieser D (1997) Cryptosporidiosis after CD34-selected autologous peripheral blood stem cell transplantation (PBSCT). Treatment with paromomycin, azithromycin and recombinant human interleukin-2. Bone Marrow Transplant 19(12):1261–1263. doi:10.1038/sj.bmt.1700826
Nair P, Mohamed JA, DuPont HL, Figueroa JF, Carlin LG, Jiang ZD, Belkind-Gerson J, Martinez-Sandoval FG, Okhuysen PC (2008) Epidemiology of cryptosporidiosis in North American travelers to Mexico. Am J Trop Med Hyg 79(2):210–214
Naumova EN, Egorov AI, Morris RD, Griffiths JK (2003) The elderly and waterborne Cryptosporidium infection: gastroenteritis hospitalizations before and during the 1993 Milwaukee outbreak. Emerg Infect Dis 9(4):418–425. doi:10.3201/eid0904.020260
Neill MA, Rice SK, Ahmad NV, Flanigan TP (1996) Cryptosporidiosis: an unrecognized cause of diarrhea in elderly hospitalized patients. Clin Infect Dis 22(1):168–170
Newman RD, Zu SX, Wuhib T, Lima AA, Guerrant RL, Sears CL (1994) Household epidemiology of Cryptosporidium parvum infection in an urban community in northeast Brazil. Ann Intern Med 120(6):500–505
Newman RD, Sears CL, Moore SR, Nataro JP, Wuhib T, Agnew DA, Guerrant RL, Lima AA (1999) Longitudinal study of Cryptosporidium infection in children in northeastern Brazil. J Infect Dis 180(1):167–175. doi:10.1086/314820
Nichols G (2008) Epidemiology. In: *Cryptosporidium* and cryptosporidiosis, 2nd edn. CRC Press, Boca Raton, pp 79–118
Oberhuber G, Lauer E, Stolte M, Borchard F (1991) Cryptosporidiosis of the appendix vermiformis: a case report. Z Gastroenterol 29(11):606–608
Ok UZ, Cirit M, Uner A, Ok E, Akcicek F, Basci A, Ozcel MA (1997) Cryptosporidiosis and blastocystosis in renal transplant recipients. Nephron 75(2):171–174

Okhuysen PC, Chappell CL, Sterling CR, Jakubowski W, DuPont HL (1998) Susceptibility and serologic response of healthy adults to reinfection with Cryptosporidium parvum. Infect Immun 66(2):441–443

Okhuysen PC, Chappell CL, Crabb JH, Sterling CR, DuPont HL (1999) Virulence of three distinct Cryptosporidium parvum isolates for healthy adults. J Infect Dis 180(4):1275–1281. doi:10.1086/315033

Okhuysen PC, Robinson P, Nguyen MT, Nannini EC, Lewis DE, Janecki A, Chappell CL, White AC Jr (2001) Jejunal cytokine response in AIDS patients with chronic cryptosporidiosis and during immune reconstitution. AIDS 15(6):802–804

Okhuysen PC, Rich SM, Chappell CL, Grimes KA, Widmer G, Feng X, Tzipori S (2002) Infectivity of a Cryptosporidium parvum isolate of cervine origin for healthy adults and interferon-gamma knockout mice. J Infect Dis 185(9):1320–1325. doi:10.1086/340132

Palmer SR, Biffin A, Public Health Laboratory Service Study Group (1990) Cryptosporidiosis in England and Wales: prevalence and clinical and epidemiological features. Br Med J 300:774–777

Pantenburg B, Dann SM, Wang HC, Robinson P, Castellanos-Gonzalez A, Lewis DE, White AC Jr (2008) Intestinal immune response to human Cryptosporidium sp. infection. Infect Immun 76 (1):23–29. doi:10.1128/IAI.00960-07

Pedraza-Diaz S, Amar CF, McLauchlin J, Nichols GL, Cotton KM, Godwin P, Iversen AM, Milne L, Mulla JR, Nye K, Panigrahl H, Venn SR, Wiggins R, Williams M, Youngs ER (2001) Cryptosporidium meleagridis from humans: molecular analysis and description of affected patients. J Infect 42(4):243–250. doi:10.1053/jinf.2001.0839

Pereira SJ, Ramirez NE, Xiao L, Ward LA (2002) Pathogenesis of human and bovine Cryptosporidium parvum in gnotobiotic pigs. J Infect Dis 186(5):715–718. doi:10.1086/342296

Perryman LE, Mason PH, Chrisp CE (1994) Effect of spleen cell populations on resolution of Cryptosporidium parvum infection in SCID mice. Infect Immun 62(4):1474–1477

Perryman LE, Kapil SJ, Jones ML, Hunt EL (1999) Protection of calves against cryptosporidiosis with immune bovine colostrum induced by a Cryptosporidium parvum recombinant protein. Vaccine 17(17):2142–2149

Phillips AD, Thomas AG, Walker-Smith JA (1992) Cryptosporidium, chronic diarrhoea and the proximal small intestinal mucosa. Gut 33(8):1057–1061

Pollok RC, Farthing MJ, Bajaj-Elliott M, Sanderson IR, McDonald V (2001) Interferon gamma induces enterocyte resistance against infection by the intracellular pathogen Cryptosporidium parvum. Gastroenterology 120(1):99–107

Pozio E, Rezza G, Boschini A, Pezzotti P, Tamburrini A, Rossi P, Di Fine M, Smacchia C, Schiesari A, Gattei E, Zucconi R, Ballarini P (1997) Clinical cryptosporidiosis and human immunodeficiency virus (HIV)-induced immunosuppression: findings from a longitudinal study of HIV-positive and HIV-negative former injection drug users. J Infect Dis 176 (4):969–975

Preidis GA, Wang HC, Lewis DE, Castellanos-Gonzalez A, Rogers KA, Graviss EA, Ward HD, White AC Jr (2007) Seropositive human subjects produce interferon gamma after stimulation with recombinant Cryptosporidium hominis gp15. Am J Trop Med Hyg 77(3):583–585

Riggs MW (2002) Recent advances in cryptosporidiosis: the immune response. Microbes Infect 4 (10):1067–1080

Robinson P, Okhuysen PC, Chappell CL, Lewis DE, Shahab I, Janecki A, White AC Jr (2001) Expression of tumor necrosis factor alpha and interleukin 1 beta in jejuna of volunteers after experimental challenge with Cryptosporidium parvum correlates with exposure but not with symptoms. Infect Immun 69(2):1172–1174. doi:10.1128/IAI.69.2.1172-1174.2001

Robinson P, Okhuysen PC, Chappell CL, Weinstock JV, Lewis DE, Actor JK, White AC Jr (2003) Substance P expression correlates with severity of diarrhea in cryptosporidiosis. J Infect Dis 188(2):290–296. doi:10.1086/376836

Robinson P, Martin P Jr, Garza A, D'Souza M, Mastrangelo MA, Tweardy D (2008) Substance P receptor antagonism for treatment of cryptosporidiosis in immunosuppressed mice. J Parasitol 94(5):1150–1154. doi:10.1645/GE-1458.1

Roche JK, Martins CA, Cosme R, Fayer R, Guerrant RL (2000) Transforming growth factor beta1 ameliorates intestinal epithelial barrier disruption by Cryptosporidium parvum in vitro in the absence of mucosal T lymphocytes. Infect Immun 68(10):5635–5644

Rossi P, Rivasi F, Codeluppi M, Catania A, Tamburrini A, Righi E, Pozio E (1998) Gastric involvement in AIDS associated cryptosporidiosis. Gut 43(4):476–477

Rossignol JF (2006) Nitazoxanide in the treatment of acquired immune deficiency syndrome-related cryptosporidiosis: results of the United States compassionate use program in 365 patients. Aliment Pharmacol Ther 24(5):887–894. doi:10.1111/j.1365-2036.2006.03033.x

Rossignol JF, Hidalgo H, Feregrino M, Higuera F, Gomez WH, Romero JL, Padierna J, Geyne A, Ayers MS (1998) A double-'blind' placebo-controlled study of nitazoxanide in the treatment of cryptosporidial diarrhoea in AIDS patients in Mexico. Trans R Soc Trop Med Hyg 92 (6):663–666

Rossignol JF, Ayoub A, Ayers MS (2001) Treatment of diarrhea caused by Cryptosporidium parvum: a prospective randomized, double-blind, placebo-controlled study of Nitazoxanide. J Infect Dis 184(1):103–106. doi:10.1086/321008

Roy SL, DeLong SM, Stenzel SA, Shiferaw B, Roberts JM, Khalakdina A, Marcus R, Segler SD, Shah DD, Thomas S, Vugia DJ, Zansky SM, Dietz V, Beach MJ (2004) Risk factors for sporadic cryptosporidiosis among immunocompetent persons in the United States from 1999 to 2001. J Clin Microbiol 42(7):2944–2951. doi:10.1128/JCM.42.7.2944-2951.2004

Sagodira S, Iochmann S, Mevelec MN, Dimier-Poisson I, Bout D (1999) Nasal immunization of mice with Cryptosporidium parvum DNA induces systemic and intestinal immune responses. Parasite Immunol 21(10):507–516

Sallon S, Deckelbaum RJ, Schmid II, Harlap S, Baras M, Spira DT (1988) Cryptosporidium, malnutrition, and chronic diarrhea in children. Am J Dis Child 142(3):312–315

Sallon S, el-Shawwa R, Khalil M, Ginsburg G, elTayib J, ElEila J, Green V, Hart CA (1994) Diarrhoeal disease in children in Gaza. Ann Trop Med Parasitol 88(2):175–182

Samson VE, Brown WR (1996) Pneumatosis cystoides intestinalis in AIDS-associated cryptosporidiosis. More than an incidental finding? J Clin Gastroenterol 22(4):311–312

Sarabia-Arce S, Salazar-Lindo E, Gilman RH, Naranjo J, Miranda E (1990) Case–control study of Cryptosporidium parvum infection in Peruvian children hospitalized for diarrhea: possible association with malnutrition and nosocomial infection. Pediatr Infect Dis J 9(9):627–631

Sears CL, Guerrant RL (1994) Cryptosporidiosis: the complexity of intestinal pathophysiology. Gastroenterology 106(1):252–254

Seydel KB, Zhang T, Champion GA, Fichtenbaum C, Swanson PE, Tzipori S, Griffiths JK, Stanley SL Jr (1998) Cryptosporidium parvum infection of human intestinal xenografts in SCID mice induces production of human tumor necrosis factor alpha and interleukin-8. Infect Immun 66 (5):2379–2382

Sharpstone DR, Rowbottom AW, Nelson MR, Lepper MW, Gazzard BG (1996) Faecal tumour necrosis factor-alpha in individuals with HIV-related diarrhoea. AIDS 10(9):989–994

Sharpstone D, Neild P, Crane R, Taylor C, Hodgson C, Sherwood R, Gazzard B, Bjarnason I (1999) Small intestinal transit, absorption, and permeability in patients with AIDS with and without diarrhoea. Gut 45(1):70–76

Shirley DA, Moonah SN, Kotloff KL (2012) Burden of disease from cryptosporidiosis. Curr Opin Infect Dis 25(5):555–563. doi:10.1097/QCO.0b013e328357e569

Sidhu S, Flamm S, Chopra S (1994) Pneumatosis cystoides intestinalis: an incidental finding in a patient with AIDS and cryptosporidial diarrhea. Am J Gastroenterol 89(9):1578–1579

Smith H (2008) Diagnostics. In: *Cryptosporidium* and cryptosporidiosis. CRC Press, Boca Raton, pp 173–208

Snijders F, van Deventer SJ, Bartelsman JF, den Otter P, Jansen J, Mevissen ML, van Gool T, Danner SA, Reiss P (1995) Diarrhoea in HIV-infected patients: no evidence of cytokine-mediated inflammation in jejunal mucosa. AIDS 9(4):367–373

Sodemann M, Jakobsen MS, Molbak K, Martins C, Aaby P (1999) Episode-specific risk factors for progression of acute diarrhoea to persistent diarrhoea in west African children. Trans R Soc Trop Med Hyg 93(1):65–68

Sonea IM, Palmer MV, Akili D, Harp JA (2002) Treatment with neurokinin-1 receptor antagonist reduces severity of inflammatory bowel disease induced by Cryptosporidium parvum. Clin Diagn Lab Immunol 9(2):333–340

Sprinz E, Mallman R, Barcellos S, Silbert S, Schestatsky G, Bem David D (1998) AIDS-related cryptosporidial diarrhoea: an open study with roxithromycin. J Antimicrob Chemother 41 (Suppl B):85–91

Sreedharan A, Jayshree RS, Sridhar H (1996) Cryptosporidiosis among cancer patients: an observation. J Diarrhoeal Dis Res 14(3):211–213

Stine KC, Harris JS, Lindsey NJ, Cho CT (1985) Spontaneous remission of cryptosporidiosis in a child with acute lymphocytic leukemia. Clin Pediatr 24(12):722–724

Stockdale HD, Spencer JA, Blagburn BL (2008) Prophylaxis and chemotherapy. In: Cryptosporidium and cryptosporidiosis, 2nd edn. CRC Press, Boca Raton, pp 255–287

Taghi-Kilani R, Sekla L, Hayglass KT (1990) The role of humoral immunity in Cryptosporidium spp. infection. Studies with B cell-depleted mice. J Immunol 145(5):1571–1576

Tanyuksel M, Gun H, Doganci L (1995) Prevalence of Cryptosporidium sp. in patients with neoplasia and diarrhea. Scand J Infect Dis 27(1):69–70

ten Hove R, Schuurman T, Kooistra M, Moller L, van Lieshout L, Verweij JJ (2007) Detection of diarrhoea-causing protozoa in general practice patients in The Netherlands by multiplex real-time PCR. Clin Microbiol Infect 13(10):1001–1007. doi:10.1111/j.1469-0691.2007.01788.x

Theodos CM, Sullivan KL, Griffiths JK, Tzipori S (1997) Profiles of healing and nonhealing Cryptosporidium parvum infection in C57BL/6 mice with functional B and T lymphocytes: the extent of gamma interferon modulation determines the outcome of infection. Infect Immun 65 (11):4761–4769

Travis WD, Schmidt K, MacLowry JD, Masur H, Condron KS, Fojo AT (1990) Respiratory cryptosporidiosis in a patient with malignant lymphoma. Report of a case and review of the literature. Arch Pathol Lab Med 114(5):519–522

Trevino-Perez S, Luna-Castanos G, Matilla-Matilla A, Nieto-Cisneros L (1995) Chronic diarrhea and Cryptosporidium in diabetic patients with normal lymphocyte subpopulation. 2 case reports. Gac Med Mex 131(2):219–222

Tumwine JK, Kekitiinwa A, Nabukeera N, Akiyoshi DE, Rich SM, Widmer G, Feng X, Tzipori S (2003) Cryptosporidium parvum in children with diarrhea in Mulago Hospital, Kampala, Uganda. Am J Trop Med Hyg 68(6):710–715

Tzipori S (1998) Cryptosporidiosis: laboratory investigations and chemotherapy. Adv Parasitol 40:187–221

Tzipori S, Rand W, Griffiths J, Widmer G, Crabb J (1994) Evaluation of an animal model system for cryptosporidiosis: therapeutic efficacy of paromomycin and hyperimmune bovine colostrum-immunoglobulin. Clin Diagn Lab Immunol 1(4):450–463

Tzipori S, Rand W, Theodos C (1995) Evaluation of a two-phase scid mouse model preconditioned with anti-interferon-gamma monoclonal antibody for drug testing against Cryptosporidium parvum. J Infect Dis 172(4):1160–1164

Uip DE, Lima AL, Amato VS, Boulos M, Neto VA, Bem David D (1998) Roxithromycin treatment for diarrhoea caused by Cryptosporidium spp. in patients with AIDS. J Antimicrob Chemother 41(Suppl B):93–97

Ungar BL, Kao TC, Burris JA, Finkelman FD (1991) Cryptosporidium infection in an adult mouse model. Independent roles for IFN-gamma and CD4+ T lymphocytes in protective immunity. J Immunol 147(3):1014–1022

Urban JF Jr, Fayer R, Chen SJ, Gause WC, Gately MK, Finkelman FD (1996) IL-12 protects immunocompetent and immunodeficient neonatal mice against infection with Cryptosporidium parvum. J Immunol 156(1):263–268

Vajro P, di Martino L, Scotti S, Barbati C, Fontanella A, Pettoello Mantovani M (1991) Intestinal Cryptosporidium carriage in two liver-transplanted children. J Pediatr Gastroenterol Nutr 12 (1):139

Vakil NB, Schwartz SM, Buggy BP, Brummitt CF, Kherellah M, Letzer DM, Gilson IH, Jones PG (1996) Biliary cryptosporidiosis in HIV-infected people after the waterborne outbreak of cryptosporidiosis in Milwaukee. N Engl J Med 334(1):19–23. doi:10.1056/NEJM199601043340104

Ventura G, Cauda R, Larocca LM, Riccioni ME, Tumbarello M, Lucia MB (1997) Gastric cryptosporidiosis complicating HIV infection: case report and review of the literature. Eur J Gastroenterol Hepatol 9(3):307–310

Warren CA, Guerrant RL (2008) Clinical disease and pathology. In: Fayer R, Xiao L (eds) Cryptosporidium and cryptosporidiosis, 2nd edn. CRC Press, Boca Raton

Webster KA, Smith HV, Giles M, Dawson L, Robertson LJ (1996) Detection of Cryptosporidium parvum oocysts in faeces: comparison of conventional coproscopical methods and the polymerase chain reaction. Vet Parasitol 61(1–2):5–13

White C (2010) Cryptosporidium species. In: Mandell GL, Bennett JE, Dolin R (eds) Principles and practice of infectious diseases, 7th edn. Churchill Livingstone, Elsevier, Philadelphia

White AC Jr, Chappell CL, Hayat CS, Kimball KT, Flanigan TP, Goodgame RW (1994) Paromomycin for cryptosporidiosis in AIDS: a prospective, double-blind trial. J Infect Dis 170(2):419–424

Wittenberg DF, Miller NM, van den Ende J (1989) Spiramycin is not effective in treating cryptosporidium diarrhea in infants: results of a double-blind randomized trial. J Infect Dis 159(1):131–132

Wyatt CR, Brackett EJ, Perryman LE, Rice-Ficht AC, Brown WC, O'Rourke KI (1997) Activation of intestinal intraepithelial T lymphocytes in calves infected with Cryptosporidium parvum. Infect Immun 65(1):185–190

Xiao L, Bern C, Limor J, Sulaiman I, Roberts J, Checkley W, Cabrera L, Gilman RH, Lal AA (2001) Identification of 5 types of Cryptosporidium parasites in children in Lima, Peru. J Infect Dis 183(3):492–497. doi:10.1086/318090

Chapter 10
Immunology of Cryptosporidiosis

Guoku Hu, Yaoyu Feng, Steven P. O'Hara, and Xian-Ming Chen

Abstract *Cryptosporidium* spp. infect the gastrointestinal epithelium of vertebrate hosts. Intestinal species typically cause self-limiting diarrhea in immunocompetent individuals, suggesting an efficient host immune defense to eliminate the infection. Both innate and adaptive immunity are involved in host anti-parasite defense. Because of the "minimally invasive" nature of *Cryptosporidium* infection, mucosal epithelial cells are critical to the host's anti-*Cryptosporidium* immunity. Epithelial cells not only provide the first and rapid defense against *Cryptosporidium* infection, but also mobilize immune effector cells to the infection site to activate adaptive immunity. Attachment to the apical cell surface by *Cryptosporidium*, as well as molecules inserted into host cells after attachment, can activate host cell signal pathways and thereby alter cell function. Pathogen recognition receptors (e.g., Toll-like receptors) in epithelial cells recognize *Cryptosporidium* and initiate downstream signaling pathways (e.g., NF-kappaB) which trigger a series of antimicrobial responses and activate adaptive immunity. Non-coding RNAs are critical regulators of mucosal immunity to infection, while release of exosomes from epithelial cells may be a relatively unexplored, important component of mucosal anti-parasite defense. Conversely, it appears that *Cryptosporidium* has also developed strategies of immune evasion to escape host immunity, at least at the early stage of infection.

G. Hu • X.-M. Chen (✉)
Department of Medical Microbiology and Immunology, Creighton University School of Medicine, Omaha, NE 68178, USA
e-mail: guokuhu@creighton.edu; xianmingchen@creighton.edu

Y. Feng
School of Resources and Environmental Engineering, East China University of Science and Technology, Shanghai 200237, China
e-mail: yyfeng@ecust.edu.cn

S.P. O'Hara
Division of Gastroenterology and Hepatology, Department of Internal Medicine, Mayo Clinic College of Medicine, Rochester, MN 55905, USA
e-mail: ohara.steven@mayo.edu

S.M. Cacciò and G. Widmer (eds.), *Cryptosporidium: parasite and disease*,
DOI 10.1007/978-3-7091-1562-6_10, © Springer-Verlag Wien 2014

Immune responses contribute to the pathophysiologic features of cryptosporidiosis. A better understanding the immunology of cryptosporidiosis will provide a framework for the potential development of novel therapeutic strategies.

10.1 Introduction

Host immune defense is critical to eliminate *Cryptosporidium* spp. infection. The self-limiting nature of *Cryptosporidium* infection in immunocompetent subjects suggests that the host activates an efficient immune response to eliminate the infection. After entry into host epithelial cells, the parasite resides within a unique intracellular but extracytoplasmic niche, separating the parasite from a direct interaction with other cell types. Therefore, *Cryptosporidium* spp. are classified as "minimally invasive" mucosal pathogens (Chen et al. 2002). Evidence from in vitro and in vivo studies indicates that both innate and adaptive immunity are involved in the resolution of cryptosporidiosis and resistance to infection (Chen et al. 2005b; Pantenburg et al. 2008). However, it appears that *Cryptosporidium* can survive the host innate immune attack during early stages of infection (Pantenburg et al. 2008; Zhou et al. 2009). Complete elimination of infection requires adaptive immune responses, particularly those mediated by $CD4^+$ T cells (Pantenburg et al. 2010). Interferon (IFN)-γ, mainly released by activated $CD4^+$ T cells (Boehm et al. 1997), is also critical in the control of cryptosporidiosis (Pollok et al. 2001).

Because of the "minimally invasive" nature of *Cryptosporidium* parasites, innate immune responses by epithelial cells are critical to the host's defense against the infection. Upon *Cryptosporidium* infection, epithelial cells quickly initiate a series of innate immune reactions including production of antimicrobial peptides (e.g., β-defensins) and release of inflammatory chemokines/cytokines, such as interleukin-8 (IL-8) (Hu et al. 2010; Zhou et al. 2012). Production and secretion of antimicrobial peptides (e.g., β-defensin 2) and nitric oxide (NO) can kill *Cryptosporidium* or inhibit parasite growth (Zhou et al. 2012). These chemokines/cytokines of epithelial cell origin can mobilize and activate immune effector cells (e.g., lymphocytes, macrophages and neutrophils) to the infection sites (Blikslager et al. 2007). Recent evidence implicates that the release of epithelial exosomes may be an additional element of epithelial anti-*Cryptosporidium* defense (Hu et al. 2013). In addition, the molecular mechanisms of parasite-epithelial interactions in triggering epithelial defense and how immune responses contribute to the pathology of *Cryptosporidium* infection have been vigorously investigated (Pantenburg et al. 2008). Nevertheless, how parasite antigens are presented through the mucosal epithelium and how adaptive immunity is subsequently activated remain unclear. The potential strategies of *Cryptosporidium* immune evasion have recently been recognized and experimentally tested (Choudhry et al. 2009). Both transcriptional and post-transcriptional mechanisms are involved in the regulation of mucosal anti-*Cryptosporidium* defense. This chapter will highlight recent advances in the immunology of cryptosporidiosis, particularly those pertaining to the human pathogenic *Cryptosporidium* species, *C. parvum*, and to the role of host epithelial anti-*C. parvum* defense. It will only briefly describe the immune responses from T cells and B cells and readers are referred to recent detailed reviews on this topic (Borad and Ward 2010; Petry et al. 2010).

10.2 Parasite-Epithelial Cell Interactions

The initial interaction between intestinal epithelia and the parasite is essential for parasitism, and hence for immune and pathological sequelea. *C. parvum* occupies a unique intracellular, yet extracytoplasmic, niche in the host cell, which may help protect the parasite against host immune responses as well as antimicrobial drugs. All developmental stages of *C. parvum* (asexual and sexual) occur in a single host. Upon ingestion of an oocyst, four infective sporozoites are released into the intestinal lumen. Immediately following this process (excystation), the sporozoites contact and attach to host derived mucins and exhibit gliding motility, a method of motility conserved in all apicomplexans (Chen et al. 2002; Bhat et al. 2007). Gliding requires the parasite cytoskeleton and the release of adhesive glycoproteins from micronemes, which are apical secretory organelles (Tomley and Soldati 2001). Ultimately, the apical pole of the parasite adheres to an epithelial cell, representing the first step of internalization (Lumb et al. 1988; Aji et al. 1991). A few characterized *C. parvum* secretory proteins have been implicated in the invasion process. These include GP900 (Petersen et al. 1992), the circumsporozoite-like (CSL) glycoprotein p23 (Perryman et al. 1996; Riggs et al. 1997), the Cpgp40/15 gene (Cevallos et al. 2000b), thrombospondin related proteins TRAP-C1 and TSP2-TSP12 (Spano et al. 1998; Deng et al. 2002; Putignani et al. 2008), and mucin-like antigens CpMuc4 and CpMuc5 (O'Connor et al. 2009). In addition, Gal/GalNAc-specific lectins may be essential for the initial attachment to host epithelial cells, including the Gal/GalNAc-specific lectin p30, that has been identified in both *C. parvum* and *C. hominis*, and mediates attachment to and invasion of epithelial cells in vitro (Bhat et al. 2007).

The invasion process, and consequently, the morphological and functional alterations of host epithelia, requires intimate contact between host plasma membrane and the apical pole of the parasite (Chen et al. 2004, 2005b). Early investigations demonstrated that *C. parvum* infection induced actin polymerization at sites of infection utilizing the actin branching and nucleation machinery of the Arp2/3 complex of proteins (Elliott et al. 2001). Multiple signaling pathways have been identified that modulate *C. parvum*-induced actin reorganization and parasite internalization, suggesting redundancy in these processes (Chen et al. 2003b, 2004). The identified signaling axes include: PI3-kinase-dependent activation of the small GTPase, CDC42; c-Src-dependent activation of cortactin; and Ca^{++}- dependent activation of the serine/threonine kinases PKCα and/or PKCβ (Hashim et al. 2006; Perez-Cordon et al. 2011). Hence, actin dynamics and actin-dependent membrane protrusion events are likely central to the formation of the intracellular/ extracytoplasmic niche occupied by *C. parvum*. Moreover, in *C. parvum*-infected biliary epithelial cells, the Na+/Glucose cotransporter, SGLT1, and aquaporin 1, a channel protein selective for the movement of water, accumulate at invasion sites, and participate in membrane protrusion events induced by the parasite (Chen et al. 2005a). The localization of these proteins is dependent on the localized contractility of myosin IIB (O'Hara et al. 2010a). Interestingly, SGLT1 may also protect host epithelial cells against pathogen-induced apoptosis (Yu et al. 2005, 2006a, 2008). Further research is needed to clarify whether these epithelial alterations reflect a protective innate host response, and/or a process that ensures efficient parasite entry into a viable cell.

The host-cell signaling cascades culminate in the rearrangement of the host actin cytoskeleton, and parasite internalization within a parasitophorous vacuole membrane. The parasitophorous vacuole membrane envelops the parasite on the luminal surface of the cell in a parasite-modified host membrane (Robert et al. 1994; Bonnin et al. 1995; McDonald et al. 1995; O'Hara et al. 2004). As the invasion process ensues, a unique electron dense structure forms within the host, adjacent to a network of microfilaments. The adjacent parasite membrane becomes highly invaginated. Collectively, these structures are referred to as the feeder or attachment organelle. Hence, the parasite occupies a niche that is selectively isolated from the intestinal lumen and host cytoplasm.

10.3 Activation of Pattern Recognition Receptors (PRRs) in Epithelial Cells

The gastrointestinal epithelium, in addition to providing a natural barrier that limits infection, also plays a critical role in the initial recognition of parasites and the triggering of adaptive immunity. Pathogen-associated molecular pattern molecules (PAMPs) derived from microorganisms can be recognized by PRR-bearing cells (Medzhitov 2007). When ligated by the PAMPs, these PRRs trigger cytokine and costimulatory signals that initiate both innate and adaptive cellular responses. Activation of downstream signaling cascades of PRRs regulate many biological processes that are critical to immune response to infection. Epithelial cells along the gastrointestinal and biliary tracts express a variety of PRRs, such as the Toll-like receptors (TLRs) and nucleotide binding NOD-like receptors, RIG-I-like receptors, and oligomerization domain-like receptors, which recognize pathogens or PAMPs (Nasu and Narahara 2010).

Studies on the activation of PRRs and associated downstream signaling cascades in epithelial cells following *C. parvum* infection have focused on the TLR4/NF-κB pathway. The NF-κB pathway controls many genes involved in immune responses, inflammation and cell differentiation (Zhou et al. 2009; Oeckinghaus et al. 2011). Activation of the NF-κB pathway has been demonstrated in epithelial cells following *C. parvum* infection in a variety of studies using both in vivo and in vitro models. Activation of NF-κB signaling in epithelia cells following *C. parvum* infection appears to be a "late" but persistent response, compared with the early and transient NF-κB activity in epithelial cells following stimulation of lipopolysaccharide (LPS) and tumor necrosis factor-α (TNF-α) (Zhou et al. 2009). It is also of interest that *C. parvum*-induced NF-κB activation requires direct parasite-epithelial cell interaction. Indeed, it appears that parasite-induced NF-κB activation is limited to infected epithelial cells, at least in the early stage of infection. Hence, *C. parvum*-induced TLR4/NF-κB activation, likely requires the process of parasitic invasion (Chen et al. 2001, 2005b).

While the precise molecular mechanisms underlying *C. parvum*-induced activation of TLR4/NF-κB signaling are still unclear, it is possible that parasite-derived molecules act as PAMPs and are recognized by host cells as a TLR4 ligand during the invasion process of the parasite. However, the potential PAMPs from *C. parvum* required for TLR4 activation are not known. Indeed, the *C. parvum* genome lacks enzymes associated with endotoxin synthesis (Abrahamsen et al. 2004), hence, *C. parvum* does not express LPS or LPS-like molecules. Moreover, *C. parvum* sporozoites do not initiate limulus amebocyte lysate activity (unpublished data). Recruitment of TLR4 and TLR2 to the attachment/invasion site was reported in H69 human biliary epithelial cells after exposure to *C. parvum* in vitro (Chen et al. 2005b). Whether this recruitment is required for the activation of TLR4 signaling is unclear. Furthermore, whether infection can trigger activation of other PRRs and downstream signaling pathways, in particular, the inflammasome activation pathways (Newton and Dixit 2012), is yet to be explored. However, knockdown of TLR4 and its adaptor protein MyD88 can block *C. parvum*-induced NF-κB activation (Chen et al. 2005b), suggesting an integral role of TLR4 initiated signaling. Importantly, as discussed in detail below, activation of TLR/NF-κB signaling may account for the majority of epithelial anti-parasite defense.

10.4 Epithelial Defense Response

10.4.1 Epithelial Cell-Derived Effector Molecules

Human gastrointestinal epithelial cell lines increase the expression and production of the potent neutrophil chemoattractants interleukin-8 (IL-8) and chemokine (C-X-C motif) ligand 1 (CXCL1) following *C. parvum* infection. IL-8 and CXCL1 were shown to be predominantly released from the basolateral surface of infected cultures using polarized epithelial monolayers cultured in Transwell chambers (Ojcius et al. 1999). Cultured human colon epithelial cells infected with *C. parvum* increased prostaglandin E2 (PGE2) production as much as 50-fold and prostaglandin F2alpha (PGF2α) production by up to tenfold compared with uninfected cells. Increases were observed by 12 h after infection, were maximal by 36 h and were sustained over a 72 h culture period (Kagnoff and Eckmann 1997). In addition, CXCL10 is highly upregulated in intestinal epithelial cells of AIDS patients with active cryptosporidiosis (Wang et al. 2007). Induction of CX3CL1 by *C. parvum* has been reported in cultured human intestinal and biliary epithelial cells (Harada and Nakanuma 2012). *C. parvum* infected epithelial cells respond by upregulating certain C, C–C, and C–X–C class chemokines as shown in neonate mice (Lacroix-Lamande et al. 2002). Epithelial cells are also a source of IL-18 (IFN-γ-inducing factor) (Okazawa et al. 2004). The IL-18 gene is upregulated in ilea of mice in response to infection (Tessema et al. 2009), and IL-18 mRNA and protein are upregulated in vitro upon *C. parvum* infection of intestinal epithelial cells (McDonald et al. 2006).

Release of these cytokines and chemokines from epithelial cells following *C. parvum* infection will attract inflammatory infiltration, and facilitate mucosal antimicrobial defense (Fig. 10.1). Besides cytokine/chemokine release, *C. parvum* infection also induces expression of adhesion molecules in epithelial cells. Intercellular adhesion molecule (ICAM-1) is long known for its importance in stabilizing cell-cell interactions and is critical for the firm arrest and transmigration of leukocytes out of blood vessels into tissues (Muller 2009). Upregulation of ICAM-1 transcription and protein expression was detected in human gastrointestinal epithelial cells following *C. parvum* infection in vitro (Gong et al. 2011). Expression of adhesion molecules, such as ICAM-1, on infected epithelial cell surfaces may facilitate recruitment of lymphocytes to the site of infection (Gong et al. 2011). Indeed, expression of ICAM-1 on infected epithelial cells facilitated epithelial adherence of co-cultured Jurkat T cells (Gong et al. 2011).

Several molecules synthesized in epithelial cells following *Cryptosporidium* infection reveal anti-*C. parvum* capacity, and thus play a direct role in antimicrobial defense (Fig. 10.1). Nitric oxide (NO) is produced from arginine by the enzyme nitric oxide synthase (NOS), which exists in three isoforms. One isoform, the inducible NOS (iNOS), is known to be inducibly expressed in human intestinal epithelial cells (Ricciardolo et al. 2004), and NO has been shown to have a broad range of antimicrobial properties. NO synthesis by the epithelium is significantly increased following *C. parvum* infection; and, in the absence or inhibition of iNOS, infection and oocyst shedding are significantly exacerbated (Leitch and He 1999; Gookin et al. 2005, 2006). L-arginine administered enterically significantly reduced fecal oocyst shedding in athymic nude mice chronically infected with *C. parvum* (Leitch and He 1994). Our recent studies further demonstrate that *C. parvum* infection induced NO production in host epithelial cells in a NF-κB-dependent manner, with the involvement of the stabilization of iNOS mRNA (Zhou et al. 2012).

In addition, α- and β-defensins, as well as other low-molecular mass antimicrobial peptides, comprise part of the arsenal of innate defenses present in the gastrointestinal tract. Some of these peptides have been shown to have anti-cryptosporidial activity in vitro (Dommett et al. 2005). *C. parvum* induces expression of β-defensin-2 in human biliary epithelial cells via TLR-mediated NF-κB activation, whereas β-defensin-1 appears to be constitutively expressed. Both

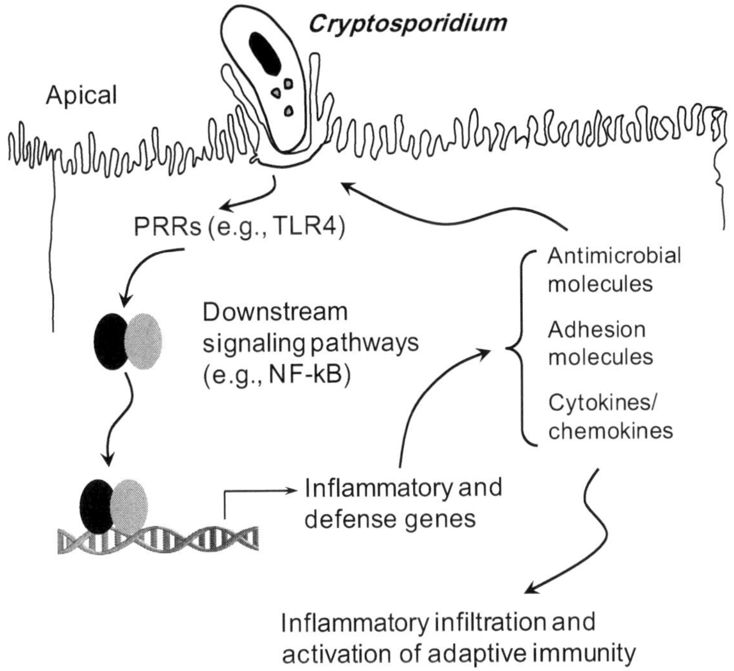

Fig. 10.1 Production and release of epithelial cell-derived effector molecules is essential for the activation of mucosal anti-*Cryptosporidium* immunity. *C. parvum* infection activates pathogen recognition receptors (e.g., TLR4) and their downstream signaling pathways (e.g., NF-κB). This activation induces the expression of cytokine and chemokine inflammatory genes and antimicrobial defense genes. Release of these cytokines and chemokiens from epithelial cells following infection will attract inflammatory infiltration, resulting in activation of adaptive immunity. Several molecules synthesized in epithelial cells following infection, including LL-37, β-defensins and NO, reveal anti-*C. parvum* capacity and play a direct role in anti-microbial defense

β-defensin-1 and β-defensin-2 have anti-*C. parvum* activity as evidenced by a decreased sporozoite viability following incubation with recombinant β-defensin-1 or 2 (Zaalouk et al. 2004). One β-defensin, enteric β-defensin, was shown to be highly expressed in epithelial cells in bovine distal small intestinal crypts and colon (Tarver et al. 1998). Moreover, expression of enteric β-defensin mRNA was upregulated by five to tenfold following *C. parvum* infection of calves, further supporting a possible role of β-defensins in the innate immune response to *C. parvum* infection (Tarver et al. 1998).

Secretion of mucins in the gastrointestinal tract is an important element of mucosal defense. Nevertheless, their role in host defense to *Cryptosporidium* infection is still unclear. Mucins are a family of high molecular weight, heavily glycosylated proteins (glycoconjugates) produced by epithelial tissues in most metazoans (Marin et al. 2008). Although some mucins are membrane-bound due to the presence of a hydrophobic membrane-spanning domain that favors retention in the plasma membrane, most mucins are secreted onto mucosal surfaces. It has been speculated that epithelial mucin expression could protect the host against further infection by extracellular stages of *C. parvum*. Notably, several mucin-like molecules have been identified in *C. parvum* which may mediate the parasite attachment to or invasion of the host cells (Cevallos et al. 2000a).

Moreover, recent studies indicate that mucosal epithelial cells along the gastro-intestinal and urogenital tracts can secrete several specific cytokines, so called epithelial-derived cytokines, such as IL-25, IL-33, IL-34, and thymic stromal lymphoietin (TSLP), upon extracellular stimuli (Maynard et al. 2012). Release of these epithelial-derived cytokines plays a critical role in mucosal immunity against mucosal infection (Taylor et al. 2009). Additional epithelial cell-derived molecules such as serum amyloid A 3 (Saa3) and regenerating islet-derived 3 gamma (Reg3g) have also recently been demonstrated to display anti-microbial activity (Reigstad et al. 2009; Choi et al. 2013). Their potential involvement in *Cryptosporidium* infection of gastrointestinal epithelium has yet to be investigated.

10.4.2 MicroRNAs (miRNAs)

miRNAs are short strands of non-coding RNAs of 18–24 nucleotides in length (Bartel 2009). These RNA molecules are emerging as key mediators of many biological processes and impact gene expression at the post-transcriptional level (Bartel 2009). Similar to other RNA molecules, most miRNAs are initially transcribed as primary transcripts (termed pri-miRNAs) by RNA polymerase II and processed by the RNase III Drosha (in the nucleus) and a second RNase III Dicer (in the cytoplasm) to generate mature miRNA molecules (Lee et al. 2003). Mature miRNAs are incorporated as single-stranded RNAs into a ribonucleoprotein complex, known as the RNA-induced silencing complex (RISC). The RISC identifies target mRNAs based on perfect or nearly perfect complementarity between the miRNA and the target mRNA 3′-untranslated region (3′UTR). The interaction between the RISC and mRNA causes either mRNA cleavage and/or translational suppression, resulting in gene suppression at the post-transcriptional level (Bartel 2009). More importantly, it has become clear that miRNAs have diverse expression patterns and play essential roles in biological processes that include development, maintenance of genome stability and viral adaptive defense mechanisms (Lewis et al. 2005). It has been predicted that miRNAs control 20–30 % of human genes (Lewis et al. 2005; Bartel 2009).

miRNAs are involved in regulation of epithelial responses to *C. parvum* infection. Differential mature miRNA expression profiles in biliary epithelial cells following *C. parvum* infection were reported using an in vitro model of human cryptosporidiosis. Transcription of a subset of miRNA genes was found to be regulated through NF-κB activation in human biliary epithelial cells following *C. parvum* infection (Zhou et al. 2009). Specifically, inhibition of NF-κB activation by SC-514, an IKK2 inhibitor, blocked LPS- and *C. parvum*-induced upregulation of a subset of pri-miRNAs, including pri-miR-125b-1, pri-miR-21, pri-miR-23b-27b-24-1 and pri-miR-30b. Moreover, direct binding of NF-κB p65 subunit to the promoter elements of *mir-125b-1*, *mir-21*, *mir-23b-27b-24-1* and *mir-30b* genes was identified by chromatin immunoprecipitation analysis and confirmed by a luciferase reporter assay using constructs covering the potential promoter elements of these miRNA genes (Zhou et al. 2009, 2010). In addition to the upregulation of some miRNAs, NF-κB signaling may be also involved in the downregulation of miRNA genes. Transcription of the *let-7i* gene in biliary epithelial cells in response to LPS stimulation and *C. parvum* infection has been reported to be suppressed through promoter binding by NF-κB subunit p50 along with CCAAT/enhancer-binding protein beta (C/EBPβ) (O'Hara et al. 2010b). Importantly, functional inhibition of selected NF-κB-responsive miRNAs in epithelial cells increased *C. parvum* burden (Zhou et al. 2009). Thus, upregulation of miRNAs through NF-κB signaling may be relevant to the regulation of epithelial anti-microbial defense.

NF-κB-responsive miRNAs may modulate epithelial anti-parasite responses at every step of the innate immune network (Fig. 10.2). Elements of the TLR/NF-κB pathway are potential targets for the NF-κB-responsive miRNAs, thus, providing feedback regulation of TLR/NF-κB signaling. For example, TLR4 is targeted by *let-7* miRNAs and suppression of *let-7* facilitates the synthesis of TLR4 protein in epithelial cells following *Cryptosporidium* infection, which promotes epithelial innate defense (Chen et al. 2007). Targeting the suppressors of cytokine signaling family of proteins (CIS/SOCS) by miR-98/*let-7* may also exert a positive feedback regulation of the innate immune response in infected cells. *C. parvum* infection induces CIS/SOCS expression in human biliary epithelial cells through TLR/NF-κB-suppressed expression of miR-98 and *let-7* (Hu et al. 2009, 2010). Induction of CIS expression enhances IκBα degradation promoting NF-κB activation (Hu et al. 2009). On the other hand, upregulation of miR-21 in infected cells may result in a negative feedback regulation of the TLR4/NF-κB signaling through targeting of PDCD4 (Sheedy et al. 2010). It was reported that LPS decreases expression of PDCD4 through induction of miR-21, resulting in subsequent inhibition of NF-κB signaling activity and promotion of IL-10 expression in human peripheral blood mononuclear cells (Sheedy et al. 2010). Similarly, miR-21 targeting of PDCD4 was shown to influence TNF-induced activation of NF-κB.

NF-κB-responsive miRNAs may play a role in the expression of many NF-κB-regulated immune or inflammatory genes. Based on bioinformatics analysis, approximately 29 % of cytokine/chemokine mRNAs have potential target sites for miRNAs (Asirvatham et al. 2008). Many of these cytokine/chemokine mRNAs are transcripts of NF-κB-regulated genes. Expression of IFN-β, the main type I IFN cytokine important to the initiation of innate responses in response to infection including *Cryptosporidium*, is finely controlled by various miRNAs. It has been experimentally confirmed that miR-26a, -34a, -145, and *let-7b* directly regulate IFN-β production by targeting IFN-β 3′UTR (Witwer et al. 2010).

Intriguingly, miRNAs may target RNA-binding proteins to regulate mRNA stability. Several RNA-binding proteins, including the KH-type splicing regulatory protein (KSRP, also known as KHSRP), tristetraprolin (TTP) and Hu antigen R

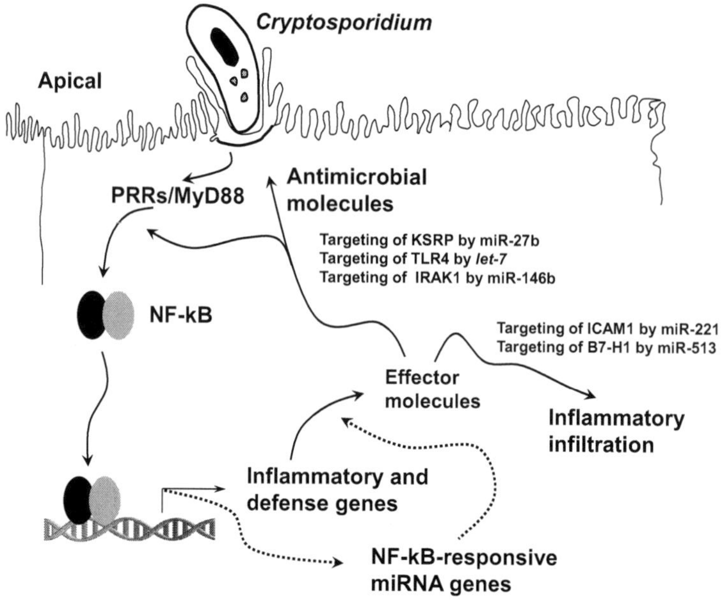

Fig. 10.2 Post-transcriptional gene suppression mediated by miRNAs is involved in the fine-tuning of epithelial anti-*Cryptosporidium* defense. *C. parvum* infection alters the expression profile of miRNA genes in infected epithelial cells. A panel of miRNA genes is trans-activated or trans-suppressed through the activation of the NF-κB pathway. These NF-κB-responsive miRNAs may modulate epithelial anti-parasite responses at every step of the innate immune network, including feedback regulation to the TLR/NF-κB signaling, post-transcriptional regulation of many of NF-κB-regulated immune or inflammatory genes, and production of anti-microbial effector molecules

(HuR), recognize AU-rich elements (AREs) within the 3′UTRs of mRNAs and diminish their half-life in the cytoplasm (Chen and Shyu 1995; Dean et al. 2004; Anderson 2008). For example, many cytokine mRNAs have AREs in their 3′UTRs and usually exhibit a very short half-life (in the range of 10–30 min) in resting cells, effectively preventing cytokine production in inactive cells. In activated cells, cytokine mRNA half-lives are in the range of several hours, accounting for a significant increase in protein production (Stoecklin and Anderson 2006; Anderson 2008). In this regard, KSRP interacts with these mRNAs that have the AREs within their 3′UTRs, including mRNAs for iNOS, IL-8 and cyclooxygenase-2 (COX-2), and is a key mediator of mRNA decay (Winzen et al. 2007; Subramaniam et al. 2008; Pautz et al. 2010). Thus, the level of KSRP can potentially coordinate mRNA stability in response to extracellular stimuli (Winzen et al. 2007). Indeed, induced expression of IL-8, NO (synthesized by iNOs) and PGE_2 (catalyzed by COX-2) has been implicated in the response to a variety of pathogens, including *C. parvum* (Stoecklin and Anderson 2006; Anderson 2008). We recently demonstrated that *C. parvum* infection of epithelial cells increases transcription of the *mir-23b-27b-24-1* gene in an NF-κB-dependent manner. Moreover, KSRP was downregulated in epithelial cells following *C. parvum* infection. Importantly, miR-27b targeted the 3′UTR of KSRP mRNA and suppressed its translation. Hence, *C. parvum* infection resulted in decreased KSRP expression through upregulation of miR-27b. Functional manipulation of KSRP or miR-27b caused reciprocal alterations in iNOS mRNA stability in infected cells. Additionally, forced expression of KSRP and inhibition of miR-27b resulted in an increased *C. parvum* burden (Zhou et al. 2012). Therefore, downregulation of KSRP and upregulation of miR-27b stabilize iNOs/IL-8/COX-2 mRNAs, contributing to epithelial anti-*C. parvum* defense (Fig. 10.2). Notably, no complementarity to IFN-γ or anti-microbial peptide mRNAs (such as LL-37 and β-defensins) has been identified for these NF-κB-responsive miRNAs.

The involvement of NF-κB-responsive miRNAs in the expression of co-stimulatory molecules, in the epigenetic impact and in the shuttling of miRNAs through release of exosomes in *Cryptosporidium* infection will be discussed in the following sections. In addition, miRNAs have been identified in both mammalian cells and non-mammalian cells including virus and parasites (Bartel 2004; Scaria et al. 2006; Winter et al. 2009). Expression of miRNAs in *C. parvum* has not yet been examined and whether *C. parvum*-derived miRNAs, if there are any, can be localized in infected host cells is unknown. The concept that a pathogen encodes mRNAs targeted by host miRNAs has recently emerged as an important mechanism of anti-viral defense (Pedersen et al. 2007). Likewise, it is of interest to test the possibility that host cell miRNAs target the internalized parasite mRNAs and silence genes of the pathogen. The direct *C. parvum*-host cell cytoplasmic tunnel-connection (Huang et al. 2004) could mediate exchange of molecules, including miRNAs, between the host cells and the internalized parasite.

10.4.3 Exosomes

Exosomes represent a specific subtype of secreted membrane vesicles that are about 30–100 nm in size and are formed inside secreting cells in endosomal compartments called multi-vesicular bodies (MVBs) (Thery 2011). Exosomes are produced by a variety of cells (e.g., reticulocytes, epithelial cells, neurons, tumor cells) and have been found in bronchoalveolar lavage, urine, serum, bile, and breast milk (Thery 2011; Hu et al. 2012a). The composition of exosomes is heterogenous, depending on their cellular origin. Exosomes do not contain a random array of intracellular proteins, but a specific set of protein families arising from the plasma membrane, the endocytic pathway, and the cytosol, especially those of endosomal origin, such as CD63, ICAM-1, and MHC molecules (Bobrie et al. 2011). These vesicles also mediate the secretion of a wide variety of other molecules, such as lipids, mRNAs, and miRNAs, and thereby traffic molecules from the cytoplasm and membranes of one cell to other cells or extracellular spaces (Smalheiser 2007; Valadi et al. 2007). There is increasing evidence that secreted vesicles play an important role in normal physiological processes, development, viral infection, and other human diseases (Liegeois et al. 2006; Yu et al. 2006b; Kolotuev et al. 2009; Thery et al. 2009).

We recently demonstrated that luminal release of exosomes from the biliary and intestinal epithelium is increased following infection by *C. parvum* (Hu et al. 2013). Secretion of exosomes is regulated by various stimuli, including the activation of P2X receptor by ATP on monocytes and neutrophils, thrombin receptor on platelets, and TLR4 by LPS on dendritic cells (Bhatnagar and Schorey 2007). Formation of exosomes within MVBs and targeting of transmembrane proteins involve a complex intracellular sorting network, including the endosomal sorting complex required for transport (ESCRT) machinery (van Niel et al. 2006). Fusion of MVBs with plasma membrane is an exocytic process that requires the association of v-SNAREs (from the vesicles) and t-SNAREs (at the membrane) to form a ternary SNARE (SNAP receptor) complex. The SNARE complex brings the two membranes in apposition, a necessary step in overcoming the energy barrier required for membrane fusion (Sudhof and Rothman 2009; Kennedy and Ehlers 2011). Several Rab family proteins, including Rab11 and Rab27b, are key regulators of the exosome secretion pathway and are involved in MVB docking at the plasma membrane (Ostrowski et al. 2010). We found that *C. parvum*-stimulated release of exosomes involves TLR4/IKK2 activation and the SNAP23-associated vesicular exocytic process (Hu et al. 2013). Whereas a basal level of exosomal luminal release exists in cultured biliary epithelial monolayers and in the murine biliary tract, a TLR4-dependent increase in luminal release of epithelial exosomes was detected following *C. parvum* infection. We demonstrated a significant co-localization of Rab11 and Rab27b with MVBs in the cytoplasm of epithelial cells following *C. parvum* infection. *C. parvum* infection also increases expression and enhances phosphorylation of SNAP23 in infected cells. Activation of TLR4 may contribute to both events: TLR4 signaling increases SNAP23 protein expression through diminished *let-7*- mediated SNAP23 translational repression, and stimulates SNAP23 phosphorylation through activation of IKK2 (Suzuki and Verma 2008).

Intriguingly, released exosomes contain antimicrobial peptides with anti-*C. parvum* activity, thus contributing to mucosal anti-*C. parvum* defense (Fig. 10.3). The anti-*C. parvum* activity of β-defesin-2 and LL-37 was previously reported in intestinal and biliary cryptosporidiosis (Zaalouk et al. 2004; Chen et al. 2005b). Both β-defesin-2 and LL-37 were identified as components of apical exosomes released from human biliary epithelial cell H69 monolayers (Hu et al. 2013). Moreover, apical exosomes from infected cells demonstrated an increase in β-defesin-2 and LL-37 content compared to exosomes from uninfected cells, while inhibition of TLR4 signaling diminished their presence in exosomes (Hu et al. 2013). Because biliary epithelial cell derived-exosomes display a protein profile different from the whole cell lysate, targeting of cytoplasmic proteins to the MVBs within epithelial cells, such as β-defesin-2 and LL-37, must be controlled through highly selective and regulated mechanisms. Moreover, exposure of *C. parvum* sporozoites to released exosomes decreases their viability and infectivity both in vitro and ex vivo (Hu et al. 2013). Anti-*C. parvum* activity of apical exosomes released

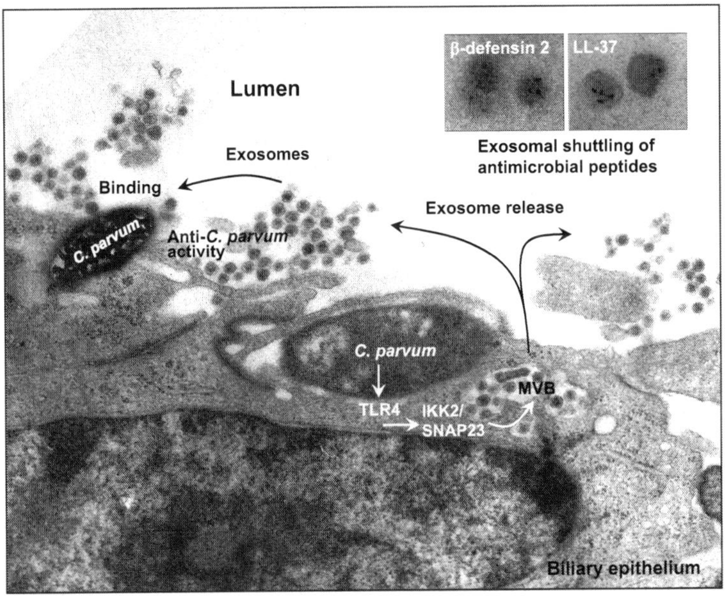

Fig. 10.3 Luminal release of exosomes from epithelium contributes to TLR4-mediated mucosal anti-*Cryptosporidium* defense. Apical exosomal vesicles are secreted by biliary epithelial cells through activation of the TLR4 signaling pathway following *C. parvum* infection. Released exosomal vesicles shuttle antimicrobial peptides, such as cathelicidin-37 and β-defensin 2, and activation of TLR4 signaling increases exosomal shuttling of antimicrobial peptides. Functionally, released exosomal vesicles can directly bind to the *C. parvum* sporozoite surface and display antimicrobial activity

from the epithelium may involve direct binding to the *C. parvum* sporozoite surface. Direct binding of exosomes to the *C. parvum* sporozoite surface was evident by scanning and transmission EM. Confocal microscopy analysis with exosomal labeling confirmed the binding of exosomes to *C. parvum* sporozoites after incubation. Interestingly, treatment of exosomes with Gal/GalNAc-specific lectin arachis hypogaea (PNA, peanut agglutinin) diminished their anti-*C. parvum* activity, suggesting that these molecules may be involved, at least partially, in exosomal binding to the *C. parvum* sporozoite surface. The lifecycle of *C. parvum*, both in vitro and in vivo, has extracellular stages (i.e., sporozoites, merozoites, and microgametocytes), and they are likely vulnerable to exosomal binding/targeting.

Release of epithelial cell-derived exosomes to the basolateral domain has also been demonstrated following *Cryptosporidium* infection (Hu et al. 2013). Although these basolateral exosomes don't exhibit anti-*C. parvum* activity as the apical exosomes, they do shuttle MHC proteins (Hu et al. 2013) and therefore, may contribute to antigen presentation (as described below). In addition, the basolateral exosomes have been shown to modulate lymphocyte immune responses during other mucosal infections (Mallegol et al. 2007). Intestinal epithelial cell-derived exosomes containing αvβ6 integrin and food antigen induced the generation of tolerogenic dendritic cells in a model of tolerance induction (Chen et al. 2011). The presence of these intestinal epithelial cell-derived exosomes impacted the development of antigen-specific T regulatory cells (Chen et al. 2011). Intriguingly, exosome-shuttled miRNA molecules can be delivered to other cell types through exosomal uptake (Hu et al. 2012b). Therefore, exosome-mediated transport of miRNAs may provide a novel mechanism of gene regulation between cells. Ohshima et al. showed that there is the enrichment of *let-7* miRNA family in the exosomes from AZ-P7a cells (Ohshima et al. 2010). Given the importance of miRNAs in epithelial innate immune responses, it would be interesting to determine if exosomes from epithelial cells also carry miRNAs and thus modulate epithelial-immune cell interactions and epithelial antimicrobial defense, via exosomal delivery of miRNAs.

10.5 Antigen Presentation

C. parvum resides at the apical surface of intestinal epithelial cells and does not invade deeper layers of the human gastrointestinal mucosa. The mechanisms by which infection with a pathogen, like *C. parvum*, that is confined to epithelial cells, result in activation of adaptive immunity is a subject of considerable interest. Microscopic analyses identified the presence of *C. parvum* in M cells in AIDS-related intestinal cryptosporidiosis (McCole et al. 2000). Infiltration of various inflammatory cells to the lumen of the biliary tract has been observed in liver tissues from AIDS patients with biliary cryptosporidiosis (Lumadue et al. 1998; Chen and LaRusso 1999). Therefore, it is possible that direct interaction between parasite and immune cells, including macrophages or dendritic cells (DCs), may present parasite antigens and activate the adaptive immunity. Until now, there is relatively little work directly addressing the role of DCs in the immune response to *C. parvum* (McDonald et al. 2013). Bone-marrow derived mouse dendritic cells exposed to *C. parvum* sporozoites or parasite antigens expressed IFN-α and IFN-β within a few hours (Barakat et al. 2009b). Soluble sporozoite antigens or recombinant parasite antigens induced maturation of these dendritic cells and stimulated production of IL-12, IL-1β and IL-6 (Barakat et al. 2009b). Human monocytic dendritic cells also increased production of IL-12 after exposure to soluble sporozoite antigens or live sporozoites (Barakat et al. 2009b; Bedi and Mead 2012).

Several pieces of evidence suggest that epithelial cells may play a role in antigen presentation during infection. The machinery required for antigen processing and presentation include major histocompatibility complex (MHC) class I and class II molecules and co-stimulatory molecules (Hershberg and Mayer 2000). As in other tissues, macrophage phagocytosis of apoptotic cells containing the pathogens may contribute to antigen presentation. Apoptotic cell death of gastrointestinal epithelial cells during *C. parvum* infection has been reported both in vivo and in vitro (Lacroix et al. 2001; Liu et al. 2008). Additionally, induction of MHC-I, MHC-II and the co-stimulatory molecule B7-H1 has been reported in biliary and intestinal epithelial cells following *Cryptosporidium* infection (Gong et al. 2010). Moreover, as described above, release of epithelial-derived exosomes has recently been implicated in antigen presentation during mucosal infection. Our recent studies demonstrate basolateral release of epithelial-derived exosomes using in vitro models of intestinal and biliary cryptosporidiosis. Of importance, those released exosomes carry both MHC-I and MHC-II (Hu et al. 2013). Coupled with the fact that epithelial cells release an array of cytokines/chemokines and adhesion molecules to attract immune cell infiltration, epithelial cells may play an active role in the antigen presentation during *Cryptosporidium* infection.

10.6 IFNs, NK Cells, Macrophages, and the Complement System

A critical role for NK cells and IFN-γ in anti-*Cryptosporidium* immunity has been documented in T and B cell-deficient mice (Barakat et al. 2009a; McDonald et al. 2013). In a recent study using mice lacking T-cells and NK-cells (SCIDbg mice) and SCIDbg mice additionally lacking macrophages and neutrophils (SCIDbgMN mice), *C. parvum* developed an acute infection with high mortality in SCIDbgMN mice but not in SCIDbg mice (Takeuchi et al. 2008). NK cells may carry out their anti-*Cryptosporidium* function through release of IFN-γ, because the resistance of SCID mice deficient in T- and B- lymphocytes to *C. parvum* infection is IFN-γ-dependent (Ungar et al. 1991; Chen et al. 1993; McDonald and Bancroft 1994). Splenic NK-cells from SCID mice which had been activated in vitro by *C. parvum* antigen produce IFN-γ (Chen et al. 1993). IFN-γ may also be released by cells other than NK-cells, including macrophages (McDonald et al. 2013). In support of this, a diligent study by Barakat et al. (2009a), demonstrated that in mice deficient in B-, T-, and NK-cells ($Rag2^{-/-}\gamma c^{-/-}$), oocyst excretion is higher than in $Rag2^{-/-}$ mice that are deficient in B- and T-cells only. Although lacking T- and NK-cells, the level of infection can be exacerbated in $Rag2^{-/-}$ $\gamma c^{-/-}$ mice by antibody neutralization of IFN-γ, suggesting that an additional intestinal cell type may be an important source of IFN-γ.

The complement system, a part of the humoral innate immune system, can be activated and can participate in host anti-*Cryptosporidium* defense. Petry et al (2008) analyzed the in vitro binding and activation of the human and mouse complement systems and tested the susceptibility to infection in complement-deficient mouse strains (Petry et al. 2008). Their results show that *C. parvum* can activate both the classical and lectin pathways, leading to the deposition of C3b on the parasite. Using real-time PCR, parasite development could be demonstrated in adult mice lacking mannan-binding lectin ($MBL\text{-}A/C^{-/-}$) but not in mice lacking complement factor C1q ($C1qA^{-/-}$) or in wild type C57BL/6 mice. MBL belongs to the C-type lectin superfamily and can recognize carbohydrate patterns of micro-organisms, resulting in activation of the lectin pathway of the complement system

(Worthley et al. 2005). The contribution of the complement system and the lectin pathway in particular to the host defense against cryptosporidiosis may become apparent in situations of immunodeficiency such as HIV infection or in early childhood.

10.7 T Cells and B Cells

The role of acquired immunity in resistance to and clearance of *C. parvum* infection has been studied most extensively in murine models in which B- and T-cell populations, or both, can be selectively depleted. Both humoral and cellular immunity play a role in the control of this infection, but the latter plays the major role, mainly in the intestinal mucosa.

Acquired resistance to cryptosporidial infection in humans is dependent upon T-cells with the αβ-form T-cell receptor; in addition, the $CD4^+$ T-cell subset has a protective role whereas $\gamma\delta^+$ or $CD8^+$ subpopulations of T-cells appear irrelevant or subordinate (Mombaerts et al. 1993; Ladel et al. 1995). The ability to clear acute and chronic *C. parvum* infection in humans correlates closely with host CD4 T-cell levels (Takeuchi et al. 2008). Similarly, studies in congenitally athymic (nude) mice and reconstitution studies in immunodeficient SCID mice demonstrate a critical role for T cells in resistance to, and clearing of, infection (Ruittenberg and van Noorle Jansen 1975). Both αβ T cells and γδ T cells are important for host defense to *C. parvum* in mice (Eichelberger et al. 2000). As in humans, CD4, but not CD8 T cells, were shown to be most important for clearing infection in mice (Wong and Pamer 2003).

Cytokines of Th1 (e.g., IFN-γ, IL-2, and IL-12), Th2 (e.g., IL-4, IL-5, IL-6, and IL-10), and Th17 (e.g., IL-23) may all play a role in host anti-*Cryptosporidium* immunity (Ehigiator et al. 2005; Weaver et al. 2006). It has been suggested that the capacity to produce all Th1, Th2 and Th17 cytokines, rather than the presence of Th2 cytokines alone, determines the effective immune response against *C. parvum* infection. The susceptibility or resistance to *C. parvum* infection depends on a delicate balance between the production of Th1 cytokines, needed to control parasite growth, and Th2 cytokines, to limit pathology. IL-18 appears to be involved in the regulation of the Th1/Th2 responses in *C. parvum*, because neutralization of IL18 resulted in a cytokine imbalance with upregulation of systemic (spleen) Th2 cytokine genes, notably IL-4 and IL-13 (Tessema et al. 2009).

C. parvum infection is accompanied by an antibody-mediated Th2 response in infected individuals with production of parasite-specific immunoglobulin (Ig) of all major classes (Ungar et al. 1986; Williams 1987; Hill et al. 1990; Peeters et al. 1992). IgG, IgM and IgA have been detected in serum and mucosa of humans and animals following infection. High levels of these immunoglobulins have been detected in AIDS patients with chronic cryptosporidiosis (Ungar et al. 1986; Cozon et al. 1994; Reperant et al. 1994; Jakobi and Petry 2008). Mucosal IgA may assist the cellular immune response by coating surfaces of the infective stages and preventing attachment to the epithelial cells. However, studies employing B-cell-deficient lMT$^{-/-}$ mice indicate that B-cells are not essential for resistance to and recovery from *C. parvum* infections in mice (Chen et al. 2003a). Therefore, the antibody response observed during *C. parvum* infection may be of minor importance as it is raised mainly to the exposed antigens of the extracellular stages of the parasite. This remains the major obstacle to the use of classical vaccines for prevention of cryptosporidiosis.

10.8 Evasion of Host Immunity and Defense

Host cell death following invasion by intracellular pathogens is an ancient defense mechanism (Velmurugan et al. 2007), which involves infiltration and activation of various immune effector cells such as macrophages, neutrophils and lymphocytes. Many intracellular pathogens have developed strategies to inhibit the induction of host cell death (Luder et al. 2001; Ashida et al. 2011). The intricate interaction between host and *C. parvum* is exemplified by the modulation of epithelial cell apoptosis during the course of cryptosporidiosis (Fig. 10.4). *C. parvum* has the ability to either induce or inhibit enterocyte apoptosis. Completion of the *C. parvum* life cycle requires viable host cells and detachment or apoptosis of infected cells in vitro leads to decreased parasite numbers (Widmer et al. 2000). Therefore, it is proposed that apoptosis may play a role in limiting and/or clearing the infection.

Early studies demonstrated that *C. parvum* infection of intestinal epithelial cells correlates with the activation of NF-κB target genes (Laurent et al. 1997, 1998; Seydel et al. 1998). NF-κB activation has been shown to prevent apoptosis in several experimental models (Beg and Baltimore 1996; Orange et al. 2005) and *Cryptosporidium*-infected cells activate anti-apoptotic pathways (Chen et al. 1998; McCole et al. 2000). During later stages of the infection cycle, pro-apoptotic mediators dominate and the host cell dies (McCole et al. 2000). Indeed, studies found that *C. andersoni* (a *Cryptosporidium* species predominantly in post-weaned and adult cattle) increased enterocyte apoptosis in vitro (Buret et al. 2003). Additionally, microarray analysis of gene expression of *C. hominis* and *C. parvum* infected ileal tissue explants revealed that the NF-κB-regulated osteoprotegerin (OPG) gene is upregulated (Castellanos-Gonzalez et al. 2008). OPG, a member of the tumor necrosis receptor superfamily, is released from intestinal epithelial cells and functions as a soluble decoy receptor, preventing the pro-apototic effects of tumor necrosis factor-related apoptosis inducing ligand (TRAIL) (O'Hara and Chen 2011). Treatment of a *Cryptosporidium*-infected cultured intestinal cell line with TRAIL induced apoptosis and decreased parasite numbers. However, the addition of recombinant OPG antagonized TRAIL-induced parasite reduction, suggesting both TRAIL-mediated regulation of parasite viability and a novel role for OPG in modulating apoptosis during the early phase of infection (Castellanos-Gonzalez et al. 2008). As described above, *Cryptosporidium* activation of TLR4 and TLR2 on the surface of biliary epithelial are required for NF-κB activation, which induces

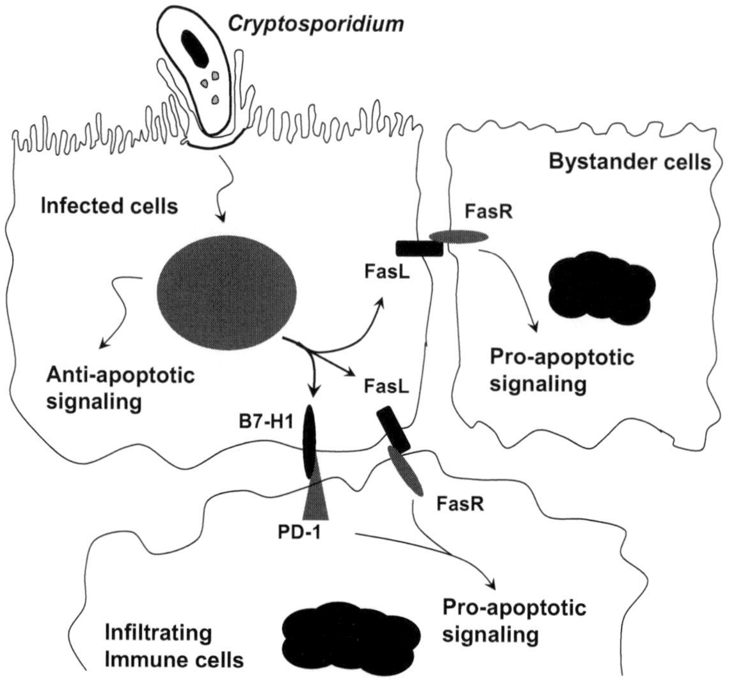

Fig. 10.4 Modulation of pro- and anti-apoptotic signals in various cell types at infection sites during *Cryptosporidium* infection. *C. parvum* reveals the ability to either induce or inhibit enterocyte apoptosis. The activation of NF-κB and the antagonism between anti- and pro-apoptotic mechanisms both promote early parasite propagation, but is a likely mechanism for parasite eradication. Induction of both FasL and B7-H1 may be involved in *C. parvum*-induced apoptotic cell death in bystander un-infected cells, including infiltrating inflammatory cells, and contributes to parasite evasion of host immune and defense

anti-apoptotic factors, but also induces the expression of proinflammatory mediators (Laurent et al. 1997; Chen et al. 2005b) and β-defensins (Tarver et al. 1998), an antimicrobial peptide that is likely to be cytopathic to zoites (Chen et al. 2005b). Therefore, evidence supports that the activation of NF-kB and the antagonism between anti- and pro-apototic mechanisms both promote early parasite propagation, but may function as a mechanism for parasite eradication.

Clearing the pathogen infected cells through apoptosis is usually facilitated by infiltrating inflammatory cells. *C. parvum* may induce apoptotic resistance in infected host epithelial cells through upregulation of B7-H1, thus preventing killing by infiltrating inflammatory cells (Fig. 10.4). B7-H1 is a key member of the B7 family of costimulatory molecules which provide signals for both stimulating and inhibiting T cell activation (Dong et al. 1999; Gong et al. 2010). Upregulation of B7-H1 has been reported in a variety of infectious diseases, including parasitic diseases (Dong and Chen 2006; Joshi et al. 2009). Recent studies indicate that B7-H1 itself acts as a receptor and can respond to PD-1 and B7-1 (Azuma et al. 2008; Keir et al. 2008). B7-H1 has been demonstrated as a ubiquitous anti-apoptotic receptor on cancer cells: B7-H1 on cancer cells receives a signal from PD-1 to rapidly induce resistance against T cell-mediated killing (Azuma et al. 2008). Interestingly, *C. parvum* induces B7-H1 expression in host epithelial cells both in vitro and in vivo (Gong et al. 2010). Whether infiltration of inflammatory cells contributes to apoptotic resistance in infected epithelial cells through interactions between B7-H1 and PD-1 is yet to be investigated.

It is important to note that *C. parvum*-induced apoptosis in vitro and in vivo may include uninfected bystander cells, in particular, during the early stage of infection (Liu et al. 2009). The Fas (APO-1 and CD95)/Fas ligand (FasL) system is an important cellular pathway that regulates the induction of apoptosis in a variety of tissues (Matter et al. 2006; Juric et al. 2009). Fas is a widely expressed, 45-kDa type I membrane protein of the TNF-nerve growth factor family of cell surface receptors. By activating a variety of downstream effector cellular proteases, including members of the caspase family, Fas induces apoptosis (Cryns and Yuan 1998). The biological importance of the Fas/FasL system has been extensively studied in T cells, where it plays a critical role in the clonal deletion of autoreactive T cells and in the activation-induced suicide of T cells (Cryns and Yuan 1998). It was reported, using an in vitro model of biliary cryptosporidiosis, that *C. parvum* stimulates FasL membrane surface translocation, increases FasL protein expression in infected biliary epithelia, and induces a marked increase of soluble FasL in supernatants from infected cells. Infected epithelial cells can induce apoptotic cell death in co-cultured Fas-sensitive Jurkat cells, which can be blocked by a neutralizing antibody to FasL and a metalloprotease inhibitor (Chen et al. 1999). Besides FasL, induction of B7-H1 in infected epithelial cells may also play a role in apoptosis in the bystander un-infected cells. PD-1, the receptor for B7-H1, is a CD28/CTLA-4 like molecule expressed on T cells, B cells and macrophages (Ishida 1992; Freeman 2000). As described above, *C. parvum* induces B7-H1 expression in host epithelial cells both in vitro and in vivo. Infected human biliary epithelial cells can induce apoptotic cell death in co-cultured activated human T cells, which can be partially blocked by a neutralizing antibody to B7-H1 (Gong et al. 2010). Therefore, induction of both FasL and B7-H1 may be involved in *C. parvum*-induced apoptotic cell death in bystander un-infected cells, including infiltrating inflammatory cells, and contributes to parasite evasion of host defense (Fig. 10.4).

Intestinal epithelial cells infected with *C. parvum* were shown to be less susceptible to activation by IFN-γ, at least in part due to parasite induced depletion of the signal transducer and activator of transcription 1 (STAT1) protein (but not STAT1 mRNA depletion) (Choudhry et al. 2009). Moreover, the depletion of STAT1 did not involve protein ubiquitination (Choudhry et al. 2009). Whether miRNA-mediated post-transcriptional suppression plays a role in STAT1 depletion is unclear. Of note, miR-146a targets the 3′UTR of STAT1 mRNA and has been demonstrated to regulate levels of STAT1 in various cell types in response to LPS stimulation (Tang et al. 2009). However, expression of miR-146a remains unchanged in biliary epithelial cells following *C. parvum* infection (Zhou et al. 2009). Given that the IFN-γ response is critical for host defense against *C.*

parvum, the parasite could benefit from the suppression of activation of epithelial cells in response to IFN-γ in infected cells.

10.9 Immunopathology of Cryptosporidiosis

10.9.1 Pathogenic Mechanisms

Intestinal epithelia serve as an absorptive/secretory surface required for nutrient acquisition, and electrolyte and water transport. Also, the epithelium functions as a protective barrier between the underlying tissues and the potentially pathogenic contents of the lumen. If either of these functions is altered, pathological consequences ensue; both are altered during cryptosporidiosis. The intestinal epithelium has a remarkable capacity for fluid and electrolyte absorption under the tight control of neurotransmitters, hormones, inflammatory mediators, and intraluminal factors (Berkes et al. 2003). Chloride secretion, involving the concerted effort of several transporters (Berkes et al. 2003), is the principal determinant of luminal hydration. With chloride secretion, paracellular movement of sodium follows providing an osmotic gradient for the diffusion of water. Chloride hypersecretion has been reported in cryptosporidiosis, and represents one pathophysiological event responsible for diarrhea (Argenzio et al. 1994; Troeger et al. 2007). Compounding this effect, cryptosporidia cause malabsorption of sodium, nutrients and water, and accelerate intestinal transit (Argenzio et al. 1990; Buret et al. 1992; Deselliers et al. 1997).

Epithelial cells maintain the intestinal barrier by forming apical junctional and adherent complexes between enterocytes. These junctions result from the association of various transmembrane and cytosolic plaque proteins, and several cytosolic regulatory proteins (Gonzalez-Mariscal et al. 2003; Turksen and Troy 2004). Intact epithelial junctions are critical to normal physiological function and prevention of diseases. The disruption of intestinal tight junctions results in a "leaky" tight junction barrier, allowing the paracellular translocation of toxic luminal substances, food antigens, and luminal microbes (Baker et al. 2003; O'Hara and Buret 2008). Indeed, impaired intestinal barrier function has been implicated in the pathophysiology of a variety of enteric infections, and chronic gastrointestinal disorders (Farquhar and Palade 1963; Ukabam et al. 1983; Hollander 1988; Furuse et al. 1996; Clayburgh et al. 2004; Dokladny et al. 2006; O'Hara and Buret 2008).

Cryptosporidiosis impairs epithelial barrier function. Epithelial tight junctions are transiently disrupted during cryptosporidial infection (Zhang et al. 2000; Buret et al. 2003; Huang and White 2006). Further research is needed to assess whether these alterations in vivo are initiated by parasitic products, the invasion process itself, or host inflammatory responses to *Cryptosporidium*. However, in the absence of other host cells, infection of bovine and human epithelial cell lines with *C. andersoni* altered the distribution of tight junctional zonula occludens 1 protein, and promoted apoptosis, consistent with a direct pathogenic effect of the parasite (Buret et al. 2003). Treatment of intestinal cells with TGF-β1 inhibits *C. parvum*-induced permeability (Roche et al. 2000).

At the structural level, cryptosporidiosis is associated with crypt hyperplasia, villous atrophy, and a diffuse shortening or loss of brush border microvilli (Argenzio et al. 1990; Farthing 2000; Huang and White 2006). Hence, the absorptive capacity of the villi surface is diminished. This imbalance of absorption and secretion is likely a major contributor to disease manifestation. While the precise mechanism of diffuse villus blunting remains elusive, it is possible, given the tissue alterations, that activated $CD8^+$ T-cells play a significant role as seen with giardiasis (Faubert 2000; Eckmann 2003; Scott et al. 2004). Physiologically, cryptosporidiosis is associated with reduced glucose-NaCl absorption, increased chloride anion secretion (Cl^-), and decreased epithelial barrier function, all of which contribute to dehydration, diarrhea, and malabsorption associated with cryptosporidiosis. Of significant interest, expression of substance P, a neuropeptide that localizes to areas of inflammation and stimulates active Cl^- secretion from intestinal epithelial cells, is elevated in experimentally infected human subjects, and in AIDS patients with uncontrolled cryptosporidiosis. Expression levels of this neuropeptide seem to correlate with severity of diarrhea (Robinson et al. 2003). Furthermore, a molecular antagonist of the substance P receptor (NK1) reduced both ion secretion and corrected the *Cryptosporidium*-induced glucose malabsorption in a macaque intestinal explant model of cryptosporidiosis (Hernandez et al. 2007).

10.9.2 *Chromatin Dynamics and* Cryptosporidium *Infection*

Epigenetic regulation refers to reversible modifications introduced by various chromatin modifying proteins on DNA and histones to regulate gene expression via modulating chromatin structures (Graeff et al. 2011). In general, DNA methylation is associated with transcriptional repression while histone acetylation is related to transcriptional activation. Depending on the lysine residue methylated, histone methylation can either activate or repress gene transcription (Cloos et al. 2008). Dynamic epigenetic regulation participates in modulating almost all important biological processes including cellular differentiation. Gene expression in eukaryotes is regulated by histone acetylation/deacetylation, an epigenetic process mediated by histone acetyltransferases (HATs) and histone deacetylases (HDACs) whose opposing activities are tightly regulated. The acetylation of histones by HATs increases DNA accessibility and promotes gene expression, whereas the removal of acetyl groups by HDACs has the opposite effect. Epigenetic modifications protect against the excessive response and autotoxicity of acute pro-inflammatory gene products (McCall et al. 2010). Evidence has accumulated showing that HDACs have immunomodulatory activity and are important to regulation of anti-microbial defense (Roger et al. 2011). Multiple reports have shown that HDAC inhibitors possess suppressive effects on the induction of immune response genes (Bode et al. 2007; Roger et al. 2011). Therefore, it is not surprising that distinct epigenetic mechanisms regulate the inflammatory responses and contribute to the pathogenetic features in many infectious diseases (Paschos and Allday 2010).

Epigenetic mechanisms may be involved in *C. parvum*-induced alteration of miRNA transcription in infected cells. Among those miRNAs suppressed in host epithelial cells following *C. parvum* infection, the *let-7i* gene is suppressed through histone-H3 deacetylation (O'Hara et al. 2010b). Analysis of the mechanism revealed that microbial infection promotes the formation of an NF-κB p50-C/EBPβ silencer complex in the *let-7i* regulatory sequence. Chromatin immunoprecipitation assays demonstrated that the repressor complex binds to the *let-7i* promoter following *C. parvum* infection and promotes histone-H3 deacetylation. Suppression of the *mir-424-503* gene following *C. parvum* infection also involves epigenetic mechanisms (Zhou et al. 2013). *C. parvum*-induced transcriptional suppression of the *mir-424-503* gene is associated with an increased recruitment of NF-κB p50 to the promoter. Intriguingly, promoter recruitment of several HDACs, including HDAC1, HDAC2 and Sirt1, was also detected in infected cells. Such promoter recruitment of NF-κB p50 and HDAC complex appears to be associated with a decrease of H3 acetylation and account for the repression of the *mir-424-503* gene in infected cells. Functionally, induction of CX3CL1 may be associated with downregulation of mature miR-424 and miR-503, both of which target the CX3CL1 3′UTR, suppress its translation and induce RNA degradation (Zhou et al. 2013).

Our understanding of epigenetic regulation of the host response to infection is advancing, but studies on the role of epigenetic modifications during *Cryptosporidium* infection are still very limited. Of particular interest, epigenetic processes may impact the metabolic functions in enterocytes and thus play a role in the pathogenicity of *Cryptosporidium* infection. Epigenetic changes may also impact mucosal immunity and determine the severity of infection.

10.10 Conclusion and Perspectives

While we have begun to recognize the critical role for mucosal epithelial cells in the immune defense against *Cryptosporidium* infection, how epithelial cells are activated to mount an efficient innate and adaptive anti-parasite immune response is still unclear. The parasite-epithelial cell interactions are very complex. PRRs expressed in epithelial cells, such as TLRs, and their downstream signaling pathways, appear to be the key effector regulators accounting for epithelial cell defense, including production and release of a panel of effector molecules of anti-microbial activity or attracting the infiltration of inflammatory immune cells to infection sites. Both transcriptional and post-transcriptional mechanisms are involved in the regulation of such epithelial responses, in particular, miRNA-mediated post-transcriptional gene suppression. *Cryptosporidium* parasites have also developed strategies of immune evasion to escape host defense, at least at the early stage of infection. Despite significant advances in understanding mucosal anti-*Cryptosporidium* immunity, how parasite antigens are presented through the mucosal epithelium and how adaptive immunity is regulated remains unclear.

Identification of parasite molecules involved in the activation of signaling pathways in infected host cells would provide new insights into the molecular mechanisms of immune activation following *Cryptosporidium* infection. Evaluating the secretion and function of these newly-identified cytokines secreted by epithelial cells will provide additional insights into how epithelial cells mediate anti-*Cryptosporidium* defense. Cell-cell communications mediated by exosomes may have a much broader impact on the innate and adaptive immune network. In particular, these mechanisms may recruit infiltrating inflammatory cells to infection sites, and may represent a new target for therapeutic intervention. The development of epithelial cell-specific knockouts will greatly advance our understanding of epithelial-mediated immune responses in vivo. Unraveling the regulatory functions of ncRNAs in anti-parasite immunity is still in its infancy, but will likely yield new insights into our understanding of the immunobiology and immunopathology of cryptosporidiosis. Given the importance of adaptive immunity to clear infection and the intramembranous localization of the parasite in epithelial cells after invasion, modulation of epithelial innate immune defense through lympho-epithelial interactions may be critical for anti-*Cryptosporidium* immunity, an understudied area of *Cryptosporidium* research. Better understanding of the impact of epigenetic modifications on the immune response will help us to evaluate the potential impact of cryptosporidial infection on malnutrition and possible co-infections identified in young children.

Acknowledgments The authors thank the members of the Chen lab, particularly Rui Zhou, Ai-Yu Gong, Jun Liu, Dongqing Chen, Grace Yang and Alex N. Eischeid for their contributions towards understanding *Cryptosporidium*-epithelial cell interactions. Support from the National Institutes of Health (AI071321 and AI095532) and the Nebraska Biomedical Research Program (LB692) is gratefully acknowledged.

References

Abrahamsen MS, Templeton TJ, Enomoto S et al (2004) Complete genome sequence of the apicomplexan, *Cryptosporidium parvum*. Science 304:441–445

Aji T, Flanigan T, Marshall R et al (1991) Ultrastructural study of asexual development of *Cryptosporidium parvum* in a human intestinal cell line. J Protozool 38:82S–84S

Anderson P (2008) Post-transcriptional control of cytokine production. Nat Immunol 9:353–359

Argenzio RA, Liacos JA, Levy ML et al (1990) Villous atrophy, crypt hyperplasia, cellular infiltration, and impaired glucose-Na absorption in enteric cryptosporidiosis of pigs. Gastroenterology 98:1129–1140

Argenzio RA, Rhoads JM, Armstrong M et al (1994) Glutamine stimulates prostaglandin-sensitive Na(+)-H+ exchange in experimental porcine cryptosporidiosis. Gastroenterology 106:1418–1428

Ashida H, Mimuro H, Ogawa M et al (2011) Cell death and infection: a double-edged sword for host and pathogen survival. J Cell Biol 195:931–942
Asirvatham AJ, Gregorie CJ, Hu Z et al (2008) MicroRNA targets in immune genes and the Dicer/Argonaute and ARE machinery components. Mol Immunol 45:1995–2006
Azuma T, Yao S, Zhu G et al (2008) B7-H1 is a ubiquitous antiapoptotic receptor on cancer cells. Blood 111:3635–3643
Baker SF, Yin Y, Runswick SK et al (2003) Peptidase allergen Der p 1 initiates apoptosis of epithelial cells independently of tight junction proteolysis. Mol Membr Biol 20:71–81
Barakat FM, McDonald V, Di Santo JP et al (2009a) Roles for NK cells and an NK cell-independent source of intestinal gamma interferon for innate immunity to *Cryptosporidium parvum* infection. Infect Immun 77:5044–5049
Barakat FM, McDonald V, Foster GR et al (2009b) *Cryptosporidium parvum* infection rapidly induces a protective innate immune response involving type I interferon. J Infect Dis 200:1548–1555
Bartel DP (2004) MicroRNAs: genomics, biogenesis, mechanism, and function. Cell 116:281–297
Bartel DP (2009) MicroRNAs: target recognition and regulatory functions. Cell 136:215–233
Bedi B, Mead JR (2012) *Cryptosporidium parvum* antigens induce mouse and human dendritic cells to generate Th1-enhancing cytokines. Parasite Immunol 34:473–485
Beg AA, Baltimore D (1996) An essential role for NF-kappaB in preventing TNF-alpha-induced cell death. Science 274:782–784
Berkes J, Viswanathan VK, Savkovic SD et al (2003) Intestinal epithelial responses to enteric pathogens: effects on the tight junction barrier, ion transport, and inflammation. Gut 52:439–451
Bhat N, Joe A, PereiraPerrin M et al (2007) *Cryptosporidium* p30, a galactose/N-acetylgalactosamine-specific lectin, mediates infection in vitro. J Biol Chem 282:34877–34887
Bhatnagar S, Schorey JS (2007) Exosomes released from infected macrophages contain mycobacterium avium glycopeptidolipids and are proinflammatory. J Biol Chem 282:25779–25789
Blikslager AT, Moeser AJ, Gookin JL et al (2007) Restoration of barrier function in injured intestinal mucosa. Physiol Rev 87:545–564
Bobrie A, Colombo M, Raposo G et al (2011) Exosome secretion: molecular mechanisms and roles in immune responses. Traffic 12:1659–1668
Bode KA, Schroder K, Hume DA et al (2007) Histone deacetylase inhibitors decrease Toll-like receptor-mediated activation of proinflammatory gene expression by impairing transcription factor recruitment. Immunology 122:596–606
Boehm U, Klamp T, Groot M et al (1997) Cellular responses to interferon-gamma. Ann Rev Immunol 15:749–795
Bonnin A, Gut J, Dubremetz JF et al (1995) Monoclonal antibodies identify a subset of dense granules in *Cryptosporidium parvum* zoites and gamonts. J Eukaryot Microbiol 42:395–401
Borad A, Ward H (2010) Human immune responses in cryptosporidiosis. Future Microbiol 5:507–519
Buret A, Hardin JA, Olson ME et al (1992) Pathophysiology of small intestinal malabsorption in gerbils infected with *Giardia lamblia*. Gastroenterology 103:506–513
Buret AG, Chin AC, Scott KG (2003) *Infection of human and bovine epithelial cells with Cryptosporidium andersoni* induces apoptosis and disrupts tight junctional ZO-1: effects of epidermal growth factor. Int J Parasitol 33:1363–1371
Castellanos-Gonzalez A, Yancey LS, Wang HC et al (2008) *Cryptosporidium* infection of human intestinal epithelial cells increases expression of osteoprotegerin: a novel mechanism for evasion of host defenses. J Infect Dis 197:916–923
Cevallos AM, Bhat N, Verdon R et al (2000a) Mediation of *Cryptosporidium parvum* infection in vitro by mucin-like glycoproteins defined by a neutralizing monoclonal antibody. Infect Immun 68:5167–5175
Cevallos AM, Zhang X, Waldor MK et al (2000b) Molecular cloning and expression of a gene encoding *Cryptosporidium parvum* glycoproteins gp40 and gp15. Infect Immun 68:4108–4116

Chen XM, LaRusso NF (1999) Human intestinal and biliary cryptosporidiosis. World J Gastroenterol 5:424–429

Chen CY, Shyu AB (1995) AU-rich elements: characterization and importance in mRNA degradation. Trends Biochem Sci 20:465–470

Chen WX, Harp JA, Harmsen AG (1993) Requirements for CD4+ cells and gamma-interferon in resolution of established *Cryptosporidium parvum* infection in mice. Infect Immun 61:3928–3932

Chen XM, Levine SA, Tietz P et al (1998) *Cryptosporidium parvum* is cytopathic for cultured human biliary epithelia via an apoptotic mechanism. Hepatology 28:906–913

Chen XM, Gores GJ, Paya CV et al (1999) *Cryptosporidium parvum* induces apoptosis in biliary epithelia by a Fas/Fas ligand-dependent mechanism. Am J Physiol 277:G599–G608

Chen XM, Levine SA, Splinter PL et al (2001) *Cryptosporidium parvum* activates nuclear factor kappaB in biliary epithelia preventing epithelial cell apoptosis. Gastroenterology 120:1774–1783

Chen XM, Keithly JS, Paya CV et al (2002) Cryptosporidiosis. Engl J Med 346:1723–1731

Chen WX, Harp JA, Harmsen AG (2003a) *Cryptosporidium parvum* infection in gene-targeted B cell-deficient mice. J Parasitol 89:391–393

Chen XM, Huang BQ, Splinter PL et al (2003b) *Cryptosporidium parvum* invasion of biliary epithelia requires host cell tyrosine phosphorylation of cortactin via c-Src. Gastroenterology 125:216–228

Chen XM, Splinter PL, Tietz PS et al (2004) Phosphatidylinositol 3-kinase and frabin mediate *Cryptosporidium parvum* cellular invasion via activation of Cdc42. J Biol Chem 279:31671–31678

Chen XM, O'Hara SP, Huang BQ et al (2005a) Localized glucose and water influx facilitates *Cryptosporidium parvum* cellular invasion by means of modulation of host-cell membrane protrusion. Proc Natl Acad Sci U S A 102:6338–6343

Chen XM, O'Hara SP, Nelson JB et al (2005b) Multiple TLRs are expressed in human cholangiocytes and mediate host epithelial defense responses to *Cryptosporidium parvum* via activation of NF-kappaB. J Immunol 175:7447–7456

Chen XM, Splinter PL, O'Hara SP et al (2007) A cellular micro-RNA, let-7i, regulates Toll-like receptor 4 expression and contributes to cholangiocyte immune responses against *Cryptosporidium parvum* infection. J Biol Chem 282:28929–28938

Chen X, Song CH, Feng BS et al (2011) Intestinal epithelial cell-derived integrin alphabeta6 plays an important role in the induction of regulatory T cells and inhibits an antigen-specific Th2 response. J Leukoc Biol 90:751–759

Choi SM, McAleer JP, Zheng M et al (2013) Innate Stat3-mediated induction of the antimicrobial protein Reg3gamma is required for host defense against MRSA pneumonia. J Exp Med 210:551–561

Choudhry N, Korbel DS, Edwards LA et al (2009) Dysregulation of interferon-gamma-mediated signalling pathway in intestinal epithelial cells by *Cryptosporidium parvum* infection. Cell Microbiol 11:1354–1364

Clayburgh DR, Shen L, Turner JR (2004) A porous defense: the leaky epithelial barrier in intestinal disease. Lab Invest 84:282–291

Cloos PAC, Christensen J, Agger K et al (2008) Erasing the methyl mark: histone demethylases at the center of cellular differentiation and disease. Genes Dev 22:1115–1140

Cozon G, Biron F, Jeannin M et al (1994) Secretory Iga antibodies to *Cryptosporidium parvum* in AIDS patients with chronic Cryptosporidiosis. J Infect Dis 169:696–699

Cryns V, Yuan JY (1998) Proteases to die for. Genes Dev 12:1551–1570

Dean JL, Sully G, Clark AR et al (2004) The involvement of AU-rich element-binding proteins in p38 mitogen-activated protein kinase pathway-mediated mRNA stabilisation. Cell Signal 16:1113–1121

Deng M, Templeton TJ, London NR et al (2002) *Cryptosporidium parvum* genes containing thrombospondin type 1 domains. Infect Immun 70:6987–6995

Deselliers LP, Tan DT, Scott RB et al (1997) Effects of *Giardia lamblia* infection on gastrointestinal transit and contractility in Mongolian gerbils. Dig Dis Sci 42:2411–2419

Dokladny K, Moseley PL, Ma TY (2006) Physiologically relevant increase in temperature causes an increase in intestinal epithelial tight junction permeability. Am J Physiol 290:G204–G212

Dommett R, Zilbauer M, George JT et al (2005) Innate immune defence in the human gastrointestinal tract. Mol Immunol 42:903–912

Dong H, Chen X (2006) Immunoregulatory role of B7-H1 in chronicity of inflammatory responses. Cell Mol Immunol 3:179–187

Dong HD, Zhu GF, Tamada K et al (1999) B7-H1, a third member of the B7 family, co-stimulates T-cell proliferation and interleukin-10 secretion. Nat Med 5:1365–1369

Eckmann L (2003) Mucosal defences against Giardia. Parasite Immunol 25:259–270

Ehigiator HN, Romagnoli P, Borgelt K et al (2005) Mucosal cytokine and antigen-specific responses to *Cryptosporidium parvum* in IL-12p40 KO mice. Parasite Immunol 27:17–28

Eichelberger MC, Suresh P, Rehg JE (2000) Protection from *Cryptosporidium parvum* infection by gamma delta T cells in mice that lack alpha beta T cells. Comp Med 50:270–276

Elliott DA, Coleman DJ, Lane MA et al (2001) *Cryptosporidium parvum* infection requires host cell actin polymerization. Infect Immun 69:5940–5942

Farquhar MG, Palade GE (1963) Junctional complexes in various epithelia. J Cell Biol 17:375–412

Farthing MJ (2000) Clinical aspects of human cryptosporidiosis. Contrib Microbiol 6:50–74

Faubert G (2000) Immune response to *Giardia duodenalis*. Clin Microbiol Rev 13:35–54

Freeman GJ, Long LA, Iwai Y et al (2000) Engagement of the PD-1 immunoinhibitory receptor by a novel B7 family member leads to negative regulation of lymphocyte activation. J Exp Med 192:1027–1034

Furuse M, Fujimoto K, Sato N et al (1996) Overexpression of occludin, a tight junction-associated integral membrane protein, induces the formation of intracellular multilamellar bodies bearing tight junction-like structures. J Cell Sci 109(Pt 2):429–435

Gong AY, Zhou R, Hu G et al (2010) *Cryptosporidium parvum* induces B7-H1 expression in cholangiocytes by down-regulating microRNA-513. J Infect Dis 201:160–169

Gong AY, Hu G, Zhou R et al (2011) MicroRNA-221 controls expression of intercellular adhesion molecule-1 in epithelial cells in response to *Cryptosporidium parvum* infection. Int J Parasitol 41:397–403

Gonzalez-Mariscal L, Betanzos A, Nava P et al (2003) Tight junction proteins. Prog Biophys Mol Biol 81:1–44

Gookin JL, Allen J, Chiang S et al (2005) Local peroxynitrite formation contributes to early control of *Cryptosporidium parvum* infection. Infect Immun 73:3929–3936

Gookin JL, Chiang S, Allen J et al (2006) NF-kappaB-mediated expression of iNOS promotes epithelial defense against infection by *Cryptosporidium parvum* in neonatal piglets. Am J Physiol 290:G164–G174

Graeff J, Kim D, Dobbin MM et al (2011) Epigenetic regulation of gene expression in physiological and pathological brain processes. Physiol Rev 91:603–649

Harada K, Nakanuma Y (2012) Cholangiopathy with respect to biliary innate immunity. Int J Hepatol 2012:793569. doi:10.1155/2012/793569

Hashim A, Mulcahy G, Bourke B et al (2006) Interaction of *Cryptosporidium hominis* and *Cryptosporidium parvum* with primary human and bovine intestinal cells. Infect Immun 74:99–107

Hernandez J, Lackner A, Aye P et al (2007) Substance P is responsible for physiological alterations such as increased chloride ion secretion and glucose malabsorption in cryptosporidiosis. Infect Immun 75:1137–1143

Hershberg RM, Mayer LF (2000) Antigen processing and presentation by intestinal epithelial cells – polarity and complexity. Immunol Today 21:123–128

Hill BD, Blewett DA, Dawson AM et al (1990) Analysis of the kinetics, isotype and specificity of serum and coproantibody in lambs infected with *Cryptosporidium-parvum*. Res Vet Sci 48:76–81

Hollander D (1988) Crohn's disease – a permeability disorder of the tight junction? Gut 29:1621–1624

Hu G, Zhou R, Liu J et al (2009) MicroRNA-98 and let-7 confer cholangiocyte expression of cytokine-inducible Src homology 2-containing protein in response to microbial challenge. J Immunol 183:1617–1624

Hu G, Zhou R, Liu J et al (2010) MicroRNA-98 and let-7 regulate expression of suppressor of cytokine signaling 4 in biliary epithelial cells in response to *Cryptosporidium parvum* infection. J Infect Dis 202:125–135

Hu G, Drescher KM, Chen XM (2012a) Exosomal miRNAs: biological properties and therapeutic potential. Front Genet 3:56. doi:10.3389/fgene.2012.00056

Hu G, Yao H, Chaudhuri AD et al (2012b) Exosome-mediated shuttling of microRNA-29 regulates HIV Tat and morphine-mediated neuronal dysfunction. Cell Death Dis 3:e381. doi:10.1038/cddis.2012.114

Hu G, Gong AY, Roth AL et al (2013) Release of luminal exosomes contributes to TLR4-mediated epithelial antimicrobial defense. PLoS Pathog 9:e1003261. doi:10.1371/journal.ppat.1003261

Huang DB, White AC (2006) An updated review on Cryptosporidium and Giardia. Gastroenterol Clin North Am 35:291–314

Huang BQ, Chen XM, LaRusso NF (2004) *Cryptosporidium parvum* attachment to and internalization by human biliary epithelia in vitro: a morphologic study. J Parasitol 90:212–221

Ishida YAY, Shibahara K, Honjo T (1992) Induced expression of PD-1, a novel member of the immunoglobulin gene superfamily, upon programmed cell death. EMBO J 11:3887–3895

Jakobi V, Petry F (2008) Humoral immune response in IL-12 and IFN-gamma deficient mice after infection with *Cryptosporidium parvum*. Parasite Immunol 30:151–161

Joshi T, Rodriguez S, Perovic V et al (2009) B7-H1 blockade increases survival of dysfunctional CD8(+) T cells and confers protection against *Leishmania donovani* infections. PLoS Pathog 5: e1000431. doi:10.1371/journal.ppat.1000431

Juric V, Chen CC, Lau LF (2009) Fas-mediated apoptosis is regulated by the extracellular matrix protein CCN1 (CYR61) in vitro and in vivo. Mol Cell Biol 29:3266–3279

Kagnoff MF, Eckmann L (1997) Epithelial cells as sensors for microbial infection. J Clin Invest 100:S51–S55

Keir ME, Butte MJ, Freeman GJ et al (2008) PD-1 and its ligands in tolerance and immunity. Annu Rev Immunol 26:677–704

Kennedy MJ, Ehlers MD (2011) Mechanisms and function of dendritic exocytosis. Neuron 69:856–875

Kolotuev I, Apaydin A, Labouesse M (2009) Secretion of Hedgehog-related peptides and WNT during *Caenorhabditis elegans* development. Traffic 10:803–810

Lacroix S, Mancassola R, Naciri M et al (2001) *Cryptosporidium parvum*-specific mucosal immune response in C57BL/6 neonatal and gamma interferon-deficient mice: role of tumor necrosis factor alpha in protection. Infect Immun 69:1635–1642

Lacroix-Lamande S, Mancassola R, Naciri M et al (2002) Role of gamma interferon in chemokine expression in the ileum of mice and in a murine intestinal epithelial cell line after *Cryptosporidium parvum* infection. Infect Immun 70:2090–2099

Ladel CH, Blum C, Dreher A et al (1995) Protective role of gamma/delta T-cells and alpha/beta T-cells in tuberculosis. Eur J Immunol 25:2877–2881

Laurent F, Eckmann L, Savidge TC et al (1997) *Cryptosporidium parvum* infection of human intestinal epithelial cells induces the polarized secretion of C-X-C chemokines. Infect Immun 65:5067–5073

Laurent F, Kagnoff MF, Savidge TC et al (1998) Human intestinal epithelial cells respond to *Cryptosporidium parvum* infection with increased prostaglandin H synthase 2 expression and prostaglandin E2 and F2alpha production. Infect Immun 66:1787–1790

Lee Y, Ahn C, Han J et al (2003) The nuclear RNase III Drosha initiates microRNA processing. Nature 425:415–419

Leitch GJ, He Q (1994) Arginine-derived nitric oxide reduces fecal oocyst shedding in nude mice infected with *Cryptosporidium parvum*. Infect Immun 62:5173–5176

Leitch GJ, He Q (1999) Reactive nitrogen and oxygen species ameliorate experimental cryptosporidiosis in the neonatal BALB/c mouse model. Infect Immun 67:5885–5891

Lewis BP, Burge CB, Bartel DP (2005) Conserved seed pairing, often flanked by adenosines, indicates that thousands of human genes are microRNA targets. Cell 120:15–20

Liegeois S, Benedetto A, Garnier JM et al (2006) The V0-ATPase mediates apical secretion of exosomes containing Hedgehog-related proteins in *Caenorhabditis elegans*. J Cell Biol 173:949–961

Liu J, Enomoto S, Lancto CA et al (2008) Inhibition of apoptosis in *Cryptosporidium parvum*-infected intestinal epithelial cells is dependent on survivin. Infect Immun 76:3784–3792

Liu J, Deng M, Lancto CA et al (2009) Biphasic modulation of apoptotic pathways in *Cryptosporidium parvum*-infected human intestinal epithelial cells. Infect Immun 77:837–849

Luder CGK, Gross U, Lopes MF (2001) Intracellular protozoan parasites and apoptosis: diverse strategies to modulate parasite-host interactions. Trends Parasitol 17:480–486

Lumadue JA, Manabe YC, Moore RD et al (1998) A clinicopathologic analysis of AIDS-related cryptosporidiosis. AIDS 12:2459–2466

Lumb R, Smith K, O'Donoghue PJ et al (1988) Ultrastructure of the attachment of Cryptosporidium sporozoites to tissue culture cells. Parasitol Res 74:531–536

Mallegol J, Van Niel G, Lebreton C et al (2007) T84-intestinal epithelial exosomes bear MHC class II/peptide complexes potentiating antigen presentation by dendritic cells. Gastroenterology 132:1866–1876

Marin F, Luquet G, Marie B et al (2008) Molluscan shell proteins: primary structure, origin, and evolution. Curr Top Dev Biol 80:209–276

Matter CM, Chadjichristos CE, Meier P et al (2006) Role of endogenous Fas (CD95/Apo-1) ligand in balloon-induced apoptosis, inflammation, and neointima formation. Circulation 113:1879–1887

Maynard CL, Elson CO, Hatton RD et al (2012) Reciprocal interactions of the intestinal microbiota and immune system. Nature 489:231–241

McCall CE, Yoza B, Liu TF et al (2010) Gene-specific epigenetic regulation in serious infections with systemic inflammation. J Innate Immun 2:395–405

McCole DF, Eckmann L, Laurent F et al (2000) Intestinal epithelial cell apoptosis following *Cryptosporidium parvum* infection. Infect Immun 68:1710–1713

McDonald V, Bancroft GJ (1994) Mechanisms of innate and acquired-resistance to *Cryptosporidium parvum* infection in SCID mice. Parasite Immunol 16:315–320

McDonald V, McCrossan MV, Petry F (1995) Localization of parasite antigens in *Cryptosporidium parvum*-infected epithelial cells using monoclonal antibodies. Parasitology 110:259–268

McDonald V, Pollok RC, Dhaliwal W et al (2006) A potential role for interleukin-18 in inhibition of the development of *Cryptosporidium parvum*. Clin Exp Immunol 145:555–562

McDonald V, Korbel DS, Barakat FM et al (2013) Innate immune responses against *Cryptosporidium parvum* infection. Parasite Immunol 35:55–64

Medzhitov R (2007) Recognition of microorganisms and activation of the immune response. Nature 449:819–826

Mombaerts P, Arnoldi J, Russ F et al (1993) Different roles of alpha-beta and gamma-delta T-cells in immunity against an intracellular bacterial pathogen. Nature 365:53–56

Muller WA (2009) Mechanisms of transendothelial migration of leukocytes. Circ Res 105:223–230

Nasu K, Narahara H (2010) Pattern recognition via the toll-like receptor system in the human female genital tract. Mediators Inflamm 2010:976024. doi:10.1155/2010/976024

Newton K, Dixit VM (2012) Signaling in innate immunity and inflammation. Cold Spring Harb Perspect Biol 4:a006049. doi:10.1101/cshperspect.a006049

O'Connor RM, Burns PB, Ha-Ngoc T et al (2009) Polymorphic mucin antigens CpMuc4 and CpMuc5 are integral to *Cryptosporidium parvum* infection in vitro. Eukaryot Cell 8:461–469
O'Hara JR, Buret AG (2008) Mechanisms of intestinal tight junctional disruption during infection. Front Biosci 13:7008–7021
O'Hara SP, Chen XM (2011) The cell biology of cryptosporidium infection. Microbes Infect 13:721–730
O'Hara SP, Yu JR, Lin JJ (2004) A novel *Cryptosporidium parvum* antigen, CP2, preferentially associates with membranous structures. Parasitol Res 92:317–327
O'Hara SP, Gajdos GB, Trussoni CE et al (2010a) Cholangiocyte myosin IIB is required for localized aggregation of sodium glucose cotransporter 1 to sites of *Cryptosporidium parvum* cellular invasion and facilitates parasite internalization. Infect Immun 78:2927–2936
O'Hara SP, Splinter PL, Gajdos GB et al (2010b) NFkappaB p50-CCAAT/enhancer-binding protein beta (C/EBPbeta)-mediated transcriptional repression of microRNA let-7i following microbial infection. J Biol Chem 285:216–225
Oeckinghaus A, Hayden MS, Ghosh S (2011) Crosstalk in NF-kappaB signaling pathways. Nat Immunol 12:695–708
Ohshima K, Inoue K, Fujiwara A et al (2010) Let-7 microRNA family is selectively secreted into the extracellular environment via exosomes in a metastatic gastric cancer cell line. PLoS One 5:e13247. doi:10.1371/journal.pone.0013247
Ojcius DM, Perfettini JL, Bonnin A et al (1999) Caspase-dependent apoptosis during infection with *Cryptosporidium parvum*. Microbes Infect 1:1163–1168
Okazawa A, Kanai T, Nakamaru K et al (2004) Human intestinal epithelial cell-derived interleukin (IL)-18, along with IL-2, IL-7 and IL-15, is a potent synergistic factor for the proliferation of intraepithelial lymphocytes. Clin Exp Immunol 136:269–276
Orange JS, Levy O, Geha RS (2005) Human disease resulting from gene mutations that interfere with appropriate nuclear factor-kappaB activation. Immunol Rev 203:21–37
Ostrowski M, Carmo NB, Krumeich S et al (2010) Rab27a and Rab27b control different steps of the exosome secretion pathway. Nat Cell Biol 12:19–31
Pantenburg B, Dann SM, Wang HC et al (2008) Intestinal immune response to human Cryptosporidium sp. infection. Infect Immun 76:23–29
Pantenburg B, Castellanos-Gonzalez A, Dann SM et al (2010) Human CD8(+) T cells clear *Cryptosporidium parvum* from infected intestinal epithelial cells. Am J Trop Med Hyg 82:600–607
Paschos K, Allday MJ (2010) Epigenetic reprogramming of host genes in viral and microbial pathogenesis. Trends Microbiol 18:439–447
Pautz A, Art J, Hahn S et al (2010) Regulation of the expression of inducible nitric oxide synthase. Nitric Oxide 23:75–93
Pedersen IM, Cheng G, Wieland S et al (2007) Interferon modulation of cellular microRNAs as an antiviral mechanism. Nature 449:919–922
Peeters JE, Villacorta I, Vanopdenbosch E et al (1992) *Cryptosporidium parvum* in calves – kinetics and immunoblot analysis of specific serum and local antibody-responses (immunoglobulin a [Iga], Igg, and Igm) after natural and experimental infections. Infect Immun 60:2309–2316
Perez-Cordon G, Nie W, Schmidt D et al (2011) Involvement of host calpain in the invasion of *Cryptosporidium parvum*. Microbes Infect 13:103–107
Perryman LE, Jasmer DP, Riggs MW et al (1996) A cloned gene of *Cryptosporidium parvum* encodes neutralization-sensitive epitopes. Mol Biochem Parasitol 80:137–147
Petersen C, Gut J, Doyle PS et al (1992) Characterization of a > 900,000-M(r) *Cryptosporidium parvum* sporozoite glycoprotein recognized by protective hyperimmune bovine colostral immunoglobulin. Infect Immun 60:5132–5138
Petry F, Jakobi V, Wagner S et al (2008) Binding and activation of human and mouse complement by *Cryptosporidium parvum* (Apicomplexa) and susceptibility of C1q- and MBL-deficient mice to infection. Mol Immunol 45:3392–3400

Petry F, Jakobi V, Tessema TS (2010) Host immune response to *Cryptosporidium parvum* infection. Exp Parasitol 126:304–309

Pollok RC, Farthing MJ, Bajaj-Elliott M et al (2001) Interferon gamma induces enterocyte resistance against infection by the intracellular pathogen *Cryptosporidium parvum*. Gastroenterology 120:99–107

Putignani L, Possenti A, Cherchi S et al (2008) The thrombospondin-related protein CpMIC1 (CpTSP8) belongs to the repertoire of micronemal proteins of *Cryptosporidium parvum*. Mol Biochem Parasitol 157:98–101

Reigstad CS, Lunden GO, Felin J et al (2009) Regulation of serum amyloid A3 (SAA3) in mouse colonic epithelium and adipose tissue by the intestinal microbiota. PLoS One 4:e5842. doi:10.1371/journal.pone.0005842

Reperant JM, Naciri M, Iochmann S et al (1994) Major antigens of *Cryptosporidium parvum* recognized by serum antibodies from different infected animal species and man. Vet Parasitol 55:1–13

Ricciardolo FL, Sterk PJ, Gaston B et al (2004) Nitric oxide in health and disease of the respiratory system. Physiol Rev 84:731–765

Riggs MW, Stone AL, Yount PA et al (1997) Protective monoclonal antibody defines a circumsporozoite-like glycoprotein exoantigen of *Cryptosporidium parvum* sporozoites and merozoites. J Immunol 158:1787–1795

Robert B, Antoine H, Dreze F et al (1994) Characterization of a high molecular weight antigen of *Cryptosporidium parvum* micronemes possessing epitopes that are cross-reactive with all parasitic life cycle stages. Vet Res 25:384–398

Robinson P, Okhuysen PC, Chappell CL et al (2003) Substance P expression correlates with severity of diarrhea in cryptosporidiosis. J Infect Dis 188:290–296

Roche JK, Martins CA, Cosme R et al (2000) Transforming growth factor beta1 ameliorates intestinal epithelial barrier disruption by *Cryptosporidium parvum* in vitro in the absence of mucosal T lymphocytes. Infect Immun 68:5635–5644

Roger T, Lugrin J, Le Roy D et al (2011) Histone deacetylase inhibitors impair innate immune responses to Toll-like receptor agonists and to infection. Blood 117:1205–1217

Ruittenberg EJ, van Noorle Jansen LM (1975) Effect of *Corynebacterium parvum* on the course of a Listeria monocytogenes infection in normal and congenitally athymic (nude) mice. Zentralbl Bakteriol Orig A 231:197–205

Scaria V, Hariharan M, Maiti S et al (2006) Host-virus interaction: a new role for microRNAs. Retrovirology 3:68

Scott KG, Yu LC, Buret AG (2004) Role of CD8+ and CD4+ T lymphocytes in jejunal mucosal injury during murine giardiasis. Infect Immun 72:3536–3542

Seydel KB, Zhang T, Champion GA et al (1998) *Cryptosporidium parvum* infection of human intestinal xenografts in SCID mice induces production of human tumor necrosis factor alpha and interleukin-8. Infect Immun 66:2379–2382

Sheedy FJ, Palsson-McDermott E, Hennessy EJ et al (2010) Negative regulation of TLR4 via targeting of the proinflammatory tumor suppressor PDCD4 by the microRNA miR-21. Nat Immunol 11:141–147

Smalheiser NR (2007) Exosomal transfer of proteins and RNAs at synapses in the nervous system. Biol Direct 2:35

Spano F, Putignani L, Naitza S et al (1998) Molecular cloning and expression analysis of a *Cryptosporidium parvum* gene encoding a new member of the thrombospondin family. Mol Biochem Parasitol 92:147–162

Stoecklin G, Anderson P (2006) Posttranscriptional mechanisms regulating the inflammatory response. Adv Immunol 89:1–37

Subramaniam D, Ramalingam S, May R et al (2008) Gastrin-mediated interleukin-8 and cyclooxygenase-2 gene expression: differential transcriptional and posttranscriptional mechanisms. Gastroenterology 134:1070–1082

Sudhof TC, Rothman JE (2009) Membrane fusion: grappling with SNARE and SM proteins. Science 323:474–477

Suzuki K, Verma IM (2008) Phosphorylation of SNAP-23 by IkappaB kinase 2 regulates mast cell degranulation. Cell 134:485–495

Takeuchi D, Jones VC, Kobayashi M et al (2008) Cooperative role of macrophages and neutrophils in host antiprotozoan resistance in mice acutely infected with *Cryptosporidium parvum*. Infect Immun 76:3657–3663

Tang YJ, Luo XB, Cui HJ et al (2009) MicroRNA-146a contributes to abnormal activation of the type I interferon pathway in human lupus by targeting the key signaling proteins. Arthritis Rheum 60:1065–1075

Tarver AP, Clark DP, Diamond G et al (1998) Enteric beta-defensin: molecular cloning and characterization of a gene with inducible intestinal epithelial cell expression associated with *Cryptosporidium parvum* infection. Infect Immun 66:1045–1056

Taylor BC, Zaph C, Troy AE et al (2009) TSLP regulates intestinal immunity and inflammation in mouse models of helminth infection and colitis. J Exp Med 206:655–667

Tessema TS, Schwamb B, Lochner M et al (2009) Dynamics of gut mucosal and systemic Th1/Th2 cytokine responses in interferon-gamma and interleukin-12p40 knock out mice during primary and challenge *Cryptosporidium parvum* infection. Immunobiology 214:454–466

Thery C (2011) Exosomes: secreted vesicles and intercellular communications. F1000 Biol Rep 3:15. doi:10.3410/B3-15

Thery C, Ostrowski M, Segura E (2009) Membrane vesicles as conveyors of immune responses. Nat Rev Immunol 9:581–593

Tomley FM, Soldati DS (2001) Mix and match modules: structure and function of microneme proteins in apicomplexan parasites. Trends Parasitol 17:81–88

Troeger H, Epple HJ, Schneider T et al (2007) Effect of chronic *Giardia lamblia* infection on epithelial transport and barrier function in human duodenum. Gut 56:328–335

Turksen K, Troy TC (2004) Barriers built on claudins. J Cell Sci 117:2435–2447

Ukabam SO, Clamp JR, Cooper BT (1983) Abnormal small intestinal permeability to sugars in patients with Crohn's disease of the terminal ileum and colon. Digestion 27:70–74

Ungar BLP, Soave R, Fayer R et al (1986) Enzyme-immunoassay detection of immunoglobulin-M and immunoglobulin-G antibodies to Cryptosporidium in immunocompetent and immunocompromised persons. J Infect Dis 153:570–578

Ungar BLP, Kao TC, Burris JA et al (1991) Cryptosporidium infection in an adult-mouse model – independent roles for Ifn-gamma and Cd4+ lymphocytes-T in protective immunity. J Immunol 147:1014–1022

Valadi H, Ekstrom K, Bossios A et al (2007) Exosome-mediated transfer of mRNAs and microRNAs is a novel mechanism of genetic exchange between cells. Nat Cell Biol 9:654–659

van Niel G, Porto-Carreiro I, Simoes S et al (2006) Exosomes: a common pathway for a specialized function. J Biochem 140:13–21

Velmurugan K, Chen B, Miller JL et al (2007) *Mycobacterium tuberculosis* nuoG is a virulence gene that inhibits apoptosis of infected host cells. PLoS Pathog 3:e110. doi:10.1371/journal.ppat.0030110

Wang HC, Dann SM, Okhuysen PC et al (2007) High levels of CXCL10 are produced by intestinal epithelial cells in AIDS patients with active cryptosporidiosis but not after reconstitution of immunity. Infect Immun 75:481–487

Weaver CT, Harrington LE, Mangan PR et al (2006) Th17: an effector CD4 T cell lineage with regulatory T cell ties. Immunity 24:677–688

Widmer G, Corey EA, Stein B et al (2000) Host cell apoptosis impairs *Cryptosporidium parvum* development in vitro. J Parasitol 86:922–928

Williams RO (1987) Measurement of class specific antibody against Cryptosporidium in serum and feces from experimentally infected calves. Res Vet Sci 43:264–265

Winter J, Jung S, Keller S et al (2009) Many roads to maturity: microRNA biogenesis pathways and their regulation. Nat Cell Biol 11:228–234

Winzen R, Thakur BK, Dittrich-Breiholz O et al (2007) Functional analysis of KSRP interaction with the AU-rich element of interleukin-8 and identification of inflammatory mRNA targets. Mol Cell Biol 27:8388–8400

Witwer KW, Sisk JM, Gama L et al (2010) MicroRNA regulation of IFN-beta protein expression: rapid and sensitive modulation of the innate immune response. J Immunol 184:2369–2376

Wong P, Pamer EG (2003) CD8 T cell responses to infectious pathogens. Annu Rev Immunol 21:29–70

Worthley DL, Bardy PG, Mullighan CG (2005) Mannose-binding lectin: biology and clinical implications. Intern Med J 35:548–555

Yu LC, Flynn AN, Turner JR et al (2005) SGLT-1-mediated glucose uptake protects intestinal epithelial cells against LPS-induced apoptosis and barrier defects: a novel cellular rescue mechanism? FASEB J 19:1822–1835

Yu LC, Turner JR, Buret AG (2006a) LPS/CD14 activation triggers SGLT-1-mediated glucose uptake and cell rescue in intestinal epithelial cells via early apoptotic signals upstream of caspase-3. Exp Cell Res 312:3276–3286

Yu X, Harris SL, Levine AJ (2006b) The regulation of exosome secretion: a novel function of the p53 protein. Cancer Res 66:4795–4801

Yu LC, Huang CY, Kuo WT et al (2008) SGLT-1-mediated glucose uptake protects human intestinal epithelial cells against *Giardia duodenalis*-induced apoptosis. Int J Parasitol 38:923–934

Zaalouk TK, Bajaj-Elliott M, George JT et al (2004) Differential regulation of beta-defensin gene expression during *Cryptosporidium parvum* infection. Infect Immun 72:2772–2779

Zhang Y, Lee B, Thompson M et al (2000) Lactulose-mannitol intestinal permeability test in children with diarrhea caused by rotavirus and cryptosporidium. Diarrhea Working Group, Peru. J Pediatr Gastroenterol Nutr 31:16–21

Zhou R, Hu G, Liu J et al (2009) NF-kappaB p65-dependent transactivation of miRNA genes following *Cryptosporidium parvum* infection stimulates epithelial cell immune responses. PLoS Pathog 5:e1000681. doi:10.1371/journal.ppat.1000681

Zhou R, Hu G, Gong AY et al (2010) Binding of NF-kappaB p65 subunit to the promoter elements is involved in LPS-induced transactivation of miRNA genes in human biliary epithelial cells. Nucleic Acids Res 38:3222–3232

Zhou R, Gong AY, Eischeid AN et al (2012) miR-27b targets KSRP to coordinate TLR4-mediated epithelial defense against *Cryptosporidium parvum* infection. PLoS Pathog 8:e1002702. doi:10.1371/journal.ppat.1002702

Zhou R, Gong AY, Chen D et al (2013) Histone deacetylases and NF-kB signaling coordinate expression of CX3CL1 in epithelial cells in response to microbial challenge by suppressing miR-424 and miR-503. PLoS One 8:e65153. doi:10.1371/journal.pone.0065153

Chapter 11
Treatment of Cryptosporidiosis

Jan R. Mead and Michael J. Arrowood

Abstract *Cryptosporidium* species are protozoan parasites that infect the epithelial cells of the gastrointestinal tract. In humans, infections occur primarily in the small intestine resulting in diarrheal illness. *Cryptosporidium* has a worldwide distribution and is considered an emerging zoonosis. Despite intensive efforts to develop workable experimental models, and the evaluation of nearly 1,000 chemotherapeutic agents, adequate therapies to clear the host of these parasites are still lacking. The reasons for the lack of drug efficacy are probably manifold and may include the unusual location of the parasite in the host cell, distinct structural and biochemical composition of drug targets, or its ability to either block import or rapidly efflux drug molecules. Understanding the basic mechanisms by which drugs are transported to the parasite and identifying unique targets are important steps in developing effective therapeutic agents.

11.1 Introduction

Therapeutic agents and treatment strategies for cryptosporidiosis have been pursued for over 35 years since *Cryptosporidium* was first identified in humans (Nime et al. 1976). Even though promising targets and lead compounds have been identified, robust therapies that clear the host of these parasites are still lacking. Particular groups would benefit from an effective therapy against cryptosporidiosis:

J.R. Mead (✉)
Atlanta Veterans Medical Center and Department of Pediatrics, Emory University, 1670 Clairmont Road, Decatur, GA 30033, USA
e-mail: jmead@emory.edu

M.J. Arrowood (✉)
Centers for Disease Control and Prevention, National Center for Emerging and Zoonotic Infectious Diseases, Division of Foodborne, Waterborne and Environmental Diseases, 1600 Clifton Road, Mailstop D66Atlanta, GA 30329, USA
e-mail: mja0@cdc.gov

S.M. Cacciò and G. Widmer (eds.), *Cryptosporidium: parasite and disease*,
DOI 10.1007/978-3-7091-1562-6_11, © Springer-Verlag Wien 2014

children, the elderly, and immunocompromised individuals including organ transplant recipients, patients undergoing cancer chemotherapy, patients with congenital or disease-induced immunodeficiencies, and especially HIV-infected individuals. Even if not fully curative, many patients would benefit if drug therapy only reduced the parasite burden. Chemotherapy would also be useful among immunocompetent populations in outbreak situations to curb the spread of disease. While long recognized as one of the most ubiquitous and frequent cause of protozoal diarrhea in humans (Huston and Petri 2001), the recent Global Enteric Multicenter Study (GEMS) of children under 5 years old in developing countries found *Cryptosporidium* to be among the top 4 causes of moderate-to-severe diarrhea and that such diarrhea is a "high risk factor for linear growth faltering and death" (Kotloff et al. 2013).

Among the more important groups in need of therapy are individuals infected with human immunodeficiency virus (HIV). While cryptosporidiosis can be serious in immunocompetent people, it can be devastating to those with AIDS. In these individuals, symptoms may include chronic or protracted diarrhea that can become life threatening. Infections among HIV-infected individuals may also become extra-intestinal, spreading to other sites including the gall bladder, biliary tract, pancreas and pulmonary system (Lopez-Velez et al. 1995). Drug development should take into consideration the location of the parasite within the host (intestinal vs. extraintestinal sites) and the unique relationship between the parasite and host including location in the host cell and development of asexual and sexual life cycle stages.

The introduction and widespread use of highly-active anti-retroviral therapy (HAART) in AIDS patients has resulted in a decrease in opportunistic infections, however many adults and children living with HIV/AIDS in sub-Saharan Africa are currently not being treated with ART. Prevalence of *Cryptosporidium* among HIV-positive children with diarrhea has been reported to range between 13 % and 74 % in sub-Saharan Africa (Mor and Tzipori 2008). An early study of the impact of HAART on AIDS-defining illnesses in HIV-infected patients noted a 60 % decrease in the incidence of cryptosporidiosis (Ives et al. 2001). It is not known, however, how long these effective treatments will persist since multi-drug resistance and severe side effects associated with protease inhibitors may result in rebounding viral loads and, ultimately, increases in opportunistic infections. Despite intensive efforts, workable experimental models and testing of hundreds of chemotherapeutic agents, adequate therapies to clear the host of these parasites are lacking. Unique characteristics of *Cryptosporidium* species and the cryptosporidial life cycle in the host may account for the failure of most drug treatments tested to date. These include unique biochemical pathways (drug targets), lack of potential drug targets, unique or selective transporters functioning in the parasitophorous vacuole, or other features unique to the location of the parasite in the host cells (Chap. 2, Fig. 2.1) that facilitates sequestration of the parasite from drugs or other therapies.

11.2 Clinically-Evaluated Therapeutics

Controlled treatment trials have provided useful information about the activity of potential anti-cryptosporidial agents, but have not resulted in the identification of a robustly effective therapeutic agent. While many compounds have demonstrated anti-cryptosporidial activity using in vitro assays, fewer therapeutic agents have demonstrated significant potency using animal models. Further, many case reports describe attempts to treat cryptosporidiosis, but provide limited statistical value in aggregate analyses. The therapies most widely tested in animal models, case reports, and clinical trials are summarized below.

11.2.1 Macrolide Antibiotics

Some of the first drugs to be tested against *Cryptosporidium* species were macrolide antibiotics. Spiramycin, evaluated in the mid to late 1980s, produced encouraging results in pilot studies but was found to have little efficacy in larger, controlled trials and exhibited potentially unacceptable toxicity (Portnoy et al. 1984; Saez-Llorens et al. 1989; Wittenberg et al. 1989). Indeed, escalating high dose i.v. spiramycin treatment of cryptosporidiosis in two AIDS patients led to worsening of intestinal symptoms without significantly impacting cryptosporidial infection. Upon discontinuation of treatment one patient continued to deteriorate, while the second showed worsening diarrhea although gross and histological improvement was observed in upper and lower intestinal biopsies (Weikel et al. 1991).

A newer macrolide, azithromycin, was evaluated with mixed results. Azithromycin treatment resulted in symptomatic relief in immunosuppressed children and in patients with chronic cryptosporidiosis (Russell et al. 1998; Vargas et al. 1993). When given at a dose of 500 mg daily to patients with AIDS it was reported to be both effective (Kadappu et al. 2002) or ineffective (Blanshard et al. 1997). Extraintestinal infections (respiratory) in immunocompromised (HIV) patients responded positively to azithromycin alone or in combination with paromomycin (Palmieri et al. 2005; Tali et al. 2011). Apparent parasitologic cure was achieved in a liver transplant patient with cryptosporidial cholangitis after combination treatment with azithromycin and paromomycin, despite ongoing immunosuppression with tacrolimus (Denkinger et al. 2008). Combination treatment using azithromycin and paromomycin of an ileum transplant patient infected with *C. hominis* resulted in transient disappearance of cryptosporidia in biopsy samples and a cessation of clinical signs. However, immunosuppressive therapy was maintained and the patient became infected with *C. parvum* within 3 months and, despite additional treatment regimens including pyrimethamine-sulphadiazine and paromomycin, intermittently shed oocysts over the next 3 years (Pozio et al. 2004). Repeated treatment with azithromycin and paromomycin failed to eradicate cryptosporidiosis in a patient with X-linked hyper-IgM syndrome (an immunodeficiency disease resulting from a mutation in the CD40L gene) (Rahman et al. 2012).

Clarithromycin demonstrated modest efficacy as a prophylactic drug. The prevalence of cryptosporidiosis in HIV-infected patients was reduced when patients received clarithromycin daily, suggesting a protective effect (Holmberg et al. 1998; Jordan 1996). In a contrasting study, rifabutin, but not clarithromycin, decreased the risk of developing cryptosporidiosis in HIV-infected patients (Fichtenbaum et al. 2000). Therapeutic efficacy was reported with clarithromycin (Fujikawa et al. 2002) and with roxithromycin (Uip et al. 1998; Sprinz et al. 1998). Symptomatic improvement and resolution of oocyst shedding in an AIDS patient followed a 3 week course of clarithromycin (800 mg/day), although $CD4^+$ counts concomitantly rose from 152 to 180/μL perhaps indicating immune responsiveness played an important role (Fujikawa et al. 2002). In an uncontrolled study, patients received oral roxithromycin (300 mg b.i.d.) for 4 weeks. Out of the 22 patients that completed the study, 15 patients (68 %) were considered to be cured, six patients (27 %) improved, and one patient failed treatment (5 %) (Uip et al. 1998). In another uncontrolled study of 24 AIDS patients with cryptosporidiosis, treatment with roxithromycin (300 mg b.i.d.) for 4 weeks produced symptomatic improvement of diarrhea in 79 % of cases, with 50 % of patients achieving complete response (Sprinz et al. 1998).

11.2.2 Aminoglycoside Antibiotics

Paromomycin has been one of the most widely used agents to treat cryptosporidial infections in AIDS patients. It is a poorly absorbed aminoglycoside that is related to neomycin and kanamycin and achieves high concentrations in the gut, in part due to poor bio-availability. Paromomycin has shown efficacy in animal models (Healey et al. 1995; Tzipori et al. 1995) and as a prophylactic treatment in neonatal calves (Fayer and Ellis 1993b), lambs (Viu et al. 2000) and goats (Johnson et al. 2000; Mancassola et al. 1995).

Patients have reported positive responses to oral paromomycin treatment with doses ranging from 1,500 to 2,000 mg/day (Flanigan and Soave 1993). The drug was considered partially effective in that decreases in symptoms (frequency of stools) were noted but the parasite was not eradicated. From 1990 to 1996, 11 studies reported an overall response rate of 67 % (Hewitt et al. 2000). In a placebo-controlled, double blind study, treatment was found to be partially effective (White et al. 1994). Ten patients with AIDS and cryptosporidiosis were randomized to paromomycin (500 mg t.i.d. or q.i.d.) or placebo in a double-blind trial. After 14 days, patients were switched to the other treatment for 14 additional days. During the paromomycin treatment phase, oocyst excretion decreased, suggesting clinical improvement in the treated group. However, the parasite load remained high in some patients with disseminated biliary tract infections. In another study in which AIDS patients were treated with paromomycin (500 mg q.i.d.), 19 out of

25 people had improvement in clinical symptoms but in only six cases was the parasite eliminated (Stefani et al. 1996). An NIAID multi-site study was initiated in 1995. Thirty-five adults with $CD4^+$ cell counts of $\leq 150/\mu L$ were enrolled through the AIDS Clinical Trials Group. Initially, 17 patients received paromomycin (500 mg q.i.d.) and 18 received matching placebo for 21 days. Then all patients received paromomycin (500 mg q.i.d.) for an additional 21 days. The results of this study were disappointing, showing no significant difference between the treated and placebo groups (Hewitt et al. 2000). Three paromomycin recipients (17.6 %) versus two placebo recipients (14.3 %) responded completely. Rates of combined partial and complete responses in the paromomycin arm (8 out of 17, 47.1 %) and the placebo arm (5 out of 14, 35.7 %) of the study were similar ($P = 0.72$). Consequently, many patients initially respond to paromomycin treatment with a decrease in diarrhea but then relapse. A statistical meta-analysis of data from two studies (Hewitt et al. 2000; White et al. 1994) found no evidence that paromomycin was effective in reducing the frequency of diarrhea nor in achieving parasitological cure (Abubakar et al. 2007). Nevertheless, paromomycin as a therapeutic agent is routinely used as a positive control in animal studies and in vitro (Graczyk et al. 2011; Benitez et al. 2007; Umejiego et al. 2008; Klein et al. 2008).

11.2.3 Benzeneacetonitrile Antibiotics

The benzeneacetonitrile derivatives diclazuril and letrazuril have been evaluated in clinical trials. Diclazuril was evaluated in a double-blind, placebo-controlled study in which patients were given 50–800 mg/kg daily (Soave et al. 1990). No significant differences were observed between the diclazuril and placebo-treated groups. Letrazuril, a derivative of diclazuril, showed promise in a pilot study. At a dosage of 150–200 mg daily, letrazuril was associated with an improvement in symptoms in 40 % of patients treated and cessation of excretion of cryptosporidial oocysts in the stool in 70 % (Blanshard et al. 1997). However, biopsies from these patients remained positive. Letrazuril also demonstrated modest activity in clinical trials (Harris et al. 1994). Of 14 patients evaluated, five showed a major response (symptomatic improvement and eradication of cryptosporidial oocysts from the stool), two had a minor response (symptomatic improvement with persistence of oocysts in stool), and seven had no response to therapy with letrazuril. A 1997 review concluded diclazuril and letrazuril had modest anticryptosporidial activities and were not very promising since patients experienced little clinical improvement (Blagburn and Soave 1997).

11.2.4 Immunotherapeutic Treatment (Colostrum)

There has been a great deal of interest in the potential of immunotherapy for cryptosporidiosis over the years. The earliest reports of immunotherapy involved individual cases where treatment with immune or hyperimmune bovine colostrum was associated with both success as well as failure (Jenkins 2004; Kelly 2003; Crabb 1998). These included a child with hypogammaglobulinemia (Tzipori et al. 1986), and patients with either hypogammaglobulinemia (Saxon and Weinstein 1987) or AIDS (Nord et al. 1990; Saxon and Weinstein 1987; Ungar et al. 1990). Reduction in parasite load was observed in animals given immune colostrum (Fayer et al. 1989; Perryman and Bjorneby 1991). In an observational study in Nigeria, HIV-associated diarrhea was alleviated by a 4 week treatment regimen with a commercial bovine colostrum product (Floren et al. 2006). Among a 30 patient group with diarrhea, 20 had cryptosporidiosis. The study did not differentiate which patients, on average, benefited most from colostrum treatment, nor whether parasite shedding ceased, but average daily bowel movements decreased from 7.0 ± 2.7 (s.d.) to 1.6 ± 0.9 (s.d.) after 4 weeks of treatment (week 5 of the study). Interestingly, $CD4^+$ cell counts increased from an average of 153 ± 62 (s.d.) cells/μL to 310 ± 106 (s.d.) cells/μL at week 7 of the study which the authors attribute, in part, to the colostrum product since ART was not initiated until the end of week 7.

Several open label studies using hyperimmune bovine colostrum (HBC) to treat AIDS patients with cryptosporidiosis have been reported. Patients treated with 48 enteric-coated capsules (40 total grams) per day over a 21-day period showed decreases in mean stool weight and stool frequency. Unfortunately, parasite load was not evaluated in this study, so HBC's direct effect upon parasite growth could not be determined (Greenberg and Cello 1996).

HBC was also used in a placebo-controlled, double-blind, one-way cross-over study in which AIDS patients were treated for 1–2 weeks with 20 g/day followed by increasing doses to 80 g/day in some patients. No statistical differences were observed in clinical symptoms but a reduction in oocyst shedding was reported (Fries et al. 1994). The prophylactic effect of hyperimmune bovine anti-*Cryptosporidium* colostrum immunoglobulin (BACI) was evaluated in healthy adults challenged with *C. parvum* (Okhuysen et al. 1998). Sixteen volunteers were randomized to receive either a (1) BACI prior to *C. parvum* challenge and a nonfat milk placebo 30 min later, (2) BACI prior to and 30 min after challenge (reinforced BACI group), or (3) nonfat milk placebo prior to and 30 min after challenge. Subjects received BACI (10 g) or nonfat milk placebo three times a day for a total of 5 days and were monitored for clinical symptoms and oocyst excretion for 30 days. Subjects receiving BACI or nonfat milk placebo had a 100-fold reduction in oocyst excretion as compared with excretion in the baseline group, however no difference was observed between the BACI and nonfat milk placebo treatment groups. In terms of clinical symptoms, no significant differences were observed in the duration of disease, time to onset of diarrhea, and severity of the disease among groups.

11.2.5 Thiazolide Antibiotics

Nitazoxanide (NTZ), a thiazolide drug with reported broad antiparasitic activities, is currently the only FDA-approved drug for use against cryptosporidiosis in immunocompetent patients. Early studies demonstrated moderate efficacy in cell culture systems and in neonatal mice and gnotobiotic piglets (Blagburn et al. 1998; Gargala et al. 2000; Theodos et al. 1998). Duration of oocyst shedding and severity of diarrhea in experimentally-infected newborn calves was reduced following treatment with NTZ (1,500 mg b.i.d.) (Ollivett et al. 2009). Calves received colostrum on day 1, were infected between days 2 and 3 and NTZ treatment began between days 5 and 7 (to allow infections to become established) and continued for 5 days.

Although early clinical trials in the U.S. were not encouraging, one uncontrolled open study in Mali suggested that the drug was efficacious against the parasite in stage 4 AIDS patients with "light" cryptosporidial infections, but failed in "heavily infected" stage 4 AIDS patients (Doumbo et al. 1997). In a double-blind placebo-controlled study, patients were randomly treated orally with either 500 or 1,000 mg (b.i.d.) NTZ or placebo for 14 days (Rossignol et al. 1998). Patients on NTZ then crossed over to placebo while the placebo patients crossed over to NTZ therapy at either the high or low dose depending on their randomization. Patients were considered 'cured' if oocysts were not found in post-treatment fecal examinations. Both doses of NTZ produced parasitological cure rates superior to the placebo responses (12/19 [63 %, $P = 0.016$] for patients receiving 1 g/d and 10/15 [67 %, $P = 0.013$] for those receiving 2 g/d). It should be noted that 25 % (5/20) of placebo-treated patients were parasitologically "cured" and 50 % (10/20) had complete resolution of clinical diarrhea before crossing over to nitazoxanide therapy.

An HIV positive patient with persistent diarrhea was sequentially treated with paromomycin, azithromycin, paromomycin with octreotide, and finally nitazoxanide, all of which failed to adequately control the cryptosporidial infection (Giacometti et al. 1999). Interestingly, seven samples of the patient's cryptosporidial isolate(s) were collected separately over a nearly 1.5 year period and tested for susceptibility to paromomycin, azithromycin and nitazoxanide in an in vitro cultivation system. Although some activity was noted for each drug, none completely inhibited parasite growth. Maximum inhibition for azithromycin was 26.5 % at 8 mg/L (10.7 μM), for paromomycin 63.4 % at 1 mg/L (1.6 μM), and for nitazoxanide 67.2 % at 10 mg/L (33.0 μM). It was also noted that nitazoxanide was toxic for the cell cultures at the maximum concentration tested (100 mg/L (330 μM)). In a subsequent study by the same researchers, nitazoxanide alone at 8 mg/L (26 μM) yielded an inhibition of parasite growth in vitro of approximately

55 %, but inhibition increased to approximately 80–84 % when combined with rifabutin at 8 mg/L (9.5 μM) or azithromycin at 8 mg/L (10.7 μM), respectively (Giacometti et al. 2000).

In a prospective, randomized, double-blind, placebo-controlled study, 50 adults and 49 children from the Nile delta of Egypt, were treated with nitazoxanide or placebo for diarrhea caused by *C. parvum*. Nitazoxanide was administered in 500 mg doses b.i.d. for 3 days in adults and adolescents, in 200 mg doses b.i.d. for 3 days in children aged 4–11 years, and in 100-mg doses b.i.d. for 3 days in children aged 1–3 years (Rossignol et al. 2001). Seven days after initiation of therapy, diarrhea resolved in 39 (80 %) of the 49 patients in the nitazoxanide treatment group, compared with 20 (41 %) of 49 in the placebo group ($P < 0.0001$). Diarrhea resolved in most patients receiving nitazoxanide within 3 or 4 days of treatment initiation. Nitazoxanide treatment reduced the duration of diarrhea ($P < 0.0001$) and oocyst shedding ($P < 0.0001$). When children were excluded, 72 % (18/25) of nitazoxanide-treated adults and adolescents resolved diarrhea while 44 % (11/25) of placebo-treated adults and adolescents resolved diarrhea; differences were not statistically significant.

Additionally, a prospective, randomized, double-blind, placebo-controlled study was performed in children with diarrhea. The efficacy and safety of nitazoxanide compared with placebo was assessed in the treatment of diarrhea caused by *C. parvum* in HIV-infected and HIV-uninfected Zambian children, most of whom were younger than 3 years and malnourished. A 3-day course of nitazoxanide significantly improved the resolution of diarrhea, parasitological eradication, and mortality in HIV-seronegative children. However, HIV-seropositive children did not benefit from nitazoxanide treatment (Amadi et al. 2002).

In an in vitro study, NTZ and tizoxanide were reportedly more active against sporozoites while tizoxanide glucuronide appeared to be more active against intracellular stages (Gargala et al. 2000). To improve the efficacy of this class of drugs, second generation compounds were generated. Among a group of 39 thiazolide and thiadiazolide compounds evaluated for efficacy compared to NTZ, several were identified as more potent than NTZ (Gargala et al. 2010). One of these compounds, RM-4865, demonstrated efficacy in a gerbil model and was equally as potent as NTZ in reducing oocysts in this model while RM-5038 was significantly more active than NTZ (Gargala et al. 2013).

The mode of action for NTZ against *Cryptosporidium* has not been determined and the following in vitro data are worth noting when interpreting the anticryptosporidial efficacy. These observations should be kept in mind when testing any agent for anticryptosporidial activity. NTZ has been reported to cause dose-dependent toxicity in cell cultures (Theodos et al. 1998; Giacometti et al. 1999; Gargala et al. 2000). A clear association between mammalian cell cycle and thiazolide toxicity (via induction of apoptosis) has also been reported (Müller et al. 2008). In the latter study, growth inhibition was pronounced for proliferating Caco2 cells (transformed) [IC_{50} 6.8 ± 1.3 μM] and human foreskin fibroblasts (primary cell culture) [IC_{50} 18.9 ± 1.2 μM], but much less so for confluent, non-proliferating cells. The proposed mode of action (toxicity) against the cell lines is via inhibition of glutathione-*S*-transferase P1. Since the mode of action for NTZ against *Cryptosporidium* is unknown, the toxicity/apoptosis exhibited by NTZ against rapidly proliferating cells (e.g. intestinal epithelium) may partially explain the capacity to reduce parasite numbers independent of a direct mode of action against the parasite itself.

11.2.6 Combination Therapy

There have been a number of studies over the years that used combination chemotherapy (some mentioned above). These include in vitro studies (Giacometti et al. 1996, 1999, 2001; You et al. 1998), in vivo studies (Fayer and Ellis 1993a; Kimata et al. 1991) and treatment given to a patient with disseminated cryptosporidiosis (Giacometti et al. 1999).

In an open-label, combination study by Smith and colleagues (Smith et al. 1998), patients with AIDS, chronic cryptosporidiosis, and low $CD4^+$ cell counts (<100 $CD4^+$ cells/μL) were given paromomycin (1.0 g b.i.d.) plus azithromycin (600 mg once daily) for 4 weeks, followed by paromomycin alone for 8 weeks. In 11 patients, median stool frequency decreased from 6.5/day (baseline) to 4.9/day (week 4) and 3.0/day (week 12). Median reductions in 24 h oocyst excretion were 84 %, 95 %, and >99 % at 2, 4, and 12 weeks, respectively. Of five survivors at 12–30 months follow-up, three remained asymptomatic off medications, and two had chronic, mild diarrhea. Both partial clinical and parasitological responses were observed, as treatment of cryptosporidiosis with azithromycin and paromomycin was associated with significant reduction in oocyst excretion and some clinical improvement. Unfortunately, no placebo group was included in the study.

The use of potent highly active anti-retroviral therapy (HAART) in patients with advanced HIV infection can improve symptoms or lead to the clearance of *C. parvum* oocysts from stools. Patients treated with double anti-retroviral therapy or protease inhibitors have demonstrated excellent responses and sustained therapeutic effects after follow-up (Ives et al. 2001; Miao et al. 2000). In other studies it was found that combination antiretroviral therapy that included a protease inhibitor restored immunity to *C. parvum* in HIV-1 infected individuals, resulting in complete clinical, microbiological, and histological responses (Carr et al. 1998). HAART-associated resolution of diarrhea is linked to increased $CD4^+$ cell counts rather than to modulation of the viral load. Four patients had recurrent diarrhea at 7–13 months (one with positive stool microscopy), associated with declining $CD4^+$ counts.

Although HAART is not thought to have a direct effect on *C. parvum*, there is evidence that the protease inhibitor, indinavir (IND) can reduce *C. parvum* infection in both in vitro and in vivo models (Bobin et al. 1998; Hommer et al. 2003; Mele et al. 2003; Miao et al. 2000).

A retrospective cohort study spanning 1999 through 2009 of 1,243 HIV^+ patients in Morocco receiving HAART identified 91 deaths (Sodqi et al. 2012). Average

$CD4^+$ counts were 95 cells/μL at enrollment and initiation of HAART. Cryptosporidiosis was the second most common cause of death (19 %) among these patients (following tuberculosis at 35 %). Of those patients that died, HAART duration averaged 9 months and 63 % died within the first year of treatment. At the time of death, ~53 % had $CD4^+$ counts above 200 cells/μL. Although some patients succumbed to cryptosporidiosis, it appears that prompt initiation of HAART and effective immune reconstitution had the most efficacious impact on surviving opportunistic infections, including cryptosporidiosis.

Looking historically at the range of drugs tested clinically against *Cryptosporidium*, it appears several drugs may have had promising initial assessments, not because the drugs were efficacious against the parasite itself, but rather because of their impact on the gut microflora and/or the epithelial (host) cells of the gastrointestinal tract.

11.3 Innate Drug Resistance

Why are so many therapies ineffective against this parasite? It seems that *C. parvum* has an innate resistance to drug therapy. Several factors may contribute to this lack of efficacy. These include: (1) lack of specific targets or differences in targets either at the molecular or structural levels compared to the host, (2) differences in biochemical pathways, (3) the parasite's unique location in the host cell which may affect drug concentration (transported from the lumen or host cell cytoplasm across to the parasite), and (4) existence of transport proteins or efflux pumps that move drugs out of the parasite to the lumen or back into the host cell.

11.3.1 Genomic, Proteomic, Metabolic Characteristics

The establishment of large scale sporozoite-expressed sequence tag (EST) and genome sequencing projects has aided the identification of *C. parvum* genes and in understanding phylogenetic similarities and diversities within the genome. It appears that *C. parvum* is more divergent and less related to other coccidia. Comparisons of small-subunit ribosomal RNA gene sequences first demonstrated that the gregarine/*Cryptosporidium* clade is separated from the other major apicomplexan lineages (Carreno et al. 1999; Leander et al. 2003). Another study compared six different proteins along with the SSU rRNA and concluded that *Cryptosporidium* is an early emerging branch of the Apicomplexa (Zhu et al. 2000a). Full genome sequencing of *C. parvum* demonstrated significant differences between cryptosporidia and coccidia (Abrahamsen et al. 2004), and though it reinforced the relatedness to gregarines, there are significant differences between cryptosporidia and gregarines. In particular, the streamlined metabolism of cryptosporidia is represented by the absence of the shikimate pathway, synthesis of select amino acids, and *de novo* pyrimidine synthesis, all of which are commonly found in apicomplexans (Templeton et al. 2010). Nevertheless, cryptosporidia contain a number of metabolic targets in common with coccidia (and gregarines), including Type I fatty acid and polyketide synthesis enzymes (see Chap. 8).

Other structural differences have been noted. Unlike most of the Apicomplexa, *C. parvum* appears to lack a plastid genome (Riordan et al. 1999; Zhu et al. 2000b; Abrahamsen et al. 2004). Primers based upon the highly conserved plastid small- or large-subunit rRNA (SSU/LSU rRNA) and the tufA-tRNAPhe genes of other members of the phylum Apicomplexa failed to amplify products from intracellular stages of *C. parvum* grown in human adenocarcinoma (HCT-8) cells (Zhu et al. 2000b). The plastid (apicoplast) in other apicomplexan parasites has been shown to code for ribosomal proteins and tRNAs (Wilson and Williamson 1997) as well as metabolic pathways involved in Type II fatty acid synthesis and isoprenoid metabolism (Templeton et al. 2010). The activities of clindamycin and other macrolide antibiotics may be achieved through inhibition of protein synthesis within the plastid (Beckers et al. 1995; Fichera and Roos 1997). Clindamycin, an effective anti-*Toxoplasma* drug, expressed some in vitro anticryptosporidial activity (Woods et al. 1996) but was not efficacious in vivo (Rehg 1991). In general, macrolides have not demonstrated consistent anticryptosporidial efficacy even after long-term administration. While this lack of efficacy may or may not be related to the "missing" plastid, drug development targeting a plastid genome or metabolic pathways associated with it, may not be useful.

Despite the "missing" plastid and, presumably, the associated isoprenoid biosynthetic pathway (Clastre et al. 2007; Templeton et al. 2010), a nonspecific polyprenyl pyrophosphate synthase (*Cp*NPPPS) was identified in *C. parvum* that is distinctly different from similar isoprenoid synthesis enzymes in other organisms (Artz et al. 2008). At least three prenyl synthase enzymes are coded for in the cryptosporidial genome. While downstream enzymes are present in *Cryptosporidium* that can utilize substrates in the isoprenoid pathway, no evidence exists that cryptosporidia have either the mevalonate (MVA, e.g. human) or methylerythritol phosphate (MEP, see *Plasmodium* and *Toxoplasma*) pathways to generate these medium-length isoprenoid substrates (Clastre et al. 2007). It is unknown whether *Cryptosporidium* relies on an uncharacterized enzyme pathway to produce these substrates or whether they are acquired by a salvage pathway from the host cells. Interestingly, *Cp*NPPPS was inhibited by sub-nanomolar concentrations of nitrogen-containing bisphosphonates and significant inhibition of in vitro parasite growth in MDCK cells was observed (Artz et al. 2008). Of the bisphosphonates tested, ibandronate showed the highest anticryptosporidial activity ($IC_{50} = 3.0\ \mu M$) in cell culture.

Evidence suggests that *C. parvum* has a putative mitochondrion and mitochondrion-associated enzymes markedly different in structure from those of its nearest relatives (Riordan et al. 1999) but lacks much of the electron transport chain (Templeton et al. 2010). This may, in part, explain why atovaquone (targets cytochrome C) is not effective against *Cryptosporidium* (Rohlman et al. 1993; Giacometti et al. 1996; Mather et al. 2007).

Some parasite targets (e.g. enzymes, structural proteins) are different from those of other related parasites. For example, the dihydrofolate reductase (DHFR) gene of *C. parvum* differs from the homologous *Plasmodium* gene. Sequencing of the DHFR gene has suggested that the enzyme may be intrinsically resistant to 2,4-diaminopyrimidine inhibitors (Vasquez et al. 1996). The *C. parvum* DHFR active site contains novel residues at several positions analogous to those at which point mutations have been shown to produce antifolate resistance in other DHFRs (Vasquez et al. 1996). This may, in part, explain why *C. parvum* is resistant to clinically used antibacterial and anti-protozoal antifolates. An analysis of the crystal structure of *C. hominis* DHFR led to the development of novel trimethoprim (TMP) analogs with significantly enhanced (368-fold) inhibition of DHFR (TMP $IC_{50} = 14\ \mu M$ versus 38 nM for analog # 37) in a cell free assay system (Pelphrey et al. 2007). Recent drug design efforts utilizing "virtual screening" of a non-active site pocket in the cryptosporidial DHFR linker region identified a noncompetitive inhibitor, flavin mononucleotide (FMN) (Martucci et al. 2009). As a first-generation lead compound, FMN exhibited an $IC_{50} = 55\ \mu M$ versus 14 μM for TMP.

11.3.2 Location and Physiology

A factor that may contribute to the ineffectiveness of many drugs is the unusual location of the parasite in the host cell. The parasite has a unique niche inside the cell that is considered "intracellular" but "extracytoplasmic." The organism resides in a parasitophorous vacuole (PV) surrounded, in part, by host cell membrane but is outside the cytoplasm. It has been postulated that the PV basal membranes, including the "feeder organelle", modulate the transport of certain drugs, so that drugs entering the cytoplasm of the host cell may not be transported across to the parasite. In cell culture studies, apical but not basolateral exposure of these drugs led to significant parasite inhibition. There is evidence to suggest that this is the case for geneticin and paromomycin, a clinically-relevant drug (Griffiths et al. 1998). Alternatively, the parasite may be able to modulate drug transport from the lumen through the host/parasite membranes into the PV, perhaps bypassing the host cytoplasm.

Another example where physiology may affect drug uptake was shown when the cryptosporidial inosine 5′-monophosphate dehydrogenase (*Cp*IMPDH) gene was introduced into *T. gondii* under the control of a *T. gondii* promoter (Gorla et al. 2012). Many drugs were effective in this *Toxoplasma* model and were due to inhibition of the *Cp*IMPDH (since the native *T. gondii* IMPDH gene was rendered inactive). However, these same drugs were generally not as active in a *C. parvum* in vitro cultivation assay suggesting reduced efficacy may be due to differences in the biology of the two parasites (Sharling et al. 2010). A phthalazinone-based inhibitor of *Cp*IMPDH with promising efficacy and selectivity was not effective in an in vivo mouse model of cryptosporidiosis (Johnson et al. 2013). Location, structure, and function of PV from these two parasites is clearly different: e.g. the PV membrane of *T. gondii* is in direct contact with the host cell cytoplasm whereas *C. parvum* remains beneath the apical membrane of the host cell and separated from the host cell cytoplasm by a feeder organelle (Sharling et al. 2010). Differences in drug efflux transporters between these parasites may also play a role in drug susceptibility.

11.3.3 ABC and Other Parasite Transporters

Transporters may also be involved in the rapid efflux of drugs as well as nutrient uptake. The existence of multi-drug resistant (MDR) transporters could facilitate drug resistance. P-glycoproteins and MDR transporters are members of the ATP-binding cassette (ABC) superfamily that are responsible for drug resistance by extruding drugs against a concentration gradient. In cancer chemotherapy, chemosensitizing agents are used to reduce efflux of chemotherapeutic agents from tumor cells by MDR pumps. To date, a set of 21 ABC genes has been identified in the fully sequenced *C. parvum* genome (Sauvage et al. 2009). Originally, three ABC transporters were identified in *C. parvum*. These include *Cp*ABCC3 (*Cp*ABC1) and *Cp*ABCC4 (*Cp*ABC2) that are similar to the multidrug resistance-associated (MRP) subfamily and *Cp*ABCB1 (CpABC3) that groups with the MDR subfamily (Zapata et al. 2002). The *Cp*ABCC3 gene shows similarity to human cystic fibrosis transmembrane receptor (CFTR) and to rat conjugate export pump protein (Strong and Nelson 2000) and is localized in sporozoites and at the boundary between host cells and the extracytoplasmic meront (Perkins et al. 1999; Zapata et al. 2002). *Cp*ABCC4 shows similarity to human CFTR and MRP-like protein from *Candida albicans* and has been identified in the apical end of sporozoites (Zapata et al. 2002). Gene expression of *Cp*ABCB1, an ABC transporter with similarity to MDR1, was detected in intracellular stages of *C. parvum* but its localization is unknown (Bonafonte et al. 2004; Zapata et al. 2002). These transporters are constitutively expressed in infected cell cultures, with transcript levels of *Cp*ABCC3 4 logs higher than either *Cp*ABCB1 or *Cp*ABCC4. In addition, MDR or MRP modulators (cyclosporine A, verapamil and probenicid) significantly affected transcript levels of all three of these ABC transporters in *C. parvum* (Bonafonte et al. 2004) and several sesquiterpenes have been shown to bind to the recombinant nucleotide-binding site of an MDR-like efflux pump when assayed on the recombinant nucleotide-binding domain of *Cp*ABCB1 (Lawton et al. 2007). While classic drug resistance has not been demonstrated with any of these ABC proteins, treatment by paromomycin, one of the most widely used agents to treat cryptosporidial infections in AIDS patients, significantly upregulates transcription of *Cp*ABCC14 (cgd7_4510) and of the *Cp*ABCB4 (cgd1_1350 or *Cp*ABC4 or *Cp*ATM1) (Benitez et al. 2007), a Pgp homolog half transporter. In order to characterize the catalytic site of this half-transporter, an extended region of the nucleotide-binding domain of *Cp*ABCB4 (H6-1350NBD) was expressed and purified as an N-terminal hexahistidine-tagged protein in *E. coli*. A dose-dependent quenching of the domain's intrinsic fluorescence was observed with the fluorescent analogue substrate TNP-ATP as well as the flavonoids quercetin and silibinin, previously shown to inhibit parasite development in a cell-based assay.

11.4 New Areas of Drug Research

A number of drug targets have been proposed over the years but have not been pursued either because the presumed drug target could not be found, toxicity of the drug was too great, or because the drug lacked efficacy in animal models or in clinical trials. Some of the following therapies remain promising because the target is selective or unique and efficacy has been demonstrated either in vitro or in vivo.

11.4.1 Flavonoids

Flavonoids, polyphenolic compounds found in plants, have demonstrated activity against several parasites and can augment the efficacy of other drugs by either increasing the uptake or decreasing the efflux of these drugs. The flavonoids apigenin and genistein inhibited *C. parvum* growth in vitro (Forney et al. 1999; Mead and McNair 2006). When the dinitroaniline drug trifluralin was tested in combination with the isoflavone genistein in an in vitro cultivation model the effective concentration that inhibited growth by 50 % (EC_{50}) dropped by 1 log (from 0.7 to 0.09 μM), suggesting that isoflavone compounds may be used alone or in combination with other moderately active drugs to increase efficacy (Mead and McNair 2006). Flavonoids such as genistein have been shown to inhibit epidermal growth factor receptor protein tyrosine kinases (EGFR PTK), which are located at the surface of target cells and may be involved in the activation of the intracellular tyrosine kinase (TK) domain, initiating downstream signaling pathways. Synthetic analogs retaining the 5,7-dihydroxyisoflavone core of genistein were found to have in vitro activity against *C. parvum* (Stachulski et al. 2006). Two of these agents with demonstrable activity in vitro decreased oocyst shedding and limited *C. parvum* intracellular development in the gut and in the biliary tract using an immunosuppressed-gerbil model of cryptosporidiosis.

11.4.2 Bisphosphonates

The isoprenoid metabolic pathway is another potential drug target in parasitic protozoa including those causing malaria, Chagas' disease, toxoplasmosis, and leishmaniasis (Ling et al. 2005; Martin et al. 2001; Moreno et al. 2001). Bisphosphonates (used in bone resorption therapy) are known to inhibit the biosynthesis of isoprenoids. In a mouse xenograft model of cryptosporidiosis (Moreno et al. 2001), the pyrophosphate analog drug risedronate was shown to inhibit *C. parvum* growth. Nitrogen-containing bisphosphonates (N-BPs) are capable of inhibiting *C. parvum* in infected MDCK cells at low micromolar concentrations (Artz et al. 2008). Crystal structures of the enzyme (nonspecific polyprenyl pyrophosphate synthase) with risedronate and zoledronate show how it could accommodate larger substrates and products.

In the process of screening a library of drug repurposing candidates (NIH Clinical Collections), researchers employing a high-throughput in vitro assay system identified itavastatin, an inhibitor of human 3-hydroxy-3-methyl-glutaryl-coenzyme A (HMG-CoA) reductase with potent in vitro activity against *C. parvum* ($IC_{50} = 0.62$ μM) (Bessoff et al. 2013). The investigators predict that the anticryptosporidial activity is mediated by the inhibition of host cell HMG-CoA, thus limiting the availability of isoprenoid precursors. *Cryptosporidium* species appear to lack the enzymes necessary for synthesis of isoprenoid precursors and depend on the host cell for these precursors. It remains to be seen if the parasite can scavenge sufficient precursors necessary for propagation under itavastatin drug pressure in an animal model or in clinical trials.

11.4.3 Polyamine Biosynthesis

It has also been determined that *C. parvum* differs fundamentally from other eukaryotes in polyamine metabolism. Mammals and most parasitic protozoa, including the coccidian parasites *E. tenella* and *P. falciparum*, convert arginine into ornithine. Ornithine is subsequently converted to putrescein via ornithine decarboxylase, and finally into spermidine and spermine (Bacchi and Narlett 1995). However, polyamine biosynthesis in *C. parvum* occurs via a pathway used by plants and certain bacteria, in which arginine is converted to agmatine by the action of arginine decarboxylase (Keithly et al. 1997). Neither arginine decarboxylase nor agmatine is found in other parasitic protozoa. Agmatine serves as the precursor for other polyamines.

C. parvum also has a reverse polyamine biosynthetic pathway not found in other protozoa that enables the interconversion of spermine, spermidine, and putrescein. Treatment with a polyamine analogue (SL-11047) prevented *C. parvum* infection in suckling TCR-alpha-deficient mice and cleared an existing infection in older mice. Treatment with putrescein, while capable of preventing infection, did not clear *C. parvum* from previously infected mice. Inhibitors of arginine decarboxylase (ADC), including difluoromethylarginine (DFMA), significantly reduced intracellular growth of *C. parvum* (Keithly et al. 1997), whereas inhibitors of ornithine had no effect upon ADC activity or upon growth of the parasite. Back-conversion of spermine to spermidine and putrescein via spermidine:spermine-N1-acetyltransferase (SSAT) was also detected. Although ADC activity was consistently detected in *C. parvum* parasites, a gene encoding an ADC homolog has not been identified in the completed genome sequence.

11.4.4 Fatty Acid Synthase

C. parvum possesses a unique Type I fatty acid synthase (*Cp*FAS1) and a putative polyketide synthase (*Cp*PKS1) encoded by 25-kb and 40-kb intronless open reading frames (ORFs), respectively (Zhu et al. 2002, 2000c) (see Chap. 8). Since *Cp*FAS1 and *Cp*PKS1 are structurally and functionally different from human Type I FAS, these enzymes may serve as novel drug targets. The fatty acid synthase gene (*Cp*FAS1) also differs from the organellar type II fatty acid enzymes identified in *Toxoplasma gondii* and *Plasmodium falciparum* (Zhu et al. 2000c). The FAS inhibitor cerulenin inhibited the growth of *C. parvum* in vitro by 96 % at 10 μg/mL, reinforcing the potential efficacy of FAS as a drug target.

11.4.5 Anti-tubulin Agents

Microtubules of various parasites differ from mammalian microtubules in a number of ways, including being relatively stabile and resistant to low temperatures. Anti-tubulin agents such as the benzimidazoles and the dinitroaniline herbicides have demonstrated efficacy against a number of parasitic agents (Roos 1997; Stokkermans et al. 1996; Traub-Cseko et al. 2001). Although benzimidazoles are widely used against helminth infections and some protozoans, *C. parvum* did not have the predicted amino acids for benzimidazole drug sensitivity in the beta-tubulin-coding gene (Edlind et al. 1994; Katiyar et al. 1994). In a subsequent study, benzimidazole treatment was not effective against *C. parvum* when evaluated in mice (Fayer and Fetterer 1995). Conversely, dinitroaniline herbicides were effective against *C. parvum* in vitro (Arrowood et al. 1996). The efficacy of this class of compounds has been determined in neonatal mice. At doses of 100 mg/kg body weight administered twice daily for 3 consecutive days, trifluralin had no statistically significant effect on the number of oocysts recovered from the excised gut of either rats or mice compared with controls, whereas at the same concentration, oryzalin caused 90 % and 79 % inhibition of oocysts recovered from mice and rats, respectively (Armson et al. 1999). Treatment with oryzalin doubled the villus/crypt ratio in the duodenum, jejunum and ileum following doses of 5 mg, 50 mg and 200 mg/kg respectively (Armson et al. 2002).

Dinitroanilines, as a rule, are poorly soluble which complicates their development as therapeutic agents. Synthetic derivatives with increased solubility were compared to parent compounds (Benbow et al. 1998; Mead et al. 2003). Acetylated glycoconjugate forms of dinitroaniline compounds were more active in vitro than non-acetylated forms, suggesting acetylation may facilitate drug transport across cell membranes (Mead et al. 2003). Although enhanced solubility did not necessarily correlate with better activity, several analogs were produced with similar anticryptosporidial activities but with toxicities lower than those of the lead compounds (Benbow et al. 1998). It may be possible that these compounds could be used at lower concentrations when combined with other efficacious drugs (Mead et al. 1999).

11.4.6 Dihydrofolate Reductase (DHFR) Inhibitors

Dihydrofolate reductase (DHFR), occurring as a bifunctional protein with thymidylate synthase (DHFR–TS) in apicomplexan protozoa, has been an established drug target for treatment of protozoal infections for many years. DHFR utilizes the cofactor NADPH to catalyze the reduction of dihydrofolate to tetrahydrofolate. In initial studies with *C. parvum*, standard dihydrofolate reductase (DHFR) inhibitors were not very effective, which was explained by nucleic acid sequences encoding resistant dihydrofolate reductase-thymidylate synthase (Vasquez et al. 1996). The sequence of *C. hominis* DHFR contains 3 amino acid mutations known to confer antifolate resistance in *P. falciparum* DHFR (*Pf*DHFR).

Several lipophilic DHFR inhibitors were reported to be active in a complementation assay but were nonselective (Brophy et al. 2000). Ninety-three lipophilic di- and tricyclic diaminopyrimidine derivatives were evaluated in a cell-free enzyme assay for the ability to inhibit recombinant DHFR cloned from human and bovine isolates of *C. parvum* (Nelson and Rosowsky 2001). The library of compounds was also tested for anticryptosporidial activity in an in vitro cultivation assay using Madin-Darby canine kidney (MDCK) cells cultured in folate-free culture medium supplemented with thymidine (10 μM) and hypoxanthine (100 μM). Cell culture assays identified 16 compounds with IC_{50}s <3 μM, of which five had IC_{50}s <0.3 μM. Anticryptosporidal activities of the latter were comparable to trimetrexate.

The crystal structure showed that the *C. hominis* DHFR domain has a unique 9-stranded rather than the 8-stranded ß-sheet (O'Neil et al. 2003). Through protein: ligand interactions it was suggested that TMP did not extend deep enough into the hydrophobic pocket occupied by the natural substrate, dihydrofolate (Popov et al. 2007) . A series of C7-TMP derivatives, designed to exploit a unique pocket in *Ch*DHFR was synthesized and evaluated. Modification of the TMP structure generated analogs with four times greater activity against *Ch*DHFR compared to TMP (Popov et al. 2006).

11.4.7 Inosine Monophosphate Dehydrogenase Inhibitors

Another potential target is the enzyme inosine 5-monophosphate dehydrogenase (*Cp*IMPDH), used for the production of guanine nucleotide. Through phylogenetic analysis, it was suggested that *Cryptosporidium* sp. obtained their IMPDH gene from an ε-proteobacterium by lateral gene transfer (Striepen et al. 2002). The bacterial origin of *Cp*IMPDH and the different kinetic properties of these enzymes compared with their human counterparts suggested *Cp*IMPDH-selective inhibitors could be obtained and, in fact, several different inhibitors active in the μM range were identified in initial high throughput screens (Umejiego et al. 2008). All of these compounds were shown to bind to the nicotinamide portion of the NAD site suggesting that they stack up against the purine ring of IMP. Subsequent modifications of these initial hits, through medicinal chemistry, resulted in increased drug potency when evaluated in a *Toxoplasma* model expressing *Cp*IMPDH (Sharling et al. 2010).

11.4.8 Pyrimidine Salvage Enzymes

Since *Cryptosporidium parvum*, unlike other protozoa, lacks de novo pyrimidine nucleotide biosynthesis, two other potential targets are the pyrimidine salvage enzymes thymidine kinase (*Cp*TK) and kinase-uracil phosphoribosyltransferase (*Cp*UK-UPRT). Both trifluoromethylthymidine (TFT) and 5-fluorodeoxyuridine (FUdR) inhibit in vitro growth of the parasite at concentrations much lower than those that inhibit host cell proliferation (20–400-fold, respectively). This cytotoxicity differential is comparable to the therapeutic index for ribavirin, and indeed, TFT treatment in an acute mouse model of cryptosporidial infection resulted in a significant decrease in oocyst shedding (Sun et al. 2010).

11.4.9 Protein Kinases

An analysis of the *C. parvum* genome identified over 70 protein kinases, which may include promising drug targets (Wernimont et al. 2010; Artz et al. 2011). Protein kinases are recognized as essential to cell cycle regulation and exhibit a high degree of conservation, especially in the ATP-binding site. While the latter may challenge drug design, the observation that almost a quarter of the *C. parvum* kinases have no known orthologues outside *Cryptosporidium* species affords opportunities for selective drug development. A comparison of kinase orthologues in *P. falciparum* and *T. gondii* revealed distinct features associated with the 'gatekeeper' amino acid that can be exploited for drug design (Wernimont et al. 2010; Ojo et al. 2010). Cell free enzyme assays of expressed kinases identified 13 pyrazolopyrimidine-based inhibitors with IC_{50} values < 10 nM (Artz et al. 2011).

A calcium-dependant protein kinase-1 was identified in *C. parvum* as well as a homologue in *T. gondii* (Wernimont et al. 2010; Murphy et al. 2010). These enzymes were targeted for drug development, in part, due to their characteristics related more to plant kinases than animal kinases. Pyrazolopyrimidine-based inhibitors showed selectivity for the protozoan kinases and exhibited low mammalian cell toxicity and significant inhibition of *T. gondii* invasion and proliferation with EC_{50}s < 1 μM and some compounds showing <100 nM (Johnson et al. 2012). Subsequent drug development based on modification of the base compound, mebendazole, identified compounds with IC_{50}s <50 nM in cell free enzyme inhibition assays and low toxicities for mammalian cells (EC_{50}s > 30 μM) (Zhang et al. 2012). However, none of the compounds showed activity against cryptosporidial or toxoplasma growth in vitro below 1 μM, although it should be noted that no details were provided for the *Cryptosporidium* in vitro growth assay (Zhang et al. 2012).

11.5 Summary Comments on Current Drug Therapy

Considering the overall pattern of drug treatment data, it is apparent that no drugs that are available against other coccidian parasites have shown a robust efficacy against cryptosporidial infections. The potent efficacy of trimethoprim-sulfamethoxazole (TMP-SMX) against *Toxoplasma*, *Isospora*, and *Cyclospora* is in stark contrast to the drugs used against *Cryptosporidium* (Goldberg and Bishara 2012; Seddon and Bhagani 2011). Even HIV-infected AIDS patients with chronic and severe diarrhea caused by *Cyclospora* or *Isospora* responded quickly and completely to TMP-SMX (Verdier et al. 2000). In a 2003–2004 study of chronic diarrhea among HIV/AIDS patients in Haiti (Dillingham et al. 2009), the occurrence of *Cyclospora* and *Isospora* at enrollment was observed to be <10 % compared to approximately 25 % of diarrheic patients in an earlier 1997–1998 study at the same Port-au-Prince clinic (Verdier et al. 2000). The lower prevalence of *Cyclospora* and *Isospora* was attributed to the widespread use of TMP-SMX as a prophylaxis or therapeutic agent in HIV/AIDS patients. This contrast is especially evident when comparing the activity of anticryptosporidial agents in immunocompetent versus immunocompromised patients. The largely modest activity of anticryptosporidial agents in immunocompetent patients combined with the essential failure of these same drugs to control or cure infections in immunocompromised patients argues that even the best drugs currently available are successful only in the context of an effective immune response, rather than by a direct and potent activity against the parasite itself.

11.6 Immunotherapy

11.6.1 Hyperimmune Colostrum

As described above, hyperimmune colostrum used as passive immunotherapy for cryptosporidiosis has been pursued as a strategy in humans since the late 1980s. In addition to the use of colostrum generated against whole antigen (Ag), several studies have evaluated colostrum or monoclonal antibodies (MAb) produced against specific antigens, many of these antigens involved in parasite attachment or invasion of host cells. For example, passive protection against cryptosporidiosis was obtained by treating immunosuppressed mice with immune colostrum generated in cows injected with recombinant pCP15/60 plasmid DNA before and after *C. parvum* infection (Jenkins et al. 1999). Immune bovine colostrum induced by immunization with *C. parvum* recombinant protein rC7, which is the C terminus of the Cp23 protein, provided substantial protection against cryptosporidiosis in neonatal calves (Perryman et al. 1999). Likewise, treatment with different polyclonal or MAbs resulted in the reduction in oocyst shedding as well as easing of clinical symptoms, although colonization still occurred, but at a considerably reduced level (Arrowood et al. 1989; Doyle et al. 1993; Perryman et al. 1990). One MAb, designated 3E2, which recognized multiple 46–770 kDa sporozoite Ags and a 1,300-kDa Ag designated CSL, was able to neutralize infectivity in vitro and control murine infection in vivo (Riggs et al. 1997). The 3E2 MAb combined with other antibodies, including anti-GP25–200 and anti-Cp23 demonstrated significant additive protection over that of the individual MAbs, reducing infection levels by 86–93 %. In addition, infection was completely prevented in up to 40 % of mice administered 3E2 alone or in combination with 3H2 and 1E10 Mabs (Schaefer et al. 2000).

Hyperimmune colostrum may also be used to reduce severity of diarrheal disease in farm animals, such as neonatal calves. This could potentially decrease transmission of *C. parvum* to animals and humans in agricultural settings or in developing countries where families and livestock live in close proximity to one another. In a recent study, cows vaccinated with rCP15/60 produced a significantly greater antibody response compared to controls and this response was strongly associated with the subsequent level of colostral antibody. Calves fed rCP15/60-immune colostrum showed a dose-dependent absorption of antibody, also associated with colostral antibody levels (Burton et al. 2011). Induction of the antibody was clearly evident but treatment efficacy was not demonstrated.

Despite the variable performance of immune colostrum in clinical trials, immunotherapy may still be useful in conjunction with conventional drug therapy or as a mechanism to decrease the severity of infection in neonatal animals or moderately immunocompromised individuals.

11.6.2 Monoclonal Antibodies

An example of MAb-based immunotherapy includes the use of a human CD40 agonist MAb, CP-870,893 to treat two X-linked hyper IgM syndrome patients with biliary cryptosporidiosis (Fan et al. 2012). The MAb activated B cells and antigen presenting cells (APCs) in vitro, restoring class switch recombination in XHM B cells and inducing cytokine secretion by monocytes. Although specific antibody responses were lacking, frequent dosing in one subject primed T cells to secrete IFN-gamma and suppressed oocyst shedding in stools. Nevertheless, oocyst shedding relapse occurred after discontinuation of therapy.

Another antibody-based immunotherapy involved the generation of an antibody-biocide fusion protein. *Cryptosporidium*-specific antibodies were fused with the antimicrobial peptide LL-37 and administered orally to neonatal mice in a prophylactic model of cryptosporidiosis (Imboden et al. 2010). Infections in treated mice were reduced by as much as 81 % in the mucosal epithelium of the gut. When administered simultaneously with oocyst inocula, several versions of antibody fusion proteins that differed in antigen specificity and in the biocide conjugate inhibited parasite growth in mouse intestinal tissue (up to 82 %), although none completely prevented infection.

11.7 Vaccines

Because of the lack of efficacious drug treatments, vaccine development that prevents disease or reduces the severity of infection is a relevant option. This is particularly true for certain groups such as immunocompromised individuals and children in developing countries since cryptosporidiosis in early childhood has been reported to be associated with subsequent impairment in growth, physical fitness, and intellectual capacity (Dillingham et al. 2002). Numerous immunogenic antigens of the *C. parvum* invasive stages involved in attachment or penetration of host cells have been identified (reviewed in (Boulter-Bitzer et al. 2007)). Several cryptosporidial antigens are immunodominant; some are surface and/or apical complex proteins that may mediate attachment and invasion. Sera from infected animals and humans recognize a number of immunodominant sporozoite antigens, including polypeptides of approximately 11, 15, 23, 44, 100, 180 and >200 (Boulter-Bitzer et al. 2007). These include the surface antigens CSL, Cp900, Cp23/27, Cp40/45, Cp15/17, Muc4 and Muc5, some of which are partially or heavily glycosylated. Antibodies developed against some of these antigens demonstrated therapeutic efficacy in mouse and animal models. Much of this work has focused on the Cp15 and Cp23 antigens.

DNA immunization has been used to induce antigen-specific B and T cell responses in various infection model systems (Hong-Xuan et al. 2005; Jenkins et al. 1995; Sagodira et al. 1999a, b). DNA vaccines expressing the Cp15/60 gene, a sporozoite surface antigen (Tilley et al. 1991), induced primarily a type-1 immune response when injected either intranasally or intramuscularly into mice (He et al. 2004; Sagodira et al. 1999b). Additionally, efficacy has been demonstrated by the generation of Cp23-specific immune responses: mice immunized with Cp23-DNA developed partial protection against *C. parvum* infection as shown by the >60 % reduction in oocyst shedding after challenge (Ehigiator et al. 2007). Administration of a DNA vaccine encoding *C. parvum* Cp15 and Cp23 resulted in induction of Th1 immune response and increased resistance to infection (Wang et al. 2010). Evaluation of a DNA vaccine comprised of P2 (CpP2), an important immunodominant marker in *C. parvum* infection (Priest et al. 2010), showed that CpP2-DNA followed with P2 protein (prime-boost), significantly increase antibody production over immunization with just the protein or CpP2-DNA alone. When challenged, reduction in oocysts production was not statistically significant, although a trend in reduction of infection was observed in the CpP2-DNA-immunized mice (Benitez et al. 2011).

Both humoral and cellular responses were elicited using a *Salmonella* strain-and-vector combination that delivered Cp23 and Cp40 fused to the C-terminal fragment of tetanus toxin (Benitez et al. 2009). In another study, three antigens, Cp15, profilin, and a *Cryptosporidium* apyrase, were delivered in a heterologous prime-boost regimen as fusions with cytolysin A (ClyA) in a *Salmonella* live vaccine vector and as purified recombinant antigens, and were found to induce specific and potent humoral and cellular immune responses (Manque et al. 2011). Profilin is a potent inducer of immune responses in mice by both *Eimeria* and *Toxoplasma* and works through the toll receptor TRL11. However, an analogous receptor (TRL11) has not been found in humans. Recently, a prime-boost immunization regimen using an intranasal route followed by oral *Salmonella* live vaccine vector of the Cp15 antigen increased immune responses but did not result in decreased infection (Roche et al. 2013). It is possible that other vectors (e.g. *Listeria*, adenovirus) may increase vaccine efficacy. Identification of other vaccine targets, multi-antigen formulations or constructs, or use of an attenuated *Cryptosporidium* strain could result in better immunological responses and protection from infection.

11.8 Conclusion

Key to the development of selective, efficacious anticryptosporidial agents is understanding cryptosporidial biochemistry and cell biology including studies into the basic mechanisms by which chemicals and drugs are transported from the host cell to the parasite (and vice versa), discerning stage-specific developmental changes, and identification of nutrient requirements and unique enzymatic pathways. The continuing increase in genome sequence data should aid in the identification and characterization of drug targets and increase our understanding of host/parasite interactions. However, the inability to cryopreserve the parasite, propagate the parasite continuously in vitro, and the lack of a genetic model hinders many avenues of chemotherapeutic research. Cryopreservation of the parasite would facilitate the establishment of standard isolates that could be used for multiple studies with consistency and reduced variability. The ability to propagate the parasite continuously in vitro could lead to establishment of a genetic model that would be useful for mutant/transfection studies and the identification of molecules as viable drug targets. Standardized positive control drugs are likewise needed when evaluating and validating potential anticryptosporidial agents in both in vitro and in vivo models. Understanding host-parasite interactions and the essential elements of immunity to *Cryptosporidium* spp. may lead to the development of effective immunotherapies or vaccines. Despite the many impediments faced by researchers working on *Cryptosporidium* parasites, active research in this area will hopefully overcome the barriers to success and more efficacious drugs and therapies will be developed to treat this potentially severe disease.

Acknowledgements "The findings and conclusions in this report are those of the author(s) and do not necessarily represent the views of the Department of Veterans Affairs or the Centers for Disease Control and Prevention."

References

Abrahamsen MS, Templeton TJ, Enomoto S, Abrahante JE, Zhu G, Lancto CA, Deng M, Liu C, Widmer G, Tzipori S, Buck GA, Xu P, Bankier AT, Dear PH, Konfortov BA, Spriggs HF, Iyer L, Anantharaman V, Aravind L, Kapur V (2004) Complete genome sequence of the apicomplexan, *Cryptosporidium parvum*. Science 304(5669):441–445

Abubakar I, Aliyu SH, Arumugam C, Hunter PR, Usman NK (2007) Prevention and treatment of cryptosporidiosis in immunocompromised patients. Cochrane Database Syst Rev (1): CD004932. doi:10.1002/14651858.CD004932.pub2

Amadi B, Mwiya M, Musuku J, Watuka A, Sianongo S, Ayoub A, Kelly P (2002) Effect of nitazoxanide on morbidity and mortality in Zambian children with cryptosporidiosis: a randomised controlled trial. Lancet 360(9343):1375–1380

Armson A, Sargent K, MacDonald LM, Finn MP, Thompson RC, Reynoldson JA (1999) A comparison of the effects of two dinitroanilines against *Cryptosporidium parvum* in vitro and in vivo in neonatal mice and rats. FEMS Immunol Med Microbiol 26(2):109–113

Armson A, Menon K, O'Hara A, MacDonald LM, Read CM, Sargent K, Thompson RC, Reynoldson JA (2002) Efficacy of oryzalin and associated histological changes in *Cryptosporidium*-infected neonatal rats. Parasitology 125(Pt 2):113–117

Arrowood MJ, Mead JR, Mahrt JL, Sterling CR (1989) Effects of immune colostrum and orally administered antisporozoite monoclonal antibodies on the outcome of *Cryptosporidium parvum* infections in neonatal mice. Infect Immun 57:2283–2288

Arrowood MJ, Mead JR, Xie L, You X (1996) In vitro anticryptosporidial activity of dinitroaniline herbicides. FEMS Microbiol Lett 136(3):245–249

Artz JD, Dunford JE, Arrowood MJ, Dong A, Chruszcz M, Kavanagh KL, Minor W, Russell RG, Ebetino FH, Oppermann U, Hui R (2008) Targeting a uniquely nonspecific prenyl synthase with bisphosphonates to combat cryptosporidiosis. Chem Biol 15(12):1296–1306

Artz JD, Wernimont AK, Allali-Hassani A, Zhao Y, Amani M, Lin YH, Senisterra G, Wasney GA, Fedorov O, King O, Roos A, Lunin VV, Qiu W, Finerty P Jr, Hutchinson A, Chau I, von Delft F, MacKenzie F, Lew J, Kozieradzki I, Vedadi M, Schapira M, Zhang C, Shokat K, Heightman T, Hui R (2011) The *Cryptosporidium parvum* kinome. BMC Genomics 12:478

Bacchi CJ, Narlett N (1995) Polyamine metabolism. In: Marr JJ, Muller M (eds) Biochemistry and molecular biology of parasites. Academic, New York, pp 119–131

Beckers CJ, Roos DS, Donald RG, Luft BJ, Schwab JC, Cao Y, Joiner KA (1995) Inhibition of cytoplasmic and organellar protein synthesis in *Toxoplasma gondii*. Implications for the target of macrolide antibiotics. J Clin Invest 95(1):367–376

Benbow JW, Bernberg EL, Korda A, Mead JR (1998) Synthesis and evaluation of dinitroanilines for treatment of cryptosporidiosis. Antimicrob Agents Chemother 42(2):339–343

Benitez AJ, McNair N, Mead J (2007) Modulation of gene expression of three *Cryptosporidium parvum* ATP-binding cassette transporters in response to drug treatment. Parasitol Res 101 (6):1611–1616

Benitez AJ, McNair N, Mead JR (2009) Oral immunization with attenuated *Salmonella enterica* serovar Typhimurium encoding *Cryptosporidium parvum* Cp23 and Cp40 antigens induces a specific immune response in mice. Clin Vaccine Immunol 16(9):1272–1278

Benitez A, Priest JW, Ehigiator HN, McNair N, Mead JR (2011) Evaluation of DNA encoding acidic ribosomal protein P2 of *Cryptosporidium parvum* as a potential vaccine candidate for cryptosporidiosis. Vaccine 29(49):9239–9245

Bessoff K, Sateriale A, Lee KK, Huston CD (2013) Drug repurposing screen reveals FDA-approved inhibitors of human HMG-CoA reductase and isoprenoid synthesis that block *Cryptosporidium parvum* growth. Antimicrob Agents Chemother 57(4):1804–1814

Blagburn BL, Soave R (1997) Prophylaxis and chemotherapy: human and animal. In: Fayer R (ed) *Cryptosporidium* and cryptosporidiosis. CRC Press, Boca Raton, pp 111–128

Blagburn BL, Drain KL, Land TM, Kinard RG, Moore PH, Lindsay DS, Patrick DA, Boykin DW, Tidwell RR (1998) Comparative efficacy evaluation of dicationic carbazole compounds, nitazoxanide, and paromomycin against *Cryptosporidium parvum* infections in a neonatal mouse model. Antimicrob Agents Chemother 42(11):2877–2882

Blanshard C, Shanson DC, Gazzard BG (1997) Pilot studies of azithromycin, letrazuril and paromomycin in the treatment of cryptosporidiosis. Int J STD AIDS 8(2):124–129

Bobin S, Bouhour D, Durupt S, Boibieux A, Girault V, Peyramond D (1998) Importance of antiproteases in the treatment of microsporidia and/or cryptosporidia infections in HIV-seropositive patients. Pathol Biol 46(6):418–419

Bonafonte MT, Romagnoli PA, McNair N, Shaw AP, Scanlon M, Leitch GJ, Mead JR (2004) *Cryptosporidium parvum*: effect of multi-drug reversing agents on the expression and function of ATP-binding cassette transporters. Exp Parasitol 106(3–4):126–134

Boulter-Bitzer JI, Lee H, Trevors JT (2007) Molecular targets for detection and immunotherapy in *Cryptosporidium parvum*. Biotechnol Adv 25(1):13–44

Brophy VH, Vasquez J, Nelson RG, Forney JR, Rosowsky A, Sibley CH (2000) Identification of *Cryptosporidium parvum* dihydrofolate reductase inhibitors by complementation in *Saccharomyces cerevisiae*. Antimicrob Agents Chemother 44(4):1019–1028

Burton AJ, Nydam DV, Jones G, Zambriski JA, Linden TC, Cox G, Davis R, Brown A, Bowman DD (2011) Antibody responses following administration of a *Cryptosporidium parvum* rCP15/60 vaccine to pregnant cattle. Vet Parasitol 175(1–2):178–181

Carr A, Marriott D, Field A, Vasak E, Cooper DA (1998) Treatment of HIV-1-associated microsporidiosis and cryptosporidiosis with combination antiretroviral therapy. Lancet 351 (9098):256–261

Carreno RA, Martin DS, Barta JR (1999) *Cryptosporidium* is more closely related to the gregarines than to coccidia as shown by phylogenetic analysis of apicomplexan parasites inferred using small-subunit ribosomal RNA gene sequences. Parasitol Res 85(11):899–904

Clastre M, Goubard A, Prel A, Mincheva Z, Viaud-Massuart MC, Bout D, Rideau M, Velge-Roussel F, Laurent F (2007) The methylerythritol phosphate pathway for isoprenoid biosynthesis in coccidia: presence and sensitivity to fosmidomycin. Exp Parasitol 116(4):375–384

Crabb JH (1998) Antibody-based immunotherapy of cryptosporidiosis. Adv Parasitol 40:121–149

Denkinger CM, Harigopal P, Ruiz P, Dowdy LM (2008) *Cryptosporidium parvum*-associated sclerosing cholangitis in a liver transplant patient. Transpl Infect Dis 10(2):133–136

Dillingham RA, Lima AA, Guerrant RL (2002) Cryptosporidiosis: epidemiology and impact. Microbes Infect 4(10):1059–1066

Dillingham RA, Pinkerton R, Leger P, Severe P, Guerrant RL, Pape JW, Fitzgerald DW (2009) High early mortality in patients with chronic acquired immunodeficiency syndrome diarrhea initiating antiretroviral therapy in Haiti: a case-control study. Am J Trop Med Hyg 80 (6):1060–1064

Doumbo O, Rossignol JF, Pichard E, Traore HA, Dembele TM, Diakite M, Traore F, Diallo DA (1997) Nitazoxanide in the treatment of cryptosporidial diarrhea and other intestinal parasitic infections associated with acquired immunodeficiency syndrome in tropical Africa. Am J Trop Med Hyg 56(6):637–639

Doyle PS, Crabb J, Petersen C (1993) Anti-*Cryptosporidium parvum* antibodies inhibit infectivity in vitro and in vivo. Infect Immun 61(10):4079–4084

Edlind T, Visvesvara G, Li J, Katiyar S (1994) *Cryptosporidium* and microsporidial beta-tubulin sequences: predictions of benzimidazole sensitivity and phylogeny. J Eukaryot Microbiol 41 (5):S 38

Ehigiator HN, Romagnoli P, Priest JW, Secor WE, Mead JR (2007) Induction of murine immune responses by DNA encoding a 23-kDa antigen of *Cryptosporidium parvum*. Parasitol Res 101 (4):943–950

Fan X, Upadhyaya B, Wu L, Koh C, Santin-Duran M, Pittaluga S, Uzel G, Kleiner D, Williams E, Ma CA, Bodansky A, Oliveira JB, Edmonds P, Hornung R, Wong DW, Fayer R, Fleisher T, Heller T, Prussin C, Jain A (2012) CD40 agonist antibody mediated improvement of chronic *Cryptosporidium* infection in patients with X-linked hyper IgM syndrome. Clin Immunol 143 (2):152–161

Fayer R, Ellis W (1993a) Glycoside antibiotics alone and combined with tetracyclines for prophylaxis of experimental cryptosporidiosis in neonatal BALB/c mice. J Parasitol 79 (4):553–558

Fayer R, Ellis W (1993b) Paromomycin is effective as prophylaxis for cryptosporidiosis in dairy calves. J Parasitol 79(5):771–774

Fayer R, Fetterer R (1995) Activity of benzimidazoles against cryptosporidiosis in neonatal BALB/c mice. J Parasitol 81(5):794–795

Fayer R, Andrews C, Ungar BLP, Blagburn B (1989) Efficacy of hyperimmune bovine colostrum for prophylaxis of cryptosporidiosis in neonatal calves. J Parasitol 75:393–397

Fichera ME, Roos DS (1997) A plastid organelle as a drug target in apicomplexan parasites. Nature 390(6658):407–409

Fichtenbaum CJ, Zackin R, Feinberg J, Benson C, Griffiths JK, Team ACTGNWCS (2000) Rifabutin but not clarithromycin prevents cryptosporidiosis in persons with advanced HIV infection. AIDS 14(18):2889–2893

Flanigan TP, Soave R (1993) Cryptosporidiosis. [Review]. Prog Clin Parasitol 3:1–20

Floren CH, Chinenye S, Elfstrand L, Hagman C, Ihse I (2006) ColoPlus, a new product based on bovine colostrum, alleviates HIV-associated diarrhoea. Scand J Gastroenterol 41(6):682–686

Forney JR, DeWald DB, Yang S, Speer CA, Healey MC (1999) A role for host phosphoinositide 3-kinase and cytoskeletal remodeling during *Cryptosporidium parvum* infection. Infect Immun 67(2):844–852

Fries L, Hillman K, Crabb JH, Linberg S, Hamer D, Griffiths J, Keusch G, Soave R, Petersen C (1994) Clinical and microbiological effects of bovine anti-*Cryptosporidium* immunoglobulin (BACI) on cryptosporidial diarrhea in patients with AIDS. In: 34th interscience conference on antimicrobial agents and chemotherapy, Orlando. p Abstract 198

Fujikawa H, Miyakawa H, Iguchi K, Nishizawa M, Moro K, Nagai K, Ishibashi M (2002) Intestinal cryptosporidiosis as an initial manifestation in a previously healthy Japanese patient with AIDS. J Gastroenterol 37(10):840–843

Gargala G, Delaunay A, Li X, Brasseur P, Favennec L, Ballet JJ (2000) Efficacy of nitazoxanide, tizoxanide and tizoxanide glucuronide against *Cryptosporidium parvum* development in sporozoite-infected HCT-8 enterocytic cells. J Antimicrob Chemother 46(1):57–60

Gargala G, Le Goff L, Ballet JJ, Favennec L, Stachulski AV, Rossignol JF (2010) Evaluation of new thiazolide/thiadiazolide derivatives reveals nitro group-independent efficacy against in vitro development of *Cryptosporidium parvum*. Antimicrob Agents Chemother 54 (3):1315–1318

Gargala G, Francois A, Favennec L, Rossignol JF (2013) Activity of halogeno-thiazolides against *Cryptosporidium parvum* in experimentally- infected immunosuppressed gerbils (*Meriones unguiculatus*). Antimicrob Agents Chemother 57(6):2821–2823

Giacometti A, Cirioni O, Scalise G (1996) In-vitro activity of macrolides alone and in combination with artemisin, atovaquone, dapsone, minocycline or pyrimethamine against *Cryptosporidium parvum*. J Antimicrob Chemother 38(3):399–408

Giacometti A, Burzacchini F, Cirioni O, Barchiesi F, Dini M, Scalise G (1999) Efficacy of treatment with paromomycin, azithromycin, and nitazoxanide in a patient with disseminated cryptosporidiosis. Eur J Clin Microbiol Infect Dis 18(12):885–889

Giacometti A, Cirioni O, Barchiesi F, Ancarani F, Scalise G (2000) Activity of nitazoxanide alone and in combination with azithromycin and rifabutin against *Cryptosporidium parvum* in cell culture. J Antimicrob Chemother 45(4):453–456

Giacometti A, Cirioni O, Del Prete MS, Barchiesi F, Fineo A, Scalise G (2001) Activity of buforin II alone and in combination with azithromycin and minocycline against *Cryptosporidium parvum* in cell culture. J Antimicrob Chemother 47(1):97–99

Goldberg E, Bishara J (2012) Contemporary unconventional clinical use of co-trimoxazole. Clin Microbiol Infect 18(1):8–17

Gorla SK, Kavitha M, Zhang M, Liu X, Sharling L, Gollapalli DR, Striepen B, Hedstrom L, Cuny GD (2012) Selective and potent urea inhibitors of *Cryptosporidium parvum* inosine 5′-monophosphate dehydrogenase. J Med Chem 55(17):7759–7771

Graczyk Z, Chomicz L, Kozlowska M, Kazimierczuk Z, Graczyk TK (2011) Novel and promising compounds to treat *Cryptosporidium parvum* infections. Parasitol Res 109(3):591–594

Greenberg PD, Cello JP (1996) Treatment of severe diarrhea caused by *Cryptosporidium parvum* with oral bovine immunoglobulin concentrate in patients with AIDS. J AIDS Hum Retrovirol 13(4):348–354

Griffiths JK, Balakrishnan R, Widmer G, Tzipori S (1998) Paromomycin and geneticin inhibit intracellular *Cryptosporidium parvum* without trafficking through the host cell cytoplasm: implications for drug delivery. Infect Immun 66(8):3874–3883

Harris M, Deutsch G, MacLean JD, Tsoukas CM (1994) A phase I study of letrazuril in AIDS-related cryptosporidiosis. AIDS 8(8):1109–1113

He H, Zhao B, Liu L, Zhou K, Qin X, Zhang Q, Li X, Zheng C, Duan M (2004) The humoral and cellular immune responses in mice induced by DNA vaccine expressing the sporozoite surface protein of *Cryptosporidium parvum*. DNA Cell Biol 23(5):335–339

Healey MC, Yang S, Rasmussen KR, Jackson MK, Du C (1995) Therapeutic efficacy of paromomycin in immunosuppressed adult mice infected with *Cryptosporidium parvum*. J Parasitol 81(1):114–116

Hewitt RG, Yiannoutsos CT, Higgs ES, Carey JT, Geiseler PJ, Soave R, Rosenberg R, Vazquez GJ, Wheat LJ, Fass RJ, Antoninievic Z, Walawander AL, Flanigan TP, Bender JF (2000) Paromomycin: no more effective than placebo for treatment of cryptosporidiosis in patients with advanced human immunodeficiency virus infection. AIDS Clinical Trial Group. Clin Infect Dis 31(4):1084–1092

Holmberg SD, Moorman AC, Von Bargen JC, Palella FJ, Loveless MO, Ward DJ, Navin TR (1998) Possible effectiveness of clarithromycin and rifabutin for cryptosporidiosis chemoprophylaxis in HIV disease. HIV Outpatient Study (HOPS) Investigators. J Am Med Assoc 279 (5):384–386

Hommer V, Eichholz J, Petry F (2003) Effect of antiretroviral protease inhibitors alone, and in combination with paromomycin, on the excystation, invasion and in vitro development of *Cryptosporidium parvum*. J Antimicrob Chemother 52(3):359–364

Hong-Xuan H, Lei C, Cheng-Min W, Kai Z, Yi T, Xi-Ming Q, Ming-Xing D (2005) Expression of the recombinant fusion protein CP15-23 of *Cryptosporidium parvum* and its protective test. J Nanosci Nanotechnol 5(8):1292–1296

Huston CD, Petri WA Jr (2001) Emerging and reemerging intestinal protozoa. Curr Opin Gastroenterol 17(1):17–23

Imboden M, Riggs MW, Schaefer DA, Homan EJ, Bremel RD (2010) Antibodies fused to innate immune molecules reduce initiation of *Cryptosporidium parvum* infection in mice. Antimicrob Agents Chemother 54(4):1385–1392

Ives NJ, Gazzard BG, Easterbrook PJ (2001) The changing pattern of AIDS-defining illnesses with the introduction of highly active antiretroviral therapy (HAART) in a London clinic. J Infect 42 (2):134–139

Jenkins MC (2004) Present and future control of cryptosporidiosis in humans and animals. Expert Rev Vaccines 3(6):669–671

Jenkins M, Kerr D, Fayer R, Wall R (1995) Serum and colostrum antibody responses induced by jet-injection of sheep with DNA encoding a *Cryptosporidium parvum* antigen. Vaccine 13 (17):1658–1664

Jenkins MC, O'Brien C, Trout J, Guidry A, Fayer R (1999) Hyperimmune bovine colostrum specific for recombinant *Cryptosporidium parvum* antigen confers partial protection against cryptosporidiosis in immunosuppressed adult mice. Vaccine 17(19):2453–2460

Johnson EH, Windsor JJ, Muirhead DE, King GJ, Al-Busaidy R (2000) Confirmation of the prophylactic value of paromomycin in a natural outbreak of caprine cryptosporidiosis. Vet Res Commun 24(1):63–67

Johnson SM, Murphy RC, Geiger JA, DeRocher AE, Zhang Z, Ojo KK, Larson ET, Perera BG, Dale EJ, He P, Reid MC, Fox AM, Mueller NR, Merritt EA, Fan E, Parsons M, Van Voorhis WC, Maly DJ (2012) Development of *Toxoplasma gondii* calcium-dependent protein kinase 1 (TgCDPK1) inhibitors with potent anti-toxoplasma activity. J Med Chem 55(5):2416–2426

Johnson CR, Gorla SK, Kavitha M, Zhang M, Liu X, Striepen B, Mead JR, Cuny GD, Hedstrom L (2013) Phthalazinone inhibitors of inosine-5′-monophosphate dehydrogenase from *Cryptosporidium parvum*. Bioorg Med Chem Lett 23(4):1004–1007

Jordan WC (1996) Clarithromycin prophylaxis against *Cryptosporidium* enteritis in patients with AIDS. J Natl Med Assoc 88(7):425–427

Kadappu KK, Nagaraja MV, Rao PV, Shastry BA (2002) Azithromycin as treatment for cryptosporidiosis in human immunodeficiency virus disease. J Postgrad Med 48(3):179–181

Katiyar SK, Gordon VR, Mclaughlin GL, Edlind TD (1994) Antiprotozoal activities of benzimidazoles and correlations with beta-tubulin sequence. Antimicrob Agents Chemother 38(9):2086–2090

Keithly JS, Zhu G, Upton SJ, Woods KM, Martinez MP, Yarlett N (1997) Polyamine biosynthesis in *Cryptosporidium parvum* and its implications for chemotherapy. Mol Biochem Parasitol 88 (1–2):35–42

Kelly GS (2003) Bovine colostrums: a review of clinical uses. Altern Med Rev 8(4):378–394

Kimata I, Uni S, Iseki M (1991) Chemotherapeutic effect of azithromycin and lasalocid on *Cryptosporidium* infection in mice. J Protozool 38(6):S232–S233

Klein P, Cirioni O, Giacometti A, Scalise G (2008) In vitro and in vivo activity of aurintricarboxylic acid preparations against *Cryptosporidium parvum*. J Antimicrob Chemother 62 (5):1101–1104

Kotloff KL, Nataro JP, Blackwelder WC, Nasrin D, Farag TH, Panchalingam S, Wu Y, Sow SO, Sur D, Breiman RF, Faruque AS, Zaidi AK, Saha D, Alonso PL, Tamboura B, Sanogo D, Onwuchekwa U, Manna B, Ramamurthy T, Kanungo S, Ochieng JB, Omore R, Oundo JO, Hossain A, Das SK, Ahmed S, Qureshi S, Quadri F, Adegbola RA, Antonio M, Hossain MJ, Akinsola A, Mandomando I, Nhampossa T, Acacio S, Biswas K, O'Reilly CE, Mintz ED, Berkeley LY, Muhsen K, Sommerfelt H, Robins-Browne RM, Levine MM (2013) Burden and aetiology of diarrhoeal disease in infants and young children in developing countries (the

Global Enteric Multicenter Study, GEMS): a prospective, case-control study. Lancet 382 (9888):209–222

Lawton P, Pelandakis M, Petavy AF, Walchshofer N (2007) Overexpression, purification and characterization of a hexahistidine-tagged recombinant extended nucleotide-binding domain 1 (NBD1) of the *Cryptosporidium parvum* CpABC3 for rational drug design. Mol Biochem Parasitol 152(1):101–107

Leander BS, Clopton RE, Keeling PJ (2003) Phylogeny of gregarines (Apicomplexa) as inferred from small-subunit rDNA and beta-tubulin. Int J Syst Evol Microbiol 53(1):345–354

Ling Y, Sahota G, Odeh S, Chan JM, Araujo FG, Moreno SN, Oldfield E (2005) Bisphosphonate inhibitors of *Toxoplasma gondi* growth: in vitro, QSAR, and in vivo investigations. J Med Chem 48(9):3130–3140

Lopez-Velez R, Tarazona R, Garcia Camacho A, Gomez-Mampaso E, Guerrero A, Moreira V, Villanueva R (1995) Intestinal and extraintestinal cryptosporidiosis in AIDS patients. Eur J Clin Microbiol Infect Dis 14(8):677–681

Mancassola R, Reperant JM, Naciri M, Chartier C (1995) Chemoprophylaxis of *Cryptosporidium parvum* infection with paromomycin in kids and immunological study. Antimicrob Agents Chemother 39(1):75–78

Manque PA, Tenjo F, Woehlbier U, Lara AM, Serrano MG, Xu P, Alves JM, Smeltz RB, Conrad DH, Buck GA (2011) Identification and immunological characterization of three potential vaccinogens against *Cryptosporidium* species. Clin Vaccine Immunol 18(11):1796–1802

Martin MB, Grimley JS, Lewis JC, Heath HT III, Bailey BN, Kendrick H, Yardley V, Caldera A, Lira R, Urbina JA, Moreno SN, Docampo R, Croft SL, Oldfield E (2001) Bisphosphonates inhibit the growth of *Trypanosoma brucei*, *Trypanosoma cruzi*, *Leishmania donovani*, *Toxoplasma gondii*, and *Plasmodium falciparum*: a potential route to chemotherapy. J Med Chem 44(6):909–916

Martucci WE, Udier-Blagovic M, Atreya C, Babatunde O, Vargo MA, Jorgensen WL, Anderson KS (2009) Novel non-active site inhibitor of *Cryptosporidium hominis* TS-DHFR identified by a virtual screen. Bioorg Med Chem Lett 19(2):418–423

Mather MW, Henry KW, Vaidya AB (2007) Mitochondrial drug targets in apicomplexan parasites. Curr Drug Targets 8(1):49–60

Mead J, McNair N (2006) Antiparasitic activity of flavonoids and isoflavones against *Cryptosporidium parvum* and *Encephalitozoon intestinalis*. FEMS Microbiol Lett 259(1):153–157

Mead JR, Benbow JW, Garmon D, Stewart J (1999) Improved efficacy of dinitroaniline analogs for use as anti-cryptosporidial drugs. J Eukaryot Microbiol 46(5):62S–63S

Mead JR, Fauq AH, Khan MA, McNair N (2003) Efficacy of glycoconjugated dinitroanilines against *Cryptosporidium parvum*. J Eukaryot Microbiol 50(Suppl):550–552

Mele R, Gomez Morales MA, Tosini F, Pozio E (2003) Indinavir reduces *Cryptosporidium parvum* infection in both in vitro and in vivo models. Int J Parasitol 33(7):757–764

Miao YM, Awad-El-Kariem FM, Franzen C, Ellis DS, Muller A, Counihan HM, Hayes PJ, Gazzard BG (2000) Eradication of cryptosporidia and microsporidia following successful antiretroviral therapy. J AIDS 25(2):124–129

Mor SM, Tzipori S (2008) Cryptosporidiosis in children in Sub-Saharan Africa: a lingering challenge. Clin Infect Dis 47(7):915–921

Moreno B, Bailey BN, Luo S, Martin MB, Kuhlenschmidt M, Moreno SN, Docampo R, Oldfield E (2001) (31)P NMR of apicomplexans and the effects of risedronate on *Cryptosporidium parvum* growth. Biochem Biophys Res Commun 284(3):632–637

Müller J, Sidler D, Nachbur U, Wastling J, Brunner T, Hemphill A (2008) Thiazolides inhibit growth and induce glutathione-S-transferase Pi (GSTP1)-dependent cell death in human colon cancer cells. Int J Cancer 123(8):1797–1806

Murphy RC, Ojo KK, Larson ET, Castellanos-Gonzalez A, Perera BG, Keyloun KR, Kim JE, Bhandari JG, Muller NR, Verlinde CL, White AC Jr, Merritt EA, Van Voorhis WC, Maly DJ (2010) Discovery of potent and selective inhibitors of calcium-dependent protein kinase 1 (CDPK1) from *C. parvum* and *T. gondii*. ACS Med Chem Lett 1(7):331–335

Nelson RG, Rosowsky A (2001) Dicyclic and tricyclic diaminopyrimidine derivatives as potent inhibitors of *Cryptosporidium parvum* dihydrofolate reductase: structure-activity and structure-selectivity correlations. Antimicrob Agents Chemother 45(12):3293–3303

Nime FA, Burek JD, Page DL, Holscher MA, Yardley JH (1976) Acute enterocolitis in a human being infected with the protozoan *Cryptosporidium*. Gastroenterology 70(4):592–598

Nord J, Ma P, Dijohn D, Tzipori S, Tacket CO (1990) Treatment with bovine hyperimmune colostrum of cryptosporidial diarrhea in AIDS patients. AIDS 4(6):581–584

O'Neil RH, Lilien RH, Donald BR, Stroud RM, Anderson AC (2003) Phylogenetic classification of protozoa based on the structure of the linker domain in the bifunctional enzyme, dihydrofolate reductase-thymidylate synthase. J Biol Chem 278(52):52980–52987

Ojo KK, Larson ET, Keyloun KR, Castaneda LJ, Derocher AE, Inampudi KK, Kim JE, Arakaki TL, Murphy RC, Zhang L, Napuli AJ, Maly DJ, Verlinde CL, Buckner FS, Parsons M, Hol WG, Merritt EA, Van Voorhis WC (2010) *Toxoplasma gondii* calcium-dependent protein kinase 1 is a target for selective kinase inhibitors. Nat Struct Mol Biol 17(5):602–607

Okhuysen PC, Chappell CL, Crabb J, Valdez LM, Douglass ET, DuPont HL (1998) Prophylactic effect of bovine anti-*Cryptosporidium* hyperimmune colostrum immunoglobulin in healthy volunteers challenged with *Cryptosporidium parvum*. Clin Infect Dis 26(6):1324–1329

Ollivett TL, Nydam DV, Bowman DD, Zambriski JA, Bellosa ML, Linden TC, Divers TJ (2009) Effect of nitazoxanide on cryptosporidiosis in experimentally infected neonatal dairy calves. J Dairy Sci 92(4):1643–1648

Palmieri F, Cicalini S, Froio N, Rizzi EB, Goletti D, Festa A, Macri G, Petrosillo N (2005) Pulmonary cryptosporidiosis in an AIDS patient: successful treatment with paromomycin plus azithromycin. Int J STD AIDS 16(7):515–517

Pelphrey PM, Popov VM, Joska TM, Beierlein JM, Bolstad ES, Fillingham YA, Wright DL, Anderson AC (2007) Highly efficient ligands for dihydrofolate reductase from *Cryptosporidium hominis* and *Toxoplasma gondii* inspired by structural analysis. J Med Chem 50 (5):940–950

Perkins ME, Riojas YA, Wu TW, Le Blancq SM (1999) CpABC, a *Cryptosporidium parvum* ATP-binding cassette protein at the host-parasite boundary in intracellular stages. Proc Natl Acad Sci U S A 96(10):5734–5739

Perryman LE, Bjorneby JM (1991) Immunotherapy of cryptosporidiosis in immunodeficient animal models. J Protozool 38(6):S98–S100

Perryman LE, Riggs MW, Mason PH, Fayer R (1990) Kinetics of *Cryptosporidium parvum* sporozoite neutralization by monoclonal antibodies, immune bovine serum, and immune bovine colostrum. Infect Immun 58:257–259

Perryman LE, Kapil SJ, Jones ML, Hunt EL (1999) Protection of calves against cryptosporidiosis with immune bovine colostrum induced by a *Cryptosporidium parvum* recombinant protein. Vaccine 17(17):2142–2149

Popov VM, Chan DC, Fillingham YA, Atom Yee W, Wright DL, Anderson AC (2006) Analysis of complexes of inhibitors with *Cryptosporidium hominis* DHFR leads to a new trimethoprim derivative. Bioorg Med Chem Lett 16(16):4366–4370

Popov VM, Yee WA, Anderson AC (2007) Towards in silico lead optimization: scores from ensembles of protein/ligand conformations reliably correlate with biological activity. Proteins 66(2):375–387

Portnoy D, Whiteside ME, Buckley E III, MacLeod CL (1984) Treatment of intestinal cryptosporidiosis with spiramycin. Ann Intern Med 101:202–204

Pozio E, Rivasi F, Caccio SM (2004) Infection with *Cryptosporidium hominis* and reinfection with *Cryptosporidium parvum* in a transplanted ileum. APMIS 112(4–5):309–313

Priest JW, Kwon JP, Montgomery JM, Bern C, Moss DM, Freeman AR, Jones CC, Arrowood MJ, Won KY, Lammie PJ, Gilman RH, Mead JR (2010) Cloning and characterization of the acidic ribosomal protein P2 of *Cryptosporidium parvum*, a new 17-kilodalton antigen. Clin Vaccine Immunol 17(6):954–965

Rahman M, Chapel H, Chapman RW, Collier JD (2012) Cholangiocarcinoma complicating secondary sclerosing cholangitis from cryptosporidiosis in an adult patient with CD40 ligand deficiency: case report and review of the literature. Int Arch Allergy Immunol 159(2):204–208

Rehg JE (1991) Anti-cryptosporidial activity of macrolides in immunosuppressed rats. J Protozool 38(6):S228–S230

Riggs MW, Stone AL, Yount PA, Langer RC, Arrowood MJ, Bentley DL (1997) Protective monoclonal antibody defines a circumsporozoite-like glycoprotein exoantigen of *Cryptosporidium parvum* sporozoites and merozoites. J Immunol 158:1787–1795

Riordan CE, Langreth SG, Sanchez LB, Kayser O, Keithly JS (1999) Preliminary evidence for a mitochondrion in *Cryptosporidium parvum*: phylogenetic and therapeutic implications. J Eukaryot Microbiol 46(5):52S–55S

Roche JK, Rojo AL, Costa LB, Smeltz R, Manque P, Woehlbier U, Bartelt L, Galen J, Buck G, Guerrant RL (2013) Intranasal vaccination in mice with an attenuated *Salmonella enterica* Serovar 908htr A expressing Cp15 of *Cryptosporidium*: impact of malnutrition with preservation of cytokine secretion. Vaccine 31(6):912–918

Rohlman VC, Kuhls TL, Mosier DA, Crawford DL, Hawkins DR, Abrams VL, Greenfield RA (1993) Therapy with atovaquone for *Cryptosporidium parvum* infection in neonatal severe combined immunodeficiency mice. J Infect Dis 168(1):258–260

Roos MH (1997) The role of drugs in the control of parasitic nematode infections: must we do without? Parasitology 114(7):S137–S144

Rossignol JF, Hidalgo H, Feregrino M, Higuera F, Gomez WH, Romero JL, Padierna J, Geyne A, Ayers MS (1998) A double-'blind' placebo-controlled study of nitazoxanide in the treatment of cryptosporidial diarrhoea in AIDS patients in Mexico. Trans R Soc Trop Med Hyg 92 (6):663–666

Rossignol JF, Ayoub A, Ayers MS (2001) Treatment of diarrhea caused by *Cryptosporidium parvum*: a prospective randomized, double-blind, placebo-controlled study of nitazoxanide. J Infect Dis 184(1):103–106

Russell TS, Lynch J, Ottolini MG (1998) Eradication of *Cryptosporidium* in a child undergoing maintenance chemotherapy for leukemia using high dose azithromycin therapy. J Pediatr Hematol Oncol 20(1):83–85

Saez-Llorens X, Odio CM, Umana MA, Morales MV (1989) Spiramycin vs. placebo for treatment of acute diarrhea caused by *Cryptosporidium*. Pediatr Infect Dis J 8(3):136–140

Sagodira S, Buzoni-Gatel D, Iochmann S, Naciri M, Bout D (1999a) Protection of kids against *Cryptosporidium parvum* infection after immunization of dams with CP15-DNA. Vaccine 17 (19):2346–2355

Sagodira S, Iochmann S, Mevelec MN, Dimier-Poisson I, Bout D (1999b) Nasal immunization of mice with *Cryptosporidium parvum* DNA induces systemic and intestinal immune responses. Parasite Immunol 21(10):507–516

Sauvage V, Aubert D, Escotte-Binet S, Villena I (2009) The role of ATP-binding cassette (ABC) proteins in protozoan parasites. Mol Biochem Parasitol 167(2):81–94

Saxon A, Weinstein W (1987) Oral administration of bovine colostrum anti-cryptosporidia antibody fails to alter the course of human cryptosporidiosis. J Parasitol 73:413–415

Schaefer DA, Auerbach-Dixon BA, Riggs MW (2000) Characterization and formulation of multiple epitope-specific neutralizing monoclonal antibodies for passive immunization against cryptosporidiosis. Infect Immun 68(5):2608–2616

Seddon J, Bhagani S (2011) Antimicrobial therapy for the treatment of opportunistic infections in HIV/AIDS patients: a critical appraisal. HIV/AIDS 3:19–33

Sharling L, Liu X, Gollapalli DR, Maurya SK, Hedstrom L, Striepen B (2010) A screening pipeline for antiparasitic agents targeting *Cryptosporidium* inosine monophosphate dehydrogenase. PLoS Negl Trop Dis 4(8):e794

Smith NH, Cron S, Valdez LM, Chappell CL, White AC Jr (1998) Combination drug therapy for cryptosporidiosis in AIDS. J Infect Dis 178(3):900–903

Soave R, Dieterich D, Kotler D, Gassyuk E, Tierney AR, Liebes L (1990) Oral diclazuril therapy for cryptosporidiosis. Paper presented at the Sixth International Conference on AIDS, San Francisco, CA, June 20–23, 1990. [Note: published December 30, 1990]
Sodqi M, Marih L, Lahsen AO, Bensghir R, Chakib A, Himmich H, El Filali KM (2012) Causes of death among 91 HIV-infected adults in the era of potent antiretroviral therapy. Presse Med 41 (7–8):e386–e390
Sprinz E, Mallman R, Barcellos S, Silbert S, Schestatsky G, Bem David D (1998) AIDS-related cryptosporidial diarrhoea: an open study with roxithromycin. J Antimicrob Chemother 41 (Suppl B):85–91
Stachulski AV, Berry NG, Lilian Low AC, Moores SL, Row E, Warhurst DC, Adagu IS, Rossignol JF (2006) Identification of isoflavone derivatives as effective anticryptosporidial agents in vitro and in vivo. J Med Chem 49(4):1450–1454
Stefani HN, Levi GC, Amato Neto V, Braz LM, Azevedo HD, Possa TA, Silva NF, de Mendonca JS, Fernandes AO (1996) The treatment of cryptosporidiosis in AIDS patients using paromomycin. Rev Soc Bras Med Trop 29(4):355–357
Stokkermans TJ, Schwartzman JD, Keenan K, Morrissette NS, Tilney LG, Roos DS (1996) Inhibition of *Toxoplasma gondii* replication by dinitroaniline herbicides. Exp Parasitol 84 (3):355–370
Striepen B, White MW, Li C, Guerini MN, Malik SB, Logsdon JM Jr, Liu C, Abrahamsen MS (2002) Genetic complementation in apicomplexan parasites. Proc Natl Acad Sci U S A 99 (9):6304–6309
Strong WB, Nelson RG (2000) Preliminary profile of the *Cryptosporidium parvum* genome: an expressed sequence tag and genome survey sequence analysis. Mol Biochem Parasitol 107 (1):1–32
Sun XE, Sharling L, Muthalagi M, Mudeppa DG, Pankiewicz KW, Felczak K, Rathod PK, Mead J, Striepen B, Hedstrom L (2010) Prodrug activation by *Cryptosporidium* thymidine kinase. J Biol Chem 285(21):15916–15922
Tali A, Addebbous A, Asmama S, Chabaa L, Zougaghi L (2011) Respiratory cryptosporidiosis in two patients with HIV infection in a tertiary care hospital in Morocco. Ann Biol Clin 69 (5):605–608
Templeton TJ, Enomoto S, Chen WJ, Huang CG, Lancto CA, Abrahamsen MS, Zhu G (2010) A genome-sequence survey for Ascogregarina taiwanensis supports evolutionary affiliation but metabolic diversity between a Gregarine and *Cryptosporidium*. Mol Biol Evol 27(2):235–248
Theodos CM, Griffiths JK, D'Onfro J, Fairfield A, Tzipori S (1998) Efficacy of nitazoxanide against *Cryptosporidium parvum* in cell culture and in animal models. Antimicrob Agents Chemother 42(8):1959–1965
Tilley M, Upton SJ, Fayer R, Barta JR, Chrisp CE, Freed PS, Blagburn BL, Anderson BC, Barnard SM (1991) Identification of a 15-kilodalton surface glycoprotein on sporozoites of *Cryptosporidium parvum*. Infect Immun 59(3):1002–1007
Traub-Cseko YM, Ramalho-Ortigao JM, Dantas AP, de Castro SL, Barbosa HS, Downing KH (2001) Dinitroaniline herbicides against protozoan parasites: the case of *Trypanosoma cruzi*. Trends Parasitol 17(3):136–141
Tzipori S, Roberton D, Chapman C (1986) Remission of diarrhoea due to cryptosporidiosis in an immunodeficient child treated with hyperimmune bovine colostrum. Br Med J 293:1276–1277
Tzipori S, Griffiths J, Theodus C (1995) Paromomycin treatment against cryptosporidiosis in patients with AIDS. J Infect Dis 171(4):1069–1070; author reply 1071
Uip DE, Lima AL, Amato VS, Boulos M, Neto VA, Bem David D (1998) Roxithromycin treatment for diarrhoea caused by *Cryptosporidium* spp. in patients with AIDS. J Antimicrob Chemother 41(Suppl B):93–97
Umejiego NN, Gollapalli D, Sharling L, Volftsun A, Lu J, Benjamin NN, Stroupe AH, Riera TV, Striepen B, Hedstrom L (2008) Targeting a prokaryotic protein in a eukaryotic pathogen: identification of lead compounds against cryptosporidiosis. Chem Biol 15(1):70–77
Ungar BLP, Ward DJ, Fayer R, Quinn CA (1990) Cessation of *Cryptosporidium*-associated diarrhea in an acquired immunodeficiency syndrome patient after treatment with hyperimmune bovine colostrum. Gastroenterology 98:486–489

Vargas SL, Shenep JL, Flynn PM, Pui CH, Santana VM, Hughes WT (1993) Azithromycin for treatment of severe *Cryptosporidium* diarrhea in two children with cancer. J Pediatr 123 (1):154–156

Vasquez JR, Gooze L, Kim K, Gut J, Petersen C, Nelson RG (1996) Potential antifolate resistance determinants and genotypic variation in the bifunctional dihydrofolate reductase-thymidylate synthase gene from human and bovine isolates of *Cryptosporidium parvum*. Mol Biochem Parasitol 79(2):153–165

Verdier RI, Fitzgerald DW, Johnson WD Jr, Pape JW (2000) Trimethoprim-sulfamethoxazole compared with ciprofloxacin for treatment and prophylaxis of *Isospora belli* and *Cyclospora cayetanensis* infection in HIV-infected patients. A randomized, controlled trial. Ann Intern Med 132(11):885–888

Viu M, Quilez J, Sanchez-Acedo C, del Cacho E, Lopez-Bernad F (2000) Field trial on the therapeutic efficacy of paromomycin on natural *Cryptosporidium parvum* infections in lambs. Vet Parasitol 90(3):163–170

Wang C, Luo J, Amer S, Guo Y, Hu Y, Lu Y, Wang H, Duan M, He H (2010) Multivalent DNA vaccine induces protective immune responses and enhanced resistance against *Cryptosporidium parvum* infection. Vaccine 29(2):323–328

Weikel C, Lazenby A, Belitsos P, Mcdewitt M, Fleming HE, Barbacci M (1991) Intestinal injury associated with spiramycin therapy of *Cryptosporidium* infection in AIDS. J Protozool 38(6): S147

Wernimont AK, Artz JD, Finerty P Jr, Lin YH, Amani M, Allali-Hassani A, Senisterra G, Vedadi M, Tempel W, Mackenzie F, Chau I, Lourido S, Sibley LD, Hui R (2010) Structures of apicomplexan calcium-dependent protein kinases reveal mechanism of activation by calcium. Nat Struct Mol Biol 17(5):596–601

White AC, Chappell CL, Hayat CS, Kimball KT, Flanigan TP, Goodgame RW (1994) Paromomycin for cryptosporidiosis in aids - a prospective, double-blind trial. J Infect Dis 170(2):419–424

Wilson RJ, Williamson DH (1997) Extrachromosomal DNA in the Apicomplexa. Microbiol Mol Biol Rev 61(1):1–16

Wittenberg DF, Miller NM, van den Ende J (1989) Spiramycin is not effective in treating *Cryptosporidium* diarrhea in infants: results of a double-blind randomized trial. J Infect Dis 159:131–132

Woods KM, Nesterenko MV, Upton SJ (1996) Efficacy of 101 antimicrobials and other agents on the development of *Cryptosporidium parvum* in vitro. Ann Trop Med Parasitol 90(6):603–615

You X, Schinazi RF, Arrowood MJ, Lejkowski M, Juodawlkis AS, Mead JR (1998) In-vitro activities of paromomycin and lasalocid evaluated in combination against *Cryptosporidium parvum*. J Antimicrob Chemother 41(2):293–296

Zapata F, Perkins ME, Riojas YA, Wu TW, Le Blancq SM (2002) The *Cryptosporidium parvum* ABC protein family. Mol Biochem Parasitol 120(1):157–161

Zhang Z, Ojo KK, Johnson SM, Larson ET, He P, Geiger JA, Castellanos-Gonzalez A, White AC Jr, Parsons M, Merritt EA, Maly DJ, Verlinde CL, Van Voorhis WC, Fan E (2012) Benzoylbenzimidazole-based selective inhibitors targeting *Cryptosporidium parvum* and *Toxoplasma gondii* calcium-dependent protein kinase-1. Bioorg Med Chem Lett 22 (16):5264–5267

Zhu G, Keithly JS, Philippe H (2000a) What is the phylogenetic position of *Cryptosporidium*? Int J Syst Evol Microbiol 50(Pt 4):1673–1681

Zhu G, Marchewka MJ, Keithly JS (2000b) *Cryptosporidium parvum* appears to lack a plastid genome. Microbiology 146(2):315–321

Zhu G, Marchewka MJ, Woods KM, Upton SJ, Keithly JS (2000c) Molecular analysis of a Type I fatty acid synthase in *Cryptosporidium parvum*. Mol Biochem Parasitol 105(2):253–260

Zhu G, LaGier MJ, Stejskal F, Millership JJ, Cai X, Keithly JS (2002) *Cryptosporidium parvum*: the first protist known to encode a putative polyketide synthase. Gene 298(1):79–89

Part IV
Cryptosporidium and Water

Chapter 12
Cryptosporidium Oocysts in Drinking Water and Recreational Water

Paul A. Rochelle and George D. Di Giovanni

Abstract Oocysts belonging to a wide variety of species and genotypes of *Cryptosporidium* are common in livestock, wild animals, and humans. Consequently, water is frequently contaminated through direct contact with infected animals and their waste, run-off from contaminated land, or structural and engineering failures in water conveyance, storage, or treatment facilities. Oocysts are resistant to chlorine disinfection at the concentrations typically applied during drinking water treatment but properly operated treatment plants that utilize filtration usually remove oocysts from source water with high efficiency. Nevertheless, waterborne *Cryptosporidium* continues to be a public health concern. Outbreaks have been linked to treated drinking water, but regulations enacted over the last decade, better watershed management, and operational improvements have led to a decline in drinking water related cryptosporidiosis in some countries. However, the same period has seen a marked increase in cryptosporidiosis outbreaks caused by contamination of recreational water, particularly swimming pools. This chapter reviews recent waterborne outbreaks of cryptosporidiosis, discusses oocyst prevalence in drinking water and recreational water, examines the risk of waterborne transmission, and describes the principal methods for detecting oocysts in water, including genotyping environmental oocysts.

P.A. Rochelle (✉)
Metropolitan Water District of Southern California, Water Quality Laboratory, 700 Moreno Avenue, La Verne, CA 91750, USA
e-mail: prochelle@mwdh2o.com

G.D. Di Giovanni
University of Texas School of Public Health, El Paso Regional Campus, 1101 N. Campbell CH 412, El Paso, TX 79902, USA
e-mail: George.D.DiGiovanni@uth.tmc.edu

S.M. Cacciò and G. Widmer (eds.), *Cryptosporidium: parasite and disease*, DOI 10.1007/978-3-7091-1562-6_12, © Springer-Verlag Wien 2014

12.1 *Cryptosporidium* in Water

The story of the 1993 cryptosporidiosis outbreak in Milwaukee is well known within the drinking water industry (MacKenzie et al. 1994). It remains the largest waterborne disease outbreak in U.S. history with an estimated 400,000 illnesses and approximately 150 deaths linked to the contamination. However, 20 years after the Milwaukee event, outbreaks of cryptosporidiosis still occur, linked to both drinking water and recreational water, so *Cryptosporidium* in water continues to be a public health concern. The response to the Milwaukee outbreak by regulators, public health professionals, academic researchers, and water utility scientists led to a plethora of detection methods, development of viability and infectivity assays, the application of molecular tools for genotyping oocysts in water, and evaluation of various disinfection technologies. In addition, many oocyst occurrence surveys were conducted, culminating in the U.S. with a period of regulatory monitoring under the Long Term 2 Enhanced Surface Water Treatment Rule (LT2ESWTR; USEPA 2006) and an intensive 10-year finished water monitoring program in the U.K. (DWI 1999). The LT2ESWTR is designed to improve the control of *Cryptosporidium* (and other pathogens) in drinking water systems and implementation of the regulation is expected to reduce the annual incidence of cryptosporidiosis by an estimated 89,000–1,459,000 cases and prevent 20–314 related premature deaths (USEPA 2006).

Cryptosporidium is particularly problematic for water utilities because it is common in surface waters and the oocyst stage of the parasite, which is found in environmental waters and is immediately infective after being excreted from infected hosts, is resistant to chlorine disinfection at the concentrations typically applied during drinking water treatment (see Chap. 13). However, despite everything that has been learned since the Milwaukee outbreak, it is still difficult to assess the true risk to public health from waterborne *Cryptosporidium*.

In the immediate aftermath of the Milwaukee outbreak, the only concern for the drinking water industry was "*Cryptosporidium*". However, research using polymerase chain reaction (PCR)-based techniques greatly enhanced our knowledge of the species and genotypes of *Cryptosporidium* oocysts found in surface waters. We now know that human-infectious oocysts predominantly belong to just two species (*C. parvum* and *C. hominis*), with a few other species playing a minor role in human infection (see Chap. 3). Additionally, there are many other animal-associated species and genotypes that do not appear to be pathogenic to humans. At least eight species and seven genotypes have been linked to human disease but it is generally recognized that the vast majority of human cryptosporidiosis cases are caused by *C. parvum* and *C. hominis*, and it appears that only *C. hominis*, *C. parvum*, and *C. meleagridis* readily infect immune-competent humans in all age groups (McLauchlin et al. 2000; Leoni et al. 2006). A genotyping study of 2,414 human clinical specimens from England obtained between 1985 and 2000 found *C. parvum* and *C. hominis* in 99 % of samples (Leoni et al. 2006). The remaining samples were identified as containing *C. meleagridis* (0.9 %), *C. felis* (0.2 %),

C. andersoni (0.1 %), *C. canis* (0.04 %), *C. suis* (0.04 %) and the *Cryptosporidium* cervine genotype (0.04 %; now recognized as *C. ubiquitum*). Similarly, Chalmers et al. (2009a) genotyped 7,758 human clinical specimens obtained in England and Wales between 2000 and 2003, and 96 % were identified as *C. parvum* and/or *C. hominis*. The remaining 1 % of typeable samples was identified as *C. meleagridis* (0.7 %), *C. felis* (0.05 %), *C. ubiquitum* (0.05 %), *C. canis* (0.01 %), horse genotype (0.01 %), and skunk genotype (0.01 %). There have also been two documented cases of cryptosporidiosis due to the W17 chipmunk genotype (Feltus et al. 2006). So, finding *C. parvum* or *C. hominis* in drinking water sources is certainly a cause for concern and finding any of the other species that have been recovered from human infections would need investigating. However, oocysts of species and genotypes that do not usually pose a serious risk of human disease frequently contaminate surface waters. While these wildlife- and livestock-associated oocysts may be genetically distinct from the primary human-pathogenic species, they are morphologically similar and cross-react with the antibodies that are used in the most widely adopted detection methods, thus complicating risk assessments. Consequently, important issues have been raised with regard to *Cryptosporidium* risk assessment and the development of watershed management strategies. From a watershed management perspective, different strategies are needed to control human, wildlife, and livestock sources of oocysts. It is therefore reasonable to expect that detection methodologies which differentiate animal-related genotypes from the three *Cryptosporidium* species causing the majority of cryptosporidiosis in immune-competent humans (*C. hominis*, *C. parvum* and *C. meleagridis*) may meet the needs of regulatory agencies and the water industry in addressing *Cryptosporidium* risk and control.

Species of *Cryptosporidium* are common in many different types of animals and consequently oocysts belonging to various species and genotypes are common in the environment. Oocysts are readily mobilized by rainfall, and so even if they are not deposited directly into a body of water by animal defecation, they are easily transported across land into water (Miller et al. 2008). There have been few studies of oocyst survival in untreated water under realistic environmental conditions, although some studies have examined survival in livestock wastewater lagoons and soil (Jenkins et al. 1999). However, compared to most human pathogens, *Cryptosporidium* is relatively stable in environmental waters, remaining infectious for weeks or months at typical surface water temperatures. Factors that determine whether or not oocysts released into surface waters will cause outbreaks or sporadic cases of disease include: the source of contamination and hence the species; the concentration of contaminating oocysts; the length of time oocysts are in the water; environmental temperature; and the presence of effective filtration and/or effective disinfection at treatment plants. The fact that waterborne outbreaks occur demonstrates that human-pathogenic oocysts can survive in environmental waters and retain their infectivity, but the amount of attenuation due to combined environmental stressors is not known. Oocysts are very sensitive to inactivation by UV-C irradiation (Rochelle et al. 2004; Johnson et al. 2005) and it has been suggested that sunlight may be the most significant inactivating agent in environmental waters (see Chap. 13). Water temperature is also a critical factor in oocyst survival in the environment. In a study using inclusion/exclusion of fluorescent dyes to assess loss of viability, oocyst suspensions lost 70 % of their infectivity when incubated in natural mineral waters for 84 days at 20 °C (Nichols et al. 2004). Cell culture assays have also been used to measure reductions in infectivity of oocysts in natural waters. Infectivity of oocysts stored in river water at 21–23 °C decreased by 2.6–3.3 $\log_{10}$ over 84 days and was undetectable after 98 days (Pokorny et al. 2002). Infectivity of *C. parvum* oocysts suspended in raw surface water, as measured by cell culture combined with reverse transcription polymerase chain reaction (RT-PCR), was reduced by 3-$\log_{10}$ in about 120 days at 18 °C and 40 days at 26 °C (Johnson et al. 2008). These values translated to a daily inactivation rate of 0.072-$\log_{10}$/day at 18 °C. Similarly, inactivation of oocysts suspended in groundwaters and surface waters ranged from 0.009 $\log_{10}$/day at 5 °C to 0.2 $\log_{10}$/day at 30 °C (Ives et al. 2007).

12.2 Waterborne Outbreaks of Cryptosporidiosis

Cryptosporidiosis outbreaks linked to both drinking water and recreational water continue to occur around the world (Tables 12.1 and 12.2). Of the 325 water-associated protozoan disease outbreaks reported worldwide for the hundred years ending in 2004, 51 % were caused by *Cryptosporidium* spp. (Karanis et al. 2007). During the next 6 years (2004–2010) *Cryptosporidium* caused 60 % of the 199 reported outbreaks of protozoan diseases linked to all sources (Baldursson and Karanis 2011). Almost 200 waterborne outbreaks of cryptosporidiosis had been reported globally by the end of 2010 and have occurred in the U.S., U.K., Australia, New Zealand, Germany, Sweden, Japan, Spain, Norway, Denmark, France, and Ireland. It is highly likely that outbreaks also occur in most other countries but they may lack suitable reporting mechanisms or detection methods, and comprehensive monitoring data are not available for most countries.

During the period 1991–2002, *Cryptosporidium* caused 7 % of the 207 waterborne disease outbreaks in the U.S. However, due to the Milwaukee outbreak, *Cryptosporidium* caused more illnesses than all other waterborne pathogens combined during the 11 year period of 1991–2002 (Craun et al. 2006). Although the overall number of documented waterborne cryptosporidiosis outbreaks has increased substantially in the U.S. over the last 20 years (4 outbreaks in 1992 vs. 39 in 2007; Yoder et al. 2010), there have been no outbreaks linked to community surface water supplies in the last two decades (Yoder et al. 2012).[1] In fact, the majority of outbreaks are associated

[1] The 20-year period without a cryptosporidiosis outbreak linked to a municipal surface water supply in the U.S. ended recently. An outbreak in Baker City, Oregon during the summer of 2013 was linked to an unfiltered surface water supply with up to 91 oocysts/L in one supply creek (www.bakercity.com; accessed Sept. 25, 2013).

Table 12.1 Examples of cryptosporidiosis outbreaks linked to drinking water[a]

Date	Location	Source/cause	Cases[b]
2000	North West England	Ground water under the influence of surface water, heavy rainfall, inadequate treatment	58
2000	Scotland	Inadequate filtration	90
2000	Northern Ireland	Human sewage from septic tank contaminated water supply, *C. parvum*	168
2001	Saskatchewan, Canada	Treatment plant failure	1,907
2001	Northern Ireland	Structural failures at treatment plant, wastewater contamination, *C. hominis*	306
2004	Bergen, Norway	Possible sewage contamination, coincident with large giardiasis outbreak, *C. parvum*	133
2005	South East England	Public water supply, *C. hominis*	140
2005	North Wales, UK	Contaminated source water reservoir, *C. hominis*	231
2007	Galway, Ireland	Contaminated source water lake and heavy rainfall	>240
2008	Northampton, England	Contamination of treated water by rabbit feces/carcass, *C. cuniculus*	>400
2010	Ostersund, Sweden	Sewage contamination of source water lake, *C. hominis*	12,700
2011	Skellefteå, Sweden	Sewage contamination of river, *C. hominis*	147

[a]Compiled from various sources: (Baldursson and Karanis 2011; Chalmers 2012; EPA 2011; Glaberman et al. 2002; Karanis et al. 2007)
[b]Case numbers for some outbreaks are based on laboratory confirmed infections while others are based on epidemiological studies and the size of the potentially impacted population

Table 12.2 Examples of cryptosporidiosis outbreaks linked to recreational water[a]

Date	Location	Source/cause	Cases
2002	Sweden	Swimming pool, *C. parvum*	800
2003	Majorca, Spain	Hotel swimming pool, *C. parvum*	391
2004	California, USA	Treated recreational water in water park, *C. parvum*	336
2004	California, USA	Water slide in a water park	>250
2005	Nagano, Japan	Hotel swimming pool, *C. hominis*	41
2005	New York, USA	Interactive fountain	2,307
2005	Ohio, USA	Community swimming pool	523
2005	London, England	Swimming pools, *C. hominis*	>120
2005	NSW, Australia	Swimming pool	254
2006	Missouri, USA	Interactive fountain in water park	116
2007	Utah, USA	Various treated recreational water venues	1,302
2008	Texas, USA	Lake, pools, interactive fountain, *C. hominis*	2,050
2009	Australia	Swimming pools, *C. hominis*	1,141

[a]Compiled from various sources: (Baldursson and Karanis 2011; Beach 2008; Chalmers 2012; Hlavsa et al. 2011)

with recreational use of water, particularly treated water such as swimming pools (Yoder et al. 2012).

There were 149 outbreaks of cryptosporidiosis in the U.K. for the period 1983–2005 (Nichols et al. 2006). Fifty five of these outbreaks were linked to municipal drinking water, six were associated with private water supplies, and swimming pools were the source of infections in 43 outbreaks. During the period 1983–1997, there were 25 waterborne outbreaks of cryptosporidiosis in the U.K. with 3,455 reported cases of illness (Nichols 2003). As recently as May 2013, an outbreak of cryptosporidiosis with at least 13 confirmed cases in Roscommon, Ireland, led to a boil water advisory affecting over 10,000 people (Roscommon County Council 2013). While acute contamination and failures at treatment plants have often been implicated as the cause of cryptosporidiosis outbreaks, they can occur in water produced by treatment plants meeting all regulatory standards (Neira-Munoz et al. 2007).

Although much of the early research and regulatory focus was on drinking water, *Cryptosporidium* has become the leading cause of recreational water outbreaks in the U.S. The spatial and temporal proximity of contaminating individuals and "recipient" individuals in recreational water, particularly swimming pools, means that outbreaks linked to recreational water probably result from recent contamination and so oocyst survival in recreational water is not a primary issue. In the 30 years spanning 1971–2000, *Cryptosporidium* caused 15 % of outbreaks associated with recreational water (Craun et al. 2005). More recently (2007–2008), 59 out of 82 disease outbreaks with an identified infectious agent linked to treated recreational water were caused by *Cryptosporidium*. Between 1988 and 2008, 136 documented outbreaks of cryptosporidiosis linked to recreational use of water in nine countries sickened a total of 19,271 people (Beach 2008). The average number of people infected in these outbreaks was 142 but there were over 5,000 infections in one outbreak. There were 5,697 cases of illness in a statewide outbreak linked to swimming pools in Utah in 2007 (Hlavsa et al. 2011) leading to restrictions on the use of public pools by young children. Of the U.K. outbreaks linked to recreational water in the period 1983–2005, the number of cases ranged from 3 to 152 and the sources of infection were public and private swimming pools, rivers, interactive water features (e.g., splash zones), and water fountains. During a 1 year monitoring period of seven swimming pools in the Netherlands, 4.6 % of samples contained *Cryptosporidium* oocysts (Schets et al. 2004). Finally, at the time of writing this chapter, another outbreak during the first quarter of 2013 in Victoria, Australia was being linked to swimming pools (Lester 2013). Detailed reviews of outbreaks associated with recreational water are provided by Beach (2008) and Chalmers (2012).

Some of the earliest applications of genotyping tools demonstrated that in the U.K. and the U.S., 67 % of waterborne cryptosporidiosis outbreaks were caused by *C. hominis* while *C. parvum* was the causative agent in the remaining 33 % (McLauchlin et al. 2000; Sulaiman et al. 1998). Genotyping of *Cryptosporidium* isolates associated with the 1993 Milwaukee outbreak identified *C. hominis* as the etiologic agent (Zhou et al. 2003). With the exception of one recent incident, only

C. parvum and *C. hominis* have been identified as causes of waterborne outbreaks of cryptosporidiosis (Xiao and Ryan 2008). A 2008 outbreak in Northamptonshire, England was caused by a rabbit genotype, now renamed *C. cuniculus* (Chalmers et al. 2009b; Robinson et al. 2010) (Chap. 3). It is noteworthy that multilocus genotyping of the rabbit genotype indicated that it is very closely related to *C. hominis*, as opposed to typical animal-associated genotypes (Robinson et al. 2010). Reanalysis of 3,030 sporadic human cryptosporidiosis samples previously identified as containing *C. hominis* revealed that *C. cuniculus* had a prevalence of 1.2 % among clinical isolates (Chalmers et al. 2011).

12.3 Risk Assessment

There are many risk factors for acquiring cryptosporidiosis, including recreational use of treated water (swimming pools and water parks), swimming in untreated water, consuming contaminated food, person to person transmission, contact with farm animals, exposure in day care facilities, drinking contaminated water, and foreign travel (see Chap. 2). However, the potential magnitude of outbreaks from drinking water is greater than for most other sources. Consequently, there have been numerous efforts to derive estimates of the risk to public health from *Cryptosporidium* oocysts in water but the contribution of drinking water to the overall cryptosporidiosis disease burden is still not clear. Many factors contribute to the uncertainty of calculating this risk, including:

- Widely varying oocyst occurrence data in source and finished waters
- Relative prevalence of human-pathogenic and animal-associated species and genotypes
- Variable recovery efficiencies of detection methods
- Non-standardized viability and infectivity assessment methods
- Differing risk assessment models and underlying assumptions
- Varying susceptibility to infection within different human populations
- Possible protective immunity resulting from low-level endemic exposure
- Varying infectious doses for different species and strains of human-pathogenic *Cryptosporidium* spp. (Okhuysen et al. 1999; Chappell et al. 2006)
- Relative contributions of other sources of infection

According to one of the first risk assessment studies, based on an individual's average water consumption of 1.5 L/day and an acceptable daily oocyst intake of 6.5×10^{-5} (calculated from human dose response data), the theoretical maximum acceptable concentration in drinking water was 4.4×10^{-2} oocysts/1,000 L (Haas et al. 1996). This value is considerably lower than the detection limit of currently used monitoring methods. In another study, based on a finished water concentration of one oocyst/1,000 L, it was estimated that tap water was responsible for 6,000 cases of illness annually in New York City (Perz et al. 1998). More recently, it was estimated that the daily infection risk for the general immune-competent population in New York City was 3–10 cases per 100,000 people (Makri et al. 2004). However, it is well established that some sections of the community are more susceptible than others to infection by *Cryptosporidium*. For example, in a case–control study of AIDS patients in San Francisco up to 85 % of endemic cryptosporidiosis cases were attributed to drinking unboiled tap water (Aragón et al. 2003).

In contrast to the above risk assessments, some studies have found no association between drinking water and endemic cryptosporidiosis. The most significant risk factors for sporadic cryptosporidiosis among 282 immunocompetent individuals in seven U.S. states were recent international travel, contact with cattle, contact with young children with diarrhea, and swimming in untreated surface water (Roy et al. 2004). Drinking water was not associated with *Cryptosporidium* infection in six of these states but consumption of well water within Minnesota was a significant risk factor. Similarly, in a case–control study of immunocompetent individuals in the San Francisco Bay Area, there was no significant association between cryptosporidiosis and drinking water (Khalakdina et al. 2003). The major risk factor for cryptosporidiosis within this population was recent travel to another country. Similar results were obtained in the U.K. where epidemiological and case–control studies found no association between drinking water and sporadic cases of cryptosporidiosis in northwest England (Hunter et al. 2004; Hughes et al. 2004).

Nevertheless, regulatory improvements within the water industry, better watershed management, and changes in treatment practices have resulted in reduced disease incidence, confirming that drinking water is a contributing factor to cryptosporidiosis in the community. Following the introduction of new drinking water regulations in England and Wales in 1999, the annual incidence of cryptosporidiosis was reduced by 6,770 cases (Lake et al. 2007). There was a significant reduction in disease incidence during the first 6 months of each year but interestingly there was no change during the second half of the year. This disparity in seasonal disease incidence may have been due to improved land and water management practices reducing domestically acquired infections during the first part of the year. However, travel-associated illnesses starting in the summer and continuing through the second part of the year were not impacted by regulatory changes in the U.K. After installing membrane filtration at two treatment plants in the U.K., the annual incidence of 22 cryptosporidiosis cases per 100,000 people declined to less than 10 cases per 100,000 people (Goh et al. 2005). The authors concluded that drinking cold, unboiled municipal tap water was one of the greatest risk factors for sporadic cryptosporidiosis.

Regardless of the theoretical or actual risk of waterborne cryptosporidiosis, when oocysts are detected in drinking water, utilities and regulatory agencies take measures to minimize community-wide outbreaks. The usual action is to issue boil-water orders or advisories. However, a recent cost-benefit analysis determined that nine illnesses per 10,000 people was the point at which the expected monetized benefit of a boil water notice would exceed the expected costs (Ryan et al. 2013). This infection rate far exceeds the USEPA "acceptable" infection rate of 1 in 10,000, and it corresponded to a raw water concentration of 46 oocysts/L and a finished water concentration of 0.046 oocysts/L (assuming 99.9 % removal). But the costs of disease are not just economic, particularly at the level of the individual. Utilities that are routinely detecting human-pathogenic oocysts in finished water should take mitigating action regardless of the economic break-even point.

Despite the emphasis on drinking water as a potential mode of transmission, it is well established that sources such as food, contact with animals, and international travel also commonly transmit *Cryptosporidium* oocysts. In fact, some studies have suggested that food and other modes of transmission may be at least as important as drinking water and may be more likely to transmit higher oocyst doses (Frost et al. 2005). There is some indication that low level endemic exposure to oocysts in drinking water may confer protective immunity during outbreaks (Frost et al. 2005; Chappell et al. 1999). This has led to speculation that cryptosporidiosis only became a serious epidemic disease in western countries after improvements in sanitation and drinking water treatment reduced levels of low-dose exposure that had formerly maintained protective immunity in the community (Frost et al. 2005).

12.4 Detection Methods

USEPA Method 1623 is the most commonly used oocyst detection method in the U.S. and is recognized as the standard method for *Cryptosporidium* and *Giardia* in surface waters (USEPA 2005). It was used in the most recent round of regulatory *Cryptosporidium* monitoring of untreated water in the U.S. under the LT2ESWTR (USEPA 2006). It involves sample concentration by filtration and centrifugation, oocyst purification by immunomagnetic separation (IMS), and detection by immunofluorescence assay microscopy. Method 1623 includes four options for initial sample concentration: Envirochek and Envirochek HV filter capsules, Filta-Max compressed foam filters, and a portable continuous flow centrifuge. Each of these methods requires different post-concentration elution and processing prior to sample clean up by IMS. Ultrafiltration through 50–80 kDa molecular weight cut-off hollow-fiber filters has also proved efficient for recovering *Cryptosporidium* oocysts from large volumes of environmental and finished water samples (Hill et al. 2007; Kuhn and Oshima 2002; Lindquist et al. 2007). However, ultrafiltration has not yet been validated for use with Method 1623. The Envirochek filter is a pleated membrane of polyethersulfone whereas the membrane in the Envirochek HV (high volume) capsule is polyester. The minimum volume of untreated surface water required for analysis by Method 1623 is 10 L, although depending on sample turbidity more than one filter may be required to filter 10 L, and special precautions are sometimes necessary to ensure sample integrity (Fig. 12.1). The Envirochek HV filter can typically be used for up to 50 L of surface water and up to 1,000 L of finished water. However, some authors have reported difficulty in passing 10 L of turbid source waters through Envirochek and Envirochek HV filters (DiGiorgio et al. 2002). Detected organisms are verified by staining with 4’6-diamidino-2-phenylindole (DAPI) and differential interference contrast (DIC) microscopy. The method was recently updated to include pre-elution of the filter in sodium

Fig. 12.1 Special sampling precautions to ensure samples remain adequately cooled during prolonged filtration of large volumes of finished water using USEPA Method 1623 filtration equipment. In this example, finished drinking water samples were being filtered for the detection of infectious *Cryptosporidium* using a cell culture assay (Rochelle et al. 2012). The ambient temperature was 43 °C (110 °F) and samples needed to be kept cool during filtration to prevent inactivation of oocysts (Photos courtesy of J. Hernandez, City of Scottsdale Water Resources Dept. (Arizona, USA))

hexametaphosphate and adding an additional rinse to the IMS (Method 1623.1; USEPA 2012).

Until 2008, regulations in the U.K. required continuous *Cryptosporidium* monitoring of treated drinking water (DWI 1999). The method required continuous sampling at a flow rate of at least 40 L/h with the filter being changed every 24 h (≥960 L in 24 h). As with Method 1623, the U.K.-approved method involved sample collection by filtration through Envirochek HV capsules or FiltaMax compressed foam filters, secondary concentration and purification by IMS, staining with FITC-labeled anti-*Cryptosporidium* antibody, and examination by fluorescence and DIC microscopy (DWI 2005). The method used 0.5 % sodium polyphosphate for the pre-elution treatment. A slightly modified IMS procedure allowed solid-phase automated scanning cytometry to be used for slide examination.

Although numerous PCR-based methods and primers for the detection of *Cryptosporidium* in water have been described and compared, PCR has not been developed as a routine tool for monitoring or regulatory compliance within the water industry. The various PCR assays have different nucleic acid targets and include modifications for viability determination by detecting mRNA (RT-PCR), different species specificities (e.g. all *Cryptosporidium* spp. or individual species) and reported sensitivities as low as one oocyst. Amplification targets include the 18S ribosomal RNA gene, β-tubulin, heat shock protein 70, and oocyst wall proteins, among others (Leetz et al. 2007; Rochelle 2001; Xiao et al. 2006). Most PCR-based assays were developed as presence/absence methods but assays using various quantitative-PCR formats applied to oocysts in water have also been developed (Guy et al. 2003; Kishida et al. 2012; Masago et al. 2006). However, despite reports of sensitive and reproducible PCR-based detection of oocysts in water, a recent study concluded that although quantitative-PCR methods are useful for the detection and species identification of *Cryptosporidium* in environmental samples, they cannot accurately and consistently measure the low concentrations of oocysts that are typically present in sources of drinking water (Staggs et al. 2013).

C. parvum, *C. hominis*, and other species of *Cryptosporidium* are morphologically similar but can be readily differentiated by DNA sequence variations in many genes. Consequently, the real power of applying PCR-based methods to *Cryptosporidium* in water comes from the ability to analyze amplification products by restriction fragment length polymorphisms (RFLP) or sequencing. The current USEPA Method 1623 and the U.K. DWI regulatory methods are microscopy-based, as described above. While these methods are capable of enumerating low levels of oocysts, they do not determine the species or genotypes of the detected *Cryptosporidium*. Researchers have attempted to address this shortcoming by developing PCR methods for genotyping oocysts directly recovered from water samples or from microscope slides following analysis by fluorescence microscopy. While valuable information can be obtained by directly genotyping oocysts from water samples, accurate quantification of oocysts is not possible with current PCR technology. Therefore, methods focused on genotyping oocysts recovered from regulatory slides are the most useful since the overriding regulatory requirement of oocyst quantification (including empty oocysts) is satisfied. Molecular species identification or genotyping of oocysts detected by Method 1623 have not been approved by the USEPA, and it is unlikely that genotyping will be a regulatory requirement in the near future in the U.S. However, a simplified genotyping method for oocysts present on Method 1623 slides has been developed (Di Giovanni et al. 2010) and a genotyping method was developed for oocysts isolated during routine water monitoring in the U.K. (Nichols et al. 2010).

Nested polymerase chain reaction restriction fragment length polymorphism (PCR-RFLP) methods have been used for the sensitive detection and discrimination of *Cryptosporidium* species and genotypes in water (e.g., Xiao et al. 2000; Nichols et al. 2010; Ruecker et al. 2013). These methods involve two rounds of PCR amplification for increased sensitivity. The amplified DNA fragments are then subjected to restriction enzyme digestion which generates DNA fragment patterns unique to different *Cryptosporidium* species and genotypes. Although nested PCR-RFLP genotyping methods have contributed greatly to our understanding of waterborne *Cryptosporidium*, they are not practical for routine use by most water utility and water quality laboratories. Nested PCR is prone to contamination with PCR product, and the extensive manipulation of PCR products for RFLP analysis increases this risk. At present, the majority of utility and water quality laboratories have limited or no molecular experience. In response, a slide genotyping method that can readily be used by water quality laboratories to distinguish human-pathogenic (i.e. *C. hominis*, *C. parvum*, and *C. meleagridis*) from animal-associated *Cryptosporidium* oocysts was developed under Water Research Foundation Project 4099 (Di Giovanni et al. 2010). The developed method uses single-round multiplex PCR targeting the *Cryptosporidium* 18S rRNA and heat shock protein 70 (hsp70) genes and is compatible with conventional or real-time formats. The melting temperature (T_m) of PCR products (amplicons) is dictated by their DNA sequences. High resolution melt (HRM) analysis allows precise determination of amplicon T_m and further discrimination of human and animal-associated *Cryptosporidium* spp. The method is currently undergoing international multi-laboratory evaluation as part of Water Research Foundation Project 4284. Although slide genotyping will not be a regulatory requirement under the next round of LT2ESWTR monitoring (due to start in April, 2015), its use as a research tool will provide added value to the monitoring, aid development of effective watershed management plans, and provide critical data for refinement of *Cryptosporidium* human health risk assessments.

12.5 Occurrence in Source Waters

The actual occurrence of human-pathogenic *Cryptosporidium* oocysts in source and finished waters is still not clear due to natural variation in oocyst occurrence, variable method recovery efficiencies, and the effect of the water matrix on the sensitivity of the detection method. Various research surveys conducted both before the Milwaukee outbreak and in the immediate aftermath generally reported high oocyst prevalence in untreated and finished drinking waters. In the first survey of oocyst occurrence in surface waters, samples that were impacted by domestic and agricultural waste had *Cryptosporidium* concentrations as high as 5,800 oocysts/L (Madore et al. 1987). A large survey of North America spanning 1988–1993 reported that 60 % of samples were positive for *Cryptosporidium* oocysts (LeChevallier and Norton 1995) while a study in Canada demonstrated lower levels of contamination with oocysts detected in 4.5 % of raw water samples (Wallis et al. 1996). Other studies detected oocysts in 6 % of stream samples in Wisconsin (Archer et al. 1995), 63 % of river samples in Pennsylvania (States et al. 1997), and 13 % of surface waters in New Zealand (Ionas et al. 1998). More recently, a version of Method 1623 using Filta-Max filters was used to detect oocysts in water at Portuguese river beaches (Julio et al. 2012). Oocysts were detected in 82 % of samples with concentrations up to 5.3 oocysts/L. In addition, a study of raw water samples in Canada using USEPA Method 1623 (approximately 20 L grab samples), reported that >45 % of water samples contained *Cryptosporidium* oocysts (N = 1,296; Ruecker et al. 2013). In general, occurrence surveys using a variety of detection methods reported high oocyst prevalence in rivers and streams, with 100 % of samples being positive in some surveys and concentrations of less than one to hundreds of oocysts per liter (Clancy and Hargy 2008).

Regulatory monitoring programs typically demonstrate lower prevalence than most research studies. The first regulatory monitoring of *Cryptosporidium* in U.S. waters was conducted under the USEPA's Information Collection Rule (ICR). This survey of 5,838 untreated source waters throughout the U.S. reported an average occurrence of 6.8 % with a mean concentration of 0.067 oocysts/L (Messner and Wolpert 2003). In follow-up supplementary surveys of fewer facilities, 14 % of samples were positive with an average concentration of 0.053

oocysts/L. Starting in 2006, monitoring for *Cryptosporidium* was mandated in the U.S. by the LT2ESWTR (USEPA 2006). Drinking water facilities serving over 10,000 people were required to monitor for oocysts in at least 24 consecutive monthly samples of raw water using USEPA Method 1623. Out of almost 40,000 samples (mostly 10 L raw water grab samples) collected under the first round of monitoring at 1,670 sampling sites from 1,376 facilities across the U.S., 7 % contained at least one *Cryptosporidium* oocyst (Messner 2011a). Over half of the facilities (51 %) reported no positive samples at all throughout the entire monitoring period. In oocyst-positive samples, concentrations ranged from 0.1 oocysts/L (1,720 samples out of 2,895 positives) to a high of 16 oocysts/L (Ongerth 2013a). A total of 57 samples (0.14 %) had concentrations greater than one oocyst per liter. In the positive samples, a single oocyst was detected in more than half of the samples (59 %), 94 % had 1–4 oocysts, 4 % had 5–9 oocysts, and 2 % had ten or more oocysts. The locations with the highest prevalence of oocysts were on the Mississippi River downstream of the Ohio River and Missouri River confluence. Based on the LT2ESWTR monitoring data, at least 80 treatment plants in the U.S. serving over 10,000 people will be required to make improvements to increase oocyst removal or inactivation.

However, as with most microbe detection methods, recovery efficiencies for Method 1623 are rarely 100 % and can vary considerably between laboratories, between analysts, and between different sampling sites, depending on a number of factors, including the chemical and biological characteristics of water samples. Recovery efficiencies for sample sites monitored during the LT2ESWTR ranged from zero to >90 % with a mean of 39.7 % (N = 3,370; Messner 2011b). Similarly, Method 1623 recoveries ranged from 36 % in highly turbid water (99 NTU) to 75 % in lower turbidity samples (20 NTU; DiGiorgio et al. 2002) and were 3.1–90.3 % in samples from coastal streams in Australia (Ongerth 2013b). Consequently, since most oocyst occurrence surveys do not account for matrix effects and measurement of recovery efficiency, it is likely that occurrence is actually higher than most reported values. Analyzing the LT2ESWTR dataset by cumulative probability distributions and accounting for recovery efficiencies, a recent study concluded that one oocyst per liter will occur at 50 % of raw water sample sites at least 25 % of the time (Ongerth 2013a).

12.6 Species Identification and Genotyping of Oocysts in Water

Most genotyping studies of oocysts in environmental waters have demonstrated the presence of a wide variety of species and genotypes, including human infectious oocysts and species/genotypes that have not been linked to human disease (Jiang et al. 2005; Nichols et al. 2010; Ruecker et al. 2005; Wilkes et al. 2013; Xiao et al. 2000; Yang et al. 2008). PCR-RFLP targeting the 18S rRNA gene demonstrated that 93 % of storm water samples were positive for *Cryptosporidium* spp. with 12 different genotypes represented. None of the detected genotypes matched those typically found in human, farm animals, or domestic animals. However, four were identical or closely related to *C. baileyi*, and *Cryptosporidium* genotypes from opossums and snakes, indicating that wildlife was the primary source of oocyst contamination of surface water during storms (Xiao et al. 2000). The same method was also used to analyze untreated surface water and wastewater samples. *Cryptosporidium* was detected in 45.5 % of surface water samples and 24.5 % of raw wastewater samples (Xiao et al. 2001). The predominant genotypes in surface water were *C. parvum* and *C. hominis*, while *C. andersoni* was most commonly detected in wastewater. In the South Nation River Basin in Canada, 21 % of raw water samples (N = 690) contained wildlife-specific genotypes and 13 % contained livestock genotypes, while avian and human-specific genotypes were identified in 3 % and <1 % of samples, respectively (Wilkes et al. 2013).

Several research groups have developed methods for genotyping *Cryptosporidium* oocysts recovered from regulatory slides. A nested PCR-RFLP assay for the 18S rRNA gene was used to analyze 1,042 *Cryptosporidium* IFA microscopy positive slides obtained from Scottish raw and finished drinking waters under the Scottish Water Routine *Cryptosporidium* Monitoring Programme (Nichols et al. 2010). Volumes of filtered water ranged from 300 L for turbid raw waters to a maximum of 4,000 L for groundwaters. The frequency distribution was a single oocyst on 34 % of slides, 75 % of slides had 1–4 oocysts, 12 % had 5–9 oocysts, and 11 % had ten or more oocysts. Genotyping success was 45 % for single oocyst slides and overall genotyping success was 62 % for all slides. Genotyping revealed that animal-associated *Cryptosporidium* were present in almost 90 % of the positive samples. The most frequently observed species and genotypes were *C. andersoni* (27 %), *C. ubiquitum* (20 %), human-pathogenic *C. parvum* and *C. hominis* (combined at 12 %), and muskrat genotypes (7 %).

In a similar study of 1,296 Canadian raw water samples (approximately 20 L grab samples) analyzed using USEPA Method 1623, more than 45 % of water samples tested positive for *Cryptosporidium* by IFA microscopy, yielding 601 slides for subsequent genotyping analysis (Ruecker et al. 2013). Distribution of oocysts on slides was very similar to that reported by Nichols et al. (2010), with 30 % of slides having a single oocyst, 71 % had 1–4 oocysts, 19 % had 5–9 oocysts, and 10 % had ten or more oocysts. Slides were analyzed by nested PCR-RFLP of the 18S rRNA gene for species/genotype identification (Ruecker et al. 2011). Genotyping success was 57 % for single oocyst slides and overall genotyping success was 70 % for all slides. Genotyping results were also similar to the Nichols et al. (2010) study in Scotland, with over 90 % of the samples found to contain animal-associated *Cryptosporidium* species. The most frequently observed species and genotypes were *C. andersoni* (44 %), muskrat genotypes (48 %), *C. ubiquitum* (5 %), and human-pathogenic *C. parvum* and *C. hominis* occurring in only 4 % (combined) of genotyped samples. Unfortunately, provisions for slide archival and genotyping for research purposes were not in place during the first round of the LT2ESWTR.

12.7 Oocysts in Treated Drinking Water

Oocysts are resistant to chlorine disinfection at the concentrations typically applied during drinking water treatment but correctly operating treatment plants that utilize filtration usually remove oocysts from source water. However, oocysts were reported in 3.8–40 % of treated drinking water samples at concentrations up to 0.48 oocysts/L (Rose et al. 1997). A survey of treatment plants in Wisconsin detected oocysts in 4.2 % of finished water samples (Archer et al. 1995) and another survey detected oocysts in 3.5 % of treated drinking waters (Wallis et al. 1996). In 1998, high oocyst concentrations detected in the treated water of Sydney, Australia and multiple detections in source water led to a period of on-again, off-again boil water advisories over 2 months. Although no infections were attributed to this contamination, the incident had long-ranging political, technical, operational, and managerial consequences in Australia (McClellan 1998). During the subsequent 6 years of monitoring Sydney's water supply that was precipitated by this incident, oocysts were detected in just 0.04 % of treated drinking water samples (N = 4,961; O'Keefe 2010). However, the monitoring period was characterized by an extreme multi-year drought and so low oocyst occurrence may not have been reflective of "normal" conditions. Finished water was also monitored at a treatment plant in Japan. No oocysts were detected in 365 consecutive daily samples of 180 L each (Masago et al. 2004).

Regulatory or systematic and widespread monitoring of *Cryptosporidium* in either raw or treated water has only been conducted in a few countries (Table 12.3). Until October 2008, the UK drinking water regulations included the most intensive *Cryptosporidium* monitoring program ever undertaken. The regulation required continuous monitoring of *Cryptosporidium* oocysts in finished drinking water for at least 23 h per day at a flow rate of at least 40 L per hour (DWI 1999). Although the majority of samples analyzed during this decade-long monitoring program were negative, *Cryptosporidium* oocysts were occasionally detected in finished drinking water. During the period 2000–2002, a total of 97,999 samples were analyzed (total volume = 115,303,050 L), 5.5 % were positive, and the average oocyst concentration was 0.0002 oocysts/L (Smeets et al. 2007). In the earlier years of the monitoring program, oocysts were detected at least once in the finished water from many plants. For example, in 2002, 1.9 % of samples were positive but oocysts were detected at least once from 68 % of sample sites (DWI 2002). Similarly, in 2003, 1.1 % of samples were positive with 54 % of sample locations reporting detection in at least one sample. So clearly, the public is being exposed to low levels of *Cryptosporidium* oocysts in finished drinking water. Nevertheless, the results of this extensive monitoring program allowed the DWI to conclude that treated drinking water is not a major source of exposure of the population to *Cryptosporidium* oocysts. In 2008, a total of 50,569 samples were analyzed (total volume = 46,523,480 L) from 204 plants but none of them exceeded the treatment standard of <1 oocyst/L (DWI 2008).

Table 12.3 Regulatory or required monitoring of *Cryptosporidium* in water

Country	Regulation/requirement	Monitoring dates	Type of water	Oocyst occurrence
Sydney, Australia	Outcome of Sydney Water Enquiry, 1998	1998–2004	Finished	0.04 %
England and Wales	UK Water Supply Regulation (Amended), 1999	1999–2008	Finished	2.8 %
Scotland	*Cryptosporidium* (Scottish Water) Directions, 2003	2006–2007	Finished	7 %
			Raw	26 %
United States	Long Term 2 Enhanced Surface Water Treatment Rule, 2006	2006–2008	Raw	7 %

Despite significant efforts in watershed management and treatment upgrades, *Cryptosporidium* remains a challenge to drinking water quality in Scotland. Consequently, regulatory monthly monitoring is required at all treatment plants in Scotland (The *Cryptosporidium* [Scottish Water] Directions 2003). During 22 months of monitoring finished water (2006–2007), a total of 20,249 treated water samples from 304 drinking water supplies were analyzed. Oocysts were detected in 7 % of samples with a maximum concentration of 1.6 oocysts/L (Wilson et al. 2008). Oocysts were detected at least once in 54 % of supplies. From a public health perspective, the most risky source water catchments were those that were drier than average but had occasional high rainfall events, which flushed oocysts into the water. DNA could not be amplified from all of these monitoring slides, but of those that could be amplified, 19 % were identified as *C. andersoni*, 8 % were a cervine genotype, 5 % were *C. parvum*, 0.3 % were *C. baileyi*, 0.3 % were *C. bovis*, and 0.3 % were a muskrat genotype. Similarly, *C. andersoni*, *C. parvum*, and *C. ubiquitum* were identified on 4.1 %, 4.3 %, and 12.6 % of regulatory finished water monitoring slides in Scotland (Nichols et al. 2010). Mixed species or genotypes were detected on 19 % of slides. During 2011, a total of 8,919 finished water samples were analyzed under the regulatory directive in Scotland. Of these, 378 (4.2 %) samples tested positive for oocysts, the highest percentage of positive samples since 2008. Similarly, 91 of 264 (34 %) treatment plants reported at least one positive sample (Drinking Water Quality Regulator for Scotland 2012).

12.8 Infectivity of Waterborne Oocysts

Cryptosporidium infectivity assays have typically been used to assess the efficacy of disinfection processes but they can also be used to inform risk assessment models by determining the infectivity of oocysts in environmental water samples. The infectivity of oocysts recovered by Method 1623 can be tested in cell culture assays by modification of the oocyst-magnetic bead disassociation step at the end of the IMS procedure and omitting microscopic examination of oocysts. A variety of methods have been developed for detecting infection in cell culture and assays have

Table 12.4 Prevalence of infectious *Cryptosporidium* oocysts in water

Type of water	Number of samples	Mean sample volume	Positive samples	Reference
Finished drinking water	370	943 L	0 %	Rochelle et al. (2012)
Finished drinking water	1,690	100 L	1.4 %	Aboytes et al. (2004)
Source water	560	10 L	3.9 %	LeChevallier et al. (2003)
Source water	122	10 L	4.9 %	Di Giovanni et al. (1999)
Filter backwash water	121	10 L	7.4 %	Di Giovanni et al. (1999)
Disinfected reclaimed effluent	15	400 L	40 %	Gennaccaro et al. (2003)
Raw wastewater	9	4 L	22 %	Lalancette et al. (2012)
Secondary undisinfected wastewater effluents	10	4 L	30 %	Lalancette et al. (2012)
Raw wastewater	18	1 L	33 %	Gennaccaro et al. (2003)

been used to detect infectious *Cryptosporidium* oocysts in raw wastewater, disinfected reclaimed effluent, raw source water, treatment plant filter backwash water, and finished drinking water (Table 12.4). Although cell culture-based methods for assessing *Cryptosporidium* infectivity have been standardized within individual laboratories, and detailed protocols have been developed, they have not yet been validated in large-scale multi-laboratory trials and there are no procedures approved for regulatory use. There are a variety of approaches that include different cell lines and media formulations, oocyst treatments prior to inoculation, incubation periods, and infection detection assays.

Optimized cell culture assays are very sensitive to infections with low levels of oocysts (Fig. 12.2) and have been used to assess the prevalence of infectious oocysts in finished water. A study utilizing cell culture to assess oocyst infectivity reported that 26.8 % of surface water treatment plants (N = 82) were releasing infectious oocysts in their finished water (Aboytes et al. 2004). Overall, 1.4 % of treated drinking water samples (N = 1,690) contained infectious *Cryptosporidium* oocysts but in all cases the follow-up repeat samples were negative. This detection rate translated into a calculated annual risk of 52 infections in 10,000 people, far exceeding the USEPA's 1 in 10,000 risk goal. In a study with contrasting results, treated drinking water samples from 14 treatment plants across the U.S. were analyzed using a modified version of Method 1623 coupled with a cell culture infectivity assay and immunofluorescence detection of infection (Rochelle et al. 2012). Sample volumes were 83.5–2,282 L with an average of 943 L (N = 370). None of the 370 finished water samples produced infections. This lack of infectious oocysts in a total volume of 349,053 L translated to an annual

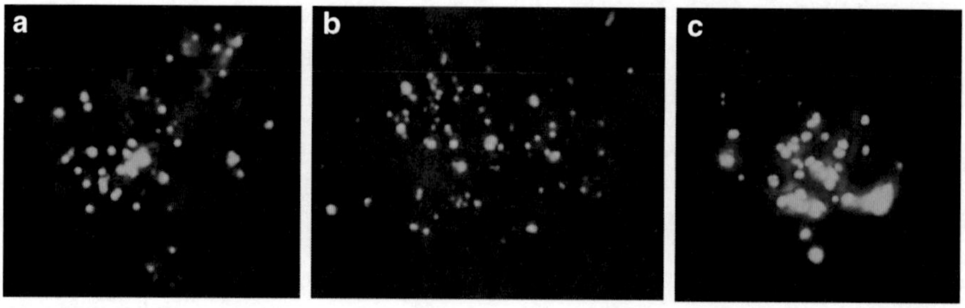

Fig. 12.2 Assay of oocyst infectivity in cultured cell monolayers. Clusters of intracellular life cycle stages (infectious foci) from infectious *C. parvum* oocysts recovered from large volume, spiked drinking water samples (**a** and **b**) and *C. hominis* from an unspiked treated wastewater sample (**c**). Human ileocecal adenocarcinoma (HCT-8) cell monolayers grown in multi-well chamber slides were inoculated with oocysts to allow the development of intracellular *Cryptosporidium* growth stages. Infected monolayers were fixed with methanol and stained with fluorescently-labeled antibodies that recognize sporozoites and other intracellular life cycle stages. Infectious foci were visualized using fluorescence microscopy

risk of less than one infection per 10,000 people. So, more research and more extensive sampling are needed to accurately determine the prevalence of human-infectious oocysts in treated drinking water and thus improve the accuracy of risk assessments.

There have been several recent methodological advances in the detection of infectious *Cryptosporidium* in water. A method for enumerating both total and infectious oocysts in individual water samples was recently described (Lalancette et al. 2010). The method, referred to as 3D-CC-IFA, relies on dual direct detection using differential immunofluorescent staining to detect both oocysts and cell culture infection foci for each sample. This method helps address issues related to splitting of water samples, which typically contain low numbers of oocysts, between oocyst enumeration and cell culture infectivity assays (LeChevallier et al. 2003). The 3D-CC-IFA method was used to quantify the proportion of total *Cryptosporidium* oocysts in raw sewage and wastewater treatment plant effluents that were potentially human-infectious (Lalancette et al. 2012). Human-infectious *Cryptosporidium* were observed in 20 % of the samples. The maximum proportion of total oocysts that were infectious was 22 % for raw sewage and 7 % for treated effluents (with one exception). Both of these values are well below commonly used quantitative microbial risk assessment values for the occurrence of infectious *Cryptosporidium* in water. Further application of the 3D-CC-IFA method will provide significant cost and labor savings and improved estimates of infectious oocyst fractions to aid risk assessments.

Numerous studies have examined the interplay between oocyst pretreatment, cell lines, media formulations, pH, infectivity periods, and other factors on the development of *Cryptosporidium* in cell culture. In particular, there are several, and sometimes conflicting reports in the literature examining the effects of bile salts and sodium taurocholate on oocyst excystation, sporozoite mobility, and susceptibility of host cells to parasite invasion (e.g., Feng et al. 2006; Gold et al. 2001; Kato et al. 2001; King et al. 2012; Woodmansee et al. 1987). However, significant increases in the in-vitro development of *Cryptosporidium* have been reported using combination of cell monolayer growth medium with a high fetal bovine serum concentration (approximately 10 %) followed by centrifugal inoculation and infection development in medium with low (approximately 1 %) fetal bovine serum (King et al. 2012; B. King, personal communication). Additional research is needed in this area to fully understand these factors and develop standardized protocols that can then be used to gain a more accurate assessment of the prevalence of infectious oocysts in treated drinking water.

12.9 Conclusions

Waterborne cryptosporidiosis was recognized as a major problem for the drinking water industry after large outbreaks in the 1980s and 1990s. More recently, recreational water, particularly treated water venues such as swimming pools, has emerged as the primary waterborne route of transmission. Remarkably, there has been no drinking water outbreak linked to treated surface water for the last 2 decades in the U.S. (Yoder et al. 2012).[2] Nevertheless, routine detection in raw waters, occasional detection in finished water, resistance to chlorine disinfection, continued transmission in the community, and the large number of potential cases in the event of an outbreak, all continue to ensure that waterborne *Cryptosporidium* remains a matter of great concern to the water industry. While still not accurately quantified, the risks of endemic disease and outbreaks are clearly greater than zero as illustrated by recent outbreaks linked to drinking water in some countries and frequent swimming pool-related outbreaks. Some of the questions that still need to be answered regarding *Cryptosporidium* in water include: the relative contribution of drinking water to the overall burden of cryptosporidiosis compared to other sources of infection; the prevalence in source waters of human-infectious species or genotypes compared to species that do not infect humans; accurate measures of the frequency of oocyst breakthrough into treated water; the relative virulence of human-infectious species in water; and the contribution to population immunity of low-level endemic exposure to waterborne oocysts.

[2] The 20-year period without a cryptosporidiosis outbreak linked to a municipal surface water supply in the U.S. ended recently. An outbreak in Baker City, Oregon during the summer of 2013 was linked to an unfiltered surface water supply with up to 91 oocysts/L in one supply creek (www.bakercity.com; accessed Sept. 25, 2013).

References

Aboytes A, Di Giovanni GD, Abrams FA, Rheinecker C, McElroy W, Shaw N, LeChevallier MW (2004) Detection of infectious *Cryptosporidium* in filtered drinking water. J Am Water Works Assoc 96(9):88–98

Aragón TJ, Novotny S, Enanoria W, Vugia DJ, Khalakdina A, Katz MH (2003) Endemic cryptosporidiosis and exposure to municipal tap water in persons with acquired immunodeficiency syndrome (AIDS): a case-control study. BMC Public Health 3:2

Archer JR, Ball JR, Standridge JH, Greb SR, Rasmussen PW, Masterson JP, Boushon L (1995) *Cryptosporidium* spp. oocyst and *Giardia* spp. cyst occurrence, concentrations and distribution in Wisconsin waters. Wisconsin Department of Natural Resources, Publication No. WR420-95

Baldursson S, Karanis P (2011) Waterborne transmission of protozoan parasites: review of worldwide outbreaks – an update 2004–2010. Water Res 45:6603–6614

Beach MJ (2008) Waterborne: recreational water. In: Fayer R, Xiao L (eds) *Cryptosporidium* and cryptosporidiosis, 2nd edn. IWA Publishing/CRC Press, Boca Raton, pp 334–369

Chalmers RM (2012) Waterborne outbreaks of cryptosporidiosis. Ann Ist Super Sanita 48:429–446

Chalmers RM, Elwin K, Thomas AL, Guy EC, Mason B (2009a) Long-term *Cryptosporidium* typing reveals the aetiology and species-specific epidemiology of human cryptosporidiosis in England and Wales, 2000 to 2003. Euro Surveill 14:1–9

Chalmers RM, Robinson G, Elwin K, Hadfield SJ, Xiao L, Ryan U, Modha D, Mallaghan C (2009b) *Cryptosporidium* sp. rabbit genotype, a newly identified human pathogen. Emerg Infect Dis 15:829–830

Chalmers RM, Elwin K, Hadfield SJ, Robinson G (2011) Sporadic human cryptosporidiosis caused by *Cryptosporidium cuniculus*, United Kingdom, 2007–2008. Emerg Infect Dis 17:536–538

Chappell CL, Okhuysen PC, Sterling CR, Wang C, Jakubowski W, Dupont HL (1999) Infectivity of *Cryptosporidium parvum* in healthy adults with pre-existing anti-*C. parvum* serum immunoglobulin G. Am J Trop Med Hyg 60:157–164

Chappell CL, Okhuysen PC, Langer-Curry R, Widmer G, Akiyoshi DE, Tanriverdi S, Tzipori S (2006) *Cryptosporidium hominis*: experimental challenge of healthy adults. Am J Trop Med Hyg 75:851–857

Clancy JL, Hargy TM (2008) Waterborne: drinking water. In: Fayer R, Xiao L (eds) *Cryptosporidium* and cryptosporidiosis, 2nd edn. IWA Publishing/CRC Press, Boca Raton, pp 305–333

Craun GF, Calderon RL, Craun MF (2005) Outbreaks associated with recreational water in the United States. Int J Environ Health Res 15:243–262

Craun MF, Craun GF, Calderon RL, Beach MJ (2006) Waterborne outbreaks reported in the United States. J Water Health 4:19–30

Cryptosporidium (Scottish Water) Directions (2003) Scottish Executive. www.scotland.gov.uk/Resource/Doc/26487/0013541.pdf. Accessed June 2013

Di Giovanni GD, Hashemi FH, Shaw NJ, Abrams FA, LeChevallier MW, Abbaszadegan M (1999) Detection of infectious *Cryptosporidium parvum* oocysts in surface and filter backwash water samples by immunomagnetic separation and integrated cell culture PCR. Appl Environ Microbiol 65:3427–3432

Di Giovanni GD, Hoffman, RM, Sturbaum GD (2010). *Cryptosporidium* genotyping method for regulatory microscope slides. Project 4099 Report, Water Research Foundation, Denver, pp 56

DiGiorgio CL, Gonzalez DA, Huitt CC (2002) *Cryptosporidium* and *Giardia* recoveries in natural waters by using Environmental Protection Agency Method 1623. Appl Environ Microbiol 68:5952–5955

Drinking Water Inspectorate (DWI) (1999) UK water supply regulations (Amended) 1999. www.dwi.gov.uk

Drinking Water Inspectorate (DWI) (2002) Drinking water 2002. www.dwi.gov.uk

Drinking Water Inspectorate (DWI) (2005) Standard operating protocol for the monitoring of *Cryptosporidium* oocysts in treated water supplies to satisfy the Water Supply (Water Quality) regulations. U.K. Department of Environment

Drinking Water Inspectorate (DWI) (2008) Drinking water 2008. www.dwi.gov.uk

Drinking Water Quality Regulator for Scotland (2012) Drinking water quality in Scotland 2011: annual report by the drinking water quality regulator for Scotland. p 151. http://www.dwqr.org.uk/technical/annual-report. Accessed 18 June 2013

Environmental Protection Agency (Ireland) (2011) EPA drinking water advice note no. 9: *Cryptosporidium* sampling and monitoring. Office of Environmental Enforcement

Feltus DC, Giddings CW, Schneck BL, Monson T, Warshauer D, McEvoy JM (2006) Evidence supporting zoonotic transmission of *Cryptosporidium* spp. in Wisconsin. J Clin Microbiol 44:4303–4308

Feng H, Nie W, Sheoran A, Zhang Q, Tzipori S (2006) Bile acids enhance invasiveness of *Cryptosporidium* spp. into cultured cells. Infect Immun 74:3342–3346

Frost F, Craun G, Mihály K, György B, Calderon R, Muller T (2005) Serological responses to *Cryptosporidium* antigens among women using riverbank-filtered water, conventionally filtered surface water and groundwater in Hungary. J Water Health 3(1):77–82

Gennaccaro AL, McLaughlin MR, Quintero-Betancourt W, Huffman DE, Rose JB (2003) Infectious *Cryptosporidium parvum* oocysts in final reclaimed effluent. Appl Environ Microbiol 69:4983–4984

Glaberman S, Moore JE, Lowery CJ, Chalmers RM, Sulaiman I, Elwin K, Rooney PJ, Millar BC, Dooley JS, Lal AA, Xiao L (2002) Three drinking water-associated cryptosporidiosis outbreaks, Northern Ireland. Emerg Infect Dis 8:631–633

Goh S, Reacher M, Casemore DP, Verlander NQ, Charlett A, Chalmers RM, Knowles M, Pennington A, Williams J, Osborn K, Richards S (2005) Sporadic cryptosporidiosis decline after membrane filtration of public water supplies, England, 1996–2002. Emerg Infect Dis 11:251–259

Gold D, Stein B, Tzipori S (2001) The utilization of sodium taurocholate in excystation of *Cryptosporidium parvum* and infection of tissue culture. J Parasitol 87:997–1000

Guy RA, Payment P, Krull UJ, Horgen PA (2003) Real-time PCR for quantification of *Giardia* and *Cryptosporidium* in environmental water samples and sewage. Appl Environ Microbiol 69:5178–5185

Haas CN, Crockett CS, Rose JB, Gerba CP, Fazil AM (1996) Assessing the risk posed by oocysts in drinking water. J Am Water Works Assoc 88(9):131–136

Hill VR, Kahler AM, Jothikumar N, Johnson TB, Hahn D, Cromeans TL (2007) Multistate evaluation of an ultrafiltration-based procedure for simultaneous recovery of enteric microbes in 100-liter tap water samples. Appl Environ Microbiol 73:4218

Hlavsa MC, Roberts VA, Anderson AR, Hill VR, Kahler AM, Orr M, Garrison LE, Hicks LA, Newton A, Hilborn ED, Wade TJ, Beach MJ, Yoder JS (2011) Surveillance for waterborne disease outbreaks and other health events associated with recreational water- United States, 2007–2008. MMWR Surveill Summ 60(ss12):1–32

Hughes S, Syed Q, Woodhouse S, Lake I, Osborn K, Chalmers RM, Hunter PR (2004) Using a geographical information system to investigate the relationship between reported cryptosporidiosis and water supply. Int J Health Geogr 3:15. doi:10.1186/1476-072X-3-15

Hunter PR, Hughes S, Woodhouse S, Syed Q, Verlander NQ, Chalmers RM, Morgan K, Nichols G, Beeching N, Osborn K (2004) Case–control study of sporadic cryptosporidiosis with genotyping. Emerg Infect Dis 10:1241–1249

Ionas G, Learmouth JJ, Keys EA, Brown TJ (1998) Distribution of *Giardia* and *Cryptosporidium* in natural water systems in New Zealand – a nationwide survey. Water Sci Technol 38:57–60

Ives RL, Kamarained AM, John DE, Rose JB (2007) Use of cell culture to assess *Cryptosporidium parvum* survival rates in natural groundwaters and surface waters. Appl Environ Microbiol 73:5968–5970

Jenkins MB, Walker MJ, Bowman DD, Anthony LC, Ghiorse WC (1999) Use of a sentinel system for field measurements of *Cryptosporidium parvum* oocyst inactivation in soil and animal waste. Appl Environ Microbiol 65:1998–2005

Jiang J, Alderisio KA, Xiao L (2005) Distribution of *Cryptosporidium* genotypes in storm event samples from three watersheds in New York. Appl Environ Microbiol 71:4446–4454

Johnson AM, Linden K, Ciociola KM, De Leon R, Widmer G, Rochelle PA (2005) UV inactivation of *Cryptosporidium hominis* as measured in cell culture. Appl Environ Microbiol 71:2800–2802

Johnson AM, Rochelle PA, Di Giovanni GD (2008) The risk of cryptosporidiosis from drinking water. In: Proceedings of the American Water Works Association Water Quality Technology Conference, American Water Works Association, Denver

Julio C, Sa C, Ferreira I, Martins S, Oleastro M, Angelo H, Guerreiro J, Tenreiro R (2012) Waterborne transmission of *Giardia* and *Cryptosporidium* in Southern Europe (Portugal). J Water Health 10(3):484–496

Karanis P, Kourenti C, Smith H (2007) Waterborne transmission of protozoan parasites: a worldwide review of outbreaks and lessons learnt. J Water Health 5:1–38

Kato S, Jenkins MB, Ghiorse WC, Bowman DD (2001) Chemical and physical factors affecting the excystation of *Cryptosporidium parvum* oocysts. J Parasitol 87:575–581

Khalakdina A, Vugia DJ, Nadle J, Rothrock GA, Colford Jr JM (2003) Is drinking water a risk factor for endemic cryptosporidiosis? A case–control study in the immunocompetent general population of the San Francisco Bay Area. BMC Public Health 3:11

King BJ, Keegan AR, Phillips R, Fanok S, Monis PT (2012) Dissection of the hierarchy and synergism of the bile derived signal on *Cryptosporidium parvum* excystation and infectivity. Parasitology 139:1533–1546

Kishida N, Miyata R, Furuta A, Izumiyama S, Tsuneda S, Sekiguchi Y, Noda N, Akiba M (2012) Quantitative detection of *Cryptosporidium* oocysts in water source based on 18S rRNA by alternately binding probe competitive reverse transcription polymerase chain reaction (ABC-RT-PCR). Water Res 46:187–194

Kuhn RC, Oshima KH (2002) Hollow-fiber ultrafiltration of *Cryptosporidium parvum* oocysts from a wide variety of 10-L surface water samples. Can J Microbiol 48:542–549

Lake IR, Nichols G, Bentham G, Harrison FC, Hunter PR, Kovats SR (2007) *Cryptosporidiosis* decline after regulation, England and Wales, 1989–2005. Emerg Infect Dis 13:623–625

Lalancette C, Di Giovanni GD, Prévost M (2010) Improved risk analysis by dual direct detection of total and infectious *Cryptosporidium* oocysts on cell culture in combination with immunofluorescence assay. Appl Environ Microbiol 76(2):566–577

Lalancette C, Genereux M, Mailly J, Servais P, Cote C, Michaud A, Di Giovanni GD, Prévost M (2012) Total and infectious *Cryptosporidium* oocyst and total *Giardia* cyst concentrations from distinct agricultural and urban contamination sources in Eastern Canada. J Water Health 10:147–160

LeChevallier MW, Norton WD (1995) *Giardia* and *Cryptosporidium* in raw and finished water. J Am Water Works Assoc 87:54–68

LeChevallier MW, Di Giovanni GD, Clancy JL, Bukhari Z, Bukhari S, Rosen JS, Sobrinho J, Frey MM (2003) Comparison of Method 1623 and cell culture-PCR for detection of *Cryptosporidium* spp. in source waters. Appl Environ Microbiol 69:971–979

Leetz AS, Sotiriadou I, Ongerth J, Karanis P (2007) An evaluation of primers amplifying DNA targets for the detection of *Cryptosporidium* spp. using *C. parvum* HNJ-1 Japanese isolate in water. Parasitol Res 101:951

Leoni F, Amar C, Nichols G, Pedraza-Diaz S, McLauchlin J (2006) Genetic analysis of *Cryptosporidium* from 2414 humans with diarrhea in England between 1985 and 2000. J Med Microbiol 55:703–707

Lester R (2013) Outbreak of cryptosporidiosis extends to regional Victoria. Chief Health Officer Alert, State Government Victoria, Department of Health

Lindquist HD, Harris S, Lucas S, Hartzel M, Riner D, Rochelle P, De Leon R (2007) Using ultrafiltration to concentrate and detect *Bacillus anthracis*, *Bacillus atrophaeus* subspecies *globigii*, and *Cryptosporidium parvum* in 100-liter water samples. J Microbiol Methods 70:484–492

MacKenzie WR, Hoxie NJ, Proctor ME, Gradus MS, Blair KA, Peterson DE, Kazmierczak JJ, Addiss DG, Fox KR, Rose JB, Davis JP (1994) A massive outbreak in Milwaukee of *Cryptosporidium* infection transmitted through the public water supply. N Engl J Med 331:161–167

Madore MS, Rose JB, Gerba CP, Arrowood MJ, Sterling CR (1987) Occurrence of *Cryptosporidium* oocysts in sewage effluents and select surface waters. J Parasitol 73:702–705

Makri A, Modarres R, Parkin R (2004) *Cryptosporidiosis* susceptibility and risk: a case study. Risk Anal 24(1):209–220

Masago Y, Oguma K, Katayama H, Hirata T, Ohgaki S (2004) *Cryptosporidium* monitoring at a water treatment plant, based on waterborne risk assessment. Water Sci Technol 50:293–299

Masago Y, Oguma K, Katayama H, Ohgaki S (2006) Quantification and genotyping of *Cryptosporidium* spp. in river water by quenching probe PCR and denaturing gradient gel electrophoresis. Water Sci Technol 54:119–126

McClellan P (1998) Sydney Water Inquiry. Fifth report. Premiers Department, New South Wales

McLauchlin J, Amar C, Pedraza-Diaz S, Nichols GL (2000) Molecular epidemiological analysis of *Cryptosporidium* spp. in the United Kingdom: results of genotyping *Cryptosporidium* spp. in 1,705 fecal samples from humans and 105 fecal samples from livestock animals. J Clin Microbiol 38:3984–3990

Messner M (2011a) LT2 Round 1: *Cryptosporidium* occurrence. http://water.epa.gov/lawregs/rulesregs/sdwa/lt2/upload/lt2round1crypto.pdf

Messner M (2011b) LT2 Round 1: *Cryptosporidium* matrix spike recovery. http://water.epa.gov/lawregs/rulesregs/sdwa/lt2/upload/lt2round1cryptomatrix.pdf

Messner MJ, Wolpert RL (2003) *Cryptosporidium* and *Giardia* occurrence in ICR drinking water sources: statistical analysis of ICR data. In: McGuire MJ, McLain JL, Obolensky A (eds) Information Collection Rule data analysis. AWWA Research Foundation and the American Water Works Association, Denver, pp 463–481

Miller WA, Lewis DJ, Pereira MD, Lennox M, Conrad PA, Tate KW, Atwill ER (2008) Farm factors associated with reducing *Cryptosporidium* loading in storm runoff from dairies. J Environ Qual 37:1875–1882

Neira-Munoz E, Okoroa C, McCarthy ND (2007) Outbreak of waterborne cryptosporidiosis associated with low oocyst concentrations. Epidemiol Infect 135:1159–1164

Nichols G (2003) Using existing surveillance-based systems. In: Hunter PR, Waite M, Ronchi E (eds) Drinking water and infectious disease: establishing the links. CRC Press, Boca Raton, pp 131–141

Nichols RA, Paton CA, Smith HV (2004) Survival of *Cryptosporidium parvum* oocysts after prolonged exposure to still natural mineral waters. J Food Prot 67:517–523

Nichols RAB, Campbell BM, Smith HV (2006) Molecular fingerprinting of *Cryptosporidium* oocysts isolated during water monitoring. Appl Environ Microbiol 72:5428–5435

Nichols RA, Connelly L, Sullivan CB, Smith HV (2010) Identification of *Cryptosporidium* species and genotypes in Scottish raw and drinking waters during a one-year monitoring period. Appl Environ Microbiol 76:5977–5986

O'Keefe B (2010) Sydney Water Inquiry: Ten Year Review. NSW Government. nsw.gov.au. Accessed June 2013

Okhuysen PC, Chappell CL, Crabb JH, Sterling CR, DuPont HL (1999) Virulence of three distinct *Cryptosporidium parvum* isolates for healthy adults. J Infect Dis 180:1275–1281

Ongerth JE (2013a) LT2 *Cryptosporidium* data: what do they tell us about *Cryptosporidium* in surface water in the United States? Environ Sci Technol 47:4029–4038

Ongerth JE (2013b) The concentration of *Cryptosporidium* and *Giardia* in water- the role and importance of recovery efficiency. Water Res 47:2479–2488

Perz JF, Ennever FK, Le Blancq SM (1998) *Cryptosporidium* in tap water: comparison of predicted risks with observed levels of disease. Am J Epidemiol 147:289–301

Pokorny NJ, Weir SC, Carreno RA, Tevors JT, Lee H (2002) Influence of temperature on *Cryptosporidium parvum* oocyst infectivity in river water samples as detected by tissue culture assay. J Parasitol 88:641–643

Robinson G, Wright S, Elwin K, Hadfield SJ, Katzer F, Bartley PM, Hunter PR, Nath M, Innes EA, Chalmers RM (2010) Re-description of *Cryptosporidium cuniculus* in man and Takeuchi, 1979 (Apicomplexa: *Cryptosporidiidae*): morphology, biology and phylogeny. Int J Parasitol 40:539–1548

Rochelle PA (2001) Detection of protozoa in environmental water samples. In: Rochelle PA (ed) Environmental molecular microbiology: protocols and applications. Horizon Scientific Press, Wymondham, p 91

Rochelle PA, Fallar D, Marshall MM, Montelone BA, Upton SJ, Woods K (2004) Irreversible UV inactivation of *Cryptosporidium* spp. despite the presence of UV repair genes. J Eukaryot Microbiol 51:553–562

Rochelle PA, Johnson AM, De Leon R, Di Giovanni GD (2012) Assessing the risk of infectious *Cryptosporidium* in drinking water. J Amer Water Works Assoc, http://dx.doi.org/10.5942/jawwa.2012.104.0063, E325–E336

Roscommon County Council (2013) http://www.roscommoncoco.ie/en/Services/Sanitation/Notice_to_Consumers_Supplied_by_the_Roscommon_CWSS/. Accessed June 2013

Rose JB, Lisle JT, LeChevallier M (1997) Waterborne cryptosporidiosis: incidence, outbreaks, and treatment strategies. In: Fayer R (ed) Cryptosporidium and *Cryptosporidiosis*. CRC Press, Boca Raton, pp 93–109

Roy SL, DeLong SM, Stenzel SA, Shiferaw B, Roberts JM, Khalakdina A, Marcus R, Segler SD, Shah DD, Thomas S, Vugia DJ, Zansky SM, Dietz V, Beach MJ, Emerging Infections Program FoodNet Working Group (2004) Risk factors for sporadic cryptosporidiosis among immunocompetent persons in the United States from 1999 to 2001. J Clin Microbiol 42:2944–2951

Ruecker NJ, Bounsombath N, Wallis P, Ong CS, Isaac-Renton JL, Neumann NF (2005) Molecular forensic profiling of *Cryptosporidium* species and genotypes in raw water. Appl Environ Microbiol 71:8991–8994

Ruecker NJ, Hoffman RM, Chalmers RM, Neumann NF (2011) Detection and resolution of *Cryptosporidium* species and species mixtures by genus-specific nested PCR-restriction fragment length polymorphism analysis, direct sequencing, and cloning. Appl Environ Microbiol 77:3998–4007

Ruecker NJ, Matsune JC, Lapen DR, Topp E, Edge TA, Neumann NF (2013) The detection of *Cryptosporidium* and the resolution of mixtures of species and genotypes from water. Infect Genet Evol 15:3–9

Ryan MO, Gurian PL, Haas CN, Rose JB, Duzinski PJ (2013) Acceptable microbial risk: cost-benefit analysis of a boil water order for *Cryptosporidium*. J Am Water Works Assoc 105:51–52

Schets FM, Engels GB, Evers EG (2004) *Cryptosporidium* and *Giardia* in swimming pools in the Netherlands. J Water Health 2:191–200

Smeets PW, van Dijk JC, Stanfield G, Rietveld LC, Medema GJ (2007) How can the UK statutory *Cryptosporidium* monitoring be used for quantitative risk assessment of *Cryptosporidium* in drinking water? J Water Health 5:107–118

Staggs SE, Beckman EM, Keely SP, Mackwan R, Ware MW, Moyer AP, Ferretti JA, Sayed A, Xiao L, Villegas EN (2013) The applicability of TaqMan-based quantitative real-time PCR assays for detecting and enumerating *Cryptosporidium* spp. oocysts in the environment. PLOS One 8(6):e66562. doi:10.1371/journal.pone.0066562

States S, Stadterman K, Ammon L, Vogel P, Baldizar J, Wright D, Conley L, Sykora J (1997) Protozoa in river water: sources, occurrence, and treatment. J Am Water Works Assoc 89:74–83

Sulaiman IM, Xiao L, Yang C, Escalante L, Moore A, Beard CB, Arrowood MJ, Lal AA (1998) Differentiating human from animal isolates of *Cryptosporidium parvum*. Emerg Infect Dis 4:681–685

U.S. Environmental Protection Agency (USEPA) (2005) Method 1623: *Cryptosporidium* and *Giardia* in water by filtration/IMS/FA. EPA 815-R-05-002. Office of Research and Development, Government Printing Office, Washington, DC

U.S. Environmental Protection Agency (USEPA) (2006) National Primary Drinking Water Regulations: Long Term 2 Enhanced Surface Water Treatment Rule; Final Rule. Fed Regist 71:654–786

U.S. Environmental Protection Agency (USEPA) (2012) Method 1623.1: *Cryptosporidium* and *Giardia* in water by filtration/IMS/FA. EPA 816-R-12-001. Office of Research and Development, Government Printing Office, Washington, DC

Wallis PM, Erlandsen SL, Isaac-Renton JL, Olson ME, Robertson WJ, van Keulen H (1996) Prevalence of *Giardia* cysts and *Cryptosporidium* oocysts and characterization of *Giardia* spp. cysts isolated from drinking water in Canada. Appl Environ Microbiol 62:2789–2797

Wilkes G, Ruecker NJ, Neumann NF, Gannon VP, Jokinen C, Sunohara M, Topp E, Pintar KD, Edge TA, Lapen DR (2013) Spatiotemporal analysis of *Cryptosporidium* species/genotypes and relationships with other zoonotic pathogens in surface water from mixed use watersheds. Appl Environ Microbiol 79:434–448

Wilson L, Anthony S, Kay D, Procter C (2008) Evaluation and development of *Cryptosporidium* risk assessment. Final Report ADA/012/07, Scottish Government

Woodmansee DB, Powell EC, Pohlenz JF, Moon HW (1987) Factors affecting motility and morphology of *Cryptosporidium* sporozoites in vitro. J Protozool 34:295–297

Xiao L, Ryan UM (2008) Molecular epidemiology. In: Fayer R, Xiao L (eds) Cryptosporidium and cryptosporidiosis. CRC Press, Boca Raton, p 119

Xiao L, Alderisio K, Limor J, Royer M, Lal AA (2000) Identification of species and sources of *Cryptosporidium* oocysts in storm waters with a small-subunit rRNA-based diagnostic and genotyping tool. Appl Environ Microbiol 66:5492–5498

Xiao L, Singh A, Limor J, Graczyk TK, Gradus S, Lal A (2001) Molecular characterization of *Cryptosporidium* oocysts in samples of raw surface water and wastewater. Appl Environ Microbiol 67:1097–1101

Xiao L, Alderisio K, Singh A (2006) Development and standardization of a *Cryptosporidium* genotyping tool for water samples. Awwa Research Foundation, Denver

Yang W, Chen P, Villegas EN, Landy RB, Kanetsky C, Cama V, Dearen T, Schultz CL, Orndorff KG, Prelewicz GJ, Brown MH, Young KR, Xiao L (2008) *Cryptosporidium* source tracking in the Potomac River watershed. Appl Environ Microbiol 74:6495–6504

Yoder JS, Harral C, Beach MJ (2010) Cryptosporidiosis surveillance: United States, 2006–2008. MMWR Surveill Summ 59:1–14

Yoder JS, Wallace RM, Collier SA, Beach MJ, Hlavsa MC (2012) Cryptosporidiosis surveillance – United States, 2009–2010. MMWR Surveill Summ 61:1–12

Zhou L, Singh A, Jiang J, Xiao L (2003) Molecular surveillance of *Cryptosporidium* spp. in raw wastewater in Milwaukee: implications for understanding outbreak occurrence and transmission dynamics. J Clin Microbiol 41:5254–5257

Chapter 13
Removal and Inactivation of *Cryptosporidium* from Water

Paul Monis, Brendon King, and Alexandra Keegan

Abstract Water is a major route of transmission for *Cryptosporidium* and oocysts commonly occur in surface and recreational waters as a consequence of fecal contamination from Wildlife or anthroponotic sources. There are many characteristics possessed by *Cryptosporidium* oocysts that allow them to persist in aquatic environments, including recreational waters, and to bypass water treatment processes. These types of events lead to outbreaks of cryptosporidiosis, caused by direct exposure to contaminated recreational water (such as swimming pools) or by drinking contaminated potable water. Previous chapters have discussed the epidemiology of *Cryptosporidium* in relation to waterborne transmission and also the sources and presence of oocysts in drinking and recreational waters. This chapter will review the processes contributing to the removal and inactivation of *Cryptosporidium* oocysts from surface waters and wastewaters, including natural processes that occur in surface waters and engineered processes used for the production of drinking water or for the treatment of wastewater.

13.1 Measurement of Inactivation

In order to review the removal and inactivation of *Cryptosporidium* in water, it is important to first define key terms, such as viability and infectivity, and to briefly review the methods used to measure inactivation. In general terms, "viability" is loosely taken to indicate that an organism is capable of living or developing under favorable conditions. Many techniques that measure viability at the level of a cell provide general indicators of cell health, such as membrane integrity or enzyme activity. Infectivity is more specifically taken to indicate that an organism is able to

P. Monis (✉) • B. King • A. Keegan
Australian Water Quality Centre, South Australian Water Corporation, GPO 1751, Adelaide, SA 5001, Australia
e-mail: Paul.Monis@sawater.com.au; brendon.king@sawater.com.au; alex.keegan@sawater.com.au

S.M. Cacciò and G. Widmer (eds.), *Cryptosporidium: parasite and disease*,
DOI 10.1007/978-3-7091-1562-6_13, © Springer-Verlag Wien 2014

survive and multiply in a host. In the case of *Cryptosporidium*, viability measurements have historically been used as surrogates for infectivity, but as will be shown below, there are cases where this assumption has been proven to be invalid. In the context of protecting public health, it is critical to understand how treatment processes affect oocyst infectivity.

Animal bioassays have been considered the "gold standard" for assessing *Cryptosporidium* oocyst infectivity. The first description of experimental infection in mice using *Cryptosporidium* oocysts from a diarrheic calf found that neonatal mice were readily infected but that 21 day old mice were only mildly infected (Sherwood et al. 1982). The study of Sherwood et al. (1982) also provided the first evidence that oocysts were susceptible to freezing but retained infectivity for 4–6 months at 4 °C. The neonatal mouse model has since been used extensively in the assessment of *Cryptosporidium parvum* oocyst inactivation by a number of mechanisms (Anderson 1985, 1986; Peeters et al. 1989; Korich et al. 1990; Finch et al. 1993a, b; Fayer 1994, 1995; Black et al. 1996; Fayer et al. 1996; Harp et al. 1996; Xunde and Brasseur 2000; Araki et al. 2001; Craik et al. 2001; Biswas et al. 2003; Li et al. 2004, 2005; Garvey et al. 2010), which will be discussed elsewhere in this chapter. Mouse infectivity has been shown to be reproducible under standardized conditions using a logistic dose response model (Korich et al. 2000). Aside from practical, cost and ethical issues around the use of animals for measuring infectivity, the mouse model has limited applicability because the most common species that infects humans, *Cryptosporidium hominis*, cannot infect mice (Morgan-Ryan et al. 2002). Gnotobiotic piglets have been described for the propagation of *C. hominis* (Widmer et al. 2000; Akiyoshi et al. 2002) and for measuring drug efficacy against *C. hominis* (Theodos et al. 1998) but such a system is even less practical and more costly than neonatal mice. More recently, an animal model using immunosuppressed gerbils has been described for the propagation of *C. parvum* and *C. hominis* (Baishanbo et al. 2005) and also for *C. muris* and *C. andersoni* (Kvac et al. 2007, 2009). The gerbil model has been used to study drug efficacy against *C. parvum* (Baishanbo et al. 2006) but does not appear to have been further used to evaluate the effectiveness of disinfectants against *C. hominis*.

Viability assays have long been pursued as surrogates for measuring infectivity, being cheaper, simpler, faster and more amenable to high-throughput quantitative analysis than using animal models of infectivity. Vital dyes were shown in an early study to be poor predictors of viability (Korich et al. 1990), although excystation correlated with mouse infectivity for measuring inactivation by oxidant-based disinfectants (Korich et al. 1990). However, according to Korich et al. (1990), reliable use of excystation required sporozoite counts, rather than oocyst counts, since high doses of disinfectants resulted in empty oocysts that could be confused with excysted oocysts. Further studies comparing excystation and vital dye staining with mouse infectivity for assessing the effect of chemical disinfectants found that the in vitro viability assays consistently underestimated inactivation (Finch et al. 1993a; Black et al. 1996). Although vital dyes such as propidium iodide have been shown to poorly correlate with infectivity, nucleic acid dye staining has been shown to correlate with mouse infectivity for chemical disinfectants

(Belosevic et al. 1997; Neumann et al. 2000). However, inter-laboratory comparison of in vitro viability assays, including dye exclusion, nucleic acid binding dyes and excystation, with mouse infectivity found that the in vitro techniques could not reliably predict inactivation following ozone treatment (Bukhari et al. 2000). While some agreement has been reported between viability techniques and animal infectivity for chemical disinfectants, the same has not been found for alternative methods of inactivation such as exposure to ultraviolet light. Excystation has been shown to greatly underestimate inactivation by UV compared with animal infectivity (Morita et al. 2002). When compared with cell culture infectivity, viability methods have also been shown to underestimate inactivation caused by oocyst ageing at 15 °C (Schets et al. 2005). Detection of ribosomal RNA using fluorescent in situ hybridization (FISH) has also been reported as a viability indicator to measure inactivation of oocysts in soil (Davies et al. 2005). The reported oocyst inactivation rate at 35 °C was only 2 $\log_{10}$ after 25 days (Davies et al. 2005), compared with 4 $\log_{10}$ within 4 days at 37 °C and 4 $\log_{10}$ after 14.6 days at 30 °C measured using cell culture infectivity (King et al. 2005), suggesting that the FISH technique is a poor surrogate for infectivity when measuring metabolic exhaustion of oocysts. While cost effective and simple to use, in vitro viability assays should be used with caution because of the inconsistent agreement with animal infectivity measures, and such assays should only be used when validated against an infectivity assay for a particular mode of inactivation.

An in vitro alternative to viability assays is the use of cell culture systems to measure oocyst infectivity. Complete development of *Cryptosporidium* was first reported using cultured human fetal lung cells and primary cultures of chicken and pig kidney cells (Current and Haynes 1984). However, while this system supported the production of sporulated oocysts it did not appear to be adopted for further use, possibly because of the requirement for primary-derived cells, with mouse infectivity being used predominantly in the 1980s. In subsequent years, a number of cell lines were evaluated for their ability to support *Cryptosporidium* development, but not as quantitative assays for measuring infectivity. For example, a canine kidney cell line (MDCK) was reported to support high rates of infection but did not produce sporulated oocysts (Gut et al. 1991), while a human endometrial carcinoma cell line was reported to support the complete development of *Cryptosporidium* in vitro (Rasmussen et al. 1993). An extensive comparison of cell lines suggested that HCT-8 cells (a human ileocecal adenocarcinoma cell line) supported the production of the highest number of parasite developmental stages (Upton et al. 1994b). The timing of sporozoite release from oocysts was also shown to be critical for successful infection of cell monolayers, with highest infection achieved when oocysts excysted when in contact with the monolayer (Upton et al. 1994a). While various cell lines have continued to be evaluated (Yang et al. 1996; Lawton et al. 1997; Deng and Cliver 1998; Lacharme et al. 2004), HCT-8 cells are currently the most commonly used for measuring in vitro infectivity.

Different end-points have been used for detecting cell infectivity. PCR-based assays have been used for end-point measurement of infection of cell monolayers either qualitatively (Rochelle et al. 1997; Di Giovanni et al. 1999) or through the use of quantitative PCR to measure the relative change in infectivity due to exposure to disinfectants or environmental exposures (Keegan et al. 2003; King et al. 2005). A chemiluminescence immunoassay has also been described for measuring *Cryptosporidium* proliferation during cell infection (You et al. 1996). Possibly the most useful end-point measurement has been the application of fluorescent antibodies to detect intracellular developmental stages of *Cryptosporidium*, allowing the detection of infectious foci (Slifko et al. 1997). In this method, a single focus of infection is caused by the localized infection of nearby cells by sporozoites released from a single oocyst followed by the subsequent infection of neighboring cells by other life-cycle stages. The focus detection method (FDM) allows quantitation of infectious oocysts in a sample (assuming each focus arises from a single oocyst). If the input number of oocysts is know the FDM also allows measurement of the proportion of infectious oocysts in a sample. For FDM to be accurate it is critical to control oocyst excystation so that sporozoites are released when the oocyst is on the monolayer to prevent dispersal of sporozoites, which could result in multiple foci from a single oocyst (King et al. 2011, 2012). An alternative approach for the detection of intracellular stages has been the use of in situ hybridization with a *Cryptosporidium*-specific oligonucleotide probe (Rochelle et al. 2001), although this technique does not appear to be in widespread use. A recent comparison of infectivity methods for both *C. parvum* and *C. hominis* has shown that the FDM method is less prone to the detection of false positives compared with a PCR end-point for infectivity detection and that FDM was better at detecting infectious oocysts spiked into water samples (Johnson et al. 2012). The robustness of cell culture infectivity using HCT-8 cells and FDM has been demonstrated in an interlaboratory trial (Bukhari et al. 2007).

Multiple comparison studies of in vitro cell culture infectivity assays with mouse infectivity have been conducted and these methods shown to be equivalent (Rochelle et al. 2002; Slifko et al. 2002; Garvey et al. 2010; Alum et al. 2011), although not for assessing the efficacy of some anti-cryptosporidial drug treatments (Theodos et al. 1998). A comparison of in vitro infectivity for three cell lines measured using a reverse-transcription PCR assay with CD-1 mouse infectivity found that HCT-8 cells gave the best correlation with the "gold standard" for oocysts exposed to UV light and ozone (Rochelle et al. 2002). A similar study compared HCT-8 infectivity using the focus detection method with BALB/c mouse infectivity for dose response and response to UV and chlorine dioxide disinfection (Slifko et al. 2002). The results were not significantly different, although it was suggested that cultured cells might allow more sensitive detection of infection compared with measuring infection in mice (Slifko et al. 2002). A HCT-8 infectivity assay using qPCR also provided comparable results with a SCID mouse assay for UV inactivation (Garvey et al. 2010). Considering that these studies used different cell culture infection measurement end-points and different mouse strains, the correlation between cell culture and mouse infectivity can be considered to be robust.

13.2 Removal and Inactivation by Natural Processes

13.2.1 The Terrestrial Environment

13.2.1.1 Heat

Oocysts can enter surface waters when infected animals defecate directly into creeks or streams or as a result of discharge from sewage systems or sewage treatment plants. However, in many instances fecal deposition is directly onto the ground, so the terrestrial environment often poses the first barrier to the entry of oocysts into surface waters. Dispersal from animal feces relies on rainfall to release oocysts from the fecal matrix and wash them into surface waters (Davies et al. 2004), and this mobilization is one of the major factors influencing the output of process-based models for oocyst fate in catchments (Ferguson et al. 2007). While in the terrestrial environment, oocysts can be exposed to stresses such as temperature and desiccation, which are particularly effective at inactivating oocysts. Experiments using moist heat demonstrated complete oocyst (presumably *C. parvum*) inactivation, measured using mouse infectivity, for exposure times between 5 and 20 min at 45 °C, although details of oocyst doses and oocyst infectivity before heat treatment were not provided (Anderson 1985). Mouse infectivity data have also shown that *C. parvum* oocysts suspended in water were completely inactivated after 1 min at 72 °C and 2 min at 64 °C (Fayer 1994). In the study by Fayer (1994), the oocyst dose administered to each mouse was 1.5×10^5 and considering that the oocysts were at most 1 month old and isolated directly from experimentally infected calves, it would be reasonable to infer a reduction in infectivity of at least 4 $\log_{10}$ units. Similar results were obtained using a pasteurization temperature of 71.7 °C for 5–15 s, resulting in complete inactivation using doses of 10^5 treated oocysts administered to mice (Harp et al. 1996). Sensitivity to high temperature appears to be common to *Cryptosporidium* species, with *C. parvum*, *C. muris* and a *Cryptosporidium* spp. isolated from a chicken all exhibiting complete inactivation after 15 s at 60 °C or 30 s at 55 °C using a dose of 10^6 oocysts into mice for *C. parvum* and *C. muris* or 2-week-old chickens for the *Cryptosporidium* spp. (Fujino et al. 2002). These temperatures can occur in animal feces exposed to sunlight, with bovine fecal material shown to exhibit temperature peaks between 40 °C and 70 °C once air temperature exceeds 25 °C (Li et al. 2005). Using diurnal temperature cycles typical of spring – autumn conditions for California rangelands, an inactivation rate of 3.27 $\log_{10}$/day was observed using a mouse infectivity model (Li et al. 2005). In these experiments, Li et al. (2005) observed that the primary loss of infectivity appeared to be due to partial or complete excystation of oocysts in the fecal matrix. The rate of loss of infectivity was much slower for winter diurnal temperatures, with an internal fecal matrix temperature of 30 °C resulting in an inactivation rate of 0.2 $\log_{10}$/day and a rate of 0.03 $\log_{10}$/day at 20 °C (Li et al. 2010).

13.2.1.2 Freezing and Desiccation

While oocysts are sensitive to heat, they are able to withstand short periods of freezing, surviving days at −10 °C and hours at −15 °C (Fayer and Nerad 1996). However, oocysts have been shown to be susceptible to freeze/thawing cycles particularly in soil (Jenkins et al. 1999; Kato et al. 2002), where soil particles cause abrasion and fragmentation of oocysts (Jenkins et al. 1999). Oocysts have been shown to be particularly sensitive to desiccation, which is significant in a terrestrial environment. Desiccation can occur within the fecal matrix, as a consequence of fecal processing by insects such as dung beetles (Ryan et al. 2011) or once oocysts have been washed from the fecal matrix into soil. Loss of infectivity from desiccation can be relatively rapid, with oocysts in calf feces dried for as little as 1 day losing infectivity for neonatal mice (Anderson 1986). Within 2 h of air-drying on slides, 97 % of oocysts were dead (as measured by vital dye staining). Mortality increased to 100 % of oocysts after 4 h (Robertson et al. 1992). Similar results were obtained when assessing oocyst survival following desiccation on stainless steel surfaces (Deng and Cliver 1999). Aside from desiccation, the physical, chemical and biological properties of soil may affect oocyst survival (Ferguson et al. 2003). Viability assays have shown that oocyst survival is affected by soil type (Jenkins et al. 2002; Davies et al. 2005), in particular soil texture and possibly low soil pH (Jenkins et al. 2002). In these studies temperature was suggested as a major factor affecting oocyst inactivation, and soil moisture appeared to have no correlation with loss of viability (Jenkins et al. 2002; Davies et al. 2005). However, a study evaluating the effect of water potential (using osmolarity as a surrogate) suggests that soil moisture content will greatly influence the rate of oocyst degradation. The study of Walker et al. (2001) used water potentials representing soil moisture contents required to support normal crop growth, the wilting point (the minimum soil moisture content required to prevent plant wilting) and extremely dry conditions. Water potential stresses were found to combine with temperature effects and freeze-thawing to enhance oocyst degradation. Degradation was likely to be more rapid in dry soil compared with calf feces or in water at low temperatures (Walker et al. 2001). A more recent study using cell culture infectivity provided further support for the findings of Walker et al. (2001), suggesting that saturated loamy soil did not affect oocyst inactivation compared with oocysts suspended in distilled water, with both resulting in 0.93 $\log_{10}$ inactivation after 10 days at 30 °C, whereas dry loamy soil had a greater effect, with 2.5 $\log_{10}$ inactivation after 10 days at 32 °C (Nasser et al. 2007). Furthermore, the study of Nasser et al. (2007) compared viability and infectivity measures for the different test conditions and found that the viability method overestimated oocyst survival by 1–2 $\log_{10}$, measuring no loss of infectivity at 30 °C and only 0.5 $\log_{10}$ inactivation at 32 °C.

13.2.1.3 Ammonia

Ammonia can be present at high levels in decomposing feces, especially in manure storages (Muck and Steenhuis 1982; Muck and Richards 1983; Patni and Jui 1991), ranging from 0.05 M, rising to 0.2 M in cattle slurries depending on storage time (Whitehead and Raistrick 1993). Oocysts have been shown to be sensitive to ammonia, initially in a study investigating gaseous disinfection of oocysts using different compounds (Fayer et al. 1996). In that study, oocysts suspended in water were exposed to one atmosphere of pure ammonia at 21–23 °C for 24 h and were rendered non-infectious to mice. Another study used excystation and vital dye staining assays to assess the effect of a wide range of ammonia concentrations (0.007–0.148 M) on oocyst viability, finding that all concentrations reduced viability and calculating that 0.06 M ammonia would inactivate 99.999 % oocysts in 8.2 days at 24 °C (Jenkins et al. 1998). The inactivation rate was temperature dependent, taking much longer (55 days) at 4 °C. Considering the sensitivity of oocysts to ammonia, it is likely that some oocyst inactivation will occur from this source in animal feces and that practices such as long-term storage of animal waste will effectively reduce *Cryptosporidium* risk, even at low temperatures (Hutchison et al. 2005).

13.2.1.4 Transport into Receiving Waters

Oocysts need to be mobilized from soil surfaces/sub-surfaces into receiving waters. There are conflicting reports of the interaction between oocysts and soil particles. Oocysts have been reported not to attach to soil particles in water (Dai and Boll 2003) and to behave as single particles, rather than aggregates (Kaucner et al. 2005; Brookes et al. 2006). However, in the context of agricultural runoff, significant attachment of oocysts was shown for clay loam and for both clay loam and sandy loam in the presence of manure (Kuczynska et al. 2005). Irrespective of adsorption/desorption, soils can greatly limit the mobility of pathogens by acting as a filter (Tufenkji et al. 2004) and for this reason the preferred method for disposal of human waste in wilderness areas is by digging small holes and burying feces (Cilimburg et al. 2000). Retention in soils has been shown to be a function of surface roughness and not to be affected by soil organic matter (Santamaria et al. 2012). A laboratory study using simulated rainfall and soil cores for three contrasting soil types (clay loam, silty loam, sandy loam) found that the distribution of *Cryptosporidium parvum* oocysts was similar for all three soils, with the majority of oocysts (73 %) in the top 2 cm of soil (Mawdsley et al. 1996). However, oocysts have been shown to rapidly pass through saturated macroporous soils, although oocyst losses through the soil columns were 90–99.9 % (Darnault et al. 2003), similar to that reported by Kuyczynska et al. (2005). The presence of the majority of oocysts near the soil surface suggests that they could be readily mobilized during a precipitation event. Removal of oocysts through soil appears to be variable and the observed transport not consistent with colloid theory (Santamaria et al. 2011). The variability in removal, particularly as a function of travel distance, has been suggested to be due to variability within the population of oocysts (Santamaria et al. 2011). The degree of agricultural land use can also affect transport, particularly the use of tile drainage, which is used to remove excess water from the soil sub-surface to improve crop growth, minimize erosion and reduce the impact of livestock on land. In the case of land with tile drainage, pathogen transport (including *Cryptosporidium*) has been shown to be predominantly via the tile drainage system and not by surface transport, except on rare occasions (Dorner et al. 2006).

Soil is not the only factor affecting oocyst mobilization, with vegetation proving to be effective at reducing *Cryptosporidium* numbers in surface runoff, causing a reduction of up to 1.44 $\log_{10}$ per meter of vegetated buffer (Tate et al. 2004). The combination of soil type and vegetated buffer influences oocyst removal, with buffers on sandy loam being less effective at removing *Cryptosporidium* oocysts (1–2 $\log_{10}$ reduction/meter) compared with vegetation on silty clay or loam (2–3 $\log_{10}$ reduction/meter) (Atwill et al. 2002). Rainfall simulation using artificial 1 kg cow fecal pats spiked with 10^7 *Cryptosporidium* oocysts on a loam soil (49 % sand, 27 % silt, 24 % clay) demonstrated that vegetation promoted infiltration (compared with a bare soil surface) and that there were fewer oocysts (1–4 $\log_{10}$ lower) in vegetated runoff compared with bare runoff. In these simulations, the number of oocysts detected in the vegetated runoff and infiltrate was much lower than that in the runoff and infiltrate from the bare soil (Davies et al. 2004). A more extensive study examining a wider range of slopes, vegetation level and rainfall conditions found that 2–5 $\log_{10}$ oocysts were retained in soil without buffer, but that 3–8.8 $\log_{10}$ oocysts were retained by soils with grassland buffers (Atwill et al. 2006). Considering that infected calves can excrete 6×10^7 oocysts per gram of feces (Uga et al. 2000) and up to 10^{10} oocysts per day for up to 14 days during the infection (Meinhardt et al. 1996), significant numbers of oocysts could still be mobilized into the aquatic environment, despite attenuation through soils and vegetation.

13.3 Surface Waters

13.3.1 Hydrological Parameters

Surface waters can provide an excellent environment for supporting the survival of oocysts, providing a thermal buffer from temperature extremes (in contrast to the terrestrial environment) and also providing a medium for transmission to and ingestion by a susceptible host. However, there are biotic and abiotic factors that can cause removal or inactivation of oocysts in an aquatic environment and also processes that control transport and distribution within a water body (Brookes et al. 2004). Key factors include particle interaction, temperature, sunlight (UV) exposure and grazing, all of which are affected by the hydrodynamics of a stream or reservoir (e.g. dispersion, dilution) (Brookes et al. 2004). Oocyst transport within lakes/reservoirs is predominantly driven by inflows, particularly those from rainfall events, and also by circulation patterns from wind-generated currents and internal waves (Brookes et al. 2004). Inflows are considered to be major sources of pathogens for reservoirs. The behavior of an inflow is controlled by the inflow density relative to that of the receiving water, with water temperature being a major factor affecting transport of oocysts. Inflow water that is warmer than the receiving water (and so less dense) will travel across the top of the receiving water surface, whereas inflow water that is colder will be more dense and flow along the bottom of the receiving water (Brookes et al. 2004). Entrainment of the receiving water by the inflow leads to dilution of oocyst concentrations. Inflow temperature, velocity, insertion depth and entrainment rate are critical factors determining the hydrodynamic distribution of oocysts in lakes and reservoirs (Brookes et al. 2004).

Sedimentation rate, along with flow velocity, is a key factor affecting the position of oocysts within a water column. Oocyst position can influence survival, oocysts at the surface can be exposed to stressors such as sunlight, whereas oocysts at depth or in sediments may be protected from sunlight but may be subjected to microbial interactions via predation or interaction with biofilms. Empirically determined oocyst sedimentation rates have been shown to closely match theoretical calculations based on oocyst size and density, being measured to be 0.27–0.35 μm/s (Medema et al. 1998; Dai and Boll 2006). Attachment to particles or entrapment within an organic matrix will greatly affect sedimentation rates (Searcy et al. 2005), but as discussed in the previous section, there have been conflicting reports on the interaction of oocysts with particles and characteristics of the water such as type and concentration of organic matter, as well as the nature of the sediment contributing particles, may influence oocyst attachment to particulates. In the context of wastewater, oocysts have been shown to readily attach to effluent particles (Medema et al. 1998), so oocysts in water resulting from sewage effluent discharge may behave differently compared with oocysts from land runoff. While the sedimentation velocity of single oocysts is too low to result in significant settling in a large water body, the rate can increase dramatically (28.9 μm/s) for oocysts attached to particles (Medema et al. 1998). A study of the oocyst loss rate in Lake Burragorang (a water storage in Sydney, Australia) measured a high sedimentation rate, 57.9–115.7 μm/s, much higher than the laboratory-measured values, and proposed that oocyst aggregates with other oocysts or particles was responsible for the accelerated removal (Hawkins et al. 2000). The results of Hawkins et al. (2000) suggest that transport under field conditions can vary dramatically from those observed under laboratory-simulated conditions. Factors that may not be accounted for are the effect of turbulence in inflows on oocyst interaction/aggregation with particles and the influence of processes within a water body, such as internal waves, which can cause rapid movement of oocysts through the water column (Brookes et al. 2004). Sediment re-suspension events, caused by turbulence from underflows or internal waves (Michallet and Ivey 1999), can remobilize oocysts, which will still pose a water quality challenge in drinking water service reservoirs. Storm events have also been shown to cause significant re-suspension of pathogens in river sediments (Dorner et al. 2006).

13.3.2 Predation and Microbial Activity

Oocysts are exposed to a variety of aquatic environments where they may come into contact with other microorganisms that are antagonistic or will reduce oocyst persistence. Despite the potential importance of biological interactions for the removal of oocysts in natural and engineered systems, relatively little has been published in this field. One of the first indications of the potential role of biological interactions was the study of Chauret et al. (1998), who observed better survival of oocysts in 0.2 μm-filtered river water compared with unfiltered river water and concluded that this was due to biological antagonism (Chauret et al. 1998). It is unclear whether the antagonism was due to predation or other microbial interactions that resulted in decreased oocyst viability (measured by excystation). Predation of oocysts by a free-living nematode (*Caenorhabditis elegans*) has been demonstrated in soils (Huamanchay et al. 2004) and interaction with *Pseudomonas aeruginosa* extracts (modeling soil exposure) led to degradation of the oocyst wall, reducing oocyst infectivity (Nasser et al. 2007). Ingestion of oocysts by *C. elegans* did not reduce the infectivity of excreted oocysts (Huamanchay et al. 2004). A variety of aquatic organisms have been shown to be capable of ingesting oocysts, including rotifers (Fayer et al. 2000; Stott et al. 2003; King et al. 2007), ciliates (Stott et al. 2001, 2003; King et al. 2007), amoebae (Stott et al. 2003; King et al. 2007), gastrotrichs (King et al. 2007) and platyhelminths (King et al. 2007). Examples are shown in Figs. 13.1, 13.2, 13.3, 13.4, 13.5, and 13.6 (reproduced from (King et al. 2007), http://www.wqra.com.au/publications/document-search/?download=81).

The effect of ingestion on oocyst infectivity has not been assessed. However, Fayer et al. (2000) noted that rotifers excreted oocysts in boluses and King et al. (2007) also identified oocyst clumping following biotic activity. As discussed above, aggregated oocysts will settle faster and so may be less mobile. Excreted oocysts, which may have been exposed to proteolytic digestion and also may be coated in other material as a consequence of the ingestion process, may also interact with particles differently, which may further increase sedimentation and decrease opportunity for further transport through a water body. In feeding experiments conducted by King et al. (2007), it was observed that ingestion by some predators resulted in the degradation of oocysts in food vacuoles (e.g. Fig. 13.1). One limitation of feeding experiments is that in general high oocyst numbers have been used (e.g., greater than 10^4 oocysts/ml) whereas in environmental waters oocyst concentrations are very low, at most 1–10 oocysts L under base-flow conditions (Brookes et al. 2004). In the environment many of the predators are size selective feeders and *Cryptosporidium* oocysts would only be an incidental

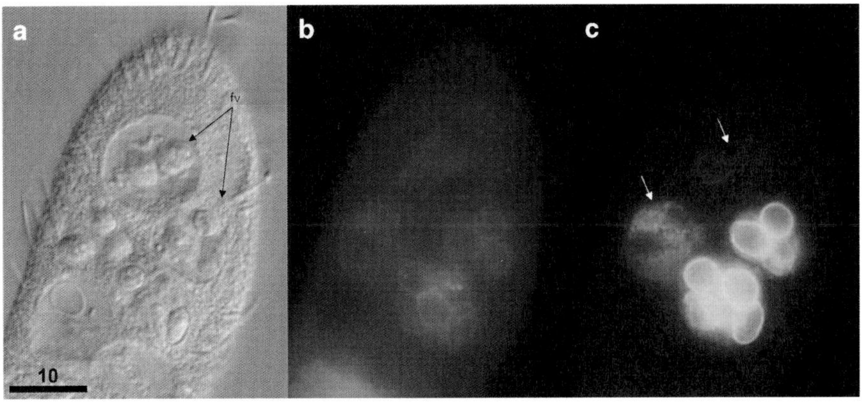

Fig. 13.1 *Paramecium* sp. enlarged to show oocysts within the food vacuoles. (**a**) DIC, (**b**) DAPI staining under UV, (**c**) oocysts labelled with a FITC antibody under *blue light*. Bar = 10 μm

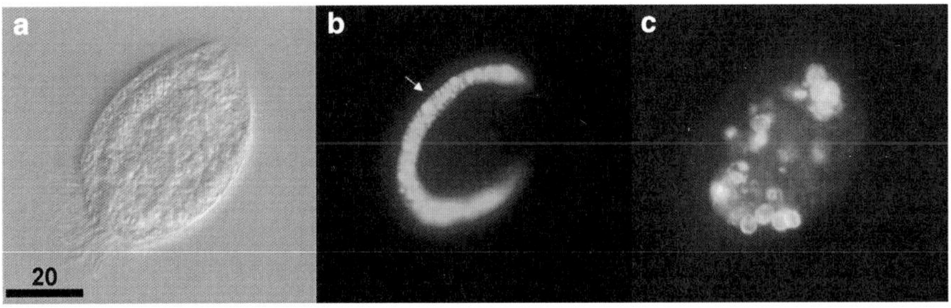

Fig. 13.2 *Euplotes* sp. containing ingested oocysts of *Cryptosporidium parvum*. The arrow indicates the *horseshoe shaped* macronucleus of the cell. (**a**) DIC, (**b**) DAPI staining under UV, (**c**) oocysts labelled with a FITC antibody under *blue light*. Bar = 20μm

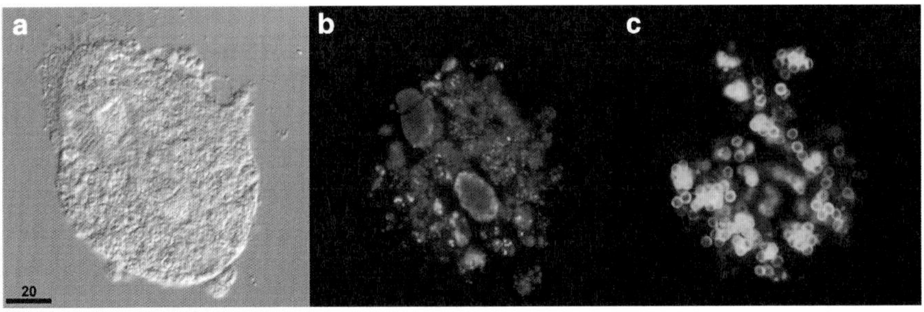

Fig. 13.3 *Oxytricha* sp. containing oocysts of *Cryptosporidium parvum*. (**a**) DIC, (**b**) DAPI staining under UV, (**c**) oocysts labelled with a FITC antibody under *blue light*. Bar = 20μm

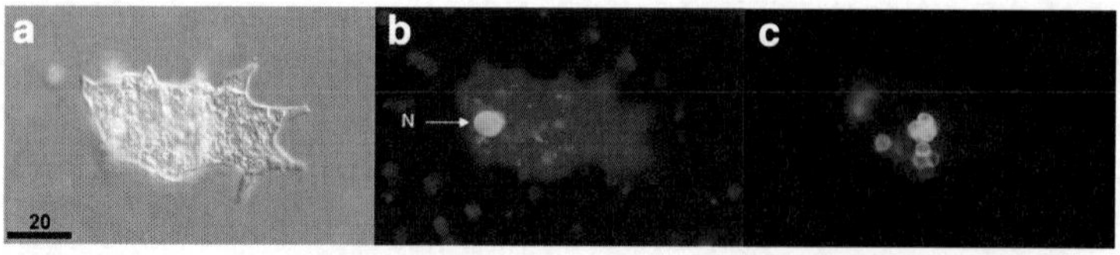

Fig. 13.4 *Mayorella* sp. containing oocysts of *Cryptosporidium parvum*. The amoeba displays conical pseudopodia (*black arrow*), which are a classifying feature of the species (**a**). The nucleus (N) of the amoeba is highlighted by the DAPI staining (**b**). (**a**) DIC, (**b**) DAPI staining under UV, (**c**) oocysts labelled with a FITC antibody under *blue light*. Bar = 20μm

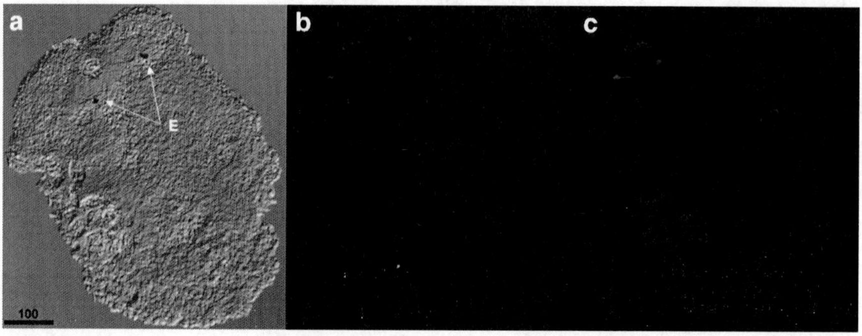

Fig. 13.5 Rhabdocoel platyhelminth containing ingested oocysts of *Cryptosporidium parvum*. Contraction and lysis of the organism occurred upon fixation, but it is still possible to determine whether the organism has ingested oocysts. (**a**) DIC, (**b**) DAPI staining under UV, (**c**) oocysts labelled with a FITC antibody under *blue light*. Bar = 100μm

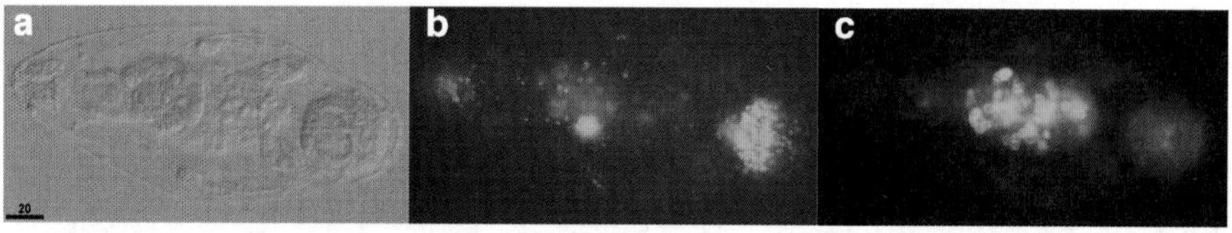

Fig. 13.6 Bdelloid rotifer (unidentified genus) containing ingested oocysts of *Cryptosporidium parvum*. (**a**) DIC, (**b**) DAPI staining under UV, (**c**) oocysts labelled with a FITC antibody under *blue light*. Bar = 20μm

food source amongst the many other organisms (such as yeast, algae and bacteria) in that size range in water. Therefore, food density would be sufficient to sustain predator numbers and oocysts in aquatic environments rich in predators (such as sediment water interfaces or biofilm surfaces) may be effectively removed.

Another mechanism for removal and release of oocysts in water is attachment to biofilms. Flow chamber experiments, using biofilms established from environmental biofilm microorganisms collected during different seasons, have been used to compare attachment and detachment of oocysts (Wolyniak et al. 2010). Oocysts were found to rapidly attach and reach a steady state with between 46 % and 77 % of oocysts attached to the biofilm during oocyst challenge, with steady detachment of oocysts once the oocyst challenge had been removed to a lower steady state of 24–65 % of oocysts retained within the biofilm (Wolyniak et al. 2010). Biofilms collected at different seasons had different levels of oocyst retention, with the spring biofilm retaining the highest proportion of oocysts (65 %) and the autumn biofilm retaining the fewest (24 %). Oocyst retention appeared to be a specific function of the microbial community harvested and was not affected by temperature (5 °C or 20 °C), biofilm thickness or water characteristics, with water harvested during autumn not reducing retention by the spring biofilm and water harvested during spring not increasing retention by the autumn biofilm (Wolyniak et al. 2010). The steady-state attachment and persistence of oocysts in natural wastewater biofilms has also been demonstrated for (Skraber et al. 2007) and drinking water and wastewater biofilms established in annular reactors (Helmi et al. 2008). The attachment and release of oocysts by natural biofilms is in contrast with earlier reports using a biofilm established with *P. aeruginosa*, which strongly retained oocysts even under high flow conditions (Searcy et al. 2006). An extension of the study by Wolyniak et al. (2010) found that seasonal differences in oocyst retention were consistent across different sites but not across different years, and that oocyst retention correlated with biofilm roughness. Challenging biofilms with increasing numbers of oocysts demonstrated that saturation could be achieved, suggesting that the number of attachment sites for oocysts is limited (DiCesare et al. 2012b). Biofilms in shallow waters may play a role in protecting oocysts from inactivation by solar radiation, with oocysts embedded at the bottom of a biofilm shown to retain infectivity up to two times better compared with oocysts at the top of a biofilm or oocysts with no biofilm in solar radiation experiments (DiCesare et al. 2012a).

Another potential source of oocyst removal in freshwaters are bivalve molluscs, which have been proposed as biological indicators of *Cryptosporidium* in river water (Graczyk et al. 2001; Izumi et al. 2006). Zebra mussels collected from a river were shown to contain significant numbers of oocysts, up to 220/g tissue (Graczyk et al. 2001). Experimental exposure of the clam *Corbicula japonica* to oocysts found that oocysts were excreted in the clam feces at high rates, with 1–3 % of the input oocysts retained in tissue (based on the challenge doses, similar numbers in tissue compared with the report of Graczyk et al. 2001) (Izumi et al. 2006). Passage through the clams had no effect on oocyst infectivity, as measured using cell culture infectivity. Clams placed in a river were able to concentrate oocysts, which were isolated and identified by genotyping to be *C. parvum* (Izumi et al. 2006). The high clearance rate reported suggests that detection of low numbers of oocysts in clam tissue may represent previous contamination events, and high numbers of oocysts representative of recent events. Infectious oocysts have been recovered from marine molluscs harvested for food (Gomez-Bautista et al. 2000). While filter-feeding molluscs may play a role in the removal of oocysts in freshwater systems, there appears to have been little research activity in this area (aside from use as a sentinel organism), with research activity predominantly in the area of oocyst contamination of commercially harvested shellfish.

13.3.3 Inactivation by Temperature

The temperatures in aquatic environments are generally more moderate compared with the high temperatures that can be encountered in a terrestrial environment. While inactivation at high temperatures is rapid, particularly at temperatures above 40 °C, it can be relatively slow at temperatures more relevant in water bodies. Oocysts have been shown capable of retaining infectivity for at least 24 weeks at 4 °C and 16–24 weeks at 15 °C (Fayer et al. 1998; Keegan et al. 2008), with at most a 1 $\log_{10}$ reduction in infectivity. Using mouse infectivity, Fayer et al. (1998) found that oocysts retained infectivity after 24 weeks stored at 20 °C and 12 weeks stored at 25 °C, but that the number of mice infected and the level of infection was greatly reduced compared with control oocysts. Quantitative measurement of inactivation using cell culture infectivity found more than 3 $\log_{10}$ inactivation of oocysts after 12 weeks stored at 20 °C, 8 weeks stored at 25 °C and 3 weeks stored at 30 °C (King et al. 2005). In cooler climates, temperature is unlikely to play a role in oocyst inactivation unless the detention time in water is extremely long. However, depending on reservoir hydrodynamics and oocyst detention times, tropical and possibly sub-tropical climates may well influence oocyst survival in the aquatic environment.

The mechanism responsible for oocyst inactivation, at least at temperatures of 37 °C or less, has been shown to be metabolic exhaustion. *Cryptosporidium* oocysts contain finite carbohydrate energy reserves, stored as amylopectin, which are required to initiate infection and are consumed due to metabolic activity at ambient temperatures (Fayer et al. 1998). Oocyst energy levels (measured as ATP) have been shown to closely correlate with retention of infectivity in cell culture for temperatures between 4 °C and 37 °C (King et al. 2005). The finite energy reserve within sporozoites is a key factor in determining host cell infection success, meaning that the timing of sporozoite release is critical and must be controlled for optimal measurement of infectivity using cell culture models (Upton et al. 1994b; King et al. 2011).

13.3.4 Inactivation by Sunlight

Sunlight is possibly the most potent environmental stressor in the aquatic environment. The most damaging components of sunlight are short-wavelength ultraviolet (UV) light, specifically UV-B (280–320 nm) and UV-A (320–400 nm), with UV-B principally causing DNA damage and UV-A causing damage to a variety of cellular components including lipids and proteins (Caldwell 1971; Friedberg et al. 1995; Malloy et al. 1997; Ravanat et al. 2001). UV exposure can be detrimental to a variety of organisms ranging from bacteria to plants in both terrestrial and aquatic environments (King and Monis 2007).

In terms of *Cryptosporidium* in water, the bulk of research appears to have focused on measuring the effectiveness of solar disinfection (SODIS) for oocyst inactivation, which combines solar irradiation with temperature inactivation and may have variable wavelengths of UV involved, depending on the UV transmissivity of the containers used. However, several studies have assessed the effectiveness of solar radiation on oocyst infectivity under environmentally relevant conditions. An early study utilized excystation to measure a 90 % reduction in oocyst viability following a 3-day exposure to sunlight in marine water (Johnson et al. 1997). More recently, solar radiation and artificial UV-B have been shown to be effective at reducing the infectivity of oocysts suspended in phosphate-buffered saline (Connelly et al. 2007), although it was not clear how inactivation was calculated in that study. A dose of 66 kJ/m^2 (measured at 320 nm) resulted in a 97.5 % reduction in infectivity compared with the dark control, and a claimed >99.99 % loss of infectivity was observed following 10 h exposure to sunlight on a mid-summer day. However, some of the calculated reductions may not be accurate, based on the data presented (triplicate samples of 5,000 oocysts used in solar radiation experiments, each replicate analyzed in triplicate using 100 oocysts per analysis, measuring infectivity from a total of 900 oocysts) At most a 3 $\log_{10}$ inactivation (99.9 %) could be measured if all of the oocysts were infectious at the start of the experiment (but at most 22 % were infectious in the controls). Additionally, there was also between two and almost tenfold difference in the infectivity measured for dark and laboratory control oocysts, suggesting either large variability in the measured infectivity, or some other effect on infectivity aside from solar radiation (temperature can be excluded because it did not exceed a maximum of 22.5 °C on any of the testing days). The authors noted the high variability in some samples and suggested it could be due to differences within the oocyst population, oocyst clumping (causing underestimation of infectious foci) or variability in the quality/density of the cell monolayer (Connelly et al. 2007). Irrespective of these possible shortcomings of the cell assay or inactivation calculations, the findings clearly demonstrated the impact of solar radiation on oocyst infectivity.

Another study assessed the effect of solar radiation on the survival of *C. parvum* oocysts suspended in tap water or environmental waters, using a cell culture infectivity assay (King et al. 2008). These experiments were conducted on days with different levels of solar insolation, measured using a pyranometer, and compared with the UV index. The UV index is a measure developed by the World Health Organization to raise public awareness of the risk of UV radiation and provides an indication of the daily danger of solar UV radiation intensity (http://www.who.int/uv/intersunprogramme/activities/uv_index/en/index.html). It is calculated at solar noon, when the sun's radiation is most intense, with each point on the index equivalent to 25 mW/m^2 of UV radiation 290 nm and 400 nm. The water quality ranged from 2.8 mg/L dissolved organic carbon (DOC), 0.23 nephelometric turbidity units (NTU) turbidity and a color of 2 Hazen units (HU) for the tap water, to maximums of 12.3 mg/L DOC, 7.42 NTU turbidity and 77 HU color for the reservoir waters. For days with UV indices of 4 or higher, reductions in infectivity of between 2.2 and 3 $\log_{10}$ were achieved within a single day of exposure (King et al. 2008). The inactivation rates on different days were compared using T_{90}, which is the time (hours) required to achieve 90 % inactivation. The T_{90} was larger for low UV index days (e.g. 6.4 for a day with a UV index of 1) and smaller for high UV index days (e.g. 0.4 for a day with a UV index of 12). Comparison of T_{90} for the different water types suggested that DOC had the largest influence on oocyst inactivation by solar UV, with waters with the T_{90} increasing as the DOC increased, and waters with similar DOC having similar T_{90}, despite differing by up to tenfold in color and turbidity (King et al. 2008). The penetration of UV in water has been shown to be highly dependent on the concentration and type of DOC (Morris et al. 1995; Jerome and Bukata 1998; Hutchison et al. 2005), providing protection from solar radiation for oocysts that are at sufficient depth to allow UV attenuation. However, particular events such as warm-water inflows into cold reservoirs will place oocysts within a zone where they can be inactivated by sunlight, which can cause >90 % inactivation after a few hours on a high UV index day.

The relative contribution of the different wavelengths of sunlight has been investigated using long-pass filters (Connelly et al. 2007; King et al. 2008). Connelly et al. (2007) used a <404 nm long pass filter to remove UV-B and UV-A. Compared with dark controls, oocysts exposed to sunlight with the UV component removed had between 42 % and 61 % reduction in infectivity, compared with reductions of 93 and 97 % for oocysts exposed to solar UV. King et al. (2008) used cut off filters to remove UV-B (<323 nm) and UV-A and UV-B (<400 nm). In these experiments, full sunlight achieved 2.5–2.7 $\log_{10}$ oocyst inactivation, whereas the removal of UV-B reduced the amount of inactivation to 0.8–1.3 $\log_{10}$ and the removal of both UV-A and UV-B abolished any reduction in infectivity compared with the dark control, which is in contrast with the finding of Connelly et al. (2007). While UV-B might be more effective in causing DNA damage and reducing infectivity (causing approximately two thirds of the observed inactivation), UV-A light has also been previously shown to cause cytotoxic and mutagenic effects, although to a smaller extent than UV-B (Ravanat et al. 2001). While UV-A causes less damage and a longer exposure time is required, it is likely to be of greater ecological significance than UV-B in the context of oocyst inactivation within a water column because longer wavelengths are better able to penetrate deeper into water.

The mechanism of solar UV inactivation has been further elucidated through the study of the effect of UV radiation on sporozoite DNA and cellular processes (King et al. 2010). One of the main mechanisms of inactivation caused by short wavelength UV irradiation is the production of DNA photoproducts, specifically cyclobutane pyrimidine dimers (CPDs) (Ravanat et al. 2001). Oocysts were exposed to varying doses of UV-C and an anti-CPD antibody was used to readily detect CPDs in oocyst DNA extracts (King et al. 2010). However, CPDs could not be detected in oocysts exposed to levels of solar UV causing 3 $\log_{10}$ inactivation, whereas CPDs could be detected in oocysts exposed to a UV-C dose causing the same level of inactivation (approx. 10 mJ/cm^2). It was possible to detect CPDs in the DNA from solar UV irradiated oocysts, but only by increasing the amount of

DNA assayed tenfold and also by increasing the exposure times for the luminescent assay used to detect antibody binding events (King et al. 2010). Other cellular targets were assessed to determine if they were affected by solar UV causing loss of infectivity. Solar radiation was found to induce membrane depolarization in sporozoites, leading to an increase in intracellular calcium and apical organelle discharge, which was apparent as a loss of internal granularity as measured by flow cytometry (King et al. 2010). The ultimate consequence of exposure to solar radiation was the accelerated depletion of ATP (compared with dark controls) and a decrease in the ability of sporozoites to attach to and invade host cells (King et al. 2010).

13.3.5 Inactivation by Other Environmental Stressors

While ammonia levels may become high enough in animal feces/manure storages to reduce oocyst viability, the impact of free ammonia on oocyst viability is likely to be negligible in drinking water reservoirs (Brookes et al. 2004). Brookes et al. (2004) estimated that levels of ammonia in lakes would typically be 1,000-fold lower than the lowest concentration tested by Jenkins et al. (1998) and that ammonia levels would be unlikely to significantly affect oocyst viability even in lakes undergoing eutrophication. However, anthropogenic activity could introduce other stressors into the aquatic environment. For example, algicides are often used to control blooms of toxic or taste and odor-producing cyanobacteria, which commonly occur in water storage reservoirs (due to eutrophication, often as a consequence of nutrient inputs from agriculture and urban development) or waste stabilization ponds and cause problems with the production of potable or reuse water. These blooms were historically treated using compounds such as copper sulfate, but due to environmental concerns, alternatives have been developed, such as stabilized hydrogen peroxide. Hydrogen peroxide has been shown to be effective at inactivating oocysts (Barbee et al. 1999; Weir et al. 2002; Quilez et al. 2005; Castro-Hermida et al. 2006), but the available studies have focused on high doses of peroxide for rapid inactivation of *Cryptosporidium*. It is not known if oocysts are also inactivated by stabilized hydrogen peroxide or if the amounts dosed to control cyanobacteria will provide a high enough concentration to be effective against *Cryptosporidium* oocysts.

13.4 Removal by Treatment Processes

Water treatment provides a key barrier to the entry of *Cryptosporidium* into drinking water supplies. For the purpose of this chapter, we will consider water treatment processes to be those engineered processes responsible for the physical removal of oocysts from water and will consider disinfection processes separately (although they are part of the treatment plants used to produce drinking water or reuse water). As discussed elsewhere in this book (see Chaps. 3 and 12), many of the largest waterborne outbreaks of cryptosporidiosis around the world have been caused by failures in water treatment systems (e.g., Hayes et al. 1989; Mac Kenzie et al. 1994; Yamamoto et al. 2000), so water treatment is a critical barrier for the protection of public health (Risebro et al. 2007).

13.4.1 Flocculation

Cryptosporidium poses a challenge to conventional water treatment plants. Most older plants utilize treatment processes designed to remove turbidity and to a certain degree natural organic matter (NOM), but have not been designed to target the removal of particles in the *Cryptosporidium* oocyst size range. The first step in conventional filtration plants is the addition of coagulants, such as alum or ferric chloride, as well as coagulant aids (cationic polyelectrolytes), to remove particulates by flocculation followed by sedimentation. Coagulation results in the aggregation of suspended particles in source water, forming a gel like precipitate (aluminium hydroxide in the case of alum). The removal of *Cryptosporidium* through flocculation can be quite effective and is the basis of a cost effective method for the concentration of oocysts in water samples for routine monitoring (Vesey et al. 1993). Poor coagulation conditions have been shown to result in poor filtration performance (Emelko 2003). A pilot plant study using alum as a coagulant removed approximately 1 $\log_{10}$ oocysts after coagulation and sedimentation (Hsu and Yeh 2003). Jar testing of river water, using alum as the coagulant and an oocyst surrogate (with similar characteristics to an oocyst – 4–6 μm in size, 1.05–1.10 specific gravity, −25 mV (pH 6.0–6.5) Zeta potential), measured 1.2–1.5 $\log_{10}$ surrogate removal by coagulation/sedimentation (Lee et al. 2007). The mechanism of oocyst removal during flocculation is dependent on the oocyst charge (Bustamante et al. 2001). A study of oocyst zeta potential in the presence of different coagulants suggests that oocysts maintain a negative charge in the presence of ferric chloride and so sweep flocculation (physical entrainment in the developing floc) is required for oocyst removal, whereas oocysts undergo charge neutralization in the presence of alum, meaning that oocysts can adsorb to alum flocs, as well as be removed by sweep flocculation (Bustamante et al. 2001). These findings have been further supported by studies of oocyst charge characteristics in the presence of different salt concentrations and coagulants (Butkus et al. 2003). The chemicals used for flocculation do not appear to reduce oocyst infectivity. Examination of filter backwash water from water treatment plants detected infectious oocysts (Di Giovanni et al. 1999) and jar testing using alum as the coagulant did not measure any reduction in oocyst infectivity. A small amount of infectivity reduction of 0.4 $\log_{10}$ also occurred in the control jars without alum, indicating an effect due to mechanical processing (Keegan et al. 2008).

13.4.2 Media Filtration

Sand filtration was first reported to have potential for removal of oocysts based on a laboratory-scale experiment using a 1 m column of pool filter sand challenged with oocysts purified from a human clinical sample (Chapman and Rush 1990). In this experiment, 10^6 oocysts were used to challenge the filter and none were detected in the filtrate, while oocysts were detected only in the top 300 mm of the filter bed (Chapman and Rush 1990). The transport of oocysts through a filter medium has been measured for wastewater treated with sand filters. Grain size was a key factor, with fine-grained sand effectively removing oocysts while coarse-grained media had more oocyst break-through, depending on hydraulic loading rates (Logan et al. 2001). Filter removal efficiency has been shown to be affected by the presence of biofilm and NOM. The removal efficiency was measured for oocysts coagulated using calcium chloride and filtered through silicon beads, finding that biofilm reduced the filtration efficiency by 50 % and the presence of 5 ppm of NOM reduced efficiency by 75 % (Dai and Hozalski 2002). The reduction in filtration efficiency was likely due to charge interactions, with improved filtration performance observed using alum as the coagulant, even in the presence of NOM (Dai and Hozalski 2002). Filtration efficiency has also been shown to be affected by ionic strength and pH (Hsu et al. 2001). However, a more recent study suggests that oocyst removal by a packed quartz sand column was not overly affected by ionic strength and that physical straining is an important capture mechanism (Tufenkji et al. 2004). Atomic force microscopy has characterized the interaction between oocysts and sand particles, suggesting that proteins on the oocyst surface contribute to electrostatic repulsion against the surface of sand particles, and that dissolved organic carbon and dissolved calcium assist with overcoming this repulsion (Considine et al. 2002). A further study of oocyst wall macromolecules suggests that they are responsible for electrostatic repulsion from surfaces, even in the presence of ionic strengths predicted to overcome any electrostatic repulsion, and that treatment of the oocysts with proteinase K to remove the macromolecules resulted in improved oocyst attachment to surfaces (Kuznar and Elimelech 2006).

A monitoring program of source water and drinking water produced by a conventional treatment plant (coagulation-flocculation, sedimentation and rapid filtration) measured a combined oocyst removal of 2.5–2.7 $\log_{10}$, depending on the method used to determine oocyst numbers (geometric mean versus 90 % observation level) (Hashimoto et al. 2001). Pilot-scale experiments using similar treatment processes measured a combined oocyst removal of 5–6 $\log_{10}$ (Gale et al. 2002). There appear to be fewer studies examining filtration performance separately from coagulation/sedimentation. Sand filter column testing, using a *Cryptosporidium* tracer (described in the *Flocculation* section) and water that had been coagulated with alum measured a 1.3–1.5 $\log_{10}$ removal of the tracer (Lee et al. 2007). Poorer performance (0.5 removal) was reported for a pilot plant sand filter receiving secondary treated effluent from an activated sludge plant, although performance was greatly increased by dosage of alum (2 $\log_{10}$ removal) (Suwa and

Suzuki 2003). Other studies have reported variable filter performance in wastewater treatment plants, ranging from 0.4 to 1.5 $\log_{10}$ depending on the filter type (Levine et al. 2008; Fu et al. 2010). The differences in performance are likely due to differences in operational parameters such as hydraulic load, chemical pretreatments and backwashing regimes (Levine et al. 2008). In contrast, laboratory-scale slow sand filters have achieved >5 $\log_{10}$ oocyst removal, a similar performance as observed for pool sand filters by Chapman et al. (1990) (Hijnen et al. 2004). This removal rate was confirmed in a later study by loading a mature pilot sand filter with oocysts, dosing for 100 days and monitoring for 150 days, demonstrating oocyst removal of 4.7 $\log_{10}$ (Hijnen et al. 2007). Oocysts did not accumulate in the filter and it was hypothesized that they were removed as a result of predation by zooplankton (Hijnen et al. 2007). Granular activated carbon (GAC), often used in filters to reduce NOM or remove organic contaminants, has also been shown to be effective at removing oocysts (1.3–2.7 $\log_{10}$) (Hijnen et al. 2010). A study on the fate of oocysts on GAC also observed oocyst loss from the filter medium due to biotic activity (Bichai et al. 2010). Zooplankton were isolated, identifying rotifers as a dominant group (previously shown to be predators of *Cryptosporidium*) and oocysts were recovered from zooplankton following release by sonication (Bichai et al. 2010).

Conventional filters using sand, gravel, anthracite, etc., are not the only types of filters in use for the production of drinking water or the treatment of wastewater. Bank filtration, which passes water through a river bank, is in use in some parts of the world as a pretreatment for drinking water or for wastewater. The removal efficiency of sandy alluvial riverbank sediment in bank filtration, based on modeled distribution profiles, has been claimed to be 23–200 $\log_{10}$/m (Faulkner et al. 2010). In the case of bank filtration, straining is not the only removal mechanism and surface charge of the medium has been suggested to play a major role, such that gravel soil can have better performance than sandy soils for removal of microorganisms (Hijnen et al. 2005; Faulkner et al. 2010). The performance of bank filtration for oocyst removal has been found to correlate with the surface coverage of the sediment grains by metal oxides, with higher coverage providing better removal (Metge et al. 2010). A novel filtration medium was developed using Zeolite treated with a quaternary ammonium chloride. This modified Zeolite exhibited an oocyst removal efficiency of 0.54, which is comparable with sand filters but of most interest it also resulted in 1.19 $\log_{10}$ inactivation of the oocysts (Abbaszadegan et al. 2006). Surface functionalization of the sand in sand filters may improve filter performance. Functionalizing silica with 3-(2-aminoethyl) aminopropyltrimethoxysilane improved the filtration performance from 1 $\log_{10}$ for the uncoated silica to 1.3–1.4 $\log_{10}$ for the functionalize silica, depending on the level of coating (Majewski and Keegan 2012).

13.4.3 Dissolved Air Flotation

Dissolved Air Flotation (DAF) treatment of surface waters and wastewaters uses coagulants to develop flocs, which are floated to the surface using aeration and then collected from the top of the reactor and disposed of as sludge (for a detailed review of this process see Edzwald 2010). The effectiveness of the DAF process, like conventional flocculation processes, is affected by turbidity and natural organic matter and has reached oocyst removals of 1.7–2.5 $\log_{10}$, increasing to $> 5.4 \log_{10}$ when combined with filtration (DAFF) for drinking water (Edzwald et al. 2001). In the case of DAFF, the operation of the plant in terms of filter backwashing is important for determining if oocysts are concentrated in the floated sludge or in the filters, with the volume of backwash water used determining the level of oocyst build-up in the plant influent (Edzwald et al. 2001).

13.4.4 Membrane Filtration

Membrane filtration systems can provide excellent removal of oocysts, through size exclusion. Most research on membrane filters in terms of *Cryptosporidium* has been in relation to the concentration of oocysts from large volume water samples (e.g. Hill et al. 2007; Lindquist et al. 2007; Liu et al. 2012), rather than validating the technology for oocyst removal. Full-scale validation of membrane filters will often measure removal of bacteria (such as *E. coli*) and use calculated $\log_{10}$ removal values (LRV) for smaller organisms to provide a conservative LRV for the much larger oocysts. In practice, the performance of filters is higher than the maximum LRV that will be credited by health regulators. As a guiding principle, health regulators (at least in Australia) will not provide a credit of more than 4 LRV for any individual treatment barrier. In the case of membranes this is partly driven by limitations in real-time detection of membrane failure, for example parameters such as turbidity may not detect failures that will allow breakthrough of small particles (such as viruses) while still removing larger particles and producing acceptable levels of turbidity in the filtrate. Ultrafiltration has been shown to be effective ($>1.84 \log_{10}$) at removal of oocysts in a full scale testing for tertiary treatment of secondary effluent (Fu et al. 2010). A pilot-scale evaluation of conventional and membrane filters has been reported, but the numbers of oocysts in the raw water were low (56 oocysts/10L) and so limited conclusions can be drawn from that work regarding oocyst removal efficiency (Hsu and Yeh 2003).

13.4.5 Activated Sludge and Other Secondary Wastewater Treatment Processes

As water becomes scarcer, increasing attention is being given to the reuse of wastewater. Most of the common conventional wastewater treatment processes, such as activated sludge process (ASP) or waste stabilization ponds, are designed to reduce nutrients to a level suitable for environmental discharge, with pathogen removal more often being achieved by disinfection prior to discharge, although some pathogen attenuation may occur during treatment. Wastewater reuse usually requires tertiary treatment to further improve the quality of the water and also to remove pathogens and protect human health. The activated sludge process is essentially a large bioreactor designed to achieve nitrification and denitrification of the wastewater to remove nitrogen from the wastewater. Being an active biological process, there is some opportunity for removal or inactivation of oocysts, either through entrapment in biological flocs or predation. An early study of lab-scale ASP using *C. parvum* oocysts demonstrated a 1 $\log_{10}$ reduction in the intensity of infection measured using mice (Villacorta-Martinez de Maturana et al. 1992). A pilot plant ASP showed an oocyst removal efficiency of 2, which was increased to 3 $\log_{10}$ by dosing with alum (Suwa and Suzuki 2003). Monitoring of a full-scale wastewater treatment plant measured a 1.52 $\log_{10}$ oocyst removal for conventional ASP and higher removals for an oxidation ditch process (2.17 $\log_{10}$) and an anaerobic-anoxic process (1.79 $\log_{10}$) (Fu et al. 2010). Another study of a full-scale ASP reported better removal of oocysts, with an LRV of almost 3 $\log_{10}$ (Neto et al. 2006). There are many different configurations for ASP, and these may have a large effect on removal efficiency. A lab-scale ASP, using return activated sludge from a full-scale ASP, was able to achieve a 2.4 $\log_{10}$ reduction in oocyst number, whereas the full-scale plant only removed 0.7 $\log_{10}$ (Wen et al. 2009).

13.4.6 Lagoons/Waste Stabilization Ponds

Wastewater stabilization ponds and constructed wetlands provide attractive low technology and low energy solutions for treating contaminated waters such as wastewaters or storm waters, especially in areas where there is sufficient space. Provided that detention times are adequate and there are reduced opportunities for disturbance of sediments, lagoons and constructed wetlands have the potential to achieve excellent removal of oocysts from wastewater or storm water. Sedimentation is a major mechanism for oocyst reduction. Analysis of oocysts in a surface flow wetland showed that oocyst numbers in sediment were 1–3 orders of magnitude higher compared with oocyst numbers in the water column (Karim et al. 2004). Monitoring of lagoon systems used for treating wastewater has shown 1–2.5 $\log_{10}$ oocyst removal (Ulrich et al. 2005), with constructed wetlands capable of $1-2$ $\log_{10}$ LRV (Quinonez-Diaz et al. 2001; Reinoso et al. 2008). As previously mentioned, some of this removal is thought to be a result of protozoan predation (Stott et al. 2001). Some variability in the performance of constructed wetlands has been noted, with a system receiving sewage effluent only removing 64 % of oocysts (Thurston et al. 2001).

High rate algal ponds (HRAP) are a variation of waste stabilization ponds, actively using energy to provide a higher level of oxidation to support algal growth, which mediates nutrient removal. The HRAP design has been shown to be effective at oocyst inactivation, causing 2–3 $\log_{10}$ reduction in mouse infectivity (Araki et al. 2001). Further study of oocyst inactivation in a HRAP using semipermeable bags suggested that the inactivation was due to a combination of ammonia, pH and sunlight exposure (Araki et al. 2001). Sunlight has also been suggested to be a major cause of inactivation of oocysts in waste stabilization ponds, with 40 % inactivation observed after 4 days exposure to sunlight (Reinoso and Becares 2008). Considering that secondary treated effluent is high in organics, it is tempting to speculate that UV-A is a primary factor causing inactivation in this scenario. A comparison of anaerobic ponds, a facultative pond and a maturation pond suggested that sunlight and water chemistry were the main factors influencing oocyst removal in the anaerobic and maturation ponds, whereas predation may have been more important in the facultative pond (Reinoso et al. 2011).

13.5 Disinfection of *Cryptosporidium* in Water

This chapter has so far reviewed natural and engineered processes that remove or inactivate *Cryptosporidium* oocysts in environmental waters or in wastewaters. The final barrier, either for the production of safe drinking water or for the production of fit for purpose reuse water, is disinfection. Many processes have been evaluated, with some now in widespread use for the management of any *Cryptosporidium* oocyst that may break through water treatment processes. Disinfection methods, which will be discussed below, include chlorine, chloramine, chlorine dioxide, ozone, mixed oxidants and ultraviolet light (UV).

13.5.1 Oxidant-Based Disinfection

The resistance of *Cryptosporidium* to disinfectants such as chlorine is one of the primary reasons *Cryptosporidium* has been an issue for the water industry and the cause of so many drinking water-related outbreaks. The ineffectiveness of chlorine and chloramine was recognized in the early 1990s, with disinfection studies showing that a contact time (Ct) of 1,600 mg.min/L using chlorine was required to achieve a 1 $\log_{10}$ reduction in oocyst infectivity, measured using mice (Korich et al. 1990). Contact time, which is the product of the disinfectant dose (mg/L) multiplied by time of exposure (minutes), provides a measure for comparing the effectiveness of different chemical disinfectants and also provides a way of ensuring delivery of an adequate dose during water treatment. In the case of the study of Korich et al. (1990), the Ct of 1,600 mg.min/L was delivered using 80 ppm (ppm = mg/L) of free chlorine with an exposure time of 20 min. Chloramine was shown to be equally ineffective in the same study. While it is not possible to deliver the required Ct for treating drinking water or reuse water, chlorine can be used to effectively inactivate oocysts in swimming pools following a fecal accident. However, there is usually insufficient contact time immediately following a fecal accident, so any pool users in the vicinity will likely become infected if they have ingested contaminated water before the pool has been treated with elevated levels of chlorine. A more recent study using two different isolates of *C. parvum* and cell culture infectivity demonstrated that the Ct required for 3 $\log_{10}$ inactivation was 10,400–15,300 mg.min/L, which is larger than the previous recommendation from the Centre for Disease Control and Prevention for the treatment of recreational waters following a fecal accident (Shields et al. 2008). Chlorine dioxide was shown to be more effective, while ozone treatment (1 ppm) was the best option for inactivation of oocysts, achieving 2 $\log_{10}$ inactivation for a Ct estimated between 5 and 10 mg.min/L and >4 $\log_{10}$ inactivation for a Ct of 10 mg.min/L (Korich et al. 1990). Similar results were observed in an independent study using 1.11 ppm ozone, achieving approximately 4 $\log_{10}$ inactivation for a Ct of 6.6 mg.min/L (Peeters et al. 1989).

Electrochemically produced mixed oxidants (chlorine plus other uncharacterized species produced by a MIOX unit) have been shown to inactivate >3 $\log_{10}$ oocysts with a Ct of 1,200 mg.min/L (5 ppm mixed oxidant with 4 h contact), whereas no inactivation was observed after exposure to chlorine for Cts of 1,200–7,200 mg.min/L (5 ppm for 4–24 h contact) (Venczel et al. 1997), although it was not clear from the study if chlorine decay was measured to properly calculate a Ct. The lack of any observed inactivation for chlorine for the nominated Ct range is suspect considering the earlier findings of Korich et al. (1990). A study of the inactivation kinetics of chlorine dioxide for oocysts compared excystation and cell culture infectivity, showing that for a Ct of 1,000 mg.min/L, the excystation method only measure 0.5 $\log_{10}$ inactivation, whereas the infectivity method measured a 2 $\log_{10}$ inactivation (Chauret et al. 2001). This result suggests that the earlier finding of Venczel et al. (1997) may have in part been affected by the method used for estimating viability. The performance of MIOX units appears to be variable. The effectiveness of MIOX could not be reproduced in a subsequent study using a cell culture infectivity assay, where the MIOX mixed oxidants and chlorine treatment performed similarly (Keegan et al. 2003). However, a different study using mouse infectivity measured inactivation for oocysts exposed to 2–5 mg/L MIOX (measured as free chlorine) but not for 5 mg/L free chlorine (Sasahara et al. 2003). An on-site evaluation of an electrolytic system (ECO, using the same system as MIOX) found that the system was not effective at inactivating oocysts, producing equivalent results to chlorine (Venczel et al. 2004).

Conflicting results have been reported for the sensitivity of oocysts as a function of age. A study using oocysts aged in dialysis cassettes suspended in a river reported no change in chlorine sensitivity, although this study used excystation to measure chlorine inactivation (Chauret et al. 1998). In contrast, oocysts stored at 4 °C for 24 weeks were shown to have approximately 1 $\log_{10}$ more inactivation for Ct 300 mg.min/L compared with fresh oocysts using an infectivity assay (Keegan et al. 2008).

13.5.2 Ultraviolet Light

Discussion within the scientific literature regarding the use of UV-C for the disinfection of potable water first began in the late 1980s/early 1990s in response to proposed changes to the US Safe Drinking Water Act (Wolfe 1990). At that time, there was no information on UV inactivation of *Cryptosporidium*, but oocysts were presumed to be more resistant than bacteria on the basis of the observed resistance of oocysts to chlorine (Wolfe 1990). There are two types of UV-C lamps used in disinfection systems, medium pressure (MP)-UV lamps that deliver UV light in the full UV-C spectrum (200–300 nm) and low pressure (LP)-UV lamps that deliver germicidal UV-C (254 nm). The first report of the effectiveness of UV against *Cryptosporidium* used mouse infectivity to demonstrate complete inactivation of oocysts following a UV dose of 15,000 mW/s for 150 min, although the dose delivery was not measured to allow comparison with more recent studies (Lorenzo-Lorenzo et al. 1993). A subsequent study used viability assays (vital dyes and excystation) to demonstrate 2–3 log 10 reduction in viability using a total low pressure LP-UV dose of 8,748 mJ/cm^2 (Campbell et al. 1995) and on the basis of this and similar studies using viability techniques it was believed for many years that *Cryptosporidium* was relatively UV resistant.

A detailed study of the dose response of *Cryptosporidium* to LP- and MP-UV, using mouse infectivity and a collimated beam apparatus for precise UV dose delivery, demonstrated inactivation of 2 $\log_{10}$ for 10 mJ/cm^2, 3 $\log_{10}$ for 25 mJ/cm^2 and 3.4 – >4.9 $\log_{10}$ for 119 mJ/cm^2 (Craik et al. 2001). Evaluation of a UV disinfection device that delivered 120 mJ/cm^2 was shown to reduce infectivity in mice by 5.4 $\log_{10}$ (Drescher et al. 2001). The most effective wavelengths of UV-C were determined to be 250–275 nm (Linden et al. 2001). A direct comparison of animal infectivity and excystation methods demonstrated that high doses of UV were required to reduce excystation, but much lower doses were required to reduce infectivity (Morita et al. 2002). Dose–response studies at low UV doses measured >2.6 $\log_{10}$ for 3 mJ/cm^2 and >3.9 $\log_{10}$ for 10 mJ/cm^2 (Shin et al. 2001), with a similar bench-scale study using cell culture infectivity demonstrating >3 $\log_{10}$ inactivation for 5–10 mJ/cm^2 doses of LP-UV (Bukhari and LeChevallier 2003). Due to ease of production in calves or mice, the vast majority of UV studies have used *C. parvum*. Using oocysts isolated from a human patient, *C. hominis* has been shown to possess a similar UV sensitivity compared with *C. parvum* (Johnson et al. 2005). The requirement for lower UV doses has translated into suitable UV reactor design, with full-scale UV systems described for *Cryptosporidium* disinfection and validated using MS2 bacteriophage as a challenge organism (Bukhari and LeChevallier 2003).

A study of ex vivo and in vivo reactivation of oocysts found no evidence for reactivation, with oocysts exposed to 60 mJ/cm^2 of MP-UV remaining inactivated, as measured using mouse infectivity (Belosevic et al. 2001). Other studies also examined photoreactivation and dark repair, finding that although *Cryptosporidium* possesses UV repair genes and that cyclobutane pyrimidine dimers were shown to be repaired, infectivity was not restored (Oguma et al. 2001; Morita et al. 2002; Rochelle et al. 2004). Based on more recent knowledge regarding the importance of energy stores for infectivity and the impact of sunlight on infectivity (King et al. 2010), it is likely that sporozoite repair of UV-induced damage results in energy depletion and loss of infectivity. Transcriptome analysis supports this, showing an upregulation of genes for DNA repair and intracellular trafficking (Zhang et al. 2012), which is consistent with the finding of King et al. (2010) in terms of UV-B causing premature exocytosis. Some studies have recommended the use of MP-UV to prevent reactivation (Kalisvaart 2004), but the available evidence suggests that LP-UV is sufficient to cause a permanent reduction in infectivity for *Cryptosporidium*.

Conventional UV reactors used for treatment of bulk water use UV lamps enclosed in a quartz sleeve and installed either in the center of a stainless steel pipe, or suspended in an open channel. Water is passed through the pipe or channel and the UV dose delivered is a function of the lamp intensity, flow rate, mixing through the system and the UV transmissivity of the water. In conventional UV systems the lamps are always on while the system is operational. An alternative system that has been evaluated more recently uses pulsed UV light, which has been shown to achieve 6 $\log_{10}$ inactivation with an equivalent UV dose of 278 mJ/cm^2 and 2 $\log_{10}$ with a dose of 15 mJ/cm^2 (Lee et al. 2008). One issue with the use of pulsed UV technology is the difficulty in measuring the UV fluence (dose) delivered to oocysts (Garvey et al. 2010).

13.5.3 Combined Processes

Sequential disinfection has been shown to improve oocyst inactivation levels. The combination of ozone followed by free chlorine increased oocyst inactivation by four to sixfold compared with ozone alone (Li et al. 2001). Similar results were obtained using a combination of ozone and monochloramine, although the increase in inactivation was only 2.5-fold (Rennecker et al. 2001). The synergistic effectiveness of ozone followed by chlorine has been demonstrated using both animal infectivity (Li et al. 2001; Biswas et al. 2005) and excystation assays (Corona-Vasquez et al. 2002). Low ozone doses have been combined with UV disinfection to allow adequate disinfection of water while reducing the formation of disinfection by-products by ozone (Meunier et al. 2006). An added advantage of this combination is that the ozone will react with organics in the water, helping to increase the

UV transmissivity and so improve UV dose delivery. In terms of the treatment of wastewater to produce reuse water, combined treatment trains such as UV and chlorine disinfection have been recommended to allow control of both chlorine resistant pathogens such as *Cryptosporidium* and UV resistant pathogens such as adenovirus (Montemayor et al. 2008).

13.5.4 Heat

As has been previously discussed in this chapter, *Cryptosporidium* are sensitive to heat, making this an effective method for disinfection. Moist heat is highly effective (Anderson 1985) and the temperatures achieved in thermophilic aerobic digestion of sewage sludge (55 °C) are sufficient for ready inactivation of oocysts (Whitmore and Robertson 1995). Pasteurization has been shown to be effective for the inactivation of oocysts in water and milk (Harp et al. 1996), but while this treatment can be used to inactivate oocysts in food (e.g. Deng and Cliver 2001), it has not been used for large scale treatment of water or wastewater.

13.5.5 Other

Photocatalytic inactivation of *Cryptosporidium* oocysts has been assessed using titanium dioxide films and a black light lamp (essentially UV-A), achieving up to 78 % oocyst inactivation after 3 h exposure (Sunnotel et al. 2010). However, this study used viability surrogates rather than infectivity and so may have underestimated the effectiveness of the treatment. The measured inactivation was not due to UV-A exposure based on comparison of a control exposed to the light source in the absence of the titanium dioxide. A fibrous ceramic titanium oxide photocatalyst, used in combination with UV-C, has also been shown to be effective at reducing oocyst numbers, suggesting that the photo byproducts produced are active at disrupting oocysts (Navalon et al. 2009). The mechanism of photocatalytic titanium oxide or ferrioxolate inactivation is the generation of the hydroxide radical, which has been reported to be 10^4–10^7 times more effective compared to ozone, chlorine dioxide or chlorine (Cho and Yoon 2008). However, another study examining LP-UV and titanium dioxide suggested that reactive oxygen species were only responsible for a relatively small amount of inactivation when used in isolation, but that when used in combination with UV produced a synergistic effect resulting in more inactivation than either exposure alone (Ryu et al. 2008).

Sonication has been shown to be effective in physically disrupting oocysts, causing oocysts to release contents, and also to reduce the infectivity of any remaining intact oocysts (Oyane et al. 2005). Using a horn-type sonicator, 60 s at 52 W resulted in 52 % inactivation, while 126 W for the same exposure time caused

94.9 % inactivation (Oyane et al. 2005). A flow through sonicator was also able to inactivate oocysts, but it was only effective at low flow rates (Oyane et al. 2005).

13.6 Conclusions

Cryptosporidium continues to be a public health concern around the world, through outbreaks caused by contaminated drinking water or recreational water. However, we have developed an increasingly better understanding of the fate and transport of oocysts through the environment and of the processes required to remove or inactivate them in drinking water or wastewater. There are many processes, starting from the deposition of oocysts into the terrestrial environment, through to overland transport into surface waters and hydraulic transport through to water treatment plants, that either cause the removal or inactivation of a significant number of oocysts. There are also processes that can be used to treat surface waters to remove or inactivate oocysts for the production of potable water. Despite these advances, outbreaks from drinking water will still occur, either due to treatment barrier failures, or because of insufficient risk characterization of catchments, resulting in inadequate treatment barriers to prevent entry of infectious oocysts into drinking water systems. Although there are advanced treatment options, outbreaks will still occur from recreational waters and swimming pools because it is not possible to prevent transmission when fecal events occur in close proximity to other water users. However, at least for drinking water, the technology is available to remove the threat of a *Cryptosporidium* outbreak, although at the cost of additional infrastructure and operating costs.

References

Abbaszadegan M, Monteiro P, Ouwens RN, Ryu H, Alum A (2006) Removal and inactivation of *Cryptosporidium* and microbial indicators by a quaternary ammonium chloride (QAC)-treated zeolite in pilot filters. J Environ Sci Health A Tox Hazard Subst Environ Eng 41(6):1201–1210

Akiyoshi DE, Feng X, Buckholt MA, Widmer G, Tzipori S (2002) Genetic analysis of a *Cryptosporidium parvum* human genotype 1 isolate passaged through different host species. Infect Immun 70(10):5670–5675

Alum A, Rubino JR, Khalid Ijaz M (2011) Comparison of molecular markers for determining the viability and infectivity of *Cryptosporidium* oocysts and validation of molecular methods against animal infectivity assay. Int J Infect Dis 15(3):e197–200

Anderson BC (1985) Moist heat inactivation of *Cryptosporidium* sp. Am J Public Health 75 (12):1433–1434

Anderson BC (1986) Effect of drying on the infectivity of cryptosporidia-laden calf feces for 3- to 7-day-old mice. Am J Vet Res 47(10):2272–2273

Araki S, Martin-Gomez S, Becares E, De Luis-Calabuig E, Rojo-Vazquez F (2001) Effect of high-rate algal ponds on viability of *Cryptosporidium parvum* oocysts. Appl Environ Microbiol 67 (7):3322–3324

Atwill ER, Hou L, Karle BM, Harter T, Tate KW, Dahlgren RA (2002) Transport of *Cryptosporidium parvum* oocysts through vegetated buffer strips and estimated filtration efficiency. Appl Environ Microbiol 68(11):5517–5527

Atwill ER, Tate KW, Pereira MD, Bartolome J, Nader G (2006) Efficacy of natural grassland buffers for removal of *Cryptosporidium parvum* in rangeland runoff. J Food Prot 69 (1):177–184

Baishanbo A, Gargala G, Delaunay A, Francois A, Ballet JJ, Favennec L (2005) Infectivity of *Cryptosporidium hominis* and *Cryptosporidium parvum* genotype 2 isolates in immunosuppressed Mongolian gerbils. Infect Immun 73(8):5252–5255

Baishanbo A, Gargala G, Duclos C, Francois A, Rossignol JF, Ballet JJ, Favennec L (2006) Efficacy of nitazoxanide and paromomycin in biliary tract cryptosporidiosis in an immunosuppressed gerbil model. J Antimicrob Chemother 57(2):353–355

Barbee SL, Weber DJ, Sobsey MD, Rutala WA (1999) Inactivation of *Cryptosporidium parvum* oocyst infectivity by disinfection and sterilization processes. Gastrointest Endosc 49 (5):605–611

Belosevic M, Guy RA, Taghi-Kilani R, Neumann NF, Gyurek LL, Liyanage LR, Millard PJ, Finch GR (1997) Nucleic acid stains as indicators of *Cryptosporidium parvum* oocyst viability. Int J Parasitol 27(7):787–798

Belosevic M, Craik SA, Stafford JL, Neumann NF, Kruithof J, Smith DW (2001) Studies on the resistance/reactivation of *Giardia muris* cysts and *Cryptosporidium parvum* oocysts exposed to medium-pressure ultraviolet radiation. FEMS Microbiol Lett 204(1):197–203

Bichai F, Barbeau B, Dullemont Y, Hijnen W (2010) Role of predation by zooplankton in transport and fate of protozoan (oo)cysts in granular activated carbon filtration. Water Res 44 (4):1072–1081

Biswas K, Craik S, Smith DW, Belosevic M (2003) Synergistic inactivation of *Cryptosporidium parvum* using ozone followed by free chlorine in natural water. Water Res 37(19):4737–4747

Biswas K, Craik S, Smith DW, Belosevic M (2005) Synergistic inactivation of *Cryptosporidium parvum* using ozone followed by monochloramine in two natural waters. Water Res 39 (14):3167–3176

Black EK, Finch GR, Taghi-Kilani R, Belosevic M (1996) Comparison of assays for *Cryptosporidium parvum* oocysts viability after chemical disinfection. FEMS Microbiol Lett 135 (2–3):187–189

Brookes JD, Antenucci J, Hipsey M, Burch MD, Ashbolt NJ, Ferguson C (2004) Fate and transport of pathogens in lakes and reservoirs. Environ Int 30(5):741–759

Brookes JD, Davies CM, Hipsey MR, Antenucci JP (2006) Association of *Cryptosporidium* with bovine faecal particles and implications for risk reduction by settling within water supply reservoirs. J Water Health 4(1):87–98

Bukhari Z, LeChevallier M (2003) Assessing UV reactor performance for treatment of finished water. Water Sci Technol 47(3):179–184

Bukhari Z, Marshall MM, Korich DG, Fricker CR, Smith HV, Rosen J, Clancy JL (2000) Comparison of *Cryptosporidium parvum* viability and infectivity assays following ozone treatment of oocysts. Appl Environ Microbiol 66(7):2972–2980

Bukhari Z, Holt DM, Ware MW, Schaefer FW 3rd (2007) Blind trials evaluating in vitro infectivity of *Cryptosporidium* oocysts using cell culture immunofluorescence. Can J Microbiol 53(5):656–663

Bustamante HA, Shanker SR, Pashley RM, Karaman ME (2001) Interaction between *Cryptosporidium* oocysts and water treatment coagulants. Water Res 35(13):3179–3189

Butkus MA, Bays JT, Labare MP (2003) Influence of surface characteristics on the stability of *Cryptosporidium parvum* oocysts. Appl Environ Microbiol 69(7):3819–3825

Caldwell MM (1971) Solar UV irradiation and the growth and development of higher plants. In: Giese AC (ed) Photophysiology, vol 6. Academic Press, NewYork, pp 131–177

Campbell AT, Robertson LJ, Snowball MR, Smith HV (1995) Inactivation of oocysts of *Cryptosporidium parvum* by ultraviolet irradiation. Water Res 29(11):2583–2586

Castro-Hermida JA, Pors I, Mendez-Hermida F, Ares-Mazas E, Chartier C (2006) Evaluation of two commercial disinfectants on the viability and infectivity of *Cryptosporidium parvum* oocysts. Vet J 171(2):340–345

Chapman PA, Rush BA (1990) Efficiency of sand filtration for removing *Cryptosporidium* oocysts from water. J Med Microbiol 32(4):243–245

Chauret C, Nolan K, Chen P, Springthorpe S, Sattar S (1998) Aging of *Cryptosporidium parvum* oocysts in river water and their susceptibility to disinfection by chlorine and monochloramine. Can J Microbiol 44(12):1154–1160

Chauret CP, Radziminski CZ, Lepuil M, Creason R, Andrews RC (2001) Chlorine dioxide inactivation of *Cryptosporidium parvum* oocysts and bacterial spore indicators. Appl Environ Microbiol 67(7):2993–3001

Cho M, Yoon J (2008) Measurement of OH radical CT for inactivating *Cryptosporidium parvum* using photo/ferrioxalate and photo/TiO2 systems. J Appl Microbiol 104(3):759–766

Cilimburg A, Monz C, Kehoe S (2000) PROFILE: wildland recreation and human waste: a review of problems, practices, and concerns. Environ Manage 25(6):587–598

Connelly SJ, Wolyniak EA, Williamson CE, Jellison KL (2007) Artificial UV-B and solar radiation reduce in vitro infectivity of the human pathogen *Cryptosporidium parvum*. Environ Sci Technol 41(20):7101–7106

Considine RF, Dixon DR, Drummond CJ (2002) Oocysts of *Cryptosporidium parvum* and model sand surfaces in aqueous solutions: an atomic force microscope (AFM) study. Water Res 36 (14):3421–3428

Corona-Vasquez B, Samuelson A, Rennecker JL, Marinas BJ (2002) Inactivation of *Cryptosporidium parvum* oocysts with ozone and free chlorine. Water Res 36(16):4053–4063

Craik SA, Weldon D, Finch GR, Bolton JR, Belosevic M (2001) Inactivation of *Cryptosporidium parvum* oocysts using medium- and low-pressure ultraviolet radiation. Water Res 35 (6):1387–1398

Current WL, Haynes TB (1984) Complete development of *Cryptosporidium* in cell culture. Science 224(4649):603–605

Dai X, Boll J (2003) Evaluation of attachment of *Cryptosporidium parvum* and *Giardia lamblia* to soil particles. J Environ Qual 32(1):296–304

Dai X, Boll J (2006) Settling velocity of *Cryptosporidium parvum* and *Giardia lamblia*. Water Res 40(6):1321–1325

Dai X, Hozalski RM (2002) Effect of NOM and biofilm on the removal of *Cryptosporidium parvum* oocysts in rapid filters. Water Res 36(14):3523–3532

Darnault CJ, Garnier P, Kim YJ, Oveson KL, Steenhuis TS, Parlange JY, Jenkins M, Ghiorse WC, Baveye P (2003) Preferential transport of *Cryptosporidium parvum* oocysts in variably saturated subsurface environments. Water Environ Res 75(2):113–120

Davies CM, Ferguson CM, Kaucner C, Krogh M, Altavilla N, Deere DA, Ashbolt NJ (2004) Dispersion and transport of *Cryptosporidium* oocysts from fecal pats under simulated rainfall events. Appl Environ Microbiol 70(2):1151–1159

Davies CM, Altavilla N, Krogh M, Ferguson CM, Deere DA, Ashbolt NJ (2005) Environmental inactivation of *Cryptosporidium* oocysts in catchment soils. J Appl Microbiol 98(2):308–317

Deng MQ, Cliver DO (1998) *Cryptosporidium parvum* development in the BS-C-1 cell line. J Parasitol 84(1):8–15

Deng MQ, Cliver DO (1999) *Cryptosporidium parvum* studies with dairy products. Int J Food Microbiol 46(2):113–121

Deng MQ, Cliver DO (2001) Inactivation of *Cryptosporidium parvum* oocysts in cider by flash pasteurization. J Food Prot 64(4):523–527

Di Giovanni GD, Hashemi FH, Shaw NJ, Abrams FA, LeChevallier MW, Abbaszadegan M (1999) Detection of infectious *Cryptosporidium parvum* oocysts in surface and filter backwash water samples by immunomagnetic separation and integrated cell culture-PCR. Appl Environ Microbiol 65(8):3427–3432

DiCesare EA, Hargreaves BR, Jellison KL (2012a) Biofilm roughness determines *Cryptosporidium parvum* retention in environmental biofilms. Appl Environ Microbiol 78(12):4187–4193
DiCesare EA, Hargreaves BR, Jellison KL (2012b) Biofilms reduce solar disinfection of *Cryptosporidium parvum* oocysts. Appl Environ Microbiol 78(12):4522–4525
Dorner SM, Anderson WB, Slawson RM, Kouwen N, Huck PM (2006) Hydrologic modeling of pathogen fate and transport. Environ Sci Technol 40(15):4746–4753
Drescher AC, Greene DM, Gadgil AJ (2001) *Cryptosporidium* inactivation by low-pressure UV in a water disinfection device. J Environ Health 64(3):31–35
Edzwald JK (2010) Dissolved air flotation and me. Water Res 44(7):2077–2106
Edzwald JK, Tobiason JE, Dunn H, Kaminski G, Galant P (2001) Removal and fate of *Cryptosporidium* in dissolved air drinking water treatment plants. Water Sci Technol 43(8):51–57
Emelko MB (2003) Removal of viable and inactivated *Cryptosporidium* by dual- and tri-media filtration. Water Res 37(12):2998–3008
Faulkner BR, Olivas Y, Ware MW, Roberts MG, Groves JF, Bates KS, McCarty SL (2010) Removal efficiencies and attachment coefficients for *Cryptosporidium* in sandy alluvial riverbank sediment. Water Res 44(9):2725–2734
Fayer R (1994) Effect of high temperature on infectivity of *Cryptosporidium parvum* oocysts in water. Appl Environ Microbiol 60(8):2732–2735
Fayer R (1995) Effect of sodium hypochlorite exposure on infectivity of *Cryptosporidium parvum* oocysts for neonatal BALB/c mice. Appl Environ Microbiol 61(2):844–846
Fayer R, Nerad T (1996) Effects of low temperatures on viability of *Cryptosporidium parvum* oocysts. Appl Environ Microbiol 62(4):1431–1433
Fayer R, Graczyk TK, Cranfield MR, Trout JM (1996) Gaseous disinfection of *Cryptosporidium parvum* oocysts. Appl Environ Microbiol 62(10):3908–3909
Fayer R, Trout JM, Jenkins MC (1998) Infectivity of *Cryptosporidium parvum* oocysts stored in water at environmental temperatures. J Parasitol 84(6):1165–1169
Fayer R, Trout JM, Walsh E, Cole R (2000) Rotifers ingest oocysts of *Cryptosporidium parvum*. J Eukaryot Microbiol 47(2):161–163
Ferguson C, Husman AMD, Altavilla N, Deere D, Ashbolt N (2003) Fate and transport of surface water pathogens in watersheds. Crit Rev Environ Sci Technol 33(3):299–361
Ferguson CM, Croke BF, Beatson PJ, Ashbolt NJ, Deere DA (2007) Development of a process-based model to predict pathogen budgets for the Sydney drinking water catchment. J Water Health 5(2):187–208
Finch GR, Black EK, Gyurek L, Belosevic M (1993a) Ozone inactivation of *Cryptosporidium parvum* in demand-free phosphate buffer determined by in vitro excystation and animal infectivity. Appl Environ Microbiol 59(12):4203–4210
Finch GR, Daniels CW, Black EK, Schaefer FW 3rd, Belosevic M (1993b) Dose response of *Cryptosporidium parvum* in outbred neonatal CD-1 mice. Appl Environ Microbiol 59 (11):3661–3665
Friedberg E, Walker G, Siede W (1995) DNA and mutagenesis. ASM Press, Washington, DC
Fu CY, Xie X, Huang JJ, Zhang T, Wu QY, Chen JN, Hu HY (2010) Monitoring and evaluation of removal of pathogens at municipal wastewater treatment plants. Water Sci Technol 61 (6):1589–1599
Fujino T, Matsui T, Kobayashi F, Haruki K, Yoshino Y, Kajima J, Tsuji M (2002) The effect of heating against *Cryptosporidium* oocysts. J Vet Med Sci 64(3):199–200
Gale P, Pitchers R, Gray P (2002) The effect of drinking water treatment on the spatial heterogeneity of micro-organisms: implications for assessment of treatment efficiency and health risk. Water Res 36(6):1640–1648
Garvey M, Farrell H, Cormican M, Rowan N (2010) Investigations of the relationship between use of in vitro cell culture-quantitative PCR and a mouse-based bioassay for evaluating critical factors affecting the disinfection performance of pulsed UV light for treating *Cryptosporidium parvum* oocysts in saline. J Microbiol Methods 80(3):267–273

Gomez-Bautista M, Ortega-Mora LM, Tabares E, Lopez-Rodas V, Costas E (2000) Detection of infectious *Cryptosporidium parvum* oocysts in mussels (*Mytilus galloprovincialis*) and cockles (*Cerastoderma edule*). Appl Environ Microbiol 66(5):1866–1870

Graczyk TK, Marcogliese DJ, de Lafontaine Y, Da Silva AJ, Mhangami-Ruwende B, Pieniazek NJ (2001) *Cryptosporidium parvum* oocysts in zebra mussels (*Dreissena polymorpha*): evidence from the St Lawrence River. Parasitol Res 87(3):231–234

Gut J, Petersen C, Nelson R, Leech J (1991) *Cryptosporidium parvum*: in vitro cultivation in Madin-Darby canine kidney cells. J Protozool 38(6):72S–73S

Harp JA, Fayer R, Pesch BA, Jackson GJ (1996) Effect of pasteurization on infectivity of *Cryptosporidium parvum* oocysts in water and milk. Appl Environ Microbiol 62(8):2866–2868

Hashimoto A, Hirata T, Kunikane S (2001) Occurrence of *Cryptosporidium* oocysts and *Giardia* cysts in a conventional water purification plant. Water Sci Technol 43(12):89–92

Hawkins PR, Swanson P, Warnecke M, Shanker SR, Nicholson C (2000) Understanding the fate of *Cryptosporidium* and *Giardia* in storage reservoirs: a legacy of Sydney's water contamination incident. J Water Supply Res Technol-Aqua 49(6):289–306

Hayes EB, Matte TD, O'Brien TR, McKinley TW, Logsdon GS, Rose JB, Ungar BL, Word DM, Pinsky PF, Cummings ML et al (1989) Large community outbreak of cryptosporidiosis due to contamination of a filtered public water supply. N Engl J Med 320(21):1372–1376

Helmi K, Skraber S, Gantzer C, Willame R, Hoffmann L, Cauchie HM (2008) Interactions of *Cryptosporidium parvum*, *Giardia lamblia*, vaccinal poliovirus type 1, and bacteriophages phiX174 and MS2 with a drinking water biofilm and a wastewater biofilm. Appl Environ Microbiol 74(7):2079–2088

Hijnen WA, Schijven JF, Bonne P, Visser A, Medema GJ (2004) Elimination of viruses, bacteria and protozoan oocysts by slow sand filtration. Water Sci Technol 50(1):147–154

Hijnen WA, Brouwer-Hanzens AJ, Charles KJ, Medema GJ (2005) Transport of MS2 phage, *Escherichia coli*, *Clostridium perfringens*, *Cryptosporidium parvum*, and *Giardia intestinalis* in a gravel and a sandy soil. Environ Sci Technol 39(20):7860–7868

Hijnen WA, Dullemont YJ, Schijven JF, Hanzens-Brouwer AJ, Rosielle M, Medema G (2007) Removal and fate of *Cryptosporidium parvum*, *Clostridium perfringens* and small-sized centric diatoms (*Stephanodiscus hantzschii*) in slow sand filters. Water Res 41(10):2151–2162

Hijnen WA, Suylen GM, Bahlman JA, Brouwer-Hanzens A, Medema GJ (2010) GAC adsorption filters as barriers for viruses, bacteria and protozoan (oo)cysts in water treatment. Water Res 44 (4):1224–1234

Hill VR, Kahler AM, Jothikumar N, Johnson TB, Hahn D, Cromeans TL (2007) Multistate evaluation of an ultrafiltration-based procedure for simultaneous recovery of enteric microbes in 100-liter tap water samples. Appl Environ Microbiol 73(13):4218–4225

Hsu BM, Yeh HH (2003) Removal of *Giardia* and *Cryptosporidium* in drinking water treatment: a pilot-scale study. Water Res 37(5):1111–1117

Hsu BM, Huang C, Pan JR (2001) Filtration behaviors of Giardia and *Cryptosporidium*–ionic strength and pH effects. Water Res 35(16):3777–3782

Huamanchay O, Genzlinger L, Iglesias M, Ortega YR (2004) Ingestion of *Cryptosporidium* oocysts by *Caenorhabditis elegans*. J Parasitol 90(5):1176–1178

Hutchison ML, Walters LD, Moore T, Thomas DJ, Avery SM (2005) Fate of pathogens present in livestock wastes spread onto fescue plots. Appl Environ Microbiol 71(2):691–696

Izumi T, Yagita K, Endo T, Ohyama T (2006) Detection system of *Cryptosporidium parvum* oocysts by brackish water benthic shellfish (*Corbicula japonica*) as a biological indicator in river water. Arch Environ Contam Toxicol 51(4):559–566

Jenkins MB, Bowman DD, Ghiorse WC (1998) Inactivation of *Cryptosporidium parvum* oocysts by ammonia. Appl Environ Microbiol 64(2):784–788

Jenkins MB, Walker MJ, Bowman DD, Anthony LC, Ghiorse WC (1999) Use of a sentinel system for field measurements of *Cryptosporidium parvum* oocyst inactivation in soil and animal waste. Appl Environ Microbiol 65(5):1998–2005

Jenkins MB, Bowman DD, Fogarty EA, Ghiorse WC (2002) *Cryptosporidium parvum* oocyst inactivation in three soil types at various temperatures and water potentials. Soil Biol Biochem 34(8):1101–1109

Jerome JH, Bukata RP (1998) Tracking the propagation of solar ultraviolet radiation: dispersal of ultraviolet photons in inland waters. J Great Lakes Res 24(3):666–680

Johnson DC, Enriquez CE, Pepper IL, Davis TL, Gerba CP, Rose JB (1997) Survival of *Giardia*, *Cryptosporidium*, poliovirus and *Salmonella* in marine waters. Water Sci Technol 35 (11–12):261–268

Johnson AM, Linden K, Ciociola KM, De Leon R, Widmer G, Rochelle PA (2005) UV inactivation of *Cryptosporidium hominis* as measured in cell culture. Appl Environ Microbiol 71 (5):2800–2802

Johnson AM, Giovanni GD, Rochelle PA (2012) Comparison of assays for sensitive and reproducible detection of cell culture-infectious *Cryptosporidium parvum* and *Cryptosporidium hominis* in drinking water. Appl Environ Microbiol 78(1):156–162

Kalisvaart BF (2004) Re-use of wastewater: preventing the recovery of pathogens by using medium-pressure UV lamp technology. Water Sci Technol 50(6):337–344

Karim MR, Manshadi FD, Karpiscak MM, Gerba CP (2004) The persistence and removal of enteric pathogens in constructed wetlands. Water Res 38(7):1831–1837

Kato S, Jenkins MB, Fogarty EA, Bowman DD (2002) Effects of freeze-thaw events on the viability of *Cryptosporidium parvum* oocysts in soil. J Parasitol 88(4):718–722

Kaucner C, Davies CM, Ferguson CM, Ashbolt NJ (2005) Evidence for the existence of *Cryptosporidium* oocysts as single entities in surface runoff. Water Sci Technol 52(8):199–204

Keegan AR, Fanok S, Monis PT, Saint CP (2003) Cell culture-Taqman PCR assay for evaluation of *Cryptosporidium parvum* disinfection. Appl Environ Microbiol 69(5):2505–2511

Keegan A, Daminato D, Saint CP, Monis PT (2008) Effect of water treatment processes on *Cryptosporidium* infectivity. Water Res 42(6–7):1805–1811

King BJ, Monis PT (2007) Critical processes affecting *Cryptosporidium* oocyst survival in the environment. Parasitology 134(Pt 3):309–323

King BJ, Keegan AR, Monis PT, Saint CP (2005) Environmental temperature controls *Cryptosporidium* oocyst metabolic rate and associated retention of infectivity. Appl Environ Microbiol 71(7):3848–3857

King BJ, Monis PT, Keegan AR, Harvey K, Saint C (2007) Investigation of the survival of *Cryptosporidium* in environmental waters. Cooperative Research Centre for Water Quality and Treatment, Salisbury, South Australia, Australia

King BJ, Hoefel D, Daminato DP, Fanok S, Monis PT (2008) Solar UV reduces *Cryptosporidium parvum* oocyst infectivity in environmental waters. J Appl Microbiol 104(5):1311–1323

King BJ, Hoefel D, Wong PE, Monis PT (2010) Solar radiation induces non-nuclear perturbations and a false start to regulated exocytosis in *Cryptosporidium parvum*. PLoS One 5(7):e11773

King BJ, Keegan AR, Robinson BS, Monis PT (2011) *Cryptosporidium* cell culture infectivity assay design. Parasitology 138(6):671–681

King BJ, Keegan AR, Phillips R, Fanok S, Monis PT (2012) Dissection of the hierarchy and synergism of the bile derived signal on *Cryptosporidium parvum* excystation and infectivity. Parasitology 139(12):1533–1546

Korich DG, Mead JR, Madore MS, Sinclair NA, Sterling CR (1990) Effects of ozone, chlorine dioxide, chlorine, and monochloramine on *Cryptosporidium parvum* oocyst viability. Appl Environ Microbiol 56(5):1423–1428

Korich DG, Marshall MM, Smith HV, O'Grady J, Bukhari Z, Fricker CR, Rosen JP, Clancy JL (2000) Inter-laboratory comparison of the CD-1 neonatal mouse logistic dose–response model for *Cryptosporidium parvum* oocysts. J Eukaryot Microbiol 47(3):294–298

Kuczynska E, Shelton DR, Pachepsky Y (2005) Effect of bovine manure on *Cryptosporidium parvum* oocyst attachment to soil. Appl Environ Microbiol 71(10):6394–6397

Kuznar ZA, Elimelech M (2006) *Cryptosporidium* oocyst surface macromolecules significantly hinder oocyst attachment. Environ Sci Technol 40(6):1837–1842

Kvac M, Kvetonova D, Salat J, Ditrich O (2007) Viability staining and animal infectivity of *Cryptosporidium andersoni* oocysts after long-term storage. Parasitol Res 100(2):213–217
Kvac M, Sak B, Kvetonova D, Secor WE (2009) Infectivity of gastric and intestinal *Cryptosporidium* species in immunocompetent Mongolian gerbils (*Meriones unguiculatus*). Vet Parasitol 163(1–2):33–38
Lacharme L, Villar V, Rojo-Vazquez FA, Suarez S (2004) Complete development of *Cryptosporidium parvum* in rabbit chondrocytes (VELI cells). Microbes Infect 6(6):566–571
Lawton P, Naciri M, Mancassola R, Petavy AF (1997) In vitro cultivation of *Cryptosporidium parvum* in the non-adherent human monocytic THP-1 cell line. J Eukaryot Microbiol 44(6):66S
Lee SH, Lee CH, Kim YH, Do JH, Kim SH (2007) Occurrence of *Cryptosporidium* oocysts and *Giardia* cysts in the Nakdong River and their removal during water treatment. J Water Health 5 (1):163–169
Lee SU, Joung M, Yang DJ, Park SH, Huh S, Park WY, Yu JR (2008) Pulsed-UV light inactivation of *Cryptosporidium parvum*. Parasitol Res 102(6):1293–1299
Levine AD, Harwood VJ, Farrah SR, Scott TM, Rose JB (2008) Pathogen and indicator organism reduction through secondary effluent filtration: implications for reclaimed water production. Water Environ Res 80(7):596–608
Li H, Finch GR, Smith DW, Belosevic M (2001) Sequential inactivation of *Cryptosporidium parvum* using ozone and chlorine. Water Res 35(18):4339–4348
Li X, Brasseur P, Agnamey P, Ballet JJ, Clemenceau C (2004) Time and temperature effects on the viability and infectivity of *Cryptosporidium parvum* oocysts in chlorinated tap water. Arch Environ Health 59(9):462–466
Li X, Atwill ER, Dunbar LA, Jones T, Hook J, Tate KW (2005) Seasonal temperature fluctuations induces rapid inactivation of *Cryptosporidium parvum*. Environ Sci Technol 39 (12):4484–4489
Li X, Atwill ER, Dunbar LA, Tate KW (2010) Effect of daily temperature fluctuation during the cool season on the infectivity of *Cryptosporidium parvum*. Appl Environ Microbiol 76 (4):989–993
Linden KG, Shin G, Sobsey MD (2001) Comparative effectiveness of UV wavelengths for the inactivation of *Cryptosporidium parvum* oocysts in water. Water Sci Technol 43(12):171–174
Lindquist HD, Harris S, Lucas S, Hartzel M, Riner D, Rochele P, Deleon R (2007) Using ultrafiltration to concentrate and detect *Bacillus anthracis*, *Bacillus atrophaeus* subspecies *globigii*, and *Cryptosporidium parvum* in 100-liter water samples. J Microbiol Methods 70 (3):484–492
Liu P, Hill VR, Hahn D, Johnson TB, Pan Y, Jothikumar N, Moe CL (2012) Hollow-fiber ultrafiltration for simultaneous recovery of viruses, bacteria and parasites from reclaimed water. J Microbiol Methods 88(1):155–161
Logan AJ, Stevik TK, Siegrist RL, Ronn RM (2001) Transport and fate of *Cryptosporidium parvum* oocysts in intermittent sand filters. Water Res 35(18):4359–4369
Lorenzo-Lorenzo MJ, Ares-Mazas ME, Villacorta-Martinez de Maturana I, Duran-Oreiro D (1993) Effect of ultraviolet disinfection of drinking water on the viability of *Cryptosporidium parvum* oocysts. J Parasitol 79(1):67–70
Mac Kenzie WR, Hoxie NJ, Proctor ME, Gradus MS, Blair KA, Peterson DE, Kazmierczak JJ, Addiss DG, Fox KR, Rose JB et al (1994) A massive outbreak in Milwaukee of *Cryptosporidium* infection transmitted through the public water supply. N Engl J Med 331(3):161–167
Majewski P, Keegan A (2012) Surface properties and water treatment capacity of surface engineered silica coated with 3-(2-aminoethyl) aminopropyltrimethoxysilane. Appl Surf Sci 258(7):2454–2458
Malloy KD, Holman MA, Mitchell D, Detrich HW 3rd (1997) Solar UVB-induced DNA damage and photoenzymatic DNA repair in antarctic zooplankton. Proc Natl Acad Sci U S A 94 (4):1258–1263
Mawdsley JL, Brooks AE, Merry RJ (1996) Movement of the protozoan pathogen *Cryptosporidium parvum* through three contrasting soil types. Biol Fertil Soils 21(1–2):30–36

Medema GJ, Schets FM, Teunis PF, Havelaar AH (1998) Sedimentation of free and attached *Cryptosporidium* oocysts and *Giardia* cysts in water. Appl Environ Microbiol 64 (11):4460–4466

Meinhardt PL, Casemore DP, Miller KB (1996) Epidemiologic aspects of human cryptosporidiosis and the role of waterborne transmission. Epidemiol Rev 18(2):118–136

Metge DW, Harvey RW, Aiken GR, Anders R, Lincoln G, Jasperse J (2010) Influence of organic carbon loading, sediment associated metal oxide content and sediment grain size distributions upon *Cryptosporidium parvum* removal during riverbank filtration operations, Sonoma County, CA. Water Res 44(4):1126–1137

Meunier L, Canonica S, von Gunten U (2006) Implications of sequential use of UV and ozone for drinking water quality. Water Res 40(9):1864–1876

Michallet H, Ivey GN (1999) Experiments on mixing due to internal solitary waves breaking on uniform slopes. J Geophys Res Oceans 104(C6):13467–13477

Montemayor M, Costan A, Lucena F, Jofre J, Munoz J, Dalmau E, Mujeriego R, Sala L (2008) The combined performance of UV light and chlorine during reclaimed water disinfection. Water Sci Technol 57(6):935–940

Morgan-Ryan UM, Fall A, Ward LA, Hijjawi N, Sulaiman I, Fayer R, Thompson RC, Olson M, Lal A, Xiao L (2002) *Cryptosporidium hominis* n. sp. (Apicomplexa: Cryptosporidiidae) from Homo sapiens. J Eukaryot Microbiol 49(6):433–440

Morita S, Namikoshi A, Hirata T, Oguma K, Katayama H, Ohgaki S, Motoyama N, Fujiwara M (2002) Efficacy of UV irradiation in inactivating *Cryptosporidium parvum* oocysts. Appl Environ Microbiol 68(11):5387–5393

Morris DP, Zagarese H, Williamson CE, Balseiro EG, Hargreaves BR, Modenutti B, Moeller R, Queimalinos C (1995) The attentuation of solar UV radiation in lakes and the role of dissolved organic carbon. Limnol Oceanogr 40(8):1381–1391

Muck RE, Richards BK (1983) Losses of manurial nitrogen in free-stall barns. Agric Wastes 7 (2):65–79

Muck RE, Steenhuis TS (1982) Nitrogen losses from manure storages. Agric Wastes 4(1):41–54

Nasser AM, Tweto E, Nitzan Y (2007) Die-off of *Cryptosporidium parvum* in soil and wastewater effluents. J Appl Microbiol 102(1):169–176

Navalon S, Alvaro M, Garcia H, Escrig D, Costa V (2009) Photocatalytic water disinfection of *Cryptosporidium parvum* and *Giardia lamblia* using a fibrous ceramic TiO(2) photocatalyst. Water Sci Technol 59(4):639–645

Neto RC, Santos LU, Franco RM (2006) Evaluation of activated sludge treatment and the efficiency of the disinfection of *Giardia* species cysts and *Cryptosporidium* oocysts by UV at a sludge treatment plant in Campinas, south-east Brazil. Water Sci Technol 54(3):89–94

Neumann NF, Gyurek LL, Gammie L, Finch GR, Belosevic M (2000) Comparison of animal infectivity and nucleic acid staining for assessment of *Cryptosporidium parvum* viability in water. Appl Environ Microbiol 66(1):406–412

Oguma K, Katayama H, Mitani H, Morita S, Hirata T, Ohgaki S (2001) Determination of pyrimidine dimers in *Escherichia coli* and *Cryptosporidium parvum* during UV light inactivation, photoreactivation, and dark repair. Appl Environ Microbiol 67(10):4630–4637

Oyane I, Furuta M, Stavarache CE, Hashiba K, Mukai S, Nakanishi JM, Kimata I, Maeda Y (2005) Inactivation of *Cryptosporidium parvum* by ultrasonic irradiation. Environ Sci Technol 39 (18):7294–7298

Patni NK, Jui PY (1991) Nitrogen concentration variability in dairy-cattle slurry stored in farm tanks. Trans ASAE 34(2):609–615

Peeters JE, Mazas EA, Masschelein WJ, Martiez V, de Maturana I, Debacker E (1989) Effect of disinfection of drinking water with ozone or chlorine dioxide on survival of *Cryptosporidium parvum* oocysts. Appl Environ Microbiol 55(6):1519–1522

Quilez J, Sanchez-Acedo C, Avendano C, del Cacho E, Lopez-Bernad F (2005) Efficacy of two peroxygen-based disinfectants for inactivation of *Cryptosporidium parvum* oocysts. Appl Environ Microbiol 71(5):2479–2483

Quinonez-Diaz MJ, Karpiscak MM, Ellman ED, Gerba CP (2001) Removal of pathogenic and indicator microorganisms by a constructed wetland receiving untreated domestic wastewater. J Environ Sci Health A Tox Hazard Subst Environ Eng 36(7):1311–1320

Rasmussen KR, Larsen NC, Healey MC (1993) Complete development of *Cryptosporidium parvum* in a human endometrial carcinoma cell line. Infect Immun 61(4):1482–1485

Ravanat JL, Douki T, Cadet J (2001) Direct and indirect effects of UV radiation on DNA and its components. J Photochem Photobiol B 63(1–3):88–102

Reinoso R, Becares E (2008) Environmental inactivation of *Cryptosporidium parvum* oocysts in waste stabilization ponds. Microb Ecol 56(4):585–592

Reinoso R, Torres LA, Becares E (2008) Efficiency of natural systems for removal of bacteria and pathogenic parasites from wastewater. Sci Total Environ 395(2–3):80–86

Reinoso R, Blanco S, Torres-Villamizar LA, Becares E (2011) Mechanisms for parasites removal in a waste stabilisation pond. Microb Ecol 61(3):684–692

Rennecker JL, Corona-Vasquez B, Driedger AM, Rubin SA, Marinas BJ (2001) Inactivation of *Cryptosporidium parvum* oocysts with sequential application of ozone and combined chlorine. Water Sci Technol 43(12):167–170

Risebro HL, Doria MF, Andersson Y, Medema G, Osborn K, Schlosser O, Hunter PR (2007) Fault tree analysis of the causes of waterborne outbreaks. J Water Health 5(Suppl 1):1–18

Robertson LJ, Campbell AT, Smith HV (1992) Survival of *Cryptosporidium parvum* oocysts under various environmental pressures. Appl Environ Microbiol 58(11):3494–3500

Rochelle PA, Ferguson DM, Handojo TJ, De Leon R, Stewart MH, Wolfe RL (1997) An assay combining cell culture with reverse transcriptase PCR to detect and determine the infectivity of waterborne *Cryptosporidium parvum*. Appl Environ Microbiol 63(5):2029–2037

Rochelle PA, Ferguson DM, Johnson AM, De Leon R (2001) Quantitation of *Cryptosporidium parvum* infection in cell culture using a colorimetric in situ hybridization assay. J Eukaryot Microbiol 48(5):565–574

Rochelle PA, Marshall MM, Mead JR, Johnson AM, Korich DG, Rosen JS, De Leon R (2002) Comparison of in vitro cell culture and a mouse assay for measuring infectivity of *Cryptosporidium parvum*. Appl Environ Microbiol 68(8):3809–3817

Rochelle PA, Fallar D, Marshall MM, Montelone BA, Upton SJ, Woods K (2004) Irreversible UV inactivation of *Cryptosporidium* spp. despite the presence of UV repair genes. J Eukaryot Microbiol 51(5):553–562

Ryan U, Yang R, Gordon C, Doube B (2011) Effect of dung burial by the dung beetle *Bubas bison* on numbers and viability of *Cryptosporidium* oocysts in cattle dung. Exp Parasitol 129(1):1–4

Ryu H, Gerrity D, Crittenden JC, Abbaszadegan M (2008) Photocatalytic inactivation of *Cryptosporidium parvum* with TiO(2) and low-pressure ultraviolet irradiation. Water Res 42 (6–7):1523–1530

Santamaria J, Quinonez-Diaz Mde J, Lemond L, Arnold RG, Quanrud D, Gerba C, Brusseau ML (2011) Transport of *Cryptosporidium parvum* oocysts in sandy soil: impact of length scale. J Environ Monit 13(12):3481–3484

Santamaria J, Brusseau ML, Araujo J, Orosz-Coghlan P, Blanford WJ, Gerba CP (2012) Transport and retention of *Cryptosporidium parvum* oocysts in sandy soils. J Environ Qual 41 (4):1246–1252

Sasahara T, Aoki M, Sekiguchi T, Takahashi A, Satoh Y, Kitasato H, Inoue M (2003) Effect of the mixed-oxidant solution on infectivity of *Cryptosporidium parvum* oocysts in a neonatal mouse model. Kansenshogaku Zasshi 77(2):75–82

Schets FM, Engels GB, During M, de Roda Husman AM (2005) Detection of infectious *Cryptosporidium* oocysts by cell culture immunofluorescence assay: applicability to environmental samples. Appl Environ Microbiol 71(11):6793–6798

Searcy KE, Packman AI, Atwill ER, Harter T (2005) Association of *Cryptosporidium parvum* with suspended particles: impact on oocyst sedimentation. Appl Environ Microbiol 71 (2):1072–1078

Searcy KE, Packman AI, Atwill ER, Harter T (2006) Capture and retention of *Cryptosporidium parvum* oocysts by *Pseudomonas aeruginosa* biofilms. Appl Environ Microbiol 72 (9):6242–6247

Sherwood D, Angus KW, Snodgrass DR, Tzipori S (1982) Experimental cryptosporidiosis in laboratory mice. Infect Immun 38(2):471–475

Shields JM, Hill VR, Arrowood MJ, Beach MJ (2008) Inactivation of *Cryptosporidium parvum* under chlorinated recreational water conditions. J Water Health 6(4):513–520

Shin GA, Linden KG, Arrowood MJ, Sobsey MD (2001) Low-pressure UV inactivation and DNA repair potential of *Cryptosporidium parvum* oocysts. Appl Environ Microbiol 67 (7):3029–3032

Skraber S, Helmi K, Willame R, Ferreol M, Gantzer C, Hoffmann L, Cauchie HM (2007) Occurrence and persistence of bacterial and viral faecal indicators in wastewater biofilms. Water Sci Technol 55(8–9):377–385

Slifko TR, Friedman D, Rose JB, Jakubowski W (1997) An in vitro method for detecting infectious *Cryptosporidium* oocysts with cell culture. Appl Environ Microbiol 63(9):3669–3675

Slifko TR, Huffman DE, Dussert B, Owens JH, Jakubowski W, Haas CN, Rose JB (2002) Comparison of tissue culture and animal models for assessment of *Cryptospridium parvum* infection. Exp Parasitol 101(2–3):97–106

Stott R, May E, Matsushita E, Warren A (2001) Protozoan predation as a mechanism for the removal of *Cryptosporidium* oocysts from wastewaters in constructed wetlands. Water Sci Technol 44(11–12):191–198

Stott R, May E, Ramirez E, Warren A (2003) Predation of *Cryptosporidium* oocysts by protozoa and rotifers: implications for water quality and public health. Water Sci Technol 47(3):77–83

Sunnotel O, Verdoold R, Dunlop PS, Snelling WJ, Lowery CJ, Dooley JS, Moore JE, Byrne JA (2010) Photocatalytic inactivation of *Cryptosporidium parvum* on nanostructured titanium dioxide films. J Water Health 8(1):83–91

Suwa M, Suzuki Y (2003) Control of *Cryptosporidium* with wastewater treatment to prevent its proliferation in the water cycle. Water Sci Technol 47(9):45–49

Tate KW, Pereira MD, Atwill ER (2004) Efficacy of vegetated buffer strips for retaining *Cryptosporidium parvum*. J Environ Qual 33(6):2243–2251

Theodos CM, Griffiths JK, D'Onfro J, Fairfield A, Tzipori S (1998) Efficacy of nitazoxanide against *Cryptosporidium parvum* in cell culture and in animal models. Antimicrob Agents Chemother 42(8):1959–1965

Thurston JA, Gerba CP, Foster KE, Karpiscak MM (2001) Fate of indicator microorganisms, *Giardia* and *Cryptosporidium* in subsurface flow constructed wetlands. Water Res 35 (6):1547–1551

Tufenkji N, Miller GF, Ryan JN, Harvey RW, Elimelech M (2004) Transport of *Cryptosporidium* oocysts in porous media: role of straining and physicochemical filtration. Environ Sci Technol 38(22):5932–5938

Uga S, Matsuo J, Kono E, Kimura K, Inoue M, Rai SK, Ono K (2000) Prevalence of *Cryptosporidium parvum* infection and pattern of oocyst shedding in calves in Japan. Vet Parasitol 94 (1–2):27–32

Ulrich H, Klaus D, Irmgard F, Annette H, Juan LP, Regine S (2005) Microbiological investigations for sanitary assessment of wastewater treated in constructed wetlands. Water Res 39(20):4849–4858

Upton SJ, Tilley M, Brillhart DB (1994a) Comparative development of *Cryptosporidium parvum* (Apicomplexa) in 11 continuous host cell lines. FEMS Microbiol Lett 118(3):233–236

Upton SJ, Tilley M, Nesterenko MV, Brillhart DB (1994b) A simple and reliable method of producing in vitro infections of *Cryptosporidium parvum* (Apicomplexa). FEMS Microbiol Lett 118(1–2):45–49

Venczel LV, Arrowood M, Hurd M, Sobsey MD (1997) Inactivation of *Cryptosporidium parvum* oocysts and *Clostridium perfringens* spores by a mixed-oxidant disinfectant and by free chlorine. Appl Environ Microbiol 63(11):4625

Venczel LV, Likirdopulos CA, Robinson CE, Sobsey MD (2004) Inactivation of enteric microbes in water by electro-chemical oxidant from brine (NaCl) and free chlorine. Water Sci Technol 50(1):141–146

Vesey G, Slade JS, Byrne M, Shepherd K, Fricker CR (1993) A new method for the concentration of *Cryptosporidium* oocysts from water. J Appl Bacteriol 75(1):82–86

Villacorta-Martinez de Maturana I, Ares-Mazas ME, Duran-Oreiro D, Lorenzo-Lorenzo MJ (1992) Efficacy of activated sludge in removing *Cryptosporidium parvum* oocysts from sewage. Appl Environ Microbiol 58(11):3514–3516

Walker M, Leddy K, Hager E (2001) Effects of combined water potential and temperature stresses on *Cryptosporidium parvum* oocysts. Appl Environ Microbiol 67(12):5526–5529

Weir SC, Pokorny NJ, Carreno RA, Trevors JT, Lee H (2002) Efficacy of common laboratory disinfectants on the infectivity of *Cryptosporidium parvum* oocysts in cell culture. Appl Environ Microbiol 68(5):2576–2579

Wen Q, Tutuka C, Keegan A, Jin B (2009) Fate of pathogenic microorganisms and indicators in secondary activated sludge wastewater treatment plants. J Environ Manage 90(3):1442–1447

Whitehead DC, Raistrick N (1993) The volatilization of ammonia from cattle urine applied to soils as influenced by soil properties. Plant and Soil 148(1):43–51

Whitmore TN, Robertson LJ (1995) The effect of sewage sludge treatment processes on oocysts of *Cryptosporidium parvum*. J Appl Bacteriol 78(1):34–38

Widmer G, Akiyoshi D, Buckholt MA, Feng X, Rich SM, Deary KM, Bowman CA, Xu P, Wang Y, Wang X, Buck GA, Tzipori S (2000) Animal propagation and genomic survey of a genotype 1 isolate of *Cryptosporidium parvum*. Mol Biochem Parasitol 108(2):187–197

Wolfe RL (1990) Ultraviolet disinfection of potable water. Environ Sci Technol 24(6):768–773

Wolyniak EA, Hargreaves BR, Jellison KL (2010) Seasonal retention and release of *Cryptosporidium parvum* oocysts by environmental biofilms in the laboratory. Appl Environ Microbiol 76(4):1021–1027

Xunde LI, Brasseur P (2000) A NMRI suckling mouse model for the evaluation of infectivity of *Cryptosporidium parvum* oocysts. Zhongguo Ji Sheng Chong Xue Yu Ji Sheng Chong Bing Za Zhi 18(2):94–96

Yamamoto N, Urabe K, Takaoka M, Nakazawa K, Gotoh A, Haga M, Fuchigami H, Kimata I, Iseki M (2000) Outbreak of cryptosporidiosis after contamination of the public water supply in Saitama Prefecture, Japan, in 1996. Kansenshogaku Zasshi 74(6):518–526

Yang S, Healey MC, Du C, Zhang J (1996) Complete development of *Cryptosporidium parvum* in bovine fallopian tube epithelial cells. Infect Immun 64(1):349–354

You X, Arrowood MJ, Lejkowski M, Xie L, Schinazi RF, Mead JR (1996) A chemiluminescence immunoassay for evaluation of *Cryptosporidium parvum* growth in vitro. FEMS Microbiol Lett 136(3):251–256

Zhang H, Guo F, Zhou H, Zhu G (2012) Transcriptome analysis reveals unique metabolic features in the *Cryptosporidium parvum* oocysts associated with environmental survival and stresses. BMC Genomics 13:647

Index

A

Activated sludge process (ASP), 536
Acyl-CoA binding protein (ACBP), 370–372
Amino acid metabolism, 366–367
Aminoglycosides
  antibiotics, 458–459
  databases, 332
Ammonia, 521
Anseriformes, 254–255
Antibiotics
  aminoglycoside, 458–459
  benzeneacetonitrile, 459
  macrolide, 457–458
  monoclonal, 475
  thiazolide, 461–463
Antimotility agents, 404
Anura
  Caudata and Gymnophonia, 238–239
  diversity, 305
  doubtful toad, 239, 240
  oocysts, 239
Apodiformes, 253
Artiodactyla
  bovids, 282–284
  camels and llamas, 278–280
  deer and moose, 280–281
  giraffe and okapi, 280
  pigs, 278, 279
ASP. *See* Activated sludge process (ASP)
Asymptomatic infection
  epidemiological traits, 50
  experimental infections, 49
  meta analysis, 50
  prevalence, 50
Azithromycin, 457

B

Badgers, 290
Banded mongoose, 291
Bears, 288
Benzeneacetonitrile antibiotics, 459
Birds
  Anseriformes, 254–255
  Apodiformes, 253
  Caprimulgiformes, 261
  Charadriiformes, 257–258
  Ciconiiformes, 257
  Columbiformes, 258
  Coraciiformes, 262
  Cuculiformes, 253
  epithelial cells, 248
  Falconiformes, 257
  Galliformes, 255–256
  Gaviiformes, 253
  GenBank accession numbers, 249–250
  Gruiformes, 258
  Opisthocomiformes, 253
  Passeriformes, 262–264
  Pelecaniformes, 256
  Phoenicopteriformes, 256
  Piciformes, 261
  Podicipediformes and Procellariiformes, 253
  Psittaciformes, 259–261
  Sphenisciformes, 256
  Strigiformes, 259, 261
  Struthioniformes, 253–254
  Tinamiformes, 253
  Trochiliformes and Trogoniformes, 253
Bone marrow and solid-organ transplantation, 396–397

S.M. Cacciò and G. Widmer (eds.), *Cryptosporidium: parasite and disease*, DOI 10.1007/978-3-7091-1562-6, © Springer-Verlag Wien 2014

Bovids, 282–284
Bovines
  cattle (*see* Cattle)
  domesticated buffalo, 152
  water buffalo (*see* Water buffalo)

C
Camelids
  alpacas and llamas
    infection associated with clinical disease, 196
    prevalence, 194, 195
    zoonotic transmission, 196
  camels and dromedaries
    infection associated with clinical disease, 194
    prevalence, 193–194
    zoonotic transmission, 196
  family and species, 193
Cancer, 396
Caprimulgiformes, 261
Carbohydrate and energy metabolisms
  amylopectin, 363
  N-glycan biosynthesis, 366
  glycolysis, 365
  inhibitor-enzyme interactions, 363
  MDH and LDH, 365
  pyruvate and acetyl-CoA, 363, 364
  replication, transcription and translation, 365
  TCA cycle and type II NADH dehydrogenase, 363, 365
  UGGP and T6PS-TP, 365–366
Carnivora
  badger, 290
  banded mongoose, 291
  cats, domestic and wild, 291
  coyotes and bears, 288
  dogs, domestic, 285–287
  ferrets, 289
  fin-footed mammals, 285
  foxes, 287
  mink and martens, 290
  otters, 289
  raccoons, 288–289
  skunk, 290
  wolves, 287–288
Cattle
  C. parvum GP60 subtypes, 159–161
  infection associated with clinical disease, 165–167
  location, age and prevalence, 153–158
  oocyst excretion and transmission, 168–169
  prevalence, 153, 159
  zoonotic transmission, 170–171
Caviomorphs, 295–296, 299
Cetaceans, 277
Charadriiformes, 257–258
Chiroptera, 281
Ciconiiformes, 257
Clarithromycin, 458
Clinical disease
  children, developing countries, 388–390
  competent individuals, 386–388
  cryptosporidiosis characteristics, 386, 387
  extraintestinal infections, 393
  gastrointestinal infections, 392
  healthy adults infection, 384, 386
  immune compromised individuals, 390–391
  incubation period, 384
  infection characteristics and clinical features, 384, 385
Columbiformes, 258
Coraciiformes, 262
Coyotes, 288
Cryptosporidium
  environmental stage, 4
  and *Giardia*, 68
  GP60 subtyping methods, 100–103
  incubation period, 4
  oocyst wall proteins (COWPs), 354–355
  relapses, 4
Cryptosporidium infection associated with clinical disease
  camelids
    alpacas and llamas, 196
    camels and dromedaries, 194
  cattle
    C. andersoni, 166
    C. bovis, 166
    C. parvum and *C. andersoni*, 165
    C. ryanae, 166
    halofuginone lactate treatment, 166–167
    vaccination, 167
  chickens, 205–206
  deer, 190–192
  ducks and geese, 207
  pigeons, 212
  pigs, 181–188
  quails and partridges, 212
  rabbits, 197–199
  ratites, 207–212
  small ruminants, 176–177

turkeys, 206–207
water buffalo, 167
Cryptosporidium oocysts, drinking and recreational water
cell culture assays, 492
detection methods (*see* Oocyst detection methods)
development, 490
human pathogens, 491–492
identification and genotyping, 501–502
industry, 490
infectivity assays (*see* Infectivity assays)
outbreaks, 490
risk assessment
contributions, 495
cost-benefit analysis, 496
infection, 496
regulations, 496
sporadic cryptosporidiosis, 496
swimming pools and water parks, 495
water consumption, 495
source waters (*see* Source waters, *Cryptosporidium* oocysts)
in treated drinking water
chlorine disinfection, 503
DWI, 503
monitoring, 503
regulation, 503, 504
samples analysis, 503
survey, 503
watershed management, 504
types, 491
waterborne outbreaks (*see* Waterborne outbreaks, cryptosporidiosis)
Cryptosporidium oocyst wall proteins (COWPs), 354–355
Cryptosporidium parvum oocysts/sporozoites
cluster analysis, 351
EST data, 349–350
gene expression patterns, 350–351
human epithelial cells, 350
transcriptional analysis, 349–350
UV irradiation, 350
Cuculiformes, 253

D
DAF. *See* Dissolved air flotation (DAF)
Dasyuromorphia, 273
Deer
infection associated with clinical disease, 190, 192
in New Zealand, 189–190
oocyst excretion and transmission, 192
prevalence, 190, 191
zoonotic transmission, 192–193
Diabetes, 397
Didelphimorphia, 274–275
Differential interference contrast (DIC) microscopy, 498
Dihydrofolate reductase (DHFR)
inhibitors, 471
and thymidylate synthase (DHFR-TS), 368, 373
Diprotodontia, 274
Disease burden, cryptosporidiosis
geographic distribution, 56–57
infection (*see* Infections)
seasonality, 55–56
Dissolved air flotation (DAF), 535
Domestic dogs, 285–287
Druggable genome, 331–332
Dysregulated intestinal secretion, 402

E
Egg-laying mammals, 265
Epidemiology, human cryptosporidiosis
asymptomatic infection, 49–51
disease burden (*see* Disease burden, cryptosporidiosis)
genotyping techniques, 48
human pathogens, 47
incubation period, 48–49
life cycle, 44–46
oocysts, 44, 47
PCR, 47
post-infectious sequelae, 54
risk factors, 51–54
symptomatic infection, 51
transmission routes (*see* Transmission routes)
zoonotic species, 44
Epithelial defense response, cryptosporidiosis
AIDS patients, 428
cytokines and chemokines release, 428, 430
α-and β-defensins, 428–429
exosomes (*see* Exosomes, cryptosporidiosis)
human gastrointestinal cells, 427
IL-8 and CXCL1, 427
miRNAs (*see* MicroRNAs (miRNAs))
mucins, 429–430
mucosal anti-*Cryptosporidium* immunity, 428, 429
regenerating islet-derived 3 gamma (Reg3g), 430
regulation, IL-18 gene, 428
serum amyloid A3 (Saa3), 430
Esophageal damage and appendicitis, 393–394

ESTs. *See* Expressed sequence tags (ESTs)
Exosomes, cryptosporidiosis
activation, 433–434
anti-*C. parvum* activity, 434–435
basolateral, 435
binding, 435
composition, 433
confocal microscopy analysis, 435
β-defesin-2 and LL-37, 434
intestinal epithelial cell-derived, 435–436
luminal release, biliary and intestinal epithelium, 434, 435
secretion, 434
size and formation, 433
Expressed sequence tags (ESTs)
cDNA libraries, 349
CryptoDB, 355
EPICDB, 356
and genome survey sequences (GSS), 350

F
Falconiformes, 257
Farmed animals, cryptosporidiosis
bovines (*see* Bovines)
camelids (*see* Camelids)
and *Cryptosporidium* species, 150–152
deer (*see* Deer)
lifecycles, 150
as livestock/domesticated animals, 150
pigs (*see* Pigs)
poultry (*see* Poultry)
rabbits (*see* Rabbits)
small ruminants (*see* Small ruminants)
transmission, 152
zoonotic nature, 150, 152
Fatty acid synthase (FAS)
acyl-CoA thioesters, 370
acyl elongation modules, 369
bacteria and yeast, ACS proteins, 371
CpLCE1, 370
CpPKS1, 369
lipoproteins, 371
phosphatidyl-ethanolamine, 371–372
SFP-PPT and ACPS-PPT, 369–370
FDM. *See* Focus detection method (FDM)
Ferrets, 289
Fin-footed mammals, 285
Flocculation, 532
Fluorescent in situ hybridization (FISH), 517
Focus detection method (FDM), 518
Foodborne cryptosporidiosis
case–control studies, 69
contamination, food, 68
Cyclospora, 68
electronic Foodborne Outbreak Reporting System, 69–70
microsporidia, 68
oocysts, 68–70
outbreaks, 121–122
public health problem, 70
Foxes, 287

G
Galliformes, 255–256
Gastrointestinal (GI), 61–62
Gaviiformes, 253
Genomics
amino acids and nucleotides, 330
apicomplexans, 328
BAC libraries, 329
characterization, genetic variation, 335
cis-regulatory elements (CREs), 336
C. muris RN66 and *C. parvum* TU114, 333
C. parvum TU114, *C. parvum* IOWA and *C. hominis* TU506, 334
differences, phenotypic behaviour, 329
DNA replication, transcription and translation, 330
druggable genome, 331–332
energy generation, 330
eukaryotes, 335–336
gastrointestinal conditions, 332–333
GenBank database and CryptoDB, 334
host-cell binding invasion, 333
host-specificity and site-selection, 335
human infections, 332
'hypothetical proteins', 335
immunological responses, 333
oligonucleotide primers, 328
ORFs and EST data, 328
PCR-based sequencing, 334, 340
Plasmodium, 330–331
shotgun Sanger sequencing approach, 328, 329
whole-genome sequencing, 340–341
Genotyping
C. hominis monkey, 91
food and water microbiology, 98
fox and rat, 91
human infections, 83–90
isolation, 83
SSU rRNA gene, 98
and subtyping (*see* Subtyping methods)
TaqMan array, 98

transport vectors, 91
GI. *See* Gastrointestinal (GI)
Global mRNA transcriptional analysis
advantages, RNA-Seq approach, 347
gene expression profiles, 346
hybridisation-based technique, microarrays, 346–347
reverse-transcription PCR (RT-PCR), 347
SOLiD sequencing platform, 347
GP60 subtyping methods
AL3531 and AL3535, 100
C. parvum and *C. hominis*, 100, 104–107
Cryptosporidium spp. and sequences, 100–103
host immune response, 108
intra-species diversity, 99
nomenclature, 99
recombinant progeny, 108
T-RFLP, 100
Gruiformes, 258

H
HAART. *See* Highly active antiretroviral therapy (HAART)
Heat
disinfection, *Cryptosporidium*, 541
terrestrial environment, 519
Heat shock proteins (HSPs), 365, 372, 373
Hedgehogs (Erinceomorpha), 276
Highly active antiretroviral therapy (HAART), 59, 456, 463–464
Host cell transcriptome, 351–352
Host immune response
antibody, 398
$CD4^+$ and and $CD8^+$ T cells, 397–398
IFN-γ, 398
HSPs. *See* Heat shock proteins (HSPs)
Human cryptosporidiosis
AIDS, 384
antimotility agents, 404
antiparasitic agents, 404
apicomplexan protozoan, 383
clinical disease (*see* Clinical disease)
complications, 393–394
diagnosis, 399
host immune response (*see* Host immune response)
infection prevention, 406–407
macrolides, 406
molecular epidemiology (*see* Molecular epidemiology, cryptosporidiosis)
nitazoxanide, 405
paromomycin, 405–406
pathogenesis, 400–401
pathology, 399, 400
restoration, immune system, 403
risk factors, 394–397
water borne, 384
Hydrological parameters
biotic and abiotic factors, 522
inflow water, 523
oocyst interaction/aggregation, 523
sedimentation rate, 523
water service reservoirs, 524
Hyracoidea, 275

I
IBS. *See* Inflammatory bowel syndrome (IBS)
Immune competent hosts
C. hominis infections, 387
gastrointestinal symptoms, 386
oocysts, 386
symptomatic infection, 388
Immune compromised hosts, 390–391
Immunodeficiency diseases, 395
Immunology, cryptosporidiosis
antigen, 436–437
complement system, 437–438
Cryptosporidium infection, 424
epithelial defense response (*see* Epithelial defense response, cryptosporidiosis)
evasion, host immunity and defense, 439–442
IFNs, 437
immunopathology, 442–444
macrophages, 437
mucosal pathogens, 424
NK cells, SCID mice, 437
parasite-epithelial cell interactions, 425–426
PRRs activation, epithelial cells, 426–427
T and B cells, 438–439
Immunomagnetic separation (IMS)
detection method, oocysts, 498–499
waterborne outbreaks, 504
Immunopathology
chromatin dynamics and infection, 443–444
pathogenic mechanisms, 442–443
Immunotherapy
hyperimmune colostrum, 474
monoclonal antibodies, 475
IMP dehydrogenase (IMPDH), 367, 368, 373
Incubation period, cryptosporidiosis
acute gastroenteritis, 49
experimental infections, 48
investigations, foodborne, 49
waterborne outbreak, 48

Infection
in children, 57–58
colonization, 60
epidemiology and clinical features, 59
HAART, 59
in vitro experiments, 60
in immunocompromised individuals, 58–59
intestinal infections, 60
liver transplantation, 61
microscopic examination, 59
prevalence, 55
primary immunodeficiencies, 60
in travellers, 61–63
XHIM, 60
Infectivity assays
cell culture, 504
3D-CC-IFA method, 506
in vitro development, 507
IMS procedure and omitting microscopic examination, 504
literature examination, 506–507
monolayers, cultured cell, 505, 506
optimized cell culture assays, 505
prevalence, 505
risk assessment, 504
water samples examinations, 505–506
Inflammatory bowel syndrome (IBS), 54
Innate drug resistance, cryptosporidiosis
ABC and parasite transporters, 467–468
Apicomplexa, 465
coding, prenyl synthase enzymes, 465
C. parvum, 464
differences, cryptosporidia and gregarines, 464–465
identification, *C. parvum* genes, 464
lack of efficacy, 464
location and physiology, 466–467
plasmodium gene, 466
plastid, 465

L

Lactate dehydrogenase (LDH), 363, 365
Lagomorpha, 292
Leukemia/lymphoma, 396
Life cycle
apical complex, 44–45
asexual and sexual reproduction, 46
excystation, 44
oocysts, 44–46
sporogony, 46
sporozoites, 45, 46
trophozoites and merozoites, 45, 46
Lipid metabolism
ACBP, 370
FAS (*see* Fatty acid synthase (FAS))
long chain fatty acyl-CoA synthetase, 370
ORPs, 370–371

M

Macrolide antibiotics
azithromycin, 406, 457
clarithromycin, 406, 458
roxithromycin, 406, 458
spiramycin, 406, 457
Malabsorption, 402
Malate dehydrogenase (MDH), 365
Mammals
egg-laying, 265
GenBank accession numbers, 264, 269–273
marsupials, 265, 273–275
oocyst morphology and infection site, 262, 265
placental (*see* Placental mammals)
SSU rDNA-based maximum likelihood tree, 264, 266–268
Marsupials
Dasyuromorphia, 273
Didelphimorphia, 274–275
Diprotodontia, 274
Paucituberculata, 273
Peramelemorphia, 274
Measurement of inactivation
animal bioassays, 516
anti-cryptosporidial drug treatments, 518
chemiluminescence immunoassay, 518
FDM, 518
FISH and excystation, 517
HCT-8 cells, 517
in vitro cell culture infectivity assays, 518
membrane integrity/enzyme activity, 515
oligonucleotide probe, 518
oocyst infectivity, 517
PCR-based assays, 517
viability assays, 516
Metabolism
ABC transporters, 362
amino acids, 366–367
anti-oxidant molecules, 373
carbohydrate and energy, 363–366
coccidia and haematozoa, 362
DNA damage, 373
GenBank and EuPathDB, 361
HSPs, 372
insensitivity, 373

ion-sulfur clusters, 362
lipids, 368–372
nucleotides, 367–368
P-type ATPases (P-ATPases), 363
stress-related pathways, 372–373
trehalose synthesis, 373
MicroRNAs (miRNAs)
biological processes and gene expression, 430
KSRP interaction, 432–433
mammalian and non-mammalian cells, 433
mature, 430
NF-κB-responses, 431
pri-miRNAs, 430
regulation, 430–431
RNA-induced silencing complex (RISC), 430
shuttling, 433
stability, 431–432
Molecular epidemiology, cryptosporidiosis
agricultural animals, 82
animal host, 117
application, 120–124
case-case examination, 126
C. hominis family, 116
C. parvum IIaA15G2R1, 118
distribution analysis, 118
DZHRGP and CP47, 120
genotyping, 83–91
global population structures, 119
HIV + patients, 117
human transmission routes, 82
IaA28R4 and IbA9G3, 117
IIaA17G1R1 and IIaA15G2R1, 119
IIdA20G1a, 118
immunocompromised people, 124–125
influencing factors, 92–93
mapping and sequencing, 126
multi-locus subtyping, 120
national surveillance data, 119
phenotyping, 83
serological markers, 125
sporadic disease, 82
sporozoite sequencing, 125
subtyping, 91–92
water catchment, 126
zoonotic transmission, 117
Multidimensional protein identification technology (MudPIT)
1-DE and 2-DE gels, 353
hydrophobic proteins, 354
and liquid chromatography (LC), 348
QSTAR QqTOF mass spectrometer, 352
Multilocus fragment type (MLFT)
microsatellite length, 108
and MLST, 108
size assignment, 116
Multilocus genotyping (MLG)
accurate analysis, 116
C. parvum and *C. hominis*, 108–115
micro-and minisatellites, 116
MLFT and MLST, 108
and SNPs, 108
Multilocus sequence type (MLST), 108

N

Natural organic matter (NOM), 532, 533
Nitazoxanide, 404
Non-human primates
clinical signs, 299, 302
identification, 299–302
SIV infection, 303
transmission, 299
Nuclear factor kappa-light-chain-enhancer of activated B cells (NF-κB), 401
Nucleotide metabolism
AMP-GMP pathway, 367
DHFR-TS, 368
IMPDH and GMP synthase, 367, 368
pyrimidine salvaging, UK-UPRT, 367–368

O

Oocyst detection methods
4'6-diamidino-2-phenylindole (DAPI), 497
DIC microscopy, 497
DNA fragment patterns, 499
IMS procedure, 498
PCR-based assays and primers, 498–499
USEPA method 1623, 497
Water Research Foundation Project 4099, 499
Water Research Foundation Project 4284, 500
Oocyst excretion and transmission
cattle, 168–169
chickens, 213
deer, 192
ducks and geese, 214
pigs, 189
rabbits, 199
small ruminants, 176–177
turkeys, 214
water buffalo, 169
zoonotic transmission, 214–215

Opisthocomiformes, 253
OSBP-related proteins (ORPs), 370–371
Otters, 289
Oxidant-based disinfection
  chlorine and chloramine, 537
  C. parvum and cell culture infectivity, 538
  infectivity assay, 539
  MIOX mixed oxidants, 538

P
Parasite-epithelial cell interactions, 425–426
Paromomycin, 405–406
Passeriformes, 262–264
Pathogenesis, human cryptosporidiosis
  cell death, 401
  diarrhea, 401
  dysregulated intestinal secretion, 402
  malabsorption, 402
  neuropeptide substance P, 401
  NF-κB, 401
  pro-inflammatory cytokines, 400
  small intestines and proximal colon, 400
Pattern recognition receptors (PRRs), 426–427
Paucituberculata, 237
PCR. *See* Polymerase chain reaction (PCR)
Pelecaniformes, 256
Peramelemorphia, 274
Perissodactyla
  equids, 284
  rhinoceros and tapirs, 284
Phenotyping, 83
Phoenicopteriformes, 256
Phosphopantetheinyl transferase (PPT), 369–370
Photocatalytic inactivation, 541
Piciformes, 261
Pigs
  exporters, 180
  infection associated with clinical disease
    developmental stages, 188
    and diarrhoea, 181, 188
    experimental infections, 188
    necropsy, 181
  oocyst excretion and transmission, 189
  prevalence, location and age, 180–187
  zoonotic transmission, 189
Placental mammals
  Artiodactyla (*see* Artiodactyla)
  Cetacea, 277
  Cingulata and Pilosa, 275
  Erinaceomorpha, 276
  Hyracoidea, 275
  Macroscelididae, Afrosordida and Tublidentata, 275
  Pholidota, Dermoptera and Scandentia, 275
  Proboscidea, 276
  Sirenia, 275–276
  Soricomorpha, 277
Pneumatosis cystoides intestinales, 393
Podicipediformes, 253
Polymerase chain reaction (PCR), 62, 63, 490
Possible oocyst wall proteins (POWPs), 355
Post-infectious sequelae, 54
Poultry
  infection associated with clinical disease
    chickens, 205–206
    ducks and geese, 207, 210–211
    pigeons, 212
    quails and partridges, 212
    ratites, 207, 212
    turkeys, 206–209
  oocyst excretion and transmission
    chickens, 213
    ducks and geese, 214
    turkeys, 214
  prevalence
    avian cryptosporidiosis, 201
    C. galli, 201–202
    chickens, 202–203
    ducks and geese, 203
    fecal examination, 202
    pigeons, 205
    quails and partridges, 205
    ratites, 204
    turkeys, 203
  world stock, 199–200
  zoonotic transmission, 214–215
POWPs. *See* Possible oocyst wall proteins (POWPs)
PPT. *See* Phosphopantetheinyl transferase (PPT)
Predation
  Bdelloid rotifers, 524, 526
  biofilms, 527
  bivalve molluscs, 527
  Corbicula japonica, 527
  Euplotes and Oxytricha sp., 524, 525
  flow chamber experiments, 526
  food density, 526
  free-living nematode, 524
  Mayorella sp., 524, 526
  microbial community, 527
  oocyst infectivity, 524
  Paramecium sp., 524, 525
  Rhabdocoel platyhelminth, 524, 526

Proboscidea, 275, 276
Procellariiformes, 253
Proteomics, *C. parvum*
 Apicomplexa, 356–357
 CryptoDB, 355–356
 EPICDB, 356
 gel electrophoresis, 348
 gene ontology (GO) and homology, 354
 LC-MS/MS and iTRAQ isobaric labelling, 352
 MALDI-TOF MS, 352
 mass spectrometry (MS), 348
 MudPIT, 348, 353, 354
 non-redundant proteins, 353
 sporozoites, sub-proteome analysis, 354–355
 workflow, 348, 349
Psittaciformes, 259

Q
Quantitative Microbiological Risk Assessment (QMRA), 67

R
Rabbits
 infection associated with clinical disease, 197, 199
 meat production, 197
 oocyst excretion and transmission, 199
 prevalence, 197, 198
 small-scale backyard cuniculture, 196–197
 zoonotic transmission, 199
Raccoon dogs, 287
Real-time PCR
 detection and quantification, 98
 filtration and immunomagnetic separation, 95
 human samples, 95–97
 molecular algorithms, 98
 semi-purification step, 98
 SSU rRNA gene, 98
 TaqMan array 388, 98
Reptiles
 crocodylians, 239
 diversity, 305
 Squamata (*see* Squamata)
 Testudines, 239, 248
Risk factors, cryptosporidiosis
 case–control studies, 51
 environmental and socioeconomic factors, 52
 host genetics, 53
 investigations, 52–53
 outbreak investigations, 51
 prevalence and surveillance, 51–53
 stunting and malaria, 53
RNA-induced silencing complex (RISC), 430
Rodentia
 beaver and gophers, 297–298
 caviomorphs, 295–296, 299
 Cricetidae, 295
 Muridae, 293–295
 squirrels and chipmunks, Sciuridae, 295
 suborders, 292

S
Sequential disinfection, 540–541
Simian immunodeficiency virus (SIV) infection, 303
Single nucleotide polymorphisms (SNPs)
 array-based techniques, 116
 DNA sequencing, 108
 micro-and minisatellites, 116
Sirenia, 275–276
Skunk, 285, 289, 290
Small ruminants
 design factors, 172
 infection associated with clinical disease
  anorexia and apathy/depression, 177
  morbidity and mortality rates, 176–177
 oocyst excretion and transmission, 177–178
 prevalence
  C. xiaoi, 176
  in goats, 172, 175
  in sheep, 172–174
 sheep and goats, 171–172
 zoonotic transmission
  C. parvum GP60 subtypes, 178–179
  C. ubiquitum, 180
  molecular tools, 178
  multi-locus genotyping (MLG), 179–180
Small subunit (SSU) rRNA
 conventional PCR assays, 95
 sequence analysis, 98
 sporozoite sequencing, 125
 SSCP mobility, 95
SNPs. *See* Single nucleotide polymorphisms (SNPs)
Solar disinfection (SODIS), 529
Sonication, 541–542
Soricomorpha (moles and shrews), 277
Source waters, *Cryptosporidium* oocysts
 recovery efficiency, 501
 human-pathogenic, 500
 microbe detection methods, 501

Source waters (*cont.*)
  Milwaukee outbreak, 500
  prevalence, 500
  regulatory monitoring programs, 500–501
  survey, 500
Species
  amphibians and reptiles, 15–18, 24–25
  in birds
    C. baileyi, 23
    C. galli, 23
    C. meleagridis, 22–23
    description, 11–14
    genotypes I-V, 23–24
  in mammals
    C. andersoni, 27
    C. bovis, 28
    C. canis, 27
    C. cuniculus, 29
    C. fayeri, 28
    C. felis, 27
    C. hominis, 27
    C. macropodum, 28
    C. muris, 26
    C. parvum, 26
    C. ryanae, 28
    C. scrofarum, 30
    C. suis, 27–28
    C. tyzzeri and *C. pestis*, 29–31
    C. ubiquitum, 29
    C. viatorum, 30
    C. wrairi, 26–27
    C. xiaoi, 28–29
    description, 19, 29
  in marine mammals and fish
    C. molnari, 20–21
    description, 6–10
    Piscicryptosporidium, 21–22
    piscine genotypes, 21
    prevalence, 21
  phylogenetic relationships, 6, 20
Sphenisciformes, 256
Squamata
  C. serpentis, 240–241
  habitats, 240
  lizards, 241, 245–247
  snake identification, 241–245
  tuataras, 239
SSU rRNA. *See* Small subunit (SSU) rRNA
Strain eruptions
  cluster detection, 124
  control and prevention, 124
  diagnostic laboratories, 123
  drinking and recreational water, 120–121
  protozoan infections, 123
  swimming pool operators, 123
Strigiformes, 259, 261
Struthioniformes, 253–254
Subtyping methods
  Cryptosporidium characterisation, stools, 99
  DNA extraction, 93–94
  GP60 (*see* GP60 subtyping methods)
  MLG (*see* Multilocus genotyping (MLG))
  real-time PCR (*see* Real-time PCR)
  species identification, conventional PCR, 94–95
Sunlight inactivation
  membrane depolarization, 531
  oocyst population and clumping, 529
  SODIS, 529
  sporozoite DNA and cellular processes, 530
  T_{90}, 530
  UV exposure, 528
  warm-water inflows, 530
  wavelengths, 530
Surface waters
  environmental stressors, 531
  hydrological parameters (*see* Hydrological parameters)
  inactivation, temperature, 528
  predation and microbial activity, 524–527
  sunlight inactivation, 528–531
Systems biology, *C. parvum* and *C. hominis*
  culturing methods, 338
  DNA sequencing technologies, 339
  genetic and genomic differences, 336
  gp60 gene, 339
  GP15/40 protein, 339–340
  high-throughput sequencing technologies, 337
  'hypothetical proteins', 338
  laser capture and microdissection technologies, 337
  limitations, 341
  454 Roche and Illumina, 336
  structural differences, 337
  subgenotypes, 339
  transcriptomic research, 337–338

T

T and B cells, 438–439
Taxonomy
  gregarines, 5
  peculiarities, 4–5
  phylum Apicomplexa, 4

species (*see* Species)
standards, 5–6
valid species, 6
TD. *See* Traveller diarrhoea (TD)
Terminal-restriction fragment length polymorphism (T-RFLP), 100
Terrestrial environment
ammonia, 521
freezing and desiccation, 520
heat, 519
transportation, receiving waters, 521–522
Testudines, 239, 248
Thiazolide antibiotics
azithromycin, 461
cell culture systems, 461
diarrhea, 461–462
double-blind and placebo control, 461, 462
nitazoxanide (NTZ) and tizoxanide, 462–463
Tinamiformes, 253
T6PS-TP. *See* Trehalose-6P synthase and trehalose phosphatase (T6PS-TP)
Transcriptional analysis
C. parvum oocysts/sporozoites, 349–351
description, 345–346
EPICDB, 356
global mRNA, 346–347
host cell responses, 351–352
visualisation and mining, CryptoDB, 355–356
Transmission routes
case–control studies, 63
C. hominis, 63, 64
C. parvum, 63, 64
Cryptosporidium infections, 63
foodborne cryptosporidiosis, 68–70
nosocomial infection, 63
waterborne cryptosporidiosis, 66–67
zoonotic transmission, 65–66
Traveller diarrhoea (TD), 61–62
Treatment, cryptosporidiosis
antibiotics (*see* Antibiotics)
anti-cryptosporidial activity, 457
anti-tubulin agents, 470–471
bisphosphonates, 468–469
chemotherapy, 463
colostrum, 460–461
control, 457
cryopreservation, 477
cytolysin A, 477
DHFR, inhibitor, 471
fatty acid synthase, 470
flavonoids, 468
genome sequence data, 476
HAART, 456, 463–464
HIV, 456
in human, 455
immunotherapy (*see* Immunotherapy)
innate drug resistance (*see* Innate drug resistance, cryptosporidiosis)
inosine monophosphate dehydrogenase, 472
mechanisms, chemicals and drugs, 476
polyamine biosynthesis, 469–470
protein kinases, 472–473
pyrimidine salvage enzymes, 472
standardized positive control drugs, 477
testing, animal models, 457
therapy, 455–456
TMP-SMX, 473
vaccines (*see* Vaccines, cryptosporidiosis)
Trehalose-6P synthase and trehalose phosphatase (T6PS-TP), 365–366
Trochiliformes, 253
Trogoniformes, 253

U

UDP-glucose/galactose pyrophosphorylase (UGGP), 365–366
Ultrafiltration, 497
Ultraviolet (UV) light, 539–540

V

Vaccines, cryptosporidiosis
development, 475
DNA, 475–476
early childhood, 475
immunogenic antigens, *C. parvum*, 475
profilin, 476
Salmonella, 476
Vertebrates
Anura, 238–239
birds (*see* Birds)
Carnivora, 285–292
Chiroptera, 281
description, 238
diversity, 303–305
knowledge and suggestions, 305
Lagomorpha, 292
mammals (*see* Mammals)

Vertebrates (*cont.*)
non-human primates, 298–303
Perissodactyla, 284
reptiles, 239–248
Rodentia, 292–298

W
Waterborne outbreaks, cryptosporidiosis
causes, U.S., 492, 494
consumption, drinking water, 68
C. parvum and *C. hominis*, 494–495
detection, 507
drinking and recreational water, 492, 493
genotyping tools, 494
Giardia and *Cryptosporidium*, 68
GI illness, 67
identification, infectious agent, 494
intestinal protozoa, 66
oocysts, 66
outbreaks, 66, 67, 122
pathogens, 66, 492
public pools, 494
QMRA, 67
1980s and 1990s, 507
U.K., 492, 494
Water buffalo
infection associated with clinical disease, 167
location, age, and prevalence, 159, 162–164
oocyst excretion and transmission, 169
prevalence, 159, 165
zoonotic transmission, 171
Water Research Foundation Project 4099, 499
Water Research Foundation Project 4284, 500
Water treatment
ASP (*see* Activated sludge process (ASP))
cryptosporidiosis, 532
DAF (*see* Dissolved air flotation (DAF))
disinfection processes, 531
flocculation, 532
lagoons/waste stabilization ponds, 536–537
media filtration, 533–534
membrane filtration, 535
Wolves, 287–288

X
Xenopus laevis, 24, 239, 241
X-linked hyper-immunoglobulin M syndrome (XHIM), 60

Z
Zebras, 284
Zeolite, 534
Zoonotic transmission
camelids
alpacas and llamas, 196
camels and dromedaries, 196
canine and feline cryptosporidiosis, 65
cattle
with drinking water, 171
foodborne, 170–171
molecular typing methods, 170
risk, 170
C. cuniculus, 66
C. parvum, 65
deer, 192–193
epidemiologic investigations, 65
oocysts, 65
pigs, 189
poultry
C. baileyi, 215
C. meleagridis, 214–215
C. parvum, 215
rabbits, 199
small ruminants, 176–180
water buffalo, 171

Printed by Publishers' Graphics LLC
LMO131224.15.18.262